Gene Activity in Early Development

Third Edition

Gene Activity in Early Development

Third Edition

Eric H. Davidson
Division of Biology
California Institute of Technology
Pasadena, California

1986

ACADEMIC PRESS, INC.
Harcourt Brace Jovanovich, Publishers
Orlando   San Diego   New York
Austin   Boston   London   Sydney
Tokyo   Toronto

ACADEMIC PRESS, INC.
Orlando, Florida 32887

United Kingdom Edition published by
ACADEMIC PRESS INC. (LONDON) LTD.
24–28 Oval Road, London NW1 7DX

Library of Congress Cataloging in Publication Data

Davidson, Eric H., Date
   Gene activity in early development.

   Bibliography: p.
   Includes index.
   1. Developmental genetics.   I. Title.
QH453.D38   1986   574.3'3   86-17341
ISBN 0—12—205161—0 (alk. paper)

PRINTED IN THE UNITED STATES OF AMERICA

86 87 88 89   9 8 7 6 5 4 3 2 1

This work is dedicated to the memory of my father, Morris Davidson (1898–1979), in his time a leading American painter. Only in recent years have I realized how much I learned from him about the ordering of complex perceptions.

Contents

III Transcription in the Embryo and Transfer of Control to the Zygotic Genomes

IV Differential Gene Function in the Embryo

V Gene Activity during Oogenesis

Acknowledgments

Though entitled the third edition of "Gene Activity in Early Development," this is a wholly new book. Perhaps this book contains more molecular and genetic evidence that is of *biological* import than did its predecessor. Certain conceptual themes, some organizational aspects, and various figures and calculations are all that survive from the second edition. The discoveries that have been achieved in the decade since the appearance of the second edition have largely transformed this domain of biological science. Were I to have succeeded in fulfilling the objectives of this endeavor, the following pages would provide an interpretive, synthetic review, presented in sufficient detail and with sufficient accuracy to serve also as a scholarly work of reference for interested students and for my scientific colleagues. I have realized that it may not be possible to write a book that truly accomplishes these objectives. The technological variety and the depth of available experimental evidence, and the rate at which new results appear, preclude real accuracy. Even more problematic is the task of constructing from the fragmentary insights that our field now affords a consistent image of the *developmental systems* that are the ultimate subject of this essay. The scope of the subject matter and the comparative approach I have chosen have denied me refuge within the narrower conceptual purviews of any one system. Thus it is my hope that despite its many deep imperfections, this book may provide an analysis of early development that is more united than those available, for example, in most multiauthored treatises or symposium volumes.

I would probably have never embarked on this project but for the urging and advice of colleagues whose views I deeply respect. At several junctions I would surely have abandoned it, had it not been for those on whom I have depended throughout for wisdom, knowledge, and criticism. I have been the beneficiary, on each of the hundreds of days on which I struggled with this manuscript, of essential support and encouragement. The manuscript was read, corrected, and in innumerable ways improved within my Caltech laboratory by my long time associate and colleague, Dr. Barbara R. Hough-Evans, and by Dr. Frank J. Calzone, at the time a Senior Research Fellow. Dr. L. Dennis Smith of Purdue University, and Dr. Roy J. Britten, Dr. Elliot Meyerowitz, and Dr. Ellen V. Rothenberg of the Caltech Division of Biology all reviewed critically the entire manuscript, a thankless and extensive effort. I am extremely grateful for their encouragement, their criticism, and in general their diverse and perspicacious insights. Various chapters or sections were also reviewed by Dr. Richard Axel of Columbia University College of Physicians and Surgeons; Dr. Gary Freeman of the University of Texas at Austin; Dr. William H. Klein of the M. D. Anderson Hospital, University of Texas Medical Center; and Drs. Edward B. Lewis, Mark Konishi, and Barbara Wold of the Caltech Division of Biology. It is my pleasure to acknowledge the debt that I owe to the scholarship and interest of these scientists, who in their individual ways contributed essentially to this work, though it is important to stress that they bear no responsibility for my oversights.

The manuscript was put together with the close collaboration of Ms. Jane Rigg. She served at once as bibliographic researcher, compiler of figures, editor, proofreader, interpreter, and advisor. Her informed intuitions, determination, indefatigable energy, and professional expertise were truly crucial at every stage. Ms. Stephanie Canada, who expertly converted my initial handwritten scrawl (which is generally considered illegible) and all the subsequent drafts of each chapter into a finished manuscript, also made this project her own. Her contributions from beginning to end were invaluable. Ms. Rene Thorf provided extensive and characteristically accurate secretarial assistance as well. I should also like to acknowledge, with sadness as well as gratitude, the role played by the late Mr. Ryo Arai, the Academic Press editor who organized the production of the second edition and initiated the commitment that led to this third edition. Mr. William Woodcock, the editor who inherited the problems of this enterprise, has been an understanding, effective, and above all intelligent agent and friend, and for his support and surveillance I am indeed grateful.

Gene Activity in Early Development

Third Edition

I

From Genome to Embryo: The Regulation of Gene Activity in Early Development

1. INTRODUCTORY COMMENTS

It has been understood for almost a century that the process of development is itself a heritable feature of the organism, and that construction of the organism from the egg is the consequence of genomic expression. Modern measurements demonstrate that early development, including both oogenesis and embryogenesis, involves a huge outlay of genomic information. By the time of fertilization the egg has been equipped with a unique set of biosynthetic capabilities, and in most forms it is endowed with structural polarities that are ultimately reflected in the spatial organization of the embryonic cell lineages. With the current formulation of the molecular mechanisms by which gene expression establishes the properties of differentiated cells, and the development of potent new experimental technologies, the process of embryogenesis has become more accessible. A fundamental explanation would require understanding at the molecular level of exactly what developmental information is encoded in the genome; how it is utilized in morphogenetic time and space; how its expression is regulated; and exactly how its products endow the differentiated cells of the embryo with their functional characteristics. As yet, though we know that they exist, most of the crucial causal links between genome and embryo remain largely undescribed.

The objective of this book, as of the two previous editions (1968, 1976), is to provide a meaningful interpretation, and critical review, of those aspects of the molecular biology of the embryo that reflect on the role of genomic information in early development. The areas considered extend from gene expression during oogenesis to the establishment of an asymmetric pattern of lineage-specific gene expression in the embryo. The depth and the variety of the relevant literature presents a formidable challenge. It is easy for the reviewer to feel that less *knowledge* has been extracted than is actually available in the mass of current experimental data.

The molecular mechanisms required to produce a differentiated animal embryo are anything but obvious. The paradigms that emerge from molecular analyses of terminal cell differentiation, or of systems that do not undergo embryonic development, provide essential information as to molecular mechanisms of animal cell function, insights into the differentiation process, and comparative points of reference. However, as more is learned it becomes increasingly evident that to understand early development it is necessary to study oocytes, eggs and embryos. Developmental problems that are intrinsic to embryogenesis include the creation of a three-dimensional cellular morphology where there was none before; the determination *ab initio* of cell lineage precursors; the requirement for active biosynthesis within an enormous mass of cytoplasm that at first contains very few nuclei; the need to produce large numbers of new cells at a rate higher than at any other time in the life cycle (at least in some organisms), and within a constant volume of cytoplasm; and so forth. A fundamental difference from later cells, at least in higher animals, is that the genomes of the early blastomeres are "naive," and in some organisms demonstrably totipotent. During the ontogeny of the lineages leading to given differentiated cell types interactions with other molecules occur that stably, if not irreversibly, distinguish genes never to be utilized in these cells from genes whose expression is or will be required. This deep-seated difference between the adult and early embryonic genomic apparatus can be shown experimentally, both by gene transfer and by nuclear transfer, as discussed later in this chapter.

Observations on four experimental systems account for most of our current knowledge of oogenesis and early development. These are *Xenopus*, sea urchins of various species, the mouse, and *Drosophila*. For certain areas other organisms have proved extremely valuable as well, such as the nematode *Caenorhabditis*, ascidians of various species, ctenophores, gastropod molluscs, etc. Each system has its strong points and its weak points as an experimental object, and it is impossible to obtain anything but a partial view of the processes of oogenesis and early development through the lens that any one system provides. The approach taken in the following review is thus comparative. By this route it becomes evident that many of the special devices utilized during early development can be interpreted in terms of their adaptive value, given the special biological constraints to which each species is exposed, for example the time available for oogenesis or embryogenesis, and the conditions in which these processes must occur. Thus in different organisms there is significant variation in the relative importance even of basic, common phenomena such as cell-cell interaction, the initial spatial organization of the egg cytoplasm, the utilization of maternal transcripts, and many other particular mechanistic features, as will become apparent in following chapters.

2. SYNOPSIS OF MAJOR THEMES

Chapter I: From Genome to Embryo: The Regulation of Gene Activity in Early Development

The modern theory that different sets of genes are active in different cells developed originally from classical studies that were focussed on the role of the genome in embryogenesis. It was concluded that early embryo nuclei are functionally equivalent, and by the first decades of the 20th century leading investigators had correctly perceived the determinative significance of genomic expression during oogenesis and embryogenesis. Current evidence regarding genomic equivalence among differentiated cells later in development includes examples in which alteration of the primary gene structure is required for expression, e.g., in the vertebrate immune system, but in most genes examined the DNA itself appears to remain unchanged during ontogeny. Evidence from cell fusion experiments, and from regeneration and other examples of transdifferentiation, suggests that genes normally destined to remain silent in differentiated cells are retained intact and can be reactivated. When reimplanted into eggs, differentiated cell nuclei of amphibians and several other groups display the ability to direct embryogenesis and the formation of many differentiated larval tissues. Particular genes are activated and others repressed on introduction of somatic nuclei into eggs and oocytes. Somatic cell and germ line transformation experiments have confirmed that developmental activation as well as inductive modulation of genes functional in highly differentiated cell types ("late genes") are processes usually mediated by interactions of *trans*-regulators with *cis* sequences located in the vicinity of the gene. Late genes are probably engaged in cell-heritable, repressive chromatin complexes early in development, and remain so in all cell lineages save those requiring their expression. Activation of the large set of genes utilized very early in embryogenesis may be mediated by factors present in egg cytoplasm, and in some organisms these genes may also be premarked during gametogenesis for expression in the early embryo.

Chapter II: The Nature and Function of Maternal Transcripts

RNAs synthesized during oogenesis support all or most of the biosynthetic activities carried out in early embryos. The utilization of these transcripts is triggered by maturation or fertilization, and they persist until replaced by zygotic transcripts emanating from the blastomere nuclei. Maternal transcripts include messenger RNAs, ribosomal RNAs, various low molecular weight RNAs, and a prominent class of large nontranslatable poly(A) RNA molecules, the significance of which is yet unknown. The complexity of genomic information represented in the maternal RNA is high, relative to that expressed in many somatic cells, and is almost the same in the eggs of many different species, irrespective of genome size. The extent of dependence on maternal transcripts both in real and in developmental time varies according to species, and to the particular genes considered. A number of instructive examples of such genes have now been cloned and the disposition of their transcripts quantitatively characterized. The developmental functions for which maternal mRNAs code are illuminated by genetic, molecular, and cytological evidence. Among these functions are the provision of specific embryo nucleoproteins, cytoskeletal proteins, cell surface proteins, and enzymes. Maternal transcripts may also code for regulatory proteins that affect blastomere lineage fate.

Chapter III: Transcription in the Embryo and Transfer of Control to the Zygotic Genomes

The genomes of early embryos characteristically express a set of sequences that are largely the same as those transcribed during oogenesis. The pattern of transcription changes as some new genomic sequences are activated and some genes expressed early are repressed. The level of expression of any given gene that is utilized in early development, i.e., the quantity of cytoplasmic message of that species, is given by the amount of surviving maternal transcript, the rate of flow into the cytoplasmic compartments of newly synthesized embryo transcripts, and their rate of turnover, which in the embryo varies sharply among different mRNA species. The stage of development at which newly synthesized embryonic gene products become dominant with respect to maternal transcripts varies according to the organism, and when closely examined this crucial switch occurs at different times for each individual gene or functionally related set of genes. The major embryonic transcript class is high complexity nuclear RNA (nRNA), >90% of which never accumulates outside the nuclear compartment. Measurements of the kinetics of nRNA biosynthesis and decay, and of nRNA population complexities, show that when considered with respect to genome size the numbers of nRNA species transcribed, the synthesis rates, the turnover rates, and the fractions of nRNA exported as processed message are very similar in sea urchin, *Drosophila*, and *Xenopus* embryos. Over 10^4 transcription units appear to be utilized in sea urchin embryos. Thus early development requires an enormous amount of genomic information, as well as a complex regulatory apparatus to direct its expression. Ribosomal RNA genes are highly repressed in invertebrate and lower vertebrate embryos that develop as closed systems, i.e., without net growth, but they function actively from early cleavage onward in mammalian embryos. Many specific genes, the expression of which is not restricted to given cell types, are in the early embryo subject to ontogenic programs of control, while later, after an adult nucleus-to-cytoplasm ratio is attained, they are controlled in response to physiological cues. The best known are the histone genes, particularly in the sea urchin, where separate sets of genes code for the histones utilized at a high rate early in development. Additional examples are afforded by other genes that in adult cells are under cell cycle regulation, e.g., those coding for ribosomal proteins; and by environmental response genes known to be active in some embryos, e.g., metallothionein and heat shock genes.

Chapter IV: Differential Gene Function in the Embryo

Biochemical and cytological evidence indicates that after an initial phase of cell division there occurs the onset of active differentiation in particular regions of the early embryo. By differentiation is meant the imposition of specialized programs of cytoplasmic biosynthesis directed by zygotic mRNAs. The biological and molecular processes leading to the appearance of differentiated cells in *Caenorhabditis*, sea urchin, *Xenopus*, and *Drosophila* embryos are considered in some detail. In *C. elegans* cell lineage is largely invariant, and with a few exceptions the fate of each cell is determined autonomously within its embryonic lineage. In some lineages all cellular progeny carry out the same function, though in most functionally asymmetric cleavages give rise to arrays of cells displaying different states of differentiation. Lineage-specific genes coding for gut proteins and muscle proteins, among others, have been cloned and their expression under various conditions demonstrates the autonomous character of ontogenic gene activation within the lineage. Mutations are known that affect various aspects of lineage determination processes, providing insight into the regulatory architecture of the genomic control system operating in the development of this organism. The sea urchin embryo contains some lineages that similarly develop in an autonomous, and fully determinate manner, e.g., that giving rise to skeletogenic mesenchyme cells, but many of the embryonic cell types require inductive interactions for their specification. Development occurs in several stages. Some cells expressing specific sets of genes appear in the blastula stage embryo, which however, also includes a large group of pluripotential blast cells that differentiate following gastrulation. A relatively large number of genes that serve as markers for the early differentiated cell types of this embryo have been isolated and cloned. These include genes that function specifically in aboral ectoderm, muscle, gut, and skeletogenic mesenchyme cells. Specification of the progenitors of each cell type occurs long before the expression of lineage-specific genes, and gene transfer experiments suggest that their initial embryonic activation is accomplished by the presentation within given lineages of *trans*-activators. In *Xenopus* embryos differentiated cells appear only after transcription is activated at the midblastula stage, when thousands of cells have already been formed. Inductive interactions are required for the spatial organization of the embryo, though there is clear evidence as well for the influence of localized maternal cytoplasmic constituents. Differentiation occurs regionally, and is not dependent on cell lineage, which is indeterminate. Neuronal differentiaton and likely that of many other cell types appears instead to be probabilistic. Several cloned genes expressed specifically in ectoderm and in somitic mesoderm provide markers for analysis of factors leading to determination. In *Drosophila* differentiation is also regional, and embryonic lineage is indeterminate. A unique aspect is the isolation and analysis of many genes that function in the organization of large elements of morphological pattern. Genes that affect neurogenesis, genes that are required for metamerization of the embryo, and genes that define segmental identity have been cloned and their spatial and temporal modes of expression correlated with the morphological effects of mutations. Some such genes are activated early in development, prior to cellularization of the blastoderm. Initially they respond to spatially localized, maternal cytoplasmic factors. They then function autonomously in the embryonic cells, and in some examples have been shown to code for proteins localized in the cell nuclei. Together with their pleiotropic effects with respect to the individual states of differentiation they affect, this observation, and the extent to which these genes appear to interact, suggest upper level regulatory functions. Comparatively, this review emphasizes the diversity of the fundamental genomic strategies utilized to build differentiated structures in embryos of the four types considered.

Chapter V: Gene Activity during Oogenesis

Maternal mRNAs, and the proteins translated from them, are the physical entities implied in the classical concept that genomic functions required for early development are expressed during oogenesis. In diverse organisms the logistic strategies utilized in the accumulation of the maternal RNA pools can be interpreted in terms of adaptive factors, such as the size of these pools and the amount of time available for oogenesis. Three modes of oogenesis can be distinguished. In meroistic oogenesis, known mainly in some insect groups, polyploid nurse cells, which are of germ line origin, provide most of the maternal transcripts ultimately stored in the oocyte. In *Drosophila* molecular evidence obtained with several cloned genes the products of which accumulate during oogenesis confirms that the nurse cells provide oocyte maternal RNAs, and genetic evidence identifies several germ line functions required for completion of the oocyte-nurse cell complex. In both vertebrate and invertebrate organisms that have eggs with relatively enormous pools of maternal RNAs, but in which the oocyte synthesizes its own RNA, all functional transcriptional units in the 4C oocyte nucleus are operated at a high rate. This results in the appearance of "lampbrush chromosomes," the loops of which are the sites of intense RNA synthesis. The evidence suggests that a primary function of lampbrush chromosomes is to provide rapid flow of maternal transcripts to the cytoplasm. Several specific gene activities have been characterized, the best understood of which is the transcription of the 5S rRNA genes. These genes are activated early in oogenesis and are controlled by a protein that binds to 5S rRNA. In organisms that have relatively small maternal RNA pools, such as mammals and sea urchins, lampbrush chromosomes are evidently absent, and the maternal cytoplasmic RNA is accumulated at a more leisurely rate.

Chapter VI: Cytoplasmic Localization

The establishment of topologically localized, diversely committed cell lineages is the major outcome of the initial processes of development. The location of specific embryonic structures is often foreshadowed in the orientation of the egg, and in the distribution of egg cytoplasmic components before or during very early cleavage. In some examples the fate of the lineages descendant from given blastomeres is predictably determined by the section of egg cytoplasm that each inherits. This phenomenon is termed *cytoplasmic localization*. This subject was extensively investigated by classical experimentalists. The major viewpoints that emerged from their work have profoundly influenced modern perceptions of the localization phenomenon. In most organisms studied, specification of some early lineages requires particular interactions between adjacent blastomeres, while the fate of other lineages in the same embryo is at least partly determined through cytoplasmic localization. It is doubtful that in the mammalian egg there exists any localization of developmental potential, with respect to the parts of the embryo proper. Localization can be perceived as an adaptive mechanism that enhances the rapid organization of a feeding, free-living larval form. Much of the evidence regarding embryonic localization phenomena has been obtained by microsurgical procedures, or has been adduced from cytological and genetic data. Some relevant molecular evidence that indicates a role for maternally stored RNAs have been obtained for insect eggs. Localization phenomena in many diverse biological systems are considered, including nematode worms, ctenophores, molluscs, ascidians, annelids, sea urchins, and amphibians, and in each group the primary embryonic axes, and the location of given embryonic cell types, are derived from the primary organization of the egg by particular processes. The current state of knowledge provides many opportunities for future molecular analyses of this little understood but fundamental aspect of embryogenesis.

3. HISTORICAL ANTECEDENTS: A VERY BRIEF SUMMARY OF THE ORIGINS OF THE VARIABLE GENE ACTIVITY THEORY OF CELL DIFFERENTIATION

Cell differentiation is today explained as the result of the regulated expression of specific sets of genes in the various cell types of an organism. The antecedents of this powerful modern theory are to be found in late 19th century attempts to understand the role of the genome in embryological development (the term *genome* is used herein to denote the total complement of genetic material in the cell nucleus). This problem, which is essentially the subject of the present volume, emerged initially upon the discovery during the early 1880's that all cells contain genetic material in their chromosomes. The evidence was derived from studies of fertilization, meiosis, and mitosis, and until the "rediscovery" of Mendel's experiments by Sutton (1903) and Boveri (1904) the chromosome theory of cellular inheritance was based primarily upon cytological observations. Among the most persuasive studies were those of Van Beneden (1883) on pronuclear fusion and formation of the zygote genome in *Ascaris*, where the process is easily observed because there are only two large chromosomes per haploid complement. Several figures from this study are reproduced in Fig. 1.1. Van Beneden's drawings demonstrate that the haploid chromosome sets contributed by the two pronuclei together constitute the diploid zygote nucleus, with two copies of each individual chromosome, and also show clearly the replication of the diploid set in the nuclei of the first two blastomeres. Within the next several years it was proposed explicitly in the writings of Weismann (1885), Hertwig (1885), Nägeli (1884), Strasburger (1884) and others that the fundamental result of fertilization is the formation of the diploid genome of each organism from the haploid parental genomes, that the genomic determinants are located in the chromosomes, and that these determinants control all the characteristic properties of the organism at the cellular level.

The chromosome theory of cellular determinants presented a new logical challenge for the interpretation of development, in that it became necessary to explain how only a portion of the heritable functions presumably encoded in the zygotic genome are ultimately expressed in each cell type. A solution immediately proposed by Roux (1883) and by Weismann (1885, 1892) was that differentiation of cell function results from the partition of qualitatively diverse genetic determinants into different cell nuclei. Thus, each cell would contain in its nucleus only those genes needed for its particular set of functional activities, so that developmental specialization would stem from the establishment of a mosaic of diverse partial genomes. Experiments designed specifically to test this proposition were carried out on sea urchin embryos by Driesch (1892), on the embryos of a marine annelid, *Nereis*, by Wilson (1896a), and later on other organisms (reviewed in Morgan, 1927). In these experiments the normal distribution of cleavage stage nuclei in the egg cytoplasm was transiently altered by forcing cleavage to occur in two dimensions under the pressure of a flat glass plate. On removal of the plate given nuclei are found to be partitioned into cells other than those normally inheriting them. Nonetheless, normal differentiating larvae could develop from these embryos. Reviewing these experiments, Wilson (1925) concluded:

> Specification of the blastomeres *cannot, therefore, be due to specific nuclear differences produced by a fixed order of qualitative nuclear divisions, but must be sought in conditions of the oöplasm* [i.e., egg cytoplasm; his italics].

The pressure plate experiments, and demonstrations that individual blastomeres and in some species partial embryos can give rise to complete larvae (see Chapter VI), showed that *any cleavage stage nucleus retains all the zygotic genes*, in a usable form.

Though logical arguments pointed strongly to the view that nuclear determinants must somehow control early development, there was little direct experimental evidence until Boveri (1902) showed that sea urchin embryos lacking a complete set of chromosomes in every cell fail to develop normally. An early theory of De Vries (1889), which except for his terminology seems amazingly familiar, held that all differentiation is due to molecular elements called *pangens* that migrate from the nucleus to the cytoplasm, where they exercise their function of determining the character of the cell. On the other hand, a number of elegant demonstrations indicated that at least in some organisms, the region of egg cytoplasm inherited by cells of the

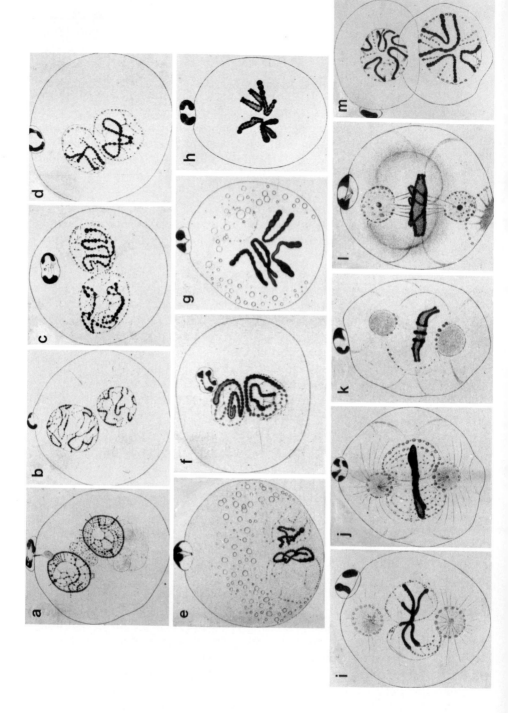

early embryo during cleavage can determine the fate of the lineage descended from it. A resolution of the paradoxical conflict between the primacy of the hereditary elements in the nucleus and the determinative role of preformed cytoplasmic components in the egg was achieved by the insight that egg cytoplasm is itself the product of prior gene expression that occurs during oogenesis (Whitman, 1895a; Wilson, 1896b; Boveri, 1910, 1918). From these origins arose the classical solution to the problem of genome function in development, which though unavoidably vague, was an important step in the right direction. As enunciated by Wilson (1896b) in a discursive review of this problem:

> If chromatin be the idioplasm [i.e., the name given by Nägeli to the genetic material] in which inheres the sum total of hereditary forces, and if it be equally distributed at every cell division, how can its mode of action so vary in different cells as to cause diversity of structure, *i.e., differentation*? [his italics] ...The central point of my own conception ... and of Driesch's (1894) views ... is as follows. All the nuclei are equivalent and all contain the same idioplasm ... Through the influence of this idioplasm the cytoplasm of the egg, or of the blastomeres derived from it, undergoes specific, progressive changes, each change then reacting upon the nucleus and thus initiating a new change. These changes differ in different regions of the egg because of pre-existing differences ... in the cytoplasmic structure ... such as the distribution of different substances in the [egg] cytoplasm.

In the succeeding decades genetic and other experimental evidence that genomic determinants indeed control the details of the developmental process became overwhelmingly convincing. Yet there seems to have been no further conceptual advance towards the basic problem of explaining *differential* cell function until 1934, when there appeared in an essay of T. H. Morgan an explicit statement of the theory that differentiation could be caused by *variation in the activity of genes in different cell types*:

> The implication in most genetic interpretations is that all the genes are acting all the time in the same way ... An alternative view would be ... *that different batteries of genes come into action as development proceeds*. ... Roux and Weismann attempted to explain development in somewhat this way, by assuming that the determinants in the chromosomes are qualitatively sorted out during development ... [but] ... there is now much evidence that is opposed to this view ... The idea that different sets of genes come into action at different times ... [requires that] some reason be given for the time relation of their unfolding. The following suggestion may meet the objections. It is known that the protoplasm of different parts of the egg is somewhat different ... and the initial differences may be supposed to affect the activity of the genes. The genes will then in turn affect the protoplasm ... In this way we can picture the gradual elaboration and differentiation of the various regions of the embryo.

Fig. 1.1.   Successive stages of pronuclear fusion, formation of the zygote genome, and first mitosis in *Ascaris*, as given by Van Beneden. The pronuclear chromosomes become visible prior to fusion (a-f). The haploid chromosome number is two, and thus four chromosomes are easily visible in the newly formed zygotic genome (g). Spindle formation and first metaphase ensue (h-k), followed by anaphase (l). The result is the creation of two blastomere nuclei, each containing the four chromosomes of the diploid complement (m). [From E. Van Beneden (1883). *Arch. Biol.* **4**, 265.]

The lasting influence of the classical interpretation of the role of the genome in development is here evident. Both Wilson (1896b) and Morgan (1934) focussed on nucleocytoplasmic *interaction* as the source of what we would today call *trans*-acting signals causing variation in gene expression, and for both the early embryo was the primary exemplar.

In its modern form the variable gene activity theory of cell differentation can be considered to date from the early 1950's, when it began to be developed in molecular terms by Brachet (1949), Sonneborn (1950), Stedman and Stedman (1950), Mirsky (1951, 1953), and others. This advance was stimulated by the new observation that within the accuracy of measurement the differentiated diploid cell types of an organism contain the same total quantity of DNA, which is just twice the content of individual sperm nuclei (Boivin *et al.*, 1948; Mirsky and Ris, 1949, 1951). The DNA content measurements implied that differentiated *adult* cells all retain the complete DNA genome, thus reinforcing the speculation that a general explanation for differentiation might be sought in the ontogenic variation of gene expression.

In this book variable regulation of genome function during oogenesis and embryogenesis is treated as the underlying causal process in early development. The development of a differentiated embryo from an undifferentiated egg is a spectacular biological process, and it is an interesting reflection that it was initially the attempt to understand the outlines of this process that led our predecessors toward the view of gene regulation that we now accept.

4. IS THERE STRUCTURAL AND FUNCTIONAL CHANGE IN THE GENOME DURING DEVELOPMENT?

(i) Ontogeny and the Structure of Specific Genes

Classical experiments established satisfactorily that the *genomes* of early blastomeres are equivalent and totipotent, even in organisms where cleavage is determinate and the cell lineages descendant from given blastomeres display invariant and diverse fates. However, these experiments demonstrate genomic equivalence only for embryonic cells, which have yet to display differentiated characteristics. Cytological evidence from studies of dipteran polytene chromosomes suggested that the complete genome may be retained in differentiated adult cells that express only a portion of the overall genetic capabilities. For example, Beerman and Clever (1964) demonstrated the same banding pattern in polytene chromosomes of four different tissues in *Chironomus*. In addition many early molecular hybridization measurements, carried out on mammalian, amphibian, and echinoderm species, showed that the overall single copy and repetitive sequence complements in the genomes of diverse cell types are indistinguishable in a given organism (reviewed in Davidson, 1976, Chapter I). In other systems, however, it is clearly not true that the genomes of all cells are equivalent, even at a gross level. For example, in *Ascaris*, as in several other species, a large fraction of the chromatin is expelled from the genomes of all somatic cell progenitors during cleavage, but not from germ line progenitors, as first shown by Boveri (1899) (see Chapter VI). The issue of genomic equivalence among fully differentiated cells is a matter of fundamental interest since ontogenic mechanisms that lead to structural alteration of expressed genes, and thus to *nonequivalent genomes*, might be quite different from those that result in genomes that contain exactly similar complements of genetic material and yet transcribe different sequences. The most direct evidence regarding this problem comes from comparisons of the structure of specific genes, in cell types where they are expressed, and in others where they are not.

Several completely distinct mechanisms have been discovered in studies focussed on the structure of genes expressed in terminally differentiated animal cells. Two classes of change observed in the primary DNA structure of differentially expressed genes are local amplification and genomic sequence rearrangement. In *Drosophila* follicle cells the genes coding for the major chorion proteins are amplified manyfold, in order to provide the requisite rates of transcription (see Chapter V for references). Amplification also occurs at a specific locus in the salivary gland chromosomes of the dipteran *Rhynchosciara angelae* (Breuer and Pavan, 1955). The best known, and most widespread example of this phenomenon is amplification of ribosomal genes, which occurs in oocytes of many different species (reviewed in Chapter V). Mechanisms that can result in amplification of local regions of the genome are apparently common, since amplification occurs randomly at a low frequency in cultured mammalian cells (Johnston *et al.*, 1983) and it occurs regularly in genes providing resistance to a variety of toxic drugs (reviewed in Brown *et al.*, 1983).

In cells of higher vertebrate immune systems activation of the genes coding for immunoglobulins (see reviews in Hood *et al.*, 1981; Leder, 1982; Honjo, 1983) and for T-cell receptors (Hedrick *et al.*, 1984; Chien *et al.*, 1984) occurs following genomic translocations, by means of which the functional genes are assembled. In the somatic cells of the immune system combinatorial translocation is utilized to generate the enormous sequence diversity required of the functional set of immunoprotein sequences. These rearrangements are also the prerequisite for maintenance of normal expression. Thus the translocations that unite variable, constant and joining regions of these genes also bring together crucial elements of the transcriptional mechanism, including the V-region promoter, and in at least some immunoglobulin genes a lymphocyte-specific enhancer located within an intron of the C-region (Gillies *et al.*, 1983; Banerji *et al.*, 1983; Mercola *et al.*, 1983; Grosschedl and Baltimore, 1985). Programmed rearrangements required for gene expression are also known in lower eukaryotes. Thus gene conversion occurs in the yeast mating type locus (reviewed in Nasmyth,

1982), and the genes coding for trypanosome coat proteins are altered during infection by a process that includes direct genomic translocation (see, e.g., Borst *et al.*, 1981).

Should ontogenic gene rearrangements and gene amplification be regarded as special devices that evolved in response to particular and unusual needs, or as *typical* mechanisms in differentiation and development? In each of the cases cited there is indeed an obvious special circumstance, such as the generation of immunological diversity in B-cell and T-cell lineages, or the provision of enormous amounts of chorion proteins within a very short time in the young *Drosophila* egg chamber. Though "special" requirements could exist in other systems as well, observations on many additional genes suggest that in most cases ontogenic gene activation is accomplished in ways that involve neither amplification nor rearrangements of the DNA sequence itself. As first pointed out by Kafatos (1972), and verified in detailed measurements for a number of genes that remain single copy throughout the life cycle, amplification is *not required* to account for the measured rates of mRNA accumulation even in terminally differentiated cells such as those dedicated to the production of ovalbumin (Palmiter, 1973), hemoglobin (Hunt, 1974), or silk fibroin (Suzuki *et al.*, 1972; Maekawa and Suzuki, 1980). With the exceptions mentioned above, in normally differentiating animal cells the copy number of expressed genes is usually not observed to increase, with respect to the same genes in other cell types where they remain silent.

Gross developmental alterations in structure such as the translocations that occur in the immunoglobulin genes are easily revealed by genome blot hybridizations, and a number of genes expressed at high levels in terminally differentiated cells have been examined in this way. For example, comparison by this method of the sequences flanking the chicken ovalbumin gene revealed no differences between oviduct DNA, where the gene is expressed, and erythrocyte or sperm DNA, where it is not (Weinstock *et al.*, 1978). The same is true for the chicken transferrin and lysozyme genes (Nguyen-Huu *et al.*, 1979; Lee *et al.*, 1980), and the *Drosophila* salivary gland glue protein genes (Garfinkel *et al.*, 1983). These genes are present in a single copy per haploid genome in all tissues, and are activated in response to steroid hormones. Similarly, the single copy mouse gene for salivary and liver α-amylase, *Amyl^a*, undergoes no amplification during differentiation, and by genome blot analysis the flanking sequences and organization of this gene remain the same in the tissues in which it is expressed as in sperm DNA (Schibler *et al.*, 1982). Nor do class I mouse transplantation antigen genes undergo rearrangement during differentiation, as shown by comparison of sperm, liver, and embryo DNAs (Steinmetz *et al.*, 1981), though the sequences of these genes identify them as members of the same extended family that includes the immunoglobulin genes (Hood *et al.*, 1982). Considered together the available observations indicate that for the genes that have been investigated neither gene rearrangement nor gene amplification are mechanisms *usually* required for expression in differentiated cells. However, such observations do not generally exclude mutational changes which if they involved only a few nucleotides, or consisted of small deletions, inversions, insertions, or conversions, would be very difficult to detect physically, except by direct sequence analysis. One example where even these kinds of minor changes can be disallowed is the *Bombyx mori* silk fibroin gene, which has been isolated from silk producing cells of the posterior silk gland, and compared in detail with the same gene isolated from cells of the middle silk gland or from pupae, neither of which synthesize fibroin. Tsujimoto and Suzuki (1984) partially purified the fibroin gene from these sources without using cloning procedures, and showed that no differences in sites of methylation are detectable with restriction enzymes in the fibroin genes, and that irrespective of source they are equally active in a cell-free transcription system. Furthermore, the primary DNA sequence throughout the known transcriptional control region, from at least 300 nt before the start of transcription, is identical in genes extracted from fibroin producing and nonproducing cells (Suzuki and Adachi, 1984).

One way in which genes can be activated in animal cells is through the specific binding of positive protein regulatory factors. Among the best known examples are the oocyte type 5S rRNA genes of *Xenopus*, which are ontogenically activated by formation of a complex with known regulatory factors (reviewed in Chapter V); the 70 kd heat shock protein of *Drosophila*, which is inductively activated in a similar manner in response to heat and certain other physiological insults (Pelham, 1982; Pelham and Bienz, 1982; Mirault *et al.*, 1982; Parker and Topol, 1984; Wu, 1984); and genes that are activated by steroid hormones (see, e.g., Payvar *et al.*, 1983; Renkawitz *et al.*, 1984; reviewed in Anderson, 1984; Yamamoto, 1985). Unlike translocational rearrangement or other changes at the DNA sequence level, ontogenic regulation that occurs by means of DNA protein interactions is not inconsistent with the concept that the genomes of differentiated cells are equivalent. Whether bound by activating factors or locked in an inactive complex with histones, the same potentially functional genes would be present in all cells. Unfortunately, precise molecular knowledge regarding the mechanism of developmental gene activation exists for only a very small fraction of the thousands of structural genes utilized during ontogeny. Furthermore, most of what we know concerns only a minority class of genes, those represented in certain cell types by extremely prevalent messages (see review in Davidson and Britten, 1979).

We have so far dealt mainly with the effects of ontogenic gene *activation*. Genes that are specifically active in given cells are repressed in other cells, and there remains the possibility that such repression involves irreversible, primary alterations in gene structure. Some evidence to the contrary comes from DNA transfection experiments reviewed later in this chapter. Thus in several instances genes cloned from the genomic DNA of cells in which they are never expressed have been shown capable of function. Here again, however, knowledge is limited to a small number of examples, and to explore this issue more generally it is necessary to turn to other forms of evidence.

(ii) Activation of Silent Genes in Differentiated Adult Cells

Several kinds of special biological circumstance elicit the activation of new sets of genes not otherwise expressed in given states of differentiation. In order of increasing abnormality, such circumstances include developmental transdifferentiation, regeneration in response to injury or experimental manipulation, and fusion between cells in diverse states of differentiation.

The term *transdifferentation* denotes the functional reorientation of already highly specialized cells in a completely new direction. Various instances of this phenomenon have been reported in insect metamorphosis. Selman and Kafatos (1974) described a clear example of transdifferentiation in the labial gland duct cells of the silk moth *Antheraea polyphemus*. The function of these cells is synthesis of a rigid extracellular cuticle during the pupal stage. At metamorphosis they redifferentiate, and become specialized for salt transport and for the secretion of large amounts of a $KHCO_3$ solution which is utilized as a solvent for the hatching enzyme cocoonase. This functional transformation occurs in nondividing polyploid cells, and Selman and Kafatos (1974) showed that the DNA synthesis associated with continued polyploidization can be suppressed without interfering with the extensive cytodifferentiation and biochemical respecialization that takes place. Another example is provided by the polytene salivary gland cells of *Drosophila melanogaster*. During the third instar the major secretory products of this gland are the glue proteins, but transcription of these genes and synthesis of these proteins ceases abruptly at pupariation, and at the prepupal stage the gland begins to synthesize and secrete a completely different set of proteins (Beckendorf and Kafatos, 1976; Sarmiento and Mitchell, 1982). Transdifferentiation also occurs frequently during organ and tissue regeneration. A famous example known since the 19th century is regeneration of the lens cells of newt eyes from pigmented iris epithelium (Wolff, 1895; earlier studies reviewed in Reyer, 1956). The cells that display this capacity on lens removal are fully differentiated, nondividing melanocytes (Yamada and McDevitt, 1974). Similarly, after amputation of the leg of a urodele, the already differentiated muscle and connective tissue cell types at the site of the injury give rise to a new population of progenitor cells from which derive a variety of internal cell types that populate the reconstructed limb (reviewed in Hay, 1968). Other examples are well known in lower metazoa. In the anthomedusan coelenterate *Podocoryne carea* differentiated striated muscle cells were shown by Schmid and Alder (1984) to transdifferentiate into smooth muscle, plus a variety of nonmuscle cell types, including gland cells, secretory cells, nematocytes, digestive cells, etc. This phenomenon can be demonstrated *in vitro*, beginning with isolated clusters of 200-300 mononucleated striated muscle cells, if these are treated with collagenase to promote partial disaggregation and disorganization of extracellular structures. At least the transition from striated to smooth muscle occurs independently of DNA synthesis.

It can be argued that transdifferentiation does not provide a general insight because all genes that might have to be activated during normal development, and in regeneration, are included in the original, ontogenically specified repertoires of the cell types that display these unusual capacities. However, this caveat does not apply to the phenomenon of transdetermination in *Drosophila* (Hadorn, 1966). When cultured for long periods in the abdomens of adult flies imaginal discs sometimes alter their state of determination, and when transplanted back into a metamorphosing host they manifest programs of differentiation other than those for which they were initially determined. Thus, for example a genital disc may give rise to cultures expressing the cuticular features of antenna, leg, wing, thorax, etc. Cell fusion experiments provide examples of the activation of specific genes in differentiated mammalian cells, the expression of which would under normal circumstances never be required. For example when fused with mouse hepatoma cells human amniocytes synthesize a variety of *human* liver-specific proteins, including serum albumin, transferrin, α-1-antitrypsin, and ceruloplasmin (Rankin and Darlington, 1979), and heterospecific fusion of hepatoma cells with lymphocytes also results in expression of the lymphocyte serum albumin gene and of various liver-specific enzymes (see, e.g., Darlington *et al.*, 1974; Brown and Weiss, 1975). Similarly, fusion of human fibroblasts with murine erythroleukemia cells results in activation of human α- and β-globin structural genes and the accumulation of these globin mRNAs on induction of the hybrid cells with dimethylsulfoxide (DMSO) (Willing *et al.*, 1979). Blau *et al.* (1985) reported the activation of a battery of different muscle-specific genes in the nuclei of human fibroblasts, hepatocytes, and other cell types, when these are fused with mouse myoblasts or differentiated myotubes. The specific activation of these genes by fusion does not require prior DNA replication. Whatever the complex, *trans* acting regulatory processes that may be involved, such results convincingly demonstrate the presence in differentiated cell nuclei of *intact* though repressed genes, which are capable of reactivation on exposure to a different cytoplasmic environment.

(iii) Developmental Capacities of Nuclei Transplanted into Eggs

The ideal experimental challenge to the proposition that differentiated animal cell genomes are intrinsically equivalent to the zygote genome is to implant the nucleus of a differentiated cell into an egg, and determine to what extent it can direct development. This approach is in principle uniquely powerful in that it is the only one that tests functionally the *whole* of the complex set of genes required for embryogenesis and differentiation, excepting those of which the maternal transcripts stored in the egg suffice. There is, however, a basic drawback in the particular difficulty of interpreting negative results. There are many reasons why a nuclear transplantation experiment might fail, i.e., terminate in the arrest of the recipient egg at an incomplete stage of development. There could indeed have occurred structural changes in some fraction of necessary genes during the differentiation of the donor cell nucleus. On the other hand the opposite could also be true, and the DNA sequence of the donor genome might remain identical to that in the zygote nucleus but have participated in reactions with tightly bound protein regulatory agents that are not easily reversible on implantation into the egg. In addition a variety of technical difficulties and incompatibilities between donor nucleus and egg cytoplasm have been uncovered that in essence have little to do with the functional competence of the differentiated cell genome. Nonetheless, a number of at least partially positive results have been achieved in nuclear transplant experiments, and from them derive general inferences that are difficult to obtain by any other route.

Transplantation of Differentiated Amphibian Cell Nuclei into Enucleated Eggs

Briggs and King (1952) demonstrated that nuclei of blastula stage *Rana pipiens* embryos can direct development to advanced stages when implanted into enucleated eggs. This basic method has since been widely utilized to test the capabilities of somatic amphibian cell genomes. From their own initial studies Briggs and King (1957) concluded that nuclei from differentiated postembryonic cells could not support extensive development. However, subsequent experiments demonstrated a significant range of potentialities in nuclei derived from late embryo, larval, and adult tissues. Gurdon (1962, 1963) showed that *Xenopus* tadpoles could be raised from enucleated eggs which had been injected with nuclei derived from differentiated tadpole intestinal cells. A significant fraction of those nuclei able to promote cleavage were also able to give rise to normal swimming larvae. Fertile adult *Xenopus* which were normal in all respects were also raised from eggs injected with intestinal cell nuclei (Gurdon and Uehlinger, 1966). At least some of the donor intestine cell genomes evidently retained the capacity to direct development of other organ systems. These positive results cannot easily be explained as the consequence of inadvertently injecting a few undifferentiated rather than differentiated larval intestine nuclei, since undifferentiated cells appear to be absent from the larval gut at the stage utilized for the cited nuclear transplantations (Marshall and Dixon, 1977).

Advanced larvae have also been obtained from enucleated eggs implanted with nuclei from a variety of *fully differentiated adult cells* (reviewed in DiBerardino *et al.*, 1984). In Fig. 1.2 is illustrated a convincing example, which demonstrates that adult skin cell nuclei retain functional genes for the construction of many normally differentiated cell types, and for overall larval morphogenesis (Gurdon *et al.*, 1975). Postneurula larvae and hatched tadpoles have been reported to develop from enucleated eggs that had been implanted with nuclei from primary cultures of adult *Xenopus* kidney, heart, lung, and testis cells (Laskey and Gurdon, 1970); from cultured melanophore cells (Kobel *et al.*, 1973); from adult *Xenopus* lymphocytes (Wabl *et al.*, 1975); from intestinal epithelium cells of late tadpoles, isolated long after feeding has begun (Marshall and Dixon, 1977); and from adult *Rana* erythrocytes (DiBerardino and Hoffner, 1983). It can be concluded that the ontogeny of the donor cell types *does not necessarily involve any loss, irreversible inactivation or permanent alteration in the sets of genes needed for gastrulation, neurulation, or the formation of many specialized cell types*, including functional muscle, nerve, hematopoietic systems, eye lens and retina, heart, pronephric tubules, etc.

There remains the difficulty of interpreting the low frequency with which advanced larvae develop when the donor nuclei are derived from late embryo, larval, or adult cells. This phenomenon has been reported in virtually every study of amphibian somatic cell nuclear transplantation. Postmetamorphosis adult frogs are obtained even more rarely in such experiments (see, e.g., Gurdon and Uehlinger, 1966; Kobel *et al.*, 1973). It is possible that the frequency of success in fact has little to do with the regulatory state of the genome, but rather reflects experimental and physiological problems that are more difficult to overcome in transfers of differentiated cell nuclei than of blastula stage embryo nuclei. Thus nuclei from differentiated cells might be damaged more easily in transfer experiments, for instance if they are more extensively engaged in cytoplasmic membranes; or they might be unable to respond quickly enough to the early cleavage requirement for very rapid cell division. There is considerable evidence to suggest that factors such as these are indeed important. For example it was found that while only a few percent of initial nuclear transfers from differentiated donor tissues result in highly developed larvae, the probability of success was increased to 24% by using a two-step procedure in which the initial transfer recipients are allowed to proceed to the blastula stage, and nuclei from these blastomeres are then transferred back into a new set of enucleated recipient eggs (Gurdon, 1963). In the experiments shown in Fig. 1.2, only 22% of the initial transfer recipients receiving presumably identical skin cell nuclei [see Fig. 1.2(a)] yielded partially cleaving embryos. Eleven *clones* of transplant embryos were then prepared by implantation of the nuclei of these partial blastulae into enucleated eggs, and four of these clones included heartbeat

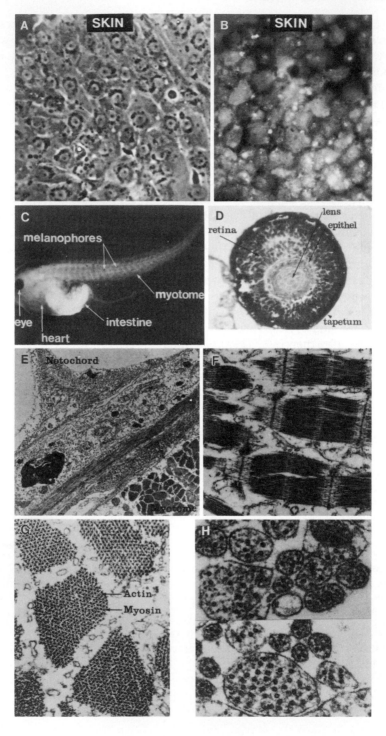

stage larvae that developed advanced differentiated structures such as shown in Fig. 1.2(c-g). It is significant that *within each clone* a wide range of results is obtained, with 70% of individual eggs again failing to cleave or producing only partial blastulae (see also Subtelny, 1965; DiBerardino and King, 1967). It follows that there exist major experimental variables that are independent of the intrinsic state of differentiation of the donor nuclei. Among the factors known to affect the level of success are the medium in which the nuclei are transferred (Hennen, 1970) and the proliferative state of the cells from which the donor nuclei are obtained. Kobel *et al.* (1973) showed that nuclei taken directly from nondividing tadpole melanophores are ineffective in promoting development compared to nuclei taken from rapidly proliferating cultures of the same melanophores, and similarly Brun (1978) found that eggs receiving nuclei of nondividing erythrocytes fail to proceed beyond gastrulation, while eggs injected with dividing erythroblast nuclei give rise to neurulae and some imperfect tadpoles as well. Further experiments with noncycling adult *Rana* erythrocyte nuclei (DiBerardino and Hoffner, 1983) demonstrated that even these nuclei can be reactivated, if injected into *oocytes* that are then induced to mature, rather than directly into already mature eggs. The blastulae and partial blastulae deriving from oocytes receiving erythrocyte nuclear transcripts were disaggregated, and individual nuclei then implanted back into enucleated mature eggs. Almost half of these eggs developed beyond the gastrula stage, compared to none

---

Fig. 1.2. Development of specialized tissues directed by nuclei derived from a differentiated adult *Xenopus* skin cell. (a) Donor skin cells in culture. Cultures were monolayer outgrowths of footweb cells from adult *Xenopus* of the 1-nucleolus (*1-nu*) genotype. (b) Fluorescence photomicrograph of same culture as in (a), after exposure to fluorescent antibody against frog keratin. Every cell binds the antibody. No fluorescence was obtained with this antibody on exposure to a primary culture of lung cells from a *1-nu* animal. (c) Tadpole with differentiated heart, melanophores, intestine, myotomes, eyes, etc., obtained from enucleated egg implanted with a nucleus derived indirectly from a cultured, keratin producing skin cell such as shown in (a) and (b). Eggs into which the skin cell nuclei were initially implanted were allowed to cleave. Five percent of these eggs cleaved completely, 25% cleaved partially, and 70% failed to cleave. One of the partially cleaving embryos was dissociated, and nuclei from individual blastomeres were reimplanted into enucleated eggs. This serial transfer procedure eliminates nuclei that are damaged or that for some other reason initially lack the ability to divide. Of the serial transfer recipient eggs, 40% failed to cleave, 30% cleaved partially, and 30% formed normal blastulae. The tadpole shown derived from a blastula of the latter category, and its descent from the original donor nucleus was authenticated by observation of the *1-nu* marker phenotye. (d) Cross-section through the eye, showing lens, lens epithelium, neural retina, and pigmented tapetum. (e)-(h) Ultrastructural evidence of cell differentiation in a skin cell nuclear transplant tadpole similar to that shown in (c). (e) Organized muscle in the myotomes. (f) Longitudinal section through a body muscle, showing characteristic striated muscle fine structure. (g) Transverse section through original tail muscle showing thick myosin filaments surrounded by actin filaments. (h) Transverse sections through nerve chord showing axons with microtubules. [From J. B. Gurdon, R. A. Laskey, and O. R. Reeves (1975). *J. Embryol. Exp. Morphol.* **34**, 93.]

when the initial transfers were made directly into mature eggs, and 10% proceeded to advanced swimming tadpole stages. This result shows clearly the significance for successful reactivation of somatic nuclei of cytoplasmic factors, e.g., the cell cycle factors that in the host oocyte determine the progression from interphase to meiotic (or mitotic) metaphase (see, e.g., Newport and Kirschner, 1984). It also serves as a reminder that in nuclear transplantation experiments only positive results can be interpreted with direct reference to developmental potentiality.

An additional common cause of failure in nuclear transfer experiments is the generation of chromosome abnormalities (reviewed in DiBerardino, 1979). Aneuploid cells and ring chromosomes, translocations, acentric fragments, etc. are often found. Chromosomal abnormalities apparently occur far more frequently in transfers involving nuclei from differentiated cells. The cell cycle phase of the donor nucleus may be important. Thus von Beroldingen (1981) found that when nuclei from cultured *Rana* cells that were in G_1 are implanted into enucleated eggs only 4% of recipients display a normal karyotype, while 25% of recipients receiving S, or G_2 nuclei retain grossly normal karyotypes. As discussed in Chapter II, blastula stage frog embryo nuclei, which on implantation produce normal embryos at relatively high frequency, are engaged in a simplified cell cycle lacking both G_1 and G_2 phases.

Conclusions that may be drawn from the amphibian nuclear transplantation experiments can be summarized as follows. Most genes needed for embryogenesis, and for the morphogenesis and cellular differentiation of advanced larvae, are clearly not altered irreversibly during somatic cell specialization. In respect to these genes differentiated cell nuclei and the zygote nucleus are indeed equivalent. It cannot be excluded that some genes required late in development are indeed irreversibly altered, accounting for the failure, other than in a very few cases, to obtain tadpoles capable of metamorphosis from eggs implanted with somatic nuclei. However, the importance of the state of proliferation of the donor cells, the high incidence of chromosome abnormalities, and the striking quantitative improvement that results from preselection of nuclei able to undergo cleavage all suggest that such a conclusion is probably unwarranted. A key observation is that the same range of developmental results is obtained on implantation of *spermatocyte* nuclei as on implantation of differentiated *somatic cell* nuclei (DiBerardino and Hoffner, 1971). Nuclei of differentiated germ cells can have undergone no irreversible changes in their unexpressed genes, and yet their totipotency is not demonstrated by nuclear transplantation. A high percentage of nuclei extracted from mitotically active primordial germ cells displays the capacity to promote development of swimming tadpoles, however (Smith, 1965). Perhaps the frequency with which advanced larvae develop from eggs implanted with somatic nuclei simply measures the improbability of complete reversal during the brief exposure to egg cytoplasm of all the specific associations with regulatory and structural proteins in which the donor genome had engaged during ontogeny. Associations of this kind might interfere in some manner with the formation of the frequent replication complexes utilized during cleavage, resulting in partially replicated genomes and broken chromosomes.

Developmental Potentiality of Transplanted Nuclei in Other Organisms

The ability of early embryonic nuclei to substitute for the zygote nucleus has been confirmed by nuclear transplantation into eggs in several urodele species (see Etkin and DiBerardino, 1983, for references) and in fish as well. Gasaryan et al. (1979) showed that blastula nuclei can promote development of feeding larvae when implanted into enucleated eggs of the teleost *Misgurnus fossilis*. A report of Tung et al. (1977) demonstrates in addition that nuclei from late gastrula stages of the ascidian *Ciona intestinalis* direct development of many highly differentiated tissues when implanted into enucleated eggs. This experiment is noteworthy because the ascidian embryo has a determinate pattern of cleavage in which each cell lineage displays a fixed fate (see Chapter VI). Thus nuclei of determined cells are shown to retain a developmental potential that in range vastly exceeds their normal fate. It follows from this result, as from the 19th century pressure plate experiments (see above), that early specification of cell fate in eggs with rigidly defined lineage assignments is determined by cytoplasmic factors acting on essentially totipotent embryo nuclei.

Early *Drosophila* embryo nuclei also display totipotency when implanted back into eggs (Illmensee, 1972; Zalokar, 1973; Okada et al., 1974a). Here the recipient egg nucleus or nuclei are not removed or destroyed as in the amphibian experiments, since both the donor and recipient nuclei can be marked genetically. These experiments show that irrespective of site of origin (and hence the embryological fate) of donor blastoderm or preblastoderm nuclei, their descendants are capable of participating in all forms of adult differentiation, including the production of fertile gametes. A series of experiments in which nuclei from already determined cellular blastoderm and gastrula stage *Drosophila* embryos were injected into unfertilized eggs was reported by Illmensee (1976). These nuclei retained the same developmental potentialities as did cleavage stage nuclei. Thus 10-30% of recipient eggs developed to the stage of organogenesis, and a smaller fraction proceeded beyond hatching. Nuclei from several cultured cell lines failed to participate even in cleavage if transplanted into unfertilized eggs, but when injected into cleavage stage embryos these nuclei also were able to direct the development of many different cell types, as shown by histochemical detection of mosaic tissues in the resulting larvae and flies (Illmensee, 1976). Thus in *Drosophila*, as in amphibians, the genomes of differentiated and committed cells retain in a potentially usable form a wide range of developmental capacities.

A very different outcome is reported for nuclear transfer experiments into mouse eggs. These eggs can be enucleated and donor nuclei introduced into them by a relatively nontraumatic method (McGrath and Solter, 1983). The donor nuclei are sucked out enclosed in a cytoplasmic vesicle, and are implanted under the extracellular coat (*zona pellucida*) of a zygote previously enucleated by the same technique. Fusion is then effected with inactivated Sendai virus. The plasma membranes of neither donor nor recipient cells are penetrated with the micropipette throughout the procedure. McGrath and Solter (1985) showed that with this method over 96% of recipient eggs develop at least to blastocyst stage when the zygote nucleus is replaced by another zygote nucleus. However, the fraction developing to blastocyst stage drops to 10% when the donor nuclei are obtained from 2-cell embryos, and to zero with donor nuclei from 4-cell or 8-cell embryos or the inner cell mass of blastocyst stage embryos. As discussed in Chapter VI, at none of these stages are the cells of the mouse embryo determined, since any cell may form any part of the future embryo. Thus the failure of the nuclear transplant eggs to develop even to early embryonic stages cannot be attributed to the existence of a committed state on the part of the donor nuclei. By the 2-cell stage the nuclei of mouse embryos have already begun to express a new pattern of genomic activity, while in the amphibian embryo there is no genomic activity at all until the blastula stage (see Chapter III), and soon after this regionally specialized patterns of gene activity begin to develop. Yet, as we have seen, nuclei from postembryonic amphibian tadpole cells as well as from differentiated adult tissues display a wide range of developmental potentialities in transplantation experiments. It thus appears unlikely that the precocious (in developmental time) expression of a new embryonic set of genes at the 2-cell stage in the mouse is what is responsible for the inability of 2-cell and later cleavage stage embryo nuclei to support development. Furthermore, McGrath and Solter (1984a) noted that 2-cell embryo nuclei successfully direct development of complete blastocysts if injected into enucleated *2-cell embryos* rather than into zygotes. The difficulty in the zygotic nuclear transfers is by this observation localized to some aspect of the recycling of the donor nuclei back through the 1-cell stage, rather than to any sort of fundamental early restriction of developmental competence. This result is similar in essence to those noted earlier, in which frog erythrocyte nuclei fail to function if transferred to enucleated eggs, but perform as do other differentiated cell nuclei when transplanted to oocytes that are then permitted to mature (DiBerardino and Hoffner, 1983). Similarly, in *Drosophila*, tissue culture cell nuclei fail to function on reimplantation into unfertilized eggs, but direct all forms of differentiation if implanted into cleaving preblastoderm embryos (Illmensee, 1976). We may conclude that if there is any fundamental restriction of developmental potential during early embryogenesis in the mouse, it has not yet been demonstrated in a way that is independent of extraneous *experimental* difficulties such as engagement of the transferred nucleus in the initial cleavage cell cycle.

5. GENE REGULATION IN DEVELOPMENT

(i) Developmental Activation of Genes by Interaction with *Trans*-Regulators

Evidence from Somatic Cell Transformation

Some of the mechanisms by which regulation of gene expression is effected in early development are common to adult differentiated cells as well, while others may be unique to the early embryo. Transformation experiments in which cloned DNA is introduced into somatic cells have provided essential insights into the regulation of genes that function in terminally differentiated cell types. Soon after introduction into the host cells, and irrespective of whether it is integrated into the genome or is present in episomal form, the transfected DNA is incorporated in chromatin complexes. Many demonstrations indicate that the configuration of such complexes is permissive for inductive interactions with diffusible, or *trans*-acting cellular regulators. An excellent example is provided by the gene for the 70 kd *Drosophila* heat shock protein. After transfection into cells of various species the regulatory region of this gene activates transcription at the adjacent promoter, on exposure of the host cells to *their* temperature of distress. This has been shown for cultured mammalian cells, where the gene is activated at 45°C (Corces *et al.*, 1981; Pelham, 1982); for cultured *Drosophila* cells, where it functions at 37°C (Di Nocera and Dawid, 1983); and for sea urchin embryos, where it functions at 25°C (McMahon *et al.*, 1984), though at 25°C the gene is quiescent in its species of origin, *Drosophila*. Additional instances include the inductive activation of the mouse metallothionein gene by heavy metal treatment, after transfection into human or mouse cultured cells (see, e.g., Mayo *et al.*, 1982), and of transfected α- and β-interferon genes by treatment with poly(I)-poly(C) or by viral infection (Canaani and Berg, 1982; Mantei and Weissmann, 1982; Ragg and Weissmann, 1983).

The somatic cell hybridization experiments mentioned earlier suggest the *continuous presence* in at least some fully differentiated cells of *trans*-regulators which specify the activity of the genes expressed particularly in those cells. This conclusion has been confirmed in DNA transformation studies in which exogenous genes are introduced into cells of the type in which they are normally expressed. For example the regulatory sequences of ovalbumin and lysozyme genes respond to endogenous cell-specific factors when intro-

duced into oviduct cells (Renkawitz *et al.*, 1982; Dean *et al.*, 1983); lens
crystallin genes are activated when introduced into lens epithelial cells, but
not other cells (Kondoh *et al.*, 1983; Piatigorsky *et al.*, 1984); the skeletal
muscle troponin gene is activated on transfection into myoblasts that are
then induced to differentiate (Konieczny and Emerson, 1985); insulin and
chymotrypsin gene regulatory sequences are activated when introduced into
pancreatic endocrine β-cells or tumor derivatives thereof, and exocrine
cells, respectively (Walker *et al.*, 1983; Episkopou *et al.*, 1984); and immunoglobulin genes are activated when transfected into lymphoid cells but
not other cells (Oi *et al.*, 1983; Banerji *et al.*, 1983; Gillies *et al.*, 1983).
Similarly, β-globin genes can be induced by DMSO treatment on transfection into erythroleukemia cells, while the same treatment does not activate
them on transfection into nonerythroid cells (Chao *et al.*, 1983; Wright *et al.*,
1983). It is clear for several genes activated by steroid hormones that interaction of diffusible hormone-receptor complexes with specific sequences of the
gene is both necessary and sufficient to cause regulated expression. Thus the
chicken ovalbumin gene is induced by estrogen on transfection into a human
estrogen-responsive mammary cell line that does not produce ovalbumin
itself (Lai *et al.*, 1983), and the mouse mammary tumor virus promoter is
transcriptionally induced in cells that contain glucocorticoid receptors (see,
e.g., Chandler *et al.*, 1983; reviewed in Yamamoto, 1985). The specific upstream sequences where the hormone-receptor complex is bound can confer
glucocorticoid regulation on other genes fused to them. An important conclusion that derives from these and many additional studies of similar import
is that *cis*-regulatory sequences with which cell type-specific *trans*-activa-
tors interact often lie no more than 1-2 kb upstream of the initiation site of
the transcript, and frequently within several hundred nucleotides of this site,
or within an intron or other internal region of the gene. Thus at least in some
of the cases examined, transfected DNA fragments that contain tissue-specific genes and their immediate environs include their own *cis*-regulatory
sequences, and they function both when integrated in many different locations in the host cell genome, and when not integrated at all. In these cases
large genomic *regions* are thus not required for the final step in ontogenic
gene activation, as might be the case were the *cis*-regulatory sequences
involved located many kilobases away from the gene itself.

Experiments in which cloned DNA containing an inducible, tissue-specific
gene is introduced into cells other than those in which it is normally expressed offer the opportunity of comparing the response of the endogenous
and the transfected genes in the same cells. The state of the endogenous
genes reflects the ontogenic "experience" undergone by the genomes of the
host cell lineage. For example, the rat genes for $\alpha_{2\mu}$ globulin, a liver-specific
protein, are activated by glucocorticoid hormone on transfection into mouse
fibroblasts, while in the same cells the endogenous $\alpha_{2\mu}$ globulin genes remain

repressed (Kurtz, 1981). The $\alpha_{2\mu}$ globulin genes are not normally utilized in the cell lineages from which fibroblasts descend, and these genes thus exist in a heritable, silent configuration that prevents the positive interactions regulating transcription of the transfected sequences. Comparisons of the normal expression of α- and β-globin genes with their expression after transfection into erythroleukemia cells, and into *non*erythroid cells, further illuminate this phenomenon (Charnay *et al.*, 1984). As noted above, human β-globin genes are transcribed actively when introduced by transfection into mouse erythroleukemia cells that are then induced to differentiate by chemical treatment, e.g., with DMSO. The endogenous (i.e., mouse) α- and β-globin genes in the same cells are also activated during the induced erythroid differentiation. In uninduced erythroleukemia cells, and in *nonerythroid cells*, e.g., HeLa cells, the transfected β-globin genes are transcribed at a low rate, while the endogenous globin genes remain silent. However, transcription of transfected β-globin genes occurs at high levels even in nonerythroid cells if a viral enhancer sequence is included in *cis* configuration in the exogenous construct (Treisman *et al.*, 1983). The natural tissue-specific interactions that lead to activation of genes such as that coding for β-globin could function by activating transcriptional enhancer elements. In their sequence characteristics and their ability to affect expression in different locations with reference to the promoter, and in opposing orientations, the *cis*-regulatory sequences of various tissue-specific genes indeed resemble viral enhancer sequences (see, e.g., Gillies *et al.*, 1983; Walker *et al.*, 1983; Chandler *et al.*, 1983; reviewed in Serfling *et al.*, 1985).

In contrast to the behavior of transfected β-globin genes, transfected α-globin genes are expressed at a high rate, both in *non*erythroid cells and in uninduced erythroleukemia cells. Thus DMSO treatment affects them relatively little (Charnay *et al.*, 1984). On the other hand, the endogenous α-globin genes of the same nonerythroid cells are inactive. One interpretation is that full expression of both α- and β-globin genes requires release from the repressed configuration during erythroid cell differentiation, but while the promoter of the α-globin gene is sufficiently active that this event alone induces a high level of transcription, the β-globin gene needs an additional *trans*-acting stimulus that is available only in erythroid cells. In nonerythroid cells *transfected* globin genes can thus be considered to represent the condition of the endogenous globin genes *after* the first of these events, i.e., the derepression that occurs during natural (or DMSO induced) erythroid differentiation. If instead of naked globin gene DNA whole human chromosomes that include the α-globin gene are introduced into mouse erythroleukemia cells by fusion with human *erythroblasts*, the human α-globin gene behaves as does the endogenous mouse α-globin gene (Deisseroth and Hendrick, 1978; Charnay *et al.*, 1984). That is, the gene is transcribed actively on chemical induction of host cell differentiation, but not otherwise. This observation confirms the implication that repression is a *cis* function of the chromatin structure of the gene, that can be transferred by passage of the chromosome on which it resides.

From these and many similar observations can be inferred the following general features of the regulatory processes undergone by genes that are expressed in particular differentiated tissues. At the beginning of development these genes are inactive, and they remain in a repressed configuration in all cell lineages except those in which they are supposed to be utilized, and even in those, until the appropriate stage of differentiation is attained. This repressed configuration can be transferred heritably from cell to cell, *in vivo* or in culture. For many tissue-specific genes the presentation of a *trans*-activator that is synthesized only in the appropriate cell type could be the sufficient cause of release from the repressed configuration as well as of active transcription. For others *a two-step process* must be necessary, first an event which is specific to the appropriate cell type, *viz.*, release of the gene from its repressed chromatin configuration, and second, the activation of high level transcription occurring as a consequence of binding a diffusible inducer (e.g., a steroid-hormone receptor complex). Such two-step activation mechanisms must be fairly general, at least for hormone responsive genes. Thus given hormones often activate different genes in different cell types, which acquire their developmental specification prior to transcriptional expression of these genes. However, signals mediating both kinds of process may coexist in some cell lineages, since it is possible to activate ontogenically repressed genes by somatic cell fusion.

Transformation of somatic cells with specifically expressed genes has yielded invaluable insights into the molecular mechanisms of inductive gene activation. However, the processes by which genes become heritably repressed during ontogeny and then derepressed in the appropriate cells, are less accessible. A powerful approach to this problem is the insertion of cloned genes into the germ line of the animal, so that their function can be investigated directly during ontogeny.

Transgenic Animals: Ontogenic Function of Exogenous Genes Introduced into the Germ Line

Exogenous genes introduced into the germ line or the zygote have been found to function during development in the correct tissues, at the proper ontogenic stages, and at least in some transformants at appropriate levels. Two systems have so far been exploited, *Drosophila* and mouse, though it appears probable that useful transfection experiments can also be carried out in the eggs of *Xenopus* (see, e.g., Etkin, 1982), *Caenorhabditis elegans* (Stinchcomb *et al.* 1985), and the sea urchin (McMahon *et al.*, 1984, 1985; Flytzanis *et al.*, 1985). In *Drosophila* the use of defective P-factor transposons as transformation vectors (Spradling and Rubin, 1982; Rubin and

Spradling, 1982) has provided an efficient means of introducing single copies of cloned DNA sequences at random locations into the germ line genome. Prokaryote vector sequences are excluded, as integration occurs at the P-factor sequence termini, which are required for transpositional insertion. Genes introduced on P-factor vectors in general display appropriate differential function in adult and larval tissues. Except for minor quantitative differences in the levels of activity the various genomic positions at which integration occurs seem to have little effect on the accuracy of ontogenic expression. Examples include the *rosy* gene, encoding xanthine dehydrogenase, which is required for eye pigment formation (Rubin and Spradling, 1982; Spradling and Rubin, 1983); genes for the salivary gland glue protein secretions (Richards *et al.*, 1983; Crosby and Meyerowitz, 1986); the DOPA decarboxylase gene (Scholnick *et al.*, 1983); and the *white* locus gene (Hazelrigg *et al.*, 1984). In the latter case, though the position of the insertion does not usually affect pigment synthesis by the enzymatic product of the *white* gene, it does affect interactions that occur with a regulatory locus, *zeste*. A clear demonstration of the tissue-specific expression of a gene inserted by P-factor transformation is shown in Fig. 1.3, from a study of Goldberg *et al.* (1984) in which the alcohol dehydrogenase gene was introduced into null mutants for this enzyme. In each of the cited examples the *cis*-regulatory elements required for appropriate developmental expression are evidently located in the vicinity of the gene, since they are included in the DNA fragments cloned into the P-factor vectors.

Transgenic mice have been constructed by microinjection of cloned DNA into the pronucleus of fertilized eggs. Following this operation, the eggs are reimplanted in foster mothers and in successful transfers, carried to term. The injected DNA integrates into the genome, usually in a single location that contains from one to many tandemly arranged copies. Transformants are generally obtained in which the exogenous genes are expressed in the correct tissues, though the quantitative level of expression is often much lower than normal, and is not related in a simple way to the number of copies

---

Fig. 1.3. (See color plates following p. 256.) Tissue-specific expression of alcohol dehydrogenase (Adh) in transgenic *Drosophila*. Embryos of a null mutant for Adh (Adh^{fn23}) were injected with a defective P-factor vector ($p\pi25.1$; Rubin and Spradling, 1982) containing an 11.8 kb DNA fragment that includes the complete Adh gene. Organs shown were prepared from Adh^{fn23} flies, and from flies of two transgenic strains in which the exogenous Adh genes had integrated at different chromosomal locations (19E on the X-chromosome and 36A on the second chromosome; the normal Adh gene is located at 35B). Adh activity was displayed by histochemical staining with a formazan dye. The left column shows organs of Adh^{fn23} flies; the middle column organs of the Adh+ strain from which the exogenous Adh gene was obtained; the right column, the same organs of the transformed strains. a, antennal disc; amg, anterior midgut; e, eye; f, fat body; fg, foregut; gc, gastric cecae; h, hypoderm; hg, hindgut; m, muscle; M, Malpighian tubules; p, paragoium; pmg, posterior midgut; s, salivary gland; sv, seminal vesicle; t, testis. [From D. A. Goldberg, J. W. Posakony, and T. Maniatis (1984). *Cell* **34**, 59. Copyright by M.I.T.]

integrated. In contrast to the results obtained in P-factor transformation of *Drosophila*, transgenic mice which display some expression of the exogenous genes in inappropriate tissues are frequently encountered as well. This could be due to integration in particular genomic locations, to the presence of multiple rather than single integrated copies of the exogenous DNA, or of vector sequence DNA, or to rearrangements or other mutational changes introduced in the exogenous DNA before or during integration. Genes that at least in some transgenic mice show correct differential activation during ontogeny include a rearranged κ immunoglobulin gene that only functions in lymphoid cells (Storb *et al.*, 1984), and a rearranged μ immunoglobulin gene (Grosschedl *et al.*, 1984; these authors reported a low level of expression in heart as well as in lymphoid tissues); β-globin genes that are expressed exclusively in erythroid cells, and in some cases at levels comparable to the endogenous globin genes (Chada *et al.*, 1985; Townes *et al.*, 1985); the transferrin gene, expressed in some though not all transformants only in liver (McKnight *et al.*, 1983); the α-fetoprotein gene, expressed in the visceral endoderm of the embryo (Krumlauf *et al.*, 1985a,b); a rat myosin light chain gene expressed specifically in skeletal muscle (Shani, 1985); and a metallothionein-growth hormone fusion gene, which is expressed preferentially in liver, just as is the endogenous metallothionein gene (Palmiter *et al.*, 1983). The rat gene for pancreatic elastase also functions specifically in transgenic mice. In a majority of transformants this gene produces more than 10^4 molecules of elastase mRNA per exocrine cell, indicating that its level of function is equivalent to that of the endogenous elastase genes (Swift *et al.*, 1984). Ornitz *et al.* (1985) showed that only 213 nt of 5' flanking sequence from the elastase gene suffices to ensure specific developmental expression in pancreatic exocrine cells. This was demonstrated by fusing the 213 nt elastase gene sequence element to the human growth hormone structural gene, the product of which can be detected by immunofluorescence. This result is illustrated in Fig. 1.4.

There may be regions of the mouse genome that are not permissive for transcription, since in many of the cited studies there are reported some transgenic animals in which no expression of the exogenous genes can be detected. However, it is clear from the many positive results achieved that as in *Drosophila*, ontogenically correct expression of a variety of genes can

---

Fig. 1.4. (See color plate following p. 256.) Tissue-specific expression of a rat elastase-human growth hormone fusion gene in transgenic mice. The fusion includes just 213 nt of 5' flanking sequence of the elastase gene (see text). Human growth hormone is evident in the exocrine acinar cells in the section of the pancreas shown, identified by bright green immunofluorescence. The endocrine cells are seen as dark islets, in which growth hormone is not synthesized. This result is in accordance with the natural expression pattern of the elastase gene, which is active only in exocrine cells. [From D. M. Ornitz, R. D. Palmiter, R. E. Hammer, R. L. Brinster, G. H. Swift, and R. J. MacDonald (1985). Reprinted by permission from *Nature* (*London*) **313**, 600. Copyright © 1985 Macmillan Journals Limited.]

occur after integration in different genomic regions. Experiments such as that reproduced in Fig. 1.4 show that relatively short proximal sequences may suffice, just as for the inductive processes observed in somatic cell transformation experiments. Thus if large chromatin "domains" are involved at all in ontogenic derepression, they must be centered on, or controlled by, interactions occurring in immediate proximity to the gene itself, and not by distant regulatory elements.

It could be argued that transfected genes do not undergo the same ontogenic derepression experienced by the endogenous tissue-specific genes, i.e., because in ectopic positions or environments they do not undergo the normal process of repression at the beginning of development, and that their correct expression thus merely reflects the eventual presence of tissue-specific *trans*-activators. An experiment that bears directly on this point has been reported by Stewart *et al.* (1984). In this study the mouse mammary tumor virus promoter was fused to an oncogene (*v-myc*), and the resulting construct injected into the egg pronucleus. Transcriptional expression of the fused gene was restricted to mammary cells in the transgenic mice, which subsequently developed mammary tumors. In contrast, the MMTV promoter is active in any *somatic* cell into which it is transfected that possesses glucocorticoid receptors, when this hormone is presented (reviewed in Yamamoto, 1985). In the transgenic mice the gene remains permanently repressed in all cells except those that have undergone the ontogenic process of mammary cell differentiation, and only then is it subject to glucocorticoid stimulation.

A general contribution of both somatic and germ line transformation experiments has been to demonstrate the reality and the ubiquity of *trans*-acting regulators in the control of gene expression in animal cells. Except for the steroid hormone receptors (reviewed in Anderson, 1984; Yamamoto, 1985), and a few examples such as the regulatory proteins that activate 5S rRNA transcription in *Xenopus* (Chapter V), little is known of the molecular nature of these *trans*-regulators. Quantitative estimates for a factor that may fall in this class, a protein that is bound tightly to sequences flanking the adult chicken β-globin gene only in cells expressing this gene, derive from a study of Emerson *et al.* (1985). There may be 10^3-10^4 molecules of this factor per erythrocyte nucleus, and the ratio of its binding constants for specific (i.e., globin gene) to nonspecific (i.e., total) DNA is $>10^4$. Several genes that apparently code for *trans*-acting regulators have been identified, for instance, the unlinked genes *raf* and *Rif*, which in the mouse determine the levels of α-fetoprotein in normal adult liver and in regenerating liver, respectively (Belayew and Tilghman, 1982). These genes also affect the levels of at least one other tissue-specific transcript which is regulated similarly though not identically to the α-fetoprotein gene (Pachnis *et al.*, 1984). Another example is the *Drosophila* gene *l(1)npr-1*, the product of which is required for transcription of three glue protein genes during the third larval instar

(Crowley *et al.*, 1984). The logical implication of differential gene regulation by means of *trans*-regulators is that the *coordinate activation* of sets of genes in given differentiated cell types could be based on *sequence homology* among the *cis*-regulatory sequences of the genes of each set (Britten and Davidson, 1969). Several examples are now known of short, homologous regulatory sequences, generally less than 30 nt in length, which occur in the 5' regions of individual unlinked members of batteries of genes that are activated in concert. Such sequences have been identified, for instance in the various heat shock genes of *Drosophila* (Pelham, 1982; Dudler and Travers, 1984; Topol *et al.*, 1985); some steroid responsive genes of chicken (Renka-witz *et al.*, 1984; Dean *et al.*, 1984); fibrinogen genes of rats (Fowlkes *et al.*, 1984); the amino acid synthetases of yeast (Donahue *et al.*, 1982); and a number of glucocorticoid responsive genes of mice and humans (Karin *et al.*, 1984; Yamamoto, 1985). Since the multiple genes of given "batteries" may thus be controlled by single *trans*-acting molecular species, the number of structural genes required to produce the characteristics of each cell type may be large compared to the number of controlling regulatory genes (Britten and Davidson, 1969, 1971).

(ii) Regulation of Gene Activity during Embryogenesis

Most of the genes considered in the last section are utilized in specific *somatic* adult or larval tissues and play no role whatsoever in early development. It is useful to regard genes of this class as "postembryonic late genes," in that during embryogenesis they are repressed, or at least not called upon to be transcribed. RNA complexity measurements indicate that these late genes constitute only a minority fraction of the genes utilized during the life cycle (reviewed in Davidson and Britten, 1979; see Chapter III), except for mammals, in which there appears to be a very extensive set of sequences represented specifically in neonatal brain RNAs (see, e.g., Hahn *et al.*, 1978, 1982). In those organisms that have been investigated a relatively large set of structural genes is represented in mRNAs of both egg and embryo. This class we refer to as the *maternal gene set*, irrespective of whether a given mRNA molecule derives from transcription in the oocyte nucleus or in an embryonic cell nucleus. A third class, of particular concern in regard to determination and differentiation in embryonic cell lineages, consists of "embryonic late genes," i.e., genes that are not significantly expressed in the oocyte but that are activated during early development, frequently in specific regions of the embryo. In well studied systems such as the sea urchin embryo it can be estimated that in terms of its diversity the latter set of sequences represents no more than about 10% of the total utilized during early development. *During embryogenesis* each of these three classes of genes must be subject to distinct regulatory processes.

Mechanism of Repression of Late Genes during Early Development; Genomic Methylation

As perceived experimentally in adult somatic cells, late genes exist in a repressed configuration, except in the tissue in which they are to be utilized. This configuration could be imposed as early as gametogenesis, or during embryogenesis. The nuclear transplantation experiments reviewed above suggest that at least in certain organisms after very early stages of development some kind of change in the state of late gene repression may occur, when nuclear totipotence becomes difficult rather than easy to demonstrate. The process of late gene repression is interesting, not only in its own right, but also because it sheds light on the mechanisms that must be required for the subsequent tissue-specific activation of these genes. Two general kinds of mechanism, that are not exclusive, have been advanced to explain late gene repression, viz., inactivation by methylation, and inactivation as a consequence of engagement in particular forms of chromatin structure. The details of this large subject lie outside the scope of the present discussion and can be reviewed only briefly here.

5-Methylcytosine occurs in CG dinucleotides in DNA of many animal species, and it has been proposed that this modification constitutes the structural basis for the cell-heritable repression and activation systems of differential gene regulation. This idea is based on three different kinds of evidence. First, there exist hemimethylases that can copy the specific pattern of methylation in a preexistent DNA strand onto a newly synthesized strand, and thus within a given lineage this pattern can be transferred from cell generation to cell generation (see, e.g., Stein et al., 1982). Thus a specific pattern of negative regulation could be heritably transferred from cell to cell in the absence of trans-acting factors. Second, it appears likely that the presence of 5-methylcytosine might alter DNA-protein interactions such as must be involved in chromatin structure and in gene regulation (see review in Razin and Riggs, 1980). Third, an extensive series of correlations between the level of methylation and the level of activity of given genes has been reported from comparisons carried out on different tissues of the same animal. Certain regions of genes are frequently found to be undermethylated, while in inactive genes they are usually methylated. However, many exceptions to this simple generalization have been noted. For example in the human $\gamma\delta\beta$-globin gene region, levels of methylation at sites that can be assayed with methyl-sensitive restriction enzymes vary from high to very low in tissues not expressing these globin genes, while a moderate level of methylation is observed in active erythroid tissues (van der Ploeg and Flavell, 1980), and in the mouse no detectable changes in methylation are observed in the α- or β-globin genes during induction of expression in erythroleukemia cells (Sheffery et al., 1982). Similarly in and around the α_2 (type I) collagen genes levels of methylation do not differ between cells that synthesize collagen and cells that do not (McKeon *et al.*, 1982). In *Xenopus*, the β_1 globin gene and the albumin gene are specifically hypomethylated in erythroid and hepatic cells respectively, but the A_1 and A_2 vitellogenin genes are heavily methylated in all adult tissues whether or not they are being transcribed (Gerber-Huber *et al.*, 1983). Similarly, though undermethylation of the 5' end of the albumin gene in rat hepatoma cell lines is highly correlated with expression, some lines that display the same undermethylation fail to synthesize albumin (Ott *et al.*, 1982). Conflicting results have also been obtained in experiments in which DNA methylated *in vitro* is introduced into cells and the effect on the level of transcriptional activity is determined. Stein *et al.* (1982) showed that on transfection into mouse fibroblasts the adenosyl-phosphoribosyltransferase gene is not transcribed if first methylated at CCGG sites, though otherwise this constitutively expressed gene is expressed in transformed cells, and Vardimon *et al.* (1982) demonstrated that a methylated adenovirus gene for a DNA binding protein cannot be transcribed on injection into the *Xenopus* oocyte nucleus, while the unmethylated gene produces adenovirus RNA. Similarly, *in vitro* methylation in the 5' region of the human γ-globin gene specifically prevents its transcription in transfected mouse fibroblasts, though methylation in the body of the gene is without effect (Busslinger *et al.*, 1983). On the other hand experiments in which specific regions of the thymidine kinase gene are methylated by synthesis *in vitro*, and the gene is then transfected, show that methylation of the 3' coding region blocks transcription even though the known locations of the *cis*-regulatory sequences by which this gene is controlled are not altered (Keshet *et al.*, 1985). This result suggests that methylation interferes with the transcription of this gene, but not with the regulatory processes by which transcription is instituted. Furthermore, methylated ribosomal gene DNA, whether prepared enzymatically *in vitro* (Pennock and Reeder, 1984) or extracted directly from sperm (Macleod and Bird, 1983), is transcribed as actively in *Xenopus* oocyte nuclei as is nonmethylated DNA. This observation suggests the need for caution in interpreting some of the other results cited, since in both rDNA studies the methylated sites include promoter sequences known to be important for regulation of these genes (see Chapter V), and since in normal development a strong correlation between synthesis of rRNA and demethylation of these sequences is observed (Bird *et al.*, 1981). A conclusion consistent with the total evidence available is that for *some genes* demethylation of specific regions may be a necessary event in activation, and methylation may accompany ontogenic repression or constitute an element of the repressed configuration. That methylation is the *causal* or *sufficient* explanation for repression, and demethylation for derepression, is so far not demonstrated. Furthermore, 5-methylcytosine is not universally present in animal DNAs. In dipteran insects, which outwardly display the same general forms of ontogenic regulatory events as do other creatures, methylated cytosines do not occur at all at a detectable level, nor do methylated adenosines (Urieli-Shoval *et al.*, 1982).

Mammalian embryos clearly possess the capacity to carry out *de novo* methylation. Jähner *et al.* (1982) showed that if Moloney murine leukemia virus (MuLV) is injected into the pronuclei of fertilized mouse eggs it is integrated into the genome and becomes heavily methylated at CG sites, while if introduced after implantation, only a small amount of methylation is observed. There is a strong negative correlation between the extent of methylation and the activity of the integrated provirus. Thus when introduced at very early stages replication of the virus is repressed until some months after birth, when viremia finally occurs, but if the virus is injected into midgestation fetuses, many cells in the embryo quickly become secondarily infected. Tissues containing hypomethylated proviral genomes derived from somatic viral infection are the only ones that display virus-specific transcripts (Stuhlmann *et al.*, 1981). In transfection tests carried out on cultured fibroblasts the methylated viral DNA isolated from the genomes of transgenic mice is noninfectious, but when demethylated by cloning, the same DNA becomes highly infectious (Harbers *et al.*, 1981). The same correlations have been established in experiments carried out with embryonal carcinoma (EC) cells infected with MuLV, which behave as do cells of the undifferentiated early embryo (Stewart *et al.*, 1982; Gautsch and Wilson, 1983). Thus the viral DNA is integrated into the genomes of these cells, where it is methylated, but cannot be expressed, even after subsequent transfection (without recloning) into a sensitive cell line. However, Gautsch and Wilson (1983) showed that methylation of the exogenous DNA does not actually occur until many days *after* integration, and that nonetheless the nonmethylated integrated copies present in the EC cells are from the beginning transcriptionally silent. Thus although methylation may interfere with proviral transcription, it is not the initial *cause* of repression. Treatment with 5-azacytidine, which when incorporated into replicating DNA inhibits the propagation of DNA methylation to the daughter cells, results in reactivation of quiescent MuLV provirus in EC cells (Stewart *et al.*, 1982), as it does of endogenous retroviral genomes in normal chicken cells (Groudine *et al.*, 1981). 5-Azacytidine also activates the metallothionein gene to a state of inducibility in a mouse cell line in which this gene is unresponsive to heavy metal induction, again an effect correlated with decrease in the level of methylation (Compere and Palmiter, 1981). Other derepressive effects reported for this agent include reactivation of the silent X-chromosome in mammalian cell hybrids (Mohandas *et al.*, 1981); activation of otherwise noninducible muscle genes in the nuclei of HeLa cells fused with mouse muscle cells (Chiu and Blau, 1985); and potentiation of muscle, cartilage, and fat cell differentiation from an embryonic stem cell line (Konieczny and Emerson, 1984). However, since the actual mode of action of 5-azacytidine is unknown (see review in Felsenfeld and McGhee, 1982) and since, like 5-methylcytidine itself, when incorporated this molecule may also affect DNA-protein interactions, we are again faced with the difficulty of distinguishing cause from correlation. Even if methylation is not the causal basis of ontogenic late gene repression, and demethylation of activation, it remains possible that the positions of some methyl groups serve to specify regions in newly replicated somatic cell DNA that in the parental cell genomes existed in a repressed configuration. The immediate causal basis of repression would then be the chromatin structure in which the inactive DNA is locked.

A vast amount of research indicates that chromatin configuration is altered in the vicinity of genes that are active, compared to the same genes in other cell types, or developmental stages where they are inactive. This can be detected by heightened sensitivity to DNase I and other nucleases (see reviews in Weisbrod, 1982; Conklin and Groudine, 1984; Weintraub, 1985a,b). A tissue-specific feature that is convincingly correlated with the state of activity or potential activity is the presence of DNase I hypersensitive sites. These are regions generally only 200-400 nt in length, where this enzyme produces double strand cuts. They usually occur in 5' flanking regions and in enhancer sequences, as well as elsewhere within active genes. In the glucocorticoid-sensitive mouse mammary tumor retrovirus promoter, for example, binding of the hormone-receptor complex causes increased hypersensitivity at the sites of the interaction (Payvar *et al.*, 1983), and analogous results have been reported for many other genes activated by steroid hormones (reviewed in Yamamoto, 1985). Similarly, the internal control region where the 5S rRNA gene is bound by its transcription factor is also marked by hypersensitive sites (Engelke *et al.*, 1980). Of particular interest in the present context is the observation that DNase I hypersensitive sites can be propagated in a dividing, differentiated cell lineage independently of transcription. In the chicken vitellogenin gene, for instance, there is a set of DNase I hypersensitive sites present only in liver, the tissue in which the gene is expressed on estrogen treatment, and some of these specific sites persist in daughter cells after the hormone is withdrawn and transcription ceases to be detectable (Burch and Weintraub, 1983). Thus it appears that ontogenic specification of the active state for this particular gene involves a permanent change in the chromatin structure, other than what is required for the act of transcription *per se*. This provides physical evidence for *a two-step process of ontogenic gene activation*. The first event alters the chromatin structures, thereby facilitating the second, interaction with the immediate *trans*-regulator.

Developmental repression of adult somatic cell late genes is extremely efficient. Ratios of at least 10^{-7} to 10^{-9} have been estimated for the level of globin (Groudine and Weintraub, 1975) and of growth hormone (Ivarie *et al.*, 1983) in cells not expressing these genes compared to cells in which they are active. As we have seen, genes present in ontogenically repressed configuration are not only transcriptionally silent but are also inaccessible or insensitive to interactions with inducers that activate transfected genes in the same cells. Evidence exists that nucleosome spacing may differ in inactive chromatin, and that repressed regions may be included in higher order structures formed by crosslinking between molecules of the nonnucleosomal histone H1, resulting in condensation of the primary DNA-nucleosome chain (see reviews in Felsenfeld and McGhee, 1982; Conklin and Groudine, 1984; Weintraub, 1985b).

In summary, whatever the detailed structures, it seems evident that at some point in early development late genes are sequestered within heritable forms of chromatin complex that are not permissive for transcription. As directly implied by the many recent observations cited above, regulation during development by sequence specific *trans*-interactions occurring at the genome level must generally be positive rather than negative, though evidence from cell fusion experiments suggests that there could exist specific negatively acting *trans*-regulators as well (see, e.g., Killary and Fournier, 1984). In general, however, any gene not otherwise specified for activity may by default be subject to repression by histones. One example of such a mechanism has emerged from the detailed experimental analysis of 5S rRNA activation and repression. These studies indicate that competition between the formation of the transcription complex on the one hand, and engagement of the 5S genes in an inactive histone complex, on the other, determines their functional state (Gargiulo *et al.*, 1984; see also Chapter V).

Evidence That Oocyte Cytoplasm Contains Factors That Can Affect Nuclear Gene Expression

Cytological observations suggest that there are factors in oocyte cytoplasm that profoundly affect the state of injected somatic nuclei, so that they rapidly assume the gross characteristics of the endogenous germinal vesicle. On injection into *Xenopus* oocyte cytoplasm, for example, HeLa cell nuclei enlarge greatly in volume, their chromatin disperses, and their prominent nucleoli diminish or disappear (Gurdon, 1976). If the host oocyte is induced to mature, the chromosomes of exogenous nuclei assume metaphase configuration synchronously with the oocyte chromosomes at the meiotic reduction divisions [reviewed in Etkin and DiBerardino, 1983; see Newport and Kirschner (1984) for discussion of cytoplasmic factors that determine chromosomal behavior during the cell cycle in *Xenopus* oocytes and eggs]. In addition, DNA synthesis is induced in nuclei injected into matured eggs, including nuclei derived from noncycling terminal cell types that normally are destined to carry out no further DNA replication, such as brain cells or erythrocytes (Graham *et al.*, 1966; Leonard *et al.*, 1982). Cytoplasmic factors acutely affect transcription in injected nuclei as well. As discussed in

Chapter V, the germinal vesicle of the amphibian oocyte is intensely active in RNA synthesis, and initiation occurs rapidly in virtually all functioning transcription units. When HeLa or other mammalian cell nuclei are injected into the cytoplasm of *Xenopus* oocytes an enormous increase in the quantity of newly synthesized nuclear RNA occurs (Gurdon *et al.*, 1976). Autoradiographic grain counts indicate that this increase is almost proportional to the change in injected nuclear volume. Most of the labeled RNA in these experiments is probably unstable heterogeneous nuclear RNA, which is normally produced at a high rate in the germinal vesicle (Chapter V).

The experiments of De Robertis and Gurdon (1977) afford a direct demonstration that the oocyte cytoplasmic environment can effect the activation of specific quiescent genes in injected somatic cell nuclei. Nuclei of cultured *Xenopus* kidney cells were injected into oocytes of the newt *Pleurodeles waltlii*, and the proteins synthesized in the injected oocytes were compared to those synthesized in control oocytes, as shown in Fig. 1.5. The somatic nuclei enlarge dramatically after injection [Fig. 1.5(a)], and after some hours new protein species appear, *all* corresponding to proteins normally synthesized in *Xenopus* oocytes. Their synthesis is abolished by the transcription inhibitor α-amanitin. Several of these oocyte proteins are not found in kidney cells [Fig. 1.5(b-e)], and none is of the class found in kidney cells but not oocytes. The experiment demonstrates that the host oocyte cytoplasm has *activated oocyte-specific genes in the somatic nuclei*, despite the phylogenetic distance separating anuran and urodele amphibians. Activation of these genes occurs in the complete absence of DNA synthesis in the injected nuclei.

The activation of the gene for an oocyte isoform of lactate dehydrogenase has been observed in an interspecific transplantation experiment in which adult liver nuclei from the salamander *Ambystoma texanum* were injected into cytoplasm of *A. mexicanum* (axolotl) oocytes (Etkin, 1976). Synthesis of alcohol dehydrogenase, which is normally found in liver but not oocytes, was not induced in the injected oocytes. A further example concerning a specific gene product is the activation of *Xenopus* 5S rRNA genes of the type normally expressed only during oogenesis. Korn and Gurdon (1981) demonstrated at least an 800-fold increase in the activity of oocyte type 5S RNA genes relative to somatic type 5S RNA genes in adult erythrocyte nuclei injected into the oocyte germinal vesicle. Evidence regarding the interactions of these genes with the specific *trans*-acting factor required for their expression is reviewed in detail in Chapter V.

Activation of the Maternal Gene Set in the Embryo

Several kinds of evidence imply that the continued expression of the maternal gene set in embryonic blastomere nuclei could be determined by

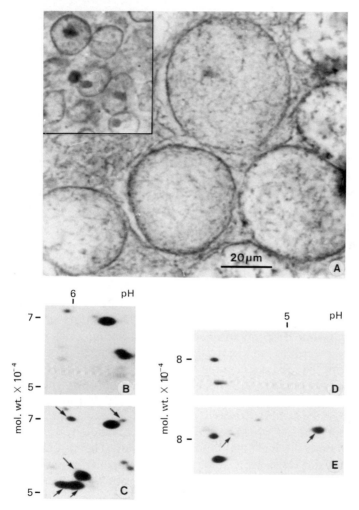

Fig. 1.5. Induction of oocyte-specific protein synthesis in somatic *Xenopus* nuclei injected into oocytes of *Pleurodeles*. (a) Stained sections showing nuclei of cultured *Xenopus* kidney cells a few minutes after microinjection into *Pleurodeles* oocyte cytoplasm (inset) and 7 days later, all photographed at the same magnification. Based on E. M. De Robertis, and J. B. Gurdon (1977). *Proc. Natl. Acad. Sci. U.S.A.* **74**, 2470. (b)-(e) Two dimensional gel electrophoretic analyses of proteins synthesized in *Pleurodeles* oocytes injected with cultured *Xenopus* kidney cell nuclei. (b)-(c) and (d)-(e) represent different regions of the gel, as indicated by the pH and molecular weight coordinates. (b), (d) Oocytes labeled immediately after injection of nuclei; (c), (e) Oocytes equivalently labeled on day 7 following injection. *Upward arrows* indicate the positions of *Xenopus* proteins synthesized only in oocytes; *downward arrows*, the positions of *Xenopus* proteins synthesized both in oocytes and in the cultured cells from which the transplanted nuclei derived. The remaining proteins are products of the endogenous *Pleurodeles* genome. [From E. M. De Robertis, and J. B. Gurdon (1977). *Proc. Natl. Acad. Sci. U.S.A.* **74**, 2470.]

factors inherited in the egg cytoplasm. This interpretation is suggested *a priori* by the large overlap between the pattern of transcription operating during oogenesis, and that which is instituted in the early embryo (reviewed in Chapters III and V), since the germinal vesicle of the oocyte and the blastomere nuclei are embedded in the same maternal cytoplasmic milieu. We have just reviewed several direct demonstrations that specific genes are activated, and others repressed, on exposure to oocyte cytoplasm. Furthermore, the entering male genome, which in the sperm is completely quiescent, is in most organisms activated soon after fertilization, and although there are exceptions (see below), in general the male genome appears to express the same pattern of gene activity as does the female genome. This is shown by the classical observation that in most forms parthenogenic eggs can complete embryogenesis normally, and by both classical and molecular studies on interspecific hybrid embryos (reviewed in Davidson, 1976; see also Chapter III).

Mechanisms other than determination by maternal cytoplasmic factors can be envisioned as well, however. Some form of control is required, since at least in the sea urchin a major fraction of the set of cytoplasmic RNAs present in the egg and embryo are not constitutively expressed sequences, in that they are absent from adult tissues (see Chapter III). The extreme form of the cytoplasmic determinant theory implies that maternal *trans*-activators specifically recognize and activate most of the genes expressed early in development. The female genome would first be exposed to these determinants during oogenesis, and the male genome on entry into the maternal cytoplasm following fertilization. A very different alternative is that the egg cytoplasm contains only factors that generally stimulate transcription, and the maternal gene set would be "marked" in advance by some process that operates in both the male and female germ lines. The specification of these genes would serve the purpose of protecting them from incorporation into the repressed configurations that involve late genes, and thereby would render the marked genes permissive with respect to *nonspecific* maternal transcription factors. The fact that *somatic* nuclei transferred into enucleated eggs can successfully direct embryogenesis provides a serious argument against the latter form of mechanism, however, at least for those organisms in which successful nuclear transfer has been demonstrated. That is, functional embryonic genomes that evidently carry out correct patterns of transcription can be derived from cells that are not of immediate germ line origin. There is some evidence for higher vertebrates, nonetheless, that specification of the maternal gene set in the early embryo is not an exclusive function of factors present in the egg cytoplasm.

In the mouse the male and female pronuclear genomes do not appear completely equivalent. Thus, e.g., the paternal X-chromosome is preferentially inactivated in extraembryonic tissues during early development (Harper *et al.*, 1982). *Both* maternal and paternal pronuclei are required for early development (Markert, 1982; Surani and Barton, 1983; McGrath and Solter, 1984b, 1985). Diploid biparental eggs containing two female pronuclei arrest at the 25 somite stage, and diploid eggs containing two male pronuclei arrest prior to implantation. Mammals (at least mice) thus differ from many other animals in which parthenotes develop normally, and even occur under natural conditions, for instance in many insect species including some of the genus *Drosophila*; and in certain lizards, urodeles, crustaceans, annelids, teleost fish, etc. (reviewed in White, 1954, 1978).

For the mouse it appears an inescapable conclusion that the prior "experience" of the male and female genomes during gametogenesis has in some way affected their respective activities during embryogenesis, so that each expresses a somewhat different set of functions. The combination constitutes the diploid pattern of transcription required for normal development, and in fact regions of maternal and paternal chromosomes 2, 6, 8, 11, and 17 have been shown to be functionally nonequivalent (Johnson, 1975; Lyon and Glenister, 1977; Searle and Beechey, 1978; Lyon, 1983; Cattanach and Kirk, 1985). Since gametogenic processes apparently affect the capacities for gene expression in the two parental genomes, such processes might be involved as well in determination of which gene sets are to be active overall. Molecular evidence that indirectly supports this possibility has been obtained by Groudine and Conkin (1985), in a study of genomic methylation occurring during spermatogenesis in the chicken. The level of methylation of many but not all genes increases in spermatocytes, and the DNA of the mature sperm in general displays a high level of methylation. However, certain specific regions remain unmethylated. Among these are sequences in the genes for thymidine kinase and for an endogenous retrovirus, both of which are expressed in the early embryo (as well as in many somatic tissues). In these genes the location of hypomethylated sequences corresponds to the DNase I hypersensitive sites present when these genes are expressed, but they remain hypomethylated in sperm, in which transcription does not occur and DNase I hypersensitive sites are absent. In contrast, the late genes coding for β-globin, ovalbumin, and vitellogenin are all heavily methylated in chicken sperm DNA. Similarly, Stein *et al.* (1983) found that in the hamster the 5' ends of the genes for adenosyl phosphoribosyl transferase and for dihydrofolate reductase are hypomethylated in sperm. These are also commonly expressed genes that are transcribed in the early embryo, as elsewhere. In contrast, late genes of the hamster, e.g., the β-globin gene, are heavily methylated in sperm DNA. These results suggest that genes destined for function in the early embryo enter the zygote already marked by the pattern of methylation in a way that distinguishes them from late genes, which are destined for embryonic repression. Of course it is possible that the hypomethylation pattern observed in sperm is simply a consequence of the transcriptional activity of these ubiquitously expressed genes *in the spermatocyte*. The implication is that genes expressed in oogenesis and in spermiogenesis would be expressed after fertilization simply because the structural residue of their transcription during gametogenesis protects them from repression. It thus remains to be determined whether *control* of expression of the maternal gene set is based primarily on factors that are internal or external to the genes of the early embryo. It is not unlikely that the solutions to this interesting problem will differ among various organisms, e.g., in mammalian systems where methylation may be important, as compared to *Drosophila,* where DNA methylation does not occur.

Activation of Embryonic Genes Not Previously Expressed

Factors that induce activation of specific genes and classes of genes in transplanted nuclei are clearly present in the cytoplasm of the early embryo. Neurula stage nuclei transplanted into enucleated eggs cease to synthesize all classes of RNA during cleavage, just as do the nuclei of normal amphibian embryos (Chapter III), and synthesis of 18S and 28S rRNA, heterogeneous nuclear RNA, tRNA, and oocyte and somatic type 5S RNAs is reactivated on the normal schedule at the blastula stage (Gurdon and Woodland, 1969; Wakefield and Gurdon, 1983). A correct developmental pattern of activation has also been demonstrated for a *Xenopus* α-actin gene in embryos derived from enucleated eggs into which nuclei of larval muscle cells had been transplanted (Gurdon *et al.*, 1984). In this experiment the donor nuclei were genetically marked with an albino mutation and with the anucleolate *1-nu* mutation to ensure the unequivocal identification of transplant recipients. The α-actin gene in the experimental embryos was regulated in a normal fashion. Thus its transcripts were absent in blastula and early gastrula stages, and they appeared at the onset of myogenesis in the late gastrula. The cytoplasmic environment to which the exogenous nuclei and their descendants were exposed thus apparently contains necessary and sufficient *trans*-acting signals to account for this ontogenic pattern of regulation.

In order to affect gene activity in implanted (or endogenous) nuclei, cytoplasmic regulatory factors must have the ability to enter the nuclear compartment. Many experiments have demonstrated that injected *nuclear proteins* will migrate into the oocyte germinal vesicle and concentrate there (reviewed in De Robertis, 1983). Proteins not of nuclear origin in general enter only if they are below a certain size, about 70 kd (Bonner, 1975a), though certain smaller proteins are specifically excluded (Bonner, 1975b). Studies on the karyophilic behavior of nucleoplasmin, an extremely prevalent acidic oocyte nuclear protein, indicate that a specific region of the polypeptide sequence contains the signal required for accumulation in the germinal vesicle (Dingwall *et al.*, 1982; Dabauvalle and Franke, 1982). Feldherr *et al.* (1984) showed that if coated with nucleoplasmin and injected into the oocyte cytoplasm colloidal gold particles will be transported into the germinal vesicle via the nuclear pores. RNAs are also transported to the nucleus, probably by means of other proteins that bind specifically to them. Thus 5S RNA injected into the cytoplasm of *Xenopus* oocytes accumulates in the nucleolus, as do injected U snRNAs in the nucleoplasm (De Robertis *et al.*, 1982). Zeller *et al.* (1983) showed that specific snRNA-binding proteins are located in the embryo cytoplasm during early *Xenopus* development, and that when they complex with the newly synthesized snRNAs appearing at blastula stage they migrate into the embryo nuclei. Many nuclear proteins stored in the oocyte germinal vesicle are distributed to the cytoplasm at germinal vesicle breakdown. During embryonic development maternal nuclear proteins reenter the various blastomere nuclei, in temporally and regionally specific ways (Dreyer and Hausen, 1983). This suggests the possibility that regulatory macromolecules, including some active during oogenesis, could be specifically targeted to sets of blastomere nuclei, and thus differentially affect the early patterns of transcription in the embryo.

Genes that are not expressed until particular stages of embryogenesis, and that are activated at specific locations, are of particular interest since they take part in the processes leading to the initial appearance of differentially functioning cell types. There are again several classes of mechanism that might reasonably explain the initial appearance of regionally localized patterns of gene activity in the embryo. As proposed by Davidson and Britten (1971) particular batteries of structural genes could be specified for expression by direct interaction with maternal *trans*-regulators located in appropriate regions of egg cytoplasm, or synthesized on localized maternal mRNAs. Or they could be specified as the result of contact with an adjacent cell type that releases specific internal activators. A third alternative that is different in emphasis, because it involves an intermediate level of internal hierarchy in the genomic regulatory system, is that the initial regional specialization of the embryo is controlled by a small number of pleiotropic developmental master genes. The crucial spatial pattern of *their* expression might be determined by a few simple polarizations in the egg. As we shall see in the following review there is evidence from different biological systems that supports, to a various extent, all of these modes of zygotic gene activation.

It is becoming possible to introduce cloned genes coding for specific proteins directly into eggs, and to observe their regulated function *during embryogenesis* (see, e.g., Steller and Pirrotta, 1984; Etkin *et al.*, 1984; McMahon *et al.*, 1984; Flytzanis *et al.*, 1985). This, and the use of transgenic animals in which exogenous genes active in the early embryo have been integrated into the germ line, can be expected to provide the means to explore the elements operative in the initial activation of embryonic late genes, whether these elements are cytoplasmic or nuclear, maternal or zygotic in origin.

The broad outlines of the process of gene regulation in development have become apparent. As already concluded long ago, the genomes of most differentiated cells are basically equivalent, though they express diverse sets of genes. The maternal gene set in expressed during gametogenesis and also in the early embryo. The postembryonic late genes whose function characterizes specific adult cell types are generally repressed early in development, and following ontogenic changes that potentiate expression, these genes are specifically activated in appropriate cells in response to *trans*-activators. Zygotically active genes that provide the initial *differential* patterns of expression in the embryo are activated in a spatially defined pattern. In the following chapters we deal first with the characteristics of the maternal gene set and then with the behavior and functions of embryonic late genes and the various mechanisms by which their differential activity is organized in diverse embryonic forms.

II

The Nature and Function of Maternal Transcripts

Many species of RNA are stored in eggs, and are inherited by the embryo. Classes of transcript identified in mature unfertilized eggs include messenger RNAs (mRNA), ribosomal RNAs (rRNA), transfer RNAs (tRNA) and other low molecular weight species, as well as mitochondrial RNAs. In addition, a class of large, nontranslatable polyadenylated transcripts distinguished by the presence at internal locations of interspersed repetitive sequences has been found in eggs of sea urchins and amphibians. The function of these "interspersed RNA" molecules, if any, is not yet known.

Maternal RNAs appear to be an intrinsic feature of development from egg and sperm. Eggs that develop outside the mother in the external environment contain sufficient cytoplasm to provision the hundreds or thousands of cells formed as the zygotic genome replicates. Net growth cannot occur until a feeding larva has been constructed, the end-product of embryogenesis. Thus at early stages the nucleus-to-cytoplasm ratio is relatively low, and the flow of newly synthesized transcripts cannot supply the biosynthetic apparatus. In some large eggs such as those of amphibians and dipterans, there is little or no transcription at all during cleavage. All embryos known require protein synthesis from the start, however, and at early stages this is carried out with maternal ribosomes and tRNAs, and is programmed with maternal mRNAs. In some organisms these components remain important far into the developmental process, though the stage at which maternal transcripts are replaced by newly synthesized zygotic transcripts varies greatly from sequence to sequence, even within a given species of embryo. Only in mammals, where the pace of early development is relatively slow, and where external nutrients are taken up almost from the beginning, is the developmental role of maternal transcripts somewhat curtailed. This provides an instructive comparison that illuminates the functional roles of maternal RNA in different strategies of development.

1. THE DISCOVERY OF MATERNAL mRNA: EARLY BIOLOGICAL AND BIOCHEMICAL EVIDENCE

The decisive demonstration of maternal mRNA in the 1960's was foreshadowed by many kinds of prior evidence. The general dominance of maternal characters early in development was illustrated convincingly in numerous classical observations on interspecific hybrid embryos (see reviews in Morgan, 1927; Davidson, 1976). Detailed examination of such embryos showed that the morphological features, as well as the rate of cleavage and of other early developmental processes, always followed the pattern of the maternal species, and the influence of the paternal genome on morphogenesis could usually not be detected until midembryogenesis, or even later. Observations of this kind were reported for sea urchins, amphibians, teleost fish, and many other forms, and the main conclusions were later confirmed in studies on the expression of individual enzymes in interspecific hybrid embryos. The discovery that enucleate egg cytoplasm is capable of carrying out cleavage pointed in the same direction. Following less convincing earlier claims, this was shown clearly by Fankhauser (1934), in experiments performed with cytoplasmic fragments of urodele eggs, and by Harvey (1936, 1940), in studies carried out on enucleate fragments of sea urchin eggs produced by centrifugation through a sucrose medium. The complete exclusion of nuclear chromatin was confirmed by the disappearance of Feulgen staining (Harvey, 1940). A decisive series of experiments demonstrating cleavage in the eggs of anuran amphibians in the absence of any functional nuclear genome was reported by Briggs *et al.* (1951). The experiment was begun by fertilizing eggs of *Rana pipiens* with irradiated sperm of *Rana catesbiana*. This interspecific hybrid combination is normally lethal, the hybrid embryos arresting by the gastrula stage (Moore, 1941). The lethal phenotype is abolished by the irradiation, and eggs fertilized with irradiated sperm develop as gynogenetic *R. pipiens* haploids. The genome of the egg was then also removed by physical enucleation, following activation with the irradiated *R. catesbiana* sperm. Extensive cleavage nonetheless occurred, and partial blastulae were even formed. An example is shown in Fig. 2.1. As can be seen there, cleavage amphiasters are formed in the blastomere cytoplasm despite the absence of functional chromosomes. Such enucleation experiments demonstrated that the process by which the egg mass is divided into many cells is dependent on maternal cytoplasmic factors rather than on the embryo genomes. It was clear that the maternal developmental systems loaded into the

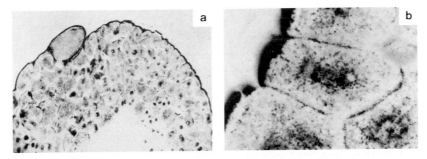

Fig. 2.1. Cleavage without nuclear genomes in the frog egg. The photographs show achromosomal partial blastulae formed by fertilizing *Rana pipiens* eggs with irradiated *Rana catesbiana* sperm and then enucleating the recipient eggs. (a) Section through cleaved animal hemisphere. Cells are intact and show well-defined boundaries throughout most of the cleaved area. X55. (b) Enlarged view of amphiastral figure in same blastula as shown in (a). The figure contains no Feulgen-positive material. X425. [From R. Briggs, E. U. Green, and T. H. King (1951). *J. Exp. Zool.* **116**, 455.]

egg cytoplasm are mechanistically very complex. Not only must the cytoplasm be able to generate or provide the components needed for the multiple cleavage mitoses, but it evidently also contains spatial information that determines their orientation.

The first clear demonstration of high molecular weight maternal RNA was accomplished by Brachet (1933), in experiments carried out on sea urchin eggs. It had already been established in many previous studies dating back to the turn of the century that pentoses are present in acid hydrolysates made from various eggs and embryos. Brachet (1933) also showed that the total quantity of RNA remains almost constant during embryonic development, in contrast to the quantity of DNA, which rises steadily. By the early 1960's it was known that the bulk of the maternal RNA stored in eggs is rRNA. An impressive demonstration of the functional importance of maternal rRNA in embryonic development was given by Brown and Gurdon (1964). They showed that homozygous anucleolate (*o-nu*) *Xenopus* embryos derived from heterozygous parents develop to a swimming tadpole stage, consisting of about 5×10^5 cells. These mutants can synthesize no rRNA, while control embryos produce significant rRNA after the neurula stage. The store of maternal ribosomes inherited in the egg cytoplasm thus suffices for the biosynthetic needs of the embryo unexpectedly far into development.

The evidence that led initially to the discovery of maternal *mRNA* came from investigations of the effect of fertilization on protein synthesis in sea urchin embryos. This material was favorable because of an unusual developmental feature, the sharp increase in the rate of protein synthesis that almost immediately follows fertilization. This occurs too rapidly to be accounted for as the consequence of new mRNA synthesis. Protein synthesis was first observed in developing sea urchin eggs by Hultin (1952), in an early radioisotope uptake experiment carried out with ^{15}N-labeled glycine and alanine. A pioneering cell-free translation study of Hultin (1961a) revealed that the remarkable increase in the protein synthesis capacity of the egg after fertilization is a property of the postnuclear fraction of the homogenate. Hultin (1961b) also showed that inhibition of protein synthesis by treatment with puromycin blocks cleavage, and this in itself implied that the egg cytoplasm contains preformed mRNA, since as Harvey (1936, 1940) had found, cleavage occurs in enucleate cytoplasm. Within the next few years several groups showed that homogenates of sea urchin eggs could translate "synthetic messenger RNAs," such as polyuridylic acid. These and other observations made with cell-free systems indicated that the maternal ribosomes and tRNAs are functional and that the egg must contain at least some of the various factors required for protein synthesis, including mRNA. Among those who contributed to this conclusion were Nemer (1962), Wilt and Hultin (1962), Brachet *et al.* (1963a), Nemer and Bard (1963), Tyler (1963), Maggio *et al.* (1964), and Monroy *et al.* (1965).

The experiments that inescapably required the proposition of maternal mRNA were those in which it was demonstrated that protein synthesis occurs in *enucleated egg cytoplasm*. Such observations were reported by Brachet *et al.* (1963b) and by Denny and Tyler (1964). Both laboratories reported that enucleated halves of unfertilized sea urchin eggs prepared by centrifugation can synthesize protein at control rates when parthenogenetically activated. Tyler (1965) also showed that the amino acid composition of the total proteins synthesized in enucleated sea urchin egg fragments is the same as that of normal eggs. Protein synthesis was demonstrated in enucleated frog eggs soon after (Smith and Ecker, 1965). At about the same time it was shown that treatment with actinomycin D, at concentrations that almost completely block embryonic transcription, fails to suppress a large fraction of the protein synthesis normally occurring in cleavage stage sea urchin embryos (Gross and Cousineau, 1963, 1964). Nor does actinomycin treatment of sea urchin eggs abolish cell division, and actinomycin-treated frog eggs may form complete blastulae (Brachet and Denis, 1963; Wallace and Elsdale, 1963; for a general review of the biological and molecular effects of actinomycin and other such drugs on embryonic development, see Davidson, 1976). At the time it was not clear to what extent mitochondrial transcripts rather than stored mRNAs synthesized in the oocyte nuclear genomes might have contributed to the protein synthesis observed in physically enucleated or actinomycin-treated eggs. As reviewed later in this Chapter, this contribution is in fact minor, as shown by the high complexity of the early embryo protein synthesis pattern, the absolute rate of protein synthesis, its insensitivity to mitochondrial inhibitors such as ethidium bromide, and ultimately by the extraction and characterization of the maternal messages themselves. Messenger RNA preparations that displayed activity in cell-free systems were extracted from unfertilized sea urchin eggs by Maggio *et al.* (1964) and Slater and Spiegelman (1966), and from *Xenopus* oocytes by Davidson *et al.* (1966) and Cape and Decroly (1969). Crude estimates from these studies based on relative translational activities suggested that at least 2% of the total egg RNA in both sea urchin and amphibian eggs is capable of function as mRNA.

The demonstration of maternal mRNA was perhaps the most remarkable initial result of the application of molecular approaches to embryonic development. The early studies mentioned here, and the many ancillary observations of which space does not permit notice, sufficed to prove at least at a qualitative level, and at least for sea urchins and amphibians, that stored transcripts originally synthesized in the nucleus of the oocyte dominate protein biosynthesis during early development. It remained to determine the amount, the complexity and the later fate of the maternal transcripts, and most importantly, to obtain direct evidence as to the functional roles played by maternal mRNAs in the embryo.

2. COMPLEXITY OF MATERNAL RNA

In all species so far examined the maternal RNA includes a very complex population of molecules, the product of thousands of diverse transcription units. *Complexity*, in the precise sense used here, denotes the length of single copy DNA sequence represented in an RNA preparation, here the RNA of the mature egg. This is measured most accurately by the method of RNA excess hybridization with a single copy genomic DNA tracer. When the reactions are carried out to kinetic termination, the fraction of the labeled single copy DNA hybridized gives directly one-half of the fraction of the genomic single copy sequence represented in the RNA, since transcrip-

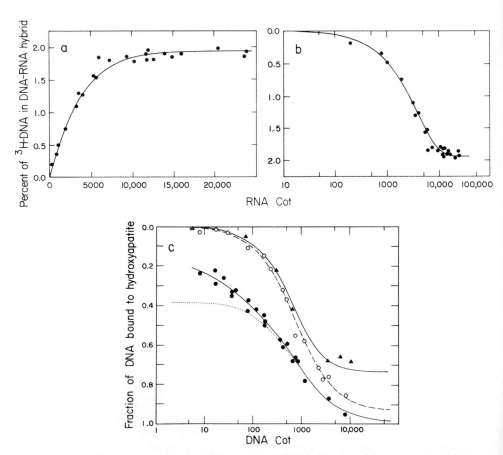

Fig. 2.2.   Sequence complexity of the maternal RNA in the egg of the sea urchin *Arbacia punctulata*. (a) A single copy sequence tracer was prepared from ³H-thymidine labeled embryo DNA by kinetic fractionation on hydroxyapatite columns, and reacted with excess total unfertilized egg RNA. The fraction of tracer hybridized (ordinate) was measured by hydroxyapatite chromatography, and is portrayed as a function of RNA C_0t, i.e., concentration (moles nt

tion is asymmetric. In practice, calculation of the RNA complexity by this means usually involves an additional small correction to take into account labeled single copy DNA fragments too small to react under the conditions used. Because the hybridizing RNA molecules are present in great sequence excess, with respect to the DNA tracer, the hybridization reaction results in no significant decrease in the concentration of the hybridizing RNA species and the reaction therefore follows pseudo-first-order kinetics. From the rate constant observed the approximate concentration in the total RNA of the hybridizing species can also be estimated. The kinetics of RNA excess hybridization reactions and the equations required for reduction of data obtained by this method are reviewed in Appendix I, part 5. Appendix I includes: a summary of the relations governing the renaturation kinetics of single copy and repetitive DNA; treatment of other forms of solution RNA-DNA hybridization reactions, and some illustrative examples; a brief discussion of the effects of various experimental variables; and appropriate references.

An example of a complexity measurement carried out on the maternal RNA of sea urchin eggs is shown in Fig. 2.2 (Anderson *et al.*, 1976). The kinetics of the hybridization reactions are illustrated in Fig. 2.2(a) and (b). In the experiment reproduced in Fig. 2.2(c) it is demonstrated that the DNA molecules that hybridized with the egg RNA were in fact very largely single

---

liter^{-1}) $\times$ time (sec). The reaction follows pseudo-first-order kinetics [equation (17) of Appendix I], and the least-squares solution assuming this form is shown by the solid line. At termination 1.93% of the single copy tracer was hybridized. (b) Same reactions as in (a) illustrated as a semilog plot, to better display kinetics. The rate constant (ln $2/C_0t_{1/2}$) is $3.0 \times 10^{-4} M^{-1}\text{sec}^{-1}$ according to the least squares fit. By reference to a kinetic standard, *viz.*, the hybridization reaction of excess ϕX174 RNA with a ϕX174 DNA tracer (Galau *et al.*, 1977b; see also Fig. AI.5 of Appendix) it can be calculated that a pure RNA of the complexity measured for this egg RNA would react with a rate constant of about $3 \times 10^{-2} M^{-1}\text{sec}^{-1}$. Thus [equation (18) of Appendix I] about 1% of the *total* egg RNA serves to drive the reaction, i.e., constitutes the complex class RNA. (c) Demonstration that the hybridizing RNAs were actually bound to single copy ^{3}H-DNA fragments. ^{3}H-DNA was recovered from a hybrid preparation obtained after incubation with egg RNA to RNA C_0t 10,000, mixed with a 20,000-fold excess of unlabeled DNA, and allowed to react to various extents (closed triangles). The kinetics with which total *Arbacia* DNA renatures are shown by the closed circles, and the single copy component of these kinetics is indicated by the dotted line [see Appendix I, Fig. AI.2, and equation (11)]. For a genome of this size (0.79 pg haploid) the second-order rate constant for the single copy component is ideally $1.3 \times 10^{-3} M^{-1}\text{sec}^{-1}$ (dotted line). Reaction with whole DNA of the ^{3}H-DNA single copy tracer preparation utilized for the experiments in (a) and (b) is shown by the open circles. The least squares solution (dashed line) indicates a rate almost identical to the theoretical rate, $1.4 \times 10^{-3} M^{-1}\text{sec}^{-1}$. The *previously hybridized* ^{3}H-DNA reacts at a very similar rate, $1.8 \times 10^{-3} M^{-1}\text{sec}^{-1}$, and there is no early reacting sequence visible such as would indicate a significant yield of hybrid molecules containing repetitive DNA sequences. There could be included a few percent of hybridized sequences present in 2 to 5 copies per haploid genome. [From D. M. Anderson, G. A. Galau, R. J. Britten, and E. H. Davidson (1976). *Dev. Biol.* **51**, 138.]

copy sequences. The validity of the measurement turns on this point, since had a significant fraction of the hybridized tracer consisted of repetitive rather than single copy sequences, the single copy complexity could not be judged from the quantity of RNA-DNA duplex obtained. From the saturation level shown in Fig. 2.2(a) it can be deduced that the complexity of the RNA in the egg of *Arbacia punctulata* is about 30×10^6 nt. This is over seven times the complexity of the whole *E. coli* genome, and is also several times the complexities of the mRNA found in many somatic cells, e.g., mouse kidney, liver, or erythroleukemia cells, or adult sea urchin cells (reviewed in Davidson and Britten, 1979). The kinetics of the reaction are illustrated most clearly in Fig. 2.2(b). As indicated in the legend it can be inferred that about 1% of the mass of the total oocyte RNA consists of the class of transcripts that account for most of the hybridization observed, henceforth referred to as "complex class RNA." About 90% of the remainder is ribosomal. The *average* sequence concentration or prevalence of each complex class RNA species is calculated from the hybridization kinetics to be about 1400 molecules per egg. In addition to complex class transcripts the sea urchin egg contains prevalent heterogeneous species (which together account for only a low fraction of the total complexity), mitochondrial RNAs, and various low molecular weight RNAs, as discussed in Section 4 of this Chapter.

Available estimates of RNA complexity for the eggs of different species are assembled in Table 2.1. All of the values listed fall within the range $35 \pm 10 \times 10^6$ nt, with the exception of the *Drosophila* measurement, which indicates an egg RNA complexity 2- to 3-fold lower. Some of the variation shown could be due to experimental error and not represent real differences,

TABLE 2.1   Comparative Summary of Maternal RNA Complexities

Organism	Complexity (nt)	RNA per egg (ng)	Approximate prevalence of hybridizing sequences[a] copies per egg	copies per ng RNA[b]
Echinoderms				
Sea urchins				
Strongylocentrotus purpuratus	37×10^{6c}	2.8^d	1.6×10^{3c}	6×10^2
Tripneustes gratilla	34×10^{6e}	1.7^e	1.9×10^{3e}	11×10^2
Lytechinus pictus	30×10^{6f}	3.9^g	4.3×10^{3h}	11×10^2
Arbacia punctulata	30×10^{6i}	2.3^j	1.4×10^{3i}	6×10^2
Starfish				
Pisaster ochraceus	43×10^{6k}	15^k	1.7×10^{4k}	11×10^2

TABLE 2.1   (continued)

Organism	Complexity (nt)	RNA per egg (ng)	Approximate prevalence of hybridizing sequences[a]	
			copies per egg	copies per ng RNA[b]
Amphibians				
Anuran				
Xenopus laevis	$27–40 \times 10^{6l}$	$4,800^{m}$	1.8×10^{6l}	4×10^{2}
Urodele				
Triturus cristatus	40×10^{6n}	$\sim 8,500^{o}$	—	—
Dipterans				
Drosophila melanogaster	12.1×10^{6p}	190^{q}	4.5×10^{4p}	2×10^{2}
Musca domestica	24.8×10^{6p}	$1,200^{p}$	7.6×10^{5p}	6×10^{2}
Echiuroid				
Urechis caupo	$31–47 \times 10^{6r}$	14^{r}	1.6×10^{3r}	1×10^{2}

[a] *Prevalence* here refers to RNA sequences of the class that dominates the complexity measurements shown in the first column. That is, highly prevalent sequences that do not contribute significantly to the diversity of the RNA population are not considered in this average. The average prevalence is calculated by comparing the observed rate of the hybridization reaction by which the complexity is measured to the rate expected for that reaction were the RNA to consist only of the hybridizing species [see equation (18) of Appendix I]. This ratio gives an estimate of the fraction of the total RNA that consists of the hybridizing species. The mass of the driver RNA in nucleotides represented in the complex class RNA fraction, e.g., per egg, is then divided by the complexity, in nucleotides, to obtain the number of times the *average* hybridizing species is present. The average estimates could in any given case be low by a factor of about two, due to various physical factors that may retard the reaction (see, e.g., Van Ness and Hahn, 1982).

[b] Number of molecules of *average* complex class transcript calculated per nanogram of total RNA (column 3).

[c] Galau *et al.* (1976); Hough-Evans *et al.* (1977).

[d] Goustin and Wilt (1981).

[e] Duncan (1978).

[f] Wilt (1977).

[g] Brandhorst (1980).

[h] Calculated by Kovesdi and Smith (note *k*), from data in notes *f* and *g*.

[i] Anderson *et al.* (1976).

[j] Calculated by Kovesdi and Smith (1982) (note *h*) from data of Whiteley (1949).

[k] Kovesdi and Smith (1982).

[l] Davidson and Hough (1971); Rosbash and Ford (1974).

[m] Taylor and Smith (1985).

[n] Rosbash *et al.* (1974).

[o] From data of Osawa and Hayashi (1953).

[p] Hough-Evans *et al.* (1980).

[q] Anderson and Lengyel (1979).

[r] Calculated by Davidson (1976), from data of Davis (1975).

but the *Drosophila* hybridization reactions were carried out together with the *Musca* reactions, and the difference in RNA complexity recorded for these two dipteran eggs cannot be explained as the consequence of differing experimental procedures (Hough-Evans *et al.*, 1980). The overall similarity of egg RNA complexities is particularly impressive when the enormous diversity in egg size and total RNA content is considered (column 3 of Table 2.1). The concentration of the complex class RNA sequences with respect to ribosomal RNA content differs little from species to species, as shown in the last column of Table 2.1. It is a remarkable observation that eggs varying in total or rRNA content by three orders of magnitude display complex class sequence concentrations that fall within a factor of three of an average value of about 500 molecules of each sequence per nanogram of total RNA. Table 2.1 thus illustrates a basic and general property of both protostome and deuterostome eggs. This is the presence of a high complexity class of maternal RNA consisting of a great many relatively rare sequences, stored in quantities that are almost proportional to the content of ribosomal RNA in the egg.

An interesting consequence of the similarity in egg RNA complexities is shown in Fig. 2.3, in which the egg RNA complexities are plotted as a function of genome size. It is clear that the amount of genomic sequence represented in egg RNA is independent of total genomic DNA content, again

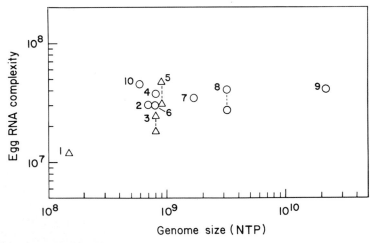

Fig. 2.3.   Complexity of egg RNA in various species, as a function of genome size. Data for egg complexities are from Table 2.1. Genome sizes (haploid) are given in original references. Two points connected by a dashed line indicate independent determinations. Both protostomes (△) and deuterostomes (○) are represented. 1, *Drosophila melanogaster*; 2, *Arbacia punctulata*; 3, *Musca domestica*; 4, *Strongylocentrotus purpuratus*; 5, *Urechis caupo*; 6, *Lytechinus pictus*; 7, *Tripneustes gratilla*; 8, *Xenopus laevis*; 9, *Triturus cristatus*; 10, *Pisaster ochraceus*. [Modified from B. R. Hough-Evans, M. Jacobs-Lorena, M. R. Cummings, R. J. Britten, and E. H. Davidson (1980). *Genetics* **95**, 81.]

with the exception of *Drosophila*, in which the low genome size could be considered to be correlated with the lower egg RNA complexity. The main implication of Table 2.1 and Fig. 2.3 is that there is a functional significance to the complex class of maternal transcripts, since in the various organisms studied this class always includes approximately the same quantity of genetic information.

3. MATERNAL TRANSCRIPTS CONTAINING INTERSPERSED REPETITIVE SEQUENCES

There has recently been discovered in sea urchin and amphibian eggs a class of poly(A) RNA molecules that is distinguished by the presence of repetitive sequence elements within regions transcribed from single copy genomic sequence. In the sea urchin egg these molecules generally exceed the average polysomal message in length. Their sequence organization resembles that of the genomic DNA, in which repetitive sequence elements only a few hundred nucleotides long are embedded in the single copy sequence at intervals of one to several kilobases (Davidson *et al.*, 1973; Graham *et al.*, 1974; Goldberg *et al.*, 1975). Though the functional significance of this class of maternal transcripts is yet unknown, it is clearly of quantitative importance, as it accounts for 65-70% of the total mass of poly(A) RNA present in the cytoplasm of fully grown *Xenopus laevis* oocytes, and of unfertilized *Strongylocentrotus purpuratus* eggs. A minor fraction of the interspersed maternal poly(A) RNA may consist of *bona fide* messages, in which the repetitive sequence elements reside in the nontranslated 3' terminal sequences. However, as demonstrated in the following summary of current evidence, these RNAs and the mature, translatable maternal message also stored in amphibian and echinoderm eggs constitute largely exclusive classes of heterogeneous maternal transcripts.

(i) Initial Evidence for Heterogeneous Repetitive Sequence Transcripts in the Egg

The existence of maternal transcripts of genomic repetitive sequences was demonstrated in the early applications of nucleic acid hybridization methods to egg RNA. The first such studies were carried out by filter hybridization procedures capable of detecting the reaction of repetitive sequences only. This limitation results from the difficulty of achieving high enough C_0t to permit reaction of single copy DNA sequences, due to the relatively low amounts of DNA that can be loaded on filters, and to insufficient RNA concentrations and reaction times. Reliable kinetics are generally not available from filter hybridization experiments in any case, since the reaction of filter-bound DNA is known to be retarded by large factors. The filter hybridization experiments nonetheless yielded interesting qualitative conclusions that foreshadowed more convincing demonstrations of maternal repetitive sequence transcripts.

RNAs extracted from *Xenopus* oocytes and uniformly labeled *in vitro* with ^{3}H-dimethyl sulfate were hybridized to filter-bound DNA by Crippa *et al.* (1967). Though undoubtedly an underestimate this experiment indicated that at least 1-2% of the total DNA sequence is represented in oocyte RNA, and thus it suggested a significant diversity of repetitive sequence transcripts. The repetitive sequence content of *Xenopus* egg RNA was measured in solution hybridization reactions with an isolated repetitive DNA tracer by Hough and Davidson (1972). The complexity of the egg repeat sequence transcripts was reported to be at least 3×10^4 nt (see Appendix I). Although further direct measurements of this nature have not been carried out on *Xenopus* egg RNA, the more recent results reviewed below suggest that the actual diversity of maternal repeat sequence transcripts in this RNA would be much higher, were low prevalence transcripts taken into account. However, it was already clear that mRNAs transcribed from known highly repeated genes, such as the histone genes, could contribute only insignificantly to the overall repetitive sequence complexity of the maternal RNA. Similar estimates of repetitive sequence representation in egg RNA were reported by Hough *et al.* (1973) for the neotropic amphibian, *Engystomops pustulosus*. Sea urchin eggs were also found to contain a diverse set of repetitive sequence transcripts, as demonstrated by Hynes and Gross (1972), using filter hybridization procedures.

Many early filter hybridization experiments on egg and embryo RNAs were carried out by the competition method. A frequent result was that the repetitive sequence hybridization observed with labeled embryonic transcripts could be competed, more or less effectively, by unlabeled egg RNA (reviewed in Davidson, 1976). In retrospect, these experiments show at least that repeat sequence transcripts are present in the eggs of many species, and that representatives of many of the same repetitive sequence families transcribed during oogenesis are also transcribed in the embryo. Among the organisms for which these conclusions may be drawn are the sand dollar *Dendraster excentricus* (Whiteley *et al.*, 1970; Mizuno *et al.*, 1974), the gastropod *Acmaea scutum* (Karp and Whiteley, 1973), the oyster *Crassostraea gigas* (McLean and Whiteley, 1974), the tunicate *Ascidia callosa* (Lambert, 1971), and several different sea urchins (Glišin *et al.*, 1966; Whiteley *et al.*, 1966, 1970; Chetsanga *et al.*, 1970; Hynes and Gross, 1972).

Some maternal transcripts that include repetitive sequences may survive far into embryogenesis. Thus Crippa *et al.* (1967) showed that a large component of the hybridization to genomic DNA of metabolically labeled *Xenopus oocyte RNA* is competed efficiently by unlabeled RNAs extracted from late *embryos*. These mesurements were carried out in solution reactions in RNA excess at RNA C_0t's of ~10-40 *M* sec. Direct evidence for the developmental persistence of maternal repeat sequence transcripts was obtained in experiments performed on embryonic RNAs of *Engystomops pustulosus*. Under appropriate laboratory conditions oogenesis occurs synchronously in this frog, and maternal RNAs labeled by injection of precursor into the body cavity of the female can later be recovered from clutches of shed eggs (Davidson and Hough, 1969). Heterogeneous maternally labeled transcripts that hybridized with repetitive DNA sequences were detected in RNAs extracted from embryos as late as the neurula and tadpole stages (Hough *et al.*, 1973).

With the advent of cloned repetitive sequence probes the representation of specific genomic repeats in the maternal RNA could be measured. A set of cloned interspersed repeat sequence elements from the genome of *Strongylocentrotus purpuratus* was constructed by Scheller *et al.* (1977). Nine such clones, derived from repetitive sequence families differing in genomic frequency, flanking sequence environment, repeat length, and primary sequence (Klein *et al.*, 1978; Anderson *et al.*, 1981; Posakony *et al.*, 1981) were reacted with total egg RNA. Transcripts of *all nine* were observed to be present in the eggs (Costantini *et al.*, 1978). A further unexpected result was that *both complementary strands* of each repetitive sequence are represented in the maternal RNA, though often at different concentrations. The prevalence of transcripts representing individual repetitive sequences was measured by hybridization kinetics in egg RNA excess, and by molecular titration, carried out in strand-separated probe excess (a summary of the quantitative basis of the titration method for measurement of transcript concentration is given in Appendix II). A typical pair of titration reactions demonstrating the reaction of both strands of a cloned repetitive sequence with sea urchin egg RNA is shown in Fig. 2.4(a). It was found that some specific repetitive sequences are represented in only a few thousand transcripts per egg, while the prevalence of other repeat sequence transcripts ranges beyond 10^5 transcripts per egg. Costantini *et al.* (1978) also measured the overall extent of genomic repetitive sequence representation in sea urchin egg RNA. The most important observation, which is reproduced in Fig. 2.4(c), is that $\geq 80\%$ of a tracer consisting of the total short genomic repetitive sequence can be hybridized on extended reaction with the maternal RNA. This result requires not only that most of the different interspersed repeat sequences in the genome are represented in the egg RNA, but also that representation of both complementary strands of each repeat sequence is general. Other experiments demonstrated that the relatively prevalent repeat sequence transcripts represent at least several hundred different repetitive sequence families. There are several thousand *distinct* families of short repetitive sequence in the *S. purpuratus* genome, and transcripts of a clear majority of these sequences are included in the maternal RNA, at a level of prevalence about equal to or greater than the average complex class single copy sequence transcript. The unfertilized sea urchin egg has only a

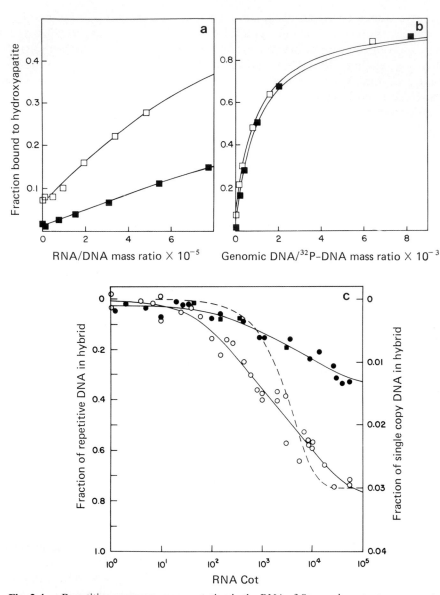

Fig. 2.4. Repetitive sequence representation in the RNA of *Strongylocentrotus purpuratus* eggs. (a) Titration of egg transcripts with cloned DNA probes representing the complementary strands (solid and open squares) of an interspersed repetitive sequence element called "2101." The length of this sequence element is about 320 nt, and it occurs about 700 times per haploid genome (Klein *et al.*, 1978). The ordinate gives the fraction of the labeled probe sequence included in RNA-DNA hybrids, as the mass ratio of egg RNA to DNA tracer is varied (abscissa). Each sample was reacted to kinetic termination, which is determined by the probe concentration, since the probe is present in the reactions in sequence excess (see Appendix II for theory and titration data reduction procedures). The prevalence of transcripts that hybridize

small pronucleus, having completed both meiotic maturation divisions prior to spawning. The high transcript prevalence measured for several of the individual repetitive sequences in experiments such as those shown in Fig. 2.4(a) thus implies a cytoplasmic location for these RNAs, and as described below, this inference has been confirmed directly for both the sea urchin and amphibian examples.

(ii) Molecular Structure of Interspersed Poly(A) RNAs in the Sea Urchin Egg

While the evidence so far reviewed demonstrates the presence of some form of repetitive sequence transcripts in the sea urchin egg, the nature of the RNA molecules containing the repeat sequence elements remained undefined. These were shown by Costantini *et al.* (1980) to be long, polyadenylated transcripts composed mainly of single copy sequences within which

---

to the two strands was calculated from this experiment as 8400 and 3000 sequences per egg, respectively. (b) Reaction of same cloned 2101 tracers with *S. purpuratus* DNA, in which both complements of the repeat sequence must be represented equally. (c) Hybridization of egg RNA with genomic repetitive sequence tracers. Sea urchin ^{3}H-DNA was sheared to an average length of 3300 nt, renatured to C_0t 40, and digested with S1 nuclease to remove single-stranded regions. The resistant fraction (21%), consisting almost wholly of repetitive DNA, was chromatographed on a Sepharose gel filtration column. The excluded fraction consisted of long repetitive sequences (≥ 1 kb). This DNA was recovered, sheared, and renatured again to remove contaminating fragments. It constituted the "long repeat" tracer. The included peak, consisting of duplex DNA fragments averaging 330 nt in length, constituted the "short repeat" tracer. About 25% of the total DNA in the *S. purpuratus* genome consists of repetitive sequence (Graham *et al.*, 1974), and of this approximately 70% is found in short repetitive sequence elements, of which there exist $\geq 5 \times 10^5$ per haploid genome. The remainder consists of long repetitive sequence elements. Hybridization kinetics for the reactions of the short repeat tracer (open circles) and the long repeat tracer (closed circles) with excess egg RNA are shown as a function of RNA C_0t (see Appendix I for kinetic treatment). Hybridization was assayed by binding to hydroxyapatite after subtraction of the small component of renatured DNA-DNA duplex in each sample. The kinetics were analyzed on the assumption of second-order rather than pseudo-first-order form, because both strands of each repeat sequence family are represented in the egg RNA (Costantini *et al.*, 1978; see text). Though neither reaction is terminated at the maximum RNA C_0t it is evident that at least 80% of the short repeat tracer and 30% of the long repeat tracer are engaged in RNA-DNA hybrids. The kinetics indicate a continuous range of transcript prevalence and this can be approximated, for the short repeat tracer reaction, as the sum of three kinetic components. The first of these includes 11% of the tracer that reacts with RNA sequences present on the average in 1.5×10^5 copies per egg; the second, 31% of the tracer, reacting with transcripts at an average prevalence of 3.3×10^4 copies per egg; and the third, 35% of the tracer, reacting with RNAs occurring on average of 1.4×10^3 times per egg. These values were calculated by reference to the kinetics of a single copy tracer reaction with egg RNA (Hough-Evans *et al.*, 1977), shown by the dashed line. The prevalence of average complex class *single copy* transcripts, according to this reaction, is about 1.6×10^3 copies per egg. [From F. D. Costantini, R. H. Scheller, R. J. Britten, and E. H. Davidson (1978). *Cell* **15**, 173. Copyright by M.I.T.]

the repeat sequences are interspersed. About 1.5% of the total egg RNA contains 3' oligo(A) sequences sufficient to promote binding to oligo(dT) cellulose. Reactions carried out with both cloned repetitive sequence tracers and genomic DNA tracers showed that the poly(A) RNA fraction includes at least 50% and possibly all of the interspersed repeat sequence transcripts (as well as most of the nonhistone maternal mRNA). The interspersed structure of these polyadenylated transcripts was demonstrated directly by separating out the molecules that include repeat sequences able to hybridize with genomic repetitive DNA, and measuring their sequence content. As a class, the interspersed poly(A) RNAs consist of about 90% single copy sequence transcript and 10% repetitive sequence transcript. Because both complements of most repeated sequences are represented in the interspersed maternal poly(A) RNAs, after partial renaturation the sequence organization of these molecules can be visualized in the electron microscope, as illustrated in Fig. 2.5. The location of the repetitive sequence elements is marked by

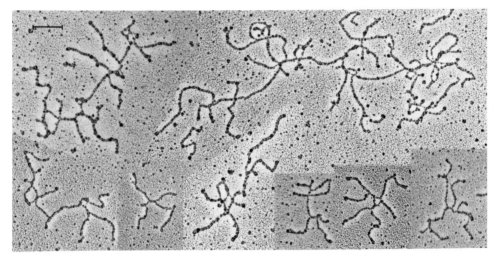

Fig. 2.5. Multimolecular structures formed by partial renaturation of sea urchin egg poly(A) RNA. Total poly(A) RNA was extracted from eggs of *Strongylocentrotus purpuratus*, denatured by brief heat treatment, and allowed to renature to an RNA C_0t of about 600 M sec, in a 58% formamide, 0.75 M NaCl medium, at 30°C. The RNA was then spread for electron microscopy. The bar represents a single strand RNA length of 1000 nt. Many multimolecular structures, including large networks, were observed. The montage shown includes several relatively simple examples. To estimate the fraction of poly(A) RNA that contains interspersed repetitive sequences, random areas of the grids were photographed and the contour lengths of all molecules recorded, including single molecules as well as those present in duplex-containing multimolecular structures. A total of 1.2×10^6 nucleotides of poly(A) RNA was scanned in this manner, of which about 65% is included in structures displaying four or more free ends. Control preparations spread immediately after denaturation ($C_0t < 10^{-3}$ M sec) included no networks, and <4% of the RNA was present in structures that could have contained single intermolecular duplexes. [From F. D. Costantini, R. J. Britten, and E. H. Davidson (1980). Reprinted by permission from *Nature* **287**, 111. Copyright © 1980 Macmillan Journal Limited.]

renatured RNA-RNA duplexes formed between complementary repeat sequences on different molecules. The average length of most of the transcribed repetitive sequences visualized in this manner is only a few hundred nucleotides. Analysis of the partially renatured egg poly(A) RNA molecules yields two important conclusions. First, as much as 65% of the total mass of poly(A) RNA is involved in multimolecular complexes containing repetitive sequence duplexes, and thus *a major fraction of the total maternal poly(A) RNA consists of interspersed transcripts*. Second, the structures observed imply that the individual repetitive sequences are not transcribed symmetrically, but rather are included within many different asymmetric transcription units oriented oppositely with respect to the repeat elements themselves. Thus the regions flanking the repetitive duplexes in structures such as shown in Fig. 2.5 are nonhomologous, and during incubation they have remained single stranded while the repeat sequences renatured. Figure 2.5 also suggests that many of the interspersed repeat elements are located in internal regions of the poly(A) RNA molecules.

Further studies have confirmed for several cloned examples these general structural features. Posakony *et al.* (1983) constructed a randomly primed cDNA library from sea urchin egg poly(A) RNA, and isolated a series of transcribed fragments, each of which included a known interspersed repetitive sequence element. As predicted, the repeat elements in these molecules were shown to be flanked by asymmetrically represented single copy sequences. The transcription units active in the synthesis of the interspersed poly(A) RNAs during oogenesis are thus proved to be unidirectional, and in general nonoverlapping.

The diverse sets of transcripts identified by the complementary strands of any given repeat sequence can be visualized in RNA gel blots, as shown in Fig. 2.6(a). Here it can be seen that the size of individual interspersed transcripts ranges from about 3 to 15 kb. The RNA gel blot experiments also reveal the persistence of most of the interspersed transcript species in late embryos, as illustrated in Fig. 2.6(b). The developmental fate of specific transcripts is displayed qualitatively by use of single copy probes isolated from regions flanking the repetitive sequence elements in the cloned maternal poly(A) RNAs. Some examples are shown in Fig. 2.6(c). Particular interspersed transcripts appear in some cases to increase in concentration in the total poly(A) RNA of late embryos, suggesting that the same transcription units are active in the embryonic nuclei. It can be seen in addition that the interspersed transcripts are discrete entities, that begin and end at defined sites.

A key issue in considering the significance of interspersed maternal poly(A) RNA is its relation to the *bona fide* maternal message. Costantini *et al.* (1980) isolated the interspersed poly(A) RNA molecules by hybridizing them with repetitive DNA, and this RNA fraction was then reacted with a single copy tracer consisting predominantly of sequences represented in egg

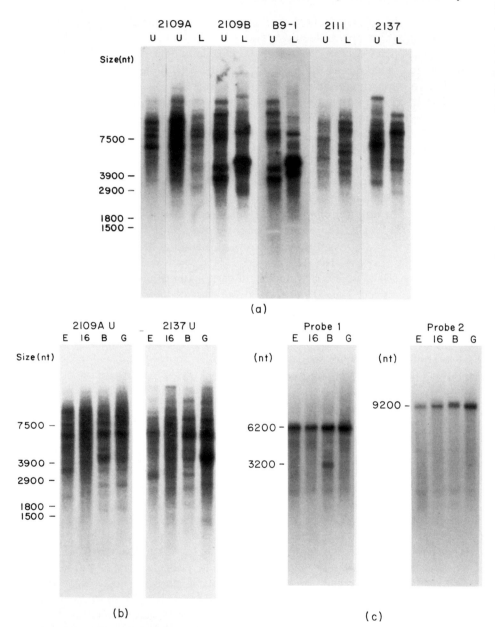

Fig. 2.6. Size distribution of interspersed maternal poly(A) RNA transcripts containing specific repetitive sequence elements. (a) Families of repeat-containing transcripts in egg RNA. *Strongylocentrotus purpuratus* egg poly(A) RNA was displayed by gel electrophoresis and transferred by blotting to nitrocellulose strips. These were hybridized individually to strand-separated cloned probes, each representing a given repetitive sequence from the sea urchin genome, and autoradiographed. The cloned repeats are indicated at the top of the Figure: 2109A is a 180 nt long sequence represented at least 900 times per haploid genome; 2109B is a 110 nt

RNA. Virtually all of this single copy tracer reacted with the interspersed poly(A) RNA. A majority of the maternal set of single copy sequences concentrated in this tracer is also represented in mRNAs associated with polysomes after fertilization (see below). On the other hand, there are many general features of the interspersed poly(A) RNA that distinguish it from *bona fide* maternal mRNA. For one thing, the size of these molecules, as shown in Fig. 2.6, is much greater than that of mRNA, the number average length of which in sea urchin embryos is about 2 to 3 kb (Slater *et al.*, 1973; Wilt, 1977; Duncan and Humphreys, 1981). In addition, the complexity of the single copy sequences included in the interspersed poly(A) RNA fraction is somewhat greater than that of the polysomal mRNA in early embryos, and hence these molecules as a class must include sequences not found in polysomal mRNA. They also display unexpectedly large differences in length when the same transcripts are compared between closely related sea urchin congeners. Posakony *et al.* (1983) found that the individual interspersed transcripts detected with cloned probes in the egg RNA of *S. purpuratus* [e.g., Fig. 2.6(a) and (c)] are *usually* of different lengths when visualized by reaction of the same probes with the egg RNA of *S. franciscanus*. Several specific examples of *intra*specific insertion/deletion polymorphism within an interspersed maternal poly(A) RNA transcription unit have been analyzed on cloned allelic DNA fragments by Calzone *et al.* (1986a). These studies all suggest that interspersed transcripts contain extensive regions that are permissive for gross mutational changes.

Primary sequences were obtained by Posakony *et al.* (1981) for a set of

---

long sequence represented at least 1000 times per haploid genome; B9-1 is another member of the 2109B family; 2111 is a 155 nt long sequence represented about 12,500 times per haploid genome; and 2137 is a 226 nt long sequence represented about 530 times per haploid genome [see Posakony *et al.* (1981) and Anderson *et al.* (1981) for primary sequences and sequence organization surrounding representatives of these repeat families]. The probe fragments were terminally labeled by the Klenow polymerase reaction. U denotes upper and L lower strand in the gel electrophoresis system utilized for separating the complements of each cloned insert. Note that the two members of the 2109B family display the same patterns of transcripts. The left-most lane shows a lower exposure of the lane hybridized to the 2109A upper strand; otherwise all exposures are the same. (b) Developmental persistence of familial sets of interspersed transcripts. As in (a), except that total poly(A) RNAs from 16-cell (16), blastula (B) and gastrula (G) stage embryos were utilized. Results are shown only for two of the probes utilized in (a), as indicated. It is not clear whether the developmental differences observed in each experiment are due to transcriptional modulation or to genomic polymorphisms amongst the animals from which the RNAs were obtained, or both. (c) Persistence and augmentation of specific interspersed transcripts identified by single copy probes. Poly(A) RNA was copied into cDNA and cloned. Interspersed transcript fragments containing the 2109A repeat sequence were selected, and single copy probes adjacent to the regions including the repetitive sequence element prepared. These were then utilized for RNA gel blot experiments as in (b). Each probe identifies a single transcript, bearing a given copy of the 2109A repeat sequence. [From J. W. Posakony, C. N. Flytzanis, R. J. Britten, and E. H. Davidson (1983). *J. Mol. Biol.* **167**, 361.]

nine cloned genomic repeats that are represented in interspersed maternal poly(A) RNAs. The large majority are blocked by translation stop signals in every reading frame. One possibility is that these repeats, and the nonconserved regions of sequence in which they are embedded, are located in long 3' tails appended to otherwise normal messages as a consequence of transcriptional readthrough, or failures of processing in the oocyte nucleus. Histone gene transcripts with this form of structure have been demonstrated in growing amphibian oocyte nuclei (Diaz *et al.*, 1981; also reviewed in Chapter V), though unlike the interspersed poly(A) RNAs the histone readthrough transcripts are confined to the germinal vesicle. However, the only complete sequence available for an interspersed maternal transcript lends no support to this interpretation. Calzone *et al.* (1986a) obtained the primary sequence of a 3.8 kb maternal transcript found in *S. purpuratus* eggs. This polyadenylated interspersed transcript appears in many ways to be typical. Thus it is colinear with genomic DNA; it is a low prevalence RNA present in only about 400 copies per egg; and it continues to be synthesized throughout embryological development. The number of these molecules per embryo has approximately doubled by pluteus stage. This transcript contains several different repetitive sequence elements as well as single copy sequence, and one of these elements, located at the 5' terminus of the RNA, is unusual in that it is represented exclusively in the same polarity in the other maternal transcription units in which it is included. The expressed copies of this repeat sequence are very highly conserved, and the single family of maternal RNAs defined by its presence could account for several percent of the total poly(A) RNA of the egg (Calzone *et al.*, 1986b). The most important feature of the particular interspersed transcript studied by Calzone *et al.* (1986a,b) is that as such it cannot be translated. It contains no significant unbroken open reading frame, though it is conceivable that translatable mRNA sequence could be obtained from it by a series of conventional splicing reactions. Sensitive nuclease protection assays have failed to reveal any such mature mRNA product in the egg or embryo, however.

Current evidence regarding the structure of interspersed maternal poly(A) RNAs in sea urchin eggs indicates that at least some of these molecules resemble heterogenous nuclear RNAs rather than mRNAs. Yet they are clearly localized in the cytoplasm, since that is where almost all the poly(A) RNA is observed in *in situ* hybridizations with ^{3}H-poly(U) (Venezky *et al.*, 1981), and since interspersed poly(A) RNAs constitute a major fraction of total poly(A) RNA mass, and certainly a large fraction of the poly(A) RNA molecules. The nuclear RNA of sea urchin embryo and adult cells is also known to display an interspersed sequence organization (Smith *et al.*, 1974; Scheller *et al.*, 1978). The interspersed sea urchin *egg poly(A) RNAs* are of similar size to sea urchin embryo *nuclear RNA*, the weight-average of which under denaturing conditions falls in the range 5 to 9 kb (Kung, 1974; Dubroff and Nemer, 1975). As does nuclear RNA, these maternal molecules contain large amounts of nontranslatable sequence, as well as mRNA sequences. An important point, however, is that the maximum aggregate complexity of the interspersed egg RNAs, about 3×10^7 nt, is probably only about 1/10th that of oocyte nuclear RNA (Chapter V), so that their deposition in the cytoplasm during oogenesis must be regarded as highly selective rather than a random process.

(iii) Interspersed Maternal Poly(A) RNA in *Xenopus* Eggs

Renatured poly(A) RNA extracted from the cytoplasm of fully grown *Xenopus* oocytes was examined in the electron microscope by Anderson *et al.* (1982). There is no further nuclear RNA synthesis after germinal vesicle breakdown until midembryogenesis in *Xenopus*, and the heterogeneous RNA present in the cytoplasm of the fully grown oocyte is generally stable. Thus it can be safely assumed that the characteristics of fully grown oocyte cytoplasmic RNA are also those of egg and early embryo cytoplasmic RNA. The same forms of intermolecular structure were observed as in the renatured sea urchin egg poly(A) RNA shown in Fig. 2.5. As reviewed in the discussion of transcriptional processes during *Xenopus* oogenesis in Chapter V, this study also showed that interspersed poly(A) RNAs can be recovered from the enucleated cytoplasm of immature as well as fully grown oocytes, and that this form of RNA is rapidly synthesized and transported to the cytoplasm during the oocyte growth phase. About 68% of the mass of the poly(A) RNA in fully grown *Xenopus* oocytes is included in transcripts that display an interspersed sequence organization, almost exactly the same fraction as found in *S. purpuratus* egg poly(A) RNA. This is an impressive correspondence, considering that as shown in Table 2.1, there are on the average a thousand times more complex class RNA molecules per egg in *Xenopus*. The sequence organization of *Xenopus* genomic DNA is similar to that of sea urchin DNA, though the genome is about four times as large. Both genomes consist largely of single copy sequences in which short repeat elements are interspersed at intervals of one to a few kilobases (Davidson *et al.*, 1973; Graham *et al.*, 1974). In both species this organization is reflected directly in the structure of the interspersed maternal poly(A) RNAs. Another similarity is that the complexity of the *interspersed poly(A) RNA* of late *Xenopus* oocytes, which must be equal to or less than that of the *total egg RNA* (Table 2.1), is again ≤10% of the complexity of germinal vesicle nRNA (see Chapter V).

The interspersed poly(A) RNA of *Xenopus* oocyte cytoplasm is largely nontranslatable. Richter *et al.* (1984) isolated this RNA fraction by use of cellulose-ethanol columns on which were trapped RNA containing double-stranded intermolecular duplexes formed by renaturation (see Fig. 5.16 for details). The translational activity of the interspersed RNA was examined in three different ways. In a cell-free protein synthesis system the denatured

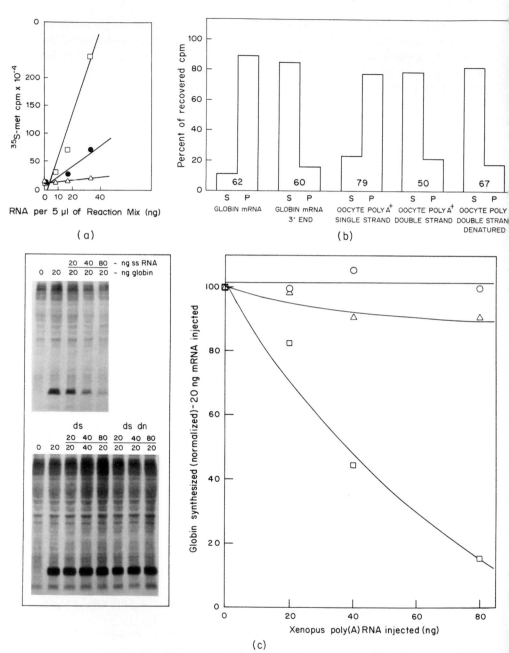

Fig. 2.7. Translational inactivity of the interspersed maternal poly(A) RNA of *Xenopus* oocytes. Total stage 6 (i.e., fully grown) *Xenopus* oocyte poly(A) RNA was renatured to C_0t 600 *M* sec in a 58% formamide, 0.75 M NaCl medium at 30°C, and then separated on an ethanol-cellulose column into RNA fractions eluting in 15% ethanol ("single-stranded" fractions) and in water ("double-stranded" fraction). The double-stranded fraction contains most of the mole-

interspersed poly(A) RNA displayed only 5-10% of the activity of the poly(A) RNA fraction lacking repetitive sequences, as illustrated in the experiment reproduced in Fig. 2.7(a). Nor does most of this RNA assemble into polysomes on injection into *Xenopus* oocyte cytoplasm, as does the noninterspersed poly(A) RNA [Fig. 2.7(b)]. A further, and perhaps more exacting test from the same study is reproduced in Fig. 2.7(c). The capacity of the *Xenopus* oocyte to support protein synthesis from exogenous mRNAs is limited, and thus injected mRNAs compete with each other, and with endogenous mRNAs for translational components. Injection of globin mRNA followed by injection of increasing amounts of *Xenopus* oocyte poly(A) RNA that lacks renaturable repetitive sequences therefore results in a progressive, competitive decrease in globin peptide synthesis. In contrast, injection of the interspersed poly(A) RNA fraction has very little effect on globin synthesis. The experiments shown in Fig. 2.7(b) and (c) thus indicate that the translational machinery of the *Xenopus* oocyte does not regard most of the interspersed poly(A) RNA molecules as usable message.

(iv) Interpretations of Interspersed Maternal Poly(A) RNA

The significance of the interspersed maternal poly(A) RNA remains an unsolved biological problem. The experiments of Richter *et al.* (1984) (Fig. 2.7) suggest that no more than about 10% of the interspersed poly(A) RNA of *Xenopus* oocytes could consist of translatable message, with or without extensive 3' terminal sequences. Nor, as we have seen, can an interspersed

---

cules that include intermolecular repetitive sequence duplexes formed by renaturation (Anderson *et al.*, 1982), while most mRNAs, including added globin mRNA, are eluted in the single-stranded fraction. (a) Comparison of translational activity of heat denatured double-stranded (△), single-stranded (□) and unfractionated poly(A) RNAs (●). The relative translational activities recorded in this experiment were: Single-stranded RNA, 1.0; double-stranded, 0.04; unfractionated, 0.25. (b) Partition of RNA fractions with polysomes (P) or supernatant (S) oocyte cytoplasmic compartments after injection. Oocytes were injected with globin ^{125}I-mRNA; 3' fragments of hydrolyzed globin ^{125}I-mRNA; single-stranded oocyte ^{125}I-poly(A) RNA; or denatured double-stranded oocyte ^{125}I-poly(A) RNA, prepared as in (a). After 12 hr incubation the oocytes were homogenized and the postmitochondrial supernatants centrifugally fractionated. The numbers indicate the percentage of total injected radioactive RNA recovered (supernatants + pellets). (c) Competition of injected oocyte poly(A) RNAs with injected globin mRNA in the *Xenopus* oocyte. Oocytes were injected with 20 ng of globin mRNA, cultured for 12 hr, and then injected with the indicated quantities of double-stranded (ds), single-stranded (ss) or denatured double-stranded (dn) poly(A) RNA, prepared as in (a). After 12 hr they were injected with ^{3}H-leucine, and 1 hr later homogenized, the newly synthesized proteins displayed by gel electrophoresis, and the gels autoradiographed (left side). The globin band is clearly evident toward the bottom of each gel, except for the control lanes labeled "0" at the left. The relative amount of globin synthesized in oocytes injected with double-stranded (o), denatured double-stranded (△), and single-stranded RNA fractions (□) was determined by densitometry and is plotted on the right. [From J. D. Richter, D. M. Anderson, E. H. Davidson, and L. D. Smith (1984). *J. Mol. Biol.* **173**, 227.]

polyadenylated transcript of the sea urchin egg be interpreted as mature mRNA. An immediate consequence is that the cytoplasmic poly(A) RNA content of the eggs of these species is more than twice their true maternal mRNA content.

Further data are required to exclude any of the possible speculative interpretations of the conclusion that the interspersed poly(A) RNAs resemble nuclear RNAs, though they are located in the egg or oocyte cytoplasm. One obvious possibility is that the interspersed poly(A) RNAs are in fact normal nuclear RNAs that "leaked" into the cytoplasm from the germinal vesicle, or that during the long period of oocyte growth were adventitiously packaged and transported to the cytoplasm along with true mRNAs. There they might be stabilized, just as are the maternal messages, but they would play no further functional role. An attempt to reconcile this interpretation with several of the facts that have been established leads to awkward, though certainly not insurmountable problems. The "leakage" hypothesis would not predict that the mass fraction of the cytoplasmic poly(A) RNA that consists of interspersed transcripts would be the same in eggs as different in size and RNA content as are those of the sea urchin and the frog. Nor is random "leakage" of transcripts from the germinal vesicle consistent with the low complexity displayed by interspersed poly(A) RNAs in both *Xenopus* and sea urchin compared to that of the respective nuclear RNAs.

An alternative possibility suggested by Costantini *et al.* (1980) is that the interspersed poly(A) RNAs are in fact pre-mRNAs that will be processed during embryogenesis, perhaps in a spatially and temporally regulated fashion. Were this a common phenomenon it would be expected that higher molecular weight forms, as well as the mature messages, would often be observed using probes for *bona fide* egg messages. However, observations on a series of specific maternal messages, e.g., the actin mRNAs (Shott *et al.*, 1984), tubulin mRNAs (Alexandraki and Ruderman, 1985a,b) and on prevalent unidentified poly(A) RNAs identified by cloned probes (see, e.g., Dworkin and Hershey, 1981; Flytzanis *et al.*, 1982; Colot and Rosbash, 1982) have so far failed to reveal the pattern of unprocessed and processed transcripts that would be predicted. However, this remains an incompletely examined hypothesis.

A feature of the interspersed maternal poly(A) RNA that is important to integrate into any proposed explanation is that the same transcripts as stored in the egg evidently continue to be synthesized in the embryo nuclei. During development the steady state *cytoplasmic* content of interspersed RNAs declines progressively. Thus Hough-Evans *et al.* (1977) found that the complexity of total *cytoplasmic* RNA exceeds that of the *polysomal* mRNA in early sea urchin embryos, but in pluteus stage embryos this difference is no longer observed, and Scheller *et al.* (1978) found that in late embryos the only prevalent repeat transcripts are nuclear. Similarly, Anderson *et al.* (1982) showed that in advanced (stage 41) tadpoles the component of cytoplasmic poly(A) RNA that is capable of forming multimolecular complexes by renaturation has largely disappeared. Experiments with several cloned interspersed maternal poly(A) RNA species in the sea urchin have demonstrated the presence of the same transcripts in the nuclei of late embryos, where they are evidently being synthesized (see, e.g., Lev *et al.*, 1980; Thomas *et al.*, 1982; Calzone *et al.*, 1986b; S. A. Johnson and E. H. Davidson, unpublished data). Thus the interspersed maternal poly(A) RNAs seem to be representative of a large class of transcription units that operate both during oogenesis and in embryogenesis. Their distinctive feature is that some of them persist in the oocyte cytoplasm, apparently in a stable form. This suggests that in the oocyte nucleus they are processed inefficiently, or perhaps not at all, since at least in *Xenopus* unprocessed transcripts may exit from the germinal vesicle if polyadenylated (see Chapter V). Thus the interspersed poly(A) RNAs could represent transcription units the products of which are processed efficiently only in somatic cells of later developmental stages, but which are transcribed along with other members of the maternal gene set in the oocyte nucleus (and in embryo nuclei). An unsatisfactory aspect of this interpretation is that it again implies a lack of function for a majority of the stored mass of *maternal* poly(A) RNA, though the suggestion is made that the homologous zygotic nuclear transcripts might indeed be functionally significant. In summary, though the biological meaning of the interspersed maternal poly(A) RNA remains mysterious, this is in a sense a reflection of our incomplete knowledge of somatic cell nuclear RNA, which the maternal interspersed transcripts much resemble. Both the sequence content and the mass of the rapidly synthesized heterogeneous RNA that is confined to the nuclei of somatic cells remain to be wholly explained (see, e.g., Wold *et al.*, 1978; Salditt-Georgieff and Darnell, 1982; cf. the non-translatable *bxd* locus transcripts of *Drosophila* described in Hogness *et al.*, 1985). The same is true of a majority of the mass of the *stable* polyadenylated RNA stored in the egg.

The remainder of the egg poly(A) is largely associated with true maternal message. Much has been learned of the role of maternal mRNA in early development, and it is to this area that the remainder of this Chapter is devoted.

4. UTILIZATION OF MATERNAL mRNA IN DEVELOPMENT

(i) Maternal mRNA and Protein Synthesis in the Early Sea Urchin Embryo

Cytoplasmic Polyadenylation of Maternal mRNA

Both polyadenylated and poly(A)-deficient maternal mRNAs are stored in sea urchin eggs. The total quantity of maternal message in the unfertilized egg can be estimated from the mass of polysomal message that can be recovered from early cleavage stage embryos. The presence of poly(A)-deficient mRNA has been demonstrated by *in vitro* translations carried out with RNA extracted from unfertilized eggs. The criterion applied is possession of a 3′ oligo(A) tract sufficient in length to promote binding to oligo(dT) cellulose (Ruderman and Pardue, 1977; Duncan and Humphreys, 1981). Nonhistone as well as histone mRNAs are included in the poly(A)-deficient class. Direct comparisons of the sequence content of poly(A) RNA and poly(A)-deficient mRNA preparations show that essentially all mRNA species except the major histone mRNAs are found in both fractions (Costantini *et al.*, 1980; Duncan and Humphreys, 1981). In addition, the same set of prominent protein species is synthesized in cell-free systems loaded with either RNA fraction (Brandhorst *et al.*, 1979). Some mRNAs are preferentially polyadenylated, however, and others are preferentially distributed to the class of transcripts deficient in terminal poly(A) tracts.

Measurements carried out on the eggs of several sea urchin species demonstrate that within the first few cleavages the poly(A) content of the maternal RNA increases about twofold (D. W. Slater *et al.*, 1972, I. Slater *et al.*, 1973; Slater and Slater, 1974; Wilt, 1973, 1977). This is due both to adenylation of previously nonadenylated RNA species, and to lengthening of preexisting poly(A) RNA tracts. As discussed below there is no significant mass contribution of newly synthesized zygotic mRNA during this early period. However, labeled adenosine is actively incorporated into the poly(A) tracts of the maternal RNA in early embryos, and from the kinetics observed it is clear that the poly(A) sequences are continuously being degraded and resynthesized (Dolecki *et al.*, 1977; Wilt, 1977). Both poly(A) turnover and the initial accretion of new terminal poly(A) tracts are cytoplasmic processes that can be demonstrated in activated enucleate egg fragments (Wilt, 1973). Nonmitochondrial poly(A) polymerase activity is present in the cytoplasm of unfertilized eggs, in enucleated merogones, and in soluble cytoplasmic fractions of early embryos (Morris and Rutter, 1976; Slater *et al.*, 1978; Egrie and Wilt, 1979). This enzyme is probably responsible for the observed synthesis of poly(A) tracts on preformed mRNA molecules. In eggs of *Lytechinus pictus* the average poly(A) tract length increases from about 45 nt to about 60 nt within 2 hr of fertilization. The fraction of the *total RNA* that can be bound by oligo(dT) cellulose also rises, from about 1.5% in the unfertilized egg to 2% in the 2-cell zygote (Wilt, 1977). Postfertilization polyadenylation appears to occur preferentially on translatable message. Thus in the unfertilized egg two-thirds of the poly(A) RNA consists of interspersed transcripts that are not mRNAs, while in early cleavage embryos 75-80% of the RNA by then bearing 3′ poly(A) tracts is found associated with the newly assembled polysomes. This result is obtained both in measurements of total poly(A) tracts carried out by hybridization with ^{3}H-poly(U), and in analyses of the distribution of newly synthesized poly(A) tracts (Slater *et al.*, 1972; Wilt, 1973; Dolecki *et al.*, 1977).

Quantity and Complexity of Maternal mRNA

The amounts of poly(A) mRNA and of poly(A)-deficient mRNA in the unfertilized egg of *Lytechinus pictus* are approximately equal, judging from the translational activity manifested by these fractions *in vitro* (Ruderman and Pardue, 1977). A larger proportion of the mRNA activity is located in the poly(A) RNA fraction of *Arbacia punctulata* eggs, but by later cleavage or early blastula stages, in both these species and in *Strongylocentrotus purpuratus* the amounts of poly(A) mRNA, and poly(A)-deficient mRNA appear to be approximately equivalent (Nemer, 1975; Ruderman and Pardue, 1977; Brandhorst *et al.*, 1979). Assuming an equal amount of poly(A) mRNA and of poly(A)-deficient mRNA in the unfertilized egg of *S. purpuratus*, the mass of maternal message can be estimated as follows. This egg contains about 2.8 ng of total RNA (Table 2.1). Thus, since 1.5% is poly(A) RNA, and about 65% of the poly(A) RNA consists of nontranslatable interspersed maternal transcript, the content of *bona fide* message would be about 30 pg per egg (0.015 × 3 ng × 0.35 × 2).

A similar estimate can be derived independently from consideration of translational rates in the early embryo. From the time course of polysome assembly determined by Goustin and Wilt (1981), and other observations reviewed below, it can be concluded that by about 4-5 hr after fertilization in *S. purpuratus* (i.e., the 8- to 16-cell stage of early cleavage) the recruitment of maternal mRNA into polysomes is in quantitative terms approaching completion, although as noted below, some maternal mRNAs continue to be utilized later in development. However, there has yet been little contribution from zygotic transcripts. Beyond the 8-cell stage the further accretion of polysomes clearly involves addition of significant quantities of newly synthesized mRNA. At this stage about 30% of the total embryo ribosomes are engaged in polysomes. The weight fraction of the total polysomal RNA that is mRNA is about 4%, as measured directly for polysomes of later *S. purpuratus* embryos (see Chapter III). The same value can also be derived from the weight average length of the polysomal mRNA in early embryos, about 2900 nt (Nemer, 1975; Goustin, 1981), and the average number of ribosomes per polysome, which is reported as about 10 in *S. purpuratus* (Brandis and Raff, 1978) and 14 in *L. pictus* (Martin and Miller, 1983) (i.e., [2900 × 350 d]/ [10 × 2.2 × 10^6 d] ≈ 4%). The *S. purpuratus* egg contains about 2.4 ng of rRNA, and thus at the 8-cell stage the content of *polysomal* mRNA would be calculated as about 29 pg per embryo (0.04 × 2.4 ng × 0.3). The close agreement that now appears between the estimated mass of translatable mRNA stored in the unfertilized egg, i.e., 30 pg, and the mass of maternal mRNA in the polysomes of the early embryo may be partly accidental, but there is a direct implication that most of the maternal message is being utilized for translation within a few hours of fertilization. Presumably the maternal mRNA content in species other than *S. purpuratus* is proportional to the total RNA content (cf. Table 2.1). Thus for example in the egg of *Lytechinus pictus*, where the total RNA content is 1.8 ng, the maternal mRNA content would be expected to be about 19 pg.

The complexity of the true maternal mRNA has been estimated for two sea urchin species, by comparison of the set of single copy DNA sequences represented in unfertilized egg RNA with that represented in early cleavage stage polysomal RNA. Hough-Evans *et al.* (1977) prepared a single copy *S. purpuratus* DNA tracer that after several cycles of prehybridization at termination with egg RNA would react with egg RNA to about 77%, compared to the initial single copy DNA reaction of 2.7%. The enriched "egg sequence" tracer reacted with total cytoplasmic RNA of 16-cell embryos to the same extent as with unfertilized egg RNA. However, only 73% of this reaction was observed with *polysomal* mRNA from 16-cell embryos. To avoid adventitious contamination with cytoplasmic RNP particles the polysomal mRNA used for this experiment was prepared by puromycin release from a centrifugal polysome fraction. Duncan and Humphreys (1981) reported a similar value in a comparison of the complexities of total egg RNA and early cleavage stage polysomal RNA in *Tripneustes gratilla* embryos. Their measurements indicate that about 80% of the egg RNA sequence set is represented in the polysomal mRNA (which, however, had not been purified by puromycin or EDTA release). In further experiments of Hough-Evans *et al.* (1977) with *S. purpuratus* RNA fractions, the same "egg sequence" tracer was reacted with total cytoplasmic and polysomal RNA fractions extracted from progressively later embryos. An example of such a reaction is illustrated in Fig. 2.8(a), and an overall summary of these data is shown in Fig. 2.8(b). As development proceeds the overall sequence complexity of the cytoplasmic RNA *declines*, while the complexity of the polysomal RNA at each stage remains clearly lower than that of the cytoplasmic RNA, until the two values converge late in embryogenesis. These measurements show that of the total maternal sequence set, the complexity of which is 37×10^6 nt (Table 2.1), the component *not* represented in polysomal RNA has a complexity of 10-14 $\times$ 10^6 nt at the cleavage to early blastula stages, respectively. Almost certainly these nonpolysomal cytoplasmic RNAs are included in the class of interspersed maternal RNAs discussed above. Thus as concluded earlier, Fig. 2.8(b) implies that these RNAs have largely disappeared from the cytoplasm, or at least decreased sharply in steady state content by late embryogenesis.

From the quantity of maternal mRNA, and its complexity, may be calculated the average number of molecules of each message present in the egg. According to cDNA hybridization kinetics reported by Duncan and Humphreys (1981), more than half the mass of maternal *polysomal mRNA* in the 3

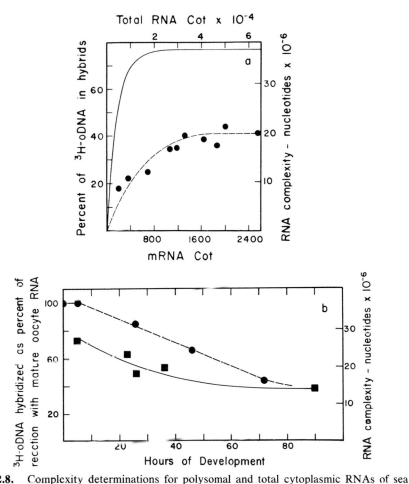

Fig. 2.8. Complexity determinations for polysomal and total cytoplasmic RNAs of sea urchin embryos. (a) Hybridization of preselected single copy ^{3}H-DNA tracer with *Strongylocentrotus purpuratus* gastrula polysomal RNA. The tracer was prepared by two cycles of prehybridization between total egg RNA and a single copy ^{3}H-DNA fraction, after each of which the hybridized DNA molecules were collected by hydroxyapatite chromatography. On reaction with egg RNA 77% of this tracer could be hybridized at kinetic termination. The polysomal RNA was prepared by puromycin release from a pelleted polysomal centrifugal fraction (Galau *et al.*, 1976). The abscissa shows the mRNA C_0t calculated on the basis that mRNA comprised 4% of the mass of polysomal RNA (see text). The left ordinate gives the percentage of the ^{3}H-DNA in hybrid. The right ordinate is calibrated in terms of RNA complexity. The solid curve represents the kinetics of the reaction of the same tracer preparation with egg RNA, as established in numerous other experiments. The dashed curve represents a least-squares solution assuming a single pseudo-first-order kinetic component with the additional assumption that the ordinate intercept is zero. At termination 41 ± 2% of the ^{3}H-DNA has reacted (errors indicate one standard deviation). (b) Persistence of the maternal sequence set in total embryo cytoplasmic RNA and in polysomal RNA. Terminal values were obtained from RNA excess hybridization reactions such as those shown in (a). The left ordinate gives these values as fractions of the reaction of the preselected tracer with the parental RNA (i.e., after normalization to 77% reaction). Absolute complexities calculated from the complexity of egg RNA (3.7×10^7 nt; see Table 2.1) are shown on the right ordinate. Total egg RNA and embryo cytoplasmic RNAs, (●); embryo polysomal RNAs (■). [From B. R. Hough-Evans, B. J. Wold, S. G. Ernst, R. J. Britten, and E. H. Davidson (1977). *Dev. Biol.* **60**, 258.]

hr (4- to 8-cell) *T. gratilla* embryo is included in the low prevalence, complex class of mRNA sequences. The more abundant sequences in the polysomal mRNA range in prevalence from 10 to 100 times the sequence concentration of the complex class transcripts, which are present on the average at about 1.6×10^3 molecules per egg (Table 2.1). The prevalence distribution of *total maternal mRNA* is similar to that of *total egg poly(A) RNA*, for *L. pictus, T. gratilla,* and *S. purpuratus* (Wilt, 1977; Duncan and Humphreys, 1981; E. H. Davidson, unpublished data). Thus given that for *S. purpuratus* about 50% of the *mRNA* mass is included in the complex class, and that its complexity is 73% of that of the total egg RNA, there would be about 1000 copies of each such sequence in the polysomes of a cleavage stage embryo (0.5×30 pg maternal mRNA $\equiv 2.6 \times 10^{10}$ nt; $[2.6 \times 10^{10}$ nt$]/[0.73 \times 37 \times 10^6$ nt$] = 960$). Considering that the number average mRNA length lies in the range 2000-3000 nt, the values available suggest that about 11,000 *different* mRNA sequences present at this low prevalence are being translated (i.e., $0.73 \times 37 \times 10^6/2.5 \times 10^3$). Such estimates are to be regarded as approximate, due to complications such as the existence of nonhomologous 3' terminal mRNA sequences appended to homologous protein coding regions, as, e.g., in the actin gene family (see Chapter IV). Nonetheless, they provide a general idea of the number of different structural genes represented in the maternal mRNA, the products of which are presumably required in the early embryo.

Analyses of the proteins synthesized in the early embryo by two-dimensional gel electrophoresis resolve at most about 10^3 diverse species (Tufaro and Brandhorst, 1979). The amount of labeled protein entering these gels, and their resolution, in general permit identification only of proteins that are coded by messages more prevalent than those included in the rare, or complex class sequence set, though certain proteins that are resolved with particular efficiency may be detected at lower levels. On this basis the complexity of the more prevalent maternal mRNA fraction would be about one-tenth that of the complex class (i.e., $10^3/10^4$), a conclusion consistent with the cDNA kinetic analyses. A population study carried out on about 200 moderately abundant egg poly(A) RNAs identified by individual cDNA clones shows that the prevalence of most such sequences falls in the range 10^4-10^5 molecules per egg (Flytzanis *et al.*, 1982), i.e., compared to slightly over 10^3 molecules per egg for complex class sequences.

Fertilization and the Activation of Protein Synthesis

In the unfertilized sea urchin egg only about 1% or less of the ribosomes are included in polysomes, and protein synthesis occurs at a relatively low rate. Within minutes of fertilization this rate begins to increase, and it continues to accelerate for the next several hours, ultimately achieving a level over 100 times that supported by the translational apparatus active in the unfertilized egg (Regier and Kafatos, 1977). Measurements of the absolute rate of protein synthesis attained in midstage embryos of *S. purpuratus* fall within the range 650 ± 250 pg hr^{-1} (Fry and Gross, 1970; Regier and Kafatos, 1977; Seale and Aronson, 1973; Goustin and Wilt, 1981), and similar values have been measured for other species (reviewed in Davidson, 1976). In these studies the rate of accumulation of precursor in embryo protein is converted to absolute protein synthesis rate by reference to the amino acid pool specific activities, or to the specific activity of aminoacyl tRNA, the most direct precursor. Various measurements made by these diverse means agree well, and are also consistent with the rate of protein synthesis predicted from polysome content (Davidson, 1976). By early blastula stage the rate of protein synthesis amounts to about 1% of the total protein content per hour. Nearly the same fraction of the newly synthesized protein, 0.8%, is lost per hour by turnover (Berg and Mertes, 1970), and the total protein content of the embryo during development thus remains almost constant.

The activation of translation is of course only one manifestation of the complex series of physiological, cytological and biochemical changes set in train by fertilization (see reviews in Vacquier, 1980; Shapiro *et al.*, 1981; Epel, 1982). The mechanism responsible for the dramatic increase in the rate of embryonic protein synthesis has been much investigated, and it appears that there are two immediate causes that function synergistically. These are a small increase in the rate of translation per active ribosome, and a 40- to 60-fold expansion in the number of active ribosomes, manifested by the rapid assembly of new polysomes during early cleavage. During this period the ribosomes and mRNA incorporated in the newly forming polysomes are almost exclusively maternal in origin, as are all other components of the translational apparatus, including initiation factors, tRNAs, aminoacyl tRNA synthetases, etc. (see review in Davidson, 1976). From measurements in which the mRNA was labeled with an exogenous precursor and the precursor pool specific activity determined, Humphreys (1971) estimated the mass of newly synthesized message that enters the polysomes during early cleavage. These experiments indicated that no more than about 10% of the polysomal mRNA present in 2 hr *Lytechinus pictus* embryos (4- to 8-cell stage at 18°C) can be accounted for as newly synthesized zygotic transcripts, the balance consisting of maternal mRNAs. The rapid assembly of polysomes following fertilization was initially observed by Monroy and Tyler (1963) and Stafford *et al.* (1964), and quantitative estimates of polysome assembly were reported by Infante and Nemer (1967), Humphreys (1971), and others. In Fig. 2.9 is reproduced a recent analysis of the relation between polysome assembly and the absolute rate of protein synthesis during the early cleavage of *S. purpuratus* embryos (Goustin and Wilt, 1981). Protein synthesis rate was determined from the incorporation of ^{3}H-lysine into embryo protein, and from the lysine pool specific activity, and the polysome

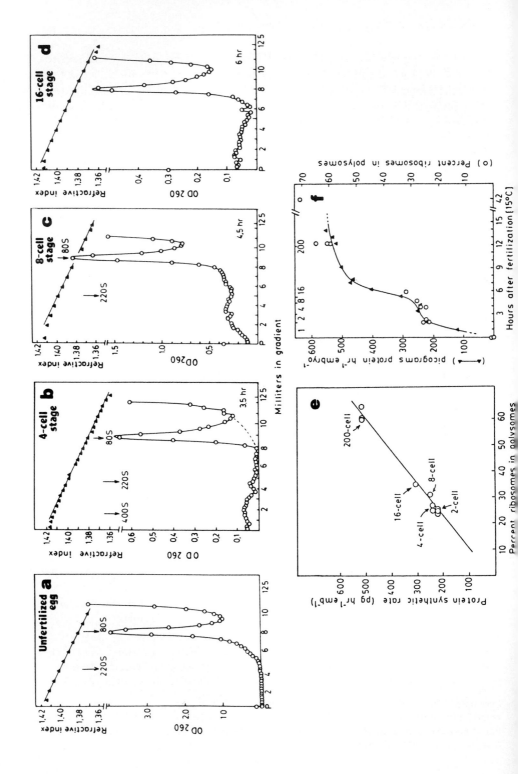

content was measured over the same time periods [Fig. 2.9(a-d)]. The data summary in Fig. 2.9(e) demonstrates that the *protein synthesis rate increases directly with polysome content.*

It can be seen in Fig. 2.9(f) that there is a pause in the assembly of polysomes around the 4- to 8-cell stage (also reported in Humphreys, 1971, for *L. pictus* embryos). As mentioned above this is most likely the point at which most of the available maternal mRNA has been incorporated into polysomes. The further increase in protein synthesis rate and polysome content after the 16-cell stage is probably due largely to incorporation of newly synthesized zygotic message. Several forms of evidence support this interpretation. Actinomycin-treated embryos in which virtually all RNA synthesis is repressed first display a decrease in histone protein synthesis compared to controls at the 16- to 32-cell stage, caused by inhibition of zygotic histone gene transcription (Kedes *et al.*, 1969). Earlier than this no significant effect of actinomycin on protein synthesis is observed (Gross, 1967). Actinomycin does not prevent assembly of maternal mRNA into polysomes in enucleate sea urchin egg fragments, nor does it materially affect the spectrum of proteins synthesized (Sargent and Raff, 1976). An independent argument derives from an electron microscope study of early cleavage stage polysomes in *L. pictus* embryos, carried out by Martin and Miller (1983). These observations provide a structural criterion for identification of the maternal mRNA molecules in the initial process of assembly into polysomes. As shown in Fig. 2.10, such polysomes are distinguished by long 3′ ends free of ribosomes, a feature that is commonly observed shortly after fertilization, but that disappears within 2 hr. By this criterion as well, assembly of previously unused maternal mRNA into polysomes approaches completion within the first several cleavages. An interesting additional observation illustrated in Fig. 2.10 is that the ribosome packing density appears maximal even in the partially loaded polysomes, implying the presence of a non-limiting supply of the components needed for translational initiation.

Though assembly of polysomes accounts quantitatively for the increase in protein synthesis rate once cleavage has begun, there is also an initial boost in the translational capacity of the embryo due to an increase in the rate of

---

Fig. 2.9. Polysome assembly and protein synthesis rate in early sea urchin embryos. (a)-(d) Increase in polysome content after fertilization. Figures show absorbance profiles (260 nm) of postmitochondrial supernatants prepared from *Strongylocentrotus purpuratus* embryos at the indicated stages: (a) Unfertilized eggs; (b) 2-cell stage; (c) 8-cell stage; and (d) 16-cell stage. (e) Absolute protein synthesis rate as a function of fraction of ribosomes in polysomes. The rate of protein synthesis was measured from ^{3}H-lysine incorporation and lysine pool specific activity, in samples of the same embryos utilized for determination of the polysome content. (f) Developmental time course described by protein synthesis rate (▲) and polysome content (O) parameters. Cell number is given at the top of the Figure. The curve shown is a free-hand fit to the numerical data. [From A. S. Goustin, and F. H. Wilt (1981). *Dev. Biol.* **82**, 32.]

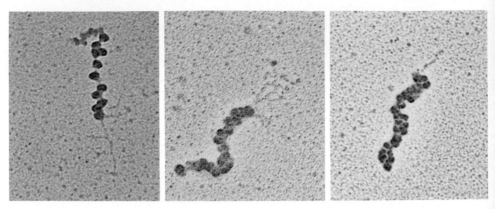

Fig. 2.10.   Electron micrographs displaying partially loaded maternal mRNAs in process of polysome assembly. *Lytechinus pictus* eggs were fertilized and homogenates prepared in a detergent solution 3 min to 60 min later. The polysomes were centrifuged onto grids through a sucrose-formalin cushion, and were shadowed before viewing. The individual plates show examples of "tailed" polysomes, in which the 5′ end of the message has been loaded with closely packed ribosomes. The average center-to-center ribosome distance is estimated to be about 135 nt. The initially loaded ribosomes had not yet completed their transit along the message at the moment of fixation, this resulting in the ribosome-free 3′ "tail" regions. These often display secondary structures visualized as loops and hairpins. [From K. A. Martin, and O. L. Miller (1983). *Dev. Biol.* **98**, 338.]

ribosome translocation stimulated by fertilization. Brandis and Raff (1978, 1979) and Hille and Albers (1979) showed that the time required for translation (and release) of the average newly synthesized protein decreases by a factor of about 2.5 within 30 min of fertilization. Fertilization does not affect the size of polysomes (Humphreys, 1969; Brandis and Raff, 1978) and therefore the higher elongation rate is accompanied by at least an equivalent increase in initiation rate. A range of values has been reported for the absolute rate of translation (codons translated per unit time) in fertilized and unfertilized sea urchin eggs. Discordance among these measurements may be due to differences in temperature at which the observations were made, possibly to species differences, and most importantly, to diverse methods of measurement (see, e.g., Regier and Kafatos, 1977; Hille and Albers, 1979; Brandis and Raff, 1978, 1979; Goustin and Wilt, 1981). Given the polysome content, measurements based on the absolute rate of protein synthesis that depend on direct precursor pool specific activity determination are most likely to be accurate. Goustin and Wilt (1981) thus calculated an average translation rate of 1.8 codons sec^{-1} at 15°C for *S. purpuratus* embryos. From the rate at which the first recruited mRNAs are loaded with ribosomes after fertilization, as visualized in the electron microscope, Martin and Miller (1983) calculated a similar translation rate of 1.5 to 1.8 codons sec^{-1} for *L. pictus* embryos at 16°C. This rate is also consistent with the fractions of tailed polysomes such as shown in Fig. 2.10, that are observed during early cleavage.

The detailed molecular mechanism(s) responsible for either polysome assembly or translation rate increase after fertilization have so far remained elusive. Spirin (1966) argued that maternal mRNA is sequestered within RNP particles that are disaggregated after fertilization, thus releasing the previously "masked message" for assembly into polysomes and translation. Interest in this hypothesis has led to extensive investigation of the properties of the maternal mRNP particles found in sea urchin eggs, generally defined as postribosomal particles containing maternal poly(A) RNA. In velocity sedimentation analyses these particles sediment at about 60-65S. Their protein content varies according to the ionic medium to which they are exposed, from about 0.25 to 1.5 times their mass of RNA (Kaumeyer *et al.*, 1978). Young and Raff (1979) showed that by buoyant density and compositional criteria the maternal poly(A) RNP particles differ from those containing newly synthesized mRNA, which have a much lower protein content. Seven major protein species ranging in molecular weight from 4×10^4 to 1.4×10^5 d, as well as additional minor proteins, are present in the maternal RNP particles, and several of the proteins are not found associated with poly(A) RNAs derived from later embryo polysomes (Moon *et al.*, 1980; Moon, 1983). Characterizations of maternal RNP particles containing poly(A) RNA are not simple to interpret, however, since much of the egg poly(A) RNA evidently consists of interspersed poly(A) RNA and not message, and the particle population is thus at the least functionally diverse. When isolated and directly tested in wheat germ and reticulocyte cell-free translation systems, the RNP particles of unfertilized *S. purpuratus* and *Arbacia punctulata* eggs are found, within a factor of two, to be as efficiently translated as are the deproteinized poly(A) RNAs extracted from them (Moon *et al.*, 1982; Moon, 1983). On the other hand Winkler *et al.* (1985) found that reticulocyte ribosomes are unable to utilize the mRNP particles in sea urchin egg lysates, or to form initiation complexes with them, though once purified the mRNA of these particles engages in initiation complexes and is freely translated. However, there is yet no direct or functional evidence that "masking" of maternal message in stable RNP particles actually accounts *in vivo* for the almost complete repression of polysome assembly prior to fertilization. Among the other possibilities that may be considered are that maternal mRNA is physically sequestered in cellular compartments other than those where polysome assembly is to occur, e.g., in some of the membrane-bound vesicles with which the egg cytoplasm abounds, or in cytoskeletal domains that are altered on fertilization (cf. Chapter VI). An alternative class of explanation is that an expansion of the capacity of the sea urchin egg protein synthesis system to accept message occurs after fertilization. Thus it was shown by Colin and Hille (1986) that in fertilized or ammonia-activated sea urchin eggs injected globin mRNA *competes* with endogenous mRNA, just as in the cytoplasm of *Xenopus* oocytes, where it is established that translational capacity is limiting (see Chapter V). Thus the globin mRNA is able to capture up to 35% of the total translational activity, but the endogenous protein synthesis is decreased concomitantly. Some element(s) of the translational apparatus other than mRNA would therefore appear to be present in limiting amounts. Free met tRNA-40S subunits are present in unfertilized eggs and in early cleavage stage *S. purpuratus* embryos in excess, with reference to the amount of mRNA being translated, and thus scarcity of activated ribosomes required for the formation of initiation complexes is not rate-limiting during this period (Hille *et al.*, 1980, 1981). Many additional factors may be involved, however. Thus within minutes of fertilization several ribosomal proteins are phosphorylated (Takeshima and Nakano, 1983; Ward *et al.*, 1983; Ballinger *et al.*, 1984), though there is again no evidence that these changes are causally significant, and there exist large differences in phosphorylation patterns among sea urchin species. Increases stimulated by fertilization have also been noted in the translational activity of isolated ribosomes when these are tested in a cell-free sea urchin system (Danilchik and Hille, 1981); and in the activity of translational elongation factor eIF-2, accompanied by release of this factor from a particulate to a soluble form (Felicetti *et al.*, 1972; Yablonka-Reuveni and Hille, 1983). Furthermore, addition of soluble factors from the reticulocyte lysate, or of eIF-2 alone, stimulates protein synthesis in sea urchin egg lysates by a factor of about two (Winkler *et al.*, 1985). A major activating trigger is undoubtedly the shift in intracellular pH from about 6.8 to 7.4 that occurs immediately after fertilization. Prior to this shift injected globin mRNA is translated very poorly (Colin and Hille, 1986), and a variety of external ionic treatments that artifically increase pH within the egg also induce activation of endogenous protein synthesis (Grainger *et al.*, 1979; Shen and Steinhardt, 1979). Furthermore, the effect of increased intracellular alkalinity can be duplicated *in vitro*, i.e., in the sea urchin cell-free translation system (Winkler and Steinhardt, 1981; Winkler *et al.*, 1985). It can be concluded that the mechanisms by which stored maternal mRNAs are mobilized and translation rate is increased are probably multiple and synergistic. They could include changes in both the functional availability of mRNA, whatever may be the mechanism of mRNA sequestration in the unfertilized egg, as well as in the activity of various components of the translational machinery, at least some pH dependent, that are still to be defined.

Maternal Histone mRNAs

The best known maternal mRNAs in the sea urchin egg are those coding for the histones. A complex series of sequential shifts in histone variants occurs during early sea urchin development. Thus the chromatin of the earliest embryonic cells is constructed with maternal histone proteins; that of later cleavage cells with newly translated histones coded by maternal histone messages; and thereafter, as described in the following chapter, with two successive sets of histone subtypes coded by mRNAs transcribed in the embryonic genome. Histones stored in the unfertilized egg and assembled into the chromatin of early cleavage cells are usually termed "CS" (cleavage stage) variants. These include CSH2a, CSH2b, H3, and CSH1 (Newrock *et al.*, 1978a; Carroll and Ozaki, 1979; Childs *et al.*, 1979; Poccia *et al.*, 1981; Salik *et al.*, 1981). A distinct maternal form of H4 that is synthesized before fertilization and utilized only in the earliest cleavage stages has not been identified. Direct measurements of Salik *et al.* (1981) indicate that there are 50-100 pg of each CS histone variant stored in the unfertilized egg, a quantity that would suffice to provide chromatin histones for at least 150 diploid genomes. The functional competence of a significant fraction of the stored CS histones was demonstrated in separate experiments in which eggs were fertilized after exposure to NH_4Cl to promote polyspermy. An average of 50 male pronuclei were thus induced to form per egg, and the chromatin of these was at least partially assembled from the maternal CS histone pool. This result was also obtained in eggs in which all new protein synthesis had been blocked with emetine. The sperm nucleus contains its own special set of histone subtypes. Immediately after fertilization the sperm H1 variant is replaced by CSH1, while CSH2a, CSH2b and H3 replace the respective sperm subtypes during the G_1 and S phases preceding the first cleavage division (Poccia *et al.*, 1981). The large amount of CS histones stored per egg constitutes something of a mystery, since so far as is known CS histones are only incorporated into embryo nuclei during the first few cleavages. From the 16-cell stage on, newly synthesized "α" or "early" histone subtypes are the dominant chromatin forms (Newrock *et al.*, 1978a; Childs *et al.*, 1979). The CS histones are stable, and they remain detectable in certain cells that form early in cleavage and then divide no further (see Chapter IV).

CS histones are synthesized both before and after fertilization. In the unfertilized egg these are the only histone subtypes being translated (Herlands *et al.*, 1982). The histone message used for translation in the unfertilized egg is newly synthesized (Brandhorst, 1980), and it follows that genes coding for the CS histone subtypes are active in the female pronucleus. Translation of CS histones from maternal mRNA continues immediately after fertilization and until the 16-cell stage, after which it can no longer be observed *in vivo* (Newrock *et al.*, 1978a; Childs *et al.*, 1979). However, mRNA for at least CSH2a can be extracted and translated *in vitro* as late as the blastula stage.

Cell-free translation experiments of Childs *et al.* (1979) reveal that the unfertilized egg contains maternal mRNA for the α-histones as well as for the CS histones. Of the total maternal histone mRNA pool about 90% is α-histone message and 10% message for the CS histone variants (DeLeon *et al.*, 1983). The quantity of maternal message coding for H3, H2b, and H1 were measured by Mauron *et al.* (1982) using the probe excess titration method (see Appendix II). There are in *S. purpuratus* eggs about 9×10^5 molecules of each of the two core nucleosomal histone mRNAs and about 8×10^4 molecules of H1 mRNA. This amounts to ~1.1 pg of total histone mRNA of all species per egg, or 3.6% of the total estimated maternal mRNA. In comparison, the amount of *newly synthesized* histone mRNA accumulated is relatively insignificant up to the 16-cell stage of cleavage. Maxson and Wilt (1981) showed in measurements of the kinetics with which newly synthesized histone mRNA accumulates, that zygotic histone transcripts amount to only about 0.06 pg by the 4-cell stage, and about 0.19 pg by the 8-cell stage. These data prove that most histone message available in early cleavage is of maternal origin, as suggested earlier by the observation that both CS and α-histone synthesis at this stage is insensitive to actinomycin (earlier references reviewed in Davidson, 1976; Newrock *et al.*, 1978a). Translation of histone occurs continuously in the cleaving embryo, displaying no temporal correlation with the phases of the cell cycle, such as is commonly observed in somatic cells (Arceci and Gross, 1977).

Unlike the CS histone subtypes, translation of the α-histone maternal message does not begin until first cleavage (Wells *et al.*, 1981; see also below). An unusual mechanism is responsible for the delayed mobilization of this maternal mRNA, relative to the overall maternal mRNA pool. The maternal α-histone mRNAs are *physically sequestered within the nuclear compartment of the zygote*, and are not released until dissolution of the nuclear membrane at first cleavage. Only at this point does the α-histone mRNA appear in polysomes and its translation begin (Woods and Fitschen, 1978; Wells *et al.*, 1981; Herlands *et al.*, 1982; Showman *et al.*, 1982). A change correlated with this event, though of unknown functional significance, is the methylation of the 5' caps of stored α-histone mRNA, which also occurs between fertilization and the 2-cell stage (Caldwell and Emerson, 1985). Venezky *et al.* (1981) and DeLeon *et al.* (1983) showed by *in situ* hybridization that before fertilization most of the mass of maternal α-histone mRNA is located within the pronucleus of the egg, in contrast to total poly(A) RNA, and to other specific maternal mRNAs which are almost completely cytoplasmic in location (Angerer and Angerer, 1981; Showman *et al.*, 1982). CS histone mRNAs are not detected by the *in situ* hybridization probes used in these studies, and since the CS variants are being translated in the unfertilized egg and pre-first cleavage zygote their mRNAs must be cytoplasmic in location. The experiments of DeLeon *et al.* (1983) reproduced in Fig. 2.11 demonstrate the release of α-histone message at breakdown of the zygote nuclear membrane (~100 min after fertilization, at 15°C). Nuclear location of histone mRNA has also been demonstrated for *Arbacia punctulata* and *S. dröbachiensis*, as well as for *S. purpuratus*, by Showman *et al.* (1982). Eggs of these species were mechanically or centrifugally separated into nucleate and enucleate fragments and 98% of the α-histone mRNA

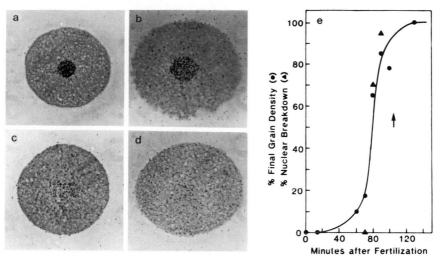

Fig. 2.11. Sequestration of maternal α histone mRNA in the zygote nucleus and its release at prophase of first cleavage in *Strongylocentrotus purpuratus* zygotes. (a)-(d) *In situ* hybridizations using a cloned radioactive probe that includes sequences of all five α-histone species. The high thermal stability of the probe-histone mRNA duplexes formed in the sections demonstrates that at the criterion conditions applied only the α-histone variants are detected. Phase contrast photographs are shown of sections of eggs seen through the autoradiographic emulsion. The eggs were fixed at (a), 70 min; (b) and (c), 80 min; and (d), 90 min postfertilization (at 15°C). (e) Correlation between amount of hybridization of *cytoplasmic* RNA to the histone gene probe and the amount of first cleavage nuclear breakdown occurring at each time point. Grain densities over the cytoplasmic areas of 30 randomly selected sections were averaged and normalized to the average plateau value at 130 min postfertilization for each point (●). Nuclear breakdown was monitored in sections of the same embryos (▲). The arrow indicates the time at which 50% of the embryos had undergone cleavage; >80% had cleaved 110-120 min postfertilization. [From D. V. DeLeon, K. H. Cox, L. M. Angerer, and R. C. Angerer (1983). *Dev. Biol.* **100**, 197.]

was recovered in the nucleate fraction. Histone mRNA is very easily released from the nucleus during cell fractionation, however. Thus mRNP particles containing maternal histone message have repeatedly been reported in postnuclear and postmitochondrial centrifugal fractions of egg homogenates, where they sediment behind the polysomes (see, e.g., Gross *et al.*, 1973; Skoultchi and Gross, 1973; Woods and Fitschen, 1978; Showman *et al.*, 1982). It is ironic that histone mRNA, which has long been considered a paradigm for maternal messages in general, in fact is unique in its localization and mobilization mechanisms. The conclusion that the maternal mRNA is located in cytoplasmic mRNPs in the unfertilized egg appears correct for other maternal messages, but is wrong for the α-histone messages.

The functional significance of the shift from CS to α-histone subtypes is not known. However, it is correlated with a clear change in internucleosome repeat length. Sea urchin sperm, in which the genome is both transcriptionally and replicationally inactive, displays the longest known repeat length, ranging from 240 to 260 nt in various species (Spadafora *et al.*, 1976; Keichline and Wassarman, 1977, 1979). In 8-cell embryo chromatin, which contains mainly CS histone subtypes, the repeat length has shortened to about 200 nt (Savić *et al.*, 1981). As the chromatin becomes loaded with α-histones the average repeat length increases again, and after the blastula-gastrula stages the internucleosome repeat length is stabilized at 210-220 nt (Keichline and Wassarman, 1977, 1979; Savić *et al.*, 1981).

Utilization of Other Specific Maternal mRNAs

Two well-supported generalities simplify interpretation of the mechanisms by which the synthesis of proteins other than the histones is activated after fertilization in sea urchin eggs. The first is that most of the resolvable protein species being translated during early cleavage are also synthesized before fertilization, though in general at manyfold lower rates (Brandhorst, 1976). Secondly, most of the nonhistone messages included in the maternal mRNA pool are mobilized with approximately similar kinetics. The near identity between the sets of proteins coded by the maternal mRNA and those actually translated shortly after fertilization has been established by *in vitro* translation experiments in which mRNA extracted from nonpolysomal mRNP particles is compared with polysomal mRNA (Infante and Heilmann, 1981; Bédard and Brandhorst, 1986a). Though there are certain specific exceptions as discussed below, the relative prevalence of the several hundred protein species that can be identified in two-dimensional electrophoretic gels is in general about the same in the products of polysomal and of nonpolysomal message translations, and for the corresponding recognizable proteins synthesized *in vivo* as well. Mobilization of the maternal message in the sea urchin egg is thus largely a quantitative, and not a qualitative, or sequence-dependent process.

Messages for a few prominent protein species are present in the egg cytoplasm immediately after fertilization, but are not translated until later in development (Bédard and Brandhorst, 1986a). These proteins are synthesized in cell-free systems loaded with mRNA extracted from the nonpolysomal RNP of 2-cell embryos, but not with message prepared from polysomes of the same stage embryos. One example is provided by a polypeptide that has been identified as a 90 kd heat shock protein (Bédard and Brandhorst, 1986a,b). After the 16-cell stage the message for this protein enters the polysomal compartment, and this protein becomes prevalent among the embryonic translational products. It may be relevant that mRNAs for these proteins are found mainly in the poly(A)-deficient message fraction of the unfertilized egg and of 2-cell stage embryos, while the same mRNAs are polyadenylated when recovered from the polysomes of later cleavage stage embryos. Exclusion of these particular mRNAs from the translational initiation processes in which other maternal mRNAs participate, and their late translational activation, would appear to indicate a sequence dependent mechanism that determines their developmental time course of expression. This is unlikely to consist simply of sequestration and release from the nucleus as with the α-histone mRNAs, since translation of these messages is not activated at first cleavage.

Only a few species of maternal mRNA have so far been identified, out of the thousands being translated in the early embryo. Among these is a set of three cytoskeletal actins which have been detected in the products of cell-free translations carried out with unfertilized egg RNAs (Merlino *et al.*, 1980; Infante and Heilmann, 1981; Crain *et al.*, 1981; Durica and Crain, 1982), and by reaction of cloned actin gene probes with unfertilized egg RNAs (Scheller *et al.*, 1981a; Crain *et al.*, 1981; Durica and Crain, 1982; Shott *et al.*, 1984). All of the members of the actin gene family of *S. purpuratus* have been cloned and characterized (J. J. Lee *et al.*, 1984), as described in Chapter IV, and the three cytoskeletal (Cy) actin genes represented in the egg RNA are identified as CyI, CyIIb, and CyIIIa (Shott *et al.*, 1984; Lee *et al.*, 1986a). Later in embryonic development transcription of all three of these genes is activated. CyI and CyIIb are initially expressed widely in the embryo, though their transcripts are later confined to certain cell lineages. The CyIIIa gene is expressed only in aboral ectoderm cells, and in the immediate precursors of these cells at the blastula stage (Angerer and Davidson, 1984; Cox *et al.*, 1986). It remains obscure why messages for this lineage-specific gene are stored in the egg (where they are not localized), or why these particular three cytoskeletal actin genes are represented in the maternal mRNA while two other cytoskeletal actin genes known to be functional later in embryogenesis are not. In any case, the prevalence of the maternal actin messages is only a few percent of their prevalence in later embryonic development. Lee *et al.* (1986a) established by probe excess titration that there are about 2100 molecules of CyI actin mRNA, about 1200 molecules of CyIII actin mRNA, and about 250 molecules of CyIIb actin mRNA per egg. These measurements place the prevalence of the maternal cytoskeletal actin mRNAs close to the average prevalence of the complex class maternal messages (see Table 2.1). Nonetheless, actin can be distinguished among the translation products of early cleavage embryos, and several isoforms are observed in two-dimensional gels, but at least some of these could represent posttranslational modifications rather than distinct products of different cytoskeletal actin mRNAs (Durica and Crain, 1982). In the egg of the sea star *Pisaster ochraceus*, which contains about five times more maternal RNA than does the egg of *S. purpuratus*, there are stored about 100 times more maternal actin mRNA molecules, and the number of these transcripts changes little throughout early development (Kovesdi and Smith, 1985). This suggests a requirement for amounts of these messages similar to those found in advanced sea urchin embryos (Chapter IV), but at early stages, before they could be supplied to the relatively large *Pisaster* egg by new transcription.

Detailed evidence also exists for several other mRNA species. Tubulins are synthesized early in sea urchin development from maternal message (Stephens, 1972; Raff and Kaumeyer, 1973; Raff, 1975; Alexandraki and Ruderman, 1985a). The amount of tubulin mRNA, identified by reaction with cloned tubulin gene probes, is low in the unfertilized egg relative to that appearing later in development (Alexandraki and Ruderman, 1981, 1985a). Messages for both α- and β-tubulin are stored, but it is not clear which of the many α-tubulin and β-tubulin genes are represented, though there is at least one α-tubulin gene that seems to function primarily in the synthesis of maternal tubulin mRNA (Alexandraki and Ruderman, 1985b). Most of the tubulin proteins used for the early cleavage mitoses and for construction of the cilia that appear initially at blastula stage are presynthesized during oogenesis. Thus only about 0.4% of the tubulins incorporated in the first mitotic apparatus are synthesized after fertilization (Bibring and Baxandall, 1977), and newly synthesized ciliary tubulins (Stephens, 1977, 1978) constitute <5% of the mass of tubulins incorporated in the initial set of cilia (Bibring and Baxandall, 1981). However, the majority of the mass of tubulin protein inherited from oogenesis is not accounted for by the amount assembled initially into mitotic spindles and cilia. New synthesis of tubulin from maternal mRNA is easily detectable in early cleavage stage embryos (Raff *et al.*, 1975). Alexandraki and Ruderman (1985b) showed that recruitment of maternal tubulin mRNA into the polysome fraction, like that of the 90 kd heat shock protein mentioned earlier, lags behind the recruitment of the bulk of the maternal message. Thus the maternal tubulin mRNA is completely engaged in translation only at late cleavage or early blastula stage.

Another protein species for which a maternal mRNA has been identified is *cyclin*, a polypeptide that during early cleavage appears and disappears in synchrony with the cell cycle. This is shown in Fig. 2.12. The synthesis of cyclin was demonstrated by Evans *et al.* (1983) in early embryos of *Arbacia punctulata* and *Lytechinus pictus*. The maternal messages for cyclin are loaded into polysomes *more rapidly* after fertilization than are most maternal messages, and cyclin is synthesized at a constant high rate during cleavage. Later its rate of synthesis declines greatly. The sharp fluctuation of cyclin levels observed in experiments such as that shown in Fig. 2.12 is caused by a selective, rapid degradation of the protein at the onset of each cleavage mitosis. The function of cyclin is not known, but as pointed out by Evans *et al.* (1983) its behavior resembles that of "maturation promoting factor," a cyclic protein to which a causal significance is attributed in the control of the cell cycle in amphibian oocyte maturation and cleavage (see, e.g., Wasserman and Smith, 1978; Newport and Kirschner, 1984; Gerhart *et al.*, 1984). In any case, the preferred assembly of the cyclins into polysomes after fertiliza-

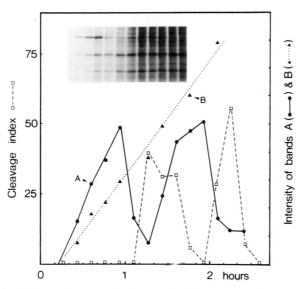

Fig. 2.12. Cyclin levels during early cleavage in *Arbacia punctulata*. The eggs were fertilized and 6 min later ^{35}S-methionine was added to the cultures. At 10 min intervals samples were removed and the protein precipitated with TCA, then dissolved in a buffer containing SDS, displayed by acrylamide gel electrophoresis, and autoradiographed (inset). Densitometric estimations of the relative quantities of two proteins are plotted. "A" is cyclin, and "B" is another protein that unlike cyclin accumulates monotonically with typical kinetics after fertilization. The percent of eggs undergoing cleavage ("cleavage index") was determined in glutaraldehyde-fixed samples removed over the same time period. Inhibitors of mitosis, such as cytochalasin, or of nuclear membrane breakdown, such as taxol, prevent the rapid proteolytic degradation of newly synthesized cyclin that normally occurs at the beginning of each cell cycle. [From T. Evans, E. T. Rosenthal, J. Youngblom, D. Distel, and T. Hunt (1983). *Cell* **33**, 389. Copyright by M.I.T.]

tion provides a further example of regulation of protein synthesis at the translational level, but one which is opposite in sign to the examples provided by the tubulins or the heat shock protein, since the latter are translated later rather than earlier than the bulk of the maternal message. Cyclin is a representative of a small set of prevalent protein species that are synthesized specifically in fertilized eggs and rapidly cleaving early embryos and are not synthesized in growing oocytes (Grainger *et al.*, 1986).

Maternal mRNAs coding for metallothionein (Nemer *et al.*, 1984), for calmodulin (Floyd *et al.*, 1986), and for two specific glycoproteins of molecular weight 65 kd and 115 kd (Lau and Lennarz, 1983) have also been identified in unfertilized sea urchin eggs. In addition, indirect evidence indicates maternal mRNA coding for several enzymes required for DNA synthesis, *viz.*, ribonucleotide reductase (Noronha *et al.*, 1972; Standart *et al.*, 1985) and thymidine kinase (De Petrocellis and Rossi, 1976). Both enzymes increase sharply in activity within a few hours of fertilization, and this increase is not blocked by actinomycin treatment. Later in development the activity of the same enzymes increases further, but this increase is sensitive to actinomycin treatment. Standart *et al.* (1985) showed that the maternal message for the small (41 kd) subunit of ribonuclease reductase is among the most prevalent of maternal mRNAs. This enzyme is required to provision the pool of deoxyribonucleotides in the embryo, which at fertilization contains sufficient precursor for the DNA of only four cells. An interesting feature of this example is that the maternal message for the 41 kd subunit is translated only after fertilization, i.e., when its products are required for DNA synthesis. Conversely, the large subunit of the enzyme is not synthesized after fertilization, and appears instead to be accumulated during oogenesis (Standart *et al.*, 1985).

The prevalence of almost all of the maternal messages discussed increases after the initial phases of development, due to new transcription. This holds true for the α-histone mRNAs, the cytoskeletal actin mRNAs, the glycoprotein mRNAs, tubulin mRNAs, calmodulin mRNA, and metallothionein mRNA. The genes utilized during oogenesis in the preparation of these maternal mRNAs are thus also utilized in the nuclei of the embryonic blastomeres, a conclusion substantiated more generally in the following chapter. Many of the specific mRNA species discussed disappear during embryogenesis or larval development. Thus they are not the products of constitutively expressed genes. For example, the α-histone mRNAs are replaced by mRNAs for the late histone subtypes during the blastula stage, and the CyIIIa cytoskeletal actin gene is utilized only during embryonic and larval life and is not expressed at all in adult tissues (Shott *et al.*, 1984).

Maternal Transcripts of Mitochondrial Origin

The sea urchin egg contains about 10^5 mitochondria (Berger, 1967). During embryonic development neither mitochondrial DNA synthesis nor mitochondrial replication occurs at a significant rate (Matsumoto *et al.*, 1974; Rinaldi *et al.*, 1979). This is due to repressive factors present initially in the zygote nucleus, since mitochondrial replication, DNA, and rRNA synthesis are all sharply activated in enucleated egg cytoplasm (Rinaldi *et al.*, 1979; Rinaldi and Giudice, 1985). Nonetheless, even under normal circumstances a large fraction of the newly synthesized RNA and of the mass of total poly(A) RNA in the unfertilized egg consists of mitochondrial transcripts. Ruderman and Schmidt (1981) estimated that >70% of ^{3}H-uridine precursor found in unfertilized egg RNA after 3 hr of labeling is resident in mitochondrial transcripts. About 50% of precursor incorporated in RNA during early cleavage is mitochondrial (Craig, 1970; Hartmann *et al.*, 1971; Chamberlain and Metz, 1972; Devlin, 1976). Identification of mitochondrial transcripts in these studies is based on hybridization with mitochondrial DNA and on comparisons with transcripts extracted directly from purified mitochondria.

However, since the relation between the amount of radioactive precursor incorporated after a given labeling period and the mass of the newly synthesized RNA depends on the molecular synthesis and turnover rates (see Chapter III and Appendix III), these estimates do not provide a measure of the fraction of egg or embryo RNA *mass* that is mitochondrial.

In *S. purpuratus* embryos about 4% of the newly synthesized 16S mitochondrial rRNA molecules is polyadenylated (Cabrera *et al.*, 1983). Probe excess titrations have established that there are about 2.2×10^6 polyadenylated 16S rRNA transcripts per egg. If the same fraction of total 16S rRNA transcripts is polyadenylated as in the newly synthesized mitochondrial 16S rRNA, there would be in all about 5×10^7 mitochondrial 16S rRNA molecules per egg. This may be compared with the number of cytoplasmic rRNA molecules per egg, $\sim 6.6 \times 10^8$ (Goustin and Wilt, 1981). The molar contribution of mitochondrial rRNA is thus about 8%. The contribution of the polyadenylated 16S rRNA with reference to total egg poly(A) RNA is perhaps more significant. From the mean length and quantity of egg poly(A) RNA (see above), it may be calculated that there are about 3.5×10^7 poly(A) RNA molecules per egg, of which about 6% would be mitochondrial poly(A) 16S rRNA. This is therefore the single most prevalent polyadenylated transcript in the egg cytoplasm. Furthermore, it appears to be incorporated into viable cDNA clones with an efficiency severalfold the average. cDNA clones representing mitochondrial 16S rRNA sequences have been described by Wells *et al.* (1982) and Cabrera *et al.* (1983). As we have seen, complex class poly(A) RNAs account for 50-60% of the total poly(A) RNA mass in the unfertilized egg, and cDNA clone libraries selected to include only very abundant sequences are likely to contain a great many copies of the mitochondrial 16S rRNA sequence. For example, Flytzanis *et al.* (1982) found that about 50% of the clones included in a sublibrary representing poly(A) RNAs present at greater than about 10^4 molecules per embryo were of mitochondrial origin. This is equivalent to about 20% of the clones in the starting unselected library.

Polyadenylated mitochondrial mRNAs have also been observed in sea urchin eggs and embryos. About eight discrete nonribosomal species have been reported (Devlin, 1976; Innis and Craig, 1978; Ruderman and Schmidt, 1981). These can be recovered from mitochondrial polysomes, and they are synthesized in the mitochondrial genome, since their transcription continues in enucleate fragments of egg cytoplasm, and in normal embryos it occurs at a constant rate throughout development, while the number of nuclei increases greatly (Devlin, 1976). However, the prevalence of the mitochondrial mRNAs is relatively low (Cabrera *et al.*, 1983), and translation of these messages accounts for an insignificant fraction of total embryo protein synthesis (Craig and Piatigorsky, 1971). Absolute synthesis and decay rate measurements were carried out by Cabrera *et al.* (1983), using cloned probes representing sea urchin mitochondrial mRNAs for subunit I of cytochrome oxidase (COI) and for a then unidentified protein (URF-1), which is now known to be a subunit of NADH dehydrogenase (Chomyn *et al.*, 1985). Cabrera *et al.* (1983) showed that the COI mRNA is present at about one molecule per mitochondrion, and the steady state level of the NADH dehydrogenase subunit mRNA is about 1/5th of this. This difference can be ascribed wholly to a difference in the turnover rates of these two mitochondrial mRNA transcripts.

Heterogeneous mitochondrial RNAs are difficult to exclude from egg and early embryo preparations, as they are released into "postmitochondrial" supernatant fractions even in the absence of detergents that lyse mitochondria. In several earlier studies agents such as ethidium bromide were used to selectively inhibit mitochondrial transcription, but these agents also affect nuclear RNA synthesis (Chamberlain and Metz, 1972; Ruderman and Schmidt, 1981). Furthermore, the distinction between mitochondrial and DNA sequences is not absolute. Thus Jacobs *et al.* (1983) found unexpectedly that the mitochondrial gene for COI is represented as well in several copies in the genomic DNA of the nucleus, where it may also be transcribed. It is important to note, however, that the aggregate *complexity* of mitochondrial species is insignificant compared to that of the maternal transcripts deriving from the oocyte nucleus. Most of the sequence length of the mitochondrial genome is represented in the known poly(A) RNA transcripts (Devlin, 1976). The maximum complexity of mitochodrial transcripts is thus only about 1.5×10^4 nt, less than 0.10% of the complexity of the cytoplasmic mRNA stored in the unfertilized egg.

(ii) Maternal mRNA in Amphibian Eggs

A large fraction of our current knowledge of maternal mRNA in amphibian material has been obtained in studies on ovarian oocytes, due to the unusual accessibility for molecular investigation of these large and easily obtained oocytes. These studies are discussed in detail in the section of Chapter V dealing with amphibian oogenesis. From them derive estimates of total maternal mRNA content, and of the approximate quantities of message for several specific proteins. Briefly, it is concluded that the fully-grown oocyte of *Xenopus laevis* contains about 21 ng of translatable mRNA, as well as a significantly larger quantity of interspersed poly(A) RNA, and about 13.5 ng of mitochondrial poly(A) RNA (see Table 5.5). Among the individual species of maternal mRNA for which some measurements exist are histone mRNAs, mRNAs for ribosomal proteins, heat shock proteins, and RNA polymerases. Table 2.1 indicates that the prevalence of the average *complex class* sequence in the unfertilized *Xenopus* egg is about 1.8×10^6 molecules per egg, and in the fully-grown oocyte the prevalences of the specific maternal mRNAs studied range from close to this average value for the heat shock mRNAs, to $3\text{-}4 \times 10^8$ molecules per egg for the individual histone maternal mRNAs (Table 5.6). One distinctive difference between fully-grown *Xenopus* oocytes and the unfertilized sea urchin eggs considered above is that the total amount of mRNA present in the *Xenopus* oocyte is only about five times the amount being translated in the oocyte polysomes, rather than 50 times as in the unfertilized sea urchin egg. From evidence obtained in studies of translation of injected mRNA in *Xenopus* oocytes (reviewed in Chapter V) it seems clear that the amount of translation occurring in the advanced oocyte is limited by the capacity of the translational apparatus, rather than by the amount of available message. While this does not exclude the possibility that the unused maternal mRNA is physically sequestered, and therefore cannot participate in translation, there is no specific evidence indicating this type of mechanism, and it is not required to explain the fivefold suppression of the possible level of protein synthesis. Another instructive conclusion is that after injection into the oocyte mRNAs may survive and continue to be translated for long periods. For example Gurdon *et al.* (1973) showed that translation of exogenous hemoglobin message occurs at 70% of the initial rate in oocytes injected with the mRNA 12-13 days earlier. Nor do fertilization and development decrease the long-term survival of injected messages (Gurdon *et al.*, 1974). Translation of injected globin mRNA continues at least to the swimming tadpole stage, 8 days after fertilization. Thus the internal milieu of the oocyte and embryo appears conducive to high stability, at least with respect to that fraction of the message being translated. This may refer as well to endogenous messages.

Oocyte maturation is stimulated hormonally in amphibians, as in other vertebrates, and it involves a complex series of physiological, biochemical, and molecular changes that are somewhat analogous to those occurring *after* fertilization in sea urchin eggs. The details of these events lie outside the scope of the present discussion (see review in Smith and Richter, 1985; other references are given in the initial section of Chapter V). During the maturation period, which occupies some hours, depending on the species, the rate of protein synthesis increases, although RNA synthesis ceases completely with germinal vesicle breakdown and meiotic metaphase chromosome condensation. The same increase in protein synthesis rate occurs in hormonally-induced maturing eggs that have been enucleated (Smith, 1975; Adamson and Woodland, 1977). The increase in absolute protein synthesis rate is about twofold in *Xenopus*, from about 17 ng hr^{-1} in the fully-grown oocyte to 35 ng hr^{-1} in the matured egg (Wasserman *et al.*, 1982; reviewed in Smith and Richter, 1985), and a similar increase is observed in maturing *Rana pipiens* oocytes (Shih *et al.*, 1978). This increase is wholly accounted for by assembly of new polysomes, as Richter *et al.* (1982) have shown that there is no change whatsoever in translational efficiency during the maturation period. Polysomes include about 2% of the ribosomes in the fully-grown oocyte and about twice this in the matured unfertilized egg (Woodland, 1974). Smith and Richter (1985) showed, furthermore, that the absolute rates of protein synthesis that obtain during maturation are consistent with the observed polysome content, given the measured values for *Xenopus* translational parameters.

In anuran amphibians zygotic RNA synthesis does not begin until the midblastula stage of embryogenesis (reviewed in Chapter III). It follows from this, and from the evidence just reviewed, that all protein synthesis occurring during maturation, and after fertilization until the midblastula stage, takes place on *maternal mRNA* templates. As shown in Fig. 2.13 the polysome content continues to rise during early *Xenopus* embryogenesis, and by the midblastula stage about 15% of the embryo (i.e., maternal) ribosomes are included in polysomes (Woodland, 1974). Thus the net increase in mRNA utilization is close to that expected were the stored, unused message of the fully-grown oocyte completely assembled into polysomes by the midblastula stage [i.e., 2% of ribosomes is included in polysomes in the oocyte and 15% in the blastula, *vs.* 21.4 ng of total maternal mRNA in the oocyte, of which about 4 ng is utilized in the oocyte polysomes (see Chapter V for data)]. Fertilization itself causes only a 1.5-fold increase in the rate of protein synthesis in either *Xenopus* or *Rana* (Ecker and Smith, 1968; Smith and Ecker, 1969; Woodland, 1974; Shih *et al.*, 1978). Thus in amphibians the process of maternal mRNA recruitment and polysome assembly is set in

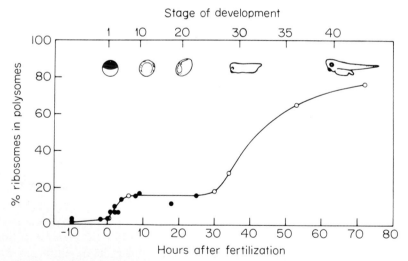

Fig. 2.13.   Percent of ribosomes in polysomes during development of *Xenopus laevis* embryos. The time of development (abscissa) applies to embryos developing at 23°C, though a lower temperature was actually used in the experiments. The upper scale shows the stage of the embryos, according to Nieuwkoop and Faber (1956). For prefertilization stages the maturation process is considered to begin 10 hr before fertilization. The value for this point was obtained by incubating normal oocytes with progesterone at 5 μg ml^{-1} *in vitro*. [From H. R. Woodland (1974). *Dev. Biol.* **40**, 90.]

train not by fertilization, but by the hormonal stimulus that initiates the events of maturation.

Maternal Histone Protein and Histone mRNAs

The mature egg of *Xenopus* contains a large complement of histone proteins, sufficient for the assembly of over 2×10^4 nuclei (Woodland, 1980). New synthesis of histone from maternal mRNA is activated during maturation and in contrast to the rate of total protein synthesis, the rate of nucleosomal histone synthesis increases 50-fold during the maturation period, to a value of about 2500 pg hr^{-1} (Adamson and Woodland, 1974, 1977). This sharp increase occurs both in enucleated and actinomycin-treated eggs (Adamson and Woodland, 1977), and it clearly results from the *selective* mobilization of stored histone mRNA, with respect to nonhistone mRNAs. By the blastula stage the rate of synthesis of the nucleosomal histones has increased further to about 5500 pg hr^{-1}. In *R. pipiens* a similar course of events obtains. Shih *et al.* (1980) showed that the rate of synthesis of histones increases tenfold during the maturation period, and a further fourfold thereafter, to a value of about 4500 pg hr^{-1} at blastula stage. The significance of these quantitative changes is illustrated by reference to the amounts of DNA present at each stage in the embryo, as shown for *Xenopus laevis* in Fig. 2.14 (Woodland, 1980). During cleavage the rate of histone synthesis,

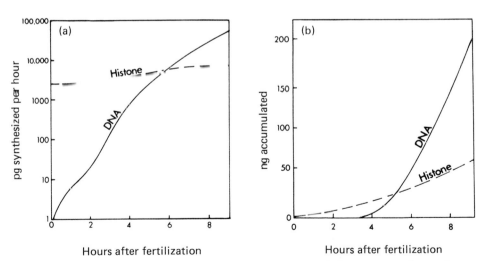

Fig. 2.14. Rates of DNA and nucleosomal histone synthesis in *Xenopus* embryos, from fertilization to early gastrula stage. At this stage (9 hr postfertilization) the embryo contains about 3×10^4 cells. (a) Rates of histone and DNA synthesis, presented as a semilog plot. (b) Linear plot of accumulated newly synthesized histone and DNA contents. The haploid genome size for *Xenopus laevis* is about 3 pg (Dawid, 1965), and there are approximately equal masses of nucleosomal histone and DNA in the nuclear chromatin. [From H. R. Woodland (1980). *FEBS Lett.* **121**, 1.]

on a mass basis, greatly exceeds the rate of DNA synthesis [Fig. 2.14(a)]. A cross-over point occurs at the blastula stage (6 hr; stage 8), after which the demand for histones is greater than the rate at which they are supplied. Calculation of the deficit, as shown in Fig. 2.14(b), indicates a shortfall of about 150 pg by the end of blastulation. This, however, is close to the quantity of stored histone protein inherited from oogenesis, estimated at about 135 pg (Adamson and Woodland, 1977). Thus *both* stored histone proteins and histone newly synthesized from maternal message are required to account for the chromatinization of the embryonic genomes formed during cleavage and blastulation. Thereafter, histone synthesis occurs on mRNAs transcribed in the embryo genomes, as described in Chapter III. In *R. pipiens* as well, both stored histone protein and newly synthesized histones translated on maternal mRNA must be utilized to construct the mass of chromatin present by the late blastula stage (Shih *et al.*, 1980). Development to an equivalent cell number in *R. pipiens* requires about 2.4 times as long as in *X. laevis*, but the genome size is about 2.4-fold greater as well. The quantitative requirements for new histone per unit time thus appear similar in these embryos. However, in *R. pipiens* the four nucleosomal histones are not synthesized at equal rates during cleavage. Synthesis of H4 is almost undetectable at this stage, and Shih *et al.* (1980) calculated that only about 2 ng of this histone would accumulate in the 24 hr required to attain stage 9 (late blastula), compared to about 20 times this value for the other nucleosomal histones. The deficit is made up by storage of a relatively large pool of H4 protein synthesized during oogenesis.

Synthesis of H1 histone from maternal mRNA appears to be regulated in a distinct manner. In early *Xenopus* embryos H1 synthesis accounts for <5% of total histone synthesis until the midblastula stage (Flynn and Woodland, 1980), though after this the fraction of total histone synthesized that is H1 equals or exceeds that attributed to each of the four nucleosomal histones. During maturation the increase in rate of synthesis of H1 lags behind that of total nonhistone proteins, while synthesis of nucleosomal histones is accelerated greatly relative to total proteins, as we have seen. Again a similar situation obtains in *R. pipiens* (Shih *et al.*, 1980). This phenomenon is not due to a deficit of H1 maternal message, since the rate of H1 synthesis during *oogenesis* is equivalent to that of the other histones. Furthermore, Ruderman *et al.* (1979) showed by analysis of the products of mRNA translations *in vitro* that similar quantities of H1 mRNA are present in the unfertilized egg and in the late blastula. Thus there is clear evidence for *sequence-specific translational preference* in the mobilization of the maternal histone mRNAs. Maternal messages for the nucleosomal histones are loaded onto the translational apparatus far more rapidly than are most maternal mRNAs, while maternal H1 histone message is specifically not mobilized until later in development.

An interesting aspect of maternal histone mRNA utilization in *Xenopus* is

the developmental change in their state of polyadenylation. About 50-70% of
the maternal histone mRNA in the oocyte is polyadenylated (Levenson and
Marcu, 1976; Ruderman and Pardue, 1977; Ruderman *et al.*, 1979), an unusual feature for animal histone mRNAs. However, the poly(A) tract is very
short, on the average <12 residues (Ballantine and Woodland, 1985). During
maturation the stored histone message is completely deadenylated, and over
90% is recovered in the poly(A)-deficient fraction of unfertilized egg or
blastula mRNA (Ruderman *et al.*, 1979; Ballantine and Woodland, 1985).
Loss of poly(A) tracts occurs at the same time as does the rapid mobilization
of nucleosomal histone message, but is unlikely either a cause or effect of the
mobilization process, since H1 histone mRNA is also deadenylated during
maturation (Woodland *et al.*, 1983).

In contrast to sea urchin embryos there is no evidence for switches in
histone subtypes during amphibian development. Though there are several
H1 variants in *Xenopus* (reviewed in Flynn and Woodland, 1980) these are
ubiquitously observed, and apparently result from genetic polymorphisms
within the H1 gene family. A similar result is reported for a urodele (Imoh,
1978). In *Xenopus borealis* there is less polymorphism and a careful developmental study of the 5' and 3' sequences of H4 mRNAs both in this species
and in *X. laevis* fails to reveal any ontogenic change in H4 subtypes (Woodland *et al.*, 1984). Nor does there appear to be switching of H3 or of H2b or
H2a subtypes (Van Dongen *et al.*, 1983a).

The developmental persistence of maternal H1 mRNA has been determined directly by use of interspecific hybrids between *Xenopus laevis* (♀)
and *X. borealis* (♂) (Woodland *et al.*, 1979). The paternal forms of H1,
which are distinguishable from those of *X. laevis*, appear only after the
midblastula stage at which general genomic activation occurs. Synthesis of
the maternal (*X. laevis*) form of H1 is exhausted shortly after the onset of
gastrulation, as seen in haploid androgenetic hybrids in which the egg nucleus had been destroyed by UV-irradiation. The rate of disappearance observed for this maternal mRNA is consistent with a half-life of only a few
hours, i.e., once it has become mobilized. It is relevant that when injected
into *X. laevis* eggs sea urchin histone mRNA also displays a half-life of about
3 hr (Woodland and Wilt, 1980a). This is in contrast to the long survival of
translated exogenous globin mRNA in *Xenopus* eggs (see above). Thus although maternal histone mRNAs may be stored for very long periods and
turn over only slowly during oogenesis (see Chapter V), once development
begins the protected status of these mRNAs is abolished.

Specific Maternal mRNAs Other than Histones

The developmental behavior of a large number of individual but unidentified *Xenopus* maternal mRNAs has been investigated by use of cDNA clone
libraries, and also in electrophoretic studies of the spectrum of proteins

synthesized in the early embryo. Clone banks in which maternal poly(A) RNAs are represented have been constructed and characterized in several laboratories (see, e.g., Dworkin and Dawid, 1980a,b; Golden *et al.*, 1980; Jacob, 1980; Dworkin and Hershey, 1981). As in the sea urchin many of the *Xenopus* cDNA clones that react with abundant poly(A) RNAs represent mitochondrial sequences (Dworkin *et al.*, 1981). These again include the most prevalent individual poly(A) RNAs in the egg, and in early embryos mitochondrial transcripts are estimated by Dworkin *et al.* (1981) to include more than 10% of the total poly(A) RNA. Dworkin and Hershey (1981) focussed on a group of 18 nonmitochondrial clones that react both with poly(A) RNAs from the unfertilized egg and with *polysomal* RNAs from late embryos, and which thus can be taken to represent *bona fide* maternal messages. The prevalence of the respective transcripts in the egg can be estimated from their data to range from around 10^6 molecules per egg, i.e., close to the average abundance of *Xenopus* poly(A) RNAs for the complex class, to over 3×10^7 molecules per egg. Most maternal poly(A) RNAs of nuclear origin studied with cDNA clones appear to remain present at constant levels over the 10 hr or so that elapse between fertilization and the early gastrula stage (Dworkin and Dawid, 1980b; Dworkin and Hershey, 1981; Colot and Rosbash, 1982). However, the distribution of such sequences between polysomal and nonpolysomal cytoplasmic compartments changes rapidly during early development, as would be predicted from the sharp increase in polysome content (Fig. 2.13). Dworkin *et al.* (1985) reported that most randomly chosen cloned maternal mRNA sequences that are predominantly nonpolysomal in late ovarian oocytes are ≥ 80% mobilized into polysomes by 2-3 hr postfertilization, i.e., by the 16-cell cleavage stage. Mobilization of about a fourth of the sample of maternal mRNAs investigated is more delayed. The rapid disappearance of several particular maternal transcripts has also been observed after fertilization. These sequences are no longer in evidence or are greatly diminished by the midblastula stage (Jacob, 1980; Dworkin and Hershey, 1981; Colot and Rosbash, 1982). A complication in interpreting specific prevalence changes is the existence of sharp developmental alterations in the extent of polyadenylation. As in the sea urchin the overall extent of polyadenylation increases after fertilization in *Xenopus* (Sagata *et al.*, 1980), though as we have already seen in considering the maternal histone messages, some mRNAs are specifically deadenylated during maturation.

Two-dimensional electrophoretic analysis reveals an interesting pattern of change in the spectrum of proteins synthesized from maternal mRNAs in early *Xenopus* embryos. A large number of new polypeptide species begin to be synthesized in the course of cleavage, and continue to be synthesized in the blastula stage, that are not synthesized in newly fertilized eggs (Brock and Reeves, 1978; Ballantine *et al.*, 1979). This cannot be due to new synthesis and appearance of their messages, since significant nuclear transcription does not take place in the embryo until the midblastula stage. Ballantine *et al.* (1979) demonstrated that essentially all protein species of this class are being translated in the ovarian oocyte, i.e., before maturation begins. Furthermore, the polypeptides not being translated in unfertilized eggs are included among the products synthesized *in vitro* from unfertilized egg mRNA. Thus a negative translation selection process, or some form of mRNA sequestration, must be operating on these messages in eggs and very early embryos, just as concluded earlier for H1 histone mRNA. Among the identified maternal nonhistone mRNAs that display this behavior are cytoskeletal β- and γ-actins, which are two of the most intensely translated proteins of oocytes, and also of later embryos. In Fig. 2.15 are reproduced experiments of Sturgess *et al.* (1980) and Ballantine *et al.* (1979) in which it is demonstrated that unfertilized eggs produce no β- or γ-actin, but that RNA extracted from these eggs contains as much β- and γ-actin message as does RNA extracted from oocytes or blastulae. Among other identified maternal mRNAs that in early embryos are mobilized with delayed kinetics relative to the majority are those coding for nuclear lamin L_{III} (Stick and Hausen, 1985), and for α-tubulin, enolase and ATP-ADP carrier protein (Dworkin *et al.*, 1985). Translation of maternal mRNA coding for fibronectin is also delayed, and the normal lag in mobilization of this maternal mRNA is observed even in unfertilized eggs that are activated by pricking rather than fertilization, and then incubated (G. Lee *et al.*, 1984). Such eggs do not carry out nuclear replication or cell division. Thus the timing mechanism responsible for the delayed mobilization of the fibronectin message, and perhaps for the other similar examples discussed, is most likely resident in some component of the egg cytoplasm.

Androgenetic haploid hybrids between *X laevis* ($\female$) and *X. borealis* ($\male$) have been utilized to study the persistence time of nonhistone maternal mRNAs, as well as of H1 histone mRNA (Woodland and Ballantine, 1980). *X. laevis* proteins produced in these hybrids are necessarily the translation products of maternal messages, and the experiment shows that such messages may be retained far into development. In the four cases investigated by Woodland and Ballantine (1980) synthesis of proteins coded by maternal mRNAs ceases in the haploid hybrids only after gastrulation is complete (stages 16-25; about 15-30 hr after fertilization). Among these are two examples that display the same delayed mobilization as do histone H1, the β- and γ- actins, and the others cited above. In summary, in amphibian embryos at least two different kinds of *sequence-specific* mechanism may be important in determining how a given maternal mRNA is utilized during the period of transcriptional quiescence extending from maturation to the midblastula stage. Certain maternal mRNAs decay rapidly, with respect to the majority, while others persist far into embryonic development; and many individual

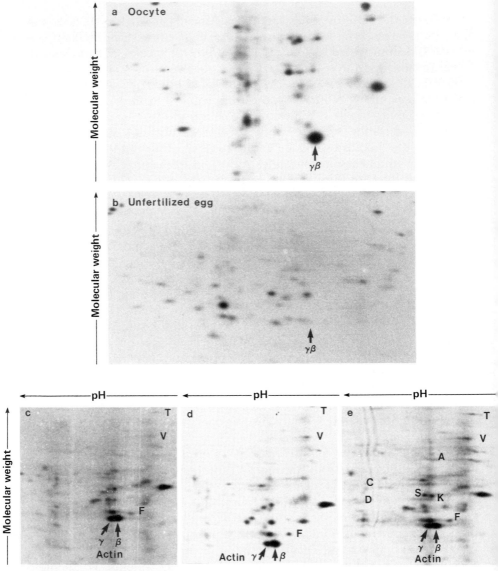

Fig. 2.15. Regulation of β- and γ-actin synthesis in oocytes and early embryos of *Xenopus laevis*. Two-dimensional gel electrophoretic displays of proteins synthesized by normal oocytes and by unfertilized eggs are shown in (a) and (b), respectively. β- and γ-actin spots are indicated. Oocytes or eggs were injected with ^{35}S-methionine, and after incubation for 3.5 hr the proteins were extracted, separated by electrophoresis, and the gels autoradiographed. The identity of the actin spots was confirmed by partial peptide mapping and other evidence. [From E. A. Sturgess, J. E. M. Ballantine, H. R. Woodland, P. R. Mohun, C. D. Lane, and G. J. Dimitriadis (1980). *J. Embryol. Exp. Morphol.* **58**, 303.] (c)-(e) Similar two-dimensional analyses of proteins synthesized *in vitro* in a rabbit reticulocyte system loaded with total RNA of oocytes (c), unfertilized eggs (d) or stage 8-9 blastulae (e). [From J. E. M. Ballantine, H. R. Woodland, and E. A. Sturgess (1979). *J. Embryol. Exp. Morphol.* **51**, 137.]

mRNAs are either specifically sequestered or are subject to positive or negative selection in the translational process. Sequence-specific alterations in the state of polyadenylation also occur, though their functional significance remains to be identified.

There are no direct measurements on early amphibian embryos of the fraction of the total poly(A) RNA sequence complexity that is represented in *polysomal mRNA*. The results of Dworkin and Dawid (1980b) cited above suggest that *for moderately abundant sequences* this fraction is high, as only 11% of such clones constructed from total stage 10 gastrula poly(A) RNA are not represented in polysomal RNA fractions. Thus a great variety of maternal mRNAs must exist. Maternal messages for various specific proteins have been qualitatively identified in *Xenopus* egg RNA by blot hybridizations with cloned probes. Aside from histones and others mentioned above, these include calmodulin (Chien and Dawid, 1984), ubiquitin (Dworkin-Rastl *et al.*, 1984), and certain ribosomal protein mRNAs (Pierandrei-Amaldi *et al.*, 1982). Maternal mRNA for a 6S form of DNA ligase has also been identified by *in vitro* translation of RNA from unfertilized axolotl (*Ambystoma mexicanum*) eggs (Thiebaud *et al.*, 1983). Maternal tubulin mRNAs were identified in *Xenopus* eggs by two-dimensional gel electrophoresis (Brock and Reeves, 1978). Maternal mRNAs coding for tubulins and for fibronectin have in addition been demonstrated by this method in eggs of the urodele *Pleurodeles waltlii*, and as might be expected translation of these proteins during cleavage is resistant to injection of actinomycin (Darribère *et al.*, 1984), just as translation of maternal fibronectin mRNA prior to renewal of transcriptional activity in *Xenopus* embryos is resistant to injection of α-amanitin (G. Lee *et al.*, 1984).

Another form of evidence regarding the functional nature of maternal transcripts in amphibian eggs is potentially available from studies of maternal mutants. Several such mutants have been discovered in the axolotl (reviewed in Malacinski and Spieth, 1979). The best known example is the "*o*" mutation (Humphrey, 1966; Briggs and Cassens, 1966). Eggs homozygous for this recessive allele arrest at gastrulation, but complete rescue can be achieved by injection of normal egg cytoplasm (Briggs, 1972). Unfortunately, except for the fact that the missing product in *o/o* eggs is (ultimately) a protein, there is little additional evidence as to its nature. Maternal effect mutants are also known in *Xenopus*, including a recessive lethal called *partial cleavage* and an egg pigmentation mutant, *periodic albinism* (Droin and Fischberg, 1984). The essential nature of neither defect is known. The maternal mutant *pale eggs* specifically affects egg pigmentation (Droin and Fischberg, 1984). This mutation is interesting in that unlike *periodic albinism* it does not interfere with melanin formation in the adult or tadpole, suggesting that at least partially nonoverlapping sets of genes of similar function are utilized for pigment formation during oogenesis, and in late larval and adult development.

(iii) Maternal mRNA in Mammalian Embryos

Quantity of Maternal Poly(A) RNA and Its Period of Persistence in Mouse Embryos

Compared to echinoderm or amphibian embryos development occurs at a very leisurely pace in preimplantation mammalian embryos. At 24 hr after fertilization the mouse embryo is at the 2-cell stage, the first cleavage occurring only at 16-20 hr, whereas by 24 hr postfertilization the sea urchin embryo consists of hundreds of cells, and depending on species and temperature, has achieved a late blastula or gastrula stage of organization. Similarly, by 24 hr the typical amphibian embryo used in the laboratory contains tens of thousands of cells, and by this time has advanced to gastrula or neurula stage of development. As illustrated in Fig. 2.16 cleavage proceeds even more slowly in the human than in the mouse embryo. An interesting comparative conclusion that follows from the data reviewed in this section is that the

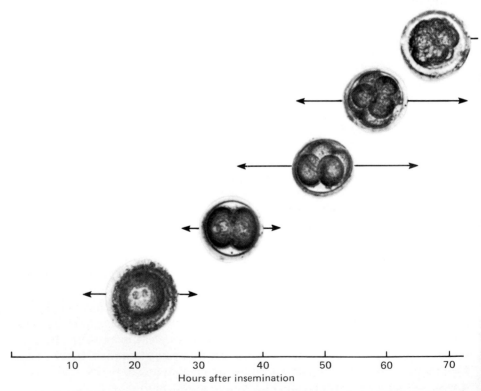

10	20	30	40	50	60	70

Hours after insemination

Fig. 2.16.   Cleavage rate in human embryos fertilized *in vitro*. Arrows indicate range observed in times at which the respective cleavages occur. [From P. R. Braude, M. H. Johnson, V. N. Bolton, and H. P. M. Pratt (1983). *In "In Vitro* Fertilization and Embryo Transfer" (P. G. Crosignani and B. L. Rubin, eds.), p. 212. Academic Press, London.]

length of the period during which functional maternal mRNA persists is in real time approximately similar in sea urchin, amphibian, and mammalian embryos. These organisms therefore differ greatly in the *stage of development* by which the maternal mRNA has disappeared or been largely replaced by new zygotic transcripts. Yet, as far as the information available indicates, the ways in which the maternal mRNA of mammalian embryos is utilized will now appear familiar. There have yet been no measurements of the complexity of maternal mRNA populations in mammalian eggs, and there so far exist few data regarding messages specifically identified by means of cloned nucleic acid probes. Nor is it known whether interspersed maternal poly(A) RNA is stored in mammalian eggs, a possibility that must be reserved in considering the following estimates of maternal mRNA content. Kaplan *et al.* (1985) reported that in mouse oocytes the concentration of RNA containing interspersed *Alu* sequences or three specific long repetitive sequences is no higher in mouse oocytes and eggs than in the cytoplasm of somatic cells. However, in somatic rodent and human cells *Alu* sequences are very extensively represented in the mRNA, where they are often embedded in the 3' trailer regions (see, e.g., Calabretta *et al.*, 1981; Tashima *et al.*, 1981; Brickell *et al.*, 1983). The form of the mouse egg RNAs containing these sequences has not been determined.

Measurements reviewed in Chapter V show that stable poly(A) RNA accumulates during oogenesis in the mouse, and that it accounts for an unusually large fraction of the total RNA in fully grown ovarian oocytes. The quantity of stored poly(A) RNA then decreases by a factor of about two during the maturation period, due to loss of terminal poly(A) tracts (Bachvarova *et al.*, 1985). From measurements of poly(A) content and tract length, and the average length of the poly(A) RNA molecules, which is about 1700 nt, Clegg and Pikó (1983a) calculated that the unfertilized mouse egg contains about 1.7×10^7 poly(A) RNAs. After fertilization this increases to 2.4×10^7 poly(A) RNA molecules, representing about 25 pg per egg, or approximately 8% of the total RNA of the egg (Bachvarova and De Leon, 1980; Pikó and Clegg, 1982; Clegg and Pikó, 1983a). The increase in poly(A) RNA content in the 1-cell embryo is due to cytoplasmic polyadenylation of preexisting poly(A)-deficient RNA molecules (Young and Sweeney, 1979; Clegg and Pikó, 1983a,b), just as in the fertilized sea urchin egg. However, Clegg and Pikó (1983b) showed that turnover and replacement of poly(A) tracts does not occur in the fertilized mouse egg. Changes in the absolute rate of protein synthesis in maturing oocytes and fertilized eggs more or less parallel those in poly(A) RNA content. Thus during the maturation period the rate decreases from about 43 to 31 pg hr^{-1} (Schultz *et al.*, 1978a), while after fertilization there is an increase of about 40%, not unlike that observed in amphibian eggs. Schultz *et al.* (1979a) determined a rate of 45 pg hr^{-1} in 1-cell embryos, and almost the same rate, 51 pg hr^{-1}, is maintained in 8-cell embryos.

Sea urchin and mouse eggs are of similar dimensions, and measurements of protein synthesis rate and maternal RNA content exist for both species. Some instructive differences appear in the comparison shown in Table 2.2. The sea urchin egg contains about ten times more total RNA, 80% of which is ribosomal, but a rather similar content of poly(A) RNA. Taking into account *both* the interspersed poly(A) RNA in the sea urchin egg, and the presence of nonpolyadenylated mRNA, the quantity of *bona fide* maternal message in the sea urchin egg was calculated above to be about 30 pg. Though this calculation cannot be performed for the mouse egg, as a *maximum* estimate the poly(A) RNA content of the 1-cell embryo can be utilized. The final column of Table 2.2 gives the surprising result that the rate of protein synthesis in the mouse egg is only about 1/5th that supported by the sea urchin egg. Among the possible (nonexclusive) explanations are that there is a large fraction of RNA that cannot function as message included in the poly(A) RNA complement (for instance, interspersed poly(A) RNA); that the basic rate of translation per polysomal ribosome is lower than in other systems, despite the favorable difference in temperature compared to the sea urchin egg; or that much of the mRNA is being underutilized. Thus it might not be assembled into polysomes, or it is included in polysomes that

TABLE 2.2   Approximate Maternal Poly(A) RNA, mRNA, and rRNA Contents, and Protein Synthesis Rates, in Early Sea Urchin and Mouse Embryos[a]

	Total RNA (pg)	Number of ribosomes	Poly(A) RNA (pg)	mRNA (pg)	Protein synthesis Rate (pg hr^{-1})
Mouse	300[b]	5.5×10^{7}[c]	25[d]	≤25[e]	45[f]
Sea Urchin	2800	6.6×10^{8}	45	30[g]	~240

[a] Data shown are for 1-cell mouse embryos, and for 8-cell cleavage stage *S. purpuratus* embryos. At this stage of sea urchin embryogenesis most of the maternal mRNA has been loaded on polysomes and is engaged in protein synthesis (Goustin and Wilt, 1981; see text) while zygotic mRNA has not yet contributed significantly to the mRNA pool. Sea urchin data are drawn from sources discussed earlier in this chapter.

[b] See Table 5.9.

[c] Pikó and Clegg (1982).

[d] Bachvarova and De Leon (1980); Pikó and Clegg (1982).

[e] Were there *no* significant nonmessage contributions from either interspersed maternal poly(A) RNA or mitochondrial poly(A) RNA, the amount of poly(A) mRNA would be 25 pg, or about 2.4×10^{7} mRNA molecules of average length (Clegg and Pikó, 1983a). About 40% of this consists of molecules polyadenylated after fertilization, which by analogy with the sea urchin case are probably *bona fide* maternal mRNAs. It is not known whether there are significant *additional* amounts of poly(A)-deficient maternal mRNAs. The mouse egg contains about 10^{5} mitochondria (Pikó and Matsumoto, 1976) but they are depleted of ribosomes and are inactive in RNA or protein synthesis until later in cleavage (Pikó and Chase, 1973).

[f] Schultz *et al.* (1979a). Amino acid pool sizes were determined for mouse eggs and embryos by Schultz *et al.* (1981b). The cited measurements were derived from methionine uptake into proteins and measurements of methionine pool specific activity.

[g] Note that both poly(A) mRNA and poly(A)-deficient mRNA are included in this estimate (see text).

are not fully loaded with ribosomes. A calculation suggests that in the fully-grown mouse oocyte ribosomes are in fact spaced more than four times more widely, on the average, than at the maximum loading densities usually observed in functional polysomes (see Chapter V). The quantity of ribosomes in the 1-cell mouse embryo (Table 2.2) is probably not rate limiting, since the measured rate of protein synthesis would be more than accounted for were all of these ribosomes cycling on and off polysomes that contain only 1/4th the number of ribosomes per message as sea urchin embryo polysomes. There is also direct evidence that the ribosomes of fertilized mouse eggs are capable of accepting newly introduced mRNA. Thus Brinster *et al.* (1980) and Ebert and Brinster (1983) showed that mouse and rabbit globin messages are translated on injection into 1-cell embryos. However, these experiments again demonstrate a limitation in the translational capacity of the egg (cf. the amphibian and sea urchin eggs discussed earlier).

Heterogenous RNA synthesis occurs in 1-cell mouse embryos, though at relatively low rates (Clegg and Pikó, 1983b; also reviewed in Chapter III). Thus in contrast to *Xenopus*, there is no transcriptionally quiescent period following fertilization. However, the fraction of late 1-cell embryo poly(A) RNA that has been newly synthesized in the zygote nucleus is calculated to amount to only about 1.5% of the total number of maternal poly(A) RNA molecules. *Translational* activity during this period is therefore almost completely dependent on maternal mRNA. A consistent and much repeated observation is that arrest of transcription by treatment with α-amanitin does not noticeably affect protein synthesis in 1-cell mouse embryos (reviewed in Davidson, 1976; Johnson, 1981).

Soon after first cleavage the quantity of maternal poly(A) RNA in the mouse embryo declines sharply, though turnover of maternal poly(A) RNA actually begins earlier (Clegg and Pikó, 1983b). During the 1-cell stage turnover is more than balanced by cytoplasmic polyadenylation and recruitment into the poly(A) fraction of previously poly(A)-deficient molecules. However, by the late 2-cell stage only about 1/3rd of the initial number of poly(A) tracts remains (Levey *et al.*, 1978; Pikó and Clegg, 1982; Clegg and Pikó, 1983a), though their average length, about 65 nt, remains unchanged. This observation almost certainly indicates degradation of the mRNA molecules themselves, as verified for some specific examples below. The disappearance by shortly after first cleavage of about 40% of both total and poly(A)-maternal RNA that had been labeled during oogenesis was demonstrated directly by Bachvarova and DeLeon (1980). Decline in the total mass of embryo RNA by the 2-cell stage was reported earlier by Olds *et al.* (1973) and confirmed by Pikó and Clegg (1982), and is reflected in an equivalent fall in the independently estimated content of ribosomes (Pikó and Clegg, 1982). Thus all the major classes of maternal transcript appear to be degraded more or less coordinately. Furthermore, Ebert *et al.* (1984) showed that on injection exogenous mRNA is degraded up to 100 times more rapidly in fertilized mouse eggs than in growing oocytes. The amount of poly(A) RNA, and of rRNA, recovers later in cleavage, due to new synthesis (Clegg and Pikó, 1983a,b; also reviewed in Chapter III). The mouse embryo is unlike non-mammalian embryos in the sequential nature of the process by which maternal RNAs are replaced by zygotic transcripts. Thus the period during which maternal mRNAs and rRNAs are required is terminated by their destruction and disappearance from the embryo, and only then do significant quantities of newly synthesized gene products accumulate.

Synthesis of Specific Proteins from Maternal mRNAs

Protein synthesis patterns have been examined extensively by two-dimensional gel electrophoresis in maturing rabbit and mouse oocytes, and in unfertilized eggs and early embryos of these species. At each stage of development specific changes in the rates of synthesis of particular resolved polypeptides have been reported, though synthesis of the large majority of polypeptide species observed in matured oocytes continues after fertilization (see Van Blerkom, 1981; Johnson, 1981, for reviews). Until the 2-cell stage in the mouse these individual changes are clearly posttranscriptional and are temporally programmed by cytoplasmic components. For example, many of the same changes in the pattern of newly synthesized protein that are observed to occur during meiotic maturation (Schultz and Wassarman, 1977; Van Blerkom and McGaughey, 1978; Schultz *et al.*, 1978b; Howlett and Bolton, 1985), also take place in enucleate oocyte fragments prepared by cytochalasin treatment (Schultz *et al.*, 1978b). Further individual alterations in the pattern of synthesis are observed in comparing fertilized and unfertilized eggs, in both mouse and rabbit. For instance Chen *et al.* (1980) found that of 95 prominent resolved polypeptides, the relative rates of synthesis of 18 were further decreased or increased in fertilized compared to unfertilized mouse eggs. At least some of these events do not depend on fertilization *per se*, but on elapsed time after ovulation, since they occur as well in unfertilized eggs cultured *in vitro* or retained in the oviduct (Van Blerkom, 1979). It is now clear, however, that a large fraction of the alterations in protein synthesis pattern detected in the two-dimensional electrophoresis studies cited actually result from *posttranslational* modifications, such as glycosylations and phosphorylations (Van Blerkom, 1981; Magnuson and Epstein, 1981; Pratt *et al.*, 1983).

There remain several specific examples for which translational selection or message sequestration must be invoked. Maternal mRNAs coding for these proteins can be demonstrated at stages when their synthesis does not occur *in vivo*, and when the majority of other maternal message species have already been mobilized for translation. The clearest examples are the "fertilization proteins" in the mouse egg, a family of prominent polypeptides of about 35 kd. Their synthesis can barely be detected in oocytes, but occurs at relatively high levels after fertilization. This phenomenon has been observed in many independent studies (see, e.g., Braude *et al.*, 1979; Cullen *et al.*, 1980; Schultz *et al.*, 1981a; Van Blerkom, 1981; Cascio and Wassarman, 1982; Howlett and Bolton, 1985). An *in vitro* translation experiment of Cascio and Wassarman (1982) demonstrating the presence of message for the fertilization proteins in ovarian oocytes is reproduced in Fig. 5.22 (Chapter V). While certain of the changes in pattern of synthesis observed for these proteins during maturation and after fertilization are also posttranslational (Pratt *et al.*, 1983), it is clear that in this case a major level of control is delay in the utilization of the stored maternal message until well after fertilization. Thus Braude *et al.* (1979) showed that approximately the same quantities of mRNA coding for the 35 kd fertilization proteins are present in unfertilized eggs as in 2-cell embryos, where their synthesis occurs at the maximal rate. As expected, the onset of translation of these proteins occurs normally in the presence of α-amanitin (Braude *et al.*, 1979).

Synthesis of several identified proteins has been reported in 1-cell mouse eggs. Among these are α- and β-tubulins, β- and γ-actins, and ribosomal proteins, all of which have been studied in growing mouse oocytes as well. Estimates of the number of mRNA molecules being utilized for the biosynthesis of these proteins in fully-grown mouse oocytes are presented in Table 5.10 (Chapter V). Schultz *et al.* (1979a) showed that the absolute rate of synthesis of the tubulins decreases by about 40% during maturation, and then returns to its previous value after fertilization. There appear to be about 2×10^5 maternal mRNA molecules engaged in tubulin synthesis per egg. The same 40% decrease in the rate of ribosomal protein synthesis is observed during maturation (LaMarca and Wassarman, 1979) [total protein synthesis during the same period declines about 25%; Schultz *et al.* (1979a)]. Absolute rates of histone synthesis have been measured in unfertilized mouse eggs by Kaye and Church (1983), and in early embryos by Kaye and Wales (1981). Rates of nucleosomal histone synthesis are about 0.17 pg hr^{-1} in the fully-grown oocyte (Wassarman and Mrozak, 1981) and are about 0.25 pg hr^{-1} in unfertilized eggs (Kaye and Church, 1983), but in contrast to the other proteins mentioned the rate in 1-cell embryos has increased to over 10 times these values (Kaye and Wales, 1981). It follows that there is a relatively large pool of stored maternal histone mRNA that is largely not utilized until after fertilization. For the nucleosomal histones this is estimated at about 6 $\times 10^5$ mRNA molecules per egg (Graves *et al.*, 1985). As in *Xenopus* and sea urchin embryos, synthesis of histone from the maternal message is uncoupled from DNA synthesis during early cleavage, and thus it continues at the same rate whether or not DNA synthesis is inhibited (Kaye and Church, 1983). Even the accelerated rate of histone synthesis in the early cleavage embryo is sufficient to provide only for the first few cell divisions. The egg also contains an amount of histone protein synthesized during oogenesis that would suffice for the chromatin of about 10 diploid cells (Wassarman and

Mrozak, 1981). It is not known how or when the newly synthesized and the stored histone complements are utilized.

The rate of histone biosynthesis in the 2-cell embryo is less than 50% of that in the fertilized egg (Kaye and Wales, 1981), and experiments with cloned histone gene probes demonstrate a precipitous decline in the mass of histone mRNA during this period, as shown in Fig. 2.17(a) (Giebelhaus *et al.*, 1983). The same is true for maternal actin mRNA [Fig. 2.17(b)]. Estimates from filter hybridization with a probe indicate that the quantity of actin mRNA in the 2-cell embryo is only about 8% of that in the unfertilized egg (Giebelhaus *et al.*, 1985; Bachvarova *et al.*, 1985) and the quantities of H3, H2a and H2b histone mRNAs about 10% of those in the unfertilized egg (Giebelhaus *et al.*, 1983; Graves *et al.*, 1985). By the 8-cell stage these mRNAs are accumulating anew, the products of synthetic activity in the zygotic genomes.

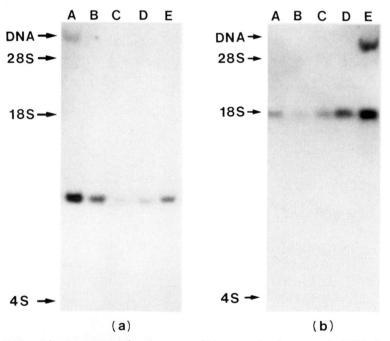

Fig. 2.17.	Disappearance and replacement of histone and actin maternal mRNAs in mouse embryos. A cloned mouse histone H3 gene probe (a) and a *Drosophila* actin gene probe (b) were reacted with RNA gel blots containing total RNA from 1,000 embryos in each lane. (a): Lane A, 4 μg mouse L cell RNA (control); lane B, unfertilized eggs; lane C, 2-cell embryos; lane D, 8-cell embryos; lane E, blastocysts. (b): Lane A, unfertilized eggs; lane B, 2-cell embryos; lane C, 8-cell embryos; lane D, blastocysts; lane E, 2 μg L cell RNA. [From D. H. Giebelhaus, J. J. Heikkila, and G. A. Schultz (1983). *Dev. Biol.* **98**, 148.]

(iv) Maternal Genetic Information in the Development of *Drosophila* and *Smittia* (Diptera)

General Characteristics of the Maternally Acting Gene Set in Drosophila

In respect to the other embryonic systems discussed in this Chapter *Drosophila* offers two unique features. These are a very different biological mode of early development, and the advantage of a genetic perspective.

Evidence reviewed in Chapter III indicates a widespread activation of zygotic nuclear transcription at the cellular blastoderm stage. However, both genetic and molecular observations demonstrate that some nuclear transcription units are functional even during the preblastoderm syncytial cleavage period. Zygotic mutants including a large deficiency and other aneuploid karyotypes that disturb development during this stage have been described (reviewed in Wright, 1970). Transcription has also been directly visualized by electron microscopy in preblastoderm nuclei by McKnight and Miller (1976) (see Fig. 3.3), and by *in situ* hybridization with cloned probes (see Chapter IV). The great majority of mRNA molecules functional prior to cellular blastoderm formation are clearly of maternal origin, however. Absolute RNA synthesis rate measurements carried out by Anderson and Lengyel (1981) show that at the onset of gastrulation over 85% of the polysomal mRNA is still of maternal origin, and even by the end of gastrulation (6 hr postoviposition) 68% of the mass of polysomal mRNA is of maternal origin. In addition, most mutations that interfere with the processes of syncytial cleavage and cellular blastoderm formation act maternally. Among the many examples that have been described are maternal mutants that fail to perform pronuclear fusion; that display mitotic defects; and that are unable to carry out normal cellularization of the blastoderm (see, e.g., Zalokar *et al.*, 1975; reviewed in Konrad *et al.*, 1985). A particularly interesting class of the latter display *regional* defects in blastoderm cellularization. For example Rice and Garen (1975) described lethal mutations in which eggs derived from homozygous mutant females carry out cleavage normally but then arrest at the cellular blastoderm stage. Eggs derived from *mat(3)3* mothers fail to form cellular blastoderm in the dorsal posterior region of the embryo, and *mat(3)1* embryos form pole cells, but otherwise do not undergo cellularization.

The biological effects of maternal mutations that interfere with known metabolic processes are instructive in that in the absence of hindsight these effects are not always easily predictable from knowledge of the biochemical lesion that they cause. A case in point is *rudimentary*, a well known mutation at a locus that encodes a multifunctional enzyme complex responsible for the first three steps in pyrimidine biosynthesis, *viz.*, carbamyl phosphate synthetase, aspartate transcarbamylase, and dihydroorotase (Jarry and Falk, 1974; Rawls and Fristrom, 1975). The *rudimentary* locus has been cloned, and it is found to encode an mRNA over 7 kb in length, from which is translated a 220 kd protein that carries out all three individual enzyme activities (Segraves *et al.*, 1983, 1984). *Rudimentary* mutants may be female sterile and they display deficiencies in wing morphogenesis. The lesion may be corrected by provision of exogenous pyrimidines. The recessive maternal effect mutations *cinnamon* (*cin*) and *maroon-like* (*mal*), which map at different locations on the X chromosome, also affect at least three enzymes, in homozygous form eliminating the activities of xanthine dehydrogenase (XDH), aldehyde oxidase, and pyridoxal oxidase (Glassman, 1965; Baker, 1973; Dickinson and Sullivan, 1975; Dickinson and Weisbrod, 1976; Browder and Williamson, 1976). However, neither the *cin* nor *mal* locus encodes the structural genes for these enzymes, which are located in two different regions of chromosome III. They may also affect other enzymes, in other pathways. *Mal* is required for the production of a molybdenum cofactor which is utilized in several other enzymes. These two loci are remarkable for their long lasting maternal effects, as observed in mutant progeny of heterozygous mothers. Thus adult *mal* and *cin* progeny of heterozygous parents have *normal* eye color and in larval stages even display increases in XDH activity, though this activity disappears at pupation in *mal* mutants and shortly after eclosion in *cin* mutants (Browder and Williamson, 1976). Marsh and Wieschaus (1977) constructed germ line mosaics by means of pole cell transplantation from *mal* embryos, and showed that the *mal*⁺ locus functions autonomously in germ line cells during oogenesis. The product of the *mal*⁺ locus accumulated in the oocyte thus suffices for the formation of eye pigment in the *adult* flies to which these eggs give rise. The *cin* mutation is in addition a maternal lethal. Most eggs of *cin* females die before hatching if fertilized by *cin* sperm, though they can be rescued by wild-type sperm. Another recessive locus that affects eye color and is also a germ line-dependent maternal lethal is *deep orange* (Marsh *et al.*, 1977). Eggs of this genotype can be rescued by injection of wild-type cytoplasm (Garen and Gehring, 1972) just as *rudimentary* eggs can be rescued by application of pyrimidines.

Systematic attempts to isolate and catalogue maternal mutants have resulted in a large collection of diverse genetic loci that display maternal effects (see, e.g., Gans *et al.*, 1975; Mohler, 1977; Nüsslein-Volhard, 1979). These include numerous genes that when mutated ultimately affect the processes of differentiation following blastoderm formation. Thus there have been identified mutations that affect the differentiation of neurons from the appropriate regions of the ventral portion of the blastoderm; that alter the postgastrular pattern of segmental identity; that interfere with anterior-posterior or dorsal-ventral assignments of cell fate; and that block germ cell determination. These and other maternal mutations affecting the pattern of differentiation in the embryo are considered specifically in Chapter VI, and some mutations that illuminate genetic functions in the process of oogenesis itself are reviewed in Chapter V. The following discussion is confined to experiments that provide insight into the size and character of the *popula-tions* of genes utilized to provide maternal developmental information.

A conclusion of general significance drawn from studies on maternal mutants as a class is that the *expression of a majority of all* Drosophila *genes that can mutate to lethality is required during oogenesis*. Mutations in some of these genes result in failure to complete oogenesis (cf. Chapter V), but when studied in such a way as to isolate their presumptive maternal effects most result in lethal arrest at embryonic stages. An experiment of Garcia-Bellido and Moscoso del Prado (1979) demonstrates the high frequency of maternal effects among lethal mutations. This study, results from which are shown in Fig. 2.18, concerns a sample of 29 different *deletions* that include from 2 to 30 salivary chromosome bands each, in total about 240 different bands or approximately 5% of the whole genome. When homozygosity for these deletions is induced by mitotic recombination in clones of somatic epidermal cells most behave as cell lethals. In Fig. 2.18 can be seen the developmental patterns of lethality resulting from crosses of heterozygous mothers and fathers (HM); from crosses of wild-type males with heterozygous deletion females (♀ HT); and from crosses of wild-type females with heterozygous deletion males (♂ HT). As expected about 25% of the offspring of the HM cross die during embryonic development. On the other hand, the embryonic lethality of ♂ HT progeny is close to that of control wild-type embryos, about 3% (Fig. 2.18), consistent with the observation that the deletions act as *recessive* lethals in epidermal cells. However, in 8 of 17 ♀ HT crosses greater embryonic lethality was observed than in the controls or in ♂ HT samples. It follows that maternal expression of the sequences deleted in these examples is required for normal development, and that a haploid complement of these sequences is inadequate for the preparation of a functional egg. Similar conclusions were derived from investigations by the same methods of the developmental effects of very large deletions that are missing from about 20 to about 60 chromosome bands, and of translocations that lead to production of aneuploid gametes (Garcia-Bellido *et al.*, 1983). For about 60% of the large deletions embryonic lethality in the ♀ HT crosses was much higher than in the ♂ HT controls. In yet another study the normal functions of 10 out of 15 lethal complementation groups from the *zeste-white* region of the X chromosome were shown to be required for the completion of oogenesis, and functions of 13 out of 15 for the production of normally developing eggs (Garcia-Bellido and Robbins, 1983). The maternal effects of 48 different recessive X chromosome lethal mutations that had been induced by EMS treatment have been examined directly by Perrimon *et al.* (1984). In this study homozygous mutant genotypes were created in dividing germ line cells by X-irradiation induced mitotic recombination, and descendants of oogonia that did not undergo the requisite recombinations were eliminated by use of stocks bearing a dominant female sterile mutation that prevents egg formation. Only 13 of the 48 loci thus investi-

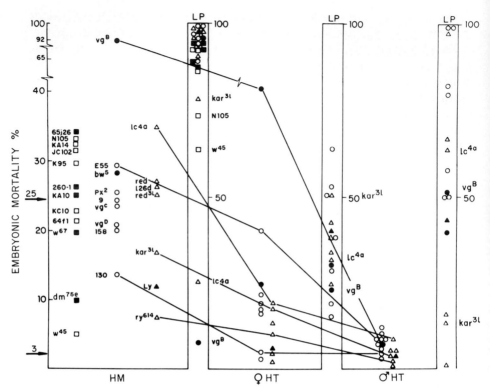

Fig. 2.18. Embryonic lethality in *Drosophila* bearing chromosomal deletions. The three panels show, respectively, lethality in progeny of crosses between males heterozygous for each deletion and females heterozygous for the same deletion (HM) (for deletions in X, males are wild type); between heterozygous females and wild-type males (♀ HT); and between heterozygous males and wild-type females (♂ HT). Open symbols represent deletions that are lethal when *homozygous* in somatic clones of epidermal cells. These deletions are viable in epidermal cells *heterozygous* for them. Closed symbols represent deletions that when homozygous are viable in epidermal cells. LP indicates lethal phase, here the percentage of embryos that reach stage 14. Deletions in X chromosome, □; in chromosome II, ○; in chromosome III, △. Arrows on ordinate represent expected mortality in HM cross (i.e., the 25% of embryos that will be homozygous for these lethals), and mortality observed in wild-type control flies (3%). [From A. Garcia-Bellido, and J. Moscoso del Prado (1979). Reprinted by permission from *Nature* (*London*) **278**, 346. Copyright © 1979 Macmillan Journals Limited.]

gated displayed no maternal effects. Development in the remainder was blocked either during oogenesis or in embryogenesis. These experiments have led to the supposition that at least 75% of all the genes that can mutate to lethality in the *Drosophila* genome are expressed during oogenesis, and exercise some necessary maternal function (Garcia-Bellido and Robbins, 1983; Konrad *et al.*, 1985).

This result is consistent with a direct interpretation of the complexity measurement for *Drosophila* egg RNA given above, about 1.2×10^7 nt

(Hough-Evans *et al.*, 1980; see also Table 2.1). If it is assumed that the complexity of the egg RNA is wholly due to maternal messages, it would suffice for about 6×10^3 different mRNA species of average length. On the other hand, the number of lethal complementation groups in the *Drosophila* genome appears to be only about 10% less than the number of cytologically visible salivary chromosome bands (Judd *et al.*, 1972; Shannon *et al.*, 1972; Hochman, 1973) of which there are about 5,000. Close molecular examination of several cloned regions of the genome each hundreds of kilobases in length suggests that the number of (easily detected) transcription units is only about a factor of two greater than the number of bands or lethal complementation groups (see, e.g., Spierer *et al.*, 1983; Spierer, 1984; Bossy *et al.*, 1984). Were the fraction of all such transcription units that are expressed during oogenesis the same as for genetic loci that can mutate to lethality, the predicted *complexity* of the maternal RNA would be very close to the measured value (i.e., 0.9×5000 bands $\times 2 \times 0.75 \times 2$ kb per mRNA, or 1.3×10^7 nt).

Are the bulk of maternal and embryonic lethals simply "housekeeping genes," i.e., genes whose products are ubiquitously required for cellular survival, mitosis, or metabolism? We have seen that most lethals in *Drosophila* display a maternal activity phase. On the other hand, tests of random collections of both point mutation lethals (Ripoll, 1977) and of deficiency lethals (Garcia-Bellido and Ripoll, 1978) have shown that in homozygous form only about 10% are *cell-lethal*, at least in abdominal epidermal cells. These estimates were made following induction of mitotic recombination in clones of somatic cells by X-irradiation of heterozygous animals. This issue has also been examined by Garcia-Bellido and Robbins (1983) in the *zeste-white* region characterized by Shannon *et al.* (1972). Less than a fourth of the lethal complementation groups in this region that are required for germ line viability turn out to be necessary for cell viability. The process of development thus requires a far greater diversity of genetic loci than does survival and multiplication of given somatic cells. As reviewed in Chapter III, in other organisms the general concept that a majority of genes active in early development are expressed only in the oocyte and/or embryo is supported directly by molecular measurements. Both these and the genetic observations reviewed here are clearly inconsistent with the view that most of the population of genes utilized in the early embryo are "housekeeping" genes.

Another conclusion indicated both by the studies of maternally acting genes cited and by direct sequence comparisons of maternal and embryo RNA sequence populations (Chapter III) is that most genes functioning in the embryo are also expressed during oogenesis. Thus as we have seen, function of the major fraction of *zygotic* lethals is also required during oogenesis (see, e.g., Garcia-Bellido and Robbins, 1983; Perrimon *et al.*, 1984). An additional analysis was carried out by Robbins (1980, 1983) on essential loci in the *zeste-white* region of the X chromosome. Of 13 potentially lethal loci, 10 were found to function during both maternal and embryonic phases of development. In these studies maternal gene function was varied by altering gene dose, and zygotic gene function by use of position effect variegation. When both maternal and zygotic gene activity were reduced lethality resulted, while normal expression in either phase of development sufficed to ensure viability. This behavior might of course be expected for constitutively active "housekeeping" genes, but the test of cell viability precludes this interpretation for a majority of the *zeste-white* lethals (Garcia-Bellido and Robbins, 1983). An impressive number of genes for which specific developmental functions are known have been shown to display both maternal and zygotic activity (reviewed in Konrad *et al.*, 1985). Molecular evidence has indicated, however, that though a given *locus* may be functional in both oogenesis and embryogenesis, its transcriptional products may be different at the two stages. For example, an unidentified locus described by Vincent *et al.* (1984) produces five distinct transcripts, generated from alternative transcription initiation sites or by alternative processing patterns, or both. Three of these transcripts are exclusively zygotic, and two accumulate during oogenesis and persist as maternal RNAs in the embryo.

In summary, genetic analyses of samples of the large population of *Drosophila* maternal effect mutants leads to three generalizations that we shall encounter again in a molecular guise in the following chapter. These are that a majority of the functional genomic loci are required for normal early development; that most of these loci are not "housekeeping" genes; and that most function *both* in oogenesis and in the embryo. A number of specific maternal mutations that provide access to certain detailed processes of determination are discussed later in this book.

Molecular Aspects of Maternal mRNA Utilization in Drosophila and Other Dipterans

The characteristics of maternal mRNA in *Drosophila*, and its mode of function, are in most general respects similar to what has been described above for other animals. The average length of the poly(A) RNA in late oocytes, preblastoderm embryos, and early blastoderm embryos is ≥ 2 kb, and according to cDNA hybridization kinetics the sequence populations of polysomal and nonpolysomal poly(A) RNA fractions from these stages are identical (Goldstein, 1978). Anderson and Lengyel (1979) reported about 3.8 ng of poly(A) RNA in very early embryos. The mass of poly(A) RNA thus accounts for about 2% of the quantity of total RNA, 180 ng. This suggests a content of about 3×10^9 molecules of maternal poly(A) mRNA per egg. As reviewed in Chapter V, at the end of oogenesis there are about 4.5×10^{10} maternal rRNA molecules per oocyte. About 50% of the ribosomes are included in polysomes during early development (Lovett and Goldstein, 1977; Anderson and Lengyel, 1984), a ratio that suggests maximal utilization of much of the available maternal poly(A) mRNA (i.e., ~7 ribosomes per polysomal mRNA). There may also be some poly(A)-deficient maternal mRNA (Zimmerman *et al.*, 1980), though this has not been examined directly in early embryos. One species that is largely poly(A)-deficient is histone mRNA, but as reviewed below this accounts for only a small percentage of the total mRNA in preblastoderm stages.

Protein synthesis varies little during preblastoderm *Drosophila* development, either qualitatively or quantitatively. Perhaps this is not surprising, given the extremely short time required to attain the cellular blastoderm stage, before which almost all synthesis occurs on maternal templates. Santon and Pellegrini (1981) reported an absolute rate of protein synthesis of 9.4 ng hr^{-1} in permeabilized 1.5-2.5 hr preblastoderm embryos, based on measurements of phenylalanine pool specific activity. There may be about a 1.5- to 2-fold increase at fertilization (Zalokar, 1976; Anderson and Lengyel, 1984). At cellular blastoderm stage, however, the rate declines to about 40% of the 1.5-2.5 hr rate (Santon and Pellegrini, 1981). This decline is probably not the consequence of exhaustion of maternal mRNA, as in the mouse, since the total poly(A) RNA content remains essentially constant (Anderson and Lengyel, 1981). Comparisons of the pattern of protein synthesis by two-dimensional gel electrophoresis in ovarian oocytes, unfertilized eggs, and 1 or 2 hr embryos reveal relatively few changes, though there are several polypeptide species not synthesized in unfertilized eggs that begin to be translated early in embryogenesis, and several other species translated only during oogenesis (Savoini *et al.*, 1981; Sakoyama and Okubo, 1981; Trumbly and Jarry, 1983). In the most detailed study Summers *et al.* (1986) identified 24 protein species, out of a total of about 1200 resolved, that first appear during the processes of nuclear migration and cellular blastoderm formation. Synthesis of most of these is not observed at later stages. Similarly, about thirty species were reported by Trumbly and Jarry (1983) to be synthesized at 1-3 hr after oviposition but not later, and an equal number were detected in embryos older than 1 hr but not earlier. Approximately half of these developmentally regulated polypeptides could be recognized among the products of *in vitro* translation, and within this set comparisons of the spectra of proteins translated from 1 hr and from 3-5 hr embryo mRNAs *in vitro* demonstrated most of the same changes as observed *in vivo*. Thus these changes in synthetic pattern reflect the appearance of new mRNAs due to early zygotic function and the disappearance of preexistent mRNAs. There are also a few species not synthesized at 1 hr *in vivo* for which the mRNAs are nonetheless demonstrable *in vitro*. Thus, regulation by translational selection or message sequestration also occurs in *Drosophila* embryos, though relatively rarely. As observed for the overall protein synthesis rate, at the cellular blastoderm stage there is a noticeable, though temporary, decrease in the rate of synthesis of many individual polypeptide species. Aside from these variations, the main conclusion from two-dimensional analyses of protein synthesis patterns in *Drosophila* embryos is that the sets of messages translated before fertilization, during the preblastoderm stages, and even later when zygotic mRNAs become prevalent, are largely though certainly not completely overlapping.

Comparable measurements have also been reported for early embryos of the chironomid midge *Smittia* by Jäckle and Kalthoff (1979) and Jäckle (1979). This egg contains only about 17 ng of total RNA, about a tenth the amount in the *Drosophila* egg. Maternal RNA was labeled by injection of precursor into the coelom of the female during oogenesis, and about 4.2% of the labeled RNA recovered in newly deposited eggs was poly(A) RNA. Assuming that the stable maternal RNA is uniformly labeled, this would indicate about 0.7 ng of poly(A) RNA. The quantity of maternally labeled poly(A) RNA then declines, and by the blastoderm stage, in *Smittia* about 9 hr after oviposition, 45% of it has disappeared. This presumably reflects the length of time that maternal mRNA is utilized in the *Smittia* embryo. Protein synthesis during the preblastoderm period depends on the store of maternal mRNA, as there is very little new RNA synthesis, and no effect of α-amanitin or actinomycin is detected until after blastoderm formation.

Maternal mRNAs for several specific proteins have been identified in *Drosophila* and *Smittia* embryos. These are largely the same proteins as discussed in respect to the other organisms considered in this chapter. Actin mRNA representing two of the six actin genes has been detected in preblastula *Drosophila* embryos by reaction with cloned probes (Sodja *et al.*, 1982; Fyrberg *et al.*, 1983). From data reviewed in Chapter V it can be estimated that the mature oocyte contains about 5×10^7 molecules of maternal actin mRNA (Table 5.1). Maternal actin mRNA has also been identified among the synthetic products of *Smittia* egg RNAs, both *in vitro* and *in vivo* (Jäckle, 1980). As expected, translation of actin in preblastoderm stages is not inhibited by α-amanitin, and the stored mRNA appears to be sequestered in maternal mRNP particles. Mature mRNAs coding for α- and β-tubulins have also been observed in *Drosophila* eggs. The mature oocyte is estimated to contain about 4×10^7 α- and β-tubulin mRNAs (see Table 5.1). There are four α- and four β-tubulin genes in the *Drosophila* genome (Kalfayan and Wensink, 1982; Mischke and Pardue, 1982; Natzle and McCarthy, 1984), and all four of the α-tubulin genes are represented in the RNA of preblastoderm embryos. One of these genes, called Tα4, is noteworthy in that it appears to be expressed only in females, and its transcript is observed in 0-2 hr eggs (Kalfayan and Wensink, 1982; Natzle and McCarthy, 1984). The exclusive function of this tubulin gene thus seems to be synthesis of a maternal mRNA. Of the four β-tubulin genes three are also represented in the maternal mRNA (Natzle and McCarthy, 1984). A third class of proteins for which maternal mRNAs have been detected in preblastoderm *Drosophila* embryos is the ribosomal proteins. This set of proteins is synthesized at an aggregate rate of about 0.42 ng hr^{-1}, or about 4.5% of total protein synthesis (Santon and Pellegrini, 1981). This suggests that the egg may contain over

10^8 molecules of mRNA coding for ribosomal proteins. Maternal rRNAs coding for proteins homologous to the vertebrate *src* proteins have also have reported in *Drosophila* (Wadsworth *et al.*, 1985). There are three genes coding for these proteins, and in adult females they are primarily active in the ovaries. Following blastoderm formation the level of *src* transcripts declines, suggesting a specifically early embryonic function.

Relatively detailed measurements exist for the *Drosophila* histone mRNAs. The histone genes of this organism have been cloned and are well characterized (Lifton *et al.*, 1978). The genome contains about 110 tandem clusters, each containing genes for all five histone species, and there is no evidence for a different set of maternal as opposed to zygotic, or early as opposed to late, histone genes (Anderson and Lengyel, 1984). Anderson and Lengyel (1980) measured the quantity of maternal histone mRNA by hybridization with a cloned histone probe that contains the whole of the histone gene cluster. In preblastoderm embryos histone mRNA accounts for about 2% of the total maternal mRNA. These data indicate a store of almost 10^8 maternal histone mRNA molecules of each species, which renders them the most prevalent maternal mRNAs in the *Drosophila* egg. After fertilization maternal histone mRNA is rapidly and specifically mobilized. Only about 25% of histone mRNA is polysomal at oviposition, but within about 2 hr the fraction that is polysomal exceeds 75% (Anderson and Lengyel, 1984). Zygotic histone mRNA synthesis is activated at the syncytial blastoderm stage, 90 min postoviposition, about an hour before cellularization. There is no evidence that the *Drosophila* egg contains histone protein synthesized during oogenesis, and the amount of stored histone mRNA suffices for synthesis of the total quantity of histone required to the blastoderm stage (Anderson and Lengyel, 1980, 1984). Thus a strategy different from both sea urchin and *Xenopus* is utilized in the early *Drosophila* embryo. The egg of *Drosophila* contains about 100 times more maternal histone mRNA of each nucleosomal species than does the egg of *S. purpuratus*. Within about 2.5 hr it must synthesize sufficient DNA and chromatin histone to supply 6,000 nuclei, each of which contains only about 0.3 pg of DNA, or a total of about 2 ng of DNA (or histone). In comparison, in the 4 hr period during which in the sea urchin embryo translation of maternal histone mRNA accounts for most histone protein biosynthesis, the amount of histone that could be produced in the sea urchin embryo assuming the translational efficiency of the *Drosophila* embryo would suffice for only about 20 sea urchin cells [(2 ng × 4/ 2.5) ÷ (1.6 pg/nucleus × 100)].

(v) Maternal mRNA in Other Organisms: Some Comparative Strategies

When considered in detail, a unique mode of maternal mRNA utilization appears characteristic of each developmental system. There are nonetheless certain invariant features, i.e., beyond the basic fact that maternal mRNA is required in the early development of every known animal egg. Among these are specific molecular aspects such as the ubiquitous presence of maternal messages coding for cytoskeletal actins and for histones. The quantitative similarity of maternal RNA complexities was discussed earlier (see Table 2.1), implying with few exceptions a basic homology in the functional role played by maternal RNA in all forms, irrespective of overall genome size. The impression that there is an underlying set of properties (not sequences) shared by the maternal gene sets of many organisms is strengthened by genetic observations. Thus analysis of large samples of maternal effect mutations in the nematode *Caenorhabditis elegans* yield many of the same conclusions as extracted from the *Drosophila* studies reviewed above, though the developmental processes in these two organisms are very different in character (cf. Chapter IV). Collections of temperature-sensitive lethals that block embryonic development in *C. elegans* have been isolated and characterized by Wood *et al.* (1980), Miwa *et al.* (1980), Cassada *et al.* (1981), and Isnenghi *et al.* (1983). Of a total of 55 different loci deriving from these assemblages that are essential for embryogenesis, maternal function of 32 has been shown to suffice for normal development; either maternal or zygotic expression is sufficient for an additional 17 loci; and four loci are required both during oogenesis and embryogenesis (Isnenghi *et al.*, 1983). Thus it can be presumed that in normal animals 53 out of the 55 loci are expressed maternally. As in *Drosophila*, mutations in most of these loci either cause defects in oogenesis that result in sterility, or disturbances in the process of early development. Among the consequences of such mutations are failure to carry out pronuclear fusion, anomalous cleavage patterns that result in early arrest or cell lineage defects, and embryonic malformations (Wood *et al.*, 1980; Denich *et al.*, 1984). In the absence of direct molecular evidence it is difficult in most cases to determine whether the essential protein products produced by the normal alleles of the embryonic lethals are translated from maternal mRNA before or after fertilization, or both, though it is clear that their transcription occurs at least during oogenesis. However, study of the temperature-sensitive periods of these mutants has identified some genes expressed before fertilization whose protein products are required during embryogenesis (Wood *et al.*, 1980; Isnenghi *et al.*, 1983).

We return in the following to consideration of differences, as well as similarities, among diverse species in their mode of utilization of maternal mRNA. These are briefly examined from the comparative vantage point afforded by some additional examples not considered earlier.

Provision of Histones in the Early Embryo

A survey of diverse embryos reveals a variety of intricate logistic solutions to the problem of supplying nucleosomal histones to the chromatin of the cleavage stage blastomeres. The initial sets of blastomere histones are variously derived from presynthesized pools of histone protein synthesized during oogenesis, from histone newly translated from maternal mRNA, or both. Whatever the source, however, the mechanisms utilized to provide early embryo histones usually include the features that in the initial phase of relatively rapid cleavage there is more histone available at any given point than is required by the number of nuclei present; that by the time the nucleus-to-cytoplasm ratio characteristic of the post-cleavage embryo has been attained, if not long before, zygotic histone mRNA has replaced the earlier sources; and that histone synthesis from maternal mRNA is uncoupled from the nuclear replication cycle. Evidence reviewed above shows that *Xenopus* and sea urchin embryos rely on both maternal mRNA and preformed histones for the chromatin of their cleavage stage blastomeres, though the relative importance of newly synthesized compared to preformed histones is greater for the sea urchin embryo. Embryos of the nonsegmented lower protostome worm *Urechis* appear to utilize mainly preformed histone protein synthesized during oogenesis, though a small amount of new histone synthesis occurs as well, probably on maternal mRNA (Franks and Davis, 1983). At fertilization the *Urechis* egg contains sufficient stored histone for 32-64 cells. Histone protein is also stored from oogenesis in the egg of the gastropod *Ilyanassa* (Mackay and Newrock, 1982), though again there is also new histone synthesized during early cleavage. The maternal origin of the histone message utilized at this time is demonstrated by the observation of Collier (1981) that histone synthesis occurs in the enucleate cytoplasm extruded into the first cleavage polar lobe. In *Urechis* and sea urchins, though not *Xenopus*, the maternal histones and the histones synthesized later in development are of different subtypes, and in the sea urchin case it is clear that these derive from distinct sets of genes. Developmental change in histone subtypes has also been observed in *Ilyanassa*, where there are oocyte-specific forms (Mackay and Newrock, 1982), and in the clam *Spisula*, where a subtype switch occurs late in cleavage (Gabrielli and Baglioni, 1977).

Several informative exceptions to these general patterns are known. The mouse embryo of course illustrates one special case, in that there is no period of rapid cleavage and hence no requirement for supply of histone at an unusually high rate. Nonetheless, as we have seen, maternal histone mRNA and presynthesized histone protein are both stored in the mouse egg, perhaps an atavistic residue of an earlier, more rapidly developing ancestor, or an inescapable requirement that follows from the general transcriptional repression prior to first cleavage. Clearly the 20 hr that elapses prior to first cleavage in the mouse would suffice for synthesis of the amount of histone message required for two diploid cells, since this is a common generation time for mammalian cells that transcribe their own histone message. The opposite extreme is illustrated by the *Drosophila* egg, where a relatively enormous store of maternal histone mRNA suffices to supply histones even for the extremely high rate of chromatin production that obtains up to the 6,000-cell stage. This solution of course depends at least in part on the relatively minute size of the *Drosophila* genome, and hence the small per nucleus mass of histone required.

The possibilities offered by the storage of either maternal histone mRNA, maternal histone protein, or both, have evidently been exploited during evolution in a very flexible manner. This element of the molecular biology of the embryo illustrates the highly adaptive nature of even such basic and early processes as the provision of materials needed for construction of blastomere chromatin. In each case the quantitative molecular details are best understood in simple logistic terms dictated by other developmental parameters, such as the rate of cleavage, the size of the egg, the length of the period of transcriptional quiescence (if any), and the size of the genome. Even in phylogenetically related forms the means that have evolved for the provision of histones may vary greatly. A good example is to be found in studies of histone gene expression in echinoderms other than the modern or "advanced" sea urchins. Though all forms investigated ultimately utilize large amounts of histone mRNA during embryogenesis, neither the eggs of the "primitive" sea urchin *Eucidaris*, and the sea cucumber *Thyone*, nor the eggs of two different genera of sea stars, *Asterias* and *Pisaster*, contain detectable *maternal* histone mRNA (Raff *et al.*, 1984; Howell *et al.*, 1986). Yet the genomes of these animals all include sets of α-histone genes. The intricate series of mechanisms by which histone mRNAs are provided during the early embryonic development of sea urchins such as *Strongylocentrotus* or *Lytechinus* thus constitute a special adaptation, the evolutionary assembly of which has required the institution of new programs of histone gene regulation.

Assembly of Stored Maternal mRNA into Polysomes

A variety of patterns is also encountered in considering the quantitative activation of maternal mRNA translation early in development. These range from the dramatic hundredfold increase in translational activity triggered by fertilization in the sea urchin egg, to the condition represented by *Drosophila*, the mouse, and the anuran amphibians, in which fertilization *per se* causes no more than a twofold change in protein synthesis rate or no change whatsoever. The end result in these diverse examples is nonetheless the same, *viz.*, engagement of most of the maternal messages and a significant fraction of the maternal ribosomes in polysomes. The distinction lies instead in the position of the translational apparatus at the time of fertilization. In the unfertilized sea urchin egg there is very little translation, a phenomenon no doubt related to the particular reproductive strategy utilized by this animal, in which matured, haploid eggs are stored in the lumen of the ovary pending appearance of the external stimulus that leads to spawning. In amphibians, where the fully-grown *ovarian oocyte* may be stored for long periods, it is the induction of ovulation rather than fertilization that leads to an increase in the translation of maternal mRNA (see above and Chapter V). Ovulation is of course ultimately a process initiated by behaviorally controlled hormonal stimulation. However different the trigger and the timing, the primary molecular mechanism of the increase in protein synthesis rate in both sea urchin and amphibian eggs is the assembly of polysomes from preformed maternal mRNA and ribosomes.

Other species in which more than a twofold increase in protein synthesis rate, polysome content, or both, follows upon fertilization are the clams *Spisula solidissima* (Bell and Reeder, 1967; Firtel and Monroy, 1970) and *Mulina lateralis* (Kidder, 1972). In *Spisula* the rate of protein synthesis in the mature oocyte is very low, and at fertilization it rapidly increases 3- to 4-fold, as does polysome content. However, in this organism germinal vesicle breakdown and maturation are stimulated by fertilization, and it is not unlikely that it is release of factors from the germinal vesicle or some other event of the maturation process, rather than fertilization *per se*, that causes the stimulation of protein synthesis (Rosenthal *et al.*, 1982). In the starfish *Asterias forbesi* the pattern of maternal mRNA mobilization during maturation also resembles that in amphibians. Maturation can be induced by a hormone, 1-methyl adenine, which like progesterone in amphibians acts at the egg surface. Soon after application of this substance the rate of protein synthesis begins to increase (Houk and Epel, 1974). The event of fertilization itself has no effect on the rate of protein synthesis. Another marine egg in which only a minor change in the rate of protein synthesis is observed after fertilization is that of *Ilyanassa obsoleta*. The rate of protein synthesis increases about 2.5-fold shortly after fertilization, and then grows slowly through gastrulation (day 3) to a final level only about three times that at fertilization. About a twofold increase in the amount of polysomes occurs between the unfertilized egg and the 8-cell cleavage stage (Collier and Schwartz, 1969; Mirkes, 1972). Similarly, fertilization results in only a twofold increase in protein synthesis rate in the eggs of a teleost fish studied by Krigsgaber and Neyfakh (1972). On the other hand a sea urchin-like pattern of rapid polysome assembly and protein synthesis increase occurs after fertilization in a different echinoid, the sand dollar *Dendraster excentricus* (Spieth and Whiteley, 1981), and probably also in a completely unrelated animal, *Urechis*. The mature oocytes of *Urechis* are stored free of accessory cells in the coelomic cavity (see Chapter V). Davis (1982) showed that only about 0.6% of ribosomes are found in polysomes in these oocytes. By the 2-cell stage (2 hr after fertilization) this fraction has increased about sevenfold, and by the 16-cell stage (3.5 hr) it has increased more than elevenfold. A rapid fivefold rise in protein synthesis rate in response to fertilization has also been reported for the polychaete annelid *Sabellaria alveolata* (Guerrier and Freyssinet, 1974). A reasonable summary of these various patterns is that where for adaptive reasons the life cycle includes a *storage phase* at the fully-grown ovarian oocyte stage, or where already matured eggs are stored, the rate of protein synthesis is low compared to that required once development is initiated. Translational activation occurs in these situations, respectively, during maturation or after fertilization, and the major mechanism is always *de novo* assembly of maternal mRNA into polysomes. In other animals where the requirement for a quiescent preembryonic phase does not exist, the rate of protein synthesis after fertilization is much the same as before.

Mobilization of stored messenger RNA can be observed in a completely different context during embryogenesis in *Artemia salina*, the brine shrimp. Under certain conditions this organism produces an encysted, dessicated gastrula in which all detectable metabolism has ceased. Upon rehydration the cryptobiotic state is relieved, and development proceeds (Clegg, 1967). Resumption of metabolism is associated with a continuous increase in protein synthesis rate (Golub and Clegg, 1968; Clegg and Golub, 1969; Hultin and Morris, 1968), and underlying this is a parallel increase in polysome content. Polysome assembly is detectable within 3 min in gastrulae prehydrated at 0°C and then raised to 30°C, or within 30 min without prehydration (Clegg and Golub, 1969). Poly(A) RNA is stored in the cryptobiotic cysts in the form of mRNP particles, from which message active in cell-free systems can be extracted (Nilsson and Hultin, 1974; Slegers and Kondo, 1977). The major features of this system are clearly reminiscent of maternal mRNA utilization during early cleavage in species such as the sea urchin.

Translational Regulation

Qualitative control of protein synthesis patterns at the level of specific mRNA sequestration or sequence-specific translational selection occurs in each of the developmental systems considered in this chapter. However, in different organisms such mechanisms are of very different significance. In the sea urchin, except for the nuclear sequestration of histone mRNAs, sequence-specific regulation of mRNA assembly into polysomes is apparently confined to a few specific proteins, and this is probably the case for *Drosophila* as well. In the mouse and *Xenopus*, such regulation may be more prominent, as is easily demonstratable by comparison of the proteins translated *in vitro* from embryo mRNA preparations with those synthesized *in vivo* in the same embryos. This form of biosynthetic regulation may be most significant where biological development proceeds for many hours in the absence of new nuclear input. In *Xenopus* such a period obtains from the time of germinal vesicle breakdown during maturation to the reactivation of transcription at the midblastula stage, and in the mouse from germinal vesicle breakdown to the end of the long interval before the sharp acceleration of transcription at the 2-cell stage.

Programmed regulation of translation can function as a strictly cytoplasmic process in early development, independent of immediate nuclear influence. This is demonstrated in various experiments with anucleate cytoplasmic fragments, some of which have been reviewed earlier. An additional example is provided by the naturally occurring preparation of anucleate egg cytoplasm available in the *Ilyanassa* polar lobe (developmental significance of this structure is discussed in detail in Chapter VI). As noted above polar lobes synthesize protein from maternal mRNA for at least 24 hr after isolation (Clement and Tyler, 1967). Geuskens (1969) showed by electron microscope autoradiography that the protein synthesis occurring in polar lobes takes place in cytoplasmic polysomes and not in mitochondria. The *Ilyanassa* egg is reported to contain about 2.5×10^8 poly(A) RNA molecules (Clark and Kidder, 1977), i.e., several times the content of such molecules in the sea urchin egg. The polar lobe includes about 14% of the total RNA (Collier, 1983), and presumably an equivalent fraction of the poly(A) RNA. Brandhorst and Newrock (1981) and Collier and McCarthy (1981) showed by two-dimensional gel electrophoresis that in normal development several changes in the species of proteins synthesized take place during the first day of development, i.e., to the early gastrula stage. Many of these same changes occur in actinomycin-treated embryos, and even in *isolated polar lobes cultured for an equivalent time* in vitro. This result is illustrated in Fig. 2.19 (Brandhorst and Newrock, 1981). The experiment implies the existence of a biologically timed mechanism that results in the delayed mobilization of preexistent maternal mRNAs.

In both the starfish *Asterias forbesii* and the clam *Spisula solidissima* protein synthesis patterns are altered sharply during the meiotic maturation period, just as in the mouse. Rosenthal *et al.* (1980, 1983) showed that in *Spisula*, where meiotic maturation *follows* fertilization, the mRNA extractable from oocytes, eggs, and early cleavage stage embryos codes for identical sets of detectable proteins, though *in vivo* the synthetic output of oocytes and fertilized eggs differs clearly. The messages not expressed *in vivo* are found in postpolysomal particles in oocytes, and their partition into polysomal *vs.* nonpolysomal compartments is correlated with the presence of a poly(A) 3′ terminal tract. It is unlikely, however, that this is causal, since there are exceptions to the correlation, and in cell-free systems no difference in translational activity could be detected between poly(A)-containing and poly(A)-deficient messages (Rosenthal *et al.*, 1983). One identified *Spisula* maternal mRNA that remains unassociated with polysomes for some hour following fertilization is that coding for α-tubulin (Tansey and Ruderman, 1983). After about 4 hr this mRNA shifts to the polysomal compartment, suggesting the presentation of some specific factor required for initiation. Similarly, the changes in protein synthesis occurring during maturation of starfish oocytes have been shown to occur by translational selection, since analyses carried out with cell-free protein synthesis systems demonstrate

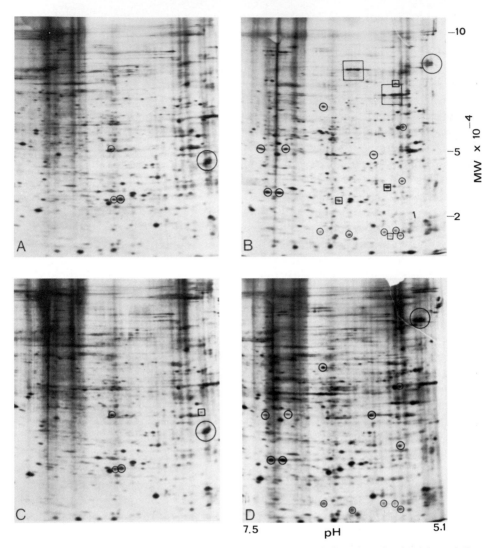

Fig. 2.19. Translation-level regulation of protein synthesis in cultured polar lobes of *Il-yanassa obsoleta*. Autoradiographs of two-dimensional gel electrophorograms displaying newly synthesized proteins are shown. (a) Embryos incubated with ³⁵S-methionine for 4 hr after first cleavage (i.e., to third cleavage); (b) embryos labeled for 4 hr, 21-25 hr later, at the gastrula stage; (c) isolated polar lobes labeled at the same time as in (a); and (d) isolated polar lobes incubated for 21 hr and then labeled at the same time as in (b). Spots in circles in (a) represent proteins the synthesis of which becomes reduced or disappears by 21-25 hr [i.e., as seen in (b)]. Spots in circles or squares in (b) are proteins not synthesized in early cleavage [i.e., as seen in (a)]. Spots in circles in (c) indicate proteins synthesized in isolated polar lobes that as in the normal embryo will cease to be synthesized by 21 hr later. Spots in circles in (d) show proteins that as in the normal embryo are synthesized in isolated polar lobes at 21-25 hr but not at 0-4 hr after first cleavage. Spots in squares in (b) represent proteins appearing in normal embryos by 21-24 hr after first cleavage, but not in isolated polar lobes. [From B. P. Brandhorst and K. M. Newrock (1981). *Dev. Biol.* **83**, 250.]

the population of maternal mRNA in oocytes before and after 1-methyl adenine treatment to be identical (Rosenthal *et al.*, 1982). In addition Martindale and Brandhorst (1984) showed that the biosynthetic alterations set in train by treatment with this hormone occur as well in anucleate fragments of oocyte cytoplasm if these are stimulated with 1-methyl adenine. The key locus of control indicated by all of these studies is sequence-specific regulation of the *extent of recruitment of certain maternal messages into polysomes*. This is clearly a prominent feature of early development, where it can be imagined as a substitute for transcriptional regulation, that provides situational flexibility in the biosynthetic output of a fixed pool of stored maternal mRNAs.

Maternal mRNA is a completely general and indispensible aspect of development from egg and sperm. Both complexity measurements and genetic analyses of maternal and zygotic mutants demonstrate that an enormous set of genes is expressed during oogenesis. This set may constitute a majority of all genomic loci engaged in the process of ontogeny. For the most part these genes continue to function in the early embryo. Seen from the vantage point of transcriptional regulation, a major initial event in embryological development is thus the establishment of the activity of the maternal gene set in the growing oocyte. As implied by the phenomenon of parthenogenesis, fertilization is in most organisms of minor immediate consequence in regard to the developmental utilization of gene products, except as a trigger. What variations in expression are at first required are regulated at the translational level. The organization of a genetic assemblage that could carry out the functions of the maternal gene set, and of mechanisms for its expression, had evidently occurred by the beginning of the Cambrian. This event was fundamental to the evolution of metazoan development from egg and sperm, and it underlies all forms of animal embryogenesis of which we are aware.

III

Transcription in the Embryo and Transfer of Control to the Zygotic Genomes

Zygotic genome activity is conveniently considered as the sum of two processes, the replacement of maternal transcripts by homologous newly synthesized embryonic transcripts, and the appearance of novel "late gene" transcripts not included in the original maternal gene set. The patterns of late gene activation indicate temporally and sometimes regionally specific regulation in the embryonic genomes, as discussed in Chapter I. Though they are of great developmental significance, late genes constitute only a numerically minor fraction of active embryonic loci. This is implied by the studies on maternal mutants reviewed earlier, and is demonstrated in a quantitative manner in the comparisons of maternal and zygotic mRNA populations considered in this Chapter. We here examine the patterns of late gene transcription, and review quantitative evidence regarding the production of messenger and other RNA species in the embryo. The expression of the histone genes and of some other specific sets of genes is then reviewed. Genes that are activated only in defined cell lineages or in particular regions of the embryo are considered in Chapter IV.

1. GENERAL QUALITATIVE ASPECTS OF NUCLEAR RNA AND MESSENGER RNA SYNTHESIS IN EMBRYOS OF VARIOUS SPECIES

(i) Zygotic Transcription in Sea Urchin Embryos

Due to fortuitous biological features such as availability, ease of handling, embryonic synchrony, and high permeability to nucleic acid precursors, transcriptional processes in the sea urchin embryo are better understood than are those of other embryos. It is useful to consider these processes initially in reference to the time course of cell division. This is illustrated for *Strongylocentrotus purpuratus* in Fig. 3.1. After a brief lag the cell number increases at a logarithmic rate until over 100 cells have appeared, and the rate of increase then declines progressively. As discussed in more detail in Chapter IV, the overall cell replication pattern described in Fig. 3.1 is the sum of the diverse programs of division operating in the different cell lineages of the embryo. In some sea urchin species cleavage occurs synchronously for the first 8 or 9 cycles and until about 6th cleavage at evenly spaced intervals, except for the micromere lineage which follows a different schedule and withdraws from mitotic activity earlier (see Chapter IV). At about the 9th or 10th cycle, the intercleavage period increases greatly, and syn-

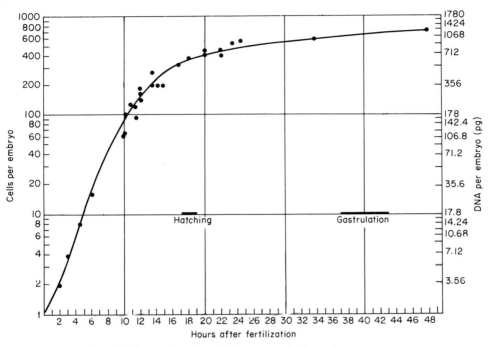

Fig. 3.1. Cell division and nuclear DNA content in *Stronglyocentrotus purpuratus* embryos. Embryos were grown at 15°C and the number of cells counted in squashes prepared with EDTA in order to form a monolayer of blastomeres on the slide (left ordinate). The right-hand ordinate gives DNA content in pg DNA per embryo, calculated from the cell number on the basis that each diploid cell contains 1.78 pg (Hinegardner, 1974). [From R. T. Hinegardner (1967). *In* "Methods in Developmental Biology" (F. H. Wilt and N. K. Wessells, eds.), p. 139. Cromwell, New York.]

chrony between lineages is lost (see, e.g., Masuda, 1979; Dan *et al.*, 1980; Masuda and Sato, 1984). Some lineages continue to proliferate in late blastula and gastrula stages, while others cease to divide, or divide only occasionally. This accounts for the overall decrease in the slope shown in Fig. 3.1 after about 20 hr. Depending on species and temperature, the period of the cell cycle during the logarithmic growth phase before about 6th cleavage falls in the range 20-90 min, and there is a slight cell cycle elongation from 6th to 8th cleavage (Fig. 3.1; see also Dan *et al.*, 1980; Masuda and Sato, 1984).

Synthesis of heterogeneous high molecular weight nuclear RNA (nRNA) is already taking place in the pronucleus of the unfertilized egg (Levner, 1974; Brandhorst, 1980), and is also active in the male pronucleus following fertilization (Poccia *et al.*, 1985). Transcription continues in the zygote nucleus and throughout cleavage (Wilt, 1963, 1964; Nemer and Infante, 1965; Rinaldi and Monroy, 1969; Kedes and Gross, 1969). This contrasts with cleavage stage *Xenopus* and *Drosophila* embryos, the genomes of which are largely quiescent until blastula and blastoderm formation respectively.

Though it is often argued that the early transcriptional quiescence in these species follows from the short length of the cell cycle, this is unlikely to be a sufficient explanation. The cell cycle length during cleavage in the more rapidly developing sea urchin species is only about a factor of two greater than in preblastoderm *Drosophila* embryos, where it is about 9 min (see below). Similarly, the period of the cell cycle in cleavage stage *Xenopus* embryos, about 35 min (Newport and Kirschner, 1982a), is the same as that of many cleavage stage sea urchin embryo cells that are active in RNA synthesis. In sea urchin embryos the rate of nRNA synthesis per nucleus increases during the first few cleavages (Wilt, 1970), and as reviewed below, by 4th-5th cleavage this rate has already attained the highest value observed at any time in embryogenesis.

The properties of the embryonic nRNA, which accounts for the large majority of the transcription products, are similar to those of nRNA in somatic mammalian cells. Thus it is DNA-like in base composition, and the newly synthesized transcripts are predominantly confined to the nucleus (see, e.g., Comb *et al.*, 1965; Gross *et al.*, 1965; Aronson and Wilt, 1969; Wilt *et al.*, 1969; Emerson and Humphreys, 1970; Aronson *et al.*, 1972; Hogan and Gross, 1972). The nuclear location probably depends on binding to intranuclear constituents, since even in demembranated nuclei greater than 90% of this class of RNA remains associated with nuclear ribonucleoprotein structures (Aronson and Wilt, 1969; Aronson *et al.*, 1972; Wilt *et al.*, 1973). Under denaturing conditions the mean size of late cleavage and blastula stage embryo nRNA lies in the range 5 to 9 kb (Kung, 1974; Dubroff and Nemer, 1975), compared to a mean of 2-3 kb for newly synthesized (nonhistone) polysomal mRNA (Kung, 1974; Nemer *et al.*, 1975; Nemer, 1975). The newly synthesized RNA appears to be smaller early in cleavage, but this could be merely a reflection of the proportionately larger contribution of mitochondrial transcripts at this stage (Dubroff, 1980). About 30% of nRNA molecules greater than about 1400 nt in length are capped (Nemer *et al.*, 1979). The cap on these molecules is exclusively of the m^7GpppX^mpYp variety. A greater fraction of the smaller nRNA molecules is also capped, and there is no significant developmental change in the size of these respective fractions. The embryo nRNA can be classified in other ways as well. Only a minor portion of the nRNA is reported to be polyadenylated (Dubroff and Nemer, 1975, 1976). Some of the nonpolyadenylated nRNA contains internal oligo(A) tracts, and oligo(U) tracts have been identified in the 3' polyadenylated nRNA class (Dubroff, 1977), as has also been reported for mammalian somatic cell nRNA (see, e.g., Molloy *et al.*, 1972). The embryo nRNA includes in addition internally complementary regions that can be isolated as double-stranded structures. In sea urchin embryos these constitute about 0.2% of the total nRNA mass (Kronenberg and Humphreys, 1972), and most probably reside within repetitive regions of the nRNA transcripts. Sea urchin embryo nRNA displays a general, interspersed sequence organization similar to that of the genomic DNA (Smith *et al.*, 1974). The repeated sequences of embryonic nRNA were studied with cloned probes, each representing a single repetitive sequence family, by Scheller *et al.* (1978). Both complements of each repetitive sequence investigated are present in the nRNA (i.e., usually in different molecules, since the nRNA is transcribed asymmetrically) just as in the interspersed egg poly(A) RNA discussed in Chapter II, and also as in HeLa cell nRNA (Fedoroff *et al.*, 1977). An additional finding of Scheller *et al.* (1978) was that a specific set of repetitive sequence families is prominently represented in the embryo nRNA, while others are predominant in adult sea urchin intestine nRNA. This observation, as well as developmental change in the fractions of nRNA containing various oligonucleotide tracts (Dubroff and Nemer, 1976), imply significant differences in the sets of primary transcripts synthesized in different cell types, a matter that we consider in more detail below.

Newly synthesized messenger RNA enters the polysomes continuously, from the earliest stages of development onward. However, as reviewed in Chapter II, during early cleavage zygotic mRNAs account for only a small fraction of the total polysomal message. After the 16-cell stage, and in the early blastula, zygotic histone message becomes the most prevalent mRNA species, as considered separately in a later section of this Chapter. The newly synthesized *non*histone messages of the early embryo include both polyadenylated and poly(A)-deficient species (Slater and Slater, 1974; Wu and Wilt, 1974; Nemer *et al.*, 1974, 1975; Nemer, 1975; Fromson and Du- chastel, 1975; Fromson and Verma, 1976). The poly(A)-deficient message fraction, i.e., generally mRNA that cannot bind to oligo(dT) cellulose, has the same molecular size and the same synthesis and turnover kinetics as does the poly(A) mRNA. Some very large newly synthesized cytoplasmic poly(A) mRNAs have also been reported in sea urchin gastrula cytoplasm, and observed under denaturing conditions they approach the mean size of the heterogeneous nRNA (Giudice *et al.*, 1974; Sconzo *et al.*, 1974). These are apparently *bona fide* messages, since they are released from polysomes by treatment with EDTA or puromycin, but their identity remains unknown. Poly(A) mRNA enters the embryo cytoplasm after a lag of 15-30 min (Wu and Wilt, 1974; Nemer *et al.*, 1975), and this lag is the same as that measured for poly(A)-deficient mRNA, and for total messenger RNA (Aronson, 1972; Brandhorst and Humphreys, 1972). Mobilization of newly synthesized mRNA into polysomes is less efficient during cleavage, and at this stage a larger fraction of newly emergent message can be recovered from nonpoly- somal mRNP particles than in blastula and later stage embryos (Dworkin and Infante, 1976). By the mesenchyme blastula stage cytoplasmic polyadenyla- tion such as occurs on preformed maternal mRNAs in early cleavage is no longer observed (see Chapter II), and instead, as in other cell types, the newly synthesized zygotic mRNAs are polyadenylated just before exit into the cytoplasm (Brandhorst and Bannet, 1978). Those newly synthesized mRNAs that are polyadenylated code for the same set of proteins as do poly(A)-deficient messages, according to two-dimensional gel electrophoretic analyses of the detectable proteins synthesized in cell-free systems loaded with these fractions (Brandhorst *et al.*, 1979). Furthermore, protein synthesis is supported equally *in vitro* by both classes of mRNA (Fromson and Verma, 1976), and both classes contain m^{7}GpppX^{m}pYp caps (Faust *et al.*, 1976; Surrey and Nemer, 1976). Experiments carried out in a cell-free system derived from sea urchin eggs indicate that the 5' cap is an absolute requirement for translation (Winkler *et al.*, 1983). Taken together, the data so far reveal no convincing distinctions of importance between newly synthesized populations of nonhistone embryo poly(A) mRNA and poly(A)-deficient mRNA, though adequate direct comparisons of the sequence contents of these classes of sea urchin embryo message have been carried out only on maternal mRNAs (Chapter II). However, the state of polyadenylation may be correlated with different individual levels of mRNA utilization, as we have seen in considering mobilization of specific maternal messages (Chapter II), and many such changes could be hidden within the overall population distribution. The fraction of the newly emergent nonhistone message that is polyadenylated increases continuously during cleavage and early blastula stages, to a level that exceeds 50-60% of the total mRNA after the midblastula stage (Nemer, 1975).

Many small RNA species are also synthesized in the developing sea urchin embryo. In addition to tRNAs (O'Melia and Villee, 1972), a number of discrete species of newly synthesized low molecular weight RNAs have been noted (Hogan and Gross, 1972; Frederiksen *et al.*, 1973; Nijhawan and Marzluff, 1979). Two of the more prominent species resemble mammalian U1 and U2 snRNAs in sequence and other characteristics (Card *et al.*, 1982; Brown *et al.*, 1985). U1 snRNA synthesis can be observed in isolated blastula stage nuclei, where it has been shown that the synthesis is carried out by polymerase II, as in other animal systems (Morris and Marzluff, 1985). These RNAs are not synthesized prior to the 64-cell stage in either *S. purpuratus* or *Lytechinus variegatus*. Earlier than this the embryo apparently relies on maternal snRNAs, which are stored in sufficient quantity to supply about 100 cells (i.e., about 10^5 molecules of each species per nucleus; Nijhawan and Marzluff, 1979). There are about 50 copies of the gene for U1 snRNA per haploid genome in each species of sea urchin, apparently arranged in a single tandem cluster with a repeat length of about 1400 nt. The genes for U2 snRNA are similarly replicate and are present in a separate cluster (Card *et al.*, 1982; Brown *et al.*, 1985). During the blastula stage the rate of snRNA synthesis is such as to require on the average synthesis of 1 to 2 molecules per gene · min, and in later development the rate of snRNA synthesis decreases about 10-fold (Nijhawan and Marzluff, 1979). This course of events implies a specific set of transcriptional regulatory controls.

A summary of the descriptive data so far reviewed is as follows. The major transcriptional product of the zygote genomes is heterogeneous nuclear RNA. Synthesis of this class of transcripts and export of new mRNAs to the cytoplasm are already active in early cleavage. There is no period of transcriptional quiescence in the sea urchin embryo, unlike some other embryos, despite the rapid, logarithmic cell division that takes place prior to the midblastula stage. In properties such as the structure of the 5' nRNA and mRNA caps, the interspersed sequence organization of the nuclear transcripts, and the synthesis of both poly(A) mRNA and poly(A)-deficient mRNA, the sea urchin embryo resembles mammalian somatic cells, except that mammalian nRNAs are reported to be somewhat longer (reviewed in Davidson and Britten, 1973). Thus from the onset of embryogenesis the zygotic genomes appear to be carrying on a "normal" process of transcription, similar to that observed in other animal cells. As shown below, this impression is substantiated on examination of the synthesis rates and the sequence content of the primary nRNA transcript population.

(ii) Transcriptional Activation in *Xenopus* Embryos

The pattern of zygotic gene activity in amphibian embryos differs sharply from that just discussed. The pronuclear genomes may be active prior to fusion, since in the axolotl a gene coding for a high molecular weight form of DNA ligase has been shown to be active in the maternal pronucleus (Signoret *et al.*, 1981; Thiebaud *et al.*, 1983). In *Xenopus* there is no further detectable nuclear RNA synthesis until 12 cleavages have occurred (about 4000 cells), when the embryo is in the midblastula stage (stage 8-8½ of Nieuwkoop and Faber, 1956), though mitochondrial RNA synthesis occurs continuously at a low rate (<0.5 pg min^{-1} per embryo; Chase and Dawid, 1972). Nuclear synthesis of many classes of RNA is then activated in concert. The sudden onset of transcription at this developmental stage, which at 23°C occurs about 7 hr postfertilization, was first noticed by Bachvarova and Davidson (1966) in an autoradiographic study, an experiment from which is reproduced in Fig. 3.2(a-d). The major class of transcripts the synthesis of which is activated in midblastula stage embryos is heterogeneous nRNA (Bachvarova *et al.*, 1966). As in sea urchin embryos this RNA is of large molecular size; it has a more or less DNA-like base composition; it hybridizes with repetitive sequences; and it turns over rapidly (see, e.g., Brown and Gurdon, 1966; Mariano and Schram-Doumont, 1965; Denis, 1966; Shiokawa and Yamana, 1968; reviewed in Davidson, 1976; Shiokawa *et al.*, 1979). Newly synthesized mRNA begins to enter the cytoplasm at this stage as well, as observed at a population level in some of the cited studies, and in specific cases reviewed below. On a per cell basis poly(A) mRNA is synthesized and transported to the cytoplasm at a severalfold higher rate at the late

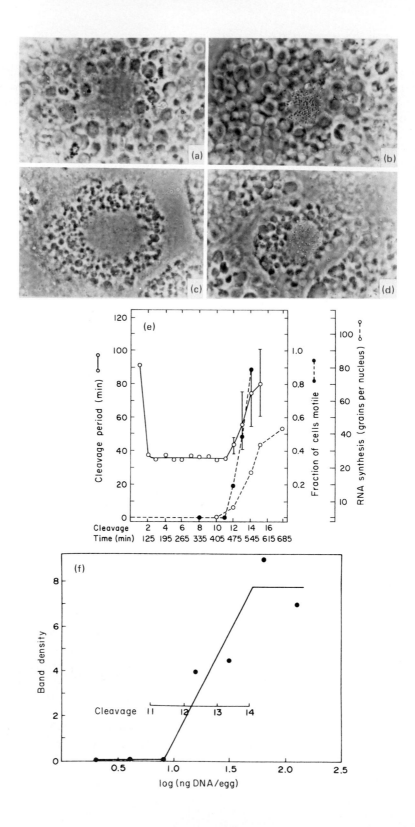

blastula stage than in more advanced embryos (Shiokawa *et al.*, 1979). The net content of poly(A) RNA does not begin to increase until the late gastrula stage, as prior to this the flow of newly synthesized poly(A) RNA into the cytoplasm is evidently balanced by degradation of surviving maternal poly(A) RNA (Sagata *et al.*, 1980). In *Xenopus* embryos only a minor fraction of the polysomal message is polyadenylated, however. Miller (1978) showed that in homozygous anucleolate mutant embryos unable to synthesize any rRNA, only about 20% of newly synthesized polysomal RNA (i.e., mRNA) in late tailbud stage embryos includes 3' poly(A) tracts. Little is known of the synthesis and accumulation of poly(A)-deficient mRNA during development.

Low molecular weight RNA synthesis is also activated at the midblastula stage (Bachvarova *et al.*, 1966; Brown and Littna, 1966a; Gurdon, 1967; Woodland and Gurdon, 1968; Shiokawa *et al.*, 1979; Newport and Kirschner, 1982a). This category includes tRNA, 5S rRNA, and various snRNAs. Among the latter are several species of U1 RNA, and U2, U4, U5, and U6 snRNAs (Newport and Kirschner, 1982a; Forbes *et al.*, 1983a, 1984; Fritz *et al.*, 1984). Two major U1 snRNA subtypes are transcribed after reinitiation

---

Fig. 3.2. Transcriptional activation at the midblastula stage in *Xenopus* embryos. (a-d) Autoradiographs of RNA synthesis in cells of *Xenopus* blastulae. Labeling is confined to the nuclei. Dorsal halves of embryos were immersed in a solution containing ^{3}H-uridine for 1 hour. The bisected embryo halves are highly permeable to this precursor at all stages. (a) Presumptive endodermal cell, stage 7. (b) endodermal cell, stage $8\frac{1}{2}$, (c) Equatorial cell, stage 8. (d) Equatorial cells, stage $8\frac{1}{2}$. Between the stages shown in (a) and (c) and those shown in (b) and (d) less than 1 hour elapses. Almost all grains above background can be removed from the sections by treatment with ribonuclease. [From R. Bachvarova and E. H. Davidson (1966). *J. Exp. Zool.* **163**, 285.] (e) Coincidence of transcriptional activation with lengthening of cleavage cell cycle and appearance of cell motility. Cell cycle period was measured from time-lapse videorecordings of dividing dissociated blastomeres. Until after cleavage 11 the cycle period is 35 ± 2 minutes. Thereafter different cells display different cycle periods, as indicated by the vertical brackets, which show the range of the period observed. RNA synthesis was measured autoradiographically as counts of grains per nucleus in 50–100 nuclei per sample. Motility was scored as the fraction of dissociated blastomeres that display either pseudopodia or blebs using videorecordings. [From J. Newport, and M. Kirschner (1982a). *Cell* **30**, 675. Copyright by M.I.T.] (f) Transcriptional activation by injection of exogenous DNA. A plasmid containing a yeast tRNA[leu] gene was injected into fertilized, uncleaved *Xenopus* eggs, and synthesis of the tRNA[leu] precursor RNA was monitored by densitometry of an autoradiograph of the low molecular weight synthesis products (ordinate). In otherwise untreated eggs the plasmid is transiently transcribed after injection, and then becomes completely inactive until the midblastula transition, when the precursor tRNA is again actively synthesized. In the experiment shown the eggs were also injected with cytochalasin to prevent cleavage, and after 2 hours, with various amounts of pBR322 DNA (abscissa), together with ^{32}P-rUTP. One hour later the RNA was extracted (i.e., 3 hours prior to the normal occurrence of the midblastula transition). The figure shows that the amounts of pBR322 DNA required to activate this premature transcription are approximately those present in the genome of the cleavage embryos, about 24 ng. [From J. Newport and M. Kirschner (1982b). *Cell* **30**, 687. Copyright by M.I.T.]

of transcription, which are different from those expressed during oogenesis (Forbes *et al.*, 1984). These derive from a tandomly reiterated set of about 500 U1 gene pairs, each containing a single copy of the genes for the two major embryonic U1 snRNAs (Lund *et al.*, 1984). An interesting aspect of both U1 and U2 snRNA accumulation is that the specific proteins to be associated with these RNAs are stored during oogenesis in the egg cytoplasm (Forbes *et al.*, 1983a), and they are apparently translocated into the nucleus only when complexed with the newly synthesized snRNAs in late blastula and early gastrula stage embryos (Zeller *et al.*, 1983; Fritz *et al.*, 1984). The translocation mediated by these snRNA binding proteins requires a particular snRNA sequence element (AU_nG, where n is 3-6), which is conserved in U1, U2, U4, and U5 snRNAs (Mattaj and De Robertis, 1985). As discussed below, synthesis of 18S and 28S rRNA is also activated in the blastula stage embryo, as is 5S rRNA transcription (regulation of 5S rRNA transcription is reviewed in Chapter V). It is clear from these studies that the process of transcriptional reactivation at the midblastula stage involves genes that are transcribed by polymerases I and III, as well as genes transcribed by polymerase II.

Transcriptional reactivation coincides with a series of other developmental changes. Newport and Kirschner (1982a) showed that after 11th cleavage the cell cycle period suddenly lengthens, losing the regular 35 min rhythm until then maintained, and at this point mitotic activity in different regions of the embryo becomes asynchronous (Satoh, 1977; reviewed in Gerhart, 1980). In urodele eggs a similar process occurs after 10th cleavage (Signoret and Lefresne, 1971; Hara, 1977). A simultaneous phenomenon is the development of cellular motility, marked by cytoplasmic blebbing and pseudopod formation (Johnson, 1976; Satoh *et al.*, 1976; Newport and Kirschner, 1982a). The coincidence of all these developmental events is shown in Fig. 3.2(e). The occurrence of multiple changes at this stage of embryogenesis has led to its designation as the "midblastula transition" (Gerhart, 1980). It is clear from several kinds of observation that the physical entities constituting the "clock" which marks the occurrence of these events are resident in the egg cytoplasm. Sawai (1979) and Sakai and Kubota (1981) showed that cyclic contractions corresponding to the cell cycle period occur in anucleate cytoplasmic fragments of amphibian eggs. Furthermore, injection of α-amanitin into the 1-cell zygote affects neither the rate nor pattern of cell division. Blastomeres attain motility on schedule in α-aminitin-treated eggs as well (Newport and Kirschner, 1982a). However, gastrulation, which normally occurs several hours later, is blocked in α-amanitin-treated eggs. Newport and Kirschner (1982a) showed that neither the elapsed time after fertilization nor the process of cleavage itself determines the onset of the midblastula transition. Thus in cytochalasin-treated eggs, where cytokinesis is totally suppressed though DNA synthesis continues at the normal rate, transcriptional activation occurs at the same time as in controls. In eggs that are ligated at the 1-cell stage so as to delay nucleation of one half for 1-3 division cycles, the retardation of the cell cycle associated with the midblastula transition occurs at different points in real time in the two halves of the embryo, taking place in the retarded half only when it too has completed 11 cleavages. The critical developmental parameter seems to be the quantity of DNA present per unit mass of cytoplasm. Thus transcriptional activation occurs prematurely in polyspermic embryos (Newport and Kirschner, 1982a) and it can be induced by injection into cytochalasin-treated eggs of the amount of DNA that has normally accumulated at 12th cleavage (Newport and Kirschner, 1982b). This experiment is shown in Fig. 3.2(f). The implication is that the mechanism of the general activation of transcription at the midblastula stage could be titration by the nuclear DNA of some general repressor stored in the egg cytoplasm, which in normal embryos blocks transcription until the quantity of DNA exceeds the critical mass attained normally at 10th cleavage. However, a different mechanism appears to control activation of the rRNA genes. Thus the time of onset of rRNA synthesis at the late blastula stage is not affected by direct alteration of the DNA to cytoplasm ratio, by means of removal of cytoplasm, or by inhibition of DNA synthesis with drugs (Takeichi *et al.*, 1985; Shiokawa *et al.*, 1986).

(iii) Timing of Zygotic Transcription in *Drosophila* and Other Insects

Cell Division and DNA Replication Rates during Cleavage

A remarkable aspect of embryogenesis in *Drosophila* is the extremely high rate of early development. Fifteen minutes after oviposition pronuclear fusion occurs, followed immediately by the initiation of zygotic nuclear replication. Within about 80 min (at 25°C) nine synchronous nuclear divisions have taken place (Zalokar and Erk, 1976). Up to this stage the embryo consists of a syncytium in which the nuclei are not sequestered within individual plasma membranes, but are distributed in the internal egg cytoplasm, each surrounded by a network of microtubules (Karr and Alberts, 1986). During 8th and 9th cleavage (256 cells) the majority of the nuclei migrate to the periphery of the embryo. Pole cells form after 9th cleavage, following the migration of about 18 nuclei into the posterior polar cytoplasm of the egg. Approximately 100 nuclei are left within the yolk. The blastoderm remains syncytial until the nuclei are enclosed in cell membranes, though each nucleus is positioned within an individual cytoskeletal domain probably constructed from elements previously resident in the cortex of the egg (Karr and Alberts, 1986). Cellularization occurs about 2.5 hr after oviposition, after the 13th synchronous cleavage, when there are approximately 6000 cells. During the preblastoderm stages the nuclear S-phase occupies just 3-4 min, and the average nuclear doubling time is only 9 min. At the syncytial blastoderm stages the nuclear doubling time slows to about 15 min (Zalokar and Erk, 1976; Anderson and Lengyel, 1981). Following cellularization there is a brief pause, and gastrulation begins at 3.5 hr, with the appearance of ventral and cephalic furrows.

Among the mechanisms underlying the extremely short S-phase during the preblastoderm stage is an unusually high frequency of replication initiation sites in the genomic DNA. The mean distance between these sites in the DNA of cleavage stage *Drosophila* embryo nuclei, as measured by electron microscopy, is about 8×10^3 ntp (Wolstenholme, 1973; Kriegstein and Hogness, 1974; Blumenthal *et al.*, 1974). In early embryos of the screw worm, a dipteran insect in which cleavage stage replication occurs even more rapidly than in *Drosophila*, a distance of only 6.9×10^3 ntp was measured between sites of replication initiation (Lee and Pavan, 1974). At the rate of replication fork progression measured in *Drosophila* embryos, about 45 ntp sec^{-1} (Blumenthal *et al.*, 1974), the mean distance between origins would be traversed by the converging forks in less than 2 min, accounting for the short S-phase observed. In *Drosophila* tissue culture cells, which provide a useful comparison, the rate of fork progression is about the same, but the distance between initiation sites is $3\text{-}6 \times 10^4$ ntp in the euchromatin portions of the genome, and may be much greater in heterochromatic regions (Blumenthal *et al.*, 1974). The same mechanism probably underlies rapid cell division during cleavage in amphibian embryos. Thus the distance between initiation sites is much shorter in embryonic amphibian cells than in cultured somatic amphibian cells (Callan, 1974). Other experiments have shown that in early *Xenopus* embryos there is no stringent preference for any particular replication origin sequence (Harland and Laskey, 1980; Méchali and Kearsey, 1984). Thus on injection into *Xenopus* eggs, all DNA sequences are replicated equally, whether of eukaryotic or prokaryotic origin, and irrespective of whether they include regions known to function as replication origins in other systems. The rate of replication of the injected DNA is simply proportional to the length of the template fragment (Méchali and Kearsey, 1984; Laskey and Harland, 1982). These data are consistent with the view that embryonic DNA replication rates are high because of a greatly relaxed stringency in the requirements for sites where initiation may occur, resulting in more closely spaced and hence more numerous replication complexes than in somatic cells. Excess quantities of the enzymes and other factors needed for replication are known to be present in eggs and early embryos (reviewed in Laskey *et al.*, 1979), and this no doubt also contributes to the high synchrony of initiation in rapidly dividing embryonic genomes (see, e.g., Blumenthal *et al.*, 1974).

Transcription

A general transcriptional activation occurs in *Drosophila* embryos at the cellular blastoderm stage (Zalokar, 1976; Anderson and Lengyel, 1979). In contrast to the superficially similar example of blastular activation in *Xenopus*, however, there is clear evidence for some prior transcriptional expression earlier in *Drosophila* development. Transcription complexes derived from the chromatin of preblastoderm cleavage stage nuclei are shown in Fig. 3.3(a)-(e), and from blastoderm stage nuclei in Fig. 3.3(f)-(g) (McKnight and Miller, 1976). Among the first identified genes to be activated are the histone genes, which begin to be expressed at a high rate, at 90 min postfertilization, early in the syncytial blastoderm stage (Anderson and Lengyel, 1980).

In some other insects a pattern similar to that observed in *Drosophila* obtains. For example Jäckle and Kalthoff (1979) were unable to detect incorporation of ^{3}H-uridine prior to blastoderm formation in embryos of the chironomid midge *Smittia*, i.e., except into DNA and mitochondrial RNA. At blastoderm stage a surge of transcriptional activity occurs, and newly synthesized mRNA enters the cytoplasm. Nor is nuclear transcription observed prior to blastoderm formation in eggs of the leafhopper *Eucelis plebejus*, as determined after injection of ^{3}H-uridine into the egg during cleavage (Schmidt and Jäckle, 1978). Autoradiographic observations also indicate initiation of nuclear RNA synthesis at blastoderm stage in *Musca* (Pietruschka and Bier, 1972) and in a bug, *Oncopeltus* (Lockshin, 1966). On the other hand in the beetle *Bruchidius*, where cellularization of the blastoderm occurs at about 4000 cells, RNA synthesis takes place in peripheral nuclei and even more intensely in the vitellophage nuclei as early as the 256 nucleus syncytial stage of development (Büning, 1980). Nuclear RNAs are also synthesized actively by preblastoderm stages of another coleopteran, *Leptinotarsa*, though labeled cytoplasmic RNAs are first detected only after migration of nuclei to the periphery (Schenkel and Schnetter, 1979). Thus a considerable range in the levels of preblastoderm transcriptional activity is encountered in different insect groups, though it is apparently a general observation that the rate of transcription per nucleus increases sharply after blastoderm cellularization.

(iv) Onset of Heterogeneous RNA Synthesis in Early Embryos of Other Species

With respect to the developmental stage at which transcription is activated, most organisms for which data exist fall somewhere between the extremes represented by the sea urchin and *Xenopus*. Transcription of heterogeneous nRNA occurs during the 20 hr prior to first cleavage in the mouse egg (Clegg and Pikó, 1983b), and during this time there is also a minor flow of newly synthesized mRNA into the cytoplasm. At the 2-cell stage transcriptional activity increases. As in other embryos the major class of RNA synthesized remains nRNA, displaying a typical heterodisperse, high molecular weight size distribution (see, e.g., Knowland and Graham, 1972; reviewed in Davidson, 1976). There is also a much augmented flow of new mRNAs to the cytoplasm from the 2-cell stage onward. Thus in the mouse the zygotic genome is not completely quiescent even at the 1-cell stage, and it displays productive transcriptional activity long in advance of the important morpho-

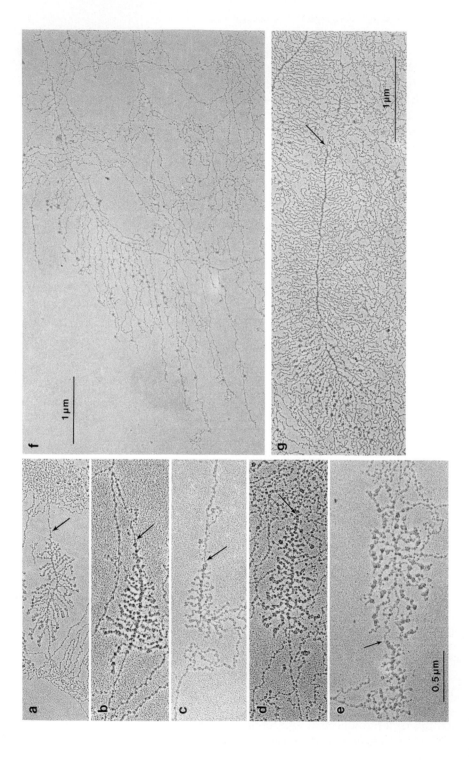

genetic events of late cleavage such as compaction and polarization at the 8-cell stage, blastocyst formation, or the initial processes of cell fate determination (see Chapter VI).

There is little convincing evidence in other organisms for an early period of transcriptional quiescence. In ascidian embryos heterogeneous RNA synthesis occurs as early as the 4-cell stage (Smith, 1967), and Meedel and Whittaker (1978) observed synthesis of high molecular weight heterodisperse transcripts 2-4 hr after fertilization, i.e., the blastula-gastrula stage in these rapidly developing organisms. Newly transcribed mRNA is present in the polysomes by 5-7 hr postfertilization. Similarly, in the gastropod *Ilyanassa obsoleta* heterogeneous nuclear RNA synthesis occurs during cleavage (Collier, 1976, 1977), and newly synthesized mRNA has been identified in polysomes of lamellibranch molluscan embryos at early cleavage stages (Firtel and Monroy, 1970; Kidder, 1972). Heterogeneous RNA is also synthesized in early cleavage stage embryos of *Urechis caupo* (Gould, 1969).

From a comparative point of view the dramatic amphibian phenomenon of transcriptional repression, followed by synchronous activation of gene expression at the midblastula stage thus appears something of a peculiarity. In the majority of species studied the zygotic genomes begin to be utilized very soon after their formation, though a significant increase in rate of transcription following the initial phases of cleavage is frequently, though certainly not invariably, observed. In some insects this occurs only after the blastoderm nuclei are separated from the mass of the interior cytoplasm, suggesting, as in *Xenopus*, the existence of a general transcriptional repressor located initially in the egg cytoplasm. Such an interpretation is unlikely to apply everywhere that transcription rate is found to increase after the initial phases of development. The mammalian embryo, for example, directs a changing program of translation during the protracted 1-cell stage (Chapter II), and the relative increase in mRNA production observed by this embryo after the 2-cell stage could be triggered by the accumulation of some of the newly synthesized protein(s). The mechanisms of the ontogenic "clocks" that pace early embryonic activities, transcription among them, are likely to

Fig. 3.3. Transcription complexes from nuclei of early *Drosophila* embryos. (a)-(e) Nonribosomal transcription complexes from *preblastoderm* embryos (90-120 minutes postovulation). The transcription complexes are tightly packed with nascent RNAs and are shorter than those observed in later embryos (average length about 2 μm, or $\geq$6000 nt). About 1% of the preblastoderm chromatin length is involved in transcription complexes. Arrows mark the initiation sites for each complex. In (e) two oppositely oriented transcription units are adjacent, separated by only about 0.25 μm of chromatin. (f), (g) Nonribosomal transcription units from embryos fixed 25-30 minutes after the beginning of cellular blastoderm stage. These transcription units are more prevalent and typically are longer, ranging from 3 to 6 μm of genomic DNA. As seen in (f) they are often less densely packed with nascent fibrils. [From S. L. McKnight and O. L. Miller (1976). *Cell* **8**, 305. Copyright by M.I.T.]

be as diverse as are other aspects of the molecular biology of early development.

2. RATES OF TRANSCRIPTION IN EMBRYONIC GENOMES AND ACCUMULATION KINETICS FOR nRNA, mRNA, and rRNA

(i) The Number of Active Transcription Units and Their Level of Expression

A quantitative measure of the expression of the zygotic genomes during embryogenesis is the rate at which they synthesize their primary gene product, heterogeneous nuclear RNA (nRNA). Combined with knowledge of the nRNA complexity, the transcription rate per nucleus also can be used to calculate the rate of initiation on the average functioning gene, and combined with the turnover rate, it provides the prevalence, or steady state level, of the nuclear transcripts. Most measurements of primary trancription rate have been based on the kinetics with which a labeled precursor is incorporated into the nRNA, and on estimates of precursor pool specific activity obtained over the same time period. The mathematical statements required to analyze kinetic incorporation data of this nature are summarized in Appendix III, which also contains a brief review of the behavior of the intracellular precursor pools from which RNA is synthesized. Two cases are there considered. In the simpler of these the pool specific activity remains essentially constant, while in the more general situation the pool specific activity changes during the experiment. As described in Appendix III [equations (1-8)], from the incorporation data are extracted the *synthesis rate constant,* in the following denoted k_s when expressed in molecular terms (here molecules min^{-1}), or k_s' when expressed in mass terms (here pg min^{-1}); and the *decay rate constant,* k_d (min^{-1}). It can be seen from the derivations given in Appendix III that the steady state content of the newly synthesized transcripts is the ratio of the synthesis to the decay rate contents, i.e., k_s/k_d [equation (6)], and that the half-life of the RNA, here denoted $t_{1/2}$, is given by $ln2/k_d$ [equation (4)].

Before embarking on the results of transcription rate estimates it is necessary to consider an important experimental issue that arises for any measurement based on the observed precursor pool specific activity. For such measurements to be valid it is necessary that the precursor pools not be compartmentalized. That is, the observed specific activity of the chemically defined precursor extracted from the embryo cells must be that of the precursor pool actually utilized within the nucleus for RNA synthesis. The greatest amount of evidence exists for sea urchin embryos, for which there are three independent observations that together preclude significant compartmentalization of RNA precursor pools. First, absolute synthesis rates

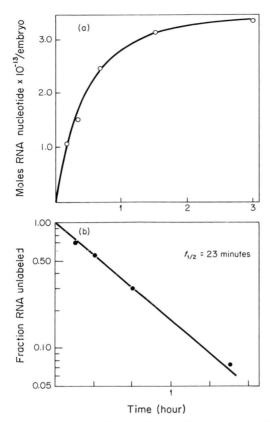

Fig. 3.4. Incorporation and decay kinetics for total heterogeneous RNA in sea urchin mesenchyme blastulae. Most of the label is present in nuclear RNA. The newly synthesized RNA was labeled with [15]N- and ^{13}C-containing nucleosides. The medium also contained all four ^{3}H-nucleosides as tracer for the exogenous precursor. Newly synthesized RNA was separated from other cellular RNAs by isopycnic centrifugation in cesium formate gradients. (a) Molar accumulation of newly synthesized RNA in *S. purpuratus* embryos over the first 3 hours of labeling. The quantities of nucleotides incorporated were calculated from the density shift observed, compared to the maximum possible density shift had the newly synthesized RNA derived wholly from the [13]C- and ^{15}N-labeled exogenous precursor. This ratio provides the dilution factor for the precursor in the newly synthesized RNA, and from the amount of radioactivity incorporated, and the specific activity of the tracer initially added to the medium, the absolute quantity of newly synthesized RNA at each time point was thus calculated. The curve is of the form given by equation (5) of Appendix III. At the plateau or steady state level approached by the accumulation curve, the embryo contains about 0.3 pmole of heterogeneous RNA nt, or approximately 100 pg. (b) Decay kinetics for total heterogeneous RNA, based on the data shown in (a), according to equation (3) of Appendix III. The slope of this line gives the first-order decay rate constant k_d which is here evaluated at 0.03 min^{-1}. The half-life ($t_{1/2}$) is 23 minutes [Equation (4) of Appendix III]. The rate of synthesis (k_s') per embryo of the heavy labeled RNA in this experiment is 3.2 pg min^{-1}. As indicated in the Notes to Table 3.1 independent observations by different methods yield very similar estimates of the absolute synthesis rate for these embryos. [From R. M. Grainger and F. H. Wilt (1976). *J. Mol. Biol.* **104**, 589.]

that agree closely have been obtained using measurements of the total chemical UTP, ATP, or GTP pool specific activities (see below, and the more extensive review in Davidson, 1976). These pools are of different sizes and during development they display different turnover rates, and with reference to the external medium, different specific activities (see Appendix III, Fig. A3.1). The close similarity in synthesis rates measured from incorporation of these different precursors requires either that none of the triphosphate pools are compartmentalized, or improbably, that despite their diverse labeling kinetics, all are compartmentalized to the same extent. Secondly, Grainger and Wilt (1976) measured the synthesis rate by a completely different procedure and showed that the results are nonetheless the same as those of measurements based on precursor pool specific activity. The embryos were permitted to incorporate ^{13}C and ^{15}N nucleosides, and the newly synthesized RNA was separated by isopycnic methods. The fraction of nucleotides incorporated from the exogenous precursor could thus be calculated independently of any pool specific activity measurements, and thereby the amount of newly synthesized RNA at each time point. Results from this study are summarized in Fig. 3.4. Third, it was shown by Galau *et al.* (1977a) that the absolute specific activity of embryo mRNA equals that of the total embryo GTP pool when both molecular populations have achieved kinetic steady state. Measurements of Clegg and Pikó (1983b) also lead to the conclusion that in early mouse embryos the ATP pool is not compartmentalized.

nRNA Transcription Rates

The rate of heterogeneous nRNA transcription has been measured directly in *Drosophila*, mouse, and sea urchin embryos, and can be inferred from available data for *Xenopus* embryos as well. These measurements have been collected in Table 3.1, which also includes the derived or assumed nRNA half-lives, and where available, the nRNA complexities, in most cases established by RNA excess hybridization with single copy DNA. We now consider some of the many interesting molecular and biological implications of the results summarized in this Table.

As might be expected from the very different numbers of cells in the embryos compared in Table 3.1, the rates of synthesis and steady state contents of heterogeneous nRNA vary enormously when expressed on a per embryo basis (columns 3 and 7, respectively). The largest part of this variation is abolished on calculation of synthesis rates and steady state content per embryonic cell (columns 4 or 5, and 8, respectively). Either as measured directly, e.g., in the experiment shown in Fig. 3.4, or as inferred from indirect observations or measurements on other cells of the same species, the turnover rates for heterogeneous nRNA are similar throughout. Thus the significant variable in determining steady state content of nRNA per cell is the transcription rate. Table 3.1 indicates that when considered in reference to *genome size* (column 12), the rates at which nRNA is produced per cell in these four species of embryo are amazingly similar (column 13). Here it can be seen that 1-3% of the genome is transcribed per minute in each embryonic nucleus, on the average. The steady state contents of nRNA, normalized to genome size, are also similar. Thus for example the value of the ratio [mass of nRNA (column 8)]/[haploid DNA content (column 12)] lies between 25 and 35% for sea urchin, *Drosophila*, and *Xenopus*, though it is severalfold higher in one and two cell mouse embryos. Despite the great differences in mode of development represented by these four creatures, the genomes of their embryonic cells are evidently transcribed at very similar levels of intensity.

Total Complexity and Average Prevalence of Embryo nRNA Species

The sequence complexity measurements for nRNA shown in Table 3.1 (column 9) provide a direct estimate of the length of genomic single copy sequence transcribed. The data are most secure for *Drosophila* and sea urchin embryos, and in both examples 30 ± 3% of the genomic single copy sequence is represented in the embryo nRNA [about 75% of sea urchin genomic DNA (Graham *et al.*, 1974) and about 80% of *Drosophila* genomic DNA (Manning *et al.*, 1975; Crain *et al.*, 1976) consist of single copy sequence]. Comparison of the complexities with the kinetic estimates of steady state nRNA contents shows that the latter are only slightly larger (i.e., column 8 versus column 9 of Table 3.1). This difference is due to that portion of the nRNA mass attributable to repetitive sequence transcripts, and to some nRNA species that are relatively prevalent. About 10% of sea urchin embryo nRNA (Smith *et al.*, 1974) and 20% of *Xenopus* embryo nRNA (Knöchel and Bladauski, 1980) consists of repetitive sequence transcript. That the complexity almost equals the steady state mass for those nRNAs considered in Table 3.1 means that most nRNA species are present in little more than one copy per nucleus, as indicated in column 11. In light of this result it is interesting to consider the length of time required for synthesis of a typical nRNA molecule, with respect to the nRNA half-lives (column 6). For the sea urchin Aronson and Chen (1977) measured a polymerase II translocation rate of 6-9 nt sec^{-1}. Therefore to synthesize an average nRNA 5-9 kb in length would require 10-25 min, which is essentially the same as the half-life. For *Xenopus* the translocation rate is 10-15 nt sec^{-1} (Anderson and Smith, 1978) and the nRNA is about the same size (Shiokawa *et al.*, 1979), so that the same conclusion holds, and this is also true of *Drosophila*, where the translocation rate (for polymerase I) is estimated at about 9 nt sec^{-1} (Anderson and Lengyel, 1979). Thus there is about one initiation event on each transcription unit per average nRNA half-life. In the sea urchin embryo the rate of nRNA synthesis, taken together with the translocation rate, suggests that there are about 1.9×10^4 polymerase molecules active per nucleus (8.6×10^6 nt min^{-1}, or 1.4×10^5 nt sec^{-1}/7.5 nt sec^{-1} per polymerase). These

TABLE 3.1 Average Nuclear RNA Transcription Rates per Cell and per Gene for Various Embryos

1 Species	2 Stage (cell no.)	3 k_s' per embryo (pg min⁻¹)	4 Synthesis rates[a] k_s' per cell (pg min⁻¹)	5 k_s' per cell (nt min⁻¹)	6 Half-life[b] $(t_{1/2})$ (min)	7 Steady state content[c] per embryo (pg)	8 per cell (nt)	9 Complexity[d] (nt)	10 Av. initiation period[e] (min)	11 Av. prevalence per transcript species per cell[f] (molecules)	12 Haploid genome size (ntp)	13 Per cell synthesis rate / Haploid genome size (min⁻¹)
S. purpuratus[g]	Cleavage (16–64 cells)	0.52[h]	0.017	2.9×10^7								
	Blastula-gastrula (av. 500 cells)	2.6[i]	0.0052	8.6×10^6	23[j]	86	2.9×10^8	1.7×10^{8}[k]	20	1.7	8.1×10^{8}[l]	0.011
D. melanogaster[m]	Blastoderm (2.5 hr, 6000 cells)[n]	47	0.0078[o]	1.3×10^7								
	Gastrula (7 hr, ~60,000 cells)[n]	90	0.0015[o]	2.6×10^6	12[p]	1.6×10^3	4.3×10^7	4.1×10^{7}[q]	16	1.1	1.6×10^{8}[r]	0.016
X. laevis	Late gastrula-neurula (stage 14, ~65,000 cells)[s]	980	0.015[t]	2.6×10^7	~20[t]	2.8×10^4	7.4×10^8	($\geq 3.5 \times 10^{8}$[u])	(≥ 11)	(≤ 2)	2.7×10^{9}[v]	0.010
Mouse	1-cell embryo	0.05[w]	0.05	8.6×10^7	~20[w]	1.5	2.6×10^9	—			3.2×10^{9}[x]	0.026
	2-cell embryo	0.1[w]	0.05	8.6×10^7	~20[w]	3.0	2.6×10^9	—				

[a] Synthesis rate is defined in Appendix III in terms of the kinetic expressions needed to resolve precursor incorporation data. The values given in this table are for k_s', i.e., synthesis rate in units of mass per unit time, pg min⁻¹ or nt min⁻¹, and for k_s, in molecular units, molecules min⁻¹. Notes indicating particulars of the following measurements are keyed to the primary sources, from which the other synthesis rate estimates listed have been calculated, i.e., either *per embryo*, or *per cell*.

[b] $t_{1/2} = ln2/k_d$, where k_d is the first order decay rate constant, in units of min⁻¹ (see Appendix III). In several of the cited cases the decay kinetics have been explicitly demonstrated to follow first order form, as illustrated, e.g., in Fig. 3.4(b).

[c] The steady state content is the ratio of the synthesis to the decay rate constants (i.e., k_s'/k_d). See Appendix III, equations (1)–(6) for derivation.

[d] Complexity is the length of single copy genomic DNA sequence represented in the nRNA. See Appendix I, section 1 for definitions, and section 5 for method of measurement by RNA excess hybridization with single copy tracers.

[e] Initiation period is the interval (in minutes) between successive transcriptional initiations on the average genomic transcription unit. This parameter is the quotient of the *complexity* (in nt) divided by the *synthesis rate per cell* (i.e., in nt min⁻¹).

[f] Prevalence is the number of molecules of the average transcript species per nucleus. This parameter is the quotient of the *steady state content* per cell (in nt) divided by the *complexity* (in nt).

[g] Though the data listed are for S. purpuratus similar measurements exist for *L. pictus* (reviewed in Davidson, 1976).

[h] Data averaged from measurement of Roeder and Rutter (1970) based on UTP pool specific activity, and carried out at the 32- to 64-cell stage, which yielded 1×10^{-2} pg min^{-1} per cell; and a measurement of Wilt (1970) based on GTP pool specific activity, and carried out on 16- to 32-cell stage embryos, which yielded 2.4×10^{-2} pg min^{-1} per cell. Satisfactory measurements of nRNA complexity and turnover rates are not available for cleavage stages, though both parameters are probably similar to those listed for the blastula–gastrula stage (Ernst et al., 1979).

[i] Many measurements have been made at blastula–gastrula stages (reviewed in Davidson, 1976). Data shown are the average of four measurements of RNA synthesis rate per cell (pg min^{-1}) as follows: 6.7×10^{-3}, hatched blastula, based on UTP pool (Roeder and Rutter, 1970); 4.5×10^{-3}, mesenchyme blastula, UTP pool, ibid.; 5.1×10^{-3}, mesenchyme blastula, based on the measurements with ^{14}C- and ^{15}N-labeled precursors also shown in Fig. 3.4 (Grainger and Wilt, 1976); and 3.5×10^{-3}, gastrula, UTP pool (Roeder and Rutter, 1970). Measurements carried out with L. pictus embryos also show that in late embryogenesis the synthesis rate per cell is only about 1/3rd that observed earlier, though the decline is not seen until after blastula stage (Wu and Wilt, 1974; Brandhorst and Humphreys, 1971, 1972). In S. purpuratus the rate of synthesis per cell observed in postgastrula stage embryos, 3.9×10^{-3} pg min^{-1}, is about the same as in gastrula stage (Roeder and Rutter, 1970). Furthermore, an almost identical rate, 2.6×10^{-3} pg min^{-1}, was measured by Nemer et al. (1979) for capped nRNA in S. purpuratus gastrulae, i.e., if it is assumed that the nRNA averages about 5 kb in length (see text).

[j] This turnover rate was measured directly by Grainger and Wilt (1976), by application of equation (3) of Appendix III.

[k] Hough et al. (1975). An identical value was obtained for embryos of T. gratilla by Kleene and Humphreys (1977).

[l] Hinegardner (1974).

[m] Data also exist for the chironomid midge Smittia (Jäckle and Kalthoff, 1979). Assuming that the internal pool in permeabilized eggs has the same specific activity as the external medium, RNA is synthesized at a rate about 1/3rd that shown for Drosophila at the blastoderm stage (7500 cells). This represents at least a 50-fold increase in synthesis rate compared to preblastoderm stages.

[n] Cell numbers were calculated from kinetic measurements of DNA synthesis and accumulation by Anderson and Lengyel (1981).

[o] Anderson and Lengyel (1981). Values based on measurements of UTP pool specific activities in embryos permeabilized by treatment with octane (Limbourg and Zalokar, 1973; Anderson and Lengyel, 1979). The value measured for blastoderm stage is at least 175 times higher than the rate of synthesis per nucleus before blastoderm formation (Anderson and Lengyel, 1979). It is not known whether the nRNA initially synthesized at blastoderm stage has the same turnover and complexity characteristics as later in development.

[p] Assumed by Anderson and Lengyel (1979) from prior measurements on tissue culture cells (Levis and Penman, 1977).

[q] Zimmerman et al. (1982). This measurement was made on nuclear RNA of third instar larvae. Earlier estimates of Turner and Laird (1973) provided a slightly lower value for larval RNA, but also showed that within their system of measurement the complexity of late embryo total RNA and of larval total RNA is the same. Total RNA complexity can be taken as nRNA complexity, since nRNA is by far the most complex class of transcript in the cell.

[r] Rasch et al. (1971).

[s] DNA content at stage 14 is about 400 ng per embryo, and the haploid genome size is 3 pg (Dawid, 1965).

[t] Estimated by Davidson (1976) from steady state measurements of Brown and Littna (1966b), assuming $t_{1/2} \cong 20$ min, as for other organisms. There are no direct measurements on whole embryos. However, a measurement of total nuclear poly(A) RNA accumulation kinetics carried out on isolated disaggregated cells of stage 23 embryos (tailbud) suggests a $t_{1/2}$ of about 45 min for the major unstable component (plus a more slowly decaying component; Shiokawa et al., 1979).

[u] Calculated from cDNA hybridization data of Knöchel and Bladauski (1981). In one of the reactions reported in this study cDNA prepared from nuclear poly(A) RNA of gastrulae was reacted with the parental RNA, and the kinetics of the reaction measured. The authors did not present a kinetic analysis of this reaction, and it is not clear whether kinetic termination had been achieved. A minimum estimate of complexity from the data is shown assuming a $R_0 t_{1/2}$ pure for the final component of the reaction of about 500 M sec. This is an approximate calculation only and could be an underestimate by a factor ≥ 2. This value and those dependent on it are therefore listed in parentheses.

[v] Dawid (1965).

[w] Estimated by Clegg and Pikó (1983b) from incorporation kinetics and ATP pool specific activity. Behavior of both UTP and ATP pool specific activities in mouse embryos were described by Clegg and Pikó (1977). The unstable nRNA synthesis rates cited were calculated from absolute steady state nRNA content per nucleus on the assumption of a 20 min half-life, based on extensive measurements of Brandhorst and McConkey (1974) on mouse L cells. In an early study of Pikó (1970) a half-life of about 10 min was estimated for mouse blastocyst cells from the rate of disappearance of labeled RNA on addition of actinomycin. No complexity data are available for mouse embryos. Clegg and Pikó (1977) concluded that the total rate of RNA synthesis per cell increases about 4-fold to the 16-32-cell morula stage, compared to the 2-cell stage, but that most of the additional transcription is accounted for by rRNA synthesis. The rate of heterogeneous nRNA synthesis per nucleus may remain fairly constant throughout early development.

[x] Based on a haploid genome size of 3.5 pg (mammalian genome size data are reviewed in Britten and Davidson, 1971).

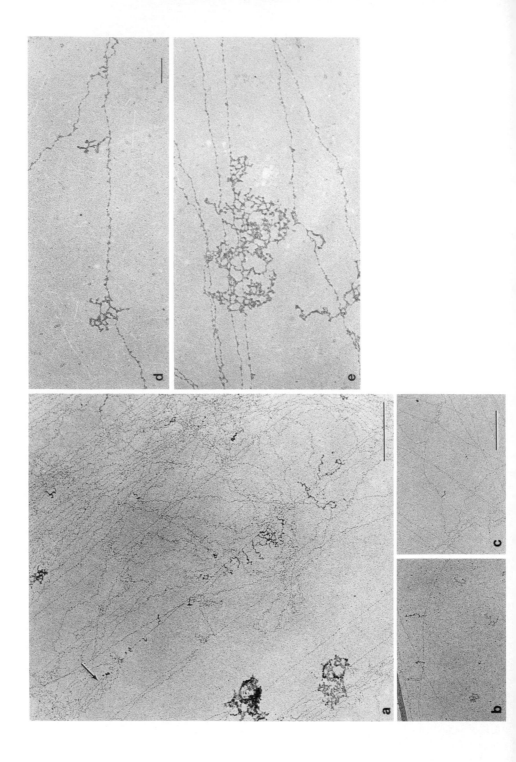

polymerase molecules must be largely distributed in separate transcription units, to account for the complexity of the nRNA (i.e., 1.7×10^8 nt complexity/1.9×10^4 polymerases = 9.3×10^3 nt per polymerase, which is only 1-2 times the length of the average nRNA molecule). Similar calculations obtain for the other organisms. In summary, the picture that emerges is that *most active transcription units are represented at any one time by only a single nascent transcript*, though as discussed below there are a minority of more heavily loaded transcription units as well.

This image of transcriptional dynamics within the nuclei of embryo cells is confirmed in an independent manner by visualization of transcription complexes in sea urchin and mammalian embryo chromatin. The distribution of transcription complexes of various degrees of nascent transcript packing in gastrula stage *S. purpuratus* embryos (i.e., the same stage and species as considered in Table 3.1) was reported by Busby and Bakken (1979). As illustrated in Fig. 3.5(a)-(c) both single transcripts and multiply loaded transcription units were observed. However, single transcripts were found on 82% of all the transcribed regions detected, and 89% of these regions contained no more than one or two RNP fibrils. Similarly, Cotton *et al.* (1980) showed that in early rabbit embryos singly occurring fibrils and pairs of fibrils account for 62-69% of all transcribed regions observed at 2-4-cell, 16-cell, and late cleavage stages [Fig. 3.5(c)-(d)]. Possibly the greater frequency of more densely packed transcripts reflects the somewhat higher steady state content of nRNA seen in early mammalian embryos, relative to genome size (Table 3.1, column 8).

Analyses of the 11% more densely packed transcription units observed in the sea urchin embryo nuclei indicate that as a class these would maintain an average steady state concentration of about 3-12 nascent nRNA molecules per nucleus, and that their aggregate complexity is about 3.5×10^6 nt (Busby and Bakken, 1979). This is only 2% of the total sea urchin embryo nRNA complexity (Table 3.1, column 9). However, only about 1/5th the expected number per nucleus of active polymerase molecules is observed altogether in this study, a calculation based on the overall frequency of nascent transcripts per unit length of chromatin. Transcripts fixed too soon after initiation would have been invisible because of their short length, and

---

Fig. 3.5. Transcription units of sea urchin and rabbit embryos. (a)-(c) Visualization of transcription units in early gastrula stage *Strongylocentrotus purpuratus* embryos. (a) Multiple fibril array on a transcription unit of the more intensely transcribed class that includes 11% of all transcribed regions observed. (b), (c) Double and single RNP fibrils illustrative of 89% of all observed transcribed regions (see text). Note the nucleosomal configuration of all the chromatin strands, both in active and wholly inactive regions. Bars = 1 μm. [(a)-(c) From S. Busby and A. Bakken (1979). *Chromosoma* **71**, 249.] (d) Two isolated transcription units from same stage rabbit embryo nucleus. (e) Active transcription unit displaying a densely packed matrix of nascent RNP fibrils from a nucleus of a 16-cell rabbit embryo. Bar = 0.5 μm. [(d) and (e) From R. W. Cotton, C. Manes, and B. A. Hamkalo (1980). *Chromosoma* **79**, 169.]

some may have been destroyed in preparation. However, the most direct interpretation of this result is that for many individual transcription units the initiation period (i.e., as in column 10, Table 3.1) is actually longer than the transcription time, and hence that in any given nucleus fixed at any given time no nascent transcript will be observed in many regions that earlier or later would have been scored as active. In any case it is clear both from the kinetic analyses summarized in Table 3.1 and from the electron microscope measurements that most embryonic transcription units are initiated at a low absolute frequency, at intervals that in many transcription units must exceed 20 min. The definition of "active" and "inactive" for this majority class of nRNA transcription units in any terms other than sequence representation in the nRNA is thus not a trivial problem. In a minority of transcription units initiation occurs much more frequently. The initiation periods shown in column 10 of Table 3.1 are averages over these two classes. The nuclear initiation rates of course set the upper limit at which the mRNA products may enter the embryo cytoplasm. Furthermore, frequency of initiation is clearly a major control point, in that this parameter varies within the same nuclei by large ratios, as indicated visually in the comparisons illustrated in Fig. 3.5.

There are two assumptions implicit in the analysis presented. First, it has been assumed in comparing nRNA complexities with per cell nRNA synthesis rates and steady state contents that the nRNA populations of the diverse cell types within a given organism are largely similar. This assumption is supported by direct measurements carried out on various sea urchin nRNAs. The complexities and sequence contents of whole embryo nRNAs at blastula, gastrula and pluteus stages are within error identical, despite the great differences in cellular function that appear as development progresses (Hough *et al.*, 1975; Kleene and Humphreys, 1977). A very similar complexity has been measured for the nRNA of adult sea urchin intestine (Wold *et al.*, 1978), which relative to a pluteus stage embryo contains few cell types. However, the nRNA populations of intestine and those of the embryo nuclei are not completely synonymous. These nRNAs include distinct distributions of interspersed repeat sequence elements (Scheller *et al.*, 1978), and Ernst *et al.* (1979) demonstrated a small component of the total nRNA complexity that is represented in adult intestine nRNA, but not in gastrula nRNA. On the assumption of typical repetitive sequence interspersion distances in the nuclear RNA, this component could account for the differences observed in the sets of repeat sequences represented in the two nRNAs, implying a specific association of certain repeat sequences with nRNA transcription units that are tissue-specific. With reference to gastrula nRNA, only about 18% of the total intestine nRNA sequence complexity is specific to that nRNA (i.e., 3.5×10^7 nt, out of 1.9×10^8 nt). In a similar vein, Knöchel and Bladauski (1980) showed that cDNA synthesized from either gastrula or tadpole stage *Xenopus* embryo nuclear poly(A) RNA largely cross reacts with the opposing nRNA. Furthermore, 90% of a set of cDNA clones representative of the more prevalent nRNA species in these *Xenopus* embryo stages cross react with the nonparental nRNAs (Knöchel and Bladauski, 1981). These data all indicate that important errors in Table 3.1 due to large qualitative distinctions in cellular nRNA sequence sets are very unlikely, at least for sea urchin and *Xenopus* embryos. A second assumption, required for the calculations shown in columns 10 and 11 of Table 3.1, is that the rapidly decaying nRNA is the same population of molecules the complexity of which is listed in column 9 of Table 3.1. This issue was resolved by Hough *et al.* (1975) in experiments carried out on the nRNA of gastrula stage sea urchin embryos. It was demonstrated that the complex class of nRNA recovered from hybrids formed with single copy DNA in the presence of a large excess of nRNA has the same synthesis and turnover kinetics as does total nRNA. Furthermore, the quantity of the complex class nRNA estimated from hybridization kinetics [equation (18) of Appendix II] was found to be close to the steady state quantity of unstable nRNA given in Table 3.1, as required on the independent grounds considered in the foregoing analysis.

Developmental Change in nRNA Synthesis Rates

Table 3.1 also lists nRNA synthesis rate comparisons for early cleavage versus blastula-gastrula stage sea urchin embryos, and for blastoderm versus late gastrula stage *Drosophila* embryos. These measurements confirm that in both organisms the absolute transcription rate per cell *decreases* sharply as development proceeds. It remains to be established in quantitative terms whether this is a general effect on the average initiation period in the sense of column 10 of Table 3.1, or is due primarily to decreases in the activity of certain specific sets of genes. However, the latter seems a likely factor. Thus in an electron microscope comparison of 64-cell and gastrula stage sea urchin embryo transcription complexes, Busby and Bakken (1980) noticed in the cleavage stage nuclei a distinct class of short, clustered, transcription units displaying densely packed nascent RNP fibril arrays, which would account for a significant fraction of the total transcriptional activity. These are almost certainly the early histone genes, which are active at the 64-cell stage and almost silent at the gastrula stage, as reviewed below. Otherwise the distributions of RNP fibril densities in the nuclei of the two embryonic stages are nearly identical. Similarly, in *Drosophila* embryos histone synthesis is most prominent at the earlier of the stages of development compared (Anderson and Lengyel, 1984). However, the absolute histone mRNA synthesis rate measurements reviewed later in this chapter show that in neither organism does histone gene transcription account for a sufficient fraction of the total instantaneous nRNA synthesis rate to explain the 3- to 5-fold decline in overall transcriptional activity per cell observed during early development (Table 3.1, columns 4 and 5). Histone gene expression is thus probably a contributing, but not a complete, explanation for the relatively high transcription rates observed in the early embryo.

(ii) Zygotic mRNA Synthesis

Cytoplasmic Entry Rates for Newly Synthesized mRNA

The rate at which mRNA enters the embryo cytoplasm is only a small fraction of the rate at which heterogeneous nRNA is synthesized. This is shown for *Drosophila*, mouse, *Xenopus*, and sea urchin embryos in Table 3.2. On an absolute scale the average cellular rates of mRNA entry shown in

TABLE 3.2   Rates of mRNA Entry Compared to nRNA Synthesis Rates in Various Embryos

1 Species	2 Stage	3 mRNA entry rate per cell (pg min^{-1})	4 nRNA synthesis rate per cell[a] (pg min^{-1})	5 mRNA entry rate nRNA synthesis rate
S. purpuratus	Blastula-gastrula (~500 cells)	2.6×10^{-4}[b]	5.2×10^{-3}	0.05
D. melano- *gaster*	Gastrula[c] (4.5 hr, ~30,000 cells)	$\geq 6 \times 10^{-5}$[d]	1.5×10^{-3}	≥ 0.04
X. laevis	Neurula[e] (stage 15-16, ~75,000 cells)	1.5×10^{-3}[f]	1.5×10^{-2}	0.10
Mouse	2-cell	4.2×10^{-3}[g]	5×10^{-2}	0.08

[a] Data from Table 3.1.

[b] Galau *et al.* (1977a). This rate was measured for total polysomal mRNA in embryos labeled for 14 to 40 hr (in different experiments) beginning at mesenchyme blastula stage (25 hr after fertilization). See Fig. 3.6. The average rate was 0.13 pg min^{-1} per embryo, and this rate is determined in the first few hours of the experiment, when the embryo contains about 500 cells.

[c] Cell number estimate from Anderson and Lengyel (1981).

[d] Anderson and Lengyel (1979). The measurement listed is for poly(A) RNA entering the cytoplasm. This is an underestimate of *total* mRNA entry rate, since poly(A)-deficient polysomal mRNA has also been reported in these embryos. Zimmerman *et al.* (1980) proposed that over 90% of the mRNA is nonpolyadenylated, but data of Lovett and Goldstein (1977) indicate that at least 50% of polysomal mRNA is polyadenylated [i.e., assuming that the usual 4% by mass of total polysomal RNA is mRNA (see Chapter II)]. A study of Brogan and Goldstein (1985) confirms for nine cloned sequences represented predominantly in moderately abundant polysomal mRNA that only about half of the cytoplasmic transcripts of each species are polyadenylated, though for some sequences this value falls during development. On the other hand, Winkles and Grainger (1985b) concluded from cell-free translation experiments carried out with poly(A) RNA and poly(A)-deficient mRNA that *most* abundant messages are

column 3 of Table 3.2 vary by large factors from species to species. However, expressed as a fraction of the basic nRNA synthesis rate the mRNA entry rates appear very similar (column 5). Thus about 5-10% of the newly synthesized heterogeneous nRNA is exported as stable message in all four species. The 10-20-fold "excess" of newly synthesized nRNA mass can perhaps be accounted for as intron sequence, and as regions of 3' terminal sequence that are discarded during mRNA processing and chain termination. However, this remains to be quantitatively demonstrated, and there is evidence to suggest that in addition to mRNA precursors there is a large class of heterogeneous transcripts in any given nucleus no portion of which appears as stable cytoplasmic message. Nucleus-confined RNAs of this nature evidently account for more than half of the newly synthesized primary transcripts in HeLa cells (Salditt-Georgieff *et al.*, 1981; Salditt-Georgieff and Darnell, 1982), and they have also been detected in mouse liver cells (Schibler *et al.*, 1983). In addition, Wold *et al.* (1978) showed that the nRNA of adult sea urchin cells includes a significant set of single copy sequences that are absent from the cytoplasm of these cells, though they are represented in polysomal mRNA of blastula stage embryos. A similar conclusion was drawn independently by Shepherd and Nemer (1980), and examples of

---

polyadenylated. If it is assumed that 50% of the total mRNA is poly(A)-deficient the value in column 3 would be 1.2×10^{-4} and that in column 5, 0.08, which would not materially change the significance of the comparison.

[e] The embryo at this stage contains about 450 ng of DNA and thus 75,000 cells. Interpolated from data of Dawid (1965).

[f] The value shown was calculated from the cell number and the per embryo synthesis rate for poly(A) RNA as follows. Shiokawa *et al.* (1979) showed that 1/6th of poly(A) RNA accumulated in labeling periods of 1.5-3 hr is cytoplasmic in dissociated cells of neurula stage embryos. Under these labeling conditions, based on UTP pool specific activities, Sagata *et al.* (1978) calculated an absolute poly(A) RNA synthesis rate of 8 ng hr^{-1} per embryo. We assume 1/6th represents polysomal poly(A) mRNA, since newly synthesized poly(A) mRNA is mobilized rapidly into polysomes (Shiokawa *et al.*, 1981a). However, only 20% of total polysomal mRNA is polyadenylated (Miller, 1978). Thus an approximate estimate of the per cell rate (in pg min^{-1}) for total polysomal mRNA synthesis is given by [$(8 \times 10^3$ pg hr$^{-1} \times 5) \div (60$ min hr$^{-1} \times 6 \times 7.5 \times 10^4$ cells)]. Shiokawa *et al.* (1981c) reported estimates for synthesis of stable RNA bearing type I caps (i.e., presumably, mainly mRNA) that are consistent with the listed value: 2.5×10^{-3} pg min^{-1} per cell for stage 10 gastrula and 1.5×10^{-3} pg min^{-1} per cell for stage 19 neurula. For comparison, the rate of appearance of stable *mitochondrial* RNAs at this stage, mainly mitochondrial rRNA, is only about 3 pg min^{-1} per embryo, or 4×10^{-5} pg min^{-1} per cell (Chase and Dawid, 1972). This rate is nonetheless about six times higher than observed in blastula and cleavage stages, indicating a transcriptional stimulation, since no mitochondrial DNA synthesis is observed until tadpole stages (Chase, 1970; Chase and Dawid, 1972).

[g] Clegg and Pikó (1983b). This measurement is based on the ATP pool specific activity in labeled embryos, and it includes both poly(A) mRNA and poly(A)-deficient mRNA. These are distinguished from nRNA by their stable accumulation kinetics. For the 2-cell embryo the rate of accumulation of these classes of zygotic transcript are given as 0.3 pg per embryo-hour for the putative poly(A)-deficient mRNA, and 0.2 pg per embryo-hour for the poly(A) mRNA.

cloned sea urchin sequences that appear in egg cytoplasm but later in life are confined to the nucleus were reported by Lev *et al.* (1980), Thomas *et al.* (1982), and Calzone *et al.* (1986b). These observations suggest the possibility that many genes active in the early embryo continue to be transcribed in adult cells, but that their products are either not processed, or are not stabilized in the cytoplasm (see discussion of the regulation of the maternal gene set in Chapter I). Whether or not such a phenomenon contributes to the high value of the ratio of nuclear RNA synthesis to mRNA entry rate, Table 3.2 shows the observation to be general, and to apply to embryos of animals that have both large and small genomes and that develop in diverse ways.

Complexities and Average Entry Rates for Newly Synthesized Embryo mRNA Populations

Table 3.3 presents mRNA entry rates per cell in molecular terms, for midstage *Drosophila* and sea urchin embryos (column 6). The nucleus of the

TABLE 3.3. **Complexities and Average Cytoplasmic Entry Rates per Sequence for Classes of Sea Urchin and *Drosophila* Embryo mRNA**

1 Embryo (stage)	2 Class of mRNA	3 Complexity mRNA (nt)	4 $\frac{\text{mRNA}^a}{\text{nRNA}}$	5 Cytoplasmic entry rates per cell[b] k'_s (nt min^{-1})	6 k_s (molecules min^{-1})	7 Av. entry period[c] (min)
Drosophila (gastrula)	Poly(A) mRNA[d]	$6.7 \times 10^{6\,d}$	0.16	$1 \times 10^{5\,e}$	70[b]	(67)
S. purpuratus blastula- gastrula)	Total mRNA	$1.7 \times 10^{7\,f}$	0.10	$4.5 \times 10^{5\,e}$	225	(38)
	Low abundance[g]	$1.5 \times 10^{7\,g}$		$1.8 \times 10^{5\,g}$	90[b]	83
	Moderate abundance[g]	$1.5 \times 10^{6\,g}$		$2.7 \times 10^{5\,g}$	135[b]	5.6

[a] nRNA complexity data from Table 3.1.

[b] k'_s is the synthesis rate constant expressed in mass terms (here nt min^{-1}) and k_s, the synthesis rate constant expressed in molecular terms (see Appendix III). $k_s = k'_s/\bar{L}$, where $\bar{L}$ is the mean length of the mRNA in nt. For *Drosophila* mRNA $\bar{L}$ is taken as 1500 nt. A mean length of 1250 nt was obtained by methylmercury gel electrophoresis by Zimmerman *et al.* (1980), while Anderson and Lengyel (1979) reported a weight average length of 1700 nt, obtained by velocity sedimentation under nondenaturing conditions. In an earlier study Lamb and Laird (1976) reported that cytoplasmic poly(A) RNA of *Drosophila* embryos has a mean length of about 1200 nt and a median length of about 1800 nt. The mean length of sea urchin embryo polysomal mRNA, both poly(A) and poly(A)-deficient, is under denaturing conditions about 2000-2500 nt (see Chapter II; see also Nemer *et al.*, 1975). The value 2000 nt is used in the calculations shown. The values of k'_s were from the entry rate measurements given in Table 3.2.

[c] Entry period is given by [complexity (nt)/entry rate (nt min^{-1})]. This is the number of minutes, on the average, between the entrance of successive molecules of given species into the cytoplasm of each

average *Drosophila* gastrula cell exports about 70 molecules of newly synthesized mRNA per minute, and the equivalent figure for the sea urchin is about 225 molecules of mRNA per minute (*S. purpuratus* embryos are cultured at 7-12°C lower temperatures than customary for laboratory *D. melanogaster*). Considered as overall systems, the gastrula stage *Drosophila* embryo generates about 2.1×10^6 molecules and the blastula-gastrula stage sea urchin embryo about 1.1×10^5 molecules of cytoplasmic message per min.

From the complexities shown in column 3 of Table 3.3 the number of diverse polysomal message species in the mRNA classes listed may also be calculated. The complexity of polysomal message in gastrula stage sea urchin embryos (Galau *et al.*, 1974) indicates the presence of some 8500 diverse message species, assuming a mean length of 2000 nt (see note *a* of Table 3.3). The measurement of mRNA complexity for *Drosophila* larvae (Zimmerman *et al.*, 1980) suggests about 5000 diverse message species, assuming a mean length of 1500 nt (note *b* of Table 3.3). Considering the several independent confirmations of these measurements in other studies (e.g., for *Drosophila*, Izquierdo and Bishop, 1979; for sea urchins, Galau *et al.*, 1976; Hough-Evans *et al.*, 1977; Duncan and Humphreys, 1981), the 2.5-fold difference between the complexities of *Drosophila* and sea urchin embryo mRNAs (Table 3.3) is undoubtedly real, and a similar difference was discussed in relation to the respective maternal RNA complexities of these

---

cell. Values in parentheses are essentially the weighted averages of entry periods for the rare and prevalent mRNAs summed together, as shown explicitly for the sea urchin case, where these mRNA classes can be separately treated. mRNA entry rates are from Table 3.2.

d Measurement of Zimmerman *et al.* (1980), obtained by mRNA excess hybridization with single copy tracers. The value shown is in reasonable agreement with estimates from cDNA hybridization kinetics (Izquierdo and Bishop, 1979). This measurement is for third instar larvae. However, the mRNA sequence sets of eggs, gastrulae and larvae show very high fractions of overlap, and similar overall complexities (Levy and McCarthy, 1975; Izquierdo and Bishop, 1979; Arthur *et al.*, 1979). A complexity for larval poly(A)-deficient polysomal mRNA of about 1.5×10^7 nt was also reported by Zimmerman *et al.* (1980). However, no comparable synthesis rate data for this unexpectedly high complexity class of mRNA are available.

e Calculated from the value of k'_s given in Table 3.2.

f Complexity of total polysomal mRNA was measured by mRNA excess hybridization with a single copy tracer by Galau *et al.* (1974).

g Low and moderate abundance classes are defined arbitrarily as <10 molecules of mRNA per cell and greater than this (see text). Statistical estimates made with cDNA clone libraries (Lasky *et al.*, 1980; Flytzanis *et al.*, 1982) indicate that about 40% of the mass of mRNA in sea urchin embryos is of the low abundance class. Sources of the complexity estimate for the moderate abundance class of mRNA are given in text. A reasonable assumption is that about 10% of the overall mRNA complexity can be attributed to moderately prevalent mRNA species. After blastula stage (i.e., excluding early histone mRNAs) there are no *very highly* prevalent mRNA species in the embryo other than mitochondrial poly(A) RNA (i.e., $>10^3$ molecules per average cell; Lasky *et al.*, 1980; Flytzanis *et al.*, 1982). Of course some sequences could be very prevalent in a subset of cells, and as discussed in Chapter IV some specific examples are known.

organisms (see Table 2.1). In both species, however, the mRNA complexities are a small fraction of the respective nRNA complexities (Table 3.1). Thus just as most nucleotides assembled into nRNA never enter the cytoplasm as mRNA (Table 3.2), 80-90% of *nRNA sequence information* is also confined to the nucleus.

Comparison of the *complexity* of each mRNA population with its *entry rate* per cell shows that most of the mRNA species in the population are exported intermittently at a low rate (Table 3.3, column 7). The mean value for the period between successive entries of any given mRNA species is over an hour in average cells of the *Drosophila* embryo, and is about 40 min in the sea urchin embryo. These, however, are averages over the whole mRNA populations. To obtain a more realistic image of mRNA entry kinetics it is useful to consider separately the prevalent and rare mRNA abundance classes. As initially pointed out by Galau *et al.* (1974) from considerations of the kinetics with which sea urchin embryo mRNA hybridizes with single copy DNA, it is clear that a significant fraction of the total polysomal message population of the embryo consists of low abundance sequences, each present in only a few copies per cell, that together account for most of the observed complexity. Prevalence distributions for sea urchin embryo mRNA populations have been derived from cDNA hybridization kinetics, and also from the reaction of sets of cDNA clones with labeled cDNA transcribed from embryo mRNA (see note *g* of Table 3.3 for references and arguments). About 40% of the mass of the mRNA consists of low abundance sequences, defined arbitrarily as sequences present at less than ten copies per *average* embryo cell. This is a convenient demarcation because it is about the minimum prevalence that can be detected without extraordinary measures in cDNA clone colony screen hybridizations (Lasky *et al.*, 1980). Furthermore, the products of many mRNAs more prevalent than this can be resolved in two-dimensional gel electrophoretic assays of newly synthesized embryo proteins (see, e.g., Brandhorst, 1976; Tufaro and Brandhorst, 1979; Bédard and Brandhorst, 1986a). Estimates based on cDNA clone colony hybridization indicate that the more prevalent nonmitochondrial poly(A) RNA sequences in sea urchin embryos range up to several hundred copies per average cell (Lasky *et al.*, 1980; Flytzanis *et al.*, 1982). From the kinetics of cDNA hybridization the complexity of this class of more prevalent cytoplasmic transcripts may be supposed to be about 10% of that of the whole mRNA population (McColl and Aronson, 1978; E. H. Davidson, unpublished data), and this value is consistent with the observation of about 10^3 diverse, newly synthesized polypeptides in the two-dimensional gel electrophoresis experiments cited.

The cytoplasmic rates of mRNA entry are unlikely to be limited by processing and export kinetics, but rather by the rate of transcription. Thus cytoplasmic entry rates were shown to be approximately equal to nuclear synthesis rates for several cloned sea urchin embryo sequences (Cabrera *et al.*, 1984), and this has been confirmed for actin nuclear and messenger

RNAs as well (J. J. Lee *et al.*, 1986b). In Table 3.3 the prevalence distribution of the mRNA is utilized to resolve the overall entry rate for newly synthesized embryo mRNA into entry rates for low abundance and higher abundance message classes (columns 5 and 6). It can then be seen (columns 3 and 7) that the period between successive entries of a given low abundance sequence is over 1 hr, while for a moderate abundance sequence it is only a few minutes. Most of the mRNA in sea urchin embryos derives from single copy genes (Goldberg *et al.*, 1973; Davidson *et al.*, 1975). By way of comparison, a single copy gene functioning at the measured polymerase translocation rates (for sea urchin embryo 6-9 nt sec^{-1}; Aronson and Chen, 1977) could produce up to 5.4 transcripts per min (assuming 100 nt center-to-center polymerase spacing), and thus the minimum entry period in a diploid cell would be about 11 sec. The 5.6 min value shown in Table 3.3 for the more prevalent mRNA class suggests that, just as indicated by the structure of even the more densely packed transcription units visualized in the electron microscope [e.g., Fig. 3.5(a)], statistically few of the embryonic genes are operating at the theoretically maximal rates.

This analysis confirms our earlier inference that the *rate of transcriptional initiation* is a major variable in determining the prevalence of a cytoplasmic mRNA species. Thus for sequences present at typical low abundances only one initiation event or less per hour is required, while for more prevalent sequences at least ten times this rate must obtain, on the average. As described in a following section, initiation in certain genes indeed occurs at rates that are severalfold higher than the average for prevalent sequences shown in Table 3.3. In the following section it is demonstrated that the *rate of turnover* is also a crucial variable in determining the prevalence of newly synthesized mRNA, i.e., the level of zygotic gene expression.

(iii) Replacement of Maternal by Zygotic Transcripts

There is no specific stage of embryonic development during which maternal transcripts are replaced by zygotic transcripts. This occurs at diverse stages for different sequences within a given embryo, though in some species it is evident when maternal mRNA has ceased to be of major importance. In *Xenopus* the net amount of poly(A) RNA and the net content of polysomal mRNA (i.e., poly(A) mRNA plus poly(A)-deficient mRNA) both increase about 6-8-fold between gastrula and swimming tadpole stages (Woodland, 1974; Sagata *et al.*, 1980; Shiokawa *et al.*, 1981a; Weiss *et al.*, 1981). Thus even if some maternal mRNAs survive until late in development, the large majority of the message is obviously of zygotic origin at postgastrular stages. In the mouse, as discussed in Chapter II, the maternal mRNA is actively degraded at the 2-cell stage, and thereafter the mRNA content increases rapidly. Clegg and Pikó (1983a) showed that by early blastocyst stage there are about 3.4×10^7 poly(A) mRNA molecules per embryo, compared to the

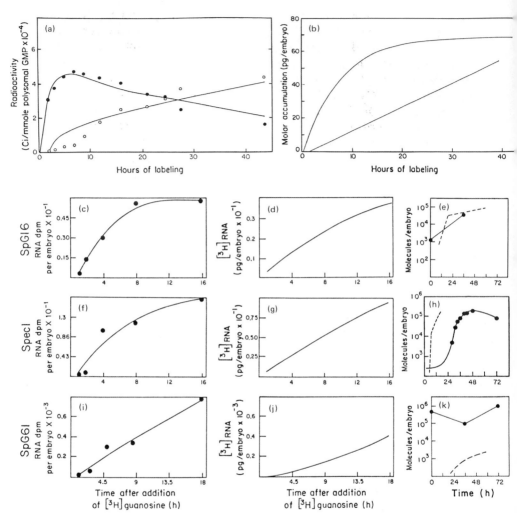

Fig. 3.6. Measurements of synthesis and turnover rates for sea urchin embryo mRNAs. The measurements shown in this figure were all carried out on *S. purpuratus* embryos which were cultured in sea water containing ³H-guanosine, beginning 24–26 hours postfertilization (time 0 on the abscissae). The data have been reduced by the kinetic methods discussed in Appendix III, where the behavior of the GTP pool in sea urchin embryos is also described (Fig. A3.1). Values for k'_s, the rate of entry of the transcript into the cytoplasm, and for k'_d, the decay rate, were obtained from least-squares analysis, assuming the form of equation (8) of Appendix III. (a) displays the kinetics of incorporation of ³H-GMP in total embryo mRNA, and (c), (f), (i) into specific poly(A) RNA transcripts. Solid lines indicate the least-squares solutions from which the rate constants were extracted, and solid or open circles the individual data points. (b), (d), (g), and (j) indicate the mass accumulation kinetics of the newly synthesized RNAs measured in (a), (c), (f), and (i) respectively. These curves were generated from the calculated values of k_s and k_d, by use of equation (5) of Appendix III. (e), (h) and (k) compare independent (i.e., nonkinetic) measurements of transcript prevalence, indicated by solid circles, with the calculated accumulation of the newly synthesized RNA, indicated by the dashed lines. It is

minimum of 0.7×10^7 at the late 2-cell stage, and 2.4×10^7 in the 1-cell zygote. Thus here too there can be no question as to the quantitative dominance of zygotic transcripts, i.e., after the earliest cleavage stages.

In the sea urchin embyro as in the *Drosophila* embryo, there is little net change in mRNA content during the period when the maternal message is being replaced by new transcripts. In both these organisms, as discussed below, most of the maternal transcript species are ultimately replaced by similar quantities of *homologous* zygotic transcripts. To determine the amount of newly synthesized mRNA accumulated at any particular stage, while total (i.e., zygotic + maternal) mRNA content remains constant, it is necessary to take into account both the cytoplasmic entry and the turnover rates of the zygotic transcripts. Galau *et al.* (1977a) calculated an average half-life of 5.7 hr for newly synthesized polysomal mRNA of blastula-gastrula stage *S. purpuratus* embryos. This was obtained from the kinetics of label incorporation in total polysomal message, by the methods described in Appendix III. A representative measurement of both the entry and the decay rates is reproduced by Fig. 3.6(a)-(b). A similar half-life estimate of 5.3 hr

---

assumed for these calculations that the rate of synthesis per nucleus remains constant, and the accumulation functions shown are derived from the measured entry and turnover rates, and the increase in number of embryonic cells due to cell division during the labeling period [see equation (9), Appendix III]. For the calculations shown in (e) and (k), and sources of independent prevalence estimates in (e), (h) and (k), see Cabrera *et al.* (1984). (a) and (b) Total polysomal mRNA (●) and ribosomal RNA (O). The cytoplasmic entry rate per embryo derived from this experiment (k_s') is 0.17 pg min^{-1} or per cell about 290 mRNA molecules min^{-1}, and the decay rate (k_d) is 2.2×10^{-3} min^{-1} ($t_{1/2} = 5.3$ hr). For rRNA the cytoplasmic entry rate is 0.027 pg min^{-1} or per cell about 16 molecules min^{-1}, and k_d approaches zero. The early lag in appearance of mature rRNA is due to the time required for processing and loading into polysomes from which the rRNA was extracted. [(a) and (b) From G. A. Galau, E. D. Lipson, R. J. Britten, and E. H. Davidson (1977a). *Cell* **10**, 415. Copyright by M.I.T.] (c)-(k) Specific sequences measured by hybridization with cloned probes. (c)-(e) Unidentified moderately prevalent transcript that displays measurable cytoplasmic turnover. For this sequence k_d is 1.3×10^{-3} min^{-1} ($t_{1/2} = 8.8$ hours), and the steady state content is about 60 molecules per average cell. Panel (e) shows that the measured accumulation of this sequence could probably be accounted for if the gene is expressed constitutively. Thus the calculated increase in transcript quantity per embryo shown may be due solely to cell multiplication. (f)-(h) Spec1 mRNA. This message codes for a Ca-binding protein and the gene is expressed only in aboral ectoderm (see Chapter IV). Per cell, k_s is 1.8 molecules min^{-1}, and the sequence is very stable ($k_d < 10^{-3}$ min^{-1}). For the calculation shown in (h) it is assumed that only about *half* the newly formed cells [i.e., those in the future aboral ectoderm of the embryo (see Chapter IV)] express the gene. This calculation shows that were the Spec1 gene *constitutively* transcribed in half the cells the rise in Spec1 transcripts would occur significantly earlier than actually observed (Lynn *et al.*, 1983). Therefore the gene must be transcriptionally activated after the main phase of cell division has been completed. (i)-(k) Unidentified stable embryonic sequence that, while fairly prevalent, is synthesized at a very low rate. k_s per cell is only 2×10^{-3} molecules min^{-1}, and $k_d < 10^{-3}$ min^{-1}. Panel (k) shows that zygotic gene activity cannot possibly account for over 90% of the transcripts of this species present in the embryo. Therefore, most maternal transcripts of this species must survive throughout embryogenesis. [(c)-(k) From C. V. Cabrera, J. J. Lee, J. W. Ellison, R. J. Britten, and E. H. Davidson (1984). *J. Mol. Biol.* **174**, 85.]

was derived by Nemer *et al.* (1975) from a "pulse-chase" procedure. On the other hand, the half-life of *total* newly synthesized *cytoplasmic poly(A) RNA* is only about 2-2.2 hr (Cabrera *et al.*, 1984). This discrepancy suggests the possibility that there is a rapidly turning over class of newly emergent polyadenylated transcript that is not polysomal, and that could consist of either the same or different sequences than those assembled into polysomes. In any case, the measurements of total polysomal mRNA entry and turnover rates in the blastula-gastrula stage sea urchin embryo show that the steady state quantity of newly synthesized mRNA, i.e., k_s/k_d, is close to the total mass of polysomal mRNA. In embryos of this stage the latter is about 58 pg per embryo, calculated on the basis that it is 4% of the mass of polysomal RNA [60% of the ribosomes are in polysomes at this stage (Infante and Nemer, 1967; Galau *et al.*, 1977a; Goustin and Wilt, 1981); hence the mRNA content is $0.04 \times 2.4 \times 10^3$ pg (rRNA/embryo) $\times 0.6$]. From the measurements of Galau *et al.* (1977a) the steady state content of newly synthesized mRNA is about 64 pg per embryo [i.e., (see Table 3.3) k_s/k_d is 0.13 pg min^{-1}/ 2.03×10^{-3} min^{-1}]. Therefore, by the blastula-gastrula stage *most if not all of the mass of maternal message has been supplanted by newly synthesized zygotic messages.*

A similar series of measurements was carried out on *Drosophila* embryos by Anderson and Lengyel (1981). Calculation from precursor incorporation kinetics of the absolute amount of zygotic mRNA accumulated as development progresses shows that at the gastrula stage 70% of the total mRNA is still maternal in origin, and >40% is still maternal even after organogenesis is underway (9 hr postoviposition). Thus maternal mRNAs remain quantitatively significant until more advanced developmental stages in *Drosophila* than in the other organisms considered.

Overall kinetic measurements of course cannot reveal any sequence-specific variations that may exist in entry rates, turnover rates, and the timing of replacement of maternal by zygotic transcripts. To examine this in the sea urchin embryo, Cabrera *et al.* (1984) determined the entry and turnover rates for a series of cloned sequences of various prevalence. Representative results are reproduced in Fig. 3.6(c)-(k). Here it can be seen that different sequences in fact display very different cytoplasmic entry and turnover kinetics. The sequence studied in Fig. 3.6(c) and (d) is unstable, with a half-life of about 9 hr [other sequences investigated by Cabrera *et al.* (1984) decay more rapidly than this]. The sequences used for the experiments in Fig. 3.6(f) and (i) are stable. These sequences of course accumulate, while the unstable ones establish a steady state level of prevalence [see Fig. 3.6(d), (g), and (j)]. As indicated in the legend the entry rates observed in these experiments differ by a factor >100. Thus k_s and k_d both vary greatly within the same embryos, and both are of crucial significance in determining the prevalence of newly synthesized transcripts. The overall mRNA turnover rates reflect mainly the more rapidly decaying species, since the amount of labeled precursor flowing into these for a given steady state quantity of transcript is proportional to the rate of decay [equation (6), Appendix III]. Cabrera *et al.* (1984) reported that all the low abundance sequences investigated are unstable, while as illustrated in Fig. 3.6 some higher prevalence sequences are stable while others are unstable. In the sea urchin embryo the zygotic geromes thus may acquire control of the population of rare single copy sequence transcripts relatively early in development. However, this may not occur at all during embryogenesis for some abundant maternal sequences. Thus Fig. 3.6(k) provides an example of a sequence prevalent in late pluteus stage embryos that is synthesized during embryogenesis at too low a rate to account for more than a small fraction of the total embryonic transcript mass. A final point illustrated in Fig. 3.6 is that the sharp rise in the quantities of certain newly synthesized transcripts observed per embryo could be due largely to increase in the number of cells producing the transcript at a constant rate per cell, rather than to "gene activation," that is, developmental increase in the rate of synthesis per cell [e.g., see Fig. 3.6(e)]. This illustration demonstrates that increase in the number of synthetically active cells alone provides a developmental mechanism by which prevalent zygotic transcripts can be accumulated in large quantities. In examples such as that shown in Fig. 3.6(h), however, the increase in transcript concentration occurs at a different time than if the accumulation kinetics were due solely to cell number increase. The implication is that the gene coding for this particular transcript is indeed activated during development, as is confirmed for this particular case by additional evidence reviewed in Chapter IV.

In summary, in the sea urchin embryo the total number of molecules of each cytoplasmic transcript species at any given time is the sum of the surviving maternal RNAs and the accumulated zygotic transcripts of that species. For each sequence the amount of zygotic mRNA depends on the entry and turnover kinetics, and both parameters differ from sequence to sequence. There could be a "default" decay rate near the average observed for total newly synthesized message that obtains for all transcripts except those that have special structural features endowing them with extra stability. In any case it seems evident that regulatory control of the cytoplasmic level of expression of each gene is a *particular property of that gene, and of its transcripts*, rather than, for instance, a general property of a given developmental stage.

With the exception of some specific examples noted below, equivalent kinetic measurements do not exist for particular messages in other embryos. However, the available data suggest similar conclusions. An experiment of Winkles and Grainger (1985a) in which *Drosophila* embryonic mRNAs still present at larval stages were physically separated from density labeled larval mRNAs, showed that the half-lives of individual embryonic message species varies from a few hours to at least 30 hours. The survival of these species was estimated from two-dimensional displays of the *in vitro* translation products of embryonic mRNA fractions extracted at successive stages. Embryonic histone mRNAs were found to be unusally unstable (see below), while mRNAs coding for actins, tubulin, and tropomyosin were more stable than the average. Similarly, Kidder and Pedersen (1982) showed that in the mouse embryo there are both long-lived mRNAs and unstable mRNAs that turn over with a half-life of about 6 hr. This can also be inferred from some of the protein synthesis studies reviewed in Chapter II, which indicate survival of some mRNAs in the mouse embryo much longer than of others after α-amanitin treatment. Thus in both the *Drosophila* and the mammalian embryo, as in the sea urchin embryo, control of the rate of mRNA decay appears to be an important and specific means of affecting the level of zygotic gene expression.

(iv) Synthesis of rRNA

Measurements of ribosomal RNA synthesis rate are collated in Table 3.4, expressed as molecules synthesized per embryonic cell (column 5) and per

TABLE 3.4   Ribosomal RNA Synthesis in Various Embryos

			4	5
			RNA synthesis rates	
1 Species	2 Maternal rRNA (no. ribosomes)	3 Stage	per cell (molecules min^{-1})	per gene (molecules min^{-1})
S. purpuratus	$6.6 \times 10^{8\,a}$	Early blastula	39^b	0.39^c
		Late gastrula	40^b	0.40^c
		Blastula-gastrula	$12\text{-}18^d$	$0.12\text{-}0.18^c$
L. pictus	$4 \times 10^{8\,a}$	Early blastula	16^b	0.031^c
		Late gastrula	23^b	0.044^c
T. gratilla	$6.7 \times 10^{8\,e}$	Blastula-gastrula	65^f	0.65^c
		Early pluteus	52^f	0.52^c
		Growing oocyte	1830^f	9.2^g
D. melanogaster	$4.5 \times 10^{10\,h}$	Gastrula	130^i	0.27^j
X. laevis	$1 \times 10^{12\,a}$	Gastrula	425^k	0.47^l
		Neurula	660^k	0.73^l
Mouse	$5.5 \times 10^{7\,m}$	2-cell^n	6700^n	11^p
		8-cell^o	5200°	9^p
		32-cell^o	5200°	9^p

[a] Data from Chapter II.
[b] Surrey *et al.* (1979). These measurements are based on the rate of incorporation into rRNA methyl groups from labeled S-adenosylmethionine precursor and on the S-adenosylmethionine pool specific

rRNA gene (column 6). It is immediately evident that the embryos for which data are listed fall into two classes. In the mouse embryo the ribosomal genes are relatively active. Thus if every gene is equally functional, the per gene rate of synthesis, about 10 molecules min⁻¹, is only a small factor short of the maximum possible rate (i.e., 10 molecules min⁻¹ requires a translocation rate of about 15 nt sec⁻¹ and a new initiation on every gene every 6 sec). In contrast, the ribosomal genes in the other embryos considered are relatively inactive. In the sea urchin embryo the rates measured for different species range from 0.03 molecules min⁻¹ to about 0.60 molecules min⁻¹ per rRNA gene, and the average value is only about 0.3 molecules min⁻¹. A comparison is shown for the *oocyte* of *Tripneustes gratilla*. In sea urchins there is no rDNA amplification during oogenesis (see Chapter V), but the genomic rRNA genes are probably operating at maximum efficiency (Griffith *et al.,* 1981). Assuming that all the rRNA genes are equally active the translocation rate implied for the oocyte is about 15 nt sec⁻¹, which is higher than the maximum rate measured for *S. purpuratus* polymerase II [9 nt sec⁻¹

---

activity. Data are treated by the method described in equations (7) and (8) of Appendix III (see Fig. A3.1 for behavior of the S-adenosylmethionine pool).

[c] There are 260 rRNA genes per haploid genome in *Lytechinus variegatus* (Patterson and Stafford, 1971), and this is taken as the rRNA gene content of *L. pictus* as well. More recent measurements of Griffith and Humphreys (1979) establish the number of rRNA genes in *T. gratilla* as 50 per haploid genome. The same number is assumed for *S. purpuratus*.

[d] Galau *et al.* (1977a). Measurement based on rate of incorporation into rRNA of labeled GMP and on GTP precursor pool specific activity, according to equations (7) and (8) of Appendix III (see Fig. A3.1 for behavior of this pool). A sample of these measurements is shown in Fig. 3.6(a).

[e] *Tripneustes gratilla* eggs contain about 1.7 ng of rRNA (Kleene and Humphreys, 1977) and the weights of the mature rRNA subunits are 1.45 × 10⁶ d and 0.7 × 10⁶ d (Griffith and Humphreys, 1979).

[f] Griffith and Humphreys (1979); Griffith *et al.* (1981). Measurement based on GTP and ATP pool specific activities over relatively brief labeling periods.

[g] The oocyte has a 4C content of ribosomal genes (i.e., 200 genes). No rDNA amplification occurs in the sea urchin during oogenesis (see Chapter V).

[h] The amount of rRNA per egg is 160 ng and the weight of *Drosophila* rRNA is 2.15 × 10⁶ d (Anderson and Lengyel, 1979).

[i] Anderson and Lengyel (1979) measured a rate of production of mature rRNA in 4.5 hr gastrulae of 0.46 × 10⁻³ pg min⁻¹ per cell, based on UTP pool specific activity over brief labeling periods in permeabilized embryos.

[j] The haploid rRNA gene number is about 235 (Tartof, 1973).

[k] Shiokawa *et al.* (1981a); measurements carried out on dissociated cells. The gastrula stage embryos (Nieuwkoop and Faber stage 10-11) are assumed to have contained 3 × 10⁴ cells. The neurula stage embryos (stages 15-16) are assumed to have contained 7.5 × 10⁴ cells.

[l] There are 450 ribosomal genes per haploid genome (Brown and Weber, 1968).

[m] Table 2.2.

[n] Clegg and Pikó (1983b) measured a rate per embryo of 0.4 pg hr⁻¹ based on incorporation into rRNA in 2-cell embryos, and the ATP pool specific activity, during a 320 min labeling period.

[o] Pikó and Clegg (1982); calculated from incorporation and pool specific activity data of Pikó (1970) and Clegg and Pikó (1977).

[p] There are 300 rRNA genes per haploid genome in the mouse (Marzluff *et al.,* 1975).

(Aronson and Chen, 1977); however, the *T. gratilla* measurements were made at 24°C and the *S. purpuratus* measurements at 15°C]. Thus it can be concluded that in the sea urchin embryo, depending on species, the rRNA gene set is operating at about 1/20 to 1/200 of the maximal possible rate (i.e., about 0.5 molecules per minute for *T. gratilla*, to about 0.04 molecules per min for *L. pictus*, versus 9.2 molecules per minute for the oocyte). The values shown in column 6 of Table 3.4 for *Drosophila* and *Xenopus* indicate that the same is true for these embryos. Though they indeed synthesize rRNA they do so at only a very low rate compared to that of which their rRNA genes are capable, or compared to the rate that would be necessary to significantly affect the total content of rRNA in the embryo. The size of the maternal rRNA pool for each of these species is given in column 2 of Table 3.4. Here it can be seen that at the embryonic rates listed it would require many days to synthesize an amount of rRNA equal to that stored in the egg at fertilization, i.e., except for the case of the mouse embryo. Ribosomal RNA synthesis in this embryo begins at the 2-cell stage (Knowland and Graham, 1972; Pikó, 1970; Clegg and Pikó, 1977, 1983b). At the listed synthesis rate the 8-cell mouse embryo could resynthesize the complete maternal rRNA pool in about 20 hr, and in fact the number of ribosomes increases from a low of 4.2×10^7 per embryo at the 2-cell stage to 9.9×10^7 during the 8-cell stage (Pikó and Clegg, 1982). After this it continues to rise. Ribosomal RNA synthesis is directly related to net growth in animal cells, and the values shown in Table 3.4 demonstrate directly the salient difference between sea urchin, amphibian and insect embryos on the one hand, and the mammalian embryos on the other. Only mammalian embryos engage in *net growth* during early development, making use of the fallopian and uterine environments to provide nutrients, while the other embryos considered are essentially closed systems. The latter embryos do not grow until the larva is sufficiently differentiated to begin feeding, and hence they store much larger quantities of maternal rRNA. It was shown directly by Humphreys (1973) that the rate of rRNA synthesis in sea urchin larvae accelerates immediately upon feeding.

   A much discussed subject has been the time of onset of rRNA synthesis in sea urchin and amphibian embryos, as the rRNA genes were originally considered to provide examples of developmental gene activation. However, as pointed out previously (Davidson, 1976) and documented quantitatively in Table 3.4, this is rather a moot issue since even when "activated" the rRNA genes are barely functioning in these embryos. Furthermore, it is not clear that in the sea urchin embryo these genes are ever *not* active, i.e., at the low level observed in mid and late stage embryos. Emerson and Humphreys (1970) demonstrated rRNA synthesis as early as the blastula stage, and this has been confirmed by Surrey *et al.* (1979) for the early blastula stages of two sea urchin species in an analysis of rRNA methyl incorporation kinetics from the S-adenosyl methionine precursor (Table 3.4). In *Xenopus* the onset of rRNA synthesis shortly after the midblastula stage (Shiokawa *et al.*, 1981b) coincides with the initial appearance of nucleoli in some cells (Nakahashi and Yamana, 1976). The rate of rRNA synthesis per cell thereafter remains constant throughout embryogenesis (Shiokawa *et al.*, 1981a; see also Table 3.4). Similarly, electron microscope observations of transcription complexes in *Drosophila* show that the synthesis of rRNA begins at the cellular blastoderm stage along with transcription of most other RNA species (McKnight and Miller, 1976). The latter study also provides a molecular explanation for the low rate of rRNA synthesis that obtains during embryogenesis, relative to the possible rate. This is that only a minor fraction of the ribosomal genes in each cluster is transcriptionally active, though those that are active are fully loaded with nascent transcripts. It follows that each rRNA gene has its own regulatory apparatus, and once activated it appears to function at a maximal rate. The same seems to be true of ribosomal genes in early amphibian oocytes (see Chapter V) and thus it may be generally applied that the low embryonic rates of rRNA transcription indicate the activity of only a few of the rRNA genes.

The time course of synthesis of tRNAs in the embryo generally follows that of the large ribosomal RNAs (reviewed in Davidson, 1976; see, e.g., Anderson and Lengyel, 1979; Shiokawa *et al.*, 1979). With the exception of 5S rRNA (for which see Chapter V), it is generally not as yet possible to consider low molecular weight RNA synthesis in terms of per gene transcription rates, since few measurements exist, and neither the complexity of this RNA class nor the number of genes responsible for its synthesis is as yet known. Gelfand and Smith (1986) found that in *Xenopus* the molar rates of U1, U2 and U6 snRNA synthesis are about the same as the rate of 5S rRNA synthesis at the gastrula stage, while synthesis of U3 is not detectable, and synthesis of U4 and U5 occurs more slowly. At the neurula stage U1 and U4 snRNA synthesis have increased somewhat relative to 5S rRNA synthesis, and U3 synthesis first becomes detectable, while the relative rates of synthesis of the other snRNA species has declined slightly. These measurements show that by this point the embryo genomes are actively providing new snRNA transcripts, though it is not known whether persisting maternal stores of some of the snRNAs would still suffice for development, as is the case for the heavy rRNAs up to the swimming tadpole stage (see above).

3. ZYGOTIC AND MATERNAL GENE SETS

Among the most surprising results to emerge from molecular studies of gene function in embryogenesis is that the large majority of midstage embryo mRNA species is already represented in the maternal RNA stored in the unfertilized egg. The zygotic gene set, defined as that set of structural genes active during embryogenesis, can be considered the sum of two subcategories, those genes *also* included in the maternal gene set, and those genes from which mRNAs first accumulate significantly during *embryonic* development, as a result of transcriptional activation. Operationally, the latter class of mRNA species is identified as transcripts not detectable in unfertilized egg RNA, or present at extremely low levels compared to their ultimate prevalence in the embryo. For convenience genes producing such transcripts were referred to in the discussion of embryonic gene regulation in Chapter I as "late genes." This denotation is to be distinguished from the cliché "developmentally regulated genes," since in many cases genes that are represented by maternal transcripts are also activated (and hence regulated) during development, even if the overall prevalence of the maternal + zygotic transcripts of these genes changes little from egg to advanced embryo.

It is apparent from the genetic evidence reviewed in Chapter II that most loci active in *Drosophila* and *Caenorhabditis* embryos are also functional during oogenesis. This implies that a major portion of the zygotic gene set consists of those genes also included in the maternal gene set. What molecular evidence exists for *Drosophila* embryos conforms to this conclusion (there are as yet no relevant molecular data for *C. elegans*). Thus several studies reviewed briefly in Chapter II show that during the development of *Drosophila* embryos there are only minor changes in the pattern of protein synthesis, as assayed by two-dimensional gel electrophoresis. In addition, analyses of cDNA hybridization kinetics carried out by Arthur *et al.* (1979) revealed differences of only a few percent between *Drosophila* oocyte and gastrula stage poly(A) RNA populations. This method is of limited sensitivity for complex class sequences, however. There are also several observations on cDNA clone libraries from which can be derived at least a qualitative idea of the fraction of moderately prevalent embryonic transcripts that are *not* represented in maternal RNA. For example Scherer *et al.* (1981) reported that of 3×10^4 differentially screened cDNA clones, less than 10 reacted with poly(A) RNA extracted from late blastoderm or gastrula stage embryos but not with preblastoderm poly(A) RNA, and most of the clones that were recovered contained transcribed repetitive sequences. A similarly low frequency of exclusively zygotic transcripts was isolated from cDNA clone libraries constructed by Sina and Pellegrini (1982) and Roark *et al.* (1985). Several single copy genes expressed specifically at blastoderm stage were identified in the study of Roark *et al.* (1985) that had not been previously detected in saturation genetic screens for zygotic lethals covering the chromosomal regions where these genes map. In general, however, the available molecular and genetic data agree excellently, both suggesting that late genes constitute a very small fraction of the total number of genes active during *Drosophila* embryogenesis. Another form of evidence consistent with this conclusion derives from two-dimensional analysis of protein synthesis

pattern at various stages (see Chapter II for references). Summers *et al.*
(1986) observed, for example, that only about 6% of the ~1200 protein
species visualized appear for the first time during the gastrula and organogenesis stages.

(i) Composition of the Zygotic Gene Set in Sea Urchin Embryos

Quantitative estimates of the respective sizes of the maternal and late gene
sets have been derived from comparative analyses of sea urchin embryo
mRNA populations and also from comparisons of the populations of newly
synthesized embryo proteins. The contribution to the overall embryo polysomal mRNA complexity of sequences also represented in unfertilized egg
RNA was measured directly by Galau *et al.* (1976). These experiments were
carried out by RNA excess hybridization, using two different single copy
DNA tracer fractions, an "mDNA" (message DNA) tracer, consisting
largely of sequences represented in the gastrula stage embryo mRNA, and a
"null mDNA" tracer, consisting of the single copy DNA that had been
depleted of these sequences. As shown in Fig. 3.7(a) and (b) the mDNA
reacted to about 60% with the parent gastrula mRNA, indicating about a 50-
fold purification of the polysomal message sequences compared to their
representation in the starting single copy DNA, while the null mDNA does
not react detectably with gastrula message. Both tracer fractions were hybridized with unfertilized egg RNA, as shown in Fig. 3.7(c) and (d), respectively, and with mRNA preparations from other embryonic stages and adult
tissues, as summarized in Fig. 3.7(e). The most striking result of these experiments is that shown in Fig. 3 7(c) where it can be seen that within the
statistical error in the measurements, which was perhaps 10%, all of the
gastrula polysomal mRNA sequences are already represented in the maternal RNA. As discussed in Chapter II the egg RNA also contains a significant
sequence component not included in polysomal RNA, here measured by the
null mDNA reaction shown in Fig. 3.7(d). The null mDNA tracer reacts to
an easily detectable extent with *blastula* polysomal mRNA, indicating the
presence of an additional set of mRNA sequences at this stage [see Fig.
3.7(e)]. Some of these sequences are not represented in the maternal RNA.
In further studies Wold (1978) prepared an analogous tracer consisting of
single copy genomic DNA from which *maternal* RNA sequences had been
depleted, and showed that about 12% of the total complexity of blastula
polysomal RNA is included in transcripts absent from unfertilized egg RNA.
This direct measurement suggests that the complexity of the "late gene"
class of trancripts in the blastula is about 3.4×10^6 nt. Wold (1978) also
found a barely detectable fraction of late gene transcript species (5×10^5 nt
complexity) in gastrula mRNA. These experiments, and the results of
Hough-Evans *et al.* (1977), which were summarized earlier (Fig. 2.8), demonstrate that 85-90% of the diverse embryo mRNA sequences are also represented in maternal RNA. Therefore, there are in all probability $\leq 10^3$ late

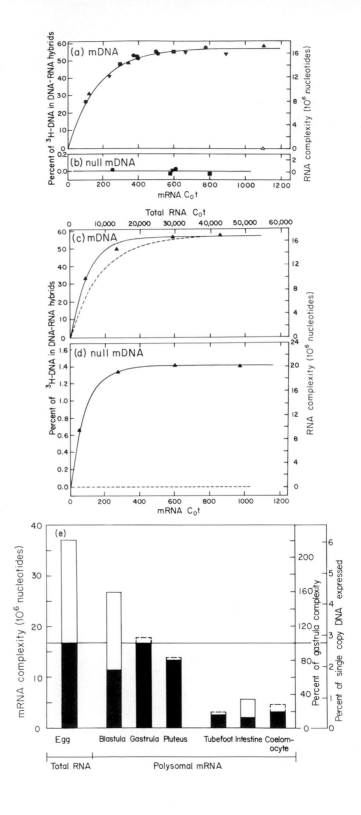

genes operative in the embryo, since as reviewed above, the polysomes of the gastrula stage embryo include *in toto* about 8500 different message species. The criterion imposed by the highly sensitive hybridization methods utilized in these experiments is a severe one, since transcript species present in only a few hundred molecules per egg (i.e., <1 molecule per the volume of an average late embryo cell) are detected, and would thus be excluded from the late gene class. This does not result in a large underestimate of the number of embryonic late genes, however. As shown below, relatively few message species increase from very low levels in the egg to moderate or high abundance in the embryo.

At the time the measurements shown in Fig. 3.7 were completed, few specific examples of low abundance messages were available, though this class of mRNA occurs ubiquitously in animal cells (Galau *et al.*, 1977b; Davidson and Britten, 1979). Message species now known to be present in only a few copies per cell in given cell types include, for example, those

---

Fig. 3.7. Hybridization of sea urchin embryo RNAs with selected single copy DNA tracers. (a) Reaction of *mDNA* tracer with four different preparations of gastrula polysomal mRNA. *mDNA* was prepared by several cycles of reaction of a single copy DNA tracer with polysomal gastrula message. Initially about 1.1% of the tracer reacted, and thus >50-fold purification of the sequences represented in the gastrula message has been achieved. The line drawn through the data is a least-squares solution assuming pseudo-first-order hybridization kinetics. The different gastrula messenger RNA preparations are indicated (●), (▲), (▼), and (■), and were obtained by puromycin or EDTA release. Also shown (△) is a reaction mixture which had been treated with RNase to destroy DNA-RNA hybrids before assay. It was demonstrated in addition that on isolation by nuclease digestion all the previously hybridized tracer is in fact single copy sequence (see, e.g., Galau *et al.*, 1974). The terminal value for the hybridization shown is 56.9 ± 1.6% (standard deviation). (b) Reaction of *null-mDNA* tracer with gastrula mRNA. *Null mDNA* is the starting single copy tracer fraction remaining after all sequences represented in the polysomal mRNA have been removed by repeated cycles of hybridization with the mRNA. No reaction is evident, as expected. (c) and (d) Hybridization of gastrula mDNA and null mDNA with total egg RNA (▲). The dashed lines of (c) and (d) are the hybridization reaction of gastrula messenger RNA with gastrula mDNA, and null mDNA, respectively, from (a) and (b). The solid lines are the least-squares solutions, assuming pseudo-first-order hybridization kinetics. Their terminal values were 57.5 ± 2.0% (c) and 1.42 ± 0.01% (d). (e) Sets of structural genes active in sea urchin embryos and adult tissues. The *closed* portion of each bar indicates the amount of single copy sequence shared between gastrula stage messenger RNA and the RNA preparations listed along the abscissa. Data were obtained from mDNA reactions such as shown in Fig. 3.7(a) and (c). The *open* bars show the amount of single copy sequence present in the various RNAs studied that are absent from gastrula stage messenger RNA. These data derive from the null mDNA reactions such as shown in Fig. 3.7(b) and (d). Dashed lines indicate the maximum amount of null mDNA reaction which could have been present and escaped detection, in terms of complexity, for cases where no apparent null mDNA reaction was observed. The overall complexity for each RNA is indicated by the total height of each bar. Complexity is calibrated in three ways along the three ordinates shown. From left to right these are: nucleotides of single copy sequence; percent of gastrula stage messenger RNA complexity; and percent of total single copy sequence. [From G. A. Galau, W. H. Klein, M. M. Davis, B. J. Wold, R. J. Britten, and E. H. Davidson (1976). *Cell* **7**, 487. Copyright by M.I.T.]

coding for α-interferon (Mantei and Weissmann, 1982; Ragg and Weissmann, 1983), and dihydrofolate reductase in normal mouse cells (Leys and
Kellems, 1981; Leys *et al.*, 1984). Leys *et al.* (1984) also showed that DHFR
genes may be transcribed at rates that are typical of those shown in Tables
3.1 and 3.3, i.e., one initiation per gene each 50 min. The mRNA for thymidine kinase in chicken cells is similarly rare (Merrill *et al.*, 1984). Maternal
mRNAs for several actin species are also stored in sea urchin eggs at a level
of prevalence equivalent to only about one molecule per average cell-volume
of cytoplasm (Chapter II), and several other examples of similarly rare maternal messages are discussed in Chapter V.

Maternal sequences are very extensively represented in the more prevalent class of sea urchin embryo transcripts as well. This has been established
most clearly in cDNA clone population studies (see, e.g., Lasky *et al.*, 1980;
Flytzanis *et al.*, 1982; Fregien *et al.*, 1983). As noted earlier the prevalence
of transcripts detected in cDNA clone colony screens using labeled cDNA
probes is with conventional procedures ≥10 transcript molecules per cell (or
per late embryo cell-volume of cytoplasm). The vast majority of sequences
of this prevalence class that are detected by colony screening in embryo
cytoplasmic RNA are also found in egg RNA. Figure 3.8, from the study of
Flytzanis *et al.* (1982), demonstrates furthermore that *the abundance of
most such transcripts is the same in the late embryo as in the maternal
RNA.* The dominant form of the population of curves illustrated in Fig. 3.8
suggests partial degradation of the maternal transcripts in early development
and subsequent restoration by zygotic synthesis (cf. Fig. 3.6). Flytzanis *et
al.* (1982) estimated that only 8% or less of (nonmitochondrial) embryonic
transcript species of the moderate and higher abundance classes are absent
from the maternal RNA, or if present, are represented only at very low
levels. Several specific examples of such late genes are discussed below, and
in Chapter IV. The cDNA clone colony measurements shown in Fig. 3.8,
taken together with the results of the RNA excess hybridizations, require the
general conclusion that *about 90% of sea urchin embryo mRNA sequences
at all prevalence levels are represented in maternal RNA.* Yet we know from
the synthesis and turnover measurements reviewed above that by midembryogenesis these transcripts are mainly if not exclusively the newly synthesized products of zygotically active genes. It follows that about 90% of
zygotically active genes are genes that were also functional in oogenesis, just
as concluded on genetic grounds for *Drosophila* and *Caenorhabditis* (cf.
Chapter II).

Essentially the same result has been derived by a completely independent
route, two-dimensional electrophoretic analysis of protein synthesis patterns during development. Thus about 80% of approximately 900 individual
polypeptide species resolved in a study of *in vivo* protein synthesis patterns
are translated at about the same rates in 2-cell embryos and in late blastula or
pluteus stage embryos (Bédard and Brandhorst, 1983). The rates of synthesis

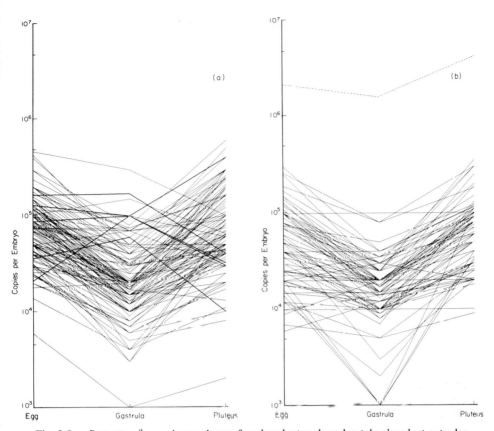

Fig. 3.8 Patterns of prevalence change for abundant and moderately abundant cytoplasmic transcripts in the sea urchin embryo. The approximate transcript prevalence per egg or embryo was estimated from the amount of ^{32}P-cDNA hybridized to matrices containing libraries of cDNA clones selected only for the property that they are represented by moderately prevalent transcripts (≥ 10 copies per average cell at late embryo stages). Data shown are derived from experiments in which ^{32}P-cDNAs synthesized from poly(A) RNA of egg, gastrula and pluteus stage embryos were reacted with the same matrices, and only clones for which complete data were available are included. Results are expressed as poly(A) RNA transcripts per egg, per gastrula (cytoplasmic compartment), or per pluteus (cytoplasmic compartment). The prevalence values were calculated from several observations on each clone at each stage, by reference to standards of known prevalence present on the same filter matrices. In (a) light solid lines represent those clones displaying the "typical" developmental pattern, i.e., more complementary transcripts in egg and pluteus than in gastrula, and the heavier lines represent those few clones which display other patterns. Prevalence patterns shown with heavy lines were those in which *either* the gastrula value was ≥ 3 times the egg value, *or* the pluteus value was 1/3rd of the gastrula value. Since only three developmental points were investigated the lines shown are not expected to represent the actual kinetics of abundance change, but are used only to connect the values obtained at the measured stages. (a) 116 clones from a two-cell cDNA clone sublibrary; (b) 85 clones from a pluteus cDNA clone sublibrary. The dashed line represents the behavior of a mitochondrial rRNA sequence which is the single most prevalent poly(A) RNA species in the embryo cytoplasm (cf. Chapter II). [From C. N. Flytzanis, B. P. Brandhorst, R. J. Britten, and E. H. Davidson (1982). *Dev. Biol.* **91**, 27.]

of only 10% of these proteins appear to alter more than 10-fold during development. Most of the changes that fall in this class consist of the new appearance of particular polypeptides, or of sharp increases in the rates of synthesis of polypeptides translated earlier at very low levels, and most occur between early blastula stage and the onset of gastrulation. Since the proteins synthesized at the 2-cell stage are all the products of maternal mRNA translation (Chapter II), these experiments indicate that the large majority of functional mRNA species are present at similar quantities in midstage embryos and in unfertilized eggs, just as implied by Fig. 3.8. The same conclusions follow from two-dimensional analysis of the populations of polypeptides translated *in vitro* from egg and embryo mRNAs (Bédard and Brandhorst, 1986a). The latter study demonstrates directly that the additions to the translation pattern observed during development reflect directly the appearance of new mRNA species, with a few exceptions accounted for by translational selection processes, as noted in Chapter II. However, Grainger *et al.* (1986) found that about 60% of prevalent proteins which after fertilization first appear during late embryogenesis are also synthesized in growing oocytes. Thus even some late genes that do not produce stored maternal mRNA are functional during both oogenesis and embryogenesis.

The large majority of the nonmitochondrial cytoplasmic transcript species found in the sea urchin embryo, both rare and prevalent, are not detectable in the cytoplasm of adult sea urchin tissues, at least those chosen for analysis, *viz.*, coelomocytes, tubefoot, and intestine. This was shown for low abundance species by Galau *et al.* (1976) as illustrated in Fig. 3.7(e), and for more prevalent species in a cDNA clone population study of Xin *et al.* (1982). Most genes active in the sea urchin embryo can therefore not be regarded as "housekeeping" genes, i.e., genes whose activity is ubiquitously required in all cells of the organism. The same conclusion is supported by evidence obtained from genetic analyses of the effects of maternal and zygotic lethals in *Drosophila*, as summarized in Chapter II, and as reviewed in the following section, this conclusion is applicable as well to the set of embryonic genes functional in *Xenopus* embryos.

(ii) Zygotic Gene Sets in Vertebrate Embryos

Comparative analyses of cDNA clone libraries representing moderately prevalent *Xenopus* embryo transcripts have yielded results equivalent to those obtained with sea urchin embryos. Dworkin and Dawid (1980b) showed that >80% of gastrula stage embryo cDNA clones are also represented in egg RNA. *Xenopus* cDNA clones that react with sufficient labeled cDNA to be accessible for study have been estimated to represent transcripts present at 10-50 molecules per cell at the 10^5 cell neurula stage (Dworkin *et al.*, 1984). New polysomal mRNA species appear mainly in the postgastrular periods, as the embryo commences organogenesis. Developmental accumulation patterns for specific unidentified transcripts have been described in several different studies (see, e.g., Golden *et al.*, 1980; Dworkin and Hershey, 1981; Colot and Rosbash, 1982; Sargent and Dawid, 1983; Dworkin *et al.*, 1984), and some known examples of *Xenopus* late genes are discussed below, and in the following Chapter. Many zygotic mRNA sequences that are represented as unfertilized egg transcripts also increase in prevalence (per embryo) after gastrulation, as indicated both by the cDNA clone experiments and by cDNA hybridization kinetic studies of Perlman and Rosbash (1978) and Rosbash (1981). A similar pattern emerges from two-dimensional analyses of protein synthesis pattern in *Xenopus* embryos. Most protein species synthesized prior to the midblastular activation of transcription continue to be translated later on, though as reviewed in Chapter II translation level selection processes are more prominent in *Xenopus* development. After gastrulation many new species of polypeptides appear (see, e.g., Brock and Reeves, 1978; Bravo and Knowland, 1979). Woodland and Ballantine (1980) showed in a species hybrid experiment in which newly synthesized *Xenopus borealis* proteins coded by the paternal nucleus could be recognized against the background of *X. laevis* proteins that synthesis of about half of the paternal species first becomes detectable by or during the gastrula stage, and the remainder during neurula stages.

Though observed relatively rarely, sequences that are present in early *Xenopus* embryos and later disappear have also been noticed, both in cDNA clone studies (see, e.g., Colot and Rosbash, 1982), and at the protein synthesis level (see, e.g., Woodland and Ballantine, 1980). However, about half the cloned embryonic sequences expressed at gastrula stage are not represented in adult *Xenopus* liver (Dworkin and Dawid, 1980b) and it is particularly striking that none of the 30 cloned transcripts represented in gastrular mRNA but not in maternal mRNA are detectable in postmetamorphic tadpoles (Sargent and Dawid, 1983; Jonas *et al.*, 1985). Thus in all of its general features the zygotic gene set of *Xenopus* embryos resembles that of sea urchin and *Drosophila* embryos. However, the available modern evidence for this organism is confined to moderately prevalent sequences, and thus to probably only about 10% of the total diversity of the transcripts utilized during early development.

Less extensive evidence has so far been obtained for mammalian embryos. However, it has been established in many two-dimensional gel electrophoretic analyses of protein synthesis patterns, carried out on mouse, rabbit, and other mammalian embryos, that most of the proteins synthesized in cleavage continue to be produced throughout preimplantation development (reviewed in Sherman, 1979; see also Chapter II). Small sets of new proteins appear at various stages as a result of the transcription of additional zygotic genes. Among these are polypeptides synthesized for the first time at the 2-cell stage, at the 8-cell stage, and following compaction and blastocyst formation.

(iii) Comparative Aspects

We may conclude that the quantitative dominance of maternally active genes within the zygotic gene set is a completely general feature, observed in *Drosophila*, *Caenorhabditis*, sea urchins, *Xenopus* and mammalian embryos. This result is the more convincing because it has been obtained by diverse methods, including genetic analyses; single copy DNA hybridization with egg and embryo RNAs; cDNA clone population studies; and two-dimensional electrophoretic analyses of protein synthesis patterns. Yet the consequence that most zygotic transcription results merely in replacement of homologous maternal transcripts seems counterintuitive. The naive view would suggest that as the biological complexity of the embryo increases so should the complexity of the transcripts utilized for protein synthesis. As Fig. 2.8 and 3.7 show, however, in the sea urchin the overall complexity of polysomal RNA actually decreases slightly during embryonic development, and in all the organisms considered newly appearing mRNA species, i.e., the products of late genes, constitute but a small minority of the total zygotic messages. It is easy to imagine the regionally or temporally specific roles of late gene products, and several illustrative examples are considered in Chapter IV. As shown there, for example, most if not all of the lineage-specific genes that operate in specific embryonic cell types belong to the late gene class. On the other hand, the functions exercised by the *major* class of zygotically expressed genes, i.e., those belonging to the maternal gene set, remain a large unanswered question. An ancillary problem is why a major fraction of this large set of genes that is transcribed in both the oocyte and the embryo, is not expressed (at least at the cytoplasmic level) in adult cells.

A speculative inference suggested by Hough-Evans *et al.* (1977) is that the maternal genes code largely for complex sets of proteins required for the morphogenesis of three-dimensional embryonic structures, and that in early development such proteins are typically translated in advance of the actual assembly of these structures. Synthesis of most embryonic proteins indeed begins in oogenesis, since the majority of maternal mRNA species are *translated* as well as transcribed long before oocyte maturation (see Chapter V). These proteins are further synthesized after fertilization from maternal mRNAs, and later, as the number of zygotic genes increases, they continue to be synthesized on newly transcribed embryo messages. Whatever their actual roles, the provision of most embryonic proteins is thus controlled very differently than in typical terminal differentiation processes, where, when a product is required, the gene that codes for it is transcriptionally activated. It is instructive to consider some prominent known proteins, such as tubulins and actin, that are synthesized both in oogenesis and in early embryogenesis (see, e.g., Grainger *et al.*, 1986), in light of the hypothesis of Hough-Evans *et al.* (1977). These are proteins that though not unique to the embryo obviously participate in the organization of multicomponent structures required for cellular morphogenesis. They display the capacity to self-

assemble into oriented functional complexes in the cytoplasm, when provided with appropriate cytoskeletal anchors and the other components with
which they interact. For example tubulin from any of several diverse
sources, on injection into echinoderm eggs, self-assembles into mitotic asters and spindles (Hamaguchi *et al.*, 1985). Oriented self-assembly appears
to be a general mechanism by which cytoskeletal organization is constructed
in differentiating cells (see, e.g., Lazarides and Moon, 1984). The best ex-
plored example of structural self-assembly in the egg is the formation of
nuclei, or nucleus-like bodies, around DNA injected into *Xenopus* eggs
(Forbes *et al.*, 1983b; Newport *et al.*, 1985). For convenience the eggs utilized for these experiments are treated with cytochalasin, which prevents
cytokinesis but permits continued nuclear division. The assembly of nuclear
structures from preformed cytoplasmic components around exogenous
DNA occurs *in vitro* as well as in injected eggs (Newport *et al.*, 1985; Stick
and Hausen, 1985). This self-assembly process involves an unknown but
clearly not simple array of protein components, and it results in the formation of authentic, functionally competent double-layered nuclear envelopes,
replete with annuli and coated with laminin on the inner surface.

Perhaps the high complexity of the genetic information required for early
embryogenesis, which is mainly a property of the maternal gene set, can be
explained on the basis of the same hypothesis. Thus three-dimensional biological structures often require an unexpectedly large diversity of proteins
for their construction, and hence of genetic information, as illustrated in
several well analyzed examples. Among these are the eukaryotic cell flagel-
lum, which includes at least 120 different proteins (Piperno *et al.*, 1977); the
chorion of the silk moth egg, in which 180 different proteins have been
identified (Regier *et al.*, 1980); the T4 phage coat, which contains over 50
different proteins (reviewed in Wood and Revel, 1976); and the fertilization
envelope of the sea urchin egg, which is assembled extracellularly from at
least seven protein species with the aid of a number of enzymes, all stored
initially in the cortical granules (reviewed in Carroll *et al.*, 1986). Morphogenesis in the early embryo might thus be an enormously complex self-
assembly process requiring the products of the maternal gene set, and organized by spatial cues positioned in the egg cytoplasm or arising through
blastomere-specific late gene functions.

4. ZYGOTIC GENE ACTIVATION: EMBRYONIC EXPRESSION OF HISTONE GENES AND OF SOME OTHER PHYSIOLOGICALLY RESPONSIVE GENES

The perspective of this review now shifts from the level of the embryonic
transcriptional *system* to that of individual genes, or sets of genes. Here
interpretation is aided by knowledge of the developmental functions of the
specific protein products. In this section we consider several sets of genes

that provide proteins required for cell division or growth, or that are induced in response to environmental factors to which the embryo is exposed. Genes that are specifically expressed in given embryonic cell types are considered in the following Chapter.

(i) "Early" and "Late" Histone Genes in the Sea Urchin Embryo

As noted in Chapter II, in the sea urchin embryo histone messages of zygotic origin become dominant as early as the 16-cell stage. During cleavage most of the newly synthesized histone messages are the product of the "early" or α-histone gene set, which also contributes maternal histone mRNA. The synthesis of α-histones increases rapidly to a peak in the early blastula stage, and then declines equally rapidly, whereafter the messages and protein products of the so-called "late" histone genes appear. However, modern measurements have shown that the conventional inference that there is a "switch" from "early" to "late" histone gene expression is partly inaccurate, since as reviewed below, some "late" histone genes begin to be transcribed as early as do the "early" histone genes.

Expression of the α-Histone Gene Set

The α-histone genes are arranged in the genomes of several sea urchin species examined in a single tandemly repeating unit, within which each gene occurs once, in the order H4-H2b-H3-H2a-H1 (reviewed in Hentschel and Birnstiel, 1981). In one species, *Lytechinus pictus*, there is in addition a second nonallelic tandem block of similarly organized histone gene repeats, but containing different spacer DNA sequences (Cohn and Kedes, 1979a,b). Both gene and spacer sequences of several species have been determined [see Schaffner *et al.* (1978) and Busslinger *et al.* (1980) for *Psammechinus miliaris*; Sures *et al.* (1978, 1980) for *S. purpuratus*; Roberts *et al.* (1984) for *L. pictus*]. The total number of genes per haploid genome coding for each early histone species is in *S. purpuratus* about 400 (Weinberg *et al.*, 1975); in *L. pictus* about 470 (Grunstein and Schedl, 1976); and in *P. miliaris* 300-600 (Hentschel and Birnstiel, 1981). The following discussion is focussed on the transcriptional expression of the α-histone genes, rather than on their structure or evolution *per se*, and on the role that the newly synthesized α-histones play in the kinetics of embryonic cell multiplication (see Fig. 3.1).

Histone synthesis was first recognized in sea urchin embryos by Nemer and Lindsay (1969) and Kedes *et al.* (1969). The histones are synthesized on a distinctive class of small polysomes that is prominent during the cleavage-early blastula period of rapid cell division. The identity of the proteins synthesized on these polysomes was initially inferred from their high relative contents of arginine and lysine, and their ultimate nuclear location, as visualized by autoradiography (Kedes *et al.*, 1969). The α-histone mRNAs range in size from 400 to 680 nt (Childs *et al.*, 1979), and they were initially isolated as a rapidly labeled 9S message fraction obtained from the small polysomes. The individual α-histone mRNAs have been identified by cell-free translation and by hybridization with their respective genes (see, e.g., Gross *et al.*, 1976; Childs *et al.*, 1979; see also earlier data reviewed in Davidson, 1976). Newly synthesized messages for the α-histones can be observed among the transcription products of the male pronucleus immediately after fertilization (Poccia *et al.*, 1985), and in the earliest cleavage stage embryos, e.g., in embryos labeled 0-2 hr after fertilization (Childs *et al.*, 1979). However, at this period most histone protein synthesis in the embryo takes place on the maternal mRNAs coding for CS and α-histones (Chapter II). Expression of α-histone genes by the 2-cell stage has also been demonstrated in a species hybrid experiment of Maxson and Egrie (1980) in which the maternal (*S. purpuratus*) and paternal (*L. pictus*) histone mRNAs could be individually distinguished. The parental α-histone gene sets are expressed equally in the hybrid embryos.

Following the 16-cell stage the absolute rate of histone protein synthesis per embryo increases sharply. It has been shown by *in situ* hybridization that during this period all of the cells in the embryo synthesize α-histone mRNAs (Angerer *et al.*, 1985). As illustrated in Fig. 3.9(a) histone biosynthesis occupies an increasing portion of the embryonic translational apparatus as cleavage progresses, accounting for over 30% of the absolute rate of total protein synthesis at the 100-200-cell stage (Moav and Nemer, 1971; Seale and Aronson, 1973; Goustin, 1981). The rise is due mainly to the accumulation of zygotic α-histone message (see below), but there is also an increase in the efficiency with which this message is loaded into polysomes. The fraction of newly synthesized histone mRNA recovered in the polysomes changes from about 20% at the 2-cell stage, to 80% at the 16-cell stage, and it later approaches 100% (Goustin, 1981). By direct measurement of the peptide elongation rate histones are translated at about 0.8 codons sec^{-1} (for *S. purpuratus* at 15°C; Goustin and Wilt, 1982), while calculation from the rate at which newly synthesized histones accumulate suggests a translation rate of about 1.35 codons sec^{-1} (Goustin, 1981). Note that these values are only about half those required to explain the rates of translation of other proteins in the same stage embryos, i.e., about 2 codons sec^{-1} (see Chapter II). At the 100-200-cell stage histones are being synthesized at close to 2 pg min^{-1} per embryo [Fig. 3.9(b)]. This is close to the rate at which DNA is accumulated during the same time period, estimated as about 1.5 pg min^{-1} (Moav and Nemer, 1971; see also Fig. 3.1). The mass of histone in chromatin is about 1.1-1.2 times the mass of DNA, and thus in the sea urchin embryo throughout the period of rapid cell division *new biosynthesis suffices for the provision of histone at just the rate required for assembly with the newly synthesized DNA*. This is in contrast to the mechanism by which histones are supplied in the cleaving *Xenopus* embryo, where as we have seen (Fig. 2.14), the large pool of maternal histone protein, plus the histone protein translated from maternal mRNAs, are utilized for the formation of thousands of cells.

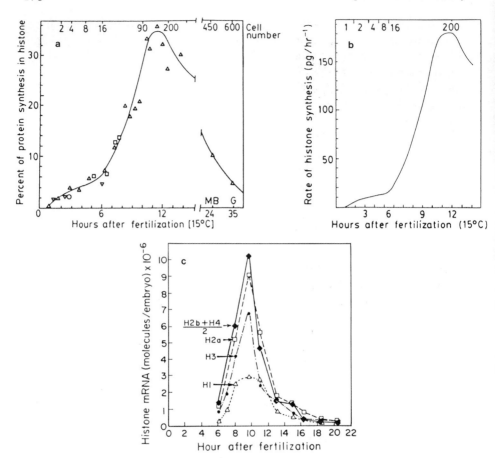

Fig. 3.9. α-Histone protein synthesis and mRNA accumulation in embryos of *Strongylocentrotus purpuratus*. (a) Fraction of total protein synthesis that is histone synthesis. Open triangles facing up represent densitometric measurements of the quantities of newly synthesized histone *mRNA* labeled *in vivo*, resolved by gel electrophoresis and then fluorographed, and expressed as a fraction of total newly synthesized polysomal mRNA. Open triangles facing down represent similar measurements on newly synthesized histone *proteins* labeled *in vivo* with ³H-leucine, and the circles and open squares represent measurements based on the amount of ³H-lysine incorporated into histone (taking into account the 12% lysine content of the histones, compared to 6% for total proteins). MB, mesenchyme blastula; G, gastrula. (b) Absolute rates of histone synthesis (at 15°C). The curve is the product of the function shown in (a) and the total protein synthesis rates measured in embryos of the same stage (data in Goustin and Wilt, 1981; see also Chapter II). [(a) and (b) From A. S. Goustin (1981). *Dev. Biol.* **87**, 163.] (c) Steady state contents of histone mRNAs, measured from the incorporation into each mRNA species of ³H-adenosine, and the ATP pool specific activities (H2b and H4 mRNAs were not separately resolved). [From R. E. Maxson and F. H. Wilt (1982). *Dev. Biol.* **94**, 435.]

For the sea urchin embryo, given the basic rates of transcription and translation, it is evident that the large number of α-histone genes per nucleus is *required* to maintain the absolute rate of cell number increase. Figure 3.9(a)-(b) also demonstrates the precipitous decline in histone synthesis rate in the midblastula stage, as cell cycles lengthen and some lineages cease dividing altogether. Only about 5% of total protein synthesis is devoted to histone synthesis in gastrula stage embryos.

The rise and decline in histone protein synthesis in early sea urchin development reflects directly the changing levels of α-histone mRNAs, as demonstrated in Fig. 3.9(c). This has been measured in several ways. Mauron *et al.* (1982) utilized single-strand probe excess titration (see Appendix II) to measure the steady state content of H3, H2b, and H1 mRNAs during development. The number of mRNA molecules for the two nucleosomal histones increases from about 8×10^5 in the unfertilized egg to 9-10 $\times 10^6$ in the early blastula stage embryo, and then declines to near the original levels. The equivalent values for H1 mRNA are 8×10^4 molecules per egg and 6×10^6 molecules in the blastula. Kinetic estimates of histone mRNA accumulation reported by Maxson and Wilt (1982) agree well with these results, as shown in Fig. 3.9(c). In these experiments the histone mRNAs were extracted from embryos that had been allowed to incorporate ^{3}H-adenosine. The histone mRNAs were separated electrophoretically, so that the mass accumulation of each species could be determined from the ATP pool specific activities. Both these and the steady state measurements show an approximately 10-fold increase in the number of mRNAs for each core nucleosomal α-histone species, comparing the maternal with the peak value at the 100-200-cell stage. The amount of H1 mRNA attains approximately half this peak level. These conclusions were confirmed in a further kinetic study of Weinberg *et al.* (1983).

The abruptness of the decline in histone mRNA content during the later blastula stage is due to the short half-life of the α-histone message. Maxson and Wilt (1982) showed that the half-life of the α-histone mRNAs measured after the 200-cell stage is only 1.5-2 hr. It is probable that this is the consequence of an increase in the decay rate, and that the half-life of the mRNA during the rising phase of the curve shown in Fig. 3.9(c) is at least 3 hr.

The rate of α-histone gene transcription has been measured directly from precursor incorporation kinetics by Maxson and Wilt (1981) and Weinberg *et al.* (1983). Both measurements indicate that on a *per cell* basis these genes are most active at the 16- to 64-cell stages, and that they are 3-6-fold less active at the 200-cell stage. Thereafter, transcription of the α-histone genes appears to be almost completely repressed, though a sensitive measurement of Childs *et al.* (1979) shows that in *S. purpuratus* a barely detectable amount of α-histone message synthesis persists as late as 40 hr postfertilization (late gastrula). Transcription measurements carried out in isolated nuclei demonstrate a high relative rate of α-histone gene transcription at the 4-cell stage, about a 2-fold lower rate per nucleus at the 100-cell stage, and no detectable activity in blastula and gastrula stage nuclei (Uzman and Wilt, 1984). At the maximum rate of transcription measured *in vivo* about 0.5 (Maxson and Wilt, 1981) to 1 (Weinberg *et al.*, 1983) molecules of histone mRNA are synthesized per gene-minute, assuming that all the histone genes are active. This may be compared to the value measured for the Spec1 genes, for example, per cell about 1.8 molecules min^{-1} or per gene, 0.9 molecules min^{-1} [Fig. 3.6(f)-(h)]. These rates are severalfold below the theoretically possible rate of transcript production, assuming 100 nt polymerase spacing and the 6-9 nt sec^{-1} translocation rate for these sea urchin embryos measured by Aronson and Chen (1977), and thus it may be assumed that the key limiting parameter is the frequency of initiation. Note, however, that the α-histone and Spec1 genes are significantly more active than the *average* gene from which moderately prevalent mRNAs arise, as calculated in Table 3.3. (i.e., $1/5.6$ min molecules^{-1} = 0.18 molecules min^{-1}).

Though they are oriented similarly, the clustered α-histone genes are transcribed monocistronically in the embryo (Kunkel *et al.*, 1978; Mauron *et al.*, 1981), and not as a single long precursor. Initiation occurs separately on each gene. Regions of 5' upstream sequence in each gene that are required for transcription are marked by DNase hypersensitive sites at developmental stages when these genes are active (Bryan *et al.*, 1983). Significant sequence elements have been identified by their conservation among histone genes in 5' regions that are otherwise divergent, and also by direct functional tests. Histone genes, like many others, are transcribed on injection into the *Xenopus* oocyte nucleus (Probst *et al.*, 1979; Etkin and Maxson, 1980), and by using this as an assay system several essential elements have been identified. These include the standard sequences often found associated with genes serviced by polymerase II, such as, TATAA, and CAAT "boxes" (Busslinger *et al.*, 1980; Grosschedl and Birnstiel, 1980a; Hentschel and Birnstiel, 1981). The most interesting discovery of this nature is that of an upstream spacer sequence element in the *P. miliaris* α-H2a histone genes, located between -111 and -165 nt with respect to the cap site. Deletion of this element or mutations within it cause large decreases in the efficiency of transcription after injection into the *Xenopus* oocyte nucleus. Two properties of this element indicate that it probably functions as a specific enhancer (Grosschedl and Birnstiel, 1980b; Grosschedl *et al.*, 1983). This sequence can act on the H2a gene to stimulate transcription in inverted as well as direct orientation, and it includes a $14/17$ nt homology with retroviral enhancer sequences. In addition Mous *et al.* (1985) reported evidence for two internal sites at which bind protein factors isolated from sea urchin embryo chromatin that stimulate transcription of an α-H2b gene 5-10-fold on injection into the *Xenopus* oocyte. One of these sequences resides within the region copied into the 5' leader of the message, and the second extends in a 3' direction from a site about 90 nt prior to the translation termination signal.

The simplest image of the process by which transcription of the early histone gene set is controlled is thus that maternal *trans*-acting factors are present in the egg, or are released or synthesized at the beginning of development, interactions with which lead to activation of these genes. The *Xenopus* oocyte nucleus may contain a weakly homologous factor, though the rate of transcription of the injected genes is only about 1% of that in the sea urchin embryo, i.e., per gene, about 1 molecule hr^{-1} (Probst *et al.*, 1979; Etkin and Maxson, 1980). During cleavage the newly replicated α-histone genes continue to become activated, but there is a gradual decline in their initiation frequency that might be due to depletion of the *trans*-acting regulatory agent. To explain the essential silencing of these genes in the later embryonic stages, it is necessary to assume further that at least a significant fraction of this regulatory agent is ultimately inactivated or trapped elsewhere. There is some experimental support for this scenario. When cloned α-histone genes are injected into sea urchin eggs they are expressed specifically in the early embryo. Thus the rate of expression is high in late cleavage, and falls sharply thereafter (Vitelli *et al.*, 1986). Furthermore, injected sperm histone genes, which are normally not expressed in the embryo, remain silent (Vitelli *et al.*, 1986). These results argue directly for the presence of a histone gene *trans*-activator that decreases to ineffective levels after the blastula stage.

Specific sequences are also required for correct 3′ processing of the primary α-histone gene transcripts. The 3′ termini of the mature α-histone mRNAs (which in the sea urchin are not polyadenylated) are produced by the endonucleolytic cleavage of longer primary transcripts (Krieg and Melton, 1984). Significant sequence homologies have been discovered in comparing the regions downstream from the mRNA termination sites of the various histone genes (reviewed in Birnstiel *et al.*, 1985). For correct processing of the primary transcripts produced by sea urchin histone genes injected into the *Xenopus* oocyte nucleus, or of the precursor RNAs transcribed from the genes *in vitro* and then injected into the oocyte nucleus, a palindromic sequence immediately preceding the mRNA terminus plus about 80 nt of adjacent spacer sequence are required (Birchmeier *et al.*, 1982, 1983, 1984). Perhaps the most interesting aspect of this discovery is the identification of another *trans*-acting regulatory macromolecule, this one, a small ($\sim$57 nt) nuclear RNA, called U7 RNA. While *Xenopus* oocytes correctly process other *P. miliaris* histone primary transcripts, they fail to terminate α-H3 primary transcripts. Correct terminal processing of this pre-mRNA can be induced, however, by prior injection of sea urchin U7 RNA (Galli *et al.*, 1983; Birnstiel *et al.*, 1985), which is present *in vivo* as an snRNP complex. The sequence of U7 RNA suggests a complementary intermolecular RNA·RNA interaction with those pre-mRNA sequences that are already known to be essential for processing, as illustrated in Fig. 3.10 (Strub *et al.*, 1984). *Trans*-acting, RNA regulatory factors that act specifically by means of complementarity with the target RNA imply that there are mechanisms for *posttranscriptional coordination of the expression of sets of*

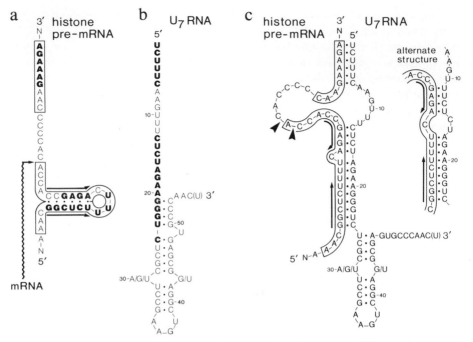

Fig. 3.10. A model for the interaction of U7 RNA with histone pre-mRNA. (a) Sequence of precursor RNA flanking the 3' end of H3 mRNA. (b) Predicted U7 RNA secondary structure. (c) Possible interaction between H3 pre-mRNA and U7 RNA. Nucleotides interacting with each other are written in boldface type. The conserved sequences also found in the other histone genes are boxed. Large arrows mark the 3' end of the mature histone mRNA, the thinner arrows indicate the nucleotides involved in the palindrome. These sequences are visualized to exist in alternative configuration, as an *intrastrand* duplex in the absence of U7 RNA (inefficient processing) and as an *interstrand* duplex in its presence (efficient processing). [From K. Strub, G. Galli, M. Busslinger, and M. L. Birnstiel (1984). *EMBO J.* **3**, 2801.]

genes, here illustrated by the diverse histone genes. As pointed out by Birchmeier *et al.* (1982), the diverse α-histone genes share almost perfectly the palindromic sequence marked by the arrows in Fig. 3.10(a).

The Late Histone Genes of Sea Urchin Embryos and Some Comparisons with Regard to Histone Gene Expression in Other Embryos

The late sea urchin histone subtypes were initially noticed as electrophoretic variants in extracts made from advanced embryos (Ruderman and Gross, 1974; Cohen *et al.*, 1975; Arceci *et al.*, 1976; Newrock *et al.*, 1978a,b). The changing pattern of histone synthesis is illustrated in Fig. 3.11(a), and it has been demonstrated by *in vitro* translation of embryo mRNAs that at each stage the embryo contains the messages coding for the specific variants observed to be synthesized *in vivo* at that stage (Newrock *et al.*, 1978b; Childs *et al.*, 1979; Hieter *et al.*, 1979). One such translation experiment is reproduced in Fig. 3.11(b). The mRNAs that code for the late histones can be distinguished by length from those coding for the α-histones (Grunstein, 1978; Childs et al., 1979), due mainly to differences in the 5' leader and 3' trailer sequences (Childs et al., 1982; Roberts et al., 1984; Busslinger and Barberis, 1985; Mohun et al., 1985). They also differ internally from the cognate early histone mRNAs in nucleotide sequence by an average of 9-14%, depending on the specific comparison. This has been shown most generally by the lowered thermal stability of mRNA·DNA hybrids formed between late histone messages and α-histone genes, compared to the respective homologous hybrids (Kunkel and Weinberg, 1978; Hieter et al., 1979). Primary sequences available for several late histone genes of P. miliaris and L. pictus (Childs et al., 1982; Roberts et al., 1984; Busslinger and Barberis, 1985) confirm the general occurrence of silent base changes within the coding regions, i.e., with respect to the α-histone genes. These studies also show that individual late histone gene sequences may vary from each other as much as they vary from the α-histone sequences. However, the proteins for which the sequenced late H3 and H4 mRNAs code are identical to those coded by the early H3 and H4 messages (Roberts et al., 1984). This accounts for the lack of detectable distinction between the H3 and H4 subtypes observed early and late in development [e.g., Fig. 3.11(b)]. There are significant differences between the amino acid sequences of two late H2b variants of P. miliaris and between either of these and the sequence of the α-H2b protein, and the same is true for three different H2a variants (Busslinger and Barberis, 1985). Most, though not all, of these differences are located at the N- and C-terminal regions of these proteins. The core regions appear by contrast highly conserved [for detailed consideration of the homologies and differences between α-histones, late histones, and the sperm histones, the reader is referred to a review of von Holt et al. (1984) and to studies of Newrock et al. (1982) and Pehrson and Cohen (1984)].

The late histone genes are not arrayed in tandem clusters as are the α-histone genes, but rather occur singly or in small sets that are dispersed and irregularly organized. Estimates from genome blots suggest that in L. pictus there are only 8-10 genes of each late histone subtype, i.e., about 2% of the number of α-histone genes (Childs et al., 1982), and 6-12 copies of the genes coding for each late histone subtype in S. purpuratus (Maxson et al., 1983). Experiments carried out with various cloned late histone gene probes have revealed complex, species-specific patterns of late histone mRNA accumulation, though the messages deriving from closely linked genes seem to appear coordinately. This has been reported for two cloned late histone gene pairs, one containing a late H2b and a late H4 gene, and the other a late H2a and a late H2b gene (Mohun et al., 1985). A major finding is that contrary to the implication of their nomenclature, the "late" histone genes are in fact expressed from the very beginning of development. Unfertilized L. pictus eggs contain about 10^4 molecules of late H3 and H4 mRNAs (Knowles and Childs, 1984), and unfertilized P. miliaris eggs about 200 molecules of late

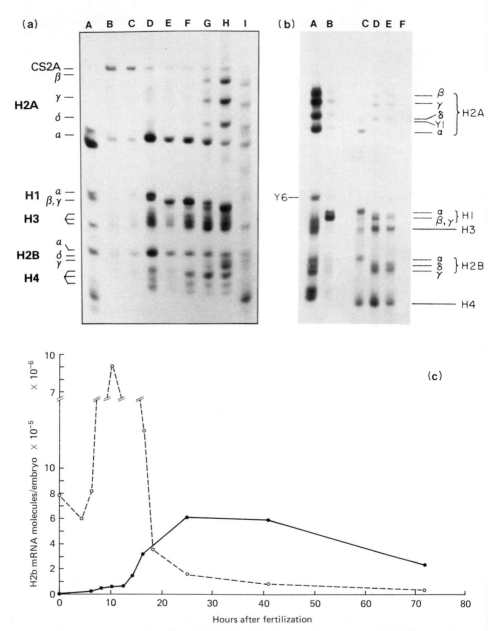

Fig. 3.11. Expression of different histone subtypes during sea urchin development. (a) and (b) Separation of *S. purpuratus* histone subtypes by gel electrophoresis in a high resolution Triton-acid-urea system. (a) Histones synthesized *in vivo*. Embryos were labeled with ³H-leucine for one hour at the indicated times. Lane A, *in vitro* translation products, 9 hr embryo mRNA; lane B, 0–100 min (cleavage); lane C, 3-4 hr (cleavage); lane D, 9–10 hr (early blastula); lanes E and F, 14½-15½ hr (blastula); lane G, 18-19 hr (hatched blastula); lane H, 23-24

H2a and H2b mRNAs (Busslinger and Barberis, 1985). Late H2b messages have also been detected in *S. purpuratus* eggs (Mohun *et al.*, 1985). Thus it is clear that at least some of the late histone genes are productively transcribed even during oogenesis, and *per gene* the late H3/H4 genes of *L. pictus* would appear to be represented by a similar number of maternal mRNAs as are the α-H3 and α-H4 genes. During cleavage and blastula formation in all three sea urchin species, the concentration of some late histone gene transcripts increases sharply, as illustrated for the H3 and H4 transcripts of *S. purpuratus* in Fig. 3.11(c). There is probably an increase in the rate of late histone gene transcription, i.e., per gene at this stage (Maxson *et al.*, 1983; Ito *et al.*, 1986), though a large portion of the rise in transcript concentration can be accounted for as the product of increase in the number of active genomes due to cell division (cf. Fig. 3.6). The late histone mRNAs attain a steady state level estimated for *S. purpuratus* at about 6×10^5 molecules per embryo at the late blastula stage (Ito *et al.*, 1986). The messages are maintained at this level for some hours, and then decline in abundance in the pluteus stage embryo. In *P. miliaris* the peak late histone mRNA level is similar, about $3\text{-}5 \times 10^5$ molecules of each histone mRNA species (Busslinger and Barberis, 1985). At the peak rate each late H3 and H4 gene is estimated from the data in Fig. 3.11(c) to produce about 0.8 stable transcripts per minute, a rate comparable to the peak rate for the α-histone genes. However, the stability of the late histone mRNAs appears to be greater than for the α-histone mRNAs (Knowles and Childs, 1984; Ito *et al.*, 1986).

The "late" histone genes are thus improperly named, since their activation in fact occurs very early in development, and they attain their maximum rate of transcription during the early blastula stage, if not before. The distinction in the pattern of expression of late and early histone genes lies more in the *cessation of α-histone gene transcription*, than in the developmen-

---

hr (mesenchyme blastula); and lane I, *in vitro* translation products of 21 hr embryo mRNA. The multiple H3 and H4 bands indicated correspond to the nonacetylated, monoacetylated and diacetylated forms of the proteins. The late histone species are designated by the letters β, δ and γ. CS2a, cleavage stage histone 2a. [Courtesy of P. Kuwabara and E. Weinberg.] (b) Histones synthesized *in vitro* in a wheat germ, cell-free translation system. Lane A, ^{3}H-leucine labeled histones made by 20-24 hr embryos *in vivo*; lane B, same except 5% perchloric acid soluble fraction of 20-24 hr embryo histones, i.e., mainly H1 histones; lane C, histones synthesized *in vitro* from mRNA extracted from small (8-14S) polysomes of 17 hr blastulae; lane D, from RNA of 26 hr mesenchyme blastulae; lane E, from mRNA of 36 hr gastrulae; and lane F, endogenous wheat germ proteins, no RNA added. Y6 is an unknown protein. Note that single forms of H3 and H4 are synthesized *in vitro* at all stages. (b) [From K. M. Newrock, L. H. Cohen, M. B. Hendricks, R. J. Donnelly, and E. S. Weinberg (1978b). *Cell* **14**, 327. Copyright by M.I.T.] (c) Accumulation of late H2b mRNA in *S. purpuratus* embryos (●). For comparison the accumulation of α-H2b mRNA is also shown (○) [see Fig. 3.9(c)]. An absolute value at the late pluteus stage was obtained by single-strand probe excess titration (see Appendix II), and this was used to convert relative measurements based on blot hybridizations to absolute estimates for the other stages. [Data from M. Ito, G. Lyons, and R. Maxson (1986). In preparation.]

tal kinetics of their respective transcriptional activations. The late histone genes continue transcription after the α-histone genes have become almost completely silent, as shown both by direct identification of newly synthesized histone mRNAs *in vivo* (see, e.g., Childs *et al.*, 1979) and by nuclear transcription *in vitro* (Knowles and Childs, 1984). Both histone gene sets function autonomously on schedule in cultures of disaggregated sea urchin embryo blastomeres (Arceci and Gross, 1980), and in embryos in which the cell cycle has been disrupted by polyspermy or drug treatment (Harrison and Wilt, 1982). Furthermore, an *in situ* hybridization study of Angerer *et al.* (1985) in which early and late transcripts could be distinguished, shows that prior to the cessation of early gene transcription all mitotically active embryonic cells produce both sets of histone message. Subsequently, they produce only late histone messages, except for those cell clusters that have withdrawn from the mitotic cycle and lack any histone mRNA. There is no evidence in this investigation for specific accumulation of late histone mRNA with respect to early histone mRNA in any particular cell lineage. In its major features the pattern of expression displayed during sea urchin development by early and late histone gene sets is analogous to that observed for oocyte and somatic sets of 5S rRNA genes in *Xenopus*. As described in Chapter V, the tandemly arrayed oocyte type 5S rRNA genes are active during oogenesis and repressed after the midblastula stage of embryogenesis, while expression of the much smaller set of somatic 5S rRNA genes continues. Somatic 5S rRNAs appear to have a higher affinity for the same *trans*-activator than do the oocyte type 5S rRNA genes, and an analogous explanation for the extended transcription of the late sea urchin histone genes may be appropriate. Maxson *et al.* (1986) demonstrated that a factor that stimulates transcription of sea urchin histone genes injected into *Xenopus* oocyte nuclei can be extracted from chromatin of late sea urchin embryos. The chromatin fraction is injected into the oocyte cytoplasm, and presumably is translocated into the germinal vesicle. This factor increases the rate of transcription of a late H2b gene about 8-fold, and the rate of transcription of an a-H2b gene in the same nuclei only about 2-fold.

A simpler picture thus emerges. The late histone gene set of the sea urchin embryo may be considered the somatic histone gene set. As are the histone genes active in most animal cells, these genes are directly subject to cell cycle regulation. The late H1 histone subtypes utilized in the embryo are in fact also found in several adult cell types (Pehrson and Cohen, 1984). Halsell *et al.* (1986) have also demonstrated late H2b and H2a mRNAs in *S. purpuratus* tubefoot, intestine and other adult cell types, while the α-H2b and α-H2a mRNAs could not be detected. During late embryogenesis the declining relative prevalence of the histone mRNAs [e.g., Fig. 3.11(c)] simply reflects the declining rate of cell division, so that in cells that have ceased dividing the late histone mRNA also disappears. Superimposed on the *cell cycle regulation* of these somatic histone genes is *ontogenic* regulation of the α-histone gene set, which can be regarded as a special genomic device utilized as a "booster" during the few hours of rapid logarithmic cell number increase. These genes then remain silent for the rest of the life cycle (except in oocytes). The unusual nature of this feature of zygotic transcriptional function in the sea urchin embryo can be demonstrated comparatively. Thus neither *Xenopus* nor *Drosophila*, both of which undergo a period of extremely rapid embryonic cell division, utilizes a special set of "early" histone genes. As reviewed in Chapter II, in both species the chromatin histones required during the rapid cell divisions of cleavage and blastula formation are derived from maternal stores of mRNA and from preformed histone protein. In *Xenopus*, the haploid genome of which contains about 40 histone genes of each type arranged in two separate clusters (Perry *et al.*, 1985), zygotic expression of these genes begins only after about 5000 cells already have formed. There are about 100 histone genes of each histone species in the *Drosophila* genome (Lifton *et al.*, 1978), and these begin transcription only at the syncytial blastoderm stage. During the last four blastular divisions and the more leisurely paced cell multiplications during and after gastrulation, the histone genes of the *Drosophila* embryo function just as in any dividing somatic cell, providing sufficient mRNA so that histone translation can keep pace with the accumulation of newly synthesized DNA (Anderson and Lengyel, 1984). This is the role also played by histone genes in late *Xenopus* and sea urchin embryos and in the mouse embryo after the 2-cell stage. The measurements of Giebelhaus *et al.* (1983) and Graves *et al.* (1985) demonstrate that after the disappearance of most maternal histone mRNA in the late 2-cell stage mouse embryo the histone mRNA content is maintained by new synthesis, at about 2×10^4 molecules per cell, a level similar to that observed in rapidly dividing cultured mouse somatic cells. In *Drosophila*, mouse, and *Xenopus* embryos, as in the sea urchin embryo, rapid turnover of histone mRNA is an essential feature of the regulation of histone translation. Thus a half-life of only a few hours is required to account not only for the decay of maternal histone messages (Chapter II) but also for the decay of zygotic histone messages when the rate of cell division falls, e.g., in blastula [Fig. 3.9 (c)] and postgastrula stage [Fig. 3.11 (c)] sea urchin embryos, or postgastrula stage *Drosophila* embryos (Anderson and Lengyel, 1984; Winkles and Grainger, 1985a).

A question emphasized by comparative considerations of embryonic histone gene expression is the functional significance of the developmental succession of histone subtypes. This has been observed not only in the sea urchin but also in the mouse (Graves *et al.*, 1985), and in several other organisms, including *Ilyanassa* and *Urechis* (Chapter II). Von Holt *et al.* (1984) have pointed out that the proteins coded by the late H2b and H2a sea urchin embryo variants display different degrees of acetylation. However, this could be due primarily to the different embryonic stages at which these variants appear rather than to the subtypes *per se*. It cannot be excluded that there is no *functional* distinction between early and late embryonic histone subtypes. The fact that there are distinct early and late H3 and H4 mRNAs in sea urchin embryos, deriving from distinct sets of genes, but coding for exactly the same proteins, suggests that there must be another explanation for the existence of these different sets of genes than the production of functionally distinct histones. Similarly, as noted in Chapter II, in neither *Xenopus* nor *Drosophila* is there variation in the subtypes of at least some histone species at any time in embryonic development. Perhaps the main significance of early and late histone genes, where these are observed, is that these gene sets are *regulated* differently. It is clear from studies of the organization of these genes in the sea urchin that histone genes regulated in concert are tandemly arrayed or closely linked, while genes that display independent patterns of expression are dispersed in the genome. The nucleotide divergence among these genes indicates that they have evolved in different ways, subject to the strong selective forces that account for the extreme conservation of the core histone protein sequences. It remains to be seen whether the sequence differences in the early and late histone proteins of the sea urchin indicate a stage-specific functional specialization. These differences could merely be the neutral consequences of a separate evolutionary history, that has followed in turn from the separation of early and late histone genes into flanking sequence environments that respond to different regulatory signals.

(ii) Growth Related Genes and Physiologically Induced Genes in Embryonic Systems

We have seen that by late embryonic stages the zygotic histone genes of all species are functioning according to cell cycle cues, as in adult somatic cells, while at the beginning of development histone genes are transcribed in a nondividing cell, the oocyte (Chapter V), and after fertilization the maternal histone mRNA is translated without regard to the cell cycle (Chapter II). This example provides an interesting paradigm for the general class of embryonically active genes that in somatic cells are *inducible*, either by internal signals such as the physiological cues that coordinate cell growth and mitosis, or by external molecules or other environmental changes. The basic rates of gene transcription are such that until the nucleocytoplasmic ratio approaches that of adult somatic cells, transcriptional *induction* is an ineffective device for low copy number genes. While this stage occurs early in cleavage in mammalian embryos, in all other forms there is a significant phase of embryogenesis when by the standards of somatic cells there is a relatively huge amount of cytoplasm for the number of nuclei (and genes) present. During this phase products that are supplied by inductive transcription in somatic cells are supplied from maternal sources in the embryo, if at all. Their provision is thus the result of an ontogenic form of gene regulation, e.g., the switching on of a gene in the course of oogenesis, rather than of a physiologically inductive form of gene regulation. In some cases certain members of given gene families function according to an ontogenic program, while the expression of others is modulated physiologically. Thus as noted in Chapter II certain specific tubulin genes are expressed only during oogenesis in the sea urchin and in *Drosophila*, and in vertebrates the oocyte type 5S rRNA genes are ontogenically activated during oogenesis (Chapter V), while later in the life cycle the somatic 5S rRNA genes are regulated as new ribosomes are required, i.e., they function under cellular growth control. Inducible somatic genes may indeed function the same way in "early" embryos as in later cells, though to little effect, as exemplified by the nearly invisible results of late histone gene expression in cleavage stage sea urchin embryo cells, or the ribosomal genes of early (nonmammalian) embryos, which account for synthesis of insignificant quantities of rRNA compared to the extant maternal rRNA pool (Table 3.4).

Ribosomal Proteins

A further example is provided by the ribosomal protein genes, the activity of which has been studied in *Xenopus* embryos, mouse embryos, and *Drosophila* embryos. In *Xenopus* these genes are actively transcribed during oogenesis, and the ribosomal proteins (rP) are assembled into the vast store of ribosomes with which the mature egg is supplied (see Chapter V). The maternal complement of rP's suffices for much of embryonic development, and for most rP species no maternal mRNA survives beyond very early cleavage (Pierandrei-Amaldi *et al.*, 1982). There are, however, four rP species that continue to be translated from maternal mRNA prior to the activation of transcription, including rRNA transcription, at the midblastula stage (the "early" rP species). Newly synthesized rP mRNA for all rP species accumulates after the gastrula stage (stage 10) (Weiss *et al.*, 1981; Pierandrei-Amaldi *et al.*, 1982), but with the exception of the four "early" rP's, which are translated continuously, the zygotic rP message is not assembled into polysomes or translated until the tailbud stage of development (stages 26-32). The translation though not the transcription of most rP messages is depressed in anucleolate mutant embryos that are unable to synthesize rRNA, probably through a mechanism that decreases the stability and hence the concentration of the rP mRNAs (Hallberg and Brown, 1969; Pierandrei-Amaldi *et al.*, 1982, 1985). On the other hand, just as the mRNAs for the "early" rP's are immune to the translation level selection process that in normal embryos accounts for the delayed loading of most zygotic rP mRNAs, the "early" rP mRNAs also continue to be translated in *o-nu* embryos. Thus, to summarize, after the tailbud stage rP gene transcription and rP translation follow the normal somatic pattern, keeping pace with the demand for new ribosomes as the embryo increases in cell number and total

RNA content. Earlier two *special* patterns of regulation are observed. All of the rP genes are ontogenically activated at stage 10 when there is but a trickle of rRNA synthesis (see Table 3.4), and the early rP's are immediately translated, while translation of most rP's is delayed, and depends on coincident rRNA synthesis.

In postblastoderm *Drosophila* embryos synthesis of rP's is also linked to rRNA synthesis. Thus the amount of rP's produced between 1 and 5 hr postoviposition is about 2 ng, while the amount of mature rRNA synthesized over the same period is about 2.6 ng, close to the ratio required for assembly of ribosomes in somatic cells (Santon and Pellegrini, 1981). As ribosome production declines after gastrulation the rate of rP synthesis also falls. However in the preblastoderm embryo, when there is no rRNA synthesis, rP production is controlled in a different way. In these embryos translation of rP's from maternal mRNAs accounts for 4.5% of total protein synthesis (compared to about 8.9% at its peak in the late blastula stage), and the rP's are either incorporated by exchange into preexisting ribosomes or assembled with preexisting rRNA to form new ribosomes (Santon and Pellegrini, 1980, 1981). In the early embryo there appears to be a selective discrimination against assembly of rP mRNA into polysomes, compared to postgastrular embryos (Al-Atia *et al.*, 1985). A contrast in the control of rP synthesis early and late in development is also observed in the mouse embryo. During cleavage, after the earliest stages, the rate of rP synthesis is coordinated with the rate of rRNA synthesis. Translation of rP's from zygotic messages is a major synthetic activity, and in the 8-cell compacted embryo, for example, it accounts for about 8% of total protein synthesis (LaMarca and Wassarman, 1979). However, in unfertilized eggs where there is no rRNA synthesis, rP's are nonetheless actively synthesized from stored maternal messages, as discussed in Chapter II. The rate of this synthesis, about 0.5 pg hr^{-1}, is about the same as the rate *per cell* in the 8-cell embryo.

Environmentally Inducible Genes

Two sets of genes that respond inductively to changing environmental circumstances have been investigated in embryos. These are the genes coding for heat shock proteins and for metallothionein, a protein utilized to chelate toxic heavy metals. Synthesis of heat shock proteins (hsp) in response to thermal stress or to other forms of physiological insult, e.g., poisoning with Na arsenite, is an ubiquitous capacity of somatic cells that is regulated at the transcriptional level (Chapter I; reviewed in Schlesinger *et al.*, 1982). However, *early embryos*, of all species studied, lack this response, and in consequence are thermosensitive, relative to late embryos (reviewed in Heikkila *et al.*, 1985a). Messenger RNA for hsp's first becomes inducible in response to heat treatment in early or late blastoderm stage

Drosophila embryos (Dura, 1981); in early or late blastula stage sea urchin embryos, depending on species (Roccheri *et al.*, 1981, 1986; Howlett *et al.*, 1983); in late blastula stage *Xenopus* embryos (Heikkila *et al.*, 1985b); and in blastocyst stage mouse and rabbit embryos (Heikkila *et al.*, 1985a). Thereafter the somatic patterns of response to heat exposure are observed. Thus the appearance of hsp's in late embryos reflects directly the inductive synthesis and accumulation of the respective messages in the stressed cells. In *Xenopus* embryos the hsp70 gene becomes inducible well before the hsp30 gene, which does not respond to heat exposure until the tadpole stage (Bienz, 1984). However, the same heat shock genes are also responsive to a completely different set of *ontogenic* regulatory controls. The *Xenopus* hsp70 gene is transcribed during oogenesis, and at this time it is not heat inducible (Bienz and Gurdon, 1982; see also Chapter V). Protection against heat stress in oocytes is instead accomplished by translational activation of the stored mRNA (Bienz and Gurdon, 1982). This mechanism is not available to the embryo, as the hsp70 mRNA is degraded after maturation, though it is apparently still present up to germinal vesicle breakdown (Baltus and Hanocq-Quertier, 1985). Hsp70 mRNA is not found at all in cleavage stage embryos (Bienz, 1984; Heikkila *et al.*, 1985a). The hsp30 gene displays a different ontogenic pattern of expression, as it is not expressed in the oocyte, with or without heat treatment. Similarly, in *Drosophila* the mRNAs for hsp26, hsp28 and hsp83 are synthesized in nurse cells during oogenesis in the absence of thermal stress, and these messages persist as maternal mRNAs in preblastoderm embryos though the mRNA for hsp70 is not found in the early embryo (Zimmerman *et al.*, 1983). Furthermore, the specific *cis*-regulatory sequences responsible for expression of the hsp26 gene in the ovary are physically distinct from those responsible for heat induced expression (Cohen and Meselson, 1985). Some hsp messages are also produced in nonstressed larvae and pupae (Sirotkin and Davidson, 1982). As already discussed in Chapter II production of the hsp90 protein during cleavage is regulated translationally in normal nonstressed *S. purpuratus* embryos. In other sea urchin species the same gene appears to be transcriptionally activated early in development in nonstressed embryos, perhaps in concert with other genes whose products are required for cell division (Bédard and Brandhorst, 1986b). Thus prior to the appearance of the induction system that controls heat shock gene expression in most somatic cells (or of sufficient nuclei so that induction can be effective), a variety of ontogenic patterns of heat shock gene expression are observed. Inductive control of heat shock gene expression occurs in late sea urchin embryos, as in adult somatic cells of all species (Chapter I), by means of interaction with *trans*-acting regulatory agents (McMahon *et al.*, 1984). Possibly the onset of stress inducibility follows the appearance of the appropriate intracellular signaling systems. Whatever the mechanism, it develops autonomously, at least in sea urchin cells. Thus Sconzo *et al.* (1983) showed that the ability to mount a molecular heat stress response occurs on the same schedule in disaggregated cultures of dividing blastomeres as in normal control embryos.

A detoxification function carried out by inducible cytochrome P-450 monooxygenases provides another instance. The metabolic products of benzo(α)pyrene formed by this enzyme causes sister chromatid exchanges in both extraembryonic and embryonic regions of postimplantation mouse embryos (Pedersen *et al.*, 1985). However this enzymatic activity is not present in morula or early blastocyst stage embryos, appearing first in late preimplantation blastocysts.

An interesting additional example is found in the regulation of the metallothionein (MT) genes during early sea urchin development. The expression of some of these genes has been studied by use of cloned cDNA probes by Nemer *et al.* (1984, 1985). MT mRNA is a relatively prevalent maternal message in *S. purpuratus* eggs, and the abundance of this mRNA remains essentially constant throughout cleavage and into the early blastula stage (9-12 hr). No change is observed in MT mRNA prevalence on presentation of heavy metals during this period. At the blastula stage the amount of MT mRNA begins to increase in untreated embryos, and as shown in Fig. 3.12(a) addition of Zn^{2+} to the medium from this time on results in an inductive increase in the levels of MT mRNA, as well as of MT protein. In later development ectoderm cells, which of course are continuously in contact with the sea water, produce a high level of MT mRNA that cannot be further augmented by exposure to Zn^{2+}, except for such prolonged exposure as to ''animalize'' the embryos, thereby increasing the fraction of ectoderm. In contrast, endoderm cells of normal embryos produce little MT mRNA, but on exposure to Zn^{2+} they display a 10-fold inductive response, as illustrated in Fig. 3.12(b). The MT_a metallothionein gene is responsible for the ectoderm Zn^{2+} binding protein while another gene, MT_b, is activated on induction within endoderm cells. The MT_a gene responds to lower levels of heavy metals (i.e., those present in normal sea water) than does the MT_b gene, which requires additional Zn^{2+} for induction (Wilkinson and Nemer, 1986a,b). For our present purposes the main point is that this cell type-specific induction appears only in the later stages of embryonic development.

Oocytes, eggs and early embryos thus lack the services of either the *internally or externally inducible* low copy number gene systems available to somatic cells. In their place we find a variety of special ontogenically regulated mechanisms. These include devices such as coordinately regulated tandem gene sets that contain a sufficiently high number of genes to compensate for the low nucleus to cytoplasm ratio of the early embryo, e.g., the α-histone genes of the sea urchin, or the oocyte 5S rRNA genes of *Xenopus*. Maternal components are also sometimes utilized to tide the embryo over this early period, e.g., metallothionein mRNAs in sea urchin eggs, but in the

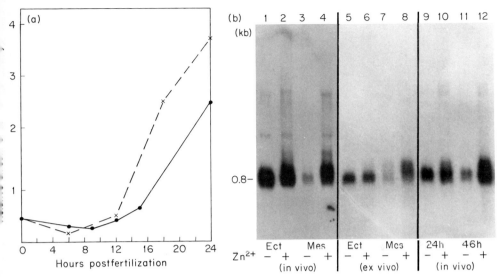

Fig. 3.12. Metallothionein (MT) induction in Zn^{2+}-treated and control *Strongylocentrotus purpuratus* embryos. (a) Accumulation of MT mRNA, visualized by densitometry of RNA gel blots carried out with a cloned MT gene probe. Total RNA was extracted from normal control embryos (●) or from embryos treated for the first 18 hours of development with 0.75 mM $ZnSO_4$ (X). Treatment with Zn^{2+} does not change the total RNA content of the embryos. (b) RNA gel blot determinations of relative MT mRNA content in Zn^{2+} treated and control ectoderm (Ect) and mesenchyme + endoderm cells (Mes). Ectoderm cells were separated from mesoderm + endoderm cells by the partial disaggregation procedure of McClay and Chambers (1978). Lanes 1-4, 72 hour pluteus stage embryos incubated for 6 hours with or without 0.75 mM $ZnSO_4$, then separated into fractions from which the RNA was extracted; lanes 5-8, 72 hour pluteus stage embryos were first separated into ectoderm and endoderm + mesoderm, and these fractions were suspended in synthetic sea water containing half the normal amount of Ca^{2+} and in the indicated samples 0.75 mM $ZnSO_4$. Lanes 9 and 10, blastulae incubated 18-24 hours after fertilization with or without 0.75 mM $ZnSO_4$; lanes 11 and 12, same, with 40-46 hour gastrulae. [From M. Nemer, E. C. Travaglini, E. Rondinelli, and J. D'Alonzo (1984). *Dev. Biol.* **102**, 471.]

case of some major heat shock proteins early embryos simply do without protection. From an adaptive point of view, the unavailability of transcription level inductive response systems in the early embryo must add to the requirement for protective developmental strategies. These have ranged from rapid development within an impermeable shell or membrane, to incubation within a living maternal environment equipped with its own inductive homeostatic mechanisms.

IV

Differential Gene Function in the Embryo

1. TIMING OF DIFFERENTIAL GENE EXPRESSION AND GENERAL ASPECTS OF GENE REGULATION IN THE EMBRYO

(i) Developmental Aspects

Sooner or later most cells in the embryo assume specialized functions and express characteristic sets of genes. As might be expected, genes that are required only in certain embryonic cell types are often genes that are not represented in the maternal RNA, or are not needed during oogenesis. That is, they are "late" genes in the sense considered earlier (see Chapter III). This seems a reasonable consequence of the principle that *activation* of given genes usually underlies the appearance of cell-type specific functions.

The developmental stage at which differentially functioning nuclei first appear in the embryo varies greatly among species. For example in *Xenopus* thousands of cells have formed before any transcriptional activity, including differential activity, is observed. In contrast, in *Caenorhabditis* and in sea urchins, differential zygotic genome function can be demonstrated in some cell lineages when there are still only a few hundred cells in the embryo. This property directly predicts the importance of *early cell lineage* in determining the fate of the descendant blastomeres. Thus, as reviewed in following sections of this Chapter, in *Caenorhabditis* and in sea urchin embryos sets of cells displaying given phenotypes are characteristically derived from one or a small number of specific "founder cells," that can be identified by their positions in the early embryo, and often by their individual cytological characteristics. In the development of these organisms, furthermore, the range of functional potentialities normally expressed by the descendants of any given founder cell is limited, though of course, many defined lineages eventually brachiate to form specialized sublineages. These statements are not generally applicable to those species in which the embryo produces a large number of cells prior to functional differentiation, however. A variety of experiments in which early embryonic cells have been marked by genetic or cytological means show that in such embryos it is difficult to demonstrate unique lineal relations between cellular phenotype and any particular cleavage stage progenitor. Thus nuclei from a given region of a predifferentiated embryo such as the *Xenopus* blastula stage embryo, the syncytial *Drosophila* embryo, or the early implantation mouse embryo may often give rise to many, and in some cases to all diversely functioning cell types among its descendants (reviewed below, and in Chapter VI). Ultimately, specialized cell lineages indeed appear, but the progenitor cells of such lineages seem to arise by a process of diversification within spatially defined regions, that consist of large assemblages of initially indifferent and equivalent embryonic cells. Though little is yet known of the mechanisms by which such processes of *regional cell specification* occur, it is clear that they require zygotic gene function, and that they involve intercellular interaction. There might exist zygotically active genes whose function is regional cell specification, for example genes that define the sets of cells among which the crucial interactions take place. In any case it seems advisable to reserve the possibility that distinct molecular processes may be involved in the genomic differentiation of embryos developing in these very different ways.

There are two general kinds of mechanism by which specification of given blastomeres early in cleavage may be influenced, and to various extents most embryos that define functional cell lineages early in development appear to utilize both. By *specification* is meant the process by which the differentiative fate of the descendants of given progenitor blastomeres is first established (whether *irreversibly* or not is in this context irrelevant). The blastomere nuclei may be affected differentially as they encounter spatially localized *maternal* molecules that can be interpreted as, or give rise to, genomic regulatory elements; or they may respond to regulatory signals that derive from spatially localized *intercellular* contacts. The cluster of problems that includes the biological mechanisms of progenitor cell specification, the role and nature of spatially localized maternal morphogenetic information, and the relative significance in diverse systems of cell-cell interactions, is the subject of Chapter VI. The present Chapter is focused on molecular and genetic evidence of *differential zygotic gene function*, both in embryos in which cell fate follows largely from the early specification of cleavage stage blastomeres, and in embryos in which cell fate is not specified until later stages, through regional organization processes. Of course this is only rarely a complete distinction. There are, for example, some cell lineages of the *Drosophila* and *Xenopus* embryos that are segregated out much earlier than are most.

The following review is arranged in order of decreasing dependence on early embryonic cell lineage decisions. The embryos considered in detail are those of *Caenorhabditis*, sea urchins, *Xenopus*, and *Drosophila*. Very little directly relevant evidence, genetic or molecular, is yet available for cell type-specific gene expression in any other embryonic systems.

(ii) Assumptions Regarding Differential Gene Activation in the Embryo from Observations on Warm-Blooded Vertebrates

Several aspects of the subject of gene regulation in development were reviewed in Chapter I. It was concluded that activation of genes in specific cells later in ontogeny is probably mediated by interaction with *trans*-regulators, and that this is likely to be the case as well for the sets of genes initially expressed in differential fashion in the early embryo. The evidence supporting such a mechanism in embryos *per se* is as yet thin, however. *Trans*-activators have been implicated for several sets of genes expression of which is not confined to specific cell types in the embryo, *viz.*, the histone genes and heat shock genes (see Chapter III), and the 5S rRNA genes (see Chapter V). There is some relevant evidence from studies on the specific expression of the α-fetoprotein (AFP) gene in the mouse embryo. This gene is active during embryogenesis in cells of the extraembryonic visceral endoderm. These cells do not contribute to the embryo proper, and thus do not provide a direct instance of specification of cell types within the embryo, as do the examples considered for other organisms in the following. Nonetheless, visceral endoderm cells are closely related in origin to the definitive embryonic cell types that later appear within the postimplantation embryo, since like the latter they derive from the inner cell mass (ICM). In the mouse the "primary endoderm" delaminates from the ICM at about day 4 after fertilization, the first and only ICM cells to visibly differentiate prior to implantation. Following implantation the primary endoderm forms the parietal, or distal, and the visceral, or proximal, extraembryonic endodermal layers of the yolk sac. The functions of these two regions are quite distinct. A major role of the parietal endoderm cells is secretion of a basement membrane (Reichert's membrane) that contains specific products including type IV collagen, and various proteoglycans and other glycoproteins (Adamson and Ayers, 1979; Hogan *et al.*, 1982). Visceral endoderm cells display the cytological organization of columnar epithelium, and they have the capability of ingesting exogenous particles and proteins (Enders *et al.*, 1978). AFP is synthesized shortly after implantation in these cells only (Dziadek and Adamson, 1978). The AFP gene is also expressed in teratocarcinoma cells induced to differentiate into cells resembling visceral endoderm, and it is not expressed when the same cells are instead induced to form parietal endoderm (Scott *et al.*, 1984). Since the activation of the AFP gene can be regarded as a specific marker of the differentiation of an early specialized cell type that derives from the ICM, the mechanism by which this occurs might provide a model for the activation of genes in the true embryonic derivatives of the ICM. As reviewed in Chapter I there is genetic and molecular evidence to indicate that expression of the AFP gene is regulated by upstream *cis*-interactions with *trans*-activators, i.e., at least in adult liver, where the AFP gene is also transcribed (Belayew and Tilghman, 1982; Pachnis *et al.*, 1984; Scott *et al.*, 1984; Krumlauf *et al.*, 1985a). However, its developmental expression in transgenic mice is apparently very sensitive to chromosomal position (Krumlauf *et al.*, 1985b). The evidence suggests, nonetheless, that at the genome level the activation of particular genes in specific embryonic cells that develop from indifferent pluripotential ICM precursors may indeed depend on the presentation, the release, or the synthesis of appropriate sequence-specific *trans*-activators.

Other than this, few identified genes that function differentially in mammalian embryos have been studied. In the mouse some genes are known that are expressed specifically in the trophectoderm but not the ICM at preimplantation stages (reviewed in Chapter VI). The cytoskeletal protein spectrin first appears at the 2-cell stage, and thus is probably a product of zygotic gene function (Sobel and Alliegro, 1985). Its pattern of localization suggests that in preimplantation development it is involved in important intercellular contacts (cf. Chapter VI). However, this protein appears by indirect immunofluorescence to be widely distributed in both the ICM and the trophectoderm. Similarly, laminin is first seen in the 16-cell stage in the compacted mouse embryo, where it appears along many intercellular contours, and type IV collagen first appears at the blastocyst stage, in the ICM (Leivo *et al.*, 1980). After implantation and differentiation of the egg cylinder types II and III collagen appear in mesodermal tissue anlagen such as the heart mesenchyme. However, no information has been reported regarding the underlying patterns of gene activity, or their control, for these or several other structural proteins that have been cytologically visualized in the mouse embryo. Those zygotically active genes that have thus far been studied at the transcript level, e.g., the genes coding for actins and histones (Chapters II and III), are not known to display specific patterns of activation in the embryo.

Among the instances of embryonic progenitor cell specification that have been investigated at the molecular level in warm-blooded vertebrates is the appearance of hemoglobin-synthesizing blood cells in the chick blastoderm. Committed lineages of cells that synthesize embryonic globins originate from hematoblast progenitors that form in the marginal zone of the blastodisc, which at this stage contains about 60,000 cells (Wilt, 1965). The hematoblast progenitor cells do not transcribe the globin genes, while their mitotic descendants do (Groudine *et al.*, 1974; Groudine and Weintraub, 1981). Activation of the embryonic globin genes occurs exclusively in this cell lineage, and it is associated with changes in the DNase I sensitivity of the chromatin and the degree of DNA methylation in the regions where these genes are located (Weintraub and Groudine, 1976; Stalder et al., 1980; cf. Chapter I). These structural alterations are an heritable characteristic of the hematopoeitic cell lineage, according to observations on virus-transformed cultured cell lines that represent various stages in chick erythroblast maturation (Weintraub et al., 1982). Groudine and Weintraub (1982) also showed that once the globin genes have been induced to become active the pattern of DNase I hypersensitive sites is propagated through many division cycles to daughter cells even in the absence of transcription of these genes. This observation was made on a cell line transformed by a temperature-sensitive virus, and grown for over 20 cycles at the temperature at which the globin genes remain transcriptionally silent.

The activation of the AFP gene in the visceral endoderm cells of the mouse embryo, and of the globin genes in the chick erythroblast lineage, imply respectively two propositions that it is reasonable to assume might be relevant generally. These are, first, that in early embryos genes are activated in previously equivalent nuclei in response to the *localized* presentation of *trans*-activators; and second, that the state of activation of lineage-specific genes can be heritably transferred to the mitotic descendants of the progenitor cells.

2. CELL LINEAGE AND DIFFERENTIAL GENE EXPRESSION IN THE EMBRYO OF *CAENORHABDITIS ELEGANS*

The complete somatic cell lineage of the free-living nematode worm *Caenorhabditis elegans* has been established (Sulston and Horvitz, 1977; Deppe et al., 1978; Kimble and Hirsh, 1979; Sulston et al., 1983). The lineage is based mainly on observations made on living embryos by Nomarski microscopy. Many partial and a few complete early cell lineages had previously been available for embryos of various nematode, molluscan, annelid, ascidian, echinoderm and other species, in some cases since the 19th century (reviewed in Chapter VI). However, classical knowledge of embryonic cell lineage rarely extended beyond the 200-cell stage, and C. elegans is the only multicellular animal for which the postembryonic lineages of all somatic cells in the larval and adult forms have been determined. During embryogenesis 671 cells are generated of which 113 (in individuals that are hermaphrodites) or 111 (in males) are programmed to die. The newly hatched larva thus consists of slightly more than 550 cells. The adult hermaphrodite contains about 850 somatic cells, plus dividing germ cells, and the mature male about 970 somatic cells, plus dividing germ cells.

Knowledge of embryonic cell lineage in C. elegans has advanced understanding of the developmental process in this creature in several important ways. The embryo is constructed from the descendants of a small number of founder cells, each of which gives rise to a defined set of differentiated cell types. The lineages by which these cell types are derived are for the most part determinate, in that the fate and functions of each cell are invariant from animal to animal. Given lineages differ in their developmental program in related species (Sternberg and Horvitz, 1982). Thus the detailed determination of cell lineage and cell fate is *prima facie* programmed in the genome. Since each founder cell inherits a fixed region of egg cytoplasm it is possible that the specification of these cells, or at least the initial stage of this process, is mediated by maternal factors localized asymmetrically in the egg cytoplasm. Some aspects of founder cell specification in *C. elegans*, and related subjects, are considered in Chapter VI. The following review is concerned with the relation between cell lineage and differentiation following founder cell specification, and then with the genetic and molecular evidence illuminating the nature of developmental gene regulation in this organism.

(i) Cell Lineage and Pathways of Differentiation in *C. elegans*

The lineage of every cell in the completed *C. elegans* embryo is shown in Fig. 4.1 (see pocket at back of book) (Sulston *et al.*, 1983). Individual cells are designated by orientations of the successive divisions in their ancestry, each of which usually produces an anterior (*a*, left on diagram) and a posterior (*p*, right on diagram) pair of daughter cells. The lineage names at the termini of the segments thus describe the unique pathway by which each cell arises (for other symbols, see legend). At the bottom of the Figure the structures to which the cells of the larva belong are indicated. Though it is by no means necessarily the case in detail, we assume implicitly in the following that all cells designated, for example, "intestine" or "body muscle," exercise similar functions, and therefore express similar sets of genes. Sophisticated classifications of the various relations that can be discerned between cell lineage and daughter cell function have been derived, to which the reader is referred for more detailed consideration (see, e.g., Sulston *et al.*, 1983; Sternberg and Horvitz, 1984). We focus here on several regions of the lineage that illustrate specific aspects of the regulatory architecture of the embryo.

The least complex relation between cell lineage and cell fate is exemplified by the lineages of the E, D, and P4 progenitor cells, illustrated toward the right of Fig. 4.1. Each of these cells is the ancestor of a *clone* of descendant cells of similar function. P4 gives rise only to cells of the germ line; D only to body muscle cells; and E only to gut cells. Thus *clonal patterns of zygotic gene function* are differentially established in these lineages. This requires the transfer from cell to daughter cell of the clonal state of specification until, at some point during the division process in each lineage, there materializes the specific pattern of gene expression. Molecular evidence of such differential function for several lineages, for example, the clone of body muscle cells descendant from the D progenitor cell, is reviewed in a following section. The number and orientation, as well as the rate (Deppe *et al.*, 1978) of the mitotic divisions in each of these clonal lineages are particular properties of the developmental programs set in train with the initial specification of the progenitor cells.

Clonal patterns of differentiation are not confined to the P4, D, and E lineages, in which all descendants of the initial progenitor cells display a single form of differentiation. Thus, for example two large, identical clones each consisting of 16 body muscle cells descend from the founder cell C. The progenitors of these clones are the 5th cleavage cells Cpp and Cap, all descendants of which become body muscle cells. However the sisters of these two cells, Cpa and Caa, give rise respectively to hypoderm cells, and to hypoderm cells and neurons. There are 81 body muscle cells in the first stage larva, and of these the muscle cell clones descendant from D, Cpp and Cap account for 52. Twenty-two additional body muscle cells arise from 6 small clones in the MS lineage, for example those descending from the progenitors MSpppp, MSppapp, and MSpapp. The remaining 6 body muscle cells of MS lineage segregate from sister cells that display other differentiated fates late in their respective lineages, either programmed cell death (e.g., cell MSpppapa) or an alternative functional specialization (e.g., cell MSppapaa). A single body muscle cell also arises from the AB lineage (ABprpppppaa), as do muscle cells of certain structures of the alimentary tract (Sulston *et al.*, 1983). Thus muscle cell differentiation occurs nonclonally as well as clonally. This distinction concerns the number of divisions *all* products of which express the same form of differentiation, with the implication that a cell-heritable regulatory state is transferred to the daughter cells at each cycle. In the body muscle cell lineages the number of such divisions is 4 or 5, for the clone descendant from the D progenitor cell, but it is zero for MSpppapp, for example. It is possible that inheritance of maternal cytoplasmic factors is responsible for the initial specification of the clonal progenitor cells when these arise very early, as do E, D, and P4, and perhaps the two C lineage muscle clone progenitors. However, this becomes an unlikely alternative when specification occurs after there are already several hundred cells in the embryo, as a terminal event in a lineage element (for instance the division that creates MSppapap, a body muscle cell, and, MSppapaa, the mother cell of two coelomocytes).

The large majority of the lineages portrayed in Fig. 4.1 are *nonclonal*. They give rise to multiple cell types, and most structures in the embryo include contributions from many different lineages. Striking examples are provided by the ring ganglion and the pharynx (see Fig. 4.1). Viewed from the vantage point of the final differentiated cell products, a very common mechanism in *C. elegans* development is the occurrence of determinate mitoses that result in the appearance of nonequivalent daughter cells expressing diverse functions. This suggests the asymmetric partition of cellular determinants associated with fixed cytoskeletal elements. Many examples of terminal cell divisions that are polarized in respect to the fate of the immediate daughter cells can be observed in Fig. 4.1, particularly within the nervous system lineages. For example programmed cell death (marked by X in Fig. 4.1) always occurs in one sister cell only, while the other sister cell differentiates, and in many cases a given cell gives rise to both a hypodermal cell and to neurons (e.g., cells ABarpppap and ABarppaap). The signals that trigger the activation of diverse regulatory programs in sister cells are unlikely to be positional in the environmental sense. Thus a series of laser ablations of single cells carried out by Sulston *et al.* (1983) demonstrated complete autonomy of cell fate, except for two specific cases in which compensation for a missing cell occurred by a change in assignment within a similar lineage late in embryogenesis. In general the missing cells are not compensated for, and the larvae developing from embryos missing given cells lack exactly the parts their lineages would normally have constructed. Examples include removal of individual muscle precursor cells of the C and D lineages, which resulted in absence of the predicted number of muscle cells in the larva; ablation of derivatives of MS, resulting in the loss of the predicted cell products but the normal differentiation of descendants from the undamaged sister cells; and a series of ablations in the AB lineage among sensory system precursors that similarly yielded no evidence of regulatory changes in cell fate. We may conclude that in lineages where the boundaries of clones formed early in cleavage do not correspond with phenotypic boundaries, cell differentiation must be controlled by *autonomous genetic programs that are activated differentially* late in the mitotic history of the lineage.

The initial events in the formation of the autonomous cell lineages of which the embryo is constructed result in partial restriction of developmental potential, since the founder cells retain the capacity to generate only certain differentiated cell types. Thus the capacity to build intestine cells is physically segregated away from AB at first cleavage, and then from other potentialities in the division of EMS; the capacity to make neurons is absent in descendants of P3; the capacity to make hypodermis is absent in descendants of D, P4, E and MS; and so forth. Three-dimensional reconstructions of early *C. elegans* embryos show that the relative locations of the various early blastomeres is positionally almost invariant (Deppe *et al.*, 1978; Schierenberg *et al.*, 1984). Though localization of fixed maternal positional cues may thus provide an explanation in principle for the initial restrictions of lineage fate, and asymmetric partition of determinants a possible mechanism for the final results, it is clear that zygotic regulatory genes must function within each lineage. This is indicated by the invariant association, in nonclonal lineages, of given *lineage patterns* with the production of given *sets of cellular phenotypes*, illustrated most impressively in the pattern of later cell fates among descendants of AB. Many different forms of neuron, and neuronal support cells, as well as gland cells, hypodermis and some late appearing specialized muscle progenitors derive from this quadrant of the embryo.

The overall lineage shown in Fig. 4.1 includes a number of *repetitive lineage patterns* that result in homologous sets of cellular phenotypes. This can be seen in the complex origin of right-left bilateral symmetry in the developing embryo (Sulston *et al.*, 1983). In the posterior end of the early embryo the initial founder cells, MS, C, and D, divide into daughters that contribute equivalent structures on the right and left sides of the embryo (Schierenberg *et al.*, 1984), and here symmetry could be regarded as an ultimate consequence of a bilateral pre-organization of the egg. However, in other regions bilateral symmetry arises only later in embryogenesis, as cells that are diverse in origin and position give rise to homologous lineage elements which produce similar or identical arrays of cell types. Such lineage elements are termed *sublineages* (Chalfie *et al.*, 1981; Sulston *et al.*, 1983; Sternberg and Horvitz, 1984). Precursor cells giving rise to bilaterally symmetrical groups of cells are marked by horizontal dashed lines in the lineage map of Fig. 4.1. For example, at the left hand end of Fig. 4.1 it can be seen that bilaterally symmetrical lineages are produced *within* as well as *between* descendants of ABalaa and ABalap. In the development of the nervous system other examples of repetitive sublineages are observed. For instance the sublineage element exemplified by the descendants of ABalaappp, which gives rise to two different inner labial neurons and to a dead sister cell, occurs six times (Sulston, 1983). Many sublineages occur repetitively in postembryonic development, and comparative observations on another species, *Panagrellus redivivus*, show that the number of repetitions is among the species-specific characters of morphogenesis (Sternberg and Horvitz, 1982). Such sublineages provide examples in which cells that express similar differentiated functions arise from initially diverse clonal origins.

Sublineage patterns of cell division and differentiation are modular, in the sense that they can be instituted as repetitive units. This implies the existence of an underlying zygotically active genetic program for each repetitive sublineage type, and thus by extension, for other sublineages that are not repetitive. Specific aspects that can be seen to be under immediate genetic control in the comparison of *C. elegans* and *P. redivivus* postembryonic lineages include the number of rounds of division in certain sublineages, and the fates of individual cells deriving from homologous lineages (Sternberg and Horvitz, 1982). Analysis of *C. elegans* mutations that affect sublineage differentiation patterns has provided particular insights into the organization of the developmental regulatory system in this organism.

(ii) Genes Required for Lineage-Dependent Morphogenetic Processes

A number of mutations that affect particular sublineages have been described (reviewed in Sternberg and Horvitz, 1984). Though in the main these alter postembryonic morphogenesis they are relevant because they demonstrate the existence of genes that exercise some of the types of regulatory function inferred from lineage analysis of embryonic development. Some genes are apparently required in multiple sublineages that are unrelated in embryonic origin, and diverse in respect to the functional cell types to which they give rise (see, e.g., Sulston and Horvitz, 1981). For example there are mutations that cause cells in different lineages to express developmental fates earlier or later in the schedule of divisions than in normal animals. Others cause widespread reversal of sexual phenotype, so that cells normally destined to produce structures such as the vulva only in hermaphrodites do so in males as well.

The effects of recessive mutations at three loci that alter both cell lineage and cell fate are illustrated in Fig. 4.2, from studies reported by Horvitz et al. (1983). The morphogenetic consequences of these mutations, lin-22, unc-86, and ced-3, are shown in Fig. 4.2(b)-(f), and the underlying sublineage changes that they induce are indicated in Fig. 4.2(g)-(j). The set of lineages here considered arise from the bilateral sets of ectoblasts V1-V6 (left and right), which in postembryonic development give rise to hypodermis and to innervated sensory structures. V1-V6 are among the blast cells marked by heavy arrowheads in Fig. 4.1, and as shown there they derive from the embryonic lineages of ABarpp (V1L, V2L, V4L, V6L, and V1R, V2R, V4R, V6R); ABplap (V5L and V3L); and ADprap (V5R and V3R). The position of these blast cells on the larva is shown diagrammatically in Fig. 4.2(a). In wild-type hermaphrodites the V5 blasts alone produce the special bilateral structures known as the "postdeirids," which include two neurons, one of which is dopaminergic [Fig. 4.2(b) and (j), wild-type], and two accessory cells. A fifth cell produced by this sublineage is programmed to die. In wild-type males V5 also gives rise to neuronal and accessory components of mechanosensory structures known as "rays." In both males and hermaphrodites V1-V4 normally generate only hypodermal syncytial nuclei (anterior daughters) and "seam cells" (final posterior daughters) that may function during the larval molts (Sulston and Horvitz, 1977). In wild-type hermaphrodites the lineage descendant from V6 is the same as those descendant from V1-V4, but in males the V6 blasts give rise to a series of five repetitive sublineages from each of which ray neurons and supporting cells are generated (Sulston et al., 1980). The wild-type hermaphrodite sublineages of V1-V6 are indicated diagrammatically in Fig. 4.2(g), and the dramatic effect of the lin-22 mutation on these ectoblast sublineages is shown in Fig. 4.2(h). In lin-22 animals V1-V4 carry out the sublineage program normally confined to V5, with the consequence that the extra dopaminergic neurons shown in Fig.

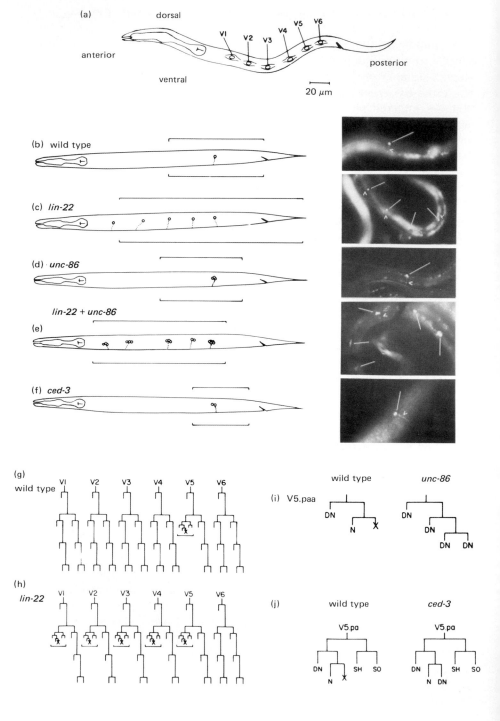

4.2(c) are produced. In *lin-22* males the fate of V1-V4 is also changed to that of V5, so that all produce rays as well as postdeirids. The appearance of supernumerary postdeirids in *unc-86* animals [Fig. 4.2(d)] is evidently due to a different alteration of the V5 sublineage, as diagrammed in Fig. 4.2(i). This effect can be regarded as a reiteration of the initial pattern of division of V5.ppa. Thus in the second and sometimes additional division cycles rather than only in the first, as in the wild-type, the anterior daughters also become dopaminergic neurons. Mutants in this gene also cause similar reiterations of the immediately *preceding* lineage pattern element in sublineages descendant from certain other neuroblasts (*viz.*, Q and T.pp), and in embryonic as well as in postembryonic development (Chalfie *et al.*, 1981; Sulston and Horvitz, 1981; Horvitz *et al.*, 1983). It is possible that these effects follow from some lesion of cytoskeletal or chromosomal structure, since *unc-86* mutants also display nondisjunction of the X chromosome during meiosis in the hermaphrodite (Hodgkin *et al.*, 1979). The reiteration of a complete pattern of differentiation in *unc-86* mutants could thus be imagined a secondary effect of failure to execute correctly a necessary asymmetric division. The effects of the *unc-86* and *lin-22* mutations are additive, so that double mutants produce ten sets of multiple dopaminergic neurons [Fig. 4.2(e)]. Another recessive mutation that causes reiterations in cell lineage, and consequent gross alterations in cell fate, is *lin-4* (Chalfie *et al.*, 1981). This mutation also affects many postembryonic somatic cell lineages, but unlike *unc-86* it does not alter neuroblast sublineages. As they are recessive, these are all probably loss of function mutations affecting the process by which the

---

Fig. 4.2. Effects of mutations on specific cell lineages in *Caenorhabditis elegans*. (a) Schematic left lateral view of a first-stage larva showing the positions of the six lateral ectoblasts V1-V6. (b)-(f) Location of dopaminergic neurons in (b), wild-type; (c), *lin-22*; (d), *unc-86*; (e), *lin-22* + *unc-86*; (f), *ced-3* animals, visualized by formaldehyde-induced fluorescence. These neurons are part of the postdeirid sensory structures, as diagrammed in (j). In certain examples, neural processes can be seen. The regions shown in the photomicrographs are indicated by brackets on the corresponding drawings. (——) cell body; (>) process. (g) Diagrams of sublineages produced in hermaphrodites by the ectoblasts V1-V6. In the wild-type only V5 generates a postdeirid structure, which is constructed from the bracketed cells. The remaining ectoblasts in hermaphrodites produce only hypodermal cells. X indicates a cell programmed to die. (h) Transformation of V1-V4 sublineages in *lin-22* mutants. These cells now produce the same sublineage as does V5 in wild-type animals, accounting for the supernumerary dopaminergic neurons seen in (c). (i) Effect of *unc-86* mutation on the V5.paa sublineage element. In the mutant animals V5.paap acts like V5.paa, as does its posterior descendant, each giving rise to an ectopic dopaminergic neuron [cf. (d)]. The effects of *unc-86* and *lin-22* are additive, as shown in (e). (j) Postdeirid lineages of wild-type and *ced-3* animals. In wild-type animals the postdeirid lineage generates a dopaminergic neuron (DN); a nondopaminergic neuron (N); two associated structural cells, the sheath cell (SH) and socket cell (SO); and a programmed cell death (X). In *ced-3* animals, this death fails to occur and a supernumerary dopaminergic neuron is generated, presumably by the lineage shown. [From H. R. Horvitz, P. W. Sternberg, I. S. Greenwald, W. Fixsen, and H. M. Ellis (1983). *Cold Spring Harbor Symp. Quant. Biol.* **48**, 453.]

proper sequence of asymmetric divisions is programmed, or perhaps more likely, executed. When this sequence is disturbed the consequence is failure of activation of the set of cell differentiations normally assigned to the affected area of the sublineage. The developmental module represented by the V5.p sublineage is generally homologous with sublineages utilized for construction of assemblages of neural cells and neuronal support cells elsewhere in the nematode, and even in insect neuroblast derivatives (Sulston and Horvitz, 1977). The regulatory and structural gene package that is required for the materialization of this characteristic sublineage is thus utilized widely, and is probably of ancient evolutionary origin.

Another genetically programmed function needed for the differentiation of the V5.p sublineage is demonstrated by the *ced-3* mutation, which causes a duplication of the dopaminergic neuron [Fig. 4.2(f)]. As indicated in Fig. 4.2(j), this genetic lesion causes failure of programmed cell death. The cell that in wild-type animals autonomously executes itself fails to do so in *ced-3* mutants, and instead it becomes a second dopaminergic neuron. In the hermaphrodite 463 neurons are generated during development, and of these 105 are sister cells slated for death soon after formation (Horvitz *et al.*, 1983). Mutants defective at the *ced-3* locus, or at several other distinct loci with similar effects (Hedgecock *et al.*, 1983), fail to carry out all of these programmed cell deaths, evidently lacking biochemical functions required for the suicides of these cells. It is interesting that, if genetically rescued, these cells may instead express another program of differentiation found nearby in the lineage, as illustrated for the V5.pa sublineage in Fig. 4.2(j).

A gene whose functions resemble those of a binary developmental switch is *lin-12*. Mutations at this locus have the effect of transforming to the same phenotype cells that normally express different phenotypes, although they occupy homologous positions in otherwise identical sublineages. Two examples from the lineages of postembryonic neural structures are diagrammed in Fig. 4.3(a) and (b) (Horvitz *et al.*, 1983). On the left side of the first stage larva the cell ABplpppaaaa is normally a motor neuron (DA9), while its homologue on the right side arising from an otherwise identical sublineage is in wild-type males a neuroblast (Y). This cell later divides to produce five neurons and six supporting cells, indicated schematically in the middle row of Fig. 4.3(a). Semidominant mutant alleles [*lin-12(d)*] cause both homologues to behave as Y neuroblasts (top row). *Revertants* of *lin-12(d)* alleles which appear to be null mutants [*lin-12(0)*] were obtained by Greenwald *et al.* (1983). These have the reciprocal effect, causing both cells to become DA9 neurons (bottom row). The *lin-12(0)* mutations are recessive, as expected for null alleles. The phenotypic effects of some *lin-12(0)* revertants can be suppressed by an amber suppressor tRNA mutation. The *lin-12* locus therefore probably codes for a protein product that is not formed in *lin-12(0)* homozygotes, and that would appear to be required for the regulatory switch by which is instituted the Y neuroblast phenotype. The effects of *lin-12(d)* mutations can thereby be interpreted as the consequence of overproduction, or of ectopic production, of this protein. An analogous set of *lin-12(d)* and *lin-12(0)* effects is shown in Fig. 4.3(b). Here, as in Fig. 4.3(a), the fates of two normally different cells, the neuroblast W and the ectoblast G2, that again occupy similar positions in homologous sublineages, are altered reciprocally by *lin-12(0)* and *lin-12(d)* mutations. Greenwald *et al.* (1983) also reported reciprocal effects of these mutations on the fates of homologous but nonidentical mesodermal and gonadal cells, and during the postembryonic development of the vulva in hermaphrodites. Several of the cells that are affected in *lin-12* mutants are known to be pluripotential in normal animals, and the mutations *alter their choice of fates*. These are among the minor number of examples in *C. elegans* in which cell fate is evidently determined by means of intercellular interactions, rather than autonomously. Thus in laser ablation experiments (Kimble, 1981; Sulston *et al.*, 1983) these particular cells express alternate functional phenotypes depending on their position, and on the presence of given neighboring cells. Examples include the G2, W pair of blast cells that is the subject of Fig. 4.3(b) (Sulston *et al.*, 1983), and a set of male ectoblasts, cells P9.p, P10.p and P11.p, shown in Fig. 4.3(c). In wild-type males P10.p gives rise to an innervated epidermal sensory structure, the "hook," located near the cloaca, and to some hypodermal nuclei, while P9.p produces syncytial hypodermal nuclei only, and P11.p produces hypodermal cells that position the hook, plus neurons. If P10.p is destroyed, P9.p differentiates the hook sensillum as P10 p would have done [Fig 4 3(c); see also Sulston, 1983]. Similarly P10.p can replace P11.p if this blast cell is ablated. These ectoblasts thus retain capacities to give rise to sublineages other than the one normally produced by each. Figure 4.3(d) demonstrates that *lin-12* can also affect their phenotype. In the absence of the *lin-12* product no hook is formed, though the P11.p sublineage is retained (bottom row). A single *lin-12(d)* allele gives two hooks, presumably by conversion of P9.p so that it produces a P10.p sublineage (second row), and a genotype containing two *lin-12(d)* alleles yields three hooks (top row). Additional evidence that developmental switches controlling alternative cell fates are affected by the quantity of *lin-12* product derives from observations of Greenwald *et al.* (1983) on the severity of defects in vulva morphogenesis produced by various *lin12(d)* alleles in animals of *lin-12(d)/lin-12(+)* genotype. Addition of a second *wild-type lin-12* gene, i.e., construction of a *lin-12(d)/(+)/(+)* genotype by use of a chromosomal duplication was found to result in more extensive defects. The *lin-12* gene has been cloned, and the sequence of a portion of the coding region displays homologies with mammalian proteins of the epidermal growth factor family, which also includes the low density lipoprotein receptor, plasminogen activator, and several other biologically active factors (Greenwald, 1985). One possible

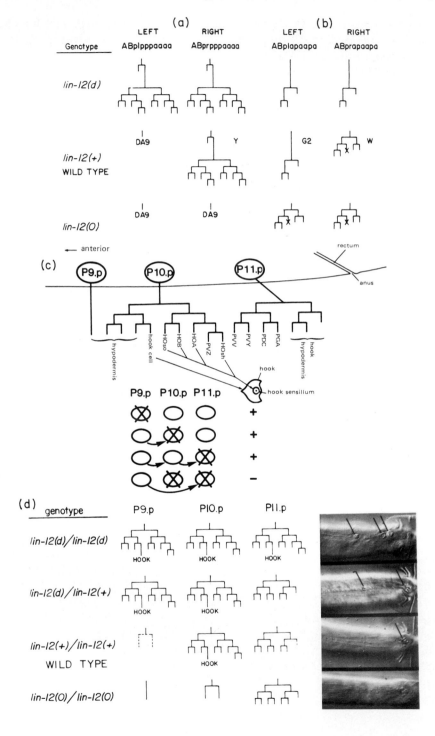

implication, since this gene affects choices of lineage fate among cells that retain multiple potentials, is that these choices are mediated by intercellular interactions in which the *lin-12* product participates.

Though few organisms are known in which embryonic cell fate is from the beginning as invariant a function of cell lineage as in *C. elegans*, both clonal patterns of differentiation and sublineages that define sets of diverse cell states occur widely. In many organisms lineage progenitors arise out of cells or groups of cells that are initially highly plastic, and that until some form of intercellular interaction or other positional instruction takes place are indeterminate in regard to their prospective fates. However *determination*, by whatever means, often results eventually in the institution of lineage-dependent patterns of differentiation similar to those discussed in this section. We consider several specific examples in sea urchin and *Drosophila* embryos in later sections, and to this class may belong a number of familiar late- and post-embryonic vertebrate developmental processes, such as the lineage-dependent differentiation pathways found in hematopoetic systems; in eye lens cell formation; and in the terminal phases of fibroblast differentiation into various connective tissue derivatives. The genetic analyses reviewed here begin to reveal aspects of lineage-dependent differentiation processes that may be under direct genetic control. The genetic programs that direct sublineage differentiation can be regarded as modular regulatory units, and the fact that single site mutations can cause the ectopic institution of a whole sublineage program would seem to require the existence of a hierarchy that

---

Fig. 4.3. Reciprocal effects of semidominant and recessive *lin-12* mutations in *C. elegans*. (a) and (b) Postembryonic fates of pairs of contralateral cells that occupy homologous positions in otherwise identical embryonic cell lineages. Top row, lineages in animals bearing a semidominant *lin-12(d)* allele; middle row, wild-type; bottom row, null *lin-12(0)* mutants. The lineage names of the contralateral homologues are shown at the top of the figure. (a) Males: Y, neuroblast; DA9, neuron. (b) Hermaphrodites and males: W, neuroblast that divides during the first larval stage; G2, ectoblast that divides during the second larval stage. [From H. R. Horvitz, P. W. Sternberg, I. S. Greenwald, W. Fixsen, and H. M. Ellis (1983). *Cold Spring Harbor Symp. Quant. Biol.* **48**, 453.] (c) Diagram showing schematically the three neuroblasts P9.p, P10.p and P11.p in the postembryonic male, the lineages to which they give rise, and the elements of the hook sensillum formed in wild-type animals from the P10.p descendants. HOB, HOA are neurons that innervate the hook, which is a sclerotized structure formed by the hook cell P10.papp. PVZ, PVV, PVY, PDC, PGA are other neurons. Below the lineage is a diagram indicating results obtained when individual neuroblasts are ablated with a laser (X). Replacement of a destroyed cell by another blast cell is indicated by arrows; e.g., when P10.p is destroyed P9.p moves into its place and generates a hook rather than syncytial hypodermal nuclei. [From J. E. Sulston (1983). *Cold Spring Harbor Symp. Quant. Biol.* **48**, 443.] (d) Cell lineages of P9.p, P10.p and P11.p in males of different *lin-12* genotypes. At the right are Nomarski photomicrographs of ventral views of adult males showing the region just anterior to the cloaca (arrowhead) with the normal and/or ectopic hooks deriving from P9.p, P10.p and/or P11.p indicated by lines. These hooks were presumably generated by the lineages shown. [From H. R. Horvitz, P. W. Sternberg, I. S. Greenwald, W. Fixsen, and H. M. Ellis (1983). *Cold Spring Harbor Symp. Quant. Biol.* **48**, 453.]

includes "upper level" control genes. Within the purview of such genes lie the "lower level" regulatory elements that directly activate the batteries of structural genes operative in the diverse cell types to which a sublineage may give rise. We also learn that, as might be expected, there are common processes required in the control of many different sublineage differentiation pathways. Thus *lin-12* affects lineages unrelated except that their development requires a choice between alternative cell fates; there are functions utilized for cell death in diverse lineages; and no doubt there are specific mechanisms of asymmetric division from which arise daughter cells of diverse fates.

Less evidence is available for the *clonal* processes of differentiation that occur early in embryogenesis in some regions of the *C. elegans* embryo. For these it is a matter of speculation whether upper level control genes are required at all. Thus the localized maternal factors that are likely to be involved in the initial specification of cells such as the E gut progenitor blastomere (see Chapter VI) could directly affect the battery of structural genes required in the early intestine cells. Alternatively, such factors might directly interact with an upper level zygotic control gene, that in turn activates this battery following the appropriate number of nuclear replication cycles.

(iii) Molecular Evidence: Genes Expressed in Particular *C. elegans* Cell Lineages

Several cytological markers of differentiation in the embryonic cell lineages of *C. elegans* are known, e.g., certain refractile fluorescent granules that appear in gut cells descendant from the E blastomere, and characteristic inclusions in the germ cell precursors (reviewed in Chapter VI). The gut cell granules are probably products of zygotically active genes expressed in the intestine precursor cells, since they appear only at about 3.5 hr postfertilization (i.e., at 25°C; Laufer *et al.*, 1980), when cells of the E lineage are within one or two cycles of completing their programmed series of divisions. However, there is yet available no direct information regarding the genes required for the synthesis of these granules. A gene for an esterase that has been purified and characterized is also expressed specifically in the gut gells (McGhee, 1986; Edgar and McGhee, 1986). This esterase activity becomes detectable when there are only 4-8 intestinal precursor cells in the E cell lineage, i.e., several cycles prior to their terminal embryonic divisions (cf. Fig. 4.1). The appearance of this lineage-specific enzyme depends on zygotic gene transcription, since it can be blocked by treatment with α-amanitin. Another early cell type-specific function for which it can be inferred that zygotic gene activity is required is programmed cell death (see discussion of *ced-3* mutants above). The earliest examples of this occur amongst the AB cell descendants in midcleavage (Fig. 4.1). Molecular evidence exists for collagen, actin, myosin and paramyosin genes, the products of which have been visualized in particular cells of early embryos by sequence-specific cytological methods.

There are four actin genes in the *C. elegans* genome, three of which are linked at distances of several kilobases, while the fourth occurs separately (Files *et al.*, 1983). Using a probe that reacts with all four of these genes, Edwards and Wood (1983) showed by *in situ* hybridization that newly fertilized eggs contain maternal actin mRNA. By 4 hr postfertilization (at 25°C), when the embryos consist of 400-450 cells, there is a noticeable increase in the quantity of actin transcripts, though this could not be assigned to any particular cell type. At this stage, however, the body muscle cells begin to express the genes for myosin and paramyosin. A large set of genes is expressed specifically in differentiated muscle cells, and mutations in over 20 different complementation groups have been identified that affect muscle cell structure in *C. elegans* (Zengel and Epstein, 1980). Myosin and paramyosin can be taken as representative. Gossett *et al.* (1982) showed by indirect immunofluorescence that during embryogenesis the number of cells in which these proteins are observed corresponds exactly to the number of body wall muscle cells at each stage, of which there are 81 in the completed embryo (see above). The rate at which these differentiated cells appear, with respect to total embryonic cell number, is shown in Fig. 4.4(a). It can be concluded from these kinetics that expression of the myosin and paramyosin genes *follows* the terminal divisions of the muscle cell precursors (Gossett *et al.*, 1982). Furthermore, the two muscle proteins appear coordinately, as also shown in Fig. 4.4(a). Yet the fate of the clone of cells descendant from the D progenitor blastomere is fixed from the beginning, and this example provides a clear demonstration of a phenomenon common in early development, *viz.*, the onset of clonal gene expression several cell cycles after the initial specification of the progenitor blastomere.

Expression of collagen genes has also been visualized by *in situ* hybridization (Edwards and Wood, 1983). Collagen is a major component of the cuticle and thus the collagen genes would be expected to be transcribed in hypodermal cell nuclei. There are over 40 collagen genes in the *C. elegans* genome (Cox and Hirsh, 1985). An experimental classification based on the pattern of transcript accumulation suggests at least four sets of collagen genes, one of which contains genes that are significantly expressed during embryogenesis. The other sets are apparently utilized sequentially at the larval molts, where they are presumably responsible for the construction of the successive larval cuticles, which differ in structure (Cox and Hirsh, 1985). Collagen mRNA is barely detectable in unfertilized eggs. However, by the 200-cell stage collagen mRNA has begun to accumulate, prior to the appearance of any other known marker of differential embryo cell function.

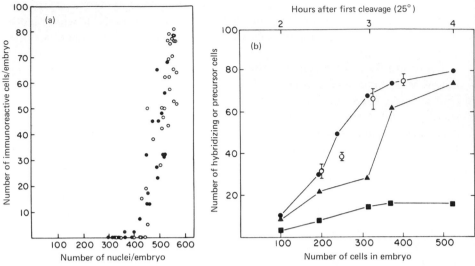

Fig. 4.4. Embryonic expression of cell type-specific molecules in *C. elegans*. (a) Appearance of paramyosin (●) and myosin (○), observed by indirect immunofluorescence in squashed preparations of cleaving embryos. The number of nuclei at each stage was counted after staining with DAPI. The antibodies used react only with the myosin and paramyosin of body wall muscles. The terminal number of cells expressing these markers is 81, the number of body wall muscle cells in the completed embryo. [From L. A. Gossett, R. M. Hecht, and H. F. Epstein (1982). *Cell* **30**, 193. Copyright by M.I.T.] (b) Appearance of collagen mRNA, assayed by *in situ* hybridization. The number of cells expressing collagen genes at each stage (O) is compared to the number of hypodermal (●), muscle (▲), and intestinal (■) precursor cells (Sulston *et al.*, 1983). The lines connecting these points are not intended to convey the exact form of the population increase within each set of precursors. [From M. K. Edwards and W. B. Wood (1983). *Dev. Biol.* **97**, 375.]

The correspondence between the number of hypoderm precursor cells and the number of cells expressing collagen genes is shown in Fig. 4.4(b). The expression of collagen genes thus indeed appears to be a hypoderm cell-specific molecular function. Activation of this function is evidently a clonal event in several of the early lineages of the embryo, e.g., those descendant from the progenitor cells Cpa, Caaa, or ABarpaap (Fig. 4.1). However, as discussed above, Fig. 4.1 shows that hypoderm differentiation, presumably including collagen gene activation, also frequently occurs in nonclonal contexts, in single cells that are the terminal products of functionally asymmetric cell divisions. For example the dorsal sister of the pair of cells arising from Caapp is a hypoderm cell, while the ventral sister is a neuron. This example suggests that where it occurs, clonal propagation of the hypodermal state of cell commitment is rather to be regarded as a useful way of distributing hypoderm cell types, than as a causal element in the process of activating hypoderm specific genes.

3. DIFFERENTIAL GENE EXPRESSION IN EMBRYONIC CELL LINEAGES OF THE SEA URCHIN

(i) Cell Lineage and Pattern of Morphogenesis

Differentiation occurs in sequential stages in sea urchin development. Specific molecular functions are already being expressed in the aboral ectoderm and the primary mesenchyme cells of blastula stage embryos, while the developmental processes that will give rise to muscle, the alimentary tract, the oral structures, neuronal assemblages, and other components of the larva do not unfold until after gastrulation. The embryonic origins of the various differentiated cell types of the completed pluteus stage embryo are indicated diagrammatically in Fig. 4.5 (see pocket at back of book). The cell lineage shown on the left is exact for the vegetal (V) half of the embryo, a region for which the lineage is known to the completion of cleavage, at the 9th-10th round of division (or earlier in some lineages). In the animal (N) half of the embryo the lineage is exactly known only to about 6th cleavage, though the number, synchrony, and rate of the subsequent several divisions has been established. This permits the canonical lineage diagram shown in Fig. 4.5 for the animal blastomere derivatives (see legend for references). The cell types of the mesenchyme stage blastula deriving from the cleavage stage lineages are indicated in the center column of Fig. 4.5. At this stage five sets of cells can be distinguished, on the basis of their current differentiated functions or their future developmental fates. These are *oral* and *aboral ectoderm*, *primary mesenchyme*, the eight *small micromeres*, and the anlagen for the gut and secondary mesenchyme derivatives, all harbored within the *vegetal plate*. The further diversification of cell types that takes place in a second phase of differentiation occurring after gastrulation has begun, and the number of cells attributed to each in the completed embryo are indicated at the right of Fig. 4.5. A third phase of differentiation and morphogenesis takes place during the several weeks of larval growth that follow the onset of feeding. This results in the development of the *imaginal rudiment* from the coelomic sacs, with contributions from ectodermal and alimentary tract elements of the mature larva. At metamorphosis the specifically larval structures initially formed during embryogenesis are destroyed or discarded, including for instance the larval skeleton, the anal, oral and esophageal structures, the ciliary band, and most of the aboral ectoderm. The juvenile sea urchin formed within emerges through the collapsed larval ectoderm (for authoritative general descriptions see Czihak, 1971, 1975; Hörstadius, 1973).

In most modern sea urchin species the whole of embryogenesis requires only a few days. It results in the formation of a free living larva that is capable of feeding, motility, and growth. In the following we shall be concerned only with this brief embryonic phase of the overall life span, which in sea urchins may last for many years. In a few sea urchin species such as

Heliocidaris erythrogramma a pelagic larva is not formed at all, and instead development proceeds directly from the blastula stage embryo to formation of the juvenile sea urchin (Mortensen, 1921; Williams and Anderson, 1975). Most of the cell types and structures shown in Fig. 4.5 never appear in *H. erythrogramma*, for example the larval skeleton, the mouth, the anus, and the oral arms. An archenteron is formed as usual by invagination but it is soon sealed off, and the juvenile is constructed directly from mesenchyme cells and bilateral coelomic sacs. This example illustrates by comparison the fact that in typical sea urchins the function of *embryogenesis* is to construct the pelagic *larva*, rather than the more complex juvenile sea urchin, and the relative simplicity of the larva of course contributes to the accessibility of embryogenesis in this organism. It suggests, furthermore, that the genetic regulatory programs utilized for embryogenesis are modular. Thus a large fraction of the slate of functions required for embryogenesis in most sea urchins, including other *Heliocidaris* species, have been deleted in *H. erythrogramma*.

Developmental Fates and Mitotic Activity in the Major Late Blastula Cell Types and Their Descendants

Following the logarithmic increase in cell number during midcleavage (see Fig. 3.1) mitotic activity slows in a lineage-specific manner. This can be seen to occur as early as 6th cleavage in some cells of the vegetal region of the embryo (*viz.*, the micromere [*M*] derivatives; Fig. 4.5). There are about 200 presumptive aboral ectoderm cells in the mesenchyme blastula stage *S. purpuratus* embryo, and this number has increased by a factor only slightly greater than two by the end of embryogenesis. Certain regions of the presumptive oral ectoderm are mitotically more active later in development, particularly during the postgastrular morphogenesis of larval arms and oral structures. The total number of oral ectoderm cells in the completed pluteus stage embryo is estimated to be about 800, and there are at least 90 neurons, which derive from the embryonic ectoderm as well (see Fig. 4.5 and legend, for sources and details).

The cessation of cleavage is marked in each ectoderm cell by the generation of a single cilium. These can be seen in the scanning electron micrograph of the vegetal pole region of an early blastula stage embryo of *Anthocidaris crassispina* reproduced in Fig. 4.6(a). Measurements reviewed in Chapter II indicate that the tubulin proteins included in blastula stage cilia are mainly if not exclusively of maternal origin, and there is little evidence from studies with inhibitors of RNA and protein synthesis that ciliation *requires* either zygotic transcripts or any newly synthesized polypeptides (Gross *et al.*, 1964; Auclair and Meismer, 1965; Auclair and Siegel, 1966; Child and Apter, 1969; Burns, 1973a,b). Figure 4.6(b)-(c), from observations of Masuda and Sato (1984) on embryos of *Temnopleurus toreumaticus*, illustrates the correlation between withdrawal from the mitotic cycle and the appearance of cilia in ectoderm cells. This usually occurs at 9th or 10th cleavage, though the exact number of division cycles varies according to species, and in some species one further division may occur following ciliogenesis (Masuda, 1979). Ciliogenesis does not take place in the eight small micromeres, as can be seen in Fig. 4.6(a), and in many species it does not occur in the presumptive mesenchyme cells either, though it does in others (Gibbons *et al.*, 1969; Raff, 1975; Masuda, 1979). Cilia-bearing ectoderm cells display cytological specializations. Their structure is polarized, with the nucleus located toward the basal (blastocoel) ends and desmosomes at the apical ends of the intercellular junctions (Wolpert and Mercer, 1963; Katow and Solursh, 1982). The apical and basal cell surfaces are also distinguished by different molecular constituents, as reviewed below.

The vital marking experiments by which the major fate assignments shown in Fig. 4.5 were initially established demonstrate that the cells giving rise to the vegetal plate are in the very early blastula positioned circumferentially around the vegetal region of the embryo (Hörstadius, 1939, 1973). These cells derive from eight specific progenitors formed at 6th cleavage (i.e., in Fig. 4.5, VAM1*l*, VAM2*l*, VOM1*l*, VOM2*l*, the two VLM1*l* and the two VLM2*l* cells). Many of the mitotically active cells of the late embryo are vegetal plate derivatives. Thus there are about 60 cells in the vegetal plate, 100 cells in the newly invaginated archenteron and about 440 cells in the completed digestive tract of the late embryo (Burke, 1980; Ettensohn, 1984). Autoradiographic observations indicate active mitosis in the gut in the late gastrula, although the initial formation of the archenteron by the inpocketing of the vegetal plate does not require cell division. The mouth is formed inductively, and is of ectodermal origin (Gustafson and Wolpert, 1967; Hörstadius, 1973). Secondary mesenchyme cells in some species begin to be released from the vegetal plate prior to invagination, and they delaminate continuously from the anterior end of the archenteron during gastrulation. These cells are best regarded as a heterogeneous set of embryonic blast cells, rather than as a single cell type. They divide actively, and include progenitors of pigment cells (Gibson and Burke, 1985), muscle cells (Gustafson and Wolpert, 1967; Ishimoda-Takagi *et al.*, 1984), and probably of the bilateral coelomic sacs from which arise the mesodermal portions of the imaginal rudiment (Gustafson and Wolpert, 1967; Czihak, 1971; Cameron and Hinegardner, 1974).

The sole known function of primary mesenchyme cells is the formation of the larval skeleton. They are distinguished from secondary mesenchyme cells by their different ancestral lineage as well as by their function. The 32 mesenchyme cells of the blastula stage embryo arise through a series of three divisions from four 5th cleavage precursor cells (in Fig. 4.5 VAM*k*, VOM*k*, and the two VLM*k* cells). As indicated in Fig. 4.5 the primary mesenchyme cells have divided once more on the average by the end of embryogenesis.

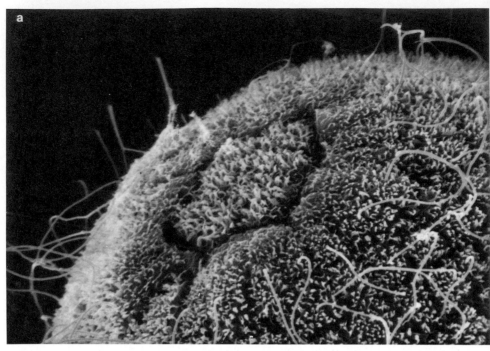

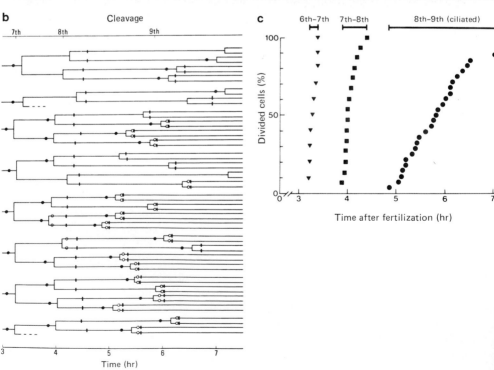

Primary mesenchyme cells are so named because in most species they enter the blastocoel prior to the formation of any secondary mesenchyme cells. However, this is not invariably the case. In certain members of the Cidaroida, the living order of sea urchins that is believed to resemble the Triassic ancestors of all modern sea urchin groups, skeletogenic mesenchyme cells appear in the blastocoel following rather than preceeding gastrular invagination (Schroeder, 1981). This is also the case in holothuroid, crinoid, and asteroid echinoderm embryos.

The four sister cells of the skeletogenic primary mesenchyme precursors (i.e., VAMs, VOMs, and the two VLMs cells) divide only once more. The eight "small micromeres" thus formed have a distinct fate. When the primary mesenchyme ingresses they remain situated in the center of the vegetal plate and thence are carried anteriorly with the invaginating archenteron. Ultimately they are distributed to the newly formed bilateral coelomic sacs, where they resume mitotic activity, and presumably contribute in some way to the imaginal rudiment (Endo, 1966; Okazaki, 1975a; Pehrson and Cohen, 1986).

The earliest morphogenetic achievement in sea urchin embryogenesis, the formation of a ciliated blastula, provides the embryo with motility. Subsequent embryonic morphogenesis creates a larva that is able to feed on planktonic microorganisms and to propel itself in an organized fashion by coordinated ciliary beat (Strathmann, 1975). The larval life style requires the development of an alimentary system equipped with neuromuscular and digestive functions, and of the larval skeleton, the ciliary bands, etc. By generating the capacities to swim, to channel food into the mouth by ciliary beat, to ingest and to digest, these larval structures in turn provide the requirements for further growth during pelagic existence. By metamorphosis the larva (in *S. purpuratus*) contains about 5×10^4 cells (Davidson *et al.*, 1982), approximately a 25-fold increase from the pluteus stage embryo. Per-

---

Fig. 4.6. Ciliogenesis and mitotic activity in ectoderm cells in blastula stage sea urchin embryos. (a) Scanning electron micrograph of vegetal pole region of early *Anthocidaris crassispina* blastula. The eight small micromeres are surrounded by a cleft separating them from the ciliated blastomeres. A single cilium can be seen protruding from the surface of each of these cells, i.e., other than the small micromeres, in the region shown clearly in the foreground. X2800. [Courtesy of S. Amemiya.] (b) Later cycles of eight ectodermal lineage elements derived from observations made on living embryos of *Temnopleurus toreumaticus* by Nomarski interference microscopy are shown. Top and bottom scales show cleavage cycle and time after fertilization (at 25°C). Short vertical lines indicate initiation of ciliogenesis, and vertical brachiations the completion of cytokinesis. Closed and open circles indicate breakdown and reformation, respectively, of complete nuclear membranes. (c) Loss of mitotic synchrony at 9th cleavage immediately precedes ciliogenesis in *T. toreumaticus*. In most of the lineages illustrated in (b) there are no further divisions for some time following ciliogenesis, though, as shown, this is not invariably the case. [(b) and (c) From M. Masuda and H. Sato (1984). *Dev., Growth Differ.* **26**, 281.]

haps two-thirds of these cells are included in the emergent juvenile, which after several weeks of morphogenesis completes its own digestive tract and begins to graze on the substratum. Thus each step in the overall developmental process changes the external biological properties of the organism. The underlying genomic programs that direct even the earliest stages of morphogenesis are therefore to be regarded as the direct product of evolutionary selective forces, acting at each stage of development.

Asymmetric Cleavage and Clonal Patterns of Differentiation in the Early Embryo

The locations of the early blastomeres, and of the embryonic structures to which they give rise are fixed with respect to the animal-vegetal axis of the unfertilized egg. However, this does not imply that the lineage shown in Fig. 4.5 describes a wholly *autonomous* pattern of blastomere differentiation. As reviewed in more detail below, and in Chapter VI, some cell lineages of the sea urchin embryo do indeed develop autonomously, but others are evidently specified for their pathways of differentiation by means of intercellular interactions or polarizations. The micromeres, for example, divide and differentiate autonomously into skeletogenic mesenchyme cells, either in culture or in ectopic positions in the embryo. On the other hand, the blastomeres of the animal half of the embryo, and the immediately subequatorial blastomere tiers remain plastic, their fates depending on the identity of their neighbors, far into cleavage. Stated another way, localized maternal components may suffice to determine the fates of the micromere lineage, while for other regions of the embryo lineage specification requires inductive interactions between blastomeres, though maternal factors may be involved as well. An aspect remarked on by early observers, particularly Hörstadius (1939; see also Fig. 6.30) is that the fates of descendants of all blastomeres in each horizontal *tier* of the 16- to 32-cell stage embryo are to some extent similar. The equivalence of the blastomeres belonging to given tiers along the animal-vegetal axis can be seen diagrammatically in Fig. 4.5. Thus, for example, the fate guides that lead to primary mesenchyme designate the progeny of a single horizontal 4th cleavage tier of cells, (*viz.*, VA*M*, the two VL*M* cells, and VO*M*; the micromere tier); those that lead to the vegetal plate designate progeny of the sisters of these cells (*viz.*, VAM, the two VLM cells and VOM; the macromere tier); and a subsequent horizontal cleavage in the latter set of lineages, i.e., the 6th, separates the progenitors of the vegetal plate (*viz.*, VAM1*l*, VAM2*l*, etc.) from the progenitors of the anal ectoderm (*viz.*, VAM1*u*, VAM2*u*, etc.). It follows that the early horizontal cleavages are largely asymmetric in regard to the fates of the daughter cell lineages. In some cases they are obviously physically asymmetric as well, for instance in the separation of micromeres (in Fig. 4.5, *M* cells) from macromeres (M cells) at 4th cleavage, and in the separation of skeletogenic mesenchyme precursors (in Fig. 4.5, k cells) from the small micromere precursors (s cells). In the vegetal half the last cleavage in which the upper and lower daughter cells have diverse fates that is known to occur is the 6th, in the M cell lineages.

Figure 4.5 shows that from the 3rd-6th cleavage on, depending on the specific lineage, the major cell types of the mesenchyme blastula differentiate *clonally*. The four clonal progenitors of the skeletogenic primary mesenchyme are separated from the clonal progenitors of the eight small micromeres at 5th cleavage, for instance (i.e., VA*M*k *vs*. VA*M*s; etc.). Czihak (1963) showed that as early as 3rd cleavage the two blastomeres on the future oral side of the embryo (i.e., No and VO) can be identified by their staining properties. Thus the oral-aboral axis is by then specified (though not irreversibly; see Chapter VI for review). This has been confirmed by microinjection of fluoresceinated dextran lineage tracers (see legend to Fig. 4.5). A large part of the oral ectoderm derives clonally from No, and similarly, a large part of the aboral ectoderm from the 8-cell blastomere at the opposite horizontal pole, Na. The progenitors that contribute to the aboral ectoderm (Fig. 4.5) each give rise to a specific region of this structure. It is not known when or how the progeny of the lateral cells NL segregate into oral and aboral progenitors, but since roughly half of the blastula stage ectoderm is included in each class, their contributions to each might be approximately equal, and this could occur as early as the meridional 4th cleavage. The ectoderm has a fixed cellular geometry and so far as is known ectoderm cell migration does not occur during embryogenesis. Clonal progeny are thus in fact contiguous.

In summary, even discounting the unknown distribution of descendants of the lateral animal blastomeres, *throughout most of the embryonic lineage clonal and phenotypic boundaries correspond*. In Fig. 4.5 the clonal progenitor cells in each lineage element are the first cells in which lower case letters appear in the lineage name (see legend). However, though they also arise in a clonal fashion, the blast cells included in the vegetal plate later differentiate in multiple directions. Similarly, in postgastrular development neuroblasts evidently arise within certain regions of the ectoderm. Evidence reviewed below indicates that only the skeletogenic mesenchyme, and the majority of the aboral ectoderm, can be regarded as *both clonally and terminally differentiated* by the mesenchyme blastula stage.

It is interesting to view the early strategy of development in the sea urchin from the vantage point of the preceding discussion of development in *C. elegans*. Though they contain about the same number of cells, the first stage *C. elegans* larva is vastly more complicated, in terms of number of diverse cell types, than is the swimming mesenchyme blastula of the sea urchin. In both organisms there are a series of early, functionally asymmetric founder cell cleavages, that could be interpreted as segregations of morphogenetically significant maternal components, though this is not required. However, simple clonal patterns of differentiation, combined with the reservation of a large set of blast cells that will differentiate later, suffice for the completion of this initial stage of sea urchin embryogenesis. Terminally asymmetric, genetically programmed cell lineages that are capable of producing many different cell types from the progeny of a single progenitor cell, are not required for the morphogenesis of the relatively simple mesenchyme blastula of the sea urchin, though they are common in *C. elegans* embryogenesis. Of course such mechanisms could well be utilized later in sea urchin development (e.g., in the construction of the larval nervous system). One consequence is the absence of the rigidity that is an innate property of the sequential and autonomous sublineage differentiation processes observed throughout embryogenesis in *C. elegans*. As noted above, many of the cell lineages that in the sea urchin embryo differentiate clonally if undisturbed, nonetheless retain the capability of differentiating in other directions. The direction in which these clones develop [e.g., oral *vs.* aboral ectoderm, or archenteron *vs.* ectoderm (see review in Chapter VI)] appears to depend on inductive signals deriving from neighboring cells. Their period of flexibility does not terminate at the point at which the clonal lineages are established (i.e., 3rd-6th cleavage). Thus many experiments reviewed by Hörstadius (1973) show that assignments along both the animal-vegetal and the oral-aboral axes can be reversed inductively in chimeric blastomere recombinants assembled later than the 32- to 64-cell stage. In addition, the effects of brief treatments with LiCl, which causes reassignment of prospective ectoderm cells to a mesenchyme-endoderm fate, show that plasticity extends at least to 7th cleavage (early blastula) in sea urchin embryos (see, e.g., Bäckström and Gustafson, 1953; Hörstadius, 1973; Mitsunaga *et al.*, 1983) and in starfish embryos (Kominami, 1984). It follows that the switches that determine nonautonomous clonal patterns of differentiation in early sea urchin embryos are not irreversibly thrown at the asymmetric division event by which the clonal progenitors are formed. On the other hand, in the autonomously differentiating micromere lineages the 5th cleavage could indeed be determinate *per se*, and the skeletogenic mesenchyme precursors may transfer to their clonal descendants completed regulatory decisions. Clonal differentiation in the sea urchin embryo is thus apparently utilized in *both autonomous and inductively controlled processes of development*.

(ii) Lineage-Specific Gene Expression in Skeletogenic Primary Mesenchyme Cells

Stages of Mesenchyme Cell Differentiation

In the early blastula the 32 presumptive primary mesenchyme cells are located in the vegetal-most region of the spherical embryo, immediately surrounding the eight small micromeres at the pole, as shown in the scanning electron micrograph reproduced in Fig. 4.7(a) [see also Fig. 4.6(a)]. The first morphological indices of differential function in the presumptive mesenchyme cells occur with the onset of the process by which these cells detach from the ectodermal wall and move into the blastocoel, termed *ingression*. The presumptive mesenchyme cells become pulsatile, assume a tear-drop shape, and protrude their nucleated basal ends into the blastocoel. The basal laminar membrane disappears from the region where ingression is to occur, though it continues to form the lining of the blastocoel in all other regions. This stage is visualized from within in Fig. 4.7(b) and in oblique views across the wall of the blastula in Fig. 4.7(c). Meanwhile the cilia (if present) disappear, and the apical surfaces of the ingressing cells lose their affinity for the external hyaline layer to which, like all other cells of the early blastula stage embryo, they had previously adhered. The lateral desmosomes joining the presumptive mesenchyme cells with their neighbors also disappear, and numerous other changes occur, including realignment of internal cytoskeletal structures and acquisition of ameboid motility. These results are summarized diagrammatically in Fig. 4.7(d) (Katow and Solursh, 1980). In some species the process of ingression is synchronous, and the mesenchyme cells migrate *en masse* into the floor of the blastocoel (e.g., *Hemicentrotus pulcherrimus*; Amemiya, 1986), while in others (e.g., *S. purpuratus*) it occurs more gradually over a period of several hours. Following completion of ingression the *vegetal plate* of the mesenchyme blastula is established as the cells of the surrounding region (i.e., presumptive archenteron and secondary mesenchyme cells) move downward, thus replacing the departed mesenchyme cells, and closing the intercellular apertures left on their withdrawal from the ectodermal wall (Okazaki, 1975a).

Once within the blastocoel the primary mesenchyme cells round up and for some time remain bunched together near the vegetal pole. They then enter a migratory phase, during which individual cells are found in many regions of the blastocoel wall. Early in gastrulation they return to the foot of the invaginating archenteron, and there they initiate skeletogenesis in two bilateral aggregations located at the future oral side of the embryo. The initial skeletal elements to be formed are small triradiate spicules, which are then extended as the mesenchyme cells move along the walls of the blastocoel, and align themselves in linear arrays within which the skeletal elements are secreted. Primary mesenchyme cell differentiation can thus be considered in three phases, which can be defined in an abbreviated manner as *ingression, migration*, and *skeletogenesis*.

Under appropriate culture conditions micromeres isolated from 4th cleavage embryos divide several times, and then display patterns of activity that correspond to each phase of their normal differentiation, including skeletogenesis (Okazaki, 1975b; Harkey and Whiteley, 1980, 1983; McCarthy and Spiegel, 1983a; Carson *et al.*, 1985). Differentiated mesenchyme cells isolated directly from late blastulae or gastrulae also carry out skeletogenesis *in*

Fig. 4.7. Ingression of primary mesenchyme cells. (a) Scanning electron micrograph of the vegetal pole of *Lytechinus pictus* blastula prior to ingression. The depression in the center of the polar region is occupied by the eight small micromeres (see Fig. 4.5), here marked by white circles. About 30 presumptive mesenchyme cells surround these. Ciliated cells are indicated by white arrows. X2430. (b) Scanning electron micrograph of *interior* of vegetal region of blastocoel, at the stage at which the apical surfaces of the ingressing mesenchyme cells have begun to intrude. White arrows indicate dividing cells in the peripheral region; white circles as above. BW, Blastula wall. X2340. [(a) and (b) From H. Katow and M. Solursh (1980). *J. Exp. Zool.* **213**, 231.] (c) Scanning electron micrograph of mesenchyme cells in process of ingression, seen in fracture of vegetal wall of blastula stage embryo of *Hemicentrotus pulcherrimus*. X2070. [(c) From S. Amemiya (1986). In preparation.] (d) Diagram of primary mesenchyme cell ingression. *1*, Morphology of primary mesenchyme cells and of adjacent blastomeres is initially identical. C, Cilia; N, nucleus; BL, basal lamina; BC, blastocoel; H, hyaline layer; B, basal body of cilium, surrounded by Golgi apparatus. *2*, Elongation of presumptive mesenchyme cell, P, into blastomere. Arrows indicate external cell processes. Basal lamina has disappeared above the ingressing cell. *3*, Detachment and further elongation. Desmosomes between primary mesenchyme cell and adjacent blastomeres have disappeared, and aligned microtubules are seen in the peripheral cytoplasm (small arrowheads). *4*, Delamination of primary mesenchyme cell. Note extracellular material between mesenchyme cell and neighboring cells. *5*, Escape into the blastocoel. [(d) From H. Katow and M. Solursh (1980). *J. Exp. Zool.* **213**, 231.]

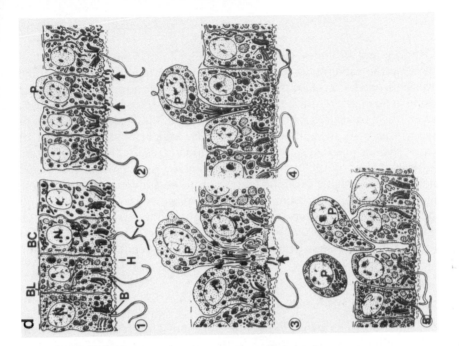

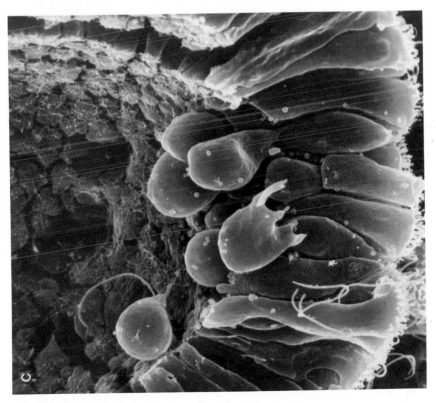

Fig. 4.7. (*Continued.*)

vitro. Harkey and Whiteley (1983) showed that in micromere cultures the cells develop the capacity to attach to plastic dishes at about the same time as ingression occurs *in vivo*; they become migratory *in vitro* at about the same time as *in vivo*; and they then begin to deposit spicules in parallel with normal or freshly cultured mesenchyme cells. Two dimensional analyses of protein synthesis in such micromere cultures indicate the appearance of many new species at the stage of differentiation corresponding to ingression *in vivo* (Harkey and Whiteley, 1983; Pittman and Ernst, 1984). Most of the mesenchyme cell-specific proteins that begin to be synthesized at this time continue to be prevalent products of mesenchyme cells at later phases of their differentiation, and conversely, *in vivo* the synthesis of most, though certainly not all late mesenchyme-specific protein species is activated around the time ingression begins (some identified examples are discussed below).

Genes Expressed in Mesenchyme Cells

Among the identified genes expressed very early in presumptive mesenchyme cell differentiation are those coding for the cytoskeletal actins CyI and CyIIb (Cox *et al.*, 1986). *In situ* hybridization demonstrates a sharp increase in the concentration of these mRNAs in the vegetal region, relative to the remainder of the embryo, prior to and immediately following primary mesenchyme cell ingression. This is shown for CyI mRNA in Fig. 4.8(a). The CyI actin message is not detected in the exact center of the vegetal pole, however, where the eight small micromeres are located. These, as indicated in Fig. 4.5, are of a lineage and fate different from those of the skeletogenic mesenchyme cells. CyI mRNAs remain prevalent in newly ingressed mesenchyme cells, but soon thereafter disappear.

The morphological observations summarized in Fig. 4.7(d) imply important molecular changes in the surface properties of mesenchyme cells during the process of ingression, and among these is loss of the specific ability to bind to the extraembryonic coat protein hyalin (McClay and Fink, 1982; Fink and McClay, 1985). This has been demonstrated *in vitro*, by measuring the force required to dislodge these cells from hyalin-coated surfaces. Primary mesenchyme cells concomitantly develop an increased affinity for the basal laminar glycoprotein fibronectin (Fink and McClay, 1985). These two characteristics, which might well be mediated at the gene level, do not occur in other embryonic cell types tested, nor in the micromere progenitors of the primary mesenchyme cell lineage. The appearance on the surfaces of primary mesenchyme cells of a new molecular species immediately following ingression has been demonstrated dramatically by indirect immunofluorescence, following reactions with monoclonal antibodies (McClay *et al.*, 1983; Wessel *et al.*, 1984; Wessel and McClay, 1985). In the example shown in Fig. 4.8(b), the antibody reacts with cells that have just completed ingression but not with cells still in the process of delamination (McClay *et al.*, 1983). The epitope recognized by this antibody is probably a carbohydrate moiety of a glycoprotein, and it is first localized in the Golgi apparatus and on the cell surfaces (Wessel and McClay, 1985; Anstrom *et al.*, 1986). It then persists on the surfaces and in the secretory vesicles of these cells throughout the skeletogenic phase of their activity. In cultured skeletogenic mesenchyme cells a different monoclonal antibody against this protein blocks spicule formation (Carson *et al.*, 1985; Grant *et al.*, 1985). Further studies have shown that this antibody specifically interferes with uptake of exogenous Ca^{2+} by the skeletogenic mesenchyme cells.

Several of the molecular species observed in primary mesenchyme cells can be specifically related to the migratory behavior these cells display in the second phase of their differentiation. Among these are fibronectin, which is concentrated in mesenchyme cells during this phase, and is also present in the basal laminar substratum on which they move within the blastocoel (Katow *et al.*, 1982; Spiegel *et al.*, 1983; Wessel *et al.*, 1984). The disappearance of fibronectin from mesenchyme cells is associated with their loss of migratory activity at the gastrula stage, when they initiate skeletogenesis (Katow and Solursh, 1982). Various additional components of the basal lamina may be required to facilitate mesenchyme cell migration, such as sulfated polysaccharides (Karp and Solursh, 1974; Solursh and Katow, 1982; Wessel *et al.*, 1984). A sulfated acid polysaccharide that binds to fibronectin *in vitro* has been isolated from sea urchin embryos, and injection into the blastocoel of a monoclonal antibody that recognizes this polysaccharide induces formation of abnormal spicules and other skeletal malformations (Iwata and Nakano, 1985). Observations on the disposition of concanavalin A (Con A) binding sites within the blastocoel provide further evidence that contact with polysaccharides, such as might be associated *in situ* with fibronectin, may be important in the control of skeletogenic mesenchyme cell migration. Thus Con A-binding sites normally disappear from the vegetal regions of the blastocoel where the primary mesenchyme cells will aggregate. Katow and Solursh (1982) showed that treatment with LiCl causes Con A-binding sites to remain uniformly distributed within the blastocoel, and in such embryos the mesenchyme cells continue in a state of migratory dispersal. They fail to aggregate or to carry out skeletogenesis. Expression of the genes coding for the cytoskeletal actin CyIIa is also correlated with migratory activity, in primary gut, and secondary mesenchyme cells (Cox *et al.*, 1986).

Skeletogenesis is the terminal differentiated function of the primary mesenchyme cell lineage. During this phase the number of skeletogenic mesenchyme cells approximately doubles (Fig. 4.5; see, e.g., Takahashi and Okazaki, 1979). The skeletal elements are secreted within syncytial "cables" attached to the cell bodies by cytoplasmic stalks, as shown for example in the electron micrographs reproduced in Fig. 4.8(d) and (e). These cables are

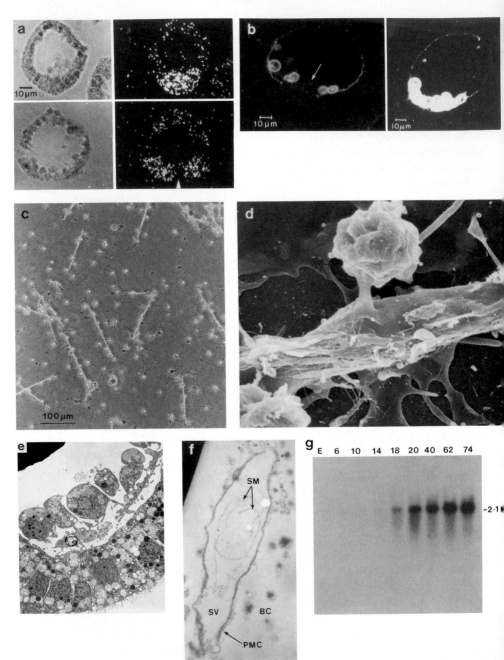

Fig. 4.8. Molecular indices of differential gene function in the primary mesenchyme cell lineage. (a) Phase micrographs and *in situ* hybridizations with sections of early mesenchyme blastula of *S. purpuratus*, using a probe that specifically recognizes the cytoskeletal actin gene CyI. The sections pass approximately along the animal-vegetal axis. Arrow indicates the gap in

formed after contacts between filopodia extended from neighboring cells, and this event is required for skeletogenesis to ensue. Thus for example, skeletogenesis does not take place in culture unless serum is supplied in the medium (Okazaki, 1975b), and McCarthy and Spiegel (1983a) showed that the specific process for which the serum component is required is extension and fusion of the filopodial networks. It is thought that the orientation of these multicellular aggregates, which determine the form of the skeletal rods, is established by means of spatial cues on the interior walls of the blastocoel (Okazaki, 1975b; Gustafson and Wolpert, 1963). The brachiations and relative lengths of the skeletal elements are species-specific characters (Okazaki, 1975b). In culture, though the mesenchyme cells freely generate skeletal rods, these are usually straight, or in any case very simple in form, lacking the specific geometry of skeletons formed *in situ* (Harkey and White-ley, 1980). A field from such a skeletogenic culture is shown in Fig. 4.8(c).

---

labeling pattern where the eight small micromeres are located. Grains are most intense in the vegetal region containing primary mesenchyme cells in process of ingression, and in some newly ingressed primary mesenchyme cells. [From K. H. Cox, L. M. Angerer, J. J. Lee, E. H. Davidson, and R. C. Angerer (1986). *J. Mol. Biol.* **188**, 159.] (b) Reactions of a monoclonal antibody named Ig8, with newly ingressed primary mesenchyme cells of *L. pictus* blastulae, visualized in sections by indirect immunofluorescence. Mesenchyme cells that have completed ingression react with the antibody while those in process of ingression but still resident in the blastocoel wall do not (arrow). [From D. R. McClay, G. W. Cannon, G. M. Wessel, R. D. Fink, and R. B. Marchase (1983). *In* "Time, Space, and Pattern in Embryonic Development" (W. R. Jeffery and R. A. Raff, eds.), p. 157. Alan R. Liss, New York.] (c) Spicule formation in *S. purpuratus* primary mesenchyme cell cultures. Cells can be observed aligned along the skeletal rods. [Courtesy of G. L. Decker and W. J. Lennarz.] (d) Scanning electron micrograph of cultured primary mesenchyme cell in process of spicule deposition. Two cell bodies can be seen attached by cytoplasmic stalks to cable within which the spicule element is deposited. X4600. [Courtesy of G. L. Decker, J. B. Morrill, and W. J. Lennarz.] (e) Transmission electron micrograph displaying spicule formation in *S. purpuratus* gastrula stage embryo. Parts of at least six primary mesenchyme cells in process of skeletogenesis are included, and the stalks by which some of these are connected to the underlying syncytial cable can be observed in several locations. At left center can be seen a cable vacuole in which spicule deposition is occurring. Below lies the ectodermal wall of the blastocoel, consisting of columnar ectoderm cells overlain with the basal laminar membrane. X1140. [Courtesy of G. Decker and W. J. Lennarz.] (f) Immunogold staining of spicule matrix proteins in transmission electron microscope section. Reaction occurs with the lamellar vacuole of the cytoplasmic cable where the spicule is depos-ited. No signal is observed after treatment with pre-immune serum. BC, blastocoel cavity; PMC, primary mesenchyme; SM, spicule matrix; SV, spicule vacuole. X5040. [From S. C. Benson, N. C. Benson, and F. Wilt (1986a). Reproduced from the journal of *J. Cell Biol.* **102,** 1878, by copyright permission of the Rockefeller University Press.] (g) Poly(A) RNA gel blot displaying accumulation of the message coding for the 50 kd major spicule matrix protein of *S. purpuratus*. Time postfertilization (hr, at 16°C) at which the poly(A) RNA samples were ex-tracted, is indicated at top of figure. Note that while this mRNA begins to accumulate as early as the initial processes of mesenchyme cell ingression, if not before, skeletogenesis is not initiated until the midgastrula stage (36-40 hr postfertilization at this temperature). [From H. M. Sucov, and E. H. Davidson, unpublished data; after Benson *et al.* (1986b).]

The mineral component of skeletal structures themselves is a 20:1 $CaCO_3:MgCO_3$ complex termed calcite. The spicules can be demineralized by treatment with EDTA or acid, leaving a proteinaceous matrix which retains the shape of the original spicule. Benson *et al.* (1986a) demonstrated that the spicule matrix consists of about ten protein species (there could be additional rare proteins) of which the most prevalent is a 47 kd glycoprotein containing about 4 kd of *N*-linked oligosaccharide. The spicule matrix proteins are secreted into the vacuole of the syncytial cable where the skeletal structure is deposited, as shown in Fig. 4.8(f) at electron microscope resolution, by the reaction of an antibody raised against these proteins. This same antibody has been utilized for the isolation of the gene coding for the major 47 kd spicule matrix protein (Sucov *et al.*, 1986). *In situ* hybridization shows that in blastula and gastrula stage embryos this gene is expressed only in primary mesenchyme cells (Benson *et al.*, 1986b). However, according both to results obtained by this method and to RNA gel blot hybridizations such as shown in Fig. 4.8(g), the transcripts of the major spicule matrix protein gene become detectable long in advance of skeletogenesis. This gene begins to be expressed at or just before ingression, though after completion of all of the divisions in the primary mesenchyme cell lineage (i.e., at about 14 hr in *S. purpuratus* embryos grown at 16°C; Benson *et al.*, 1986b). The sequence of the 47 kd spicule matrix protein includes a multiple repeating thirteen amino acid element that could be involved in structural orientation.

The observations reviewed in this section summarize an unusual breadth of information in respect to the variety of *identified structural genes* utilized in the differentiation of a single embryonic cell lineage. The functions of many of the proteins for which these genes code, e.g., the cytoskeletal actins, fibronectin, the Ca^{2+}-uptake and the spicule matrix proteins, can be reasonably related to the differentiated activities of this lineage, at each phase of its development. We have seen that many of the genes included in the primary mesenchyme gene battery are expressed as early as ingression, when the cells of this lineage first become obviously distinct from their neighbors, or even before. Yet ingression occurs long *after* the initial determination of the clonal mesenchyme cell progenitors. Since isolated micromeres (i.e., VA*M*, VO*M*, and the two VL*M* cells) divide and give rise to progeny that proceed in culture through the same processes of differentiation as *in vivo*, the determinative events accounting for the autonomous molecular differentiation of the lineage must have been set in train as early as the unequal 4th cleavage. A period of many hours ensues, in which this state of determination is propagated through four subsequent cleavages, in the first of which it is bequeathed exclusively to one of each pair of sister cells. Only after the blastula is fully formed do there appear any of the known molecular indices of primary mesenchyme cell-specific gene expression.

(iii) Differential Gene Expression in Sea Urchin Embryo Ectoderm

At least two regulatory patterns of gene expression are operating in ectoderm cells by the mesenchyme blastula stage. These are indicated respectively by genes that are expressed in all ectoderm cells (but not in mesenchyme cells), and by genes that are expressed exclusively in the *aboral*
ectoderm. There could be an oral ectoderm gene set as well, but so far no
early examples have been identified. Of course later in development elements of the oral and perhaps the aboral ectoderm as well differentiate in
additional directions, forming the ciliary band at the interface between oral
and aboral ectoderms, the mouth, and at least three distinct sets of neurons
and neuroblasts (see Fig. 4.5). Little is known of the molecular aspects of
these later differentiation processes, and the following brief review is focussed on differential zygotic gene expressions that can be observed as early
as the blastula-gastrula stages.

Some genes that in advanced embryos are expressed only in ectoderm
cells are apparently activated during cleavage. For example Nemer (1986)
demonstrated accumulation after the 16-cell stage of two unidentified ectoderm-specific poly(A) RNA sequences. These sequences are initially represented at modest levels in the maternal RNA and they attain their peak
concentrations at mesenchyme blastula stage. Their prevalence then declines precipitously. Treatment at midcleavage with LiCl, which is known to
suppress ectoderm differentiation, blocks their accumulation. These transcripts are among the earliest known products of zygotic genes that function
differentially in the embryo. At least some of the clonal ectodermal cell
lineages are already established by 4th cleavage (Fig. 4.5), and it is possible
that transcription of these sequences is activated upon segregation of the
ectoderm progenitor cells.

The gene coding for hyalin is probably expressed in all ectoderm cell
lineages. This protein is the major component of the extracellular hyaline
layer, which provides a tough, though permeable protective coat for the
embryo and later the feeding larva (Stephens and Kane, 1970; Citkowitz,
1971; McCarthy and Spiegel, 1983b; Cameron and Holland, 1985). Hyalin
can be solubilized by removal of Ca^{2+} from the medium (Herbst, 1900) and
purified from other components of the hyaline layer by repeated cycles of
Ca^{2+} precipitation (Kane, 1973). Under denaturing conditions this protein
migrates in electrophoretic gels as a polypeptide >300 kd in mass (McClay
and Fink, 1982; McCarthy and Spiegel, 1983b), though in sedimentation
analyses it behaves as a 90-100 kd polypeptide (Stephens and Kane, 1970).
Hyalin is synthesized during oogenesis, and in unfertilized eggs it is stored
within the cortical granules. This has been demonstrated by cytological immunofluorescence (McClay and Fink, 1982), as shown in Fig. 4.9(a), and at
electron microscope resolution by immunogold localization (Hylander and

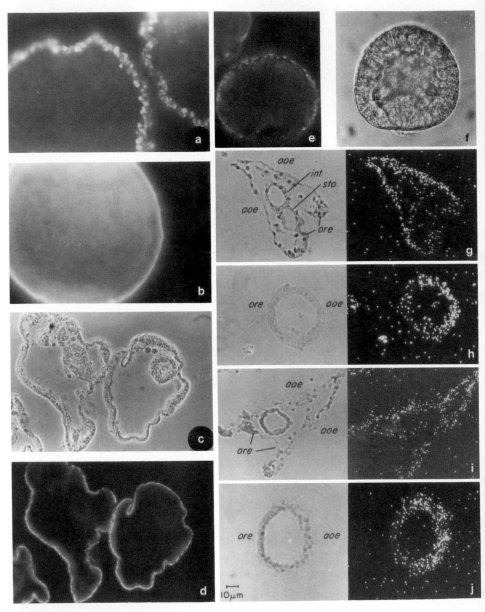

Fig. 4.9. Indices of differential gene expression in cells of the sea urchin embryo ectoderm. (a)-(f) Cytological localization of hyalin by indirect immunofluorescence. Sections of (a) an unfertilized egg, (b) a fertilized egg, and (d) a pluteus-stage *S. purpuratus* embryo that is shown by phase microscopy in (c), were reacted with a polyclonal anti-hyalin antibody, and the reaction visualized with a fluorescein-conjugated second antibody. In (a) the hyalin is localized to the cortical granules; in (b) it has been released to the external surface; and in (d) it covers the whole of the ectodermal surface, but is not observed anywhere internally. [(a)-(d) E. H. David-

Summers, 1982). At fertilization the cortical granules undergo exocytosis and the hyalin is released to the exterior [Fig. 4.9(b)]. Figures 4.9(c)-(e) show that hyalin is found exclusively on the outside surfaces of ectoderm cells, of which it is a product, and it is evidently synthesized by both oral and aboral ectoderm. It is not synthesized in archenteron or mesenchyme cells. If the maternal hyaline coat is dissolved by exposure to reduced Ca^{2+} it is regenerated as early as the blastula stage (Kane, 1973), the point at which new hyalin synthesis is first detected by immunoprecipitation (McClay and Fink, 1982). In midgastrula stage embryos stripped of their original hyaline coat it is observed that presumptive archenteron cells still in the vegetal plate lack the capacity to regenerate surface hyalin, a function carried out by the ectoderm cells covering the remainder of the embryo [Fig. 4.9(e) and (f)]. Thus expression of the hyalin gene appears confined to ectoderm cell lineages, and it does not occur in any descendants of the micromeres or of the eight 6th cleavage clonal progenitors of the vegetal plate (i.e., VAM1*l*, VAM2*l*, etc.; see Fig. 4.5).

The blastocoel of the embryo is coated with a complex of extracellular glycoproteins that form the extracellular matrix and the basal laminar membrane. These include fibronectin, laminin, heparin sulfate proteoglycan, type IV collagen, and several unknown macromolecules identified by monoclonal antibodies (Solursh and Katow, 1982; Spiegel *et al.*, 1983; Wessel *et al.*, 1984). The basal lamina later surrounds the gut as well, and thus while it is initially an ectoderm cell product it is also produced by cells of other lineages. Heparin sulfate proteoglycan, fibronectin, and laminin are also present on the surfaces of primary mesenchyme cells (Wessel *et al.*, 1984) though it is not clear which cells carry out their synthesis. Unlike hyalin, which is secreted only at the apical cell surfaces, or type IV collagen, which is seen only on the inner surfaces of the embryo (and intracellularly; Wessel *et al.*, 1984), fibronectin and laminin occur on both apical and basal ectoderm cell surfaces (Spiegel *et al.*, 1983). Cell surface glycoproteins are actively

---

son, C. N. Flytzanis, J. J. Lee, J. J. Robinson, S. J. Rose III, and H. M. Sucov (1985). *Cold Spring Harbor Symp. Quant. Biol.* **50**, 321.] (e) Localization of newly regenerated hyalin by indirect immunofluorescence on whole gastrula stage embryo of *L. pictus*, 30 minutes following removal of the original hyaline coat by treatment with low Ca^{2+} sea water. As shown in the photomicrograph (f) of an embryo of the same batch as in (e), the region that does not regenerate hyalin consists of vegetal plate cells that are soon to invaginate. [(e) From D. R. McClay and R. D. Fink (1982). *Dev. Biol.* **92**, 285. (f) Courtesy of D. R. McClay and R. D. Fink.] (g) *In situ* hybridization, right, of a section of a pluteus stage *S. purpuratus* embryo seen in the phase microscope, left, with a probe that is specific for the CyIII subfamily actin mRNAs. Only cells of the aboral ectoderm display these mRNAs; grains in other regions are at background level. (h) *In situ* hybridization of CyIII probe with section of 29 hr mesenchyme blastula; (i) section of pluteus stage embryo as in (g), but reacted with Spec1 probe; (j) same, except reacted with section of 23 hr mesenchyme blastula. [(g)-(j) From R. C. Angerer and E. H. Davidson (1984). *Science* **226**, 1153. Copyright (1984) by the AAAS.]

synthesized during development as are the key glycosyltransferases required for their synthesis (Heifetz and Lennarz, 1979; Welply et al., 1985; Grant et al., 1985; reviewed in Lennarz, 1985). Application of drugs such as tunicamycin and arylxyloside that block synthesis and assembly of the carbohydrate constituents of these proteins severely affects morphogenesis, including both primary and secondary mesenchyme cell migration and archenteron invagination (Heifetz and Lennarz, 1979; Carson and Lennarz, 1979; Akasaka et al., 1980), as does sulfate starvation. This presumably interferes with the production of the associated sulfate polysaccharides prevalent on the blastocoel wall (see, e.g., Katow and Solursh, 1981).

Though their quantity increases due to new synthesis as development proceeds, most of the basal laminar constituents discussed *are already stored in the egg at fertilization.* Wessel et al. (1984) demonstrated the colocalization in certain maternal cytoplasmic granules of fibronectin, heparin sulfate proteoglycan, laminin, collagen types I, II, and IV, and various other basal laminar antigens. It is not known whether these granules are membrane bound, and they are not obviously associated with cytoskeletal elements since they are easily stratified by low centrifugal force. Cytological immunofluorescence indicates a diffuse rather than punctate distribution of these constituents following fertilization. Later they are apparently secreted to form the extracellular matrix and basal laminar membrane of the early blastula. This provides an interesting illustration of the inference discussed in Chapter III that the embryo characteristically constructs complex three-dimensional morphological structures by self-assembly of presynthesized components. The genes coding for the known basal laminar components are also typical in that they are both maternally and zygotically active.

At least two different gene families are specifically active in the *aboral ectoderm.* These are the Spec genes, which code for Ca^{2+} binding proteins, and certain of the cytoskeletal actin genes, *viz.,* CyIIIa and CyIIIb. The CyIIIa and CyIIIb genes may be the result of a recent duplication event, as they are linked and share similar 3′ nontranslated and intron sequences which are not homologous with those of other actin genes (J. J. Lee et al., 1984; Akhurst et al., 1986). The actin protein coded by CyIIIa differs in amino acid sequence by only 1-2% from those produced by the other three functional cytoskeletal actin genes of *S. purpuratus.* The CyIII actin gene subfamily is expressed exclusively in aboral ectoderm cells, as shown in the *in situ* hybridization reproduced in Fig. 4.9(g) (Angerer and Davidson, 1984). Furthermore, these genes are utilized only in embryonic and larval life, i.e., until the disappearance of the larval aboral ectoderm at metamorphosis (Shott et al., 1984). The Spec genes code for a set of small acidic proteins, also synthesized exclusively in the aboral ectoderm cells (Bruskin et al., 1981, 1982; Lynn et al., 1983). An *in situ* hybridization demonstrating the same pattern of localization for this mRNA as for the CyIII actin mRNAs is shown in Fig. 4.9(i). Immunocytochemical localization of the Spec proteins confirms that they are present exclusively in aboral ectoderm cells (Carpenter *et al.*, 1984). The major fraction of the Spec gene family mRNA derives from the single Spec1 gene, the nucleotide sequence of which demonstrates it to be a member of the troponin C superfamily (Carpenter *et al.*, 1984; Hardin *et al.*, 1985). The special requirement of aboral ectoderm cells for these Ca^{2+} binding proteins remains unexplained, though it may be relevant that it is the induced collapse of the aboral ectoderm at metamorphosis that provides the means by which the juvenile sea urchin emerges from the larval corpus (Carpenter *et al.*, 1984).

The Spec genes and the CyIIIa and CyIIIb actin genes display similar ontogenic patterns of activation. A very small amount of maternal mRNA is present (Bruskin *et al.*, 1981; Shott *et al.*, 1984; Nemer, 1986) which in the case of the CyIIIa actin gene has been measured by single strand probe excess titration at about 1200 molecules per egg (Lee *et al.*, 1986a). Early in the blastula stage the CyIIIa gene becomes activated, as shown both by mRNA accumulation and run-off transcription in isolated nuclei (Shott *et al.*, 1984; Lee *et al.*, 1986a,b; S. A. Johnson and E. H. Davidson, unpublished data). Activation of the Spec1 gene occurs at about the same time, and the quantity of Spec1 mRNA climbs rapidly during the early blastula stage (Bruskin *et al.*, 1982; Nemer, 1986). Some hours later the CyIIIb and the Spec2 mRNAs become detectable (Bruskin *et al.*, 1982; Shott *et al.*, 1984). There are about 8×10^4 molecules of CyIIIa mRNA and about 2×10^5 molecules of Spec1 mRNA per embryo at pluteus stage, or several hundred molecules of each species per aboral ectoderm cell (Lynn *et al.*, 1983; Lee *et al.*, 1986a). Both sets of genes begin to be expressed exclusively in the *presumptive* aboral ectoderm at the blastula stage, prior to the appearance of any visible morphological distinction between future oral and aboral sides of the embryo. This can be seen in Fig. 4.9(h) and (j). The activation of these genes is evidently tightly linked to an ectoderm-specific program of differential gene activity. Thus treatment with Zn^{2+} represses ectoderm differentiation, and also delays the activation of the Spec1 gene (Nemer *et al.*, 1985). At least the central region of the aboral ectoderm consists of the clonal progeny of the eight cell blastomere, Na, though the clonal progenitors for other regions of the aboral ectoderm do not arise until subsequent cleavages (see Fig. 4.5). The early expression of the CyIII actin genes and the Spec genes serves to mark the differential specification of the genomes in the aboral ectoderm cell lineages, which either directly or otherwise may be imposed as early as 3rd cleavage.

(iv) Differential Molecular Functions in Secondary Mesenchyme, Muscle, and Archenteron Cells

The prospective secondary mesenchyme is to be regarded essentially as a pluripotential set of blast cells, as noted above, and the mechanisms leading to the appearance of the several diverse cell types that develop from this source are obscure. Various processes by which a particular pathway is chosen can be envisioned. For example a given fraction of the blast cells could spontaneously commit to one of several allowed courses of differentiation, repressing those pathways in their neighbors, an indeterminate process at the level of any given cell; or at the opposite extreme, production of given cell types could result from late, functionally asymmetric divisions, that from the beginning are programmed into the secondary mesenchyme cell lineage. There is some hint that secondary mesenchyme cells of diverse fate may be separated from one another while still resident within the vegetal plate. Thus, e.g., in some sea urchin species (though not others) the pigment cell blasts emerge into the blastocoel prior to archenteron invagination and to the release of secondary mesenchyme cells of other fates (reviewed in Gibson and Burke, 1985). Prospective pigment cells have been visualized while still within the vegetal plate in *S. purpuratus* embryos, by use of a monoclonal antibody which in the pluteus ectoderm reacts specifically with the surfaces of fully differentiated echinochrome containing pigment cells, as illustrated in Fig. 4.10(a) and (b) (Gibson and Burke, 1985). About eight pigment cell progenitors within the vegetal plate of the late mesenchyme blastula are initially visualized with this antibody. These cells begin to enter the blastocoel well before invagination of the archenteron, but many hours after ingression of the skeletogenic primary mesenchyme cells has been completed, and they continue to delaminate from the archenteron tip during the early gastrula stage [Fig. 4.10(c)-(d)]. Later in gastrulation, while secondary mesenchyme cells of other fates are still being released into the blastocoel the pigment cell blasts divide and their progeny embed themselves in the ectodermal wall of the embryo [Fig. 4.10(e)-(f)]. There are 28 ± 5 of these cells in the completed pluteus.

Another differentiated structure that derives from the secondary mesenchyme is the esophageal muscle. This is composed of contractile bands that surround the esophagus, and is required for ingestion of food into the larval alimentary tract. The muscle cells derive initially from motile, pseudopod-bearing secondary mesenchyme cells associated with the coelomic sacs (Gustafson and Wolpert, 1967). The differentiation of muscle cells follows rather than precedes most of the process of archenteron formation. Ishimoda-Takagi *et al.* (1984) demonstrated reaction of an antibody against muscle tropomyosin initially with mesenchymal precursor cells lying at the junction between the coelomic sacs and the esophagus. Similarly, as shown in Fig. 4.10(g) and (h), *in situ* hybridization with a probe specific for the muscle (M) actin isotype indicates the initial appearance of this message in a few cells located on either side of the upper archenteron at the late gastrula stage (Cox *et al.*, 1986). There are about 10-20 cells that contain M actin mRNA in the pluteus stage embryo on either side. No maternal M actin gene transcripts can be detected. The mRNAs accumulate rapidly in the small population of presumptive muscle cells, attaining concentrations of approximately

10^3 molecules per cell in the advanced pluteus stage embryo (Shott *et al.*, 1984; Lee *et al.*, 1986a,b). The muscle tropomyosin gene and the M actin gene represent a set of late activated zygotic genes that are expressed in only a small subfraction of secondary mesenchyme cell derivatives. Another gene that is activated on the same schedule as the M actin gene, and can be considered a member of the same gene battery is the sea urchin myosin heavy chain gene (S. J. Rose, III, and E. H. Davidson, unpublished data).

The secondary mesenchyme and the cells that invaginate to form the gut share a common origin. Thus it is interesting that the CyIIa cytoskeletal actin gene is expressed *both* in regions of the formed archenteron and in secondary mesenchyme cells even prior to their delamination, but in no other cells of the embryo after the late blastula stage (Cox *et al.*, 1986). This

---

Fig. 4.10. Differential molecular functions of vegetal plate derivatives in the sea urchin embryo. (a) Reaction of monoclonal antibody with surface antigen of pigment cells in late *S. purpuratus* pluteus, visualized by indirect immunofluorescence; (b) phase photomicrograph displaying location of stellate, granular pigment cells containing echinochrome in the same embryo, X200; (c) reaction of same monoclonal antibody with pigment cell blasts in the secondary mesenchyme; (d) phase photomicrograph of same embryo, X300. Some cells bearing the antigen are free within the blastocoel or are in process of delamination, while others are now embedded in the ectodermal body wall of the vegetal region. Invagination of the archenteron is about half complete. (e) Same monoclonal antibody reacted with pigment cells embedded in the ectodermal wall of an early pluteus stage embryo; (f) phase photomicrograph of same embryo, (a)-(f) are whole mounts. [(a)-(f) From A. W. Gibson and R. D. Burke (1985). *Dev. Biol.* **107**, 414.] (g) Phase photomicrograph and (h) *in situ* hybridization of muscle (M) actin probe with esophageal muscle cells of a late pluteus stage *S. purpuratus* embryo. The bilateral coelomic sacs can be seen clearly in the section and the M actin mRNAs are present in a subset of cells associated with these sacs, on either side of the esophagus. e, esophagus; i, intestine. (i) Phase photomicrograph and (j) *in situ* hybridization of cloned probe specific for the cytoskeletal actin gene CyIIa with secondary mesenchyme cells of a late gastrula stage *S. purpuratus* embryo. Five secondary mesenchyme cells containing this mRNA can be observed in the section, while the region of the archenteron included in the section is not labeled (cf. l). [(g)-(j) From K. H. Cox, L. M. Angerer, J. J. Lee, E. H. Davidson, and R. C. Angerer (1986). *J. Mol. Biol.* **188**, 159]. (k) Phase photomicrograph and (l) *in situ* hybridization of CyIIa cytoskeletal actin probe with stomach region of a pluteus stage embryo. Note that the esophagus is not labeled. [(k) and (l) From R. C. Angerer and E. H. Davidson (1984). *Science* **226**, 1153. Copyright (1984) by the AAAS.] (m) Scanning electron micrograph displaying the invaginating archenteron, with delaminating secondary mesenchyme cells, in midgastrula stage embryo of *Hemicentrotus pulcherrimus* (18 hr at 20°C). Additional secondary mesenchyme cells can be seen on the upper wall of the blastocoel, and a portion of a primary mesenchyme cell aggregate at the lower right. X800. [Courtesy of S. Amemiya.] (n) Immunofluorescent detection of an archenteron-specific antigen in the vegetal plate of a late blastula stage embryo of *Lytechinus pictus*. This antigen is not present in the egg and is probably a product of zygotic gene expression. (o) Distribution of the same antigen in the alimentary tract of a late gastrula stage embryo. Note that the cells of the esophagus, which is of the same lineage as the remainder of the gut, do not react, nor do the adjacent cells of the anal ectoderm [these are of a different lineage (cf. Fig. 4.5)]. [(n) and (o) From D. R. McClay, G. W. Cannon, G. M. Wessel, R. D. Fink, and R. B. Marchase (1983). *In* "Time, Space, and Pattern in Embryonic Development" (W. R. Jeffery and R. A. Raff, eds.), p. 157. Alan R. Liss, New York.] (See illustrations on pp. 236–237.)

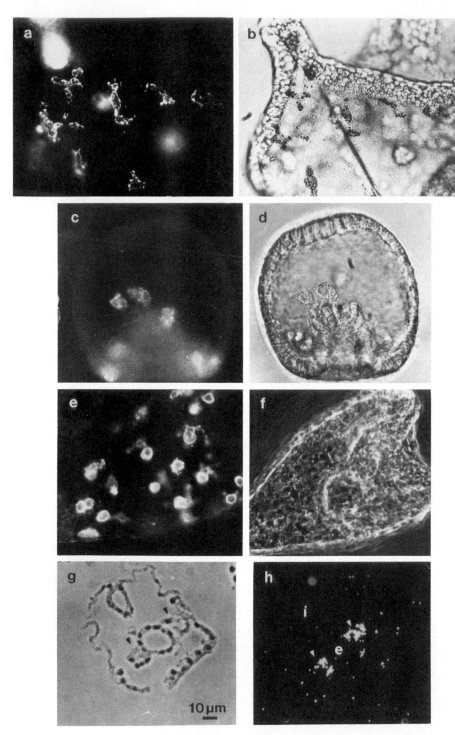

Fig. 4.10.   (See legend on p. 235.)

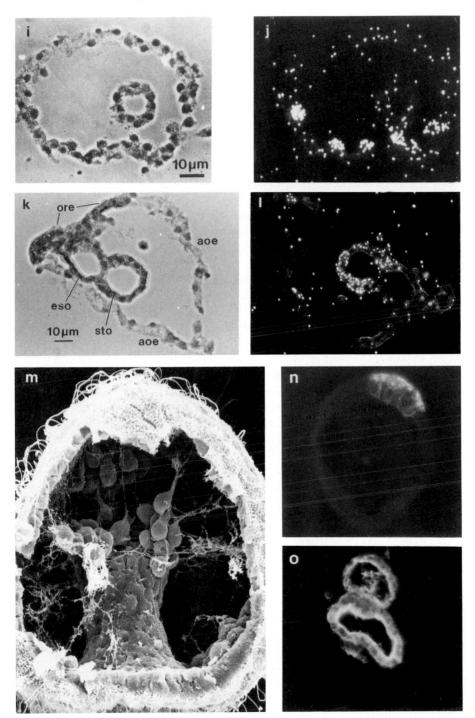

Fig. 4.10. (*Continued.*)

is shown in the *in situ* hybridizations reproduced in Fig. 4.10(j) and (l). Earlier, as noted above, this gene is also expressed transiently in primary mesenchyme cells. However, the vegetal plate and the primary mesenchyme also share a common ancestry in the four 3rd cleavage founder cells VA, VO and the two VL cells (Fig. 4.5). Perhaps a permissive derepression of the CyIIa gene occurs in these cells, with the result that the CyIIa cytoskeletal actin gene can be utilized in their progeny, but not in the progeny of other early embryonic blastomeres.

The initial phase of archenteron invagination is an autonomous function of the vegetal plate cells rather than a response to pushing or pulling forces generated by other cells (Amemiya *et al.*, 1982; Ettensohn, 1984, 1985). Thus isolated vegetal plates invaginate on schedule (Ettensohn, 1984). The invagination process has been demonstrated to involve rearrangement of cells and exchanges of the individual cells adjacent to one another (Ettensohn, 1985). Molecular changes at the cell surfaces may be inferred, given the autonomy of the initial invagination process. A late stage in the process of archenteron formation, including the delamination of secondary mesenchyme cells, can be seen in the scanning electron micrograph reproduced in Fig. 4.10(m̂). It has long been considered that the completion of invagination depends on filopodial contraction at the tip of the archenteron, exerted by mesenchymal cells that bridge to the roof of the ectoderm. Alternative explanations have been proposed, however, and in some echinoderm species invagination is completed in the absence of such filopodial contacts (Amemiya *et al.*, 1982; reviewed in Trinkaus, 1984; Ettensohn, 1985).

Though invagination can be regarded as an initial *function* of the differentiated archenteron cells of the vegetal plate, there are few known molecular markers. A surface antigen identified by a monoclonal antibody has been reported by McClay *et al.* (1983) that appears prior to invagination in the presumptive archenteron cells of the vegetal plate. As shown in Fig. 4.10(n) this antigen is evident exclusively in cells of this lineage at the mesenchyme blastula stage. After gastrulation [Fig. 4.10(o)] it is located in regions that become the stomach and intestine, but is totally absent from all but a few cells of the esophagus (McClay *et al.*, 1983; Wessel and McClay, 1985). This observation confirms that the archenteron precursor cells have begun to express specific genes prior to their morphological specialization, just as seen earlier for the primary mesenchyme cells, the aboral ectoderm, and certain of the secondary mesenchyme cells.

(v) Some General Aspects of Differential Gene Expression in Sea Urchin Embryos

Diversification of Embryonic Cell Types

The observations reviewed in the foregoing provide the most extensive and also the most direct evidence available for any species that *differential zygotic gene activation is responsible for the initial appearance of specialized cell types at the outset of development.* It is clear that many genes are expressed exclusively in one or another of the earliest appearing cell types in the embryo. Furthermore, the properties of the proteins encoded by these genes can be related more or less immediately to the functions of the embryonic cells in which they are synthesized, e.g., hyalin in ectoderm cells, the spicule matrix protein in primary mesenchyme cells, cytoskeletal actins in the cells forming the rigid aboral ectodermal wall of the larva, etc. Though yet morphologically undifferentiated the midblastula stage sea urchin embryo has already become a mosaic of diverse patterns of gene expression.

We have seen that the early differentiation patterns are largely clonal in the sea urchin embryo, and that in most of the examples considered there is a simple relation between lineage and pattern of gene expression. However, in most of the known examples, lineage-specific genes are not actually expressed until after several divisions have occurred following the appearance of the clonal progenitor cells (i.e., the first cells with lower case lineage designations in Fig. 4.5). These divisions result in expansion of the number of cells in each clonal lineage, and they generate a spatial distribution such that each clone more or less occupies a contiguous patch of the blastula wall. As pointed out by Angerer and Davidson (1984), for genes such as the CyIIIa actin gene or those genes functional in primary mesenchyme cells, the consequence is that there are a small number of nuclear "targets" at the time the lineage is specified (i.e., when the clonal progenitor cells are formed), but a large number of properly positioned, biosynthetically channelled progeny at the time expression is activated. Thus the clonal expansion that occurs throughout most of the embryo during later cleavage can be regarded as the means by which an initial set of differential biosynthetic programs is amplified and distributed spatially in the embryo, before being brought into action.

There could exist other zygotic genes that are indeed activated immediately after the establishment, by early or midcleavage, of clonal progenitor cells. However, in well studied examples such as the micromere-primary mesenchyme cell lineage there is little evidence from population level mRNA hybridization, two-dimensional protein gel electrophoresis, or analyses carried out on cDNA clones, for the earlier appearance of cell type-specific products, i.e., prior to the onset of ingression. Two unidentified ectoderm-specific transcripts that appear in cleavage were noted above, but the global blastula stage ectoderm phenotype is largely due to maternally stored proteins. Thus as we have seen, major components of the basal laminar membrane, the hyaline coat, the extracellular matrix, and the cilia have all been shown to be stored in the unfertilized egg. It should be noted that the evidence to date concerns on the one hand proteins or mRNAs that are at least moderately prevalent, and on the other, rare mRNA populations. The appearance during cleavage of a small number of newly synthesized, low abundance mRNAs, i.e., very early products of lineage-specific zygotic gene activations, would thus have escaped detection in all the investigations so far reported. Such sequences could of course be crucially significant, e.g., if they coded for *trans*-acting regulators that later activate the mesenchyme specific genes.

The now significant assemblage of identified, differentially regulated genes described for the sea urchin embryo belongs wholly to the "embryonic late gene" class discussed in Chapters I and III. Transcripts of these genes are either absent from the maternal mRNA or are present at very low levels compared to those attained in the embryonic cells in which these genes are active. Messages for the CyIIIa actin gene, for example, are present in the ~200 presumptive aboral ectoderm cells of the blastula at about 80 times the concentration per unit volume of cytoplasm as in the unfertilized egg, and messages of other lineage-specific actin genes, such as the CyIIa cytoskeletal actin gene and the M actin gene, are not represented at all in the maternal mRNA (see above). As shown in Chapter III only about 10% of all the genes represented in the embryo polysomes can be classified as late genes, and this value appears to apply to mRNAs of all prevalence levels. Therefore, there are unlikely to be more than ~10^3 lineage-specific genes utilized in embryogenesis, even including those producing very low abundance mRNAs. Probably the true number is significantly lower than this, since some late genes are expressed in many different cell lineages (e.g., the genes coding for some heat shock proteins, metallothionein, or the CyI actin protein). There is yet no evidence in the sea urchin embryo for asymmetrically localized maternal messages (cf. Chapter VI), and even in the case of the CyIIIa cytoskeletal actin gene, the *zygotic* expression of which is sharply localized, the ~2000 maternal mRNA molecules are uniformly distributed in the egg and early embryo (Cox *et al.*, 1986). Thus there are unlikely to be many lineage-specific maternal mRNAs. We may conclude that the embryo largely depends on mechanisms that activate specific genes in given lineages and cell types to produce the initial set of cellular phenotypic diversification. While this may not be a surprise, it is scarcely the only possibility, and furthermore it cannot be safely assumed the case for other embryos where the evidence available for the sea urchin embryo does not exist. An alternative that for many embryos remains to be excluded is the localization and sequestration to different lineages of maternal mRNAs that code directly for the diverse properties of the early differentiated cell types.

The Mechanisms of Gene Activation in the Early Embryo

We return here to the issue raised at the outset: What mechanism(s) account at the genome level for the differential activation of the initially equivalent genomes of the embryonic blastomeres? Totipotency and functional equivalence have been directly shown for the nuclei of early sea urchin embryos, initially by the famous pressure plate experiments of

Driesch (1892), and then by a series of classical and modern experiments demonstrating complete development from individual early cleavage blastomeres (reviewed in Chapter VI; and see discussion of nuclear equivalence in Chapter I).

Evidence from gene transfer and other experiments that implicate *trans*-activators in the control of histone gene expression in sea urchin embryos is reviewed in Chapter III. Transcriptional expression of the histone genes is in general not confined to any given embryonic cell lineage, except in the negative sense that later in development, when cells cease mitotic activity, they also cease to synthesize histone mRNA. Though relatively well known, these genes thus provide an imperfect model for genes that are activated during development exclusively in given embryonic cell lineages. Gene transfer experiments have also been carried out with the CyIIIa actin gene, which the evidence reviewed above shows to be clonally expressed in aboral ectoderm cells. These experiments were performed with a fusion gene construct, in which a sequence containing upstream CyIIIa and 5'-intron sequences is positioned in place of the promoter of the bacterial chloramphenicol acetyl transferase (CAT) gene, as shown in Fig. 4.11(a) (Davidson *et al.*, 1985; Flytzanis *et al.*, 1986). This construct was injected into unfertilized *S. purpuratus* eggs, and at various stages of development the activity of the CyIIIa sequences was measured by the production of CAT enzyme. DNA introduced in this manner is rapidly ligated into random, end-to-end concatenates, and during cleavage it is incorporated stably into nuclear compartments, where it replicates along with the embryonic nuclear DNA (McMahon *et al.*, 1985; Flytzanis *et al.*, 1985). Activation of CAT enzyme synthesis from the exogenous fusion gene appears to occur at the exact developmental stage when the CyIIIa gene is itself activated. This result is reproduced in Fig. 4.11(b). Measurements carried out at intervals show that within a few hours there occurs a greater than thousandfold increase in the rate of CAT enzyme synthesis per exogenous DNA molecule. Furthermore, *in situ* hybridizations such as shown in Fig. 4.11(d) demonstrate that CAT mRNA is synthesized only in aboral ectoderm cells. The experiment shown in Fig. 4.11(c) indicates that over a large range of exogenous CyIIIa·CAT DNA molecules per embryo only a fixed amount of CAT protein can be formed. The steady state level of CyIIIa mRNAs is about 5×10^5 molecules per embryo (Flytzanis *et al.*, 1986) or 5-6 times that of the endogenous CyIIIa mRNA (Lee *et al.*, 1986a). The data in Fig. 4.11(c) suggest competition by the excess CyIIIa·CAT sequences for a limiting amount of some specific factor(s) required for ontogenic activation of the CyIIIa gene itself. This interpretation is confirmed by experiments of Flytzanis *et al.* (1986), in which coinjection of CyIIIa·CAT together with a given molar excess of upstream CyIIIa flanking sequence was found to result in a stoichiometric, competitive decrease in CAT synthesis. Furthermore, evidence from DNA-protein binding studies carried out *in vitro* demonstrate directly the presence

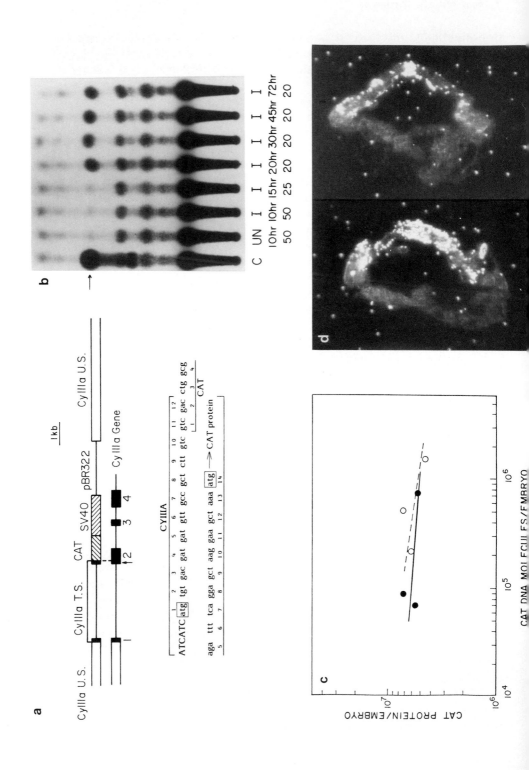

CYIIIA

ATCATC atg tgt gac gat gtt gcc gct ctt ctt gtc gtc gac ctg gcg
        1   2   3   4   5   6   7   8   9   10 11 12   CAT
                                              1   2   3   4

aga ttt tca gga gct aag gaa gct aaa atg ——→ CAT protein
5   6   7   8   9   10 11 12 13 14

of proteins that interact with certain specific regions of the 5′ flanking sequences of the CyIIIa gene (F. Calzone, unpublished data), that are also required for activity *in vivo,* according to deletion experiments (Flytzanis *et al., 1986).

Given that the experiments summarized in Fig. 4.11 indeed indicate temporally correct ontogenic expression of the CyIIIa·CAT fusion gene construct, several possible mechanisms of gene activation in the early embryo can be excluded (i.e., at least for this gene). As naked cloned DNA has been injected, specific "marking" of the gene during gametogenesis cannot be required for embryonic activation. Nor is it likely that any specific genomic location could be necessary, since the exogenous DNA, if integrated at all at this early stage, inserts ectopically (Flytzanis *et al.*, 1985). Nor can the effect observed be due to the appearance of a general (i.e., rather than sequence-

---

Fig. 4.11. Temporally and spatially correct ontogenic activation of a CyIIIa fusion gene injected into sea urchin eggs. (a) Fusion gene construct containing upstream and first intron sequences of the CyIIIa cytoskeletal actin gene of *S. purpuratus,* ligated to the bacterial chloramphenicol acetyltransferase (CAT) structural gene. T.S., transcribed sequences; U.S., upstream sequence. A diagram of the linearized construct injected into unfertilized eggs is shown in the top line. The plasmid was linearized at a restriction enzyme site 2.5 kb upstream of the 5′ end of the CyIIIa transcript. Since the molecules form a random end-to-end concatenate after injection (McMahon *et al.*, 1985), in half the cases the original 8 kb upstream sequence included in the plasmid will be regenerated. In the second line is a diagram of the CyIIIa gene (Akhurst *et al.*, 1986). Introns are denoted as thin lines, and exons are numbered. The third and fourth lines give the sequence of the junction region between the CyIIIa sequences and the bacterial CAT sequence. This region contains two in-frame ATG translation start codons, the first deriving from the CyIIIa gene and the second from the CAT gene. (b) CAT gene expression from the exogenous fusion genes. Injected eggs were allowed to develop until collected for assay of CAT activity. The enzyme activity is assayed by production of acetylated ¹⁴C-chloramphenicol, detected by thin layer chromatography (TLC) following reaction of the embryo extract with chloramphenicol in the presence of ¹⁴C-acetyl CoA. C, Control assay containing bacterial CAT enzyme. The arrow indicates the monoacetylated CAT product. Other samples contain sea urchin embryo extracts: UN, embryos derived from uninjected eggs; I, embryos derived from injected eggs. The number of hours postfertilization and the number of embryos in the samples analyzed are indicated, respectively, in the bottom two rows of numerals. [(a) and (b) From E. H. Davidson, C. N. Flytzanis, J. J. Lee, J. J. Robinson, S. J. Rose, and H. M. Sucov (1986). *Cold Spring Harbor Symp. Quant. Biol.* **50,** 321]. (c) Quantities of CAT enzyme protein (molecules of enzyme per average embryo; ordinate) produced in pooled samples of 50–100 24 hour embryos, as a function of the amount of exogenous CAT DNA that they contain (abscissa). Embryo batches containing differing amounts of exogenous DNA were prepared by injection of differing amounts of CyIIIa·CAT DNA. The injected DNA had amplified about 100-fold by 24 hours postfertilization. [(c) From C. N. Flytzanis, R. J. Britten, and E. H. Davidson (1986). *Proc. Natl. Acad. Sci. USA,* in preparation.] (d) *In situ* hybridization carried out with a section of an early pluteus state embryo, using a single-stranded antisense RNA copy of the CAT gene as probe. The figure shows consecutive oblique sections down through the face and emerging through the aboral ectoderm wall of the pluteus; upper section left. Significant labeling is observed only above the cells of the aboral ectoderm on the right side of each section, and neither the oral ectoderm nor the gut visible in the right hand section displays any CAT mRNA. [(d) From B. R. Hough-Evans and E. H. Davidson, unpublished experiment.]

specific) transcriptional activator at the stage when expression of the CyIIIa-CAT construct first occurs. As reviewed in Chapter III the highest rates of endogenous nRNA synthesis are measured at midcleavage, before the CyIIIa·CAT fusion gene becomes active. In addition, Vitelli *et al.* (1986) showed that injected α-histone genes function during midcleavage, i.e., on their expected schedule. The specificity of the response obtained from genes injected into the sea urchin egg is further demonstrated by the observation that when introduced in the same manner *sperm histone genes* remain transcriptionally silent (Vitelli *et al.*, 1986). These are of course not normally expressed in the embryo. The most direct interpretation of the experiments shown in Fig. 4.11 is thus that the exogenous CyIIIa·CAT construct encounters the same *trans*-regulators as cause expression of the endogenous CyIIIa gene, and that these regulators are presented only in the aboral ectoderm cell lineage.

The volume of maternal cytoplasm ultimately included in the aboral ectoderm cells at the time the CyIIIa gene is activated may not be greatly different from that initially included in the progenitor blastomeres of the aboral ectoderm lineages. Therefore, were these blastomeres to contain a *trans*-regulatory cytoplasmic protein, or mRNAs coding for such (either of maternal or zygotic origin), it would not be necessary to assume that the state of progenitor cell specification is transmitted to daughter cells during clonal replication at the *chromatin* level. Each newly formed CyIIIa gene could react *de novo* with the putative regulators. The significance of the *clonal* organization of the aboral ectoderm lineage, and of the unknown timing device that precludes expression of genes such as CyIIIa until clonal replication is complete, thus may not be directly regulatory. As suggested above the function of these features could simply be to ensure the proper number and location of cells carrying out the biosynthetic program needed for the construction of the aboral ectoderm.

The Actin Gene Family in Sea Urchin Development and Ontogenic Gene Batteries

The developmental patterns of actin gene transcription discussed earlier are summarized in Fig. 4.12. It is clear not only that expression of some of these genes is confined to given embryonic lineages, but also that the proteins they produce are utilized for different cellular functions. Yet as noted above the five active cytoskeletal actin genes code for proteins that differ from one another in primary amino acid sequence by only 1-2%, though the sequence of the muscle actin protein diverges about 10% from the cytoskeletal actin sequences (Cooper and Crain, 1982; Schuler *et al.*, 1983; Akhurst *et al.*, 1986). Cox *et al.* (1986) correlated the pattern of appearance of CyI and CyIIb mRNAs during embryonic development with mitotic activity, and the appearance of CyIIa mRNA with migratory activity in cells of mesenchymal

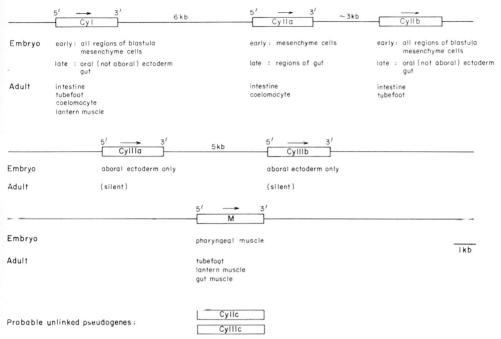

Fig. 4.12. Organization and expression of the actin gene family in *Strongylocentrotus purpuratus*. Actin genes are named M (muscle) or Cy (cytoskeletal, i.e., expressed in *non*muscle cell types). Roman numerals designate the three nonhomologous 3' nontranslated trailer sequences found in the cytoskeletal actin genes. a, b, c designate different though homologous trailer sequence variants. Linkage data are from analyses of cloned genes (J. J. Lee *et al.*, 1984; Akhurst *et al.*, 1986). Expression patterns in adult tissues were determined by RNA gel blot hybridizations, as reported by Shott *et al.* (1984). Expression of actin genes in embryonic cell types is summarized from the *in situ* hybridization study of Cox *et al.* (1986). [From E. H. Davidson, C. N. Flytzanis, J. J. Lee, J. J. Robinson, S. J. Rose, and H. M. Sucov (1985). *Cold Spring Harbor Symp. Quant. Biol.* **50**, 321.

and archenteron lineages. The presence of CyIIIa and CyIIIb mRNAs exclusively in the aboral ectoderm suggests that these actins might be constituents of rigid microfilaments. The overall import of Fig. 4.12 is in any case that though these different genes code for similar cytoskeletal actins, they are members of diverse *gene batteries* that during ontogeny are regulated differently. Thus as we have seen, in aboral ectoderm cells one such battery includes the CyIII actin genes and the Spec genes, which direct the synthesis of a family of Ca^{2+} binding proteins. Similarly the M actin gene is utilized along with the muscle tropomyosin gene, the myosin heavy chain gene and no doubt many other muscle-specific genes in the initial set of pharyngeal muscle cells. In regulatory terms the various members of the actin gene family are evidently each associated in such batteries with other kinds of genes, a particularly striking phenomenon in the case of the CyI and the

CyIIa genes. Though closely linked, these genes are utilized according to very different developmental programs (Fig. 4.12).

The evidence reviewed demonstrates the general reality of *ontogenically regulated gene batteries* in early sea urchin development. Sea urchin embryo cell types for which concrete examples of such batteries have here been reviewed include the skeletogenic mesenchyme, as well as ectoderm and pharyngeal muscle cells. A logical implication of this aspect of the regulatory architecture of the embryo is that the coordinate expression of the individual genes constituting such batteries is likely to be mediated by *trans*-acting factors which recognize homologous *cis* regulatory sequences shared amongst these genes (Britten and Davidson, 1969, 1971). As the structural gene components of some embryonic gene batteries are now identified, the molecular nature of the regulatory components has become an accessible problem. In the longer view this perspective emphasizes the importance of relating the topological distribution of the pleiotropic *trans*-regulators, and the mechanisms by which they are localized, with the relatively well defined cell fates and lineages of the sea urchin embryo.

4. REGIONAL SPECIFICATION AND TISSUE-SPECIFIC ZYGOTIC GENE EXPRESSION IN AMPHIBIAN EMBRYOS

(i) Phenomenological Aspects: Origins of Differentiated Embryonic Regions

Major Inductive Interactions

By neither morphological nor molecular criteria do *differentially functioning cells* appear in the amphibian embryo until gastrulation. This is not surprising, given the complete or near-complete absence of transcription prior to the midblastula stage (Chapter III). When gastrulation begins there are already about 10^4 cells in the *Xenopus* embryo, and more in the larger embryos of other amphibian species. During the period of transcriptional quiesence, however, there occur crucial processes that result in *specification* of the future patterns of cell differentiation. At fertilization the egg contains an animal-vegetal polarity that is interpreted morphogenetically during embryogenesis (evidence is reviewed in Chapter VI). Thus the types of cells developing from different regions of the egg can within certain limits be predicted. The animal cell cap provides ectoderm, neural tissue, and mesodermal cells of various kinds; the equatorial or marginal zone gives rise to a major portion of the mesoderm, as well as to other cell types; and the vegetal region provides endoderm, but also other cell types, including (in anuran amphibians) the germ cell progenitors. The dorsal-ventral axis is established initially by movements of preformed cytoplasmic constituents following fertilization, and prior to first cleavage (see for review Chapter VI). During cleavage the animal-vegetal and the dorsal-ventral axial specifications are transmitted to the blastomeres that inherit the respective regions of egg cytoplasm, and there is initiated the first of a series of major inductive interactions by which the larval body plan is ultimately established. These interactions are indicated in outline in the diagram shown in Fig. 4.13 (Smith and Slack, 1983). The primary inductive interaction results in specification of a fraction of the cells in the animal ectodermal cap to give rise to progeny that ultimately differentiate as mesoderm. The inducing cells are blastomeres of the vegetal region, and the inductive events by which mesodermal fate is specified in animal cap cells have probably already occurred by the beginning of the blastula stage (i.e., before stage 7 in Nieuwkoop and Faber, 1956; see, e.g., Gimlich and Gerhart, 1984). As indicated schematically in

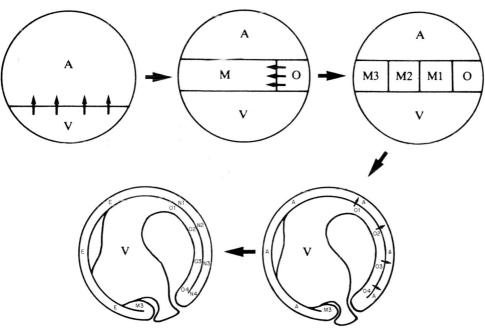

Fig. 4.13. Schematic description of major inductive interactions leading to determination of mesodermal and neural regions in the amphibian embryo. The diagram illustrates a *sequence* of inductive interactions, beginning with the maternally specified animal-vegetal polarity of the egg, and the initial establishment of the future dorsal-ventral axis (dorsal, right), which occurs by redistribution of maternal components following fertilization, as described in Chapter VI. A, animal; V, vegetal; M, mesoderm; O, organizer; M1, M2, M3, dorsal-ventral subdivisions of mesoderm; O1, O2, O3, 04, craniocaudal subdivisions of archenteron roof; E, ectoderm; N1, N2, N3, N4, craniocaudal subdivisions of neural plate. Arrows within embryos indicate inductive interactions. [From J. C. Smith, and J. M. W. Slack (1983). *J. Embryol. Exp. Morphol.* **78**, 299.]

the upper portion of Fig. 4.13, this early inductive process specifies the recipient animal cap cells as future dorsal or ventral mesoderm, in accordance with their positions with respect to the initial axes of the egg (Cooke, 1985a). Evidence considered in the following indicates that cells of the equatorial zone may in addition have the capacity to self-differentiate as dorsal and ventral mesoderm, i.e., independently of animal-vegetal induction.

The mesodermal progenitors at the most dorsal position in the late blastula (designated O in Fig. 4.13) are endowed with the special inductive properties that led to the classical designation of this region as the "organizer" (Spemann and Mangold, 1924; Spemann, 1938). These cells, which at the gastrula stage give rise to the notochord, have the capacity to induce a secondary dorsal-ventral axis if inserted in the ventral region of the embryo (Spemann and Mangold, 1924; Smith and Slack, 1983). In normal development the organizer is probably responsible for the inductive specification of a series of mesodermal fates along the dorsal-ventral axis (Keller, 1975; Slack and Forman, 1980; Smith and Slack, 1983). Thus the dorsal mesoderm contributes the notochord and somites, and in sequence toward the ventral side, the mesodermal progenitors contribute to the kidneys, the lateral plate mesoderm, and the blood islands of the postneurula stage larva. While the initial animal-vegetal induction of mesoderm occurs prior to the activation of transcription, the "dorsalizing" action of the organizer cells takes place then or shortly thereafter, i.e., in the late blastula stage (Smith and Slack, 1983). Both antecede the appearance of any overt differentiation in the embryo.

The various regions of the neural plate are organized inductively as well. Neural induction, the third of the sequence of major inductive interactions, takes place when the organizer has elongated across the archenteron roof. As indicated in Fig. 4.13 the organizer becomes regionally differentiated (O1-O4). The overlying neural plate regions (N1-N4) give rise to brain, spinal cord, etc. (Nakamura, 1978; Takaya, 1978; for reviews see Gerhart, 1980; Slack, 1983). Neural induction by implantation of a second organizer provides a spectacular demonstration of the potentiality resident even in ventral ectoderm to produce these neural structures rather than epidermis. Clear proof that the cells giving rise to secondary nervous systems in embryos receiving ectopic organizer grafts are cells that in normal embryos are destined exclusively for epidermal pathways of differentiation has been obtained by labeling donor organizer grafts, or the host cells, with cytologically detectable markers such as horseradish peroxidase or fluoresceinated dextran (Gimlich and Cooke, 1983; Smith and Slack, 1983). A great variety of chemical treatments and tissue extracts have also been shown to cause neuralization of embryonic amphibian ectoderm that would otherwise behave as epidermis. Viewed from the cellular vantage point, neural induction *in situ* might be considered a process by which is finally thrown a developmental switch that is already in a potential state of activation as a consequence of the prior ontogenic specification of the ectoderm precursor cells.

This view does not require that the "inducing principle" perform any "instructive" function (as for example, if it were a sequence-specific DNA-binding, genomic regulatory protein), but merely that it is capable of acting at the cell surface as a trigger, thereby releasing an intracellular signal. The significant feature for morphogenesis is that the inductive interaction provides the means for *spatially* designating the appropriate patches of cells to begin the differentiation of the central nervous system.

Comparisons of the *potentialities* of amphibian embryo cells when placed in ectopic situations with their actual developmental *fates in situ* have generally shown that up to the late blastula or early gastrula stages potentiality exceeds fate, as already discussed for the ectoderm. An example that has been analyzed with the aid of cell type-specific biosynthetic markers of function is dorsal-ventral mesodermal differentiation (Cleine and Slack, 1985). In early gastrula stage embryos of the axolotl, cells specified to produce mesoderm are found in a band that extends from the dorsal pole of the embryo all or most of the way around the equator. Cultured cells explanted from this band from anywhere within a 270° arc centered on the dorsal pole produce notochord-specific glycoproteins, while *in situ* only cells located within about 30° of the extreme dorsal pole do so. Similarly, though the presumptive mesoderm of the marginal zone is already dorsoventrally specified, ventral *Xenopus* explants from early gastrula stage embryos express dorsal functions if cultured in contact with dorsal axolotl explants of an equivalent stage (Slack and Forman, 1980; in the converse experiments it was found that the dorsal specification is already fixed, however, so that dorsal fate cannot be altered by combination with ventral explants).

Results demonstrating that the potentiality to carry out a given form of differentiation is spatially more widespread than actually occurs in undisturbed embryos stress the role of inductive interaction in the normal process by which cell fate is specified. On the other hand, isolates explanted from various regions of the embryo clearly display autonomous tendencies to give rise to certain cell types, suggesting "default" patterns of differentiation localized spatially from the beginning of cleavage onwards. It is reasonable to regard such default pathways as the consequence of maternal morphogenetic factors inherited from oogenesis and localized in the respective regions of the egg by first cleavage. For example, as discussed below, early cleavage cells isolated from the dorsal marginal zone of *Xenopus* embryos differentiate in isolation to give rise to muscle, as do nucleated egg fragments that include certain sectors of the egg cytoplasm (Gurdon *et al.*, 1984, 1985a); and animal pole cells isolated from cleavage stage amphibian embryos differentiate in culture mainly as epidermal ectoderm (Slack, 1984a; Jones and Woodland, 1986a; many earlier sources reviewed in Slack, 1983). A more global probe of dorsal-ventral specification as early as the 4-cell stage carried out by isolation and culture of *transverse* blastomere pairs shows that each more or less produces the structures expected of its descendants in the undisturbed embryo (Cooke, 1985b; Cooke and Webber, 1985a). Such results suggest that different regions of the egg inherit maternal factors that once distributed suffice to promote given courses of differentiation in descendants of the particular sets of blastomeres to which these factors are bequeathed. To proceed in the direction of a resolution between developmental plasticity on the one hand, and localized autonomy of differentiation, on the other, it is necessary to consider what is known of the fates and the potentialities of individual marked blastomeres.

Cell Lineage and Cell Fate in Xenopus Embryos

Modern investigations of the fates exhibited in amphibian embryos by the progeny of individual blastomeres have been carried out by injecting tracers such as horseradish peroxidase, that do not cross cell boundaries, but that can be detected in the progeny of the injected cells. In general these studies confirm the regional fate maps constructed earlier by vital staining methods (see, e.g., Vogt, 1929; Pasteels, 1942; Keller, 1975, 1976; reviewed in Gerhart, 1980). Until the beginning of gastrulation the progeny of a given labeled blastomere form a coherent patch of cells, as expected from the observation of Newport and Kirschner (1982a) that in *Xenopus* the cells do not acquire motility until 12th cleavage (see Fig. 3.2). Thereafter, however, the progeny of diverse contiguous clones display a great deal of migratory cell mixing (Cleine and Slack, 1985; Jacobson, 1985a). In the present context a major outcome of the single cell labeling studies has been the demonstration that the marked clones display somewhat variable fates. That is, in any two embryos the clonal progeny of given labeled blastomeres contribute differently in detail to the specialized structures of the postgastrular embryo. Furthermore, except for certain vegetal pole regions giving rise exclusively to endoderm, blastomeres marked as late as the early blastula stage produce ectodermal, neural, and mesodermal cell types displaying a range of diverse states of differentiation.

In Fig. 4.14(a) a projection of the midcleavage stage embryo is shown in which each cell is named (Jacobson and Hirose, 1981). Hirose and Jacobson (1979) found by injection of horseradish peroxidase that every individual blastomere up to the 16-cell stage gives rise to large populations of early tadpole stage central nervous system (CNS) cells, as well as to cells of many other tissues. Similarly, 24 out of 32, and 38 out of 64 blastomeres injected at 5th and 6th cleavage stages respectively contribute to the CNS (Jacobson and Hirose, 1981). The cells from which no CNS contribution derives are located in the vegetal pole regions [e.g., LD2.1.1.1 and LD2.2.1.1; nomenclature as in Fig. 4.14(a)]. Injection of single cells with horseradish peroxidase at 128-, 256- and 512-cell stages also results in labeled clones that in most cases include cells of many types in many organ systems, and the cell types formed by descendants of 512-cell stage blastomeres are similar to those developing from blastomeres in the same positions at earlier stages (Jacobson, 1983). Thus no progressive restriction in the potentiality of cloned progeny to give rise to *diverse differentiated cell types* has occurred by the 512-cell (i.e., early blastula) stage. Those marked clones that include CNS components always include in addition cells of the larval ectoderm and mesoderm (Jacobson, 1983). In Fig. 4.14(b) a map of all cells injected at the 512-cell stage in this series of experiments is shown, and in Fig. 4.14(c) the various CNS structures to which the progeny of these cells contribute are indicated diagrammatically, i.e., for those cells giving rise to elements of the CNS. Fig. 4.14(d) displays the *nonneuronal derivatives* of the same injected cells, i.e., their mesodermal, endodermal, and epidermal progeny. The comparison illustrates graphically the pluripotential nature of the progenitor cells in all but the extreme ventral and vegetal regions of the 512-cell blastula.

A significant developmental change occurs between the 256- and 512-cell stages in the disposition of clonal progeny of individual injected cells that contribute to the CNS. CNS descendants of 9th cleavage blastomeres are almost completely confined to one of seven distinct regions, or "compartments," of the forming nervous system, while 8th cleavage blastomeres frequently give rise to progeny distributed amongst these compartments (Jacobson, 1983, 1985b). There are three such compartments on either side of the midline, and the seventh is located at the anterior end where it bridges the midline, as indicated diagrammatically in Fig. 4.14(e). The location of 512-cell stage blastomeres whose descendants are found in each of the compartments visible in lateral view are shown by the dashed outlines in Fig. 4.14(c). Each CNS compartment is contributed by the descendants of 14-26 blastomeres of the 512-cell stage embryo, and they include neurons of many diverse phenotypes (Jacobson, 1983; Moody and Jacobson, 1983). Within the compartment the clonal progeny intermingle extensively after they acquire motility, though 98% of them are restricted to a given compartment. The mechanism underlying this strong intracompartmental preference may depend on cell surface markers, as the same preferences are observed in mixing experiments in which marked cells are allowed to mingle *in vitro* (Jacobson and Klein, 1985). From animal to animal, though the exact disposition of cells from given clones within a compartment varies, the CNS compartment for which the progeny of 512-cell stage embryo blastomeres of given regions are destined is predictable by position [Fig. 4.14(c)]. There is evidence as well that given compartments include mesodermal and ectodermal components (Jacobson, 1985b). Thus, as shown in Fig. 4.14(f), labeled cells obtained from early gastrula stage donors and grafted into 512-cell stage recipients give rise to neural, epidermal, and mesodermal progeny that all respect the same midline boundary. It is important to note, however, that there is no evidence that the compartments identified in this manner are mechanistic developmental units, in the sense that assignment to a compartment causally determines the subsequent pathways of differentiation. The

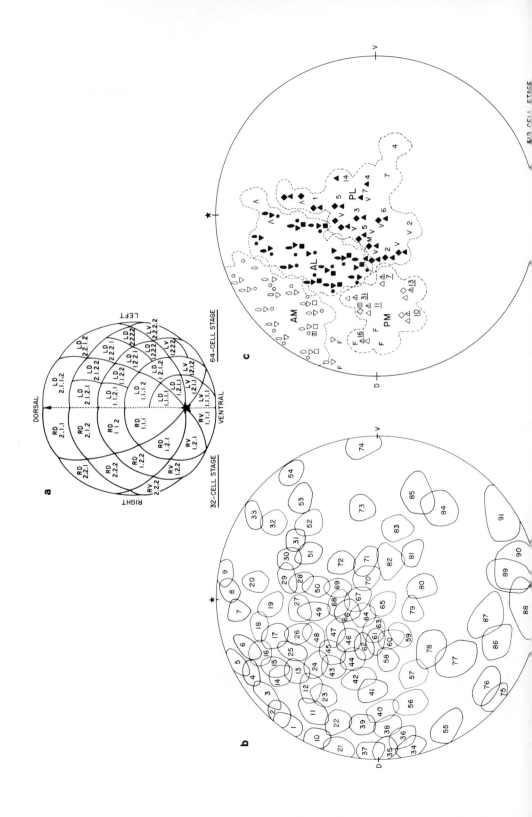

g

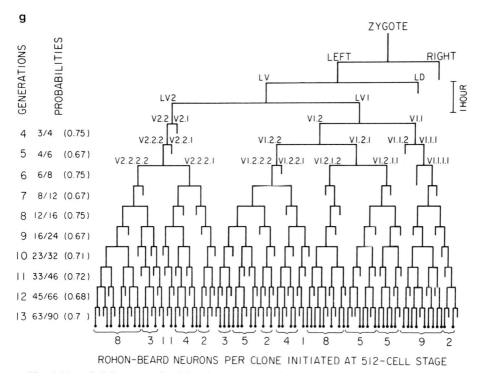

GENERATIONS	PROBABILITIES
4	3/4 (0.75)
5	4/6 (0.67)
6	6/8 (0.75)
7	8/12 (0.67)
8	12/16 (0.75)
9	16/24 (0.67)
10	23/32 (0.71)
11	33/46 (0.72)
12	45/66 (0.68)
13	63/90 (0.7)

ROHON-BEARD NEURONS PER CLONE INITIATED AT 512-CELL STAGE

Fig. 4.14. Cell lineage and cell fate in *Xenopus* embryos determined by intracellular labeling of a single blastomere. (a) Blastomere positions and designations at the 32-cell stage (left side of figure) and the 64-cell stage (right side of figure). The embryo is shown in anterior (animal) dorsal view; the star is at the animal pole, the arrow is in the dorsal midline pointing at the vegetal pole, which is not visible. Diameter of the embryo is about 1.3 mm. [From M. Jacobson and G. Hirose (1981). *J. Neurosci.* **1**, 271.] (b) Positions of blastomeres injected singly with horseradish peroxidase in individual 512-cell early blastula stage embryos. The star indicates the animal pole, and the arrow the vegetal pole. Each blastomere is denoted by a number. Fates of clonal progeny were determined by cytological detection of horseradish peroxidase in sections of early tadpole stage larvae. (c) Lateral fate map of blastomeres at the 512-cell stage that contribute to the central nervous system (CNS), indicating the regions of the CNS populated by progeny of blastomeres (shown in (b)) that had been injected with horseradish peroxidase. The positions occupied by groups of blastomeres that all contributed to the same CNS *regions* are enclosed by dashed lines. Underlined numbers in the PM (posterior medial) group are numbers of primary spinal motoneurons that originated from individual blastomeres at the positions occupied by the numbers. Numbers in the PL (posterior lateral) group are the numbers of Rohon-Beard neurons that originated from individual blastomeres at the indicated positions. F, floor plate; open ellipse, ventral telencephalon; solid ellipse, dorsal telencephalon; open circle, ventral retina; solid circle, dorsal retina; open inverted triangle, ventral diencephalon; solid inverted triangle, dorsal diencephalon; open square, ventral mesencephalon; solid square, dorsal mesencephalon; open diamond, ventral rhombencephalon; solid diamond, dorsal rhombencephalon; open triangle, ventral spinal cord; solid triangle, dorsal spinal cord; V, trunk neural crest; G, cranial neural crest. The embryo is bilaterally symmetrical with respect to the CNS fate map, and only one side is shown here. [(b) and (c) From M. Jacobson (1983). *J. Neurosci.* **3**, 1019.] [Figs. 4.14(d–f) appear in color plates following p. 256.] (d) Locations of

plasticity of the embryonic cells, as revealed by ectopic induction experiments (see above) as well as a variety of cell culture and transfer experiments utilizing molecular probes for cell-specific function (see below), argue otherwise. Compartmentalization based on adhesivity preferences could simply be a secondary manifestation of the mechanisms by which migratory cells are mobilized into sheets or layers, or assembled in appropriate numbers in given regions of the embryo, whence they are determined by other means.

The general conclusion of importance from the blastomere labeling studies reviewed is that for the large fraction of 9th cleavage blastomeres sampled the *subsequent pathways of differentiation in each lineage are indeterminate*. There are indeed certain regional restrictions of cell fate, since under normal circumstances some areas of the blastula never give rise to certain

---

mesodermal (green), epidermal (red) and endodermal (blue) progeny of same blastomeres whose CNS progeny are shown in (c). *Green*: △, head mesoderm; □, notochord; ○, ventral somitic mesoderm; ▲, head mesoderm; ◨, cardiac mesoderm; ■, lateral plate mesoderm; ●, dorsal somitic mesoderm; X, nephrotome. *Red*: +, suckers; △, ventral head; □, ventral trunk; ⊙, proctodeum; ○, ventral tail, ▲, dorsal head; ■, dorsal trunk; S, stirnorgan; P, placodes; ●, dorsal tail. *Blue*: △, ventral foregut; L, liver; □, ventral midgut and hindgut; ▲, dorsal foregut; ■, dorsal midgut and hindgut. [Data from W.-L. Xu and M. Jacobson (1986). *Dev. Biol.*, in press.] (e) Diagram showing CNS compartments in early neural plate stage embryo, shown in dorsal view. Star indicates position of animal pole. The floor plate (transverse lines) develops between the left and right posterior medial compartments (black). *Black*: The posterior medial regions (see (c)) become the posterior ventral compartments in the CNS. *Light dots*: The anterior medial region becomes the anterior ventral compartment of the CNS; *heavy dots*: the anterior lateral regions become the anterior dorsal compartments of the CNS; *crosses*: the posterior lateral regions become the posterior dorsal compartments in the CNS. [From M. Jacobson (1980). *Trends Neurosci.* **3**, 3.] (f) Cell fates observed in graft experiments, in which individual cells or very small groups of cells derived from lateral animal positions in wholly labeled stage 10 (early gastrula) embryos were implanted into equivalent positions in unlabeled recipient embryos. The donor embryos had been labeled globally by injection of horseradish peroxidase very early in development. Dorsal view: arc represents blastopore lip; star, animal pole. *Green*: CNS cell fates. Each symbol represents the results of a single graft. *Excluded* are progeny that belong to the anterior medial compartment, which as shown in (e) straddles the midline; *included* are all progeny contributing to the structures of the other compartments. *Green*: △, ventral telencephalon; ▲, dorsal telencephalon; ▽, ventral diencephalon; ▼, dorsal diencephalon, □, ventral mesencephalon; ■, dorsal mesencephalon; ◇, ventral rhombencephalon; ◆, dorsal rhombencephalon; ○, ventral spinal cord; ●, dorsal spinal cord; N, cephalic neural crest; n, trunk neural crest; R, dorsal retina; r, ventral retina; S, stirnorgan. *Red*: mesodermal progeny from same grafts: △, ventral head; □, notochord; ⊡, sclerotome; ○, ventral somite; ▲, dorsal head; ◨, heart; ■, lateral plate; ●, dorsal somite; X, nephrotome. *Blue*: epidermal progeny of same grafts: △, ventral head; □, ventral trunk, +, cement organ; ▲, dorsal head; ■, dorsal trunk; P, placodes; S, stirnorgan. [(f) From W.-L. Xu and M. Jacobson (1986). *Dev. Biol.*, in press]. (g) Lineage diagram illustrating the probability at each division, from 4th to 13th cleavage, of daughter cells that continue on pathway leading to the production of Rohon-Beard neurons. Cells are named as in (a). Though lineages are all shown to terminate at 13th cleavage, in actuality they terminate between the 11th and 14th cleavage, with the mean termination point the 13th. [(g) From M. Jacobson (1985a). *Trends Neurosci.* **8**, 151.]

cell types, but it is clear not only that the 9th cleavage blastomeres are pluripotent, but that processes yet to transpire, and factors other than their own ancestry and division pattern, ultimately determine the fates of the descendants of these blastomeres.

These features are illustrated in detailed studies carried out on the lineages from which descend several specific types of neuron, *viz.*, Rohon-Beard neurons, which are sensory neurons of the CNS that in the tadpole promote swimming in response to tactile stimuli on the skin (Clarke *et al.*, 1984); primary spinal motoneurons which innervate the axial somitic muscles; and the two bilateral Mauthner's neurons, which are involved in coordinating tadpole escape responses. The Mauthner's neurons provide a particularly straightforward case. Jacobson and Hirose (1981) showed that the clones descendant from blastomeres D2.2.2.1 or D2.2.2 [Fig. 4.14(a)] *sometimes* include a Mauthner's neuron (the frequencies are ~70% and 40% respectively). At the 512-cell stage there would appear to be an equal probability that any *one* of the ~20 blastomeres giving rise to a posterior lateral compartment [see Fig. 4.14(c)] will produce a Mauthner's neuron (Jacobson, 1983). The Rohon-Beard neurons, of which there are something less than 100 on each side in early tadpoles of stages 32-34 (Jacobson, 1981a), arise largely from two pairs of animal pole blastomeres of the 8-cell stage located contralaterally on the ventral side (V1 and V2). The lineages leading to the Rohon-Beard neurons of the left side are illustrated in Fig. 4.14(g) (Jacobson and Moody, 1984; Jacobson, 1985a). While the exact number of these neurons arising from the descendants of given blastomeres at, say, 9th cleavage, is not determinate, this diagram shows that at each cleavage after the formation of V1 and V2 there is about a 70% *probability* that the daughter cells will continue in a pathway leading to production of a Rohon-Beard neuron.

In animals in which the early cleavage planes occur at slightly variant positions, it is observed that Rohon-Beard neurons arise from other cells occupying the region normally inherited by the V1.2 and V2.2 blastomeres (Jacobson, 1981a,b). This suggests that a cytoplasmic determinant could be localized in this region of the egg. However, if V1.2, from which the majority of these neurons arise [Fig. 4.14(g)], is sucked out of the embryo, Rohon-Beard neurons are still formed in normal number (Jacobson, 1981b). Injection of horseradish peroxidase into the neighboring cell, D1.2, shows that this cell now gives rise to a majority of the Rohon-Beard neurons, and also to spinal ganglia, neither of which normally derive from D1.2. It follows that though maternal cytoplasmic factors might endow the blastomeres that inherit them with a certain probability to form Rohon-Beard neurons (i.e., on the stimulus provided by neural induction), more cells share this capacity than normally express it. A reasonable interpretation is that the failure of descendants of D1.2 to produce Rohon-Beard neurons in normal embryos is that they are inhibited from entering on this pathway of differentiation by a repressive interaction with those descendants of V1 and V2 that have done so.

Figure 4.14(c) shows that the capacity to produce Rohon-Beard neurons is distributed exclusively with respect to the capacity to produce motoneurons [the former are formed in the PL compartment (numbers), and the latter in the PM compartment (underlined numbers) of the CNS]. The motoneurons also arise in a probabilistic manner, in which the likelihood of continuation on the pathway leading to their differentiation is again about 70% throughout cleavage (Jacobson and Moody, 1984). It is not yet clear whether the reciprocal restriction in the origins of these two forms of neuron is a consequence of the different CNS compartments to which the progeny of the 9th cleavage blastomeres are assigned, i.e., in the sense that cells in the PM and PL compartments could be exposed to regionally different inductive influences, or is due to some other mechanism. However, the phenomenon of secondary CNS induction by ventral organizer implants confirms that the *capacity* to give rise to both forms of neuron is not confined to those animal hemisphere blastomeres from which they normally derive. Indeed, the final commitment to differentiation of these neuronal cell types occurs only at about the time when their lineages terminate, at about 12th-13th cleavage, just as suggested by the evidence of Fig. 4.14(g). Only at this point does permanent depletion of Rohon-Beard neurons occur when descendants of V1.2 are killed (Jacobson and Moody, 1984).

A somewhat simpler image of cell type specification emerges from marking experiments carried out on vegetal pole blastomeres. Gimlich and Gerhart (1984) showed that certain subequatorial blastomeres from the dorsal side of 32-cell embryos produce only endodermal progeny that are located after gastrulation in the floor of the archenteron. This is not true of most vegetal pole blastomeres at this stage, however, according to experiments of Heasman *et al.* (1984), which demonstrated that the large majority (though not all) of these blastomeres produce ectodermal and/or mesodermal progeny as well as endodermal progeny. As shown in Fig. 4.14(d), by the 512-cell early blastula stage the vegetal pole blastomeres sampled give rise only to endoderm of the ventral hindgut (Xu and Jacobson, 1986). Similarly, Heasman *et al.* (1984) reported exclusively endodermal fates for vegetal blastomeres labeled *in situ* at the midblastula stage (stage 8). Thus it is possible that during later cleavage there are established clonal endoderm lineages, all of the progeny in which display the same state of differentiation. The specification of such lineages remains reversible prior to the midblastular activation of transcription, however. If transplanted to the blastocoel, individual vegetal pole as well as animal pole and marginal zone blastomeres from late cleavage (stage 6-7) or midblastula (stage 8) embryos display the capacity to give rise to differentiated cells of all kinds (Heasman *et al.*, 1984, 1985). Only by stage 10 (early gastrula) have they by this test become irreversibly committed to an endodermal pathway of differentiation. It remains

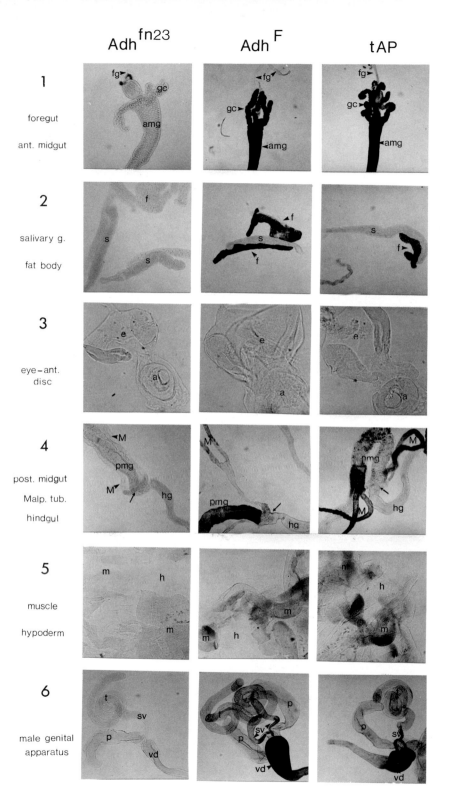

Fig. 1.3. Legend appears on page 29.

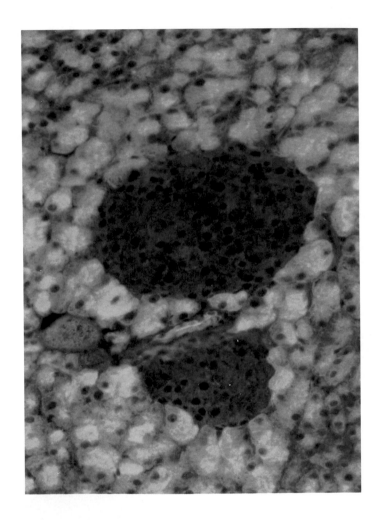

Fig. 1.4. Legend appears on page 30.

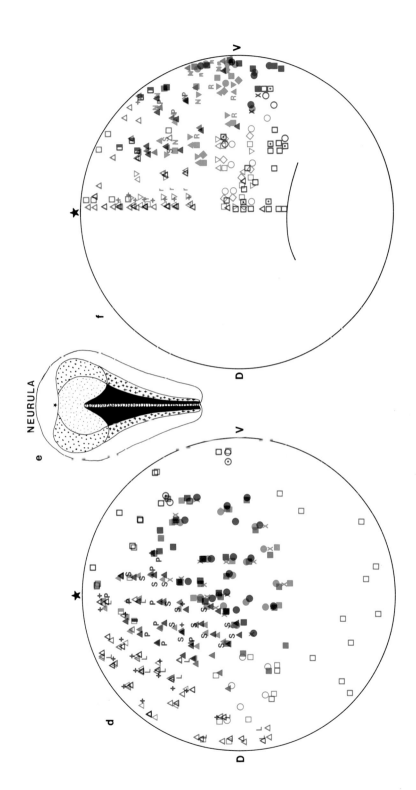

Fig. 4.14 (d–f). Legend appears on pages 253–254.

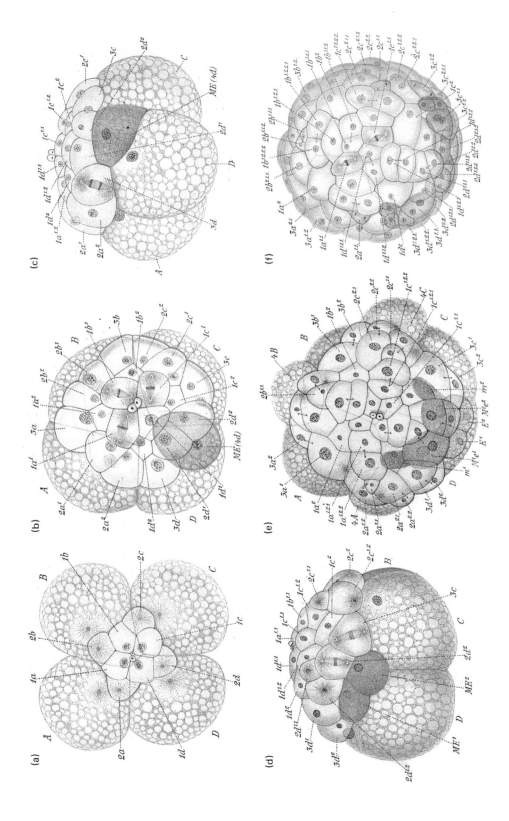

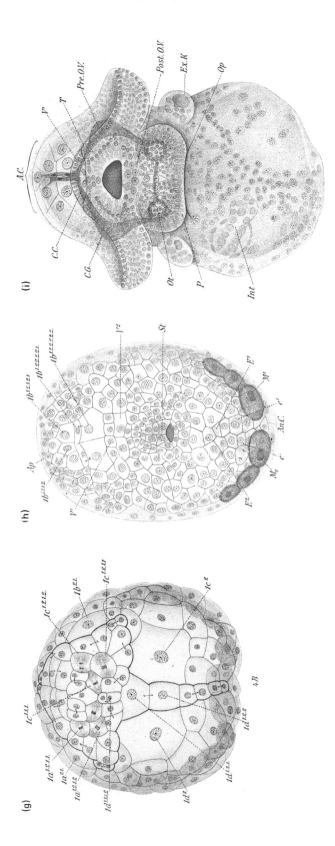

Fig. 6.2. Legend appears on page 414.

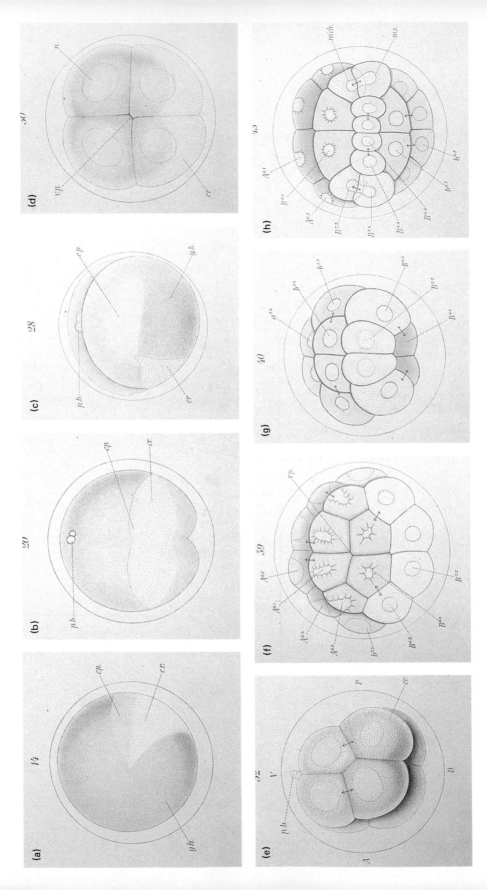

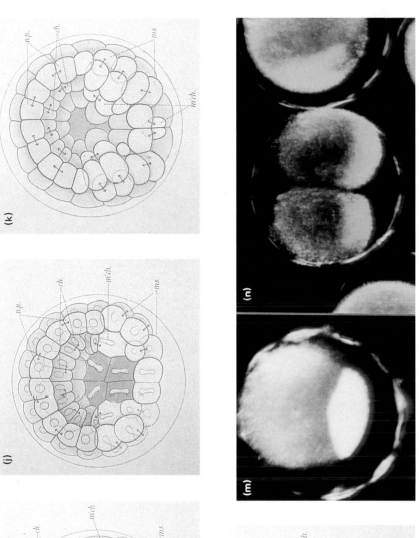

Fig. 6.3. Legend appears on page 416.

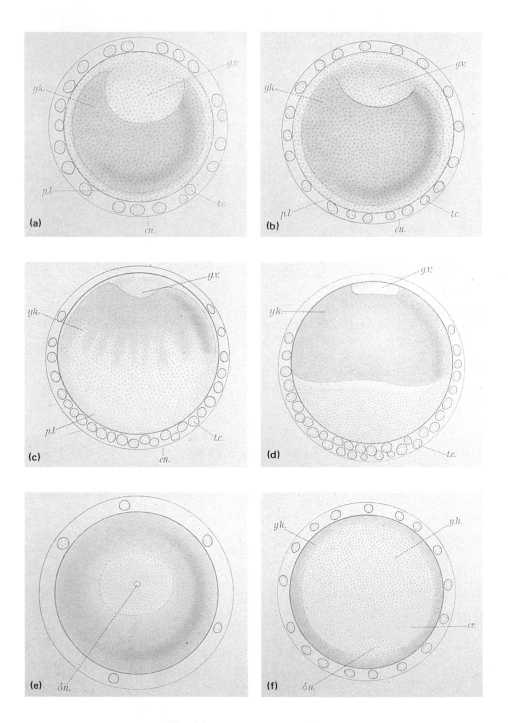

Fig. 6.4. Legend appears on page 417.

an allowable interpretation that vegetal pole blastomeres inherit cytoplasmic factors that specify an endodermal pathway of differentiation; that these are segregated into clonal endoderm progenitor cells during later cleavage; and that after transcription is initiated the progeny of these clones become irreversibly determined.

One feature of the strategy utilized in amphibian development that was noted at the outset of this chapter is its capacity to organize very large numbers of uncommitted cells into differentially functioning embryonic *regions*. We have seen that regional specification of areas of the blastula, but not their final determination, takes place prior to the onset of transcription. The migratory activity occurring at the gastrula stage is an active process probably requiring transcriptional input, that involves changes in many cytoskeletal and surface properties (see review in Trinkaus, 1984). Specification of cell fate in amphibians, in summary, can be considered to follow from the interplay of at least three factors: First, a stochastic or probabilistic likelihood that cells of given regions will embark on given courses of differentiation, perhaps in consequence of the respective regions of maternal cytoplasm that they inherit; second, the ability of cells that have begun to develop in given directions to affect the probability that nearby cells will do so; and third, the ability of certain underlying or adjacent cell layers to inductively affect cell fate, i.e., to perturb *in a spatially organized way* the first two kinds of process. In addition, as morphogenesis advances, there must continue to occur decisive interactions among cells locally, in respect to their mutual positions, in order to account for the complex spatial arrays of differentiated cell types that appear within the general regions initially specified in early development. The boundaries of the areas within which cells must interact locally to complete morphogenesis, that is of tissues and organs, are obviously much smaller than the compartment boundaries identified in early blastula single cell labeling experiments, and in general are not coincident with them.

The amphibian strategy of development is one of enormous flexibility and power. It is not restricted by lineage-specific cell division programs, requirements for exact cleavage geometry, numbers of cells (see, e.g., Cooke and Webber, 1985b), or blastomere ancestries. Nor can the initial specification processes depend on sequential cascades of genomic regulatory events. In contrast to organisms displaying invariant lineage-differentiation relationships, i.e., determinate lineages, each individual amphibian embryo is different, in the sense that different numbers of cells from given clones may have contributed to any given structure. This is also the type of strategy utilized in higher vertebrates, such as birds. Very large numbers of plastic cells are initially laid out in early avian development, here in a two-dimensional array, and after delamination of additional underlying cell layers, regional specification is completed, probably through a similar combination of mechanisms (see, e.g., Slack, 1983; Trinkaus, 1984; Jacobson, 1985b). The flexibility of the amphibian developmental mechanism is demonstrated by the observation that in certain egg-brooding hylid frogs, e.g., *Gastrotheca riobambae*, the embryo develops by a completely different morphological route, deriving from a two-dimensional disc formed by cell migration at the gastrula stage (del Pino and Elinson, 1983).

It is illuminating to compare the development of *Xenopus* and sea urchin embryos, with respect to the features here considered. Clearly there is no evidence in *Xenopus* for any *autonomously* differentiating somatic cell lineage, such as the skeletogenic mesenchyme lineage of the sea urchin embryo, in which there is a set number of cleavage divisions producing progeny that all display the same differentially expressed functions, *in vitro* or *in situ* (a possible exception in anuran embryos is the primordial germ cell lineage, discussed in Chapter VI). The endoderm in *Xenopus* may resemble lineages such as the aboral ectoderm in the sea urchin, in which a clonal pattern of differentiation obtains, but the cells remain for a long time plastic and capable of expressing different fates if placed in ectopic relationships. However, throughout most of the *Xenopus* embryo, cell lineage is clearly variable with respect to cell fate, and in addition cells remain pluripotential relatively far into development. As we have seen, the vegetal plate blast cells of the sea urchin embryo may develop in an analogous manner. Many of these cells as well may remain pluripotential until the gastrula stage, and then differentiate in various directions probabilistically. Neural differentiation in the postgastrular ectoderm of the sea urchin embryo might also occur as a probabilistic process in which intercellular interactions play a crucial role. However, neural induction by an underlying tissue layer could not take place in the sea urchin embryo as it does in the *Xenopus* embryo. Other regional inductive interactions in which cells of different prior origin participate certainly do occur in sea urchins, e.g., the archenteron-ectoderm induction resulting in mouth formation, which may be common to all deuterostomes. In many ways the earliest phases of development appear the most distinct in comparing these two forms. While both the sea urchin and the amphibian eggs display a maternal polarization along the animal-vegetal axis, and while both establish the second axis early in embryonic development, in the sea urchin embryo these localizations are interpreted in a precise fashion in terms of the early cell lineages, while in the amphibian embryo they are not.

(ii) Molecular Indices of Differential Gene Activity in the Early Amphibian Embryo

Analyses of regional protein synthesis patterns by two-dimensional gel electrophoresis first indicate differential biosynthesis at gastrulation. This conclusion obtains both for embryos bisected across the dorsal-ventral axis

(Smith and Knowland, 1984) and for embryos bisected across the animal-vegetal axis (Ballantine *et al.*, 1979). In addition Shiokawa *et al.* (1984) showed that cultured animal, vegetal, dorsal, or ventral half-embryos obtained at the 8-cell stage synthesize identical sets of (detectable) proteins through the late blastula stage, after which regional differences become apparent. These observations provide no evidence for widespread regional sequestration of maternal mRNAs (or for sequence-specific regional translational preferences). As reviewed in Chapter VI, a small number of rare maternal mRNAs have been shown by more sensitive methods to be localized along the animal-vegetal axis. However, the onset of differential patterns of protein synthesis in the gastrula stage is clearly to be attributed to the *activation of regionally-specific transcriptional programs*. Identified proteins that are synthesized in particular embryonic cell types have been particularly useful for further explorations of the pregastrular processes by which the progenitors of these cells are specified. The molecular markers so far available for embryonic amphibian cell types are so far confined mainly to the epidermal and neural ectoderm, and somitic muscle.

Specification and Gene Expression in the Embryonic Ectoderm

Proteins that are synthesized only in epidermal cells have been identified in both urodele and anuran species of amphibians. Slack (1984b) dissected the axolotl neurula into constituent structures, i.e., neural plate, lateral epidermis, endoderm, notochord, etc., and reported several newly synthesized polypeptides unique to the epidermal fraction. Cytokeratins were identified among these by immunoprecipitation, probably components of an intermediate filament network characteristic of these cells (Burnside, 1971; Slack, 1984a,b). cDNA clones for two distinct epidermal intermediate filament cytokeratins have been recovered from *Xenopus* embryos (Winkles *et al.*, 1985; Jonas *et al.*, 1985). The proteins coded by these clones were identified by their homology with a human cytokeratin. Messages for neither of these particular *Xenopus* epidermal cytokeratins, denoted XK70 and XK81, are present in the maternal mRNA, nor are they detectable after metamorphosis. Both mRNAs accumulate after the midgastrula stage, the XK70 mRNA attaining its maximum levels in tailbud stage embryos and the XK81 mRNA at the tadpole stage. Like the CyIIIa cytoskeletal actin gene in the sea urchin, these cytokeratins appear to be utilized exclusively in an *embryonic* program of epidermal differentiation. A further specific product of the embryonic epidermis in the axolotl is a high molecular weight, water-soluble surface glycoprotein of high sugar content, called epimucin (Slack, 1984b). In *Xenopus* embryos an unidentified marker recognized by a monoclonal antibody is produced by the epidermis at the late gastrula stage, when it appears synchronously over the whole external surface, except for the neural plate (Jones and Woodland, 1986a). Production of this epidermal marker requires RNA synthesis, since it is blocked by treatment with actinomycin prior to the late gastrula stage (Jones and Woodland, 1986a).

These markers have been utilized to assay at a molecular level the state of epidermal differentiation at various stages and under various conditions. As we have seen, the external cells of the animal hemisphere are pluripotent throughout the blastula stage, but by the early gastrula stage they are committed to ectodermal forms of development (i.e., neural or epidermal fate). Explants isolated from the animal pole region of early gastrula stage axolotl embryos autonomously begin to synthesize epimucin under a variety of culture conditions, at the same time as this occurs in control embryos, at the onset of neurulation (Slack, 1984a; Cleine and Slack, 1985). The epidermal *Xenopus* antigen recognized by the monoclonal antibody of Jones and Woodland (1986a) is expressed in animal but not vegetal hemisphere explants taken from embryos at all stages from 8-cell cleavage onwards, including the single-cell-deep animal cap of the stage 6 early blastula, and the outer layer of the stage 7 blastula, when these explants are cultured to the equivalent of the neurula stage. However, if animal hemisphere grafts from pregastrular donor embryos are implanted into the blastocoels of early gastrula stage recipients, their progeny usually participate in the morphogenesis of mesodermal as well as ectodermal structures (Jones and Woodland, 1986b). The donor cells were identified in these experiments by means of an injected fluorescent marker, or by use of a different *Xenopus* species which has a distinct karyotype. When the grafts were taken from early gastrula (stage $10\frac{1}{2}$) donors, only neural or epidermal progeny were obtained, and if implanted in vegetal regions, where neural induction cannot occur, the progeny of the grafted cells behave exclusively as epidermis. However, the epidermal marker is synthesized in grafts from stage $11\frac{1}{2}$ embryos even if they are implanted in the center of the forming neural plate of a recipient embryo (Jones and Woodland, 1986b), in contrast to results obtained with grafts from earlier embryos. Conversely, after stage $11\frac{1}{2}$ external cells of the dorsal ectoderm, i.e., the presumptive neural plate, fail to express the epidermal antigen even when placed in a ventral epidermal location. These experiments, in summary, confirm that in *Xenopus* prior to stage $10\frac{1}{2}$ the ability of the animal hemisphere cells to respond to mesodermal induction has disappeared, and by stage $11\frac{1}{2}$ the choice between neural and epidermal forms of differentiation has also become irreversible. A complementary study of neural differentiation has been carried out by use of another immunologically detectable marker, NCAM (neural cell adhesion molecule). Jacobson and Rutishauser (1986) showed that cultured ectodermal explants isolated from the animal pole region of early *Xenopus* gastrula fail to produce NCAM, but if combined *in vitro* with the dorsal lip organizer region the ectoderm explants express this marker of neural differentiation. These experiments also demonstrate that NCAM synthesis *follows* neural induction, occurring in undisturbed embryos about 2-3 hr after the beginning of gastrulation. In conclusion it seems evident that in early gastrula cells already committed to an ectodermal pathway, the balance between neural differentiation and the "default" epidermal form of differentiation is for a time a delicate one. It can be easily shifted in either direction, *in vivo* or *in vitro*, by exposure to ectopic cellular environments or to a range of exogenous substances that have "neuralizing activity" (see above).

As discussed earlier, the capability of explanted animal pole cells to give rise in culture to differentiated epidermis could be interpreted as the consequence of their inheritance of maternal cytoplasm initially localized at the animal pole of the fertilized egg. Evidence reviewed in Chapter VI demonstrates that in amphibians the initial polarity can be reversed by inversion of the egg, which occasions gravitational redistribution of cytoplasmic components of differing density. Jones and Woodland (1986b) showed that in inverted eggs only the *new* animal pole gives rise to cells that produce the epidermis-specific antigen. This experiment suggests a correlation between epidermal specification and maternal factors localized in the egg cytoplasm. A contrasting result that is consistent with this interpretation has been obtained by vegetal irradiation with UV light, which destroys the dorsal-ventral axial determination of the fertilized egg (reviewed in Chapter VI) without obviously affecting the animal-vegetal polarity. In embryos derived from UV-irradiated eggs the genes coding for the epidermal cytokeratins XK70 and XK81 are still normally activated (Jamrich *et al.*, 1985). However, epidermal specification in animal pole explants deprived of the vegetal inductive influences that *in situ* cause mesoderm formation may not be a completely cell-autonomous process, as might in the simplest case be supposed. Thus the late gastrula epidermal antigen is not produced in animal pole explants if they are disaggregated in Ca^{2+}-free media prior to stage 7, though the individual blastomeres remain alive and metabolically active after disaggregation (Jones and Woodland, 1986a). If the Ca^{2+} is added back prior to stage 7, the reaggregation that ensues suffices for subsequent epidermal differentiation, and disaggregation after stage 8 (midblastula) does not prevent subsequent epidermal differentiation. These results could simply reflect a requirement for certain external Ca^{2+} levels during early development, or they could indicate a special need for intercellular interaction. Another indication that epidermal differentiation requires more than simply the inheritance of animal pole egg cytoplasm is the observation that the epidermal marker is not subsequently produced if prior to stage 8 cytokinesis is blocked by treatment with cytochalasin (Jones and Woodland, 1986a). This stands in contrast to several known examples in other organisms in which early lineage-specific differentiation is known to occur autonomously, where interference with cytokinesis by cytochalasin treatment does not interfere with the manifestation on schedule of lineage-specific molecular characters (see Chapter VI for review). Treatment of whole embryos or of isolated animal pole caps with cytochalasin *after* the midblastula stage is permissive for epidermal differentiation. Thus, it is the pretranscriptional process of *ectodermal specification* which requires cell division and cell association (or certain Ca^{2+} levels). After the midblastula stage the ectodermal state of specification becomes an autonomous property of the animal pole cells, independent of these requirements; after the onset of gastrulation it becomes an irreversible property as well; and only after the midgastrular period is this state of specification manifested in a differential program of structural gene activity.

Activation of the α-Actin Gene in Differentiating Somitic Muscle

Synthesis of the muscle-specific actin (α-actin) has proved a useful molecular marker of the onset of myogenesis in the postgastrular amphibian embryo. The α-actin mRNA first appears in neurula (~stage 14) *Xenopus* embryos (Sturgess *et al.*, 1980), though mRNAs for cytoskeletal actins are prevalent in the unfertilized egg (see Chapter II). Both the cardiac and skeletal α-actin messages of *Xenopus* have been cloned (Mohun *et al.*, 1984). As shown in Fig. 4.15(a) and (b), both genes are expressed exclusively in the dorsal (somitic) mesoderm. The accumulation of these mRNAs at the neurula stage occurs long after the fate of the presumptive muscle cells has been irreversibly established, i.e., by gastrular invagination if not earlier (Cleine and Slack, 1985; reviewed in Slack, 1983). Gurdon *et al.* (1985b) utilized the α-actin probes to examine with a precise assay the process by which animal cap cells are induced to form somitic mesoderm through interaction with cells of the vegetal hemisphere. Animal and vegetal explants obtained from stage 8 blastulae were cultured separately or together, as conjugates, as indicated in the diagram shown in Fig. 4.15(c). Though according to cell lineage data the animal cap cells of these explants normally contribute to the somitic muscle [e.g., Fig. 4.14(d)], neither the isolated animal nor the vegetal explants produce α-actin mRNA, while as shown in Fig. 4.15(d) the conjugates do. About 25% of the animal cap cells in these conjugates synthesize the muscle actin, and cytological immunofluorescence shows these cells to be arranged in bilateral, myotome-like clusters. Further studies show that the inductive interaction requires at least 1.5-2.5 hr, and that the α-actin gene does not become activated until ≥5 hr after the inductive event has taken place (Gurdon *et al.*, 1985b). In no case could the animal pole cells be induced to synthesize α-actin mRNA until they are of the age when this would normally occur. It was also confirmed that vegetal cells lose their inducing ability by stage 9 (late blastula), and animal cells their ability to respond after stage 10 (early gastrula), as suggested by prior transplantation experiments (see, e.g., Nakamura *et al.*, 1970a).

These experiments demonstrate at the genome level the vegetal-animal induction of dorsal somitic mesoderm indicated diagrammatically in Fig.

4.13. Thus it is of particular interest that activation of muscle-specific actin genes also occurs *autonomously*, in cultures of blastomeres that inherit the original subequatorial dorsal cytoplasm of the fertilized egg (Gurdon *et al.*, 1984, 1985a). Thus isolated equatorial regions of stage 8 blastulae give rise to differentiated progeny that synthesize α-actin mRNA. Similarly, vegetal half embryos prepared at the 8-cell stage spontaneously activate on schedule the genes coding for α-actin, as do those fractions of 32-cell stage embryos that include subequatorial vegetal blastomeres. Ligation experiments carried out by Gurdon *et al.* (1985a) to localize the putative maternal cytoplasmic factors in the 1-cell zygote are shown in Fig. 4.15(e) and (f). It is clear that nucleated egg fragments which include the subequatorial vegetal cytoplasm, particularly that on the dorsal side, retain the autonomous capacity to differentiate muscle, though it remains possible that *within* the vegetal partial embryo some intercellular interactions are later required to potentiate α-actin gene activation.

Fig. 4.15. Expression of α-actin genes in differentiating muscle cells of the *Xenopus* embryo. (a) Transverse section of neurula (stage 18) embryo indicating regions obtained by dissection. RNA was extracted from each region and utilized for the experiments shown in (b). (b) Coordinate expression of cardiac and skeletal α-actin genes in somitic mesoderm. This figure shows autoradiographs of electrophoretic gels in which are displayed the DNA fragments derived from S1 nuclease resistant hybrids formed between the labeled α-actin probes and mRNAs present in the different embryonic regions. The *cardiac* α-actin probe yields a 250 nt hybrid species, plus a minor 120-130 nt species. The *skeletal* α-actin probe yields two prominent hybrid species 185 and 190 nt long, plus a minor fragment of 160 nt. Only the somitic mesoderm contains cardiac and skeletal α-actin messages. "$\frac{1}{5}$" denotes a lane containing one-fifth the quantity of somitic RNA as in the preceding lane. [(a) and (b) From T. J. Mohun, S. Brennan, N. Dathan, S. Fairman, and J. B. Gurdon (1984). Reprinted by permission from *Nature (London)* **311**, 716. Copyright © 1984 Macmillan Journals Limited.] (c) Diagram showing construction of animal-vegetal conjugates utilized for the experiment shown in (d). (d) Detection of cardiac α-actin mRNA in cultures of isolated embryo fragments and conjugates by S1 nuclease hybrid protection, as in (b). Eq, RNA from equatorial region that was isolated from stage 8-9 blastula and cultured as in (c) until control embryos are at the neurula stage (stage 18). An, RNA from animal region, isolated and cultured as in (c); An/Veg, RNA from conjugate, as in (c); Veg, RNA from vegetal region; Mkr, size markers. [(c) and (d) From J. B. Gurdon, S. Fairman, T. J. Mohun, and S. Brennan (1985b). *Cell* **41**, 913. Copyright by M.I.T.] (e) Diagram summarizing experiments in which fertilized egg was ligated, and the nucleated fragment indicated by the arrows cultured. Presence of cardiac actin message was then assayed by the S1 nuclease hybrid protection method (as shown in (f)). 1, Diagram of egg indicating animal-vegetal and dorsal-ventral axes, in relation to the grey crescent and sperm entry point (see Chapter VI). 2, Ligations carried out in different planes with respect to the egg axes (first column) with nomenclature utilized in (f) (second column); the fraction of resulting partial embryos in which α-actin mRNA was observed (third column); and relative strength of signal (fourth column). 3, Shows an interpretation of these results, in which the dotted region indicates localization of cytoplasmic factors that are required for autonomous activation of α-actin genes. (f) S1 nuclease assays of cardiac actin mRNA in partial embryos as indicated in (e) (part 2). RNA was extracted when the embryos had developed to a stage equivalent to stage 18-20 in controls. [(e) and (f) From J. B. Gurdon, T. J. Mohun, S. Fairman, and S. Brennan (1985a). *Proc. Natl. Acad. Sci. U.S.A.* **82**, 139.]

264

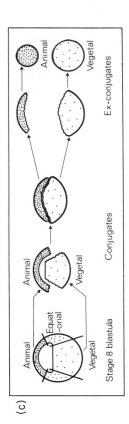

(a)

Somite
Dorsal endoderm
Neurectoderm
Notochord
Endoderm
Ventral ectomesoderm

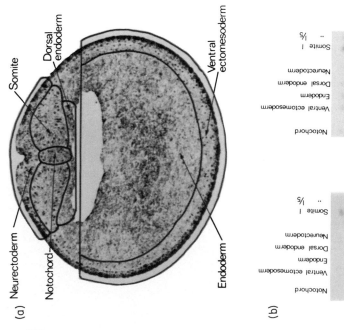

(b)

Somite I/s
Notochord
Ventral ectomesoderm
Endoderm
Dorsal endoderm
Neurectoderm
Somite I

250
120-130

185/190
160

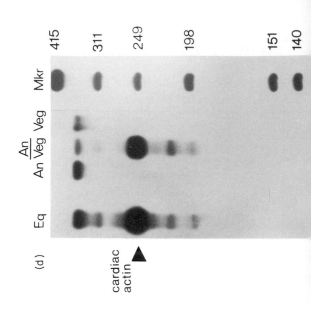

(c)

Animal
Vegetal
Ex-conjugates

Animal
Vegetal
Conjugates

Animal
Equat-orial
Vegetal
Stage 8 blastula

(d)

Mkr
Veg
An/Veg
An
Eq

415
311
249
198
151
140

cardiac actin

(e)

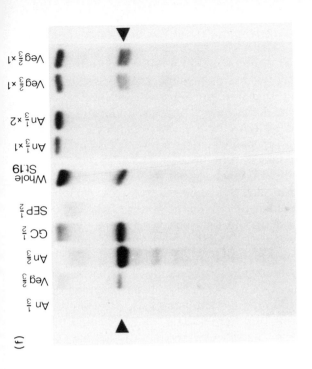

(f)

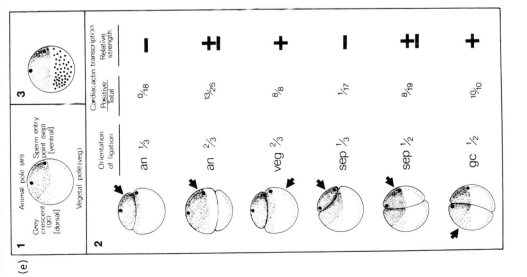

Specification of the genomic program required for differentiation of dorsal somitic muscle apparently follows two routes. In animal cap cells that do not produce muscle autonomously it depends on induction from vegetal cells, while in the vegetal subequatorial dorsal cells it appears to depend directly on the inheritance of localized maternal cytoplasmic elements. We have seen that a similar conclusion can probably be drawn for epidermal differentiation, in that this occurs autonomously only in animal cap cells, but *in situ* is apparently induced in cells of many other regions [e.g., see Fig. 4.14(d)]. However, it is not required that such data be regarded as a dichotomy. Maternal factors affecting subsequent cell fate in the first cleavage egg could be distributed so as to provide a smooth distribution of *probabilities* that cells inheriting them will differentiate in given directions. Where the probability is highest, e.g., in the subequatorial dorsal region for presumptive somitic mesoderm, the animal pole region for presumptive epidermis, or the vegetal pole region for presumptive endoderm, significant fractions of the cells will differentiate in these directions autonomously. Elsewhere, the probabilities favoring given forms of differentiation, as determined by the maternal cytoplasmic inheritance in particular regions of the egg, might require an upward boost. In normal development this would be provided inductively. Thus, for example, induction of mesoderm in the animal cap determines the incidence of cells that follow a mesodermal pathway of differentiation, while their sister cells often undergo a different fate [e.g., Fig. 4.14(d)]. Induction also provides the topological layout and the boundaries within which the specified cells of the early gastrula will differentiate, as seen most obviously in the two-dimensional, probabilistic appearance of neuroblasts in the ectoderm overlying the archenteron roof at the late gastrula stage.

5. DIFFERENTIAL GENE EXPRESSION IN THE EMBRYOGENESIS OF *DROSOPHILA*

(i) Classes of Developmentally Important Genes

The various modes of embryonic development considered in this chapter each lend themselves to certain forms of observation. The long history of analysis of mutations that affect the visible morphology of *Drosophila* has led to genetic, and in recent times molecular, identification of many loci required during development. A minor fraction of such loci when mutated cause *embryonic* arrest or malformations, though mutations at most loci cause death later in development. Among the latter are mutations that specifically block the morphogenesis of larval or adult structures, or the processes of pupariation, or cause abnormalities in the postembryonic development of imaginal disc derivatives, the development of the compound eye,

etc. These lie outside of our particular province, the formation of the embryo
per se. As noted in Chapter II most developmentally significant genes utilized in embryogenesis are expressed both maternally and zygotically. Some
maternally acting genes that are involved in establishment of the dorsoventral and anterior-posterior embryonic axes, or the specification of germ cells,
are discussed in Chapter VI, and Chapter V includes references to others
that are required for specific aspects of the process of oogenesis. The following discussion is focused on the functions of zygotically active genes that
specifically affect given *regions* of the embryo, and that are involved in
morphogenetic processes requiring differential embryonic gene expression.

Many of the developmentally important genes known in *Drosophila* differ
from any of those considered earlier in this chapter in the hierarchial level of
their function. Having initially been isolated because of the gross deformations caused by their mutation, these are often genes that directly or indirectly affect the disposition of a large number of individual differentiated cell
types within their morphological domains of action. For example, the cuticle
of the first instar larva contains a variety of diversely specialized cells, and
mutations in most of the developmentally active genes considered below
affect the spatial disposition of these cell types, and the structures they
produce. As a class such genes are often referred to as "pattern formation"
genes, and they are required for the correct localization of segmental boundaries, the morphogenesis of appropriate intrasegmental structures, symmetry, the occurrence of the gastrular invaginations, and many other processes
that are ultimately reflected in external (and internal) morphology. Amongst
the particular cell types required for the formation of the cuticular structures
are hypoderm cells that secrete the cuticle proteins of the external larval
coat; cells on the ventral side of the larva that each generate one or two
external projections, the denticles; cells on the dorsal side that produce fine
hairs or short triangular projections; cells that produce lateral sensory bristles, and other sensory cells forming the thoracic structures known as ventral pits and Keilin's organs; and cells that produce the termini of the tracheal trunks and the complex morphological specializations of the larval
head and anal regions (see, e.g., Lohs-Schardin *et al.*, 1979a, for detailed
description). The normal positions of various of these structures on the
ventral aspect of the first instar larva are shown in Fig. 4.16(a)-(c), and their
relation to the segmental organization and the structural specifications of the
adult are indicated in the diagram reproduced in Fig. 4.16(d) (Lewis, 1982).

A systematic effort to estimate by saturation mutation the number of
pattern formation genes compared to the total number of loci that when
mutated result in embryonic arrest has been reported by Nüsslein-Volhard *et
al.* (1984), Jürgens *et al.* (1984), and Wieschaus *et al.* (1984a). In these
studies the criterion applied for the recognition of pattern formation mutations is the appearance of externally detectable morphological abnormalities
in the first instar larval cuticle. While this excludes mutations that might

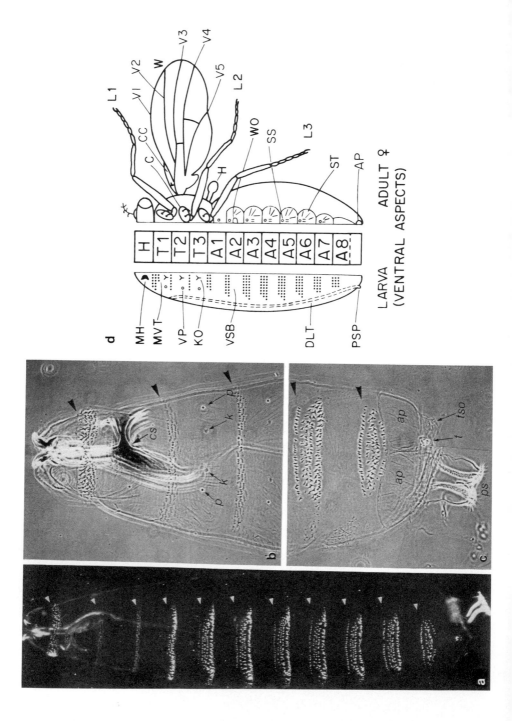

d

MH
MVT
VP
KO
VSB

DLT

PSP

LARVA
(VENTRAL ASPECTS)

| H |
| T1 |
| T2 |
| T3 |
| A1 |
| A2 |
| A3 |
| A4 |
| A5 |
| A6 |
| A7 |
| A8 |

L1
V1 V2
CC
C
W
V3
V4
V5
H
WO
SS
L2
L3
ST
AP

ADULT ♀

cs
k
p
k
p

b

ap
ap
tso
t
ps

c

a

have affected morphogenesis exclusively in internal organs, most pattern formation mutations are likely to have been included in this series of screens, which has been extended to the whole genome. Thus known homeotic mutations that act early in development were recovered, as were other known mutations that affect the differentiation of the embryonic nervous system, and previously identified mutations that disturb segmental morphogenesis. Some pattern formation mutations affect morphogenetic cell movements, e.g. *twisted gastrulation* and *folded gastrulation*, loci that in normal embryos must function within minutes of the onset of gastrulation (Zusman and Wieschaus, 1985). The main quantitative result deriving from this analysis is that pattern formation genes account for only about 3% of all zygotically active loci that when mutated cause developmental arrest prior to hatching (about 25% of all homozygous lethal mutations in animals developing from heterozygous parents cause embryonic, as opposed to postembryo-

---

Fig. 4.16. Segmental ventral structures of the first instar larval cuticle, and major segmental structures visible in ventral aspect in adult *Drosophila*. (a) Dark field view of whole wild-type larva, head at top. White arrows mark the approximate anterior boundaries of the three thoracic segments and of eight abdominal segments. The anterior boundary of each segment is marked by several rows of ventral setae or denticles. The number of rows and the disposition of the pigmented denticle hooks changes from anterior to posterior (Lohs-Schardin *et al.*, 1979a). Other characteristic segmental patterns of hairs and bristles are found on dorsal and lateral surfaces. X90. (b) Closer phase microscope view of head and thorax of larva. Anterior margins of prothorax, mesothorax, and metathorax are marked by black arrowheads. Note relatively broad denticle belt, consisting of thicker setae at anterior prothoracic border, and characteristic differences between meso- and meta-thoracic denticle belts. Each of the three thoracic segments contains three bilateral pairs of sensory organs: Keilin's organs (k); ventral pits (p); and dorsal pits (not visible). In the head segment can be seen the mouth hooks, and beneath the cuticle of the anterior thoracic region the cephalopharyngeal skeleton is visible (CS). On the dorsal and lateral sides of the head are additional specialized structures not visible here, including the maxillary and antennal sense organs and two rows of comb-like structures at either side of the mouth. X155. (c) Posterior region of first instar larva. Anterior borders of 7th and 8th abdominal segments are marked by black arrowheads. Posterior to the denticle belt of the 8th segment are the bilateral anal pads (ap), a tuft of hairs posterior to the anus (t), one of two bilateral sense organs (tso), and the posterior spiracles located more dorsally (ps). X200. The anal pads and other posterior structures may belong to reduced additional abdominal segments; their provenance in respect to segmental homology is not clear. [(a)-(c) From G. Struhl (1983). *J. Embryol. Exp. Morphol.* **76**, 297.] (d) Comparison of the ventral cuticular pattern of the first instar larval (or late embryonic) stage with that of the adult female. MH, mandibular hooks; MVT, midventral tuft; VP, ventral pits; KO, Keilin's organ; VSB, ventral setal belts; DLT, dorsal longitudinal (tracheal) trunk; PSP, posterior spiracle; H, head; T, thoracic; A, abdominal; L, leg; W, wing; H, haltere; C, coxa; CC, costal cell (of wing); V, vein; WO, Wheeler's organ; SS, sensillum (on segments A1 to A7, inclusive); ST, sternite; AP, anal plate. In the late embryo or first instar larva, the dorsal tracheal trunk terminates in an incipient anterior spiracle near the boundary of the first and second thoracic segments. In later instars, a visible anterior spiracle would be present at this point. Special sense organs on the three thoracic segments of the larval stage and on the first seven abdominal segments of the adult are depicted grossly enlarged. [(d) From E. B. Lewis (1982). *In* "Embryonic Development. Part A: Genetic Aspects" (M. M. Burger and R. Weber, eds.), p. 269. Alan R. Liss, New York.]

nic arrest). Identified pattern formation genes account in total for about 140 different loci (Jürgens *et al.*, 1984). Thus the large majority of embryonic lethals affect other processes, such as general aspects of cellular function or metabolism, or the differentiation of specific cell types in subcuticular regions of the embryo, interference with which does not induce abnormal cuticular morphogenesis. The relatively small number of pattern formation loci may even be overestimated, since there are included structural genes needed for normal cuticular biosynthesis, e.g., the gene coding for dopa decarboxylase, an enzymatic activity needed for normal cuticular pigmentation (Wright *et al.*, 1981; Nüsslein-Volhard *et al.*, 1984).

In what follows we are concerned largely with genes that fall in the pattern formation class. The identification of such genes in *Drosophila* provides a special opportunity for analysis of the genetic control of developmental processes that are pleiotropic, with respect to the range of individual cell types affected. It should be noted, however, that several *Drosophila* structural genes coding for proteins that are utilized in particular differentiated embryonic cells have also been characterized. The functions of these genes are more closely analogous to those discussed earlier, in the context of the differentiation of specific cell types of early amphibian and sea urchin embryos. Among these are several genes coding for contractile muscle proteins. The *Drosophila* genome contains six unlinked actin genes, the ontogenic expression of which has been examined with gene-specific cloned probes (Fyrberg *et al.*, 1983; Sanchez *et al.*, 1983). Two actin genes, located at chromosomal positions 5C and 42A, code for cytoskeletal actins. These genes are expressed during oogenesis and in embryonic development, as well as at certain later stages, though their individual patterns of expression during the life cycle differ significantly (cf. the sea urchin data reviewed above). Fyrberg *et al.* (1983) showed that the other four actin genes are utilized during the ontogeny of particular sets of muscles. Actin genes located at 88F and 79B are expressed at the late pupal stage, when they are required in the morphogenesis of the adult thoracic muscles and thoracic and leg muscles respectively. An actin gene located at position 57A and to a lesser extent one located at 87E encode muscle actins of the larval body wall musculature, and these genes are activated late in embryogenesis. The *Drosophila* genome contains a single myosin heavy chain gene, located in the 36B chromosome region (Rozek and Davidson, 1983; Bernstein *et al.*, 1983), and this gene is also expressed in the late embryo. The same gene is utilized in the morphogenesis of adult muscles, and it encodes alternate forms of myosin protein translated from differentially spliced mRNAs that are developmentally regulated (Falkenthal *et al.*, 1985). Another gene, already mentioned, that is utilized in a particular early embryonic cell type is the dopa decarboxylase (Ddc) gene, located in the 37C1-2 region (Wright *et al.*, 1981; Hirsh and Davidson, 1981). Ddc enzyme activity is localized in hypodermal cells where it catalyses reactions required for sclerotinization and pigment formation in the cuticle (Lunan and Mitchell, 1969). Thus the Ddc gene is expressed at periods of the life cycle when cuticle synthesis occurs, *viz.*, when the embryo hatches, at each larval molt, and at eclosion (Scholnick *et al.*, 1983). Like the muscle actin messages, the Ddc messages are not detectably represented in the maternal mRNA (Hirsh and Davidson, 1981; Beall and Hirsh, 1984), conforming to the generalization that genes expressed differentially in given embryonic cell types tend to belong to the class of genes utilized after fertilization but not during oogenesis. Scholnick *et al.* (1983) showed that an integrated exogenous Ddc gene carried on a P-factor vector undergoes accurate ontogenic activation in the transgenic host. This result demonstrates that for the Ddc gene, as for others discussed earlier (cf. Chapter I), a few kb of flanking sequences suffice to provide the genomic *cis*-regulatory elements required for specific developmental activation, and since the integrated gene functions accurately in ectopic genomic positions, its activation is likely to be mediated by diffusible *trans*-acting regulators.

(ii) Genes Required for the Embryonic Specification of Neuroblasts

Origin of Neuroblasts in Wild-Type Drosophila

The CNS of the completed embryo consists of the brain, the subesophageal ganglion, and the ventral nerve cord, in which is linked together a homologous series of segmental ganglia. These structures derive from neuroblasts that segregate from two bilateral regions of the ectoderm during gastrulation, the *ventral neurogenic ectoderm*, and the *procephalic neurogenic ectoderm* (Poulson, 1950; Campos-Ortega and Hartenstein, 1985; Hartenstein and Campos-Ortega, 1984, 1985; Hartenstein *et al.*, 1985; Technau and Campos-Ortega, 1985). As shown in Fig. 4.17(a), in each thoracic and abdominal segment (i.e., T1-A8) about 26-32 of the ~120 cells included at the cellular blastoderm stage in the ventral neurogenic ectoderm (considering both sides) are destined to become neuroblasts from which derive the ventral ganglia of the larva. The three gnathal segments [C1-C3 of Fig. 4.17(a)] contribute respectively 10, 32 and 34 neuroblasts, from which derive the subesophageal ganglion (Hartenstein and Campos-Ortega, 1984, 1985). The number of neuroblasts produced in the procephalic neurogenic region, which gives rise to the brain (Technau and Campos-Ortega, 1985), is not exactly known. In T1-A8 about one out of four of the ventral neurogenic ectoderm cells thus actually become neuroblasts, the remainder differentiating as epidermal progenitors (dermatoblasts). The basic plan by which segmental clusters of neuroblasts arise in this region of the embryonic blastoderm appears the same in hemimetabolous insects such as the grasshopper, as in holometabolous insects such as *Drosophila*, and even in distantly related crustacean arthropods (Thomas *et al.*, 1984). This homology extends as well to the later stages of the morphogenesis of the CNS. In *Drosophila* each of the segmen-

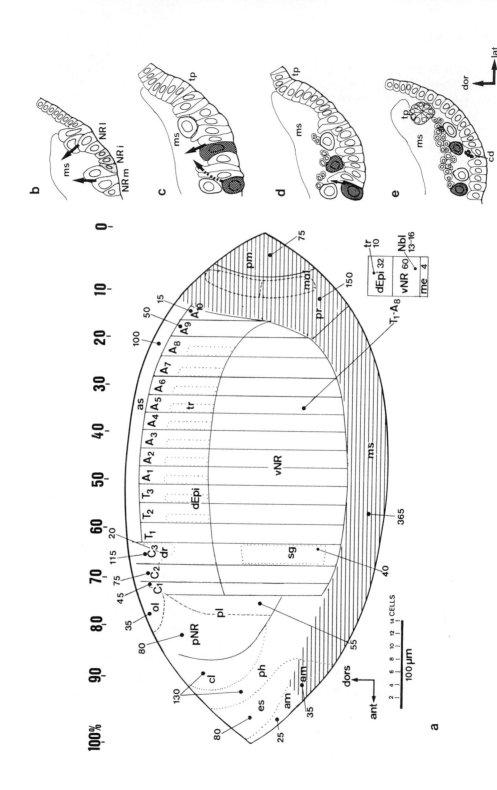

tal ganglia contains several hundred neurons, and the completed ventral nerve cord 4000-4500 ganglionic neurons (reviewed in Campos-Ortega and Hartenstein, 1985). These arise from the neuroblasts by a series of 8 or 9 subsequent divisions. The specific ganglionic progeny of individual neuroblast lineages are not known in *Drosophila*, though the initial pattern of neuroblast segregation appears to be fixed and constant (Hartenstein and Campos-Ortega, 1984). For the grasshopper *Schistocerca* however, Doe and Goodman (1985a) showed that once formed a given neuroblast generates an invariant cell lineage, giving rise to certain types of neuronal progeny according to its initial *position* in the neurogenic ectoderm. The later morphogenesis of the larval nervous system *per se* lies outside the confines of the present discussion (see, for this, Doe and Goodman, 1985a; Kuwada and Goodman, 1985; review in Campos-Ortega and Hartenstein, 1985), which is focused rather on the initial process by which occurs the specification and segregation of the embryonic neuroblasts.

Neuroblast specification in insect embryos appears to occur by a probabilistic mechanism, as in *Xenopus*, and not by a deterministic mechanism as in *Caenorhabditis* embryos. In *Drosophila* this is suggested by the observation that cells in both procephalic and ventral neurogenic ectodermal regions injected with horseradish peroxidase give rise to dermatoblasts as well as to various numbers of neuroblasts, in accord with ectodermal fate maps estab-

---

Fig. 4.17. Origin of neuroblasts in the *Drosophila* embryo. (a) Fate map of completed blastoderm. am, anterior midgut; as, amnioserosa; cl, clypeolabrum; dEpi, dorsal epidermis; dr, dorsal ridge; es, esophagus; mal, Malpighian tubes; me, mesoectodermal cells; ms, mesoderm; Nbl, neuroblast; ol, optic lobes; ph, pharynx; pl, procephalic lobe; pm, posterior midgut; pNR, procephalic neurogenic region; pr, proctodeum; sg, salivary gland; tr, trachae; vNR, ventral neurogenic region; C_1-C_3, gnathal segments; T_1-T_3, thoracic segments; A_1-A_{10}, abdominal segments. Estimated numbers of cells in each of the indicated anlagen (i.e., on one side of the bilateral embryo) are indicated. Percent egg length is shown at top. Shaded regions are destined to invaginate at gastrulation. ant, anterior; dors, dorsal. [From V. Hartenstein, G. M. Technau, and J. A. Campos-Ortega (1985). *Wilhelm Roux's Arch. Dev. Biol.* **194**, 213. For sources and details see also Hartenstein and Campos-Ortega (1985) and Technau and Campos-Ortega (1985).] (b)-(e) Successive stages of neuroblast segregation shown as transverse sections of one side of the ventral neurogenic ectoderm. (b) Longitudinal subdivisions of the neurogenic ectoderm; NRm, medial (i.e., ventral); NRi, intermediate; NRl, lateral neurogenic ectoderm cells, shown having undergone enlargement. Cells above NRl have not enlarged and do not belong to the neurogenic region; ms, mesoderm; tp, tracheal placode. Arrows indicate that neuroblasts of the initial wave will derive from NRm and NRl. (c) Neuroblasts already within are shown blank, and the second wave of neuroblasts in process of ingression from NRm and NRi are shown cross-hatched. (d) Ganglion mother cells are being given off by neuroblasts of the first and second waves (small cells), and a neuroblast of the third wave about to undergo ingression is shown in the NRm region (striped). (e) Ingression of all neuroblasts is now complete. The remaining hypodermal cells have decreased in apical-basal length. Ganglion mother cells continue to be be produced and programmed cell death (cd) also occurs in the neuroblast lineages. dor, dorsal; lat, lateral. [(b)-(e) From V. Hartenstein and J. A. Campos-Ortega (1984). *Wilhelm Roux's Arch. Dev. Biol.* **193**, 308.]

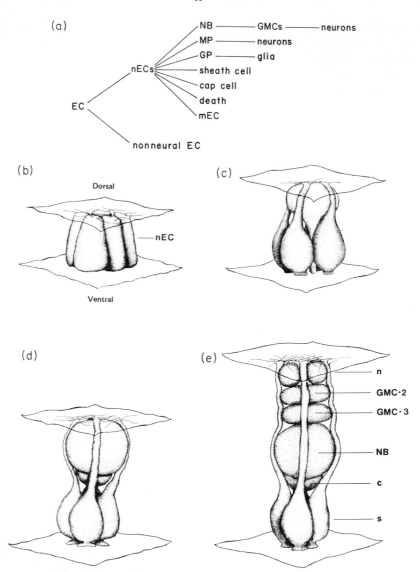

Fig 4.18.   Neuroblasts and support cells in the ventral neurogenic ectoderm of the grass-hopper *Schistocerca*. (a) Available developmental fates in the ectoderm. In this organism the lateral ectoderm cells (EC) differentiate in nonneural directions, while the midventral ectoderm is neurogenic (nEC). The nEC cells are initially equivalent, and may produce neuroblasts of two types: NB, stem cells which give rise to a series of ganglion mother cells, GMC, each of which divides to produce a pair of postmitotic neurons; and MP, which divide symmetrically to produce two postmitotic neurons directly. Or the nEC may differentiate as glial precursors, GP; as sheath cells; cap cells; as midline ectoderm cells, mEC; or undergo cell death. (b)-(e) Mor-phological differentiation of a neuroblast and supporting cell types, ventral (external) surface down; dorsal (internal) surface up. Observations are based on experiments in which individual cells were injected with a cell marker. (b) A group of contiguous EC's prior to differentiation,

lished by other means (see, e.g., Lohs-Schardin *et al.*, 1979a; Hartenstein *et al.*, 1985; Technau and Campos-Ortega, 1985). In the initial stages of the process of neuroblast delamination all of the cells in the ventral neurogenic ectoderm appear morphologically identical, though only a minor fraction become neuroblasts. The total number of cells included in the bilateral ventral neurogenic regions of the blastoderm is about 1770 (Hartenstein and Campos-Ortega, 1985), and these cells do not divide further until neuroblast ingression has begun. This process, as portrayed in Fig. 4.17(b)-(e) is not unlike the ingression of primary mesenchyme cells in the sea urchin embryo (see Fig. 4.7). Ingression of neuroblasts occurs in three waves, the neuroblasts deriving from the lateral and mainly from the medial longitudinal strips of neurogenic ectoderm on either side (Hartenstein and Campos-Ortega, 1984). A similar process occurs in the procephalic neurogenic ectoderm, except that here ingression of presumptive neuroblasts appears to be a continuous rather than periodic process.

Direct evidence that the fates of individual neurogenic ectoderm cells are not predetermined has been obtained in studies on the origins of neuroblasts in *Schistocerca* (Doe and Goodman, 1985a,b). On each side the ventral neurogenic ectoderm of this embryo consists of about 150 equivalent cells per segment, and similar to the case in *Drosophila*, about one fifth of these become neuroblasts. The remaining cells differentiate in other directions, e.g., forming neuronal support cells. The available range of cell fates is indicated in Fig. 4.18(a), and the morphological relations of the neuroblast and the support cells in Fig. 4.18(b)-(e). In *Schistocerca* as in *Drosophila* neuroblast segregation occurs in three waves, though in real time the process requires more than ten times longer in the grasshopper (Hartenstein and Campos-Ortega, 1984; Doe and Goodman, 1985b). Presumptive neuroblasts can be identified by their enlargement relative to the surrounding cells, and if these enlarged cells are destroyed by laser irradiation, an adjacent neurogenic ectoderm cell instead enlarges and replaces the ablated neuroblast (Doe and Goodman, 1985a). Thus no individual neuroectoderm cells are predetermined to become neuroblasts, and instead this capacity would appear to be distributed as a probability (i.e., ~20%) amongst all the neurogenic ectoderm cells. It follows that in the normal process of neuroblast specification, once a cell has become a neuroblast, it inhibits the proximal cells from doing so, and that the differentiation of the surrounding cells to

---

filopodia extending from dorsal surfaces of cells. (c) A single EC has begun differentiating as a neuroblast, and is in process of ingression. The contiguous cells have begun to differentiate as sheath cells. (d) NB enlarged and fully delaminated from ventral surface; sheath cells enclosing the NB, and a cap cell shown apposed to its ventral surface. (e) Mitotic T progeny of the NB, which has divided asymmetrically three times. The initally formed GMC has divided to produce two postmitotic neurons (n); two others have not yet done so; the sheath cells (s) traverse the neuroepithelium. c, cap cell. [From C. Q. Doe and C. S. Goodman (1985b). *Dev. Biol.* **111**, 193.]

become properly arranged support cells, such as shown in Fig. 4.18(b)-(e) (or to undergo cell death), is triggered through an inductive intercellular interaction occurring in the vicinity of each presumptive neuroblast within the neurogenic ectoderm. Other experiments show that the position, rather than the prior history, of a neuroblast determines the precise products of its subsequent lineage, i.e., the types of neuron to which it will give rise. Thus given neurons that are made to occupy ectopic positions by means of prior ablations differentiate to produce those specific neurons that would normally develop in those positions (Doe and Goodman, 1985a). The results of these experiments are clearly reminiscent of those demonstrating the probabilistic origins of neuroblasts in *Xenopus* (see above), suggesting an evolutionarily ancient and widespread developmental device. Note that in *Schistocerca* the initially *probabilistic* regional mechanism of neuroblast specification ultimately gives rise to *determinate cell lineages* (Doe and Goodman, 1985a,b; Kuwada and Goodman, 1985), similar to those initiated by the autonomous mechanisms of specification that prevail in organisms such as *C. elegans*.

Genetic and Molecular Evidence

A number of loci have been found that directly affect the process of neuroblast specification in *Drosophila*. Some of these act maternally, some zygotically, and some both. The first to be recognized as a gene required for embryonic neurogenesis was *Notch* (Poulson, 1937; Wright, 1970). *Notch* deficiencies result in the arrest of development at the end of embryogenesis, the larvae exhibiting a gross excess of neural cells at the expense of ventral and cephalic epidermal cells. Since in these mutants endodermal and mesodermal derivatives appear to have developed properly, the *Notch* gene product would seem to be required specifically for normal differentiation of the neurogenic ectoderm. A number of additional loci are now known that when mutated similarly cause hyperplasia of the CNS, with complementary epidermal deficiencies, *viz.*, *almondex (amx), pecanex (pcx), big brain, mastermind (mam), neuralized, Delta (Dl)* and *Enhancer of split* (Jiménez and Campos-Ortega, 1982; Lehmann *et al.*, 1983; Perrimon *et al.*, 1984), and there may be others yet unidentified (reviewed in Yedvobnick *et al.*, 1986). Use of temperature-sensitive mutants of *Dl* demonstrate that the period of development when absence of function causes the observed phenotype is during the process of neuroblast segregation. The effects of different mutations at the same locus are in some cases graded, and in others depends on the maternal genotype. Thus for example *amx, mam*, and *Notch* act maternally but mutant *amx, pcx*, and *mam* eggs can be rescued by the wild-type zygotic gene functions. Embryos homozygous for *amx* and *pcx* mutations can also be partially rescued by injection prior to cellular blastoderm stage of wild-type egg cytoplasm (LaBonne and Mahowald, 1985). Maternal *Notch* lethals cannot be rescued by wild-type sperm, however, although the

Notch[+] gene functions zygotically as well as during oogenesis (Shannon, 1972; Jiménez and Campos-Ortega, 1982; Lehmann *et al.*, 1983). When maternal effects are removed mutations at the *mam* and *Notch* loci have identical consequences. Morphological examination shows that in mutants of this class the normal number of neurogenic ectoderm cells is present in the blastoderm at the onset of gastrulation. These, however, give rise to an excess number of neuroblasts, and the fraction of dermatoblasts formed is correspondingly depressed. Examples of the resulting phenotypes are shown in Fig. 4.19(b)-(d). An important observation is that neural hyperplasia does not occur in the *dorsal* ectoderm, which also gives rise to some innervated cuticular sensory structures (see above), nor in any organ systems of mesodermal or endodermal derivation, though in extreme phenotypes it involves some regions of the lateral ectoderm that normally do not produce any neurons, e.g., the salivary gland anlage [Lehmann *et al.*, 1983; Campos-Ortega and Hartenstein, 1985; see also Fig. 4.17(a)]. Thus the effects are essentially confined to the procephalic and ventral neurogenic ectodermal regions. Furthermore, since lack of function mutations cause *neural hyperplasia*, neuroblast formation can be conceived as the default pathway of differentiation in the neurogenic ectoderm. Thus the normal function of these loci must be required for differentiation along the *epidermal* pathway normally traversed by 75% of the cells in the neurogenic ectoderm.

The extent of the neurogenic regions to which the primary effects of the loci so far discussed are mainly confined is controlled by other genes that act maternally. Two maternal effect genes that have been shown to be involved, though perhaps indirectly, in the determination of the neurogenic ectoderm domains are *dorsal* (*dl*) and *Toll* (*Tl*) (Campos-Ortega, 1983; effects of several *dl* and *Tl* mutations, and their experimental rescue, are discussed in Chapter VI). Eggs from females carrying extreme *dl* mutations develop only organs arising from the dorsal regions of the blastoderm. They lack the ventral neurogenic ectoderm altogether, as well as other ventrally derived structures such as all mesodermal products, anterior midgut, etc. [cf. Fig. 4.17(a)]. Dominant maternally acting mutations of *Tl* have the reverse effect, causing the whole blastoderm to develop as ventral structures (though recessive loss of function mutations display instead a dorsalized phenotype; Anderson *et al.*, 1985a). The *Tl*[D] phenotype includes significant hypertrophy of the ventral CNS (reviewed in Hartenstein and Campos-Ortega, 1985). Combination of *dl* or *Tl*[D] mutations with mutations such as *Notch* produces the consequences expected on the basis that the function of the normal alleles of the latter class of genes is to set the probability of forming dermatoblasts *within the neurogenic ectoderm* down to about 25%, the dimensions of the neurogenic ectoderm being determined during oogenesis by the class of genes that includes *dl* and *Tl* (cf. Chapter VI). Thus for example the entire ectoderm is hyperneuralized in *Notch* mutants derived from (heterozygous) *Tl*[D] mothers, while the ventral ectoderm of *Notch* mutants derived from

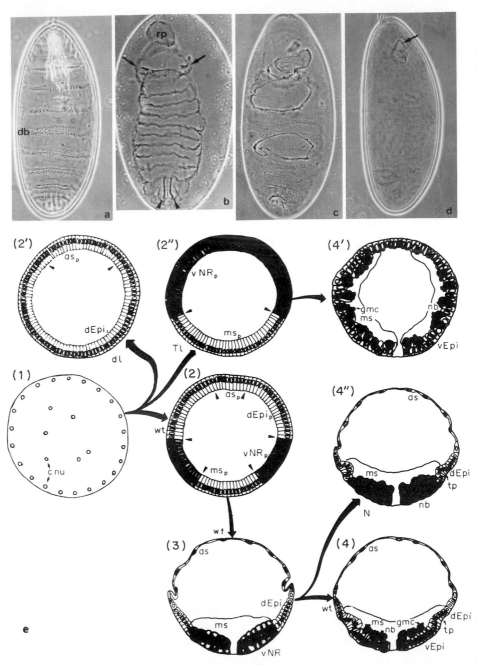

Fig. 4.19. Genetic effects on neuroblast specification in *Drosophila*. (a) Dorsal aspect of cuticle of completed wild-type embryo (for ventral aspect see Fig. 4.16). Arrowheads indicate posterior spiracles; tr, tracheal stems. (b) Cuticle of embryo mutant at the *master mind* (*mam*)

homozygous *dl* mothers produces no neural tissue whatsoever [Fig. 4.19(d); see also Campos-Ortega, 1983]. These relations are summarized in the diagram shown in Fig. 4.19(e). This analysis suggests that one process that might be affected by genes such as *Notch, mam, Dl*, etc., is the *intracellular interaction* by which differentiating neuroblasts repress neuroblast specification in neighboring cells, and promote the alternative pathway leading to epidermal differentiation.

The *Notch* gene, which is located near position 3C7 (Lindsley and Grell, 1968), has been cloned and the location of many mutations mapped at the genomic level (Artavanis-Tsakonas *et al.*, 1983, 1984; Kidd *et al.*, 1983; Grimwade *et al.*, 1985). Mutations at this complex locus also affect adult wing and eye structure, and studies with temperature sensitive mutants indicate various postembryonic periods when expression of the *Notch*[+] gene is required (reviewed in Grimwade *et al.*, 1985; Yedvobnick *et al.*, 1985). *Notch* mutations map within a region ~40 kb in length, the extremes of which include the 5′ and 3′ termini of the known *Notch* locus transcripts (Grimwade *et al.*, 1985). This observation confirms genetic evidence that the *Notch* locus is a single genetic unit, despite the diversity of the adult structures affected by different *Notch* mutations, and the several stages at which

---

locus. This particular mutation is one conferring a phenotype of intermediate severity. The cuticle consists of a segmented dorsal plate connected by epidermal bridges to remnants of the cephalic cuticle (arrows). Ventral and lateral cuticle are missing. rp, roof of pharynx. (c) Ventral aspect of cuticle of embryo mutant at the *Notch* locus; allele conferring a weak mutant phenotype. More extreme *Notch* alleles still differentiate a normal dorsal cuticle but manifest enormous hyperplasia of the CNS, and failures of hypodermal differentiation elsewhere. The head cuticle of the preparation shown is greatly reduced; the dorsal cuticle (not seen) is normal; and there are several large holes in the ventral cuticle. Denticle belts have been formed in some regions. [(a)-(c) From R. Lehmann, F. Jiménez, U. Dietrich, and J. A. Campos-Ortega (1983). *Wilhelm Roux's Arch. Dev. Biol.* **192**, 62.] (d) Cuticle preparation of an embryo mutant at both *Notch* and *Toll* loci (see text). Only a small area of cuticle over the foregut has differentiated (arrow). The remaining ectoderm over the whole circumference of the embryo, including the dorsal side, has undergone neural hyperplasia. [(d) From J. A. Campos-Ortega (1983). *Wilhelm Roux's Arch. Dev. Biol.* **192**, 317.] (e) Diagram summarizing early stages of normal process of neurogenesis and effects of two different classes of mutants. 1-4, wild-type embryos, shown in transverse section. Dark shading (2) shows vNR, ventral neurogenic ectoderm (see Fig. 4.18); at later stages, (3), (4) neuroblasts and their derivatives are shown darkly shaded, and the remaining epidermal cells are shown in light shading; as, amnioserosa; cnu, cleavage nuclei; dEpi, dorsal epidermis; gmc, ganglion mother cell; ms, mesoderm; nb, neuroblast; tp, tracheal placode. 2′, effects of the maternal mutation *dorsal* (*dl*), causing circumferential extension of dEpi; 2″, converse effect, the extension of vNR, caused by a dominant maternal mutation of the *Toll* locus; 4′, the resulting production of neuroblasts around most of the circumference of the embryo; 4″, effects of mutation causing neural hyperplasia, here *Notch, N*. [(e) Reprinted with permission from J. A. Campos-Ortega and V. Hartenstein (1985). *In* "Comprehensive Insect Physiology, Biochemistry, and Pharmacology. Nervous System: Structure and Motor Function" (G. A. Kerkut and L. I. Gilbert, eds.), Vol. 5, p. 49. Pergamon, Oxford.]

the effects of mutations are observed. The major processed *Notch*⁺ transcript, which comprises at least eight exons, is about 10.5 kb in length, and is found in eggs, early embryos, and in later stages when *Notch* gene expression is required. *In situ* hybridization indicates that *Notch* transcripts are present not only in neurogenic ectoderm cells, but also in cells of the dorsal ectoderm, mesoderm and gut (Yedvobnick *et al.*, 1985). The *Notch*⁺ transcript has been sequenced, and found to encode a 2703 amino acid protein several features of which may provide insights into the possible function of this locus (Wharton *et al.*, 1985). In the middle of the protein is a typical membrane-spanning sequence element, which might be utilized to mount on the external surface of the cell a long repetitive region that contains 36 occurrences of a sequence ~40 amino acids in length, which is homologous with mammalian epidermal growth factor (EGF). This observation could explain the wide distribution of *Notch* transcripts, since in mammalian systems EGF displays mitogenic activity and stimulates differentiation in a variety of mesodermal cell types such as fibroblasts as well as of ectodermal cell types (Carpenter and Cohen, 1979). The sequence of the *Notch* protein thus provides support for the inference drawn above, *viz.*, that the function of this gene is required for the cellular interactions that specifically induce epidermal differentiation in the neurogenic ectoderm of the *Drosophila* embryo.

(iii) Genetic Control of the Morphogenesis of Metameric Structures in the Embryo

A major characteristic of arthropod embryogenesis is the early appearance of segmental or metameric morphological organization, and many of the "pattern formation" mutations that have been isolated identify genes required for this process. The initial establishment of gross metameric units of structure can be separated from the process by which these units are individually assigned their correct developmental identities. Segmental *identity* appears to be controlled by the action of homeotic gene complexes that are also required for the correct morphogenesis of adult segmental structures. The brief discussion of such genes that follows later in this chapter is confined to their embryonic effects. Morphological and genetic aspects of the function of homeotic genes have been extensively discussed, and the reader is referred, e.g., to reviews of Ouwenweel (1976), Garcia-Bellido (1977), Lawrence (1981), Lawrence and Morata (1983), Raff and Kaufman (1983), Struhl (1983), and Lewis (1978, 1985). There has been achieved an impressive accumulation of information regarding the spatial and temporal loci of action both for the class of genes required for the initial establishment of metamerism, and for the homeotic segmental identity genes. However, it must be noted at the outset that what any of these genes actually do to produce their observed effects on multicellular morphological components of the embryo remains an unanswered question.

The Initial Establishment of Metameric Organization

The locations of the segmental epidermal primordia at the cellular blasto-derm stage are shown in Fig. 4.17(a), as demonstrated in a variety of cell lineage studies (Lohs-Schardin *et al.*, 1979b; Szabad *et al.*, 1979; Harten-stein and Campos-Ortega, 1985; Hartenstein *et al.*, 1985). The epidermal primordium of each segment consists initially of a strip 3-5 cells wide, per-pendicular to the anterior-posterior axis of the egg, and containing on each side of the embyo a total of 75-90 blastoderm cells (reviewed in Lawrence, 1981; cleavage processes leading to formation of the cellular blastoderm are reviewed in Chapter III). Soon after the onset of gastrulation ($\sim$3.75 hr post-fertilization) the presumptive epidermal cells begin to resume mitotic activ-ity, and they undergo a second wave of mitosis between 4.4 and 6 hr, with a minor fraction later dividing a third time (Hartenstein and Campos-Ortega, 1985). By the completion of embryogenesis each segment is about 10-15 cells wide, and each lateral hemisegment contains about 350 epidermal cells. The nervous system is also segmented early in postgastrular embryogenesis (Brown and Schubiger, 1981) as is the mesoderm (Martinez-Arias and Law-rence, 1985). The mesodermal anlagen of the cellular blastoderm stage embryo are positioned ventrally, as shown in Fig. 4.17(a), and prior to invag-ination the mesodermal primordium of each future segment includes about 45 cells. Following gastrulation these cells also undergo three successive waves of division (Hartenstein and Campos-Ortega, 1985).

Observations on the disposition of the progeny of clones of epidermal cells marked by X-ray induced mitotic recombination at the cellular blastoderm stage have revealed early metameric restrictions in cell fate. Thus when examined at later developmental stages the marked clones are found not to have crossed certain metameric boundaries, which define regions known as "compartments" (Garcia-Bellido *et al.*, 1973; Crick and Lawrence, 1975; Wieschaus and Gehring, 1976; reviewed in Lawrence, 1981). Lawrence (1973) demonstrated that clones marked by X-irradiation of embryos of the milkweed bug *Oncopeltus* soon after cellular blastoderm formation respect such boundaries, but clones generated from nuclei marked prior to cellular-ization overlap into adjacent compartments. Each *segment* of the *Drosoph-ila* larval and adult epidermis is believed to consist of an *anterior* and a *posterior compartment*, within which the clonal progeny of a group of pro-genitor cells intermingle extensively (reviewed in Garcia-Bellido *et al.*, 1979; Martinez-Arias and Lawrence, 1985). In adult *Drosophila* structures devel-oping from imaginal discs, particularly the wing, late developing dorsal-ventral compartmental boundaries have been described as well, but these have not been defined as unequivocally as are the anterior-posterior com-partments (see, e.g., Brower, 1985). In contrast to the "compartments" into which the majority of the descendants of 512-cell *Xenopus* blastomeres are segregated (see above), the anterior-posterior compartments of the *Drosoph-ila* embryo inferred from embryonic lineage restrictions appear to function as fundamental units of gene regulation.

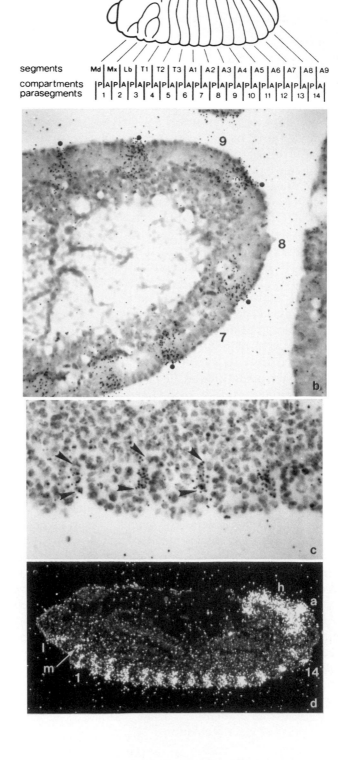

The initial morphological evidence of metamerism in the *Drosophila* embryo is the appearance of transverse grooves during the gastrula stage, at about 5 hr of development. These appear to define metameric units termed *parasegments*, that are one compartment out of register with the segments perceived in the completed larva and in the adult (reviewed in Martinez-Arias and Lawrence, 1985). The relationships among compartments, segments, and parasegments are diagrammed in Fig. 4.20(a). This morphological interpretation is supported by comparative evidence from other arthropod embryos. The organization portrayed in Fig. 4.20(a) applies to the CNS as well as the ectoderm (see, e.g., Teugels and Ghysen, 1985; Martinez-Arias and Lawrence, 1985; Beachy *et al.*, 1985). However, for the mesoderm there is no evidence for anterior and posterior compartments. Marked clones of mesodermal cells generated soon after blastoderm formation are found to extend throughout given segments when later examined, though they do not cross to adjoining segments (Lawrence, 1982). Martinez-Arias and Lawrence (1985) concluded that the metameric elements formed in the 5-6 hr embryonic mesoderm are initially aligned with the epidermal parasegments. They later shift posteriorly, assuming a segmental registration with respect to the epidermis.

The *engrailed* (*en*) locus of *Drosophila*, originally identified from the effects of a nonlethal mutation (*en¹*) on wing venation and shape, is among those required for metamerization in the embryo. Studies on mutant *en¹* flies and on normal flies bearing clones homozygous for lethal *en* mutations have shown that the defects these lesions cause in adult structures are always

---

Fig. 4.20. Parasegments as morphological and developmental units. (a) Boundaries of parasegments, segments and compartments in the *Drosophila* embryo. Parasegments are one compartment out of register with segments. Md, mandibular segment; Mx, maxillary segment; Lb, labial segment; T1-T3, thoracic segments; A1-A9, abdominal segments. [(a) From A. Martinez-Arias and P. A. Lawrence (1985). Reprinted by permission from *Nature (London)* **313**, 639. Copyright © 1985 Macmillan Journals Limited.] (b) Expression of *engrailed* gene observed by *in situ* hybridization, in an embryo at the extended germ band stage, when metamerism first becomes morphologically apparent. Anterior, left; dorsal up. A few cells immediately posterior to the ectodermal grooves (black dots) that denote the *parasegmental boundaries* express the *engrailed* gene in the longitudinal medial section shown, which displays the posterior end of the embryo. The parasegments are numbered as in (a). About one-fourth of the epidermal cells express this gene. (c) Hybridization of the *engrailed* gene to a ventral section from a later embryo at the shortened germ band stage. The epidermis now displays deep folds at the *segmental* boundaries. Three to four cells containing *engrailed* transcripts can be seen at the posterior boundary of each segment (arrowheads). The location of these cells mark the anterior-posterior compartmental boundaries between segments. (d) Medial longitudinal section of whole gastrula stage embryo, shown in dark field. Parasegments 1 and 14 are indicated. The *engrailed* gene is also expressed in a, the anal region; h, the hindgut; and in the anterior region, l, the labrum; and in a premandibular segment, m. [(b)-(d) From P. Ingham, A. Martinez-Arias, and P. A. Lawrence (1985a). Reprinted by permission from *Nature (London)* **317**, 634. Copyright © 1985 by Macmillan Journals Limited.]

confined to the posterior compartments. The absence of *engrailed* gene function in anterior cells is without detectable effect, while normal expression of the *engrailed* gene is required for development of posterior compartments in many adult segments, including proboscis, antennae, wings, legs, genitalia, etc. (Lawrence and Morata, 1976; Kornberg, 1981; Lawrence and Struhl, 1982). A cellular function for the *engrailed* gene is suggested by the observation that marked clones in the wings of *en*[1] mutant flies sometimes transgress the anterior-posterior compartment boundaries (Morata and Lawrence, 1975; Lawrence and Morata, 1976). Thus the *engrailed* gene may be considered to contribute to the mechanism by which cells of posterior compartments in all regions of the organism recognize a difference between themselves and cells of anterior compartments. A recognition property of this nature could also underlie the *embryonic* segregation of sets of clones into anterior and posterior compartments at the onset of metamerism. Lethal *en* alleles are in fact found to block the processes leading to the formation of normal segments (Nüsslein-Volhard and Wieschaus, 1980; Kornberg, 1981). In some of these mutants the posterior margins of every other segment are deleted, the arrested embryos displaying fused, paired segments of abnormal cuticular morphology.

The *engrailed* gene has been cloned and something is known of its structure and transcriptional organization (Poole *et al.*, 1985; Kuner *et al.*, 1985; Fjose *et al.*, 1985). The gene occupies a 70 kb region which includes the sites of the known *en* mutants. The initially discovered *en*[1] mutation turns out to be due to an insertion (Kuner *et al.*, 1985). Several transcripts, the major one about 2.7 kb in length, derive from a portion of the gene where the known lethal mutations occur. This transcript includes a 1700 nt long open reading frame which encodes a protein that would contain long stretches of polyglutamine, polyserine, and polyalanine, followed by a 60 amino acid sequence partially homologous with sequences shared by genes displaying homeotic function, the "homeobox" sequence (see below). The most interesting outcome from molecular isolation of *engrailed* sequences is the spatial distribution of its products in the embryo, as revealed by *in situ* hybridization of the transcripts, and cytological immunolocalization of the *en* protein. The *engrailed* gene evidently begins to function at least several division cycles *prior* to cellularization, since cleavage stage embryos bearing lethal *en* mutations display morphological abnormalities such as asymmetric location of pole cells and abnormal distribution of somatic nuclei (Karr *et al.*, 1985). *Engrailed* transcripts are observed by gel blot hybridization as early as the 12th cleavage cycle. A localized pattern of *engrailed* gene transcripts is first observed just prior to cellularization, i.e., after completion of cleavage, in a sharp transverse band, shortly followed by other bands spaced at two segment widths. After gastrulation has begun these transcripts are found sharply localized in a metameric array of 14 transverse stripes (Fjose *et al.*, 1985; Kornberg *et al.*, 1985; Ingham *et al.*, 1985a; Weir and Kornberg, 1985).

This is shown in the sections reproduced in Fig. 4.20(b)-(d). From their number and position the cells then expressing the *engrailed* gene can be seen to be present in every segmental primordium, where they are confined to the future posterior compartments. DiNardo *et al.* (1985) utilized a fusion construct to generate an antibody against the *engrailed* protein product which could then be localized by immunocytological procedures. The *engrailed* protein is also localized in the posterior compartments of every segment in gastrula stage embryos, though earlier it appears briefly only in every other segmental primordium, as do the transcripts. Figure 4.20(b) shows that at the gastrula stage the cells displaying *engrailed* transcripts are located at the posterior edges of the morphological grooves that initially delimit the parasegments, confirming that parasegments rather than segments are the developmentally primitive metameric units (Ingham *et al.*, 1985a). In later embryos transcripts of the *engrailed* locus are found in both epidermis and CNS (Ingham *et al.*, 1985a), but except for a brief period they are absent from the mesoderm and endoderm. This result is consistent with observations of Lawrence and Johnston (1984a) made on genetically mosaic flies generated by transplantation of nuclei homozygous for lethal *engrailed* alleles. While the mutant *en* cells cause visible defects in epidermis and CNS, internal organs derived from endoderm and mesoderm develop normally even if they include large clones of *engrailed*-lethal cells. *Engrailed* gene function should not be required for metamerism in the mesoderm, since anterior-posterior compartmentation is not observed there (see above). In general *engrailed* transcripts are observed in those locations affected by *en* mutations, i.e., the posterior compartments of embryonic segments and also of imaginal discs (Kornberg *et al.*, 1985). The function of the *engrailed* gene is thus to produce a property of posterior compartment cells, rather than to repress a property of anterior compartment cells. Lawrence and Morata (1976) suggested that the posterior cell property controlled by the *engrailed*$^+$ gene is one that confers physical affinity for other posterior cells, and avoidance of contact with anterior cells. This might explain the formation of straight anterior-posterior compartmental boundaries in *en*$^+$ segments, the failure of *en*$^-$ clones to respect these boundaries, and the fusion of adjacent metameric units in embryos bearing lethal *en* mutations. The sharp localization of *engrailed* transcripts right at the anterior-posterior compartmental boundaries, as shown in Fig. 4.20(d), indeed suggests that at this stage the *en* product exercises a "border marking" function. Thus early in gastrulation, a band at the parasegmental borders only a single cell in width displays *engrailed* gene transcripts and proteins (Kornberg *et al.*, 1985; Fjose *et al.*, 1985; DiNardo *et al.*, 1985). Though the various forms of evidence as to *engrailed* gene function lead to the supposition that its expression may result in alteration of external cellular properties recognized by other cells, the *en* protein itself is localized in the cell nuclei (DiNardo *et al.*, 1985). In addition regions of this protein have been shown to display DNA binding activity (Desplan *et al.*,

1985), recognizing specific sites in λ DNA, in the upstream regions of the *engrailed* gene itself, and also of the *fushi tarazu* gene, for which independent arguments suggest an interaction with the *en* gene (see below). Thus the immediate function of the *engrailed* gene may be regulatory.

Though the mechanism is unknown, the initial establishment of metamerization clearly depends on interactions between the blastoderm cells and maternal spatial information resident in the cortex of the egg. The existence of a maternal cytoplasmic organization specifying the future anterior-posterior axis, to which the segments will be perpendicularly arranged, is known from both physiological and genetic observations (reviewed in Chapter VI). Many maternal mutations affect the patterns of segmentation in the embryo (see, e.g., Nüsslein-Volhard, 1979). A well known example is afforded by the *bicaudal* mutation, which produces embryos that lack anterior structures, and at arrest instead display a mirror-image duplication of the posterior metameric cuticular pattern. A general characteristic of such maternal mutations is that they cause global disorganization of embryonic morphology, rather than specific segmental defects. It is not known whether maternal morphogenetic information is distributed along the anterior-posterior egg axis in a periodic form that is later reflected directly in metameric patterns of cell specification, or whether, as classically conceived, it is presented as a monotonic gradient extending from anterior to posterior end. In either case, however, the experimental evidence reviewed in Chapter VI indicates that the responsible maternal factors are probably associated physically with cytoskeletal elements of the egg cortex. The following discussion concerns other zygotically active genes required for the establishment of metamerism, the products of which may be thought to be utilized for the "interpretation" of maternal spatial information, whatever the global form of its distribution.

About twenty zygotically active genes that when mutated affect segmentation were detected in the saturation screens for pattern formation loci cited earlier (Nüsslein-Volhard *et al.*, 1984; Jürgens *et al.*, 1984; Wieschaus *et al.*, 1984a). These fall into three classes (Nüsslein-Volhard and Wieschaus, 1980). The first of these, which at present includes seven loci, is identified by mutations that permit the formation of a normal number of segments, but cause within every segment certain alterations of cuticular pattern, different for each locus. An example is *gooseberry*, in which every segment lacks the regions of naked cuticle that normally separate the denticle belts, so that the ventral surface of the larva is largely covered with denticles. These, however, display normal segment-specific arrangement, pigmentation, and orientation (Nüsslein-Volhard and Wieschaus, 1980). Mutations in another such gene, *fused*, which functions both maternally and zygotically, cause extensive cell death in the mesoderm and ectoderm during postgastrular development. The segmental pattern deletions and duplications caused by *fused* mutations may be secondary consequences of this (Martinez-Arias, 1985). Genes such as *gooseberry* and *fused* would appear to be involved in formation of intrasegmental structures common to all segments, rather than in the establishment of metamerism *per se*.

Mutations in a second class of zygotically active genes produce gross deletions of many contiguous segments. An example of a "gap" mutation of this kind is *Krüppel (Kr)* (Wieschaus *et al.*, 1984b). In embryos homozygous for some *Kr* alleles the thoracic and anterior abdominal segments are replaced by a partial mirror-image duplication of the abdominal segments. This phenotype is similar to that produced by maternal *bicaudal* mutations, except that *Kr* embryos display normal head development. *Kr* embryos have a normal number of cells, but the fates of some of these have been altered. The *Kr* gene has been cloned, and its transcripts are first detected prior to cellularization, at 11th cleavage (Preiss *et al.*, 1985; Knipple *et al.*, 1985). *In situ* hybridization reveals an initial accumulation of *Kr* transcripts in a transverse midregion band of 8-10 nuclei, and by cellularization, the *Kr* gene is being expressed in a strip 12-14 cells wide (Knipple *et al.*, 1985). This region is included in, but is far smaller than, the anlagen for the large set of thoracic segments missing in *Kr⁻* mutant embryos, suggesting that the morphological effects of *Kr* mutations are at least partially indirect. In later embryos a variety of specific cell types express *Kr* transcripts, including some in regions of the embryo not affected by *Kr* mutations, e.g., the posterior abdominal structures. Lethal *Kr* mutants can be at least partially rescued by injection of cloned DNA bearing a wild-type *Kr* allele, or of cytoplasm from *Kr⁺* embryos, if the injection is performed prior to blastoderm formation, and in the midregion of the egg. Furthermore, the *Kr⁻* phenotype can be induced in normal embryos by injection of antisense *Kr* RNA (Rosenberg *et al.*, 1985). One interpretation of these observations is that the *Kr* gene codes for a product that is required in thoracic and anterior abdominal cells for detection of or interaction with the maternal product of the *bicaudal⁺* gene, but this is clearly not its sole function, given the pattern of *Kr* gene expression later in embryogenesis.

A third class of zygotically active genes that affect segmentation includes nine loci that cause deletions in cuticular pattern that are repeated throughout the larva at intervals of every other segment. These are termed "pair rule" genes (Nüsslein-Volhard and Wieschaus, 1980). We have already considered one gene of this class, *engrailed*. Another locus of this class that has been studied extensively at the molecular level is *fushi tarazu (ftz)* (Weiner *et al.*, 1984; Kuroiwa *et al.*, 1984; Hiromi *et al.*, 1985). Scott *et al.* (1983) showed that the *ftz* locus is only about 20 kb from the proximal terminus of the homeotic *Antennapedia* gene, located at chromosomal band 84B1,2. Embryos homozygous for *ftz* mutant alleles display half the normal number of segments, and studies with temperature-sensitive alleles show that the activity of the *ftz⁺* gene is mainly required very early in development, between 2 and 4 hr postfertilization (Wakimoto and Kaufman, 1981; Wakimoto *et al.*, 1984). The mutant phenotype can be interpreted as the consequence of

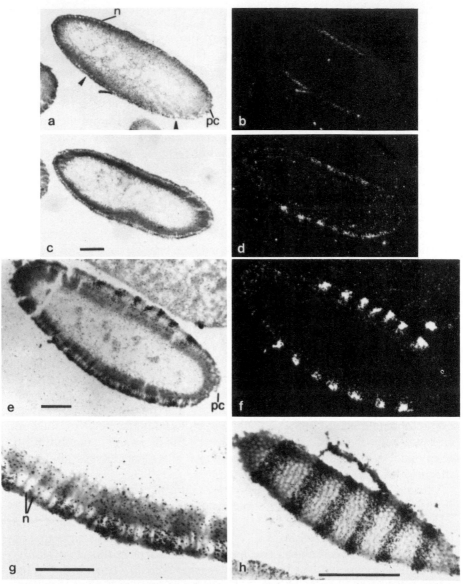

Fig. 4.21. Distribution of transcripts of *ftz*, a "pair rule" gene, in early *Drosophila* embryos. (a) and (b) Light and dark field photomicrographs of horizontal section of an 11th cleavage embryo hybridized with a cloned *ftz* probe. Arrowheads indicate region where hybridization above background is first observed. n, peripheral cleavage nuclei; pc, pole cells. (c) and (d) Light and dark field photomicrographs of longitudinal sections of a 13th cleavage syncytial blastoderm stage embryo hybridized with *ftz* probe. Periodic distribution of hybridizing cells is beginning to emerge. (e) and (f) Light and dark field view of sagittal section of embryo at cellular blastoderm stage, hybridized with *ftz* probe. Seven patches of cells expressing the *ftz* transcript can be seen on both dorsal and ventral sides. (g) Enlargement of ventral region of same section

elimination by cell death of seven metameric units that approximately coincide with alternate (i.e., the even numbered) parasegments (Martinez-Arias and Lawrence, 1985; Struhl, 1986). The *ftz* locus produces a 1.8 kb transcript that appears 2 hr after fertilization, and decreases sharply in concentration after 4 hr, though it remains detectable until late in embryogenesis (Weiner *et al.*, 1984; Kuroiwa *et al.*, 1984). As illustrated in Fig. 4.21(a)-(b), *in situ* hybridizations carried out by Hafen *et al.* (1984a) demonstrate the appearance of *ftz* transcripts in the vicinity of the peripheral syncytial blastoderm nuclei after 11th cleavage, in an area extending from about 15% to 65% of the egg length. By the final (i.e., 13th) cleavage, but even *before* cellularization of the blastoderm, a periodic distribution of *ftz* transcripts becomes evident [Fig. 4.21(c)-(d)]. This is sharply delimited to seven equally spaced regions by the time cellularization is complete [Fig. 4.21(e)-(g)]. The labeled regions consist of transverse strips each containing 3-5 cells, i.e., the width of a single segmental (or parasegmental) primordium. Cells of the ventral presumptive mesoderm are labeled as well as the presumptive ectoderm cells, and in early gastrula stage embryos the invaginated mesoderm cells also express the *ftz* gene. Hiromi *et al.* (1985) showed that in transgenic late gastrulae a fusion protein synthesized under the control of the *ftz* gene regulatory sequences also appears in neuroblasts in every segment. Different upstream sequences regulate the expression of *ftz* in the epidermis and in the neuroblasts. In the epidermis the *ftz* gene is expressed during the blastoderm-early gastrula period in regions that appear to correspond in registry to the even numbered parasegmental anlagen (Martinez-Arias and Lawrence, 1985; Hiromi *et al.*, 1985). That is, the metameric units in which the *ftz* gene is expressed are approximately equivalent to those found deleted in *ftz⁻* mutants. Metamerization thus might proceed by activation of sets of genes such as *ftz* in strips of cells that in width equal the primordium of a parasegmental (or segmental) unit, alternating with strips of cells of similar width that display other patterns of gene activity.

The earliest observed arrangement of the cells expressing *ftz* with respect to the global coordinates of the egg indicates a response, whether direct or indirect, to maternal spatial information. Thus the cells initially expressing the *ftz* gene are arranged in broad bands perpendicular to the anterior-posterior axis of the egg. When the maternal spatial coordinates are altered, the zygotic expression of pair rule genes is altered accordingly. Thus in eggs produced from *bicaudal* females half the normal number of strips of cells express the *ftz* gene (Gergen *et al.*, 1986), and a similar result has been reported for another pair rule gene, *hairy* (Ingham *et al.*, 1985b). The sharp-

shown in (e) and (f), here visualized in light micrograph. Three to five labeled cells can be seen to be spaced by an equal number of unlabeled cells. (h) Tangential section through the epithelial wall of an embryo at same stage. Bar in each figure represents 0.1 mm. [From E. Hafen, A. Kuroiwa, and W. J. Gehring (1984a). *Cell* **37**, 833. Copyright by M.I.T.]

ening of the spatial pattern of *ftz* transcript distribution late in cleavage (Weir and Kornberg, 1985) suggests that while the initial domains of *ftz* expression could reflect directly the distribution of maternal determinants, the final pattern might depend in addition on the expression of zygotic functions in the blastoderm nuclei themselves. In fact normal function of the other pair rule genes *hairy, even skipped,* and *runt,* and of several gap loci, including *Krüppel,* is required for wild-type distribution of the *ftz* protein (Carroll and Scott, 1986). This protein, detected immunocytologically, appears first at the cellular blastoderm stage in the expected pattern of seven transverse strips, and it is concentrated in the *cell nuclei,* just as is the *engrailed* protein (Carroll and Scott, 1985). Thus *ftz* could also belong to the class of genes the primary function of which is to affect the expression of other genes, here presumably genes required for the *localization of metameric patterns of genomic function.* The protein product of the *engrailed* gene begins to accumulate about 15 min after the *ftz* protein appears (DiNardo *et al.,* 1985), and the normal pattern of *engrailed* protein distribution requires wild-type *ftz* function (Howard and Ingham, 1986).

Experiments with embryos that are mosaic for pair rule mutations imply that each cell of the blastoderm assays its position independently and *autonomously,* and sets its pattern of gene expression accordingly (Gergen and Wieschaus, 1985). Studies of this kind were carried out with a lethal *runt* mutation, which in homozygous form causes specific deletions of cuticular pattern elements, again spaced at two segment intervals, and the replacement of the deleted elements with mirror-image duplications of the remaining structures (Nüsslein-Volhard and Wieschaus, 1980). In mosaic larvae *cells* of *runt* genotype were found to display the *runt* cuticular morphology, while adjacent wild-type cells produced the wild-type pattern. The only deviations from simple autonomy were observed at segmental boundaries. Similar experiments reveal autonomous expression of several additional pair rule genes that affect different cuticular pattern elements (Gergen *et al.,* 1986; Gergen and Wieschaus, 1986a). Furthermore, five of the nine known pair rule genes (*odd, prd, opa, slp,* and *en*) function autonomously with respect to *ftz,* in that mutations at these loci leave the pattern of *ftz* product expression unchanged (Carroll and Scott, 1986). Morphological analysis of the domains of action of different pair rule genes shows that the boundaries of these domains in general do not coincide with segmental, parasegmental, or compartmental boundaries, *ftz* providing an apparent exception. These domains are in fact all offset from one another in phase, though they more or less display a similar two-segment period (Gergen *et al.,* 1986). At the cellular blastoderm stage this period may be presumed to equal the width of two segment primordia, i.e., about eight cells. This interpretation is supported by *in situ* hybridizations carried out with a cloned *hairy* probe (Ingham *et al.,* 1985b). The periodic distribution of *hairy* transcripts is similar to that of *ftz* transcripts, but in a different phase, so that some cells display both, some neither, and some either *ftz* or *hairy* RNAs. Furthermore, the periodic distribution of cells expressing genes such as *ftz* is required for normal metamerization. This is demonstrated in an experiment of Struhl (1985), in which transgenic blastoderm stage embryos were induced to produce the *ftz* protein ectopically under the control of a heat shock promoter. Synthesis of the *ftz* gene product in the alternate metameric anlage where it is normally absent results in developmental failure of elements of these parasegments (i.e., the odd numbered parasegments). Thus the *overall spatial pattern of expression*, i.e., the absence as well as the presence of the *ftz* gene product, is functionally required. The same conclusion has been drawn for the *runt* gene product (Gergen and Wieschaus, 1986b). Generalizing to the whole set of pair rule genes, the *combinations* of gene activities that occur in the individual nuclei of every metameric primordium might uniquely mark the cells at each position. In fact there is evidence that the products of these genes function in a synergistic way. Thus embryos homozygous for both the *even-skipped* and *odd-skipped* pair rule mutations display some amelioration of the phenotypic effects rather than lacking *both* even and odd numbered denticle bands, as if the *ratio* of the gene products is important (Gergen *et al.*, 1986). Other observations demonstrate a graded morphological response when the number of copies of *runt*$^+$ genes is increased, so that too much *runt* product results in defects similar to those caused by mutations of other pair rule loci (Gergen *et al.*, 1986; Gergen and Wieschaus, 1986b).

Like *ftz*, other pair rule genes are also expressed very early in development (evidence for *hairy, runt, paired*, and *even-skipped* is reviewed in Gergen *et al.*, 1986), i.e., long prior to the appearance of cuticular morphological features such as denticle belts. The products of these genes, in combination, can be imagined to confer upon the cells the genomic regulatory statuses characteristic of various positions in the metameric primordia, i.e., in the sense considered earlier, the heritable state of cell specification. As the metameric width expands mitotically following gastrulation, the relative positional identities of the primordial founder cells are utilized by their progeny to generate the polarities of the mature segments. This process of *territorial designation* is clearly to be distinguished from the institution of cell differentiation and functional specialization, by which subsequently arise particular structures within the metameric units of the advanced embryo. Even so, it clearly requires early differential zygotic gene activity, and from a comparative point of view it is particularly interesting that in the *Drosophila* embryo some of the earliest differential gene activations are evidently concerned with *delimiting spatial domains* rather than with generating specialized cell types. The process by which the morphological patterns are ultimately organized within the segments (or compartments) of the larval epidermis remains unknown, and most likely involve intercellular inductive interactions that much later occur within the territories we perceive as segmental units.

Segmental Identity

The establishment of segmental identity in the embryo is controlled by a small set of homeotic genes, mutations in which may result in replacement of morphological patterns characteristic of given segments or parts of segments with those of other segments. The best known of these loci are the bithorax complex (BX-C), and the Antennapedia complex (ANT-C), both of which have been cloned. Molecular characterization indicates that each of these complexes includes more than 200 kb of genomic DNA. The BX-C controls the identity of thoracic and abdominal segments, and the ANT-C [including the *Sex combs reduced (Scr)* gene] is required for the correct development of mesothoracic, prothoracic and labial segments (Wakimoto and Kaufman, 1981; Struhl, 1983; Sato *et al.*, 1985). Additional homeotic genes may be required for determination of head and terminal posterior structures (Struhl, 1983; Sato *et al.*, 1985). The photomicrographs shown in Fig. 4.22(a) and (b) display the embryonic effects of a deficiency that removes the whole of the BX-C (Lewis, 1978). These mutants survive to late embryogenesis, and they display transformation of parasegments 5-13, plus anterior 14, toward the morphology of parasegment 4, i.e., the posterior compartment of the prothorax, and the anterior compartment of the mesothorax (Hayes *et al.*, 1984; Sánchez-Herrero *et al.*, 1985). All segments including the eight abdominal segments have produced Keilin's organs ventrally and short dorsal tracheal sections with spiracle-like structures, such as normally terminate the longitudinal tracheal trunks in the mesothorax. Both the BX-C and the ANT-C are genetically complex regions, each including several different homeotic functions, the morphological effects of which are confined to particular segmental units. The ANT-C includes several nonhomeotic complementation groups that are also required for embryonic morphogenesis. The BX-C and ANT-C interact, in that the morphological effects of mutations in either complex depend on the function of the other (see, e.g., Struhl, 1983; Hafen *et al.*, 1984b; Levine *et al.*, 1985; Sato *et al.*, 1985; evidence cited below). For reviews and genetic characterization of the ANT-C and the effects of various mutations thereof see Kaufman *et al.* (1980), Lewis *et al.* (1980a,b), and Wakimoto and Kaufman (1981); and for the BX-C, Lewis (1978, 1985), Lawrence and Morata (1983), Sánchez-Herrero *et al.* (1985), and Karch *et al.* (1985). The functional requirement for both the ANT-C and the BX-C is strictly zygotic, neither being needed for the completion of oogenesis, or the production of normally developing eggs (Lawrence *et al.*, 1983).

The left-most (proximal) portion of the BX-C including the *Ultrabithorax (Ubx)* and *postbithorax (pbx)* regions of the DNA included in the complex has been characterized in some detail. As indicated in the diagram shown in Fig. 4.22(e) (Hogness *et al.*, 1985), in the adult, mutations in the *Ubx-bxd* region alter normal morphogenesis of the structures extending from the posterior compartment of the second thoracic segment through the anterior compartment of the first abdominal segment (i.e., parasegments 5 and 6; Lewis, 1978; Struhl, 1984; Hayes *et al.*, 1984), while mutations in the abdominal domains alter posterior morphogenesis beginning with the second abdominal segment (Sánchez-Herrero *et al.*, 1985; Karch *et al.*, 1985). However, *Ubx* and *bxd* mutations also cause transformations of segmental identity within the abdominal regions of the embryo and larva. Animals bearing translocations that separate the *Ubx-bxd* region from the abdominal domains develop normally (Struhl, 1984), but within the *Ubx-bxd* region integrity of the DNA is required for normal function. Different recessive mutations within the *Ubx-bxd* region appear to define functions that affect the compartments and parasegmental units indicated in the lower portion of Fig. 4.22(e), and all of these functions may be inactivated in *cis* by *Ubx* mutations. Observations on mosaic flies show that the effects caused by mutations within the *Ubx* domain are all autonomous, both in ectodermal and mesodermal derivatives (Minana and Garcia-Bellido, 1982; Lawrence and Johnston, 1984b). That is, determination of segmental identity is mediated by the expression of the BX-C gene *in each cell*, and is not a property of groups of cells that depends on intercellular interaction.

The diagram shown in Fig. 4.22(f) relates the transcriptional structure of the 100 kb *Ubx-bxd* region to the location of mutational lesions that have been mapped at the DNA level (Bender *et al.*, 1983; Akam *et al.*, 1984; Beachy *et al.*, 1985). There are two major transcription units. These must be contiguous in the genome, and both the *Ubx* and *bxd* transcription units are required in order for normal development of parasegment 6 to occur, though

---

Fig. 4.22. Embryonic effect, genetic and molecular structure, and transcription of the *Bithorax* complex (BX-C). (a)-(d) Cuticular and tracheal patterns in arrested late embryos homozygous for a deletion of the whole BX-C, (a) and (c), and in wild-type first instar larvae, (b) and (d). (a) and (b) External ventral aspect. Arrows in (a) and (b) denote Keilin's organs, normally found only in prothoracic (PRO), mesothoracic (MS), and metathoracid (MT) segments, but in mutant embryos located on abdominal (AB) segments as well. Note also narrower thoracic-type ventral denticle belts on the abdominal segments in the embryo shown in (a). (c) and (d) Whole mounts. Mutant embryo shown in (c) has separate tracheal sections in each segment (arrows), replacing the longitudinal dorsal tracheal trunks (DLT) of the wild-type first instar larva shown in (d); MH, mandibular hooks; CP, chitinous plates. Abnormalities are also observed in the ventral nerve cord. X100. [(a)-(d). From E. B. Lewis (1978). Reprinted by permission from *Nature (London)* **276**, 565. Copyright © 1978 by Macmillan Journals Limited.] (e) Diagram describing morphological domains controlled by elements of the BX-C. The stippled region of the adult fly indicates morphological domains where mutations in the *Ubx-bxd* region alter metameric identity. Normal abdominal morphogenesis beginning with the second abdominal segment requires the functions of additional loci, denoted here "abdominal domains," as indicated by mutations that cause at least seven classes of transformation (*iab2-iab8*), located within a 200 kb right hand region of the BX-C (Karch *et al.*, 1985). The shaded portions refer to morphological effects of *Ubx* and *bxd* mutants on the abdominal segments of the *embryo* and *larvae*. The mutant effects indicated in the figure are those observed in the presence of the normal alleles of the other homeotic genes. [(e) From D. S. Hogness, H. D.

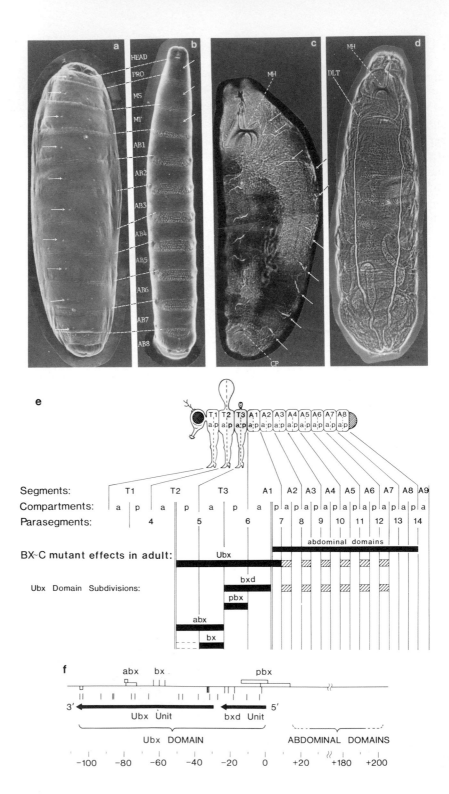

a

b

HEAD
PRO
MS
MT
AB1
AB2
AB3
AB4
AB5
AB6
AB7
AB8

c

MH

CP

d

MH
DLT

e

T1 T2 T3 A1 A2 A3 A4 A5 A6 A7 A8
a p a p a p a p a p a p a p a p a p a p a p

Segments: T1   T2   T3   A1 A2 A3 A4 A5 A6 A7 A8 A9
Compartments: a | p   a | p   a | p   a |p|a|p|a|p|a|p|a|p|a|p|a|p|a|p|a
Parasegments: | 4 | 5 | 6 | 7 | 8 | 9 | 10| 11| 12| 13| 14

                                                    abdominal domains

BX-C mutant effects in adult: Ubx

Ubx Domain Subdivisions: bxd

                                        pbx

                           abx

                           bx

f

            abx   bx   pbx

3' ◄━━━━━━━━━━━━━━━━━ ◄━━━━━━━ 5'
          Ubx Unit   bxd Unit

    └─────────────────┘   └──────────────┘
         Ubx DOMAIN   ABDOMINAL DOMAINS

  -100   -80   -60   -40   -20   0   +20 +180 +200

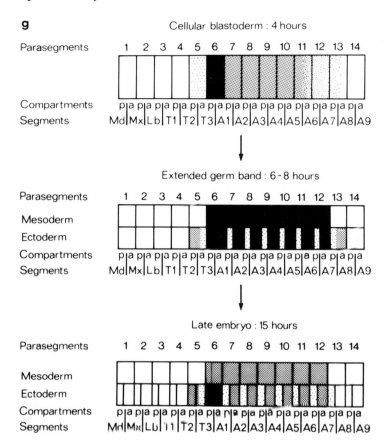

g

Cellular blastoderm : 4 hours

Parasegments 1 2 3 4 5 6 7 8 9 10 11 12 13 14

Compartments p|a p|a p|a p|a p|a p|a p|a p|a p|a p|a p|a p|a p|a p|a
Segments Md|Mx|Lb|T1|T2|T3|A1|A2|A3|A4|A5|A6|A7|A8|A9

Extended germ band : 6-8 hours

Parasegments 1 2 3 4 5 6 7 8 9 10 11 12 13 14

Mesoderm
Ectoderm
Compartments p|a p|a p|a p|a p|a p|a p|a p|a p|a p|a p|a p|a p|a p|a
Segments Md|Mx|Lb|T1|T2|T3|A1|A2|A3|A4|A5|A6|A7|A8|A9

Late embryo : 15 hours

Parasegments 1 2 3 4 5 6 7 8 9 10 11 12 13 14

Mesoderm
Ectoderm
Compartments p|a p|a p|a p|a p|a p|a n|a p|a p|a p|a p|a p|a p|a p|a
Segments Md|Mx|Lb|11|T2|T3|A1|A2|A3|A4|A5|A6|A7|A8|A9

Lipshitz, P. A. Beachy, D. A. Peattie, R. A. Saint, M. Goldschmidt-Clermont, P. J. Harte, E. R. Gavis, and S. L. Helfand (1985). *Cold Spring Harbor Symp. Quant. Biol.* **50**, 181; modified according to results of Lewis (1963) and Karch *et al.* (1985).] (f) Molecular map of the *Ubx* domain. Heavy arrows represent the *Ubx* and *bxd* transcription units (kb scale at bottom). Mapped mutation sites identified from observations on the genomic DNA of mutant strains are indicated. Breakpoints of *Ubx* chromosomal rearrangements that split the BX-C are indicated by the vertical lines in the row just above the solid arrow denoting the orientation and extent of the *Ubx* transcription unit. Four pseudo-point *Ubx* mutations (three small mutations and an insertion) are indicated by an open box (3' end) and three closely spaced vertical slashes (5' end) just below the long horizontal line representing the chromosomal DNA. The breakpoints of *bxd* rearrangements and the positions of *bxd* pseudopoint mutations are similarly indicated above the solid arrow delineating the *bxd* unit. The *abx* and *pbx* deletions are denoted by open boxes above the line whereas the sites of *bx* insertional mutations are indicated by vertical slashes. [(f) From D. S. Hogness, H. D. Lipshitz, P. A. Beachy, D. A. Peattie, R. A. Saint, M. Goldschmidt-Clermont, P. J. Harte, E. R. Gavis, and S. L. Helfand (1985). *Cold Spring Harbor Symp. Quant. Biol.* **50**, 181.] (g) Summary of *in situ* hybridization patterns with 5' exon *Ubx* probe. Darkness of shading approximately indicates intensity of labeling. [(g) From M. E. Akam and A. Martinez-Arias (1985). *EMBO J.* **4**, 689.]

the *Ubx* transcription unit alone appears sufficient for development of parasegment 5 (Bender *et al.*, 1983; Hogness *et al.*, 1985). Though spliced and
polyadenylated, the early *bxd* transcript produced during the initial establishment of larval segmental identity bears no open reading frame. Yet,
genetic observations show that sequences resident in this large region of the
BX-C DNA exercise *cis* regulatory effects on *Ubx*⁺ function (for possible
molecular interpretations see review in Hogness *et al.*, 1985). The *Ubx*
transcription unit produces *bona fide* mRNAs, however. Two 3.2 kb messages that share 5′ and 3′ exons but differ in a small internal region due to
alternative processing have been characterized (Beachy *et al.*, 1985). In
addition, a 4.3 kb mRNA that shares the same 3′ and 5′ exons is synthesized,
and a 4.7 kb nonpolyadenylated RNA that disappears late in embryogenesis
is detected as well. All of these transcript classes are observed during embryogenesis, the 4.3 kb species slightly later than the others (Hogness *et al.* ,
1985; Akam and Martinez-Arias, 1985).

Both the *Ubx* transcripts and the proteins for which the common 5′ *Ubx*
sequences code have been localized cytologically, the latter by immunofluorescent detection of proteins produced from fusion gene constructs. *In
situ* hybridization with cloned probes shows that *Ubx* transcripts appear at
the syncytial blastoderm stage, when they are localized to about 30% of egg
length in the midposterior region (Akam and Martinez-Arias, 1985). At the
cellular blastoderm stage high levels of *Ubx* transcripts are confined to the
primordium for parasegment 6, with lower concentrations in more posterior
regions. At gastrulation *Ubx* transcripts are prevalent in parasegments 6-12,
displaying a periodic distribution due to high labeling in the anterior compartments, and are present in some cells of parasegment 5 as well. Labeling
is observed in cells of the CNS and the mesoderm, as well as in epidermal
cells. These patterns are summarized in Fig. 4.22(g). Two conclusions that
can be drawn are that in general both the temporal and spatial distribution of
Ubx transcripts conform to expectation from genetic evidence, and that the
anterior ectodermal *compartments* which in the late embryo display the *Ubx*
transcripts are in fact by this stage serving as units of gene expression. Note
that anterior metameric labeling is not observed in the mesoderm [Fig.
4.22(g)], for which, as noted earlier, there is no evidence of anterior-posterior compartmentalization. *Ubx proteins* are localized in the 12 hr embryonic
CNS as would be predicted from the transcript distributions summarized in
Fig. 4.22(g) (White and Wilcox, 1984, 1985; Beachy *et al.*, 1985). Thus as
diagrammed for wild-type embryos in Fig. 4.23(a) they are detected in the
posterior compartment of segment T2, most strongly in posterior T3 and
anterior A1 compartments, and more weakly in the anterior compartments
of the abdominal segments. No *Ubx* protein is observed in embryos homozygous for a frame-shift mutation at the 5′ end of the open *Ubx* reading frame
(Fig. 4.23(a); see also Beachy *et al.*, 1985).

A direct implication of these observations is that since all the regions of the embryo requiring *Ubx* function for establishment of segmental identity display *Ubx* gene products, one or more of the protein species produced by the *Ubx* transcription unit probably mediate these functions. Like the proteins encoded by the pair rule genes discussed earlier, the *Ubx* proteins are localized in the *nuclei* of the expressing cells (Beachy *et al.*, 1985). All mutations that cause *Ubx* mutant phenotypes, including *abx*, *bx*, and *Contrabithorax* mutations, as well as *bxd* mutations, alter the specific metameric distributions of the *Ubx* proteins (White and Wilcox, 1985; White and Akam, 1985).

An example of the effects of *bxd* mutations on the distribution of *Ubx* protein is shown in Fig. 4.23(a) (Beachy *et al.*, 1985). This mutation, which alters the development of the CNS in the posterior T3 and anterior A1 compartments (Teugels and Ghysen, 1985), causes a uniform *Ubx* protein pattern extending from posterior T2 to anterior A1, in which the *Ubx* transcript concentration peak normally observed in the posterior T3-anterior A1 compartments is absent. This result provides concrete evidence for *cis* regulation of *Ubx*⁺ function by sequences of the *bxd* locus, and additional experiments of this kind reviewed by Hogness *et al.* (1985) show that the effect varies according to the region of the *bxd* locus affected by the tested mutations. The abdominal region of the BX-C also exercises an important *cis* regulatory control over the pattern of *Ubx* expression (Struhl and White, 1985). Thus embryos lacking the abdominal BX-C genes display in parasegments 7-13 the high levels of *Ubx* protein expression normally characteristic only of parasegment 6, and different gene products of the abdominal region of the BX-C gene are apparently responsible for controlling the stable level of *Ubx* expression in the various CNS cell types of the abdominal parasegments.

Molecular studies on the *Antennapedia* (*Antp*) locus of the ANT-C present a somewhat similar view. Dominant *Antp* mutations and recessive null mutations cause prothoracic cuticular features to appear in place of mesothoracic features (Wakimoto and Kaufman, 1981; Struhl, 1981, 1982). The *Antp* transcription unit is about 100 kb in length and, like the *Ubx* gene, produces a set of alternative RNAs that may be constructed by diverse patterns of splicing (Garber *et al.*, 1983; Gehring, 1984). *In situ* hybridization reveals *Antp* transcripts as early as the syncytial blastoderm stage (Levine *et al.*, 1983), and during cellularization these appear mainly in the anlagen of the thoracic segments. In later embryos, the anterior compartments of abdominal as well as thoracic segments are labeled transiently. Genetic evidence, from studies on double homeotic mutants, confirms that the *Antp* gene normally functions in the first seven abdominal segments (Struhl, 1983). In the CNS the major sites of persistent *Antp* transcription are the ganglion cells of the mesothoracic region [Fig. 4.23(b)-(c)], though at earlier stages

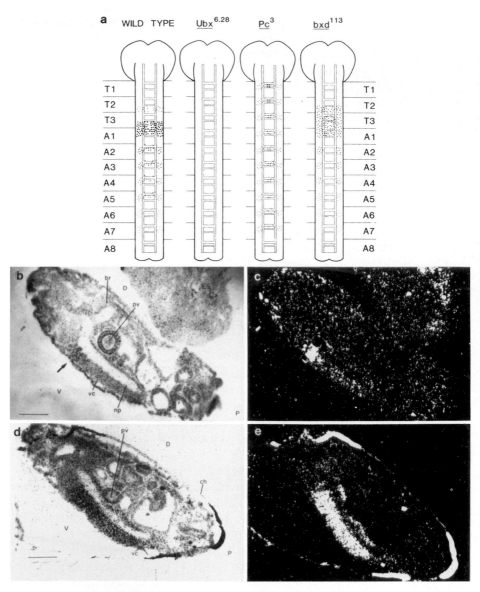

Fig. 4.23. Interactions amongst *Drosophila* genes controlling pattern formation that affect the embryonic loci of expression of these genes. (a) Diagram showing distribution of *Ubx* protein in CNS of wild-type and mutant embryos, detected by immunocytofluorescence. The number of dots in a given area is proportional to the *number of nuclei stained,* and the relative staining intensities are indicated by dot size. In the wild-type CNS the most brightly staining nuclei are located in a region comprising posterior cells of the T3 ganglia (T3p) and anterior cells of the A1 ganglia (A1a). Most cells in this region are stained. A subset of the cells in T2p is stained lightly, while the staining of T3a is so faint that it is not shown. Stained cells in segments A2-7 form a subset of the total complement of anterior cells. The number and arrangement of nuclei in these stained anterior patches is similar if not identical from segment to segment,

these transcripts are observed in all CNS segments. However, as shown dramatically in Fig. 4.23(d)-(e), abdominal synthesis of *Antp* RNAs remains high in embryos bearing a total deletion of the BX-C. That is, the abdominal segments in such embryos behave like mesothoracic segments [compare Fig. 4.22(a)-(d)]. This observation provides molecular evidence of the regulatory interactions in which the major homeotic complexes participate [Hafen *et al.*, 1984b; see also Levine *et al.* (1985) for additional *in situ* hybridization evidence of the effects of various mutations on *Ubx* and *Antp* expression patterns]. The expression of *Ubx* and *Antp* loci have many aspects in common: Both produce enormous primary transcripts from which derive different processed RNAs; both are activated at the syncytial blastoderm stage; and both are expressed mainly in those embryonic compartments transformed by mutations at the same loci, though in each case there are temporal and spatial complexities in the observed patterns of transcription that are not obviously predicted from prior evidence.

The early and from the outset *localized* transcription of *Antp* and *Ubx* genes suggests a spatially organized pattern of control by maternal factors. A maternally functioning gene, *extra sex combs* (*esc*), which is also expressed in the early embryo, has been shown to regulate the expression of major loci of the BX-C and ANT-C, including *Scr*, *Ubx*, and *Antp* (see, e.g., Struhl, 1983; Struhl and Brower, 1982). Eggs developing from *esc⁻* mothers produce embryos in which all segments are transformed to the 8th abdominal segment, the consequence of unregulated ectopic expression of the BX-C, whereas total lack of BX-C function, as shown above [see Fig. 4.22(a) (d)] leads to transformation of all segments to posterior prothoracic-anterior mesothoracic type. This interpretation is supported by the observation that expression of the *Ubx* protein in these mutants is not restricted, as in wild-type embryos, to CNS cells of parasegments 5-13, but is found in some neurons in all 14 parasegments (Struhl and White, 1985). Thus the *esc* gene contributes to the segment-specific control of the BX-C loci, which in normal

---

although the intensity of staining decreases progressively in the posterior direction (extremely faint staining observed in A8 is not depicted). $Ubx^{6.28}$ is a frame shift deletion mutation that abolishes synthesis of the protein. In the *Polycomb* mutants (Pc^3) the uniform *Ubx* protein distribution resembles that of the most posterior abdominal segments in the wild-type embryo. The bxd^{113} mutation is an inversion associated with a deletion. [From P. A. Beachy, S. L. Helfand, and D. S. Hogness (1985). *Nature (London)* **313**, 545.] (b)-(c) Phase photomicrograph and dark field photograph of *in situ* hybridization, displaying distribution of *Antp* transcripts in a wild type 18 hour embryo. A, anterior end; br, brain; D, dorsal side; np, neuropile; P, posterior end; pv, proventriculus; V, ventral side; vc, ventral cord. Scale bar, 0.1 mm. The arrow indicates strong hybridization of the *Antp* probe to the mesothoracic ganglion cells of the CNS. (d)-(e) Phase and dark field photomicrographs of a section of a 30 hour embryo homozygous for a deficiency of the BX-C (cf. Fig. 4.22(a)-(d)), hybridized with the same *Antp* probe as in (b) and (c). ch, chorion shell. Scale bar again 0.1 mm. The *Antp* transcript now accumulates heavily throughout the ventral ganglia of the metathoracic and the first seven abdominal segments. [(b)-(e) From E. Hafen, M. Levine, and W. J. Gehring (1984b). *Nature (London)* **307**, 287. Copyright © 1984 by Macmillian Journals Limited].

animals results in the production of the particular combination of BX-C products that define each segment along the anterior-posterior axis (Lewis, 1978; Struhl, 1983). In addition the *esc* gene controls *Antp* and *Scr* expression, the absence of *esc* function similarly resulting in ectopic expression of these genes. However, the *esc* gene product appears to be required for the stable pattern of expression of the segment identity genes rather than for the original institution of their spatial patterns of regulation, since the initial distribution of *Ubx* transcripts appears normal in *esc⁻* embryos (Struhl and Akam, 1985). There are a large number of zygotically active genes, perhaps as many as 40, that are also involved in the spatial *trans*-regulation of the BX-C and ANT-C (Lewis, 1978; Duncan, 1982; Duncan and Lewis, 1982; Struhl, 1983; Jürgens, 1985; Sato *et al.*, 1985; Dura *et al.*, 1985). As shown in Fig. 4.23(a) mutations in one of this class of loci, *Polycomb* (*Pc*), have been shown to produce a uniform pattern of *Ubx* protein accumulation, from T1 through A7, all of these segments in mutant *Pc* embryos displaying the relatively low level of *Ubx* protein accumulation normally characteristic of the most posterior abdominal segments (Beachy *et al.*, 1985). *Polycomb* also affects the distribution of other BX-C transcripts, and of transcripts of the homeotic genes of the ANT-C (Wedeen *et al.*, 1986). It is interesting that the effects of mutations in other genes of the *Pc* class, that do not by themselves produce such severe effects, behave *additively* (Jürgens, 1985). This suggests some sort of quantitative mechanism of control by which is regulated the expression of genes that directly determine segmental identity, and it raises the possibility that a combinatorial profile of these controlling factors provides a spatial regulatory image along the anterior-posterior axis of the embryo.

Comparative Aspects

The great complexity of the genetic program for determination of the major morphological pattern features in the *Drosophila* embryo is evident even given our present lack of understanding of how this process operates. Every aspect of the mechanism for segmental determination is polygenic. A set of genes is required that functions maternally to provide the global spatial coordinates of the egg; a set of at least nine genes is required for the transduction of these coordinates into the metameric blastoderm territories; the functions of another set of enormously complex homeotic loci directly control segmental (or parasegmental) identity; and a yet only partially identified set of zygotic genes that operate synergistically is required to regulate the spatial patterns of expression of these segment identity genes. The gene functions controlling establishment of segments are clearly separable from those determining their identity, though expression of members of both classes begins simultaneously at the blastoderm stage. Thus, for example, even in *esc⁻* or BX-C⁻ embryos, the normal number of segments are formed, and even in *ftz⁻* or *bicaudal⁻* embryos, anterior-posterior segmental identities are established. Products of the major homeotic complexes apparently affect each other's expression, as we have seen, and perhaps from this point of view it is not surprising that the *Ubx* product is nuclear in location. However, it is in no way obvious that these products regulate directly the genes that determine the actual differentiated structures and functions of the many cell types that constitute the morphological character of their domains of action. Similar differentiated structures appear in various segments, e.g., denticles, muscles, etc.; it is their *spatial disposition* that varies.

Direct evidence relevant to the genetic basis of pattern formation is not available for other than certain insect embryos, nor are comparative insights into the mechanisms by which the products of pattern formation genes might generate their effects. Some genes that affect morphological structures in *C. elegans* were discussed in the first section of this chapter. These genes control certain aspects of lineage-specific differentiation processes, and thus they provide models that are basically unrelated to pattern formation in *Drosophila*, where decisions are not clonal; lineage is not determinate; and numbers of epidermal cells exceeding the total required to construct a *C. elegans* larva are organized by *regions* to form the individual metameric units of the larva. On the other hand, in the later embryogenesis of many animals, and in postembryonic organogenesis, there commonly occur examples of morphogenetic processes in which as in the *Drosophila* ectoderm, groups of cells are set aside and then generate spatial *patterns of differentiation*. Other examples include regeneration, and perhaps even the ability of disaggregated embryos or tissues to reassemble, and carry out the reconstruction of spatially organized arrays of diverse cell types. Such pattern formation processes require many forms of intercellular interaction of which we know little. Cells must recognize and adhere to, or reject, certain other cells; they must control each other's mitotic activity, motility, and shape; and they must synthesize and mount receptors, and generate extracellular substances or structures that have the effect of triggering, or inhibiting, given forms of differentiation in their neighbors. Thus the functions of genes that affect morphogenesis of pattern must ultimately regulate such activities, and many others. Genetic analysis of the mechanisms that control the spatial distributions of veins and of chaetae in the wings of *Drosophila* suggest similar conclusions (Diaz-Benjumea *et al.*, 1986). Many genes affect these patterns, and their functions can be interpreted as effects on *intercellular interaction processes* that result either in enhancing or reducing the spatial frequency at which are established morphogenetic "centers," where formation of these structures is initiated. Whatever the mechanisms that determine spatial morphogenetic parameters, the regulatory genetic elements that immediately control the batteries of structural genes which in turn produce the individual properties of differentiated cells must lie downstream in the chain of causality. Their activation, in appropriate cells, is the end product of pattern determination.

A particularly interesting and totally new insight into the mode of function of genes controlling morphogenesis of pattern that has derived from the *Drosophila* studies summarized here is the extent of their interaction. The expression of these genes in embryonic cells at different locations, which ultimately specifies the role these cells will play in the creation of given pattern elements, appears to depend on the products of many other genes. These include products deriving both from different regions of the same homeotic complexes and from different homeotic complexes, as well as from a large number of additional loci. In an unknown way, control of pattern formation requires a multiplicity of intragenomic interactions, by which the potentialities of each cell are autonomously set. Thus in some sense genes that determine very large pattern elements are controlled differently, and function differently from the classes of cell type-specific genes with which we have largely been concerned in embryos other than *Drosophila* discussed in this chapter.

Metamerism occurs in many species, though by no means all, of both the deuterostome and protostome branches. For example neither *C. elegans* nor sea urchins display any convincing metameric organization. The *Drosophila* homeotic genes, including those of the ANT-C and BX-C, as well as some other genes, encode proteins that include a shared amino acid sequence, about 60 residues in length, that has been given the name "homeobox" (McGinnis *et al.*, 1984a; Scott and Weiner, 1984; Akam, 1984; Levine *et al.*, 1985). This relationship has led to the suggestion that genes encoding the homeobox are members of a family concerned in some way with segmentation. Sequences homologous to the homeobox have been found in amphibian, mouse, and sea urchin genomes as well, where they are expressed during embryogenesis (see, e.g., Carrasco *et al.*, 1984; McGinnis *et al.*, 1984b,c; Müller *et al.*, 1984; Hart *et al.*, 1985; Hauser *et al.*, 1985; Colberg-Poley *et al.*, 1985; Dolecki *et al.*, 1986). A distant relationship has also been reported between the *Drosophila* homeobox protein sequence and regions of yeast mating type proteins that have a DNA binding function, and this suggests a similar function for the animal homeobox sequences (Shepherd *et al.*, 1984; Laughon and Scott, 1984). It is indeed a striking fact that the protein products of several pattern formation genes that possess a homeobox sequence, i.e., *ftz, engrailed,* and *Ubx,* are all localized in the nuclei of cells that express these genes. The fact that wherever identified, genes including homeobox sequences are found to be expressed during development clearly implies that this sequence carries out a broadly useful function, required in many contexts. Whatever the role of this conserved sequence element, however, it does not seem likely that it is universally employed in a common mechanism of metamerization *per se*. The ancestors that are supposed to antedate the divergence of protostomes and deuterostomes are not thought to have been metameric (Hyman, 1955), and thus like many other features metamerism has probably arisen independently in these two great branches of the metazoa. The embryological origins of metameric structures are in fact completely different in, for instance, flies and frogs, as is the whole organization of these embryos. Thus, for example, metameric compartments display at least a polyclonal restriction in cell lineage in flies while polyclonal compartments in *Xenopus* coincide with no metameric (or any other morphological) boundaries. Furthermore, the expression of homeobox sequences in the sea urchin embryo suggests strongly the necessity of considering some other kind(s) of function. On the other hand, the same argument of biological homology implies that the mechanisms by which metameric identity are established in segmented protostomes belong to a common evolutionary series, that has progressively diversified throughout the phylogenetic development of the arthropods from segmented annelid precursors. This was originally inferred for the BX-C by Lewis (1963; see also Garcia-Bellido, 1977).

The issue of generality pervades others of the specific developmental mechanisms considered in this chapter as well. Current knowledge does not encourage the parsimonious assumption that the underlying biological processes of embryogenesis are common to all organisms, and are there to be discovered by thorough exploration of any one "model system." Each of the four species of embryo considered in some depth in this chapter is the product of a long and unique evolutionary pathway. We have seen that their biological mechanisms of development are in many ways fundamentally different, and so might be the regulatory architectures of their genomic developmental programs.

V

Gene Activity during Oogenesis

The growing oocyte can be regarded as a special form of differentiated cell, organized to carry out a unique function. It accumulates a particular collection of gene transcripts, and by the end of its growth phase has developed a complex of unusual cytological structures. We have seen that the set of genes expressed during oogenesis bears a close, largely overlapping relation to that expressed in the early embryo nuclei. Thus from the standpoint of nuclear function, *oogenesis* is the process during which the major portion of the *zygotic* pattern of gene activity is established. This pattern is also impressed on the entering male genome, and during cleavage is propagated to the blastomere nuclei. Though there are of course genes that function only in the early embryo and not during oogenesis, and *vice versa*, in quantitative terms both of these classes are minor, in respect to either the mass or the complexity of the transcripts for which they account.

This chapter is focused on gene expression during the growth phase of oogenesis, which begins following premeiotic DNA replication. Throughout this period newly synthesized transcripts are being delivered to the oocyte cytoplasm, and until the metaphase of the initial reduction division occurs at maturation, the oocyte remains in the first meiotic prophase. Thus it contains a 4C rather than a 2C genome. In the typical case transcription of the genes utilized during oogenesis begins early in the growth phase, before the onset of vitellogenesis, and continues almost until maturation. At maturation there occurs the breakdown of the germinal vesicle, or oocyte nucleus. By this time all nuclear transcription has ceased, and a new set of biosynthetic activities is instituted. The molecular biology of oocyte maturation *per se* has been reviewed extensively and is not further considered here (see, e.g., Tsafriri, 1978; Wasserman and Smith, 1978; Masui and Clarke, 1979; Wassarman *et al.*, 1979; Maller and Krebs, 1980; Porter and Whelan, 1983; Smith and Richter, 1985). For earlier events of oogenesis, such as germ cell migration, oogonial multiplication, and the many other biological and biochemical aspects not directly relevant to germ line transcription and gene expression during oogenesis, recent compilations such as those edited by Jones (1978) and Browder (1985) may be consulted. Follicle cell functions are reviewed in these works as well, and in articles by Mahowald and Kambysellis (1980); Wallace (1983); and Moor (1983). At the molecular level perhaps the best studied follicle cell activity is synthesis of the chorion or egg shell, particularly in lepidopteran (Mazur *et al.*, 1980; Jones and Kafatos, 1980; Regier *et al.*, 1982; Kafatos, 1983) and dipteran species (Kambysellis, 1974; Waring and Mahowald, 1979; Spradling and Mahowald, 1979; Kafatos, 1983). Though follicle cell function lies outside the scope of the following discussion, it is to be noted that chorion synthesis is particularly interesting in the Diptera, where it has been observed that the genes coding for the major proteins of this structure are amplified during differentiation of the ovariole (Spradling, 1981; Spradling and Mahowald, 1981; de Cicco and Spradling, 1984).

1. MEROISTIC OOGENESIS: GENE EXPRESSION IN THE NURSE CELL-OOCYTE COMPLEX

Oogenesis is always to some extent a cooperative process, involving cells other than the oocyte itself. In most, though certainly not all forms, the growing oocyte is surrounded with follicle cells of somatic origin. Follicle cells provide essential transport, hormonal and secretory functions, but with the possible rare exceptions mentioned below they do not directly feed transcription products to the oocyte. A natural division separates forms of oogenesis in which the *germ line* gives rise both to oocytes and to nurse cells that directly provide the oocyte with cytoplasmic organelles and macromolecules, including transcripts, from those in which the oocyte nucleus is alone responsible for the synthesis of its nonyolk macromolecular constituents. Oogenesis in which the egg cytoplasm is the cooperative product of both nurse cell and oocyte genomes is termed *meroistic oogenesis*. It is known mainly in insects, though it occurs as well in other protostomial invertebrates.

(i) Structural Organization and Developmental Origin of the Nurse Cell-Oocyte Complex

Meroistic oogenesis has been studied in most detail in certain holometabolous insect orders, the most prominent of which are Coleoptera, Hymenoptera, Hemiptera, Lepidoptera, and Diptera. Many other insect orders, e.g., the Orthoptera and Odonata, carry out oogenesis by the alternative route, in insects termed *panoistic* oogenesis (reviewed in King and Büning, 1985). Nurse cells are here absent, and the oocyte growth process is autonomous, except for the assistance in yolk protein and metabolite uptake provided by the follicle cells. Two types of meroistic oogenesis are diagrammed in Fig. 5.1(a) and (b), and a panoistic ovary is portrayed for comparison in Fig. 5.1(c). The chromosomes of meroistic insect oocytes typically display a condensed structure (some exceptions are noted below), and as indicated diagrammatically in Fig. 5.1, they either do not synthesize RNA at all during oocyte growth, or do so at a very low relative rate. The RNA accumulated during the growth phase in the oocyte cytoplasm is instead transported there from the nurse cells via open cytoplasmic channels. Figure 5.1 indicates the major forms of egg chamber arrangements observed. In *polytrophic* egg chambers the oocyte is directly connected to each of several nurse cells, and

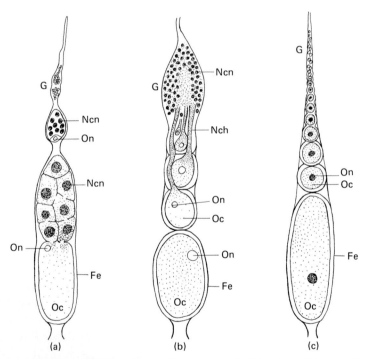

Fig. 5.1.   Diagrams of the three types of insect ovary. Polytrophic and telotrophic meroistic oogenesis are portrayed in (a) and (b), respectively, and panoistic oogenesis is portrayed in (c). In polytrophic meroistic oogenesis the oocyte is fed directly by the nurse cells via individual junctions linking the nurse cells to each other and to the oocyte. In telotrophic meroistic oogenesis the nurse cells communicate with the oocyte via a common "nutritive cord." Cell nuclei that synthesize RNA are shown in black, and those which are inactive are shown as open circles. Cytoplasmic RNA is represented by fine black dots. Fe, follicular epithelium; Ncn, nurse cell nucleus; Nch, nutritive chord; G, germarium; On, oocyte nucleus; Oc, oocyte. [After K. Bier (1967). *Naturwissenschaften* **54**, 189.]

these to one another, by discrete intercellular channels, termed *ring canals* [Fig. 5.1(a)]. In *telotrophic* egg chambers a single common duct enters each oocyte, and through this pass the combined products of a syncytial mass of nurse cell nuclei [Fig. 5.1(b)].

The role of nurse cells in feeding the oocyte was remarked upon by many classical writers (see Wilson, 1925; Telfer, 1975, for references). Early observers noted in several species that organelles as large as mitochondria pass from nurse cells to oocyte. An interesting variation exists in turbellarian flatworms, where the nurse cells, filled with yolk, are encapsulated in a cocoon along with the oocytes *after* oogenesis is completed, and the nurse cell contents are used to sustain growth of the embryos just as are *intracellular* components in other eggs. This unusual course of events draws attention to the essential aspect of nurse cell function, that of providing the egg with materials it will require for development. Wherever nurse cells are found, the functional nature of the oocyte-nurse cell interaction is evident. In certain annelids, for example, one or two nurse cells with large polyploid nuclei are applied to each oocyte, and the oocyte-nurse cell complex is released into the lumen of the ovary relatively early in oogenesis. Oocyte growth then occurs at the expense of the nurse cells, which shrink progressively until they become small compared to the relatively enormous oocytes. A general survey of accessory cell-oocyte arrangements in invertebrate oogenesis is given by Raven (1961).

Intercellular Organization and Cytoplasmic Transport in Polytrophic Egg Chambers

The structure of the nurse cell-oocyte complex of *Drosophila* as it appears at light microscope magnification is shown in Fig. 5.2(a). The five deeply staining nurse cell nuclei seen in the section are highly polyploid, while the oocyte nucleus of course contains only the 4C meiotic prophase genome. Three ring canals are shown, one connecting the oocyte with the follicle cells, and two connecting adjacent nurse cells. The growth phase of *Drosophila* oogenesis is divided into 14 stages (for a complete description see King, 1970), beginning with the formation of the oocyte-nurse cell complex and its positioning at the posterior end of the germarium [see Fig. 5.1(a)]. Oogonial replication and the establishment of new egg chambers begin during puparial life and continue in the adult. Grell and Generoso (1982) showed that the premeiotic S-phase that marks the onset of oogenesis proper occurs in the first pupal pre-oocytes within the interval 132-162 hr after oviposition. During this period the synaptonemal complexes are extended, and this is also the time at which meiotic recombination may occur. The entire 14 stages require $4\frac{1}{2}$ days in pupae (Grell and Day, 1974), and 6 days in the adult (Grell and Chandley, 1965), with the difference localized mainly to the rate at which stages 1 and 2 are traversed (Grell and Generoso, 1982). Polyploid-

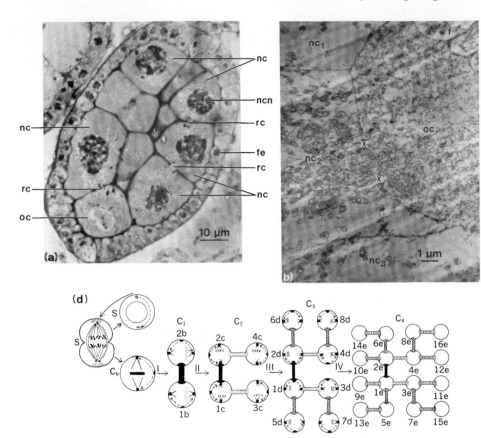

Fig. 5.2. The oocyte-nurse cell complex in the polytrophic *Drosophila* egg chamber. (a) Photomicrograph of a 2 μm section (stained by the periodic acid-Schiff procedure) through an egg chamber. At this stage a single layer of cuboidal follicle cells forms the follicular epithelium (fe). The oocyte (oc) occupies a position at the lower left corner of the chamber. Portions of 9 nurse cells [(nc); (ncn), nurse cell nucleus] can be seen in this section. The egg chamber contains 15 nurse cells in all. Three ring canals (rc) are evident, one connecting the oocyte with a nurse cell and two interconnecting nurse cells. [From E. A. Koch, P. A. Smith, and R. C. King (1967). *J. Morphol.* **121**, 55.] (b) Electron micrograph of a section through a ring canal connecting a nurse cell (nc_2) to the oocyte. Two other nurse cells (nc_1 and nc_3) and portions of several follicle cells (f) are evident. Membraneous structures bounding the ring canal can be seen at x. Mitochondria which were fixed while entering the oocyte are evident within the ring canal. [From E. H. Brown and R. C. King (1964). *Growth* **28**, 41.] (c) Diagram showing process by which the polytrophic oocyte-nurse cell complex of *Drosophila* is constructed. The cells are represented by circles lying in a single plane, and the ring canals have been lengthened for clarity. The area of each circle is proportional to the volume of the cell. The stem cell (S) divides into two daughters, one of which behaves like its parent. The other differentiates into a cystoblast (C_b) which by a series of four divisions (I-IV) produces 16 interconnected cystocytes: C_1, first; C_2, second; C_3, third; and C_4, fourth generation cystocytes. The original germ line stem cell is shown at early anaphase. Each parent-daughter pair of centrioles is attached to the plasma membrane by astral rays. The daughter stem cell receives one pair of centrioles. One remains in place while the other moves to the opposite pole. This movement is represented by the broken

ization of the nurse cell nuclei begins in stage 2, during which they increase their DNA content from 4C to 8C, and this process continues through the previtellogenic stages, i.e., stages 2-7. Together these stages require about 50 hr (King, 1970).

After the 32C stage the nurse cell DNA does not replicate uniformly, in that satellite DNA, and some other sequences, including histone gene DNA are underreplicated (Hammond and Laird, 1985). The amount of DNA present in the nurse cell genomes at stage 8, when vitellogenesis begins, is the result of 4-5 further replications of about 75% of the total DNA beyond the 32C stage. Two complete further replications have occurred by stage 10A, when the process is complete. There are then four nurse cell nuclei that contain an amount of DNA equivalent to about 1500 haploid genomes, and eleven that contain about half this amount of DNA (Hammond and Laird, 1985). The mass of oocyte constituents meanwhile increases continuously, first as the result of the flow of materials through the ring canals, and later by the pinocytotic ingestion of yolk as well. The mechanism of cytoplasmic transfer through the ring canals may be electrophoretic. Though there are no equivalent observations in *Drosophila*, a potential gradient of >1 V/cm has been demonstrated across the open ring canals of the polytrophic *Hyalophora* egg chamber, and a similar potential gradient exists in the telotrophic ovariole of the bug *Rhodnius* (Telfer, 1975; Woodruff and Telfer, 1980; Telfer *et al.*, 1981). Injection of fluorescein labeled compounds into both types of meroistic ovaries shows that negatively charged molecules are transported by intracellular electrophoresis from nurse cells to oocyte, and once inserted in the oocyte these molecules are discouraged by the polarity of the potential gradient from diffusing backwards toward the nurse cells. This is demonstrated in the experiment shown in Fig. 5.3. Here the mobility within the *Hyalophora* egg chamber of fluorescein labeled lysozyme, a positively charged protein, and of its negatively charged methylcarboxylated derivative, are compared, after injection into nurse cells (Woodruff and Telfer, 1980).

In Fig. 5.2(b) is reproduced an electron micrograph showing an oocyte-nurse cell ring canal at stage 10B in longitudinal section. Just as originally surmised by classical observers using the light microscope, the canal area is seen to be packed with mitochondria, which have apparently been caught up in the cytoplasmic streaming that carries materials from nurse cell to oocyte.

---

arrows. In the daughter cystoblast and all cystocytes the initial position of the original centriole pair is represented by a solid half-circle, whereas their final positions are represented by solid circles. The position of the future cleavage furrow is drawn as a strip of defined texture. The canal derived from the furrows is coded similarly. The oocyte in the completed chamber is cystocyte 1e and the 15 nurse cells are cystocytes 2e-16e. [From E. A. Koch, P. A. Smith, and R. C. King (1967). *J. Morphol.* **121**, 55.]

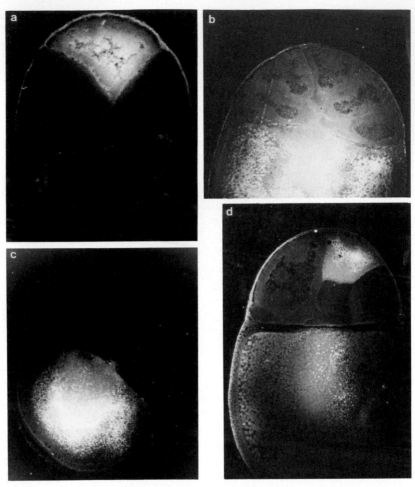

Fig. 5.3.   Intercellular electrophoresis of microinjected proteins in the polytrophic oocyte-nurse cell complex of *Hyalophora cecropia*. In the orientation shown the 7 nurse cells occupy the top half of the chamber and the oocyte the bottom half. Fixation was 1-2 hr postinjection. (a) Fluorescein-labeled lysozyme (isoelectric point 11.5) injected into a nurse cell remains confined there. (b) Injection of lysozyme into the oocyte results in transport through the intercellular ring canals from the oocyte into various nurse cells (i.e., backwards, with respect to most cytoplasmic constituents). (c) Injection of fluorescein labeled ε-amino methylcarboxy derivative of lysozyme into oocyte. The protein now has a net negative charge. It remains confined to the oocyte and does not enter the ring canals or cross into the nurse cells. (d) Injection of methylcarboxylated lysozyme into nurse cell results in transport into oocyte via the ring canals. Magnification: The width of the nurse cell cap adjacent to the oocyte is about 450 μm. [From R. I. Woodruff and W. H. Telfer (1980). Reprinted by permission from *Nature (London)* **286**, 84. Copyright © 1980 Macmillan Journals Limited.]

At this stage the flow from nurse cells to oocyte is accelerated as the result of a new process, the direct injection of nurse cell cytoplasm into the oocyte. As the injection proceeds, the cytoplasmic volume of the nurse cells decreases and that of the oocyte increases concomitantly. This phase continues for about 5 hr in *Drosophila* (stages 10B-12), until the nurse cells have relinquished most of their cytoplasmic mass. They then degenerate. The velocity of the cytoplasmic stream of the ring canal has been measured during the injection process as about 2 μm sec^{-1} (Gutzeit and Koppa, 1982), and once having entered the egg it dissipates in a circular motion that effectively distributes the entering cytoplasmic constituents. At its peak the total flow carried by the four ring canals entering the oocyte is calculated to be about 1.3×10^4 μm^3 min^{-1} (Gutzeit and Koppa, 1982). A similar rate of cytoplasmic flow into the oocyte was estimated by Telfer (1975), in the polytrophic ovaries of the moth *Hyalophora*. The end result of previtellogenic growth, followed by the direct transfer of nurse cell cytoplasm and yolk accumulation, is an enormous change in oocyte volume. In *Drosophila*, for example, the volume of the oocyte increases 90,000-fold within the few days required for oogenesis (King, 1970).

Developmental Origin of the Nurse Cells

In most Diptera the accessory cells that perform the function of feeding RNA and other cytoplasmic constituents to the growing oocyte through ring canals derive exclusively from the germ line. However an exception is known in the dipteran family Cecidomyidae, where polyploid follicle cells that are of somatic origin fuse with a single true nurse cell to form a syncytial nurse chamber that contributes cytoplasmic RNA to the oocyte (Madhavan, 1973; Mahowald and Stoiber, 1974). Careful analysis has shown that in the more typical example provided by *Drosophila* and in Lepidoptera and Hymenoptera, the nurse cells of a given egg chamber, together with the oocyte, constitute a clone, descended from a single oogonial stem cell. The development of the egg chamber was described for *Drosophila* by Koch et al. (1967), for the moth *Hyalophora cecropia* by King and Aggarwal (1965), and for the wasp *Habobracon juglandis* by Cassidy and King (1972). The structure and ontogeny of oocyte-nurse cell complexes in various insect groups are reviewed by King (1970), Telfer (1975), and King and Büning (1985). The oocyte-nurse cell complex in *Hyalophora* includes seven nurse cells and in *Drosophila* 15 nurse cells. These complexes are constructed in the terminal three and four oogonial divisions, respectively. The ring canals that connect each nurse cell to other nurse cells and/or to the oocyte are highly organized membranous structures associated with actin microfilaments (Cassidy and King, 1972; Kinderman and King, 1973; Telfer, 1975; Warn et al., 1985). The disposition of the ring canals, which originate by incomplete cytokinesis following each oogonial mitosis, indicates the order of appearance of the

nurse cells and the sequence of steps by which the egg chamber is constructed. A diagrammatic reconstruction of this process as it occurs in *Drosophila* is shown in Fig. 5.2(c) (Koch *et al.*, 1967). It can be seen that except
for that nurse cell which is formed first, the oocyte is the cell with the largest
number of ring canals (four), although oocytes with only three ring canals
have been reported (Frey *et al.*, 1984). Both the first nurse cell and the
oocyte initially form synaptonemal chromosomal complexes. The synaptonemal complexes developing in the nurse cell nucleus later disappear,
whereas in the oocyte the meiotic prophase movements proceed, and the
growth phase of oogenesis ensues.

Genetic Evidence for Particular Germ Line Functions Required in Egg Chamber Differentiation

Additional insight into the process of oocyte-nurse complex formation in
Drosophila has derived from observations on female sterile mutations that
affect egg chamber formation. Morphological effects of the mutation *fes*
[fs(2)B] were described by Johnson and King (1972). In homozygous *fes*
females cytokinesis is often *complete* rather than incomplete, with the result
that large numbers of abnormal cell clusters containing less than 16 interconnected cells are formed. The presumptive oocytes are thus deprived of their
complement of appropriately connected nurse cells. Normally the interconnected cystocytes all divide synchronously and this form of coordination is
also absent in *fes* mutants. Several mutations are known that cause disorgan-
ized cystocyte divisions, resulting in masses of ovarian cells referred to as
"tumors," including *narrow* (*nw*) and various alleles of *fused*, and of ovarian
tumor (*otu*) (reviewed in King, 1970; King and Riley, 1982; Bishop and King,
1984; Konrad *et al.*, 1985). An interesting example is the dominant mutation
fs(2)D, which in heterozygous females results in inhibition of cystocyte division (heterozygous males are fertile). Only a few egg chambers are produced, and most of these contain less than 16 cystocytes (King and Mohler,
1975). These egg chambers contain no cells identifiable as oocytes, which is
consistent with the concept that a condition for the switch to the oocyte as
opposed to the nurse cell pathway of differentiation is the presence of four
intercystocyte ring canals.

Telfer (1975) drew attention to a branched cytoskeletal structure, the *fusome*, described in detail by classical cytologists and noticed more recently
as well, which extends through the ring canals and interconnects the cystocyte complex. At division one pole of each spindle seems to be anchored in
the fusomes, which may thus serve to orient each successive cleavage and
ensure the appropriate geometry, as diagrammed in Fig. 5.2(c). The effects
of the *fs(231)* mutation (Gans *et al.*, 1975) have been interpreted by King
(1979) and King and Riley (1982), as a derangement of fusome organization.
In this mutant the cystocytes often construct long linear chains, in which

branches form less often than normally, and three dimensional reconstructions from serial electron micrograph sections show that fusomes are often not connected in the adjacent cells. *fs(231)* cystocytes frequently become polyploid and differentiate into pseudonurse cells. Incomplete cytokinesis takes place about half the time, and cells with four ring canals occur at a frequency of about 10^{-2}, while oocytes appear only at a frequency of 10^{-7}. Thus, while it may be necessary, the possession of four ring canals does not seem to be sufficient to trigger the differentiation of an oocyte. Another mutation isolated by Schüpbach (Konrad *et al.*, 1985) presumably affects this switch directly, as it causes the formation of egg chambers that contain 16 nurse cells, but no oocyte.

The development of the oocyte-nurse cell complex can be regarded as a classical example of a clonal differentiation process. The number and orientation of the divisions by which the final 16 products are produced from the stem cell, and the selection of one of the two cystocytes that contain four ring canals as the oocyte, are all seem from these examples to be subject to specific genetic controls. The implication of the genetic evidence that function of these genes occurs only in cells that are of germ line origin, is that the differentiation of this lineage involves the activation of tissue-specific genes, just as in other cell lineages.

(ii) Transcriptional Role of Nurse Cells

The structure of the meroistic egg chamber, and the observation of cytoplasmic transport from nurse cells to oocyte, suggest that the maternal transcripts present in the mature egg at fertilization were synthesized originally in the nurse cell nuclei. An indirect argument supporting this proposition was adduced by Ribbert and Bier (1969). They compared the length of time required for oogenesis in panoistic and meroistic oogenesis, and correlated this with the number of genomes putatively cooperating in the preparation of the oocyte. In the panoistic oogenesis of the cricket *Acheta domestica*, for example, oogenesis takes more than three months, while in the meroistic dipterans, such as *Drosophila* or the blowfly *Calliphora*, oogenesis is completed within a few days. The nurse cells each achieve a DNA content of 750-1500C in *Drosophila*, as we have seen, and of 256C in *Calliphora* (Ribbert and Bier, 1969). Thus there are $1\text{-}4 \times 10^3$ more genomes putatively involved in providing oocyte transcripts in these Diptera than in the 4C oocyte nucleus of the cricket. This argument indicates the great adaptive value of the meroistic method of oogenesis. A related speculation is that perhaps nurse cells are usually of germ line origin because some prior processes of germ line differentiation are required for expression of the specific set of genes needed during the growth phase of oogenesis, irrespective of whether the transcription of these genes occurs in the germinal vesicle or the nurse cell nuclei. We now consider available molecular evidence relating directly to transcription in nurse cell nuclei and its role in the provision of maternal oocyte RNAs.

Sites of Heterogeneous RNA Synthesis in the Meroistic Egg Chamber

Under normal conditions the polyploid chromosomes of most dipteran nurse cells are not *polytene*, i.e., the multiple chromatids are not aligned in register. Thus banded chromosomes such as are present in other polyploid cell types are usually not observed in nurse cells. However, the transition from polyploidy to polyteny in nurse cells seems to be fairly trivial. In some *Anopheles* species polytene chromosomes form spontaneously in nurse cells, while they are absent in close relatives (Coluzzi and Kitzmüller, 1975). Among the unexplained effects of the *fes* mutation of *Drosophila* is the appearance of polytene chromosomes in nurse cells (King, 1970), and giant polytene chromosomes also appear in nurse cells of some ovarian tumor (*otu*[7]) mutants (Bishop and King, 1984). In *Calliphora* polytene chromosomes can be induced in nurse cells merely by a mild cold treatment (Bier, 1960). Ribbert (1979) also found that in *Calliphora* reduction of polymorphism through inbreeding results in a high incidence of polyteny in nurse cell chromosomes. Cytogenetic studies on these *Calliphora* polytene chromosomes provide evidence that in this species all or most regions of the nurse cell genomes are equally replicated. Furthermore, the autoradiograph patterns obtained by Ribbert (1979) and accompanying observations on chromosomal puffing, show that transcriptional activity in nurse cell chromosomes is very widespread, and is also relatively intense. The amount of transcriptional activity was estimated as about an order of magnitude greater than in comparable polytene chromosomes of trichogen cells, and many loci were observed to remain in a puffed configuration throughout most of oogenesis. These observations imply a high complexity for nurse cell transcription, since RNA synthesis visibly occurs in a great many parts of the genome.

A classic series of autoradiographic experiments carried out by Bier (1963) demonstrated that RNA synthesized in the nurse cell nuclei is transported rapidly into the oocyte. An example, from the housefly *Musca*, is shown in Fig. 5.4. Here newly synthesized RNA can be seen localized over the polytene nurse cell nuclei after a 30 min labeling period [Fig. 5.4(a)]. Five hours later [Fig. 5.4(b)] the labeled RNA has moved into the nurse cell cytoplasm and is apparently pouring through a ring canal into the cytoplasm of the oocyte. Note that no RNA synthesis can be observed over the oocyte at 30 min, even though the film is clearly overexposed with respect to the amount of incorporation in the nurse cell nuclei. Similar autoradiographic results have been reported for other dipteran species, and also for various coleopteran, hymenopteran, and lepidopteran species, a list which includes ovaries of both telotrophic and polytrophic construction (reviewed in Telfer, 1975;

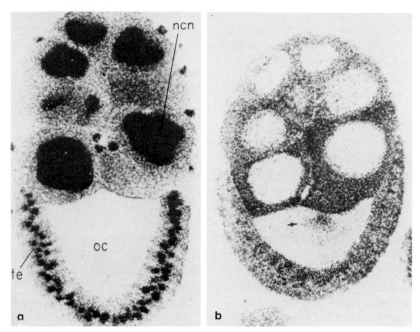

Fig. 5.4. Synthesis of RNA in nurse cell nuclei and transfer to the oocyte in the housefly. (a) Autoradiograph of an oocyte (oc), its nurse cells (ncn, nurse cell nucleus), and follicular epithelium (fe) of *Musca domestica* incubated for 30 min with ^{3}H-cytidine. (b) The same, 5 hr later. Labeled RNA can be seen entering the oocyte from an adjacent nurse cell (arrow) Densely labeled RNA is no longer present in the nurse cell nuclei, due to turnover and export, and the nurse cell cytoplasm now displays labeled RNA. [From K. Bier (1963). *J. Cell Biol.* **16**, 436.]

Muhlach and Schwalm, 1977); and in addition for the single nurse cell-oocyte complexes of a polychaete annelid (Emanuelson, 1985).

Transport of newly synthesized poly(A) RNA from nurse cell to oocyte has been directly demonstrated by Paglia *et al.* (1976a,b) in manual dissection experiments carried out on the polytrophic ovaries of the silk moths *Antheraea polyphemus* and *Actias luna*. Whole follicles were labeled *in vitro*, and ribonucleoprotein particles containing newly synthesized nurse cell poly(A) RNA were observed accumulating in the oocyte cytoplasm after several hours. Ribosomes and newly synthesized rRNA are transported as well. When further synthesis in the nurse cells was blocked with actinomycin, their content of labeled poly(A) RNA decreased more than 10-fold within 6 hr, while that of the associated oocytes increased correspondingly. No significant incorporation of precursor into vitellogenic oocyte RNA was observed on incubation of follicles from which the nurse cells had been ablated, while isolated nurse cell complexes are transcriptionally active. Furthermore, polysomes are present only in the nurse cells, and the mRNP, tRNA and ribosomes of the growing oocyte can thus be seen clearly to be stored for use later in development. By injecting ^{3}H-uridine into late pupae that were then permitted to eclose, Paglia *et al.* (1976a) showed that poly(A) RNA synthesized during vitellogenesis is indeed sequestered in mature chorionated eggs.

In some meroistic insects, particularly some species of midges, bugs and beetles, heterogeneous RNA of the oocyte cytoplasm is also contributed by the germinal vesicle, though most derives from the nurse cells. This can be inferred from autoradiographic evidence of germinal vesicle RNA synthesis (reviewed in Telfer, 1975), and has been shown more directly by other means. *In situ* hybridization with ^{3}H-poly(U) has been used to determine the source of the oocyte poly(A) RNA in the telotrophic ovary of the milkweed bug *Oncopeltus fasciatus* (Capco and Jeffery, 1979). During the growth phase of oogenesis poly(A) RNA accumulates in the oocyte cytoplasm and the newly synthesized transcripts are evidently supplied by the nurse cells. Thus these cells and the nutritive cord [see Fig. 5.1(b)] are heavily labeled by the probe, while the germinal vesicle contains no detectable poly(A) RNA. However, towards the end of oogenesis the nutritive cord is interrupted by the growth of the chorion, but thereafter a further net increase in oocyte poly(A) RNA nonetheless takes place. This is apparently due to late RNA synthesis in the germinal vesicle. Transcription of poly(A) RNA also occurs in the germinal vesicles of postvitellogenic oocytes of the bug *Dysdercus* (Winter *et al.*, 1977). In the gall midge *Wachtliella* elimination of 16 of the 20 chromosomes present in germ line cells occurs in all other cells early in embryogenesis. During oogenesis the four somatic (S) chromosomes remain condensed in the oocyte nucleus, surrounded by a concentric fibrous lamellar structure, while the remaining (E) chromosomes are dispersed (Kunz *et al.*, 1970). Autoradiography shows that the S chromosomes are transcriptionally silent in the oocyte, while the E chromosomes are active. The nucleolar organizer is located on an S chromosome and nucleolar formation and S chromosome activity are detected only in nurse cells (Kunz *et al.*, 1970). If the E chromosomes are experimentally eliminated from the germ line stem cells, oogenesis does not take place (Geyer-Duszyńska, 1966; Bantock, 1970). Though the heterogeneous RNAs produced in the oocyte by the E chromosomes may be required for oogenesis, the relative importance of nurse cell and germinal vesicle transcripts remains to be determined.

The observations reviewed so far are qualitative, and even where no germinal vesicle activity is reported it remains possible that rare maternal transcripts could be synthesized by the oocyte genomes. These might easily have escaped observation in autoradiographic experiments such as those shown in Fig. 5.4, focused as they are on the overwhelming synthetic activity of the thousand-fold polyploid nurse cell nuclei. The differences between those meroistic insects that utilize germinal vesicle transcripts, and the lepidopteran and dipteran examples that apparently do not, could be merely quantitative. On the other hand, many of the same RNA species produced in small quantities in the germinal vesicles of meroistic oocytes may also be represented in the much greater flow of nurse cell transcripts. These uncertainties should be kept in mind in considering the locus of action of germ line mutations that display maternal effects, and the origins during oogenesis of given species of transcripts, any one of which might represent only a small fraction of the total maternal RNA.

Protein Synthesis in the Drosophila Egg Chamber

Qualitative comparisons of protein synthesis patterns in isolated nurse cells and oocytes of *Drosophila* have been reported by several authors. In these studies newly synthesized proteins were labeled by injection of ^{35}S-methionine into the adult abdomen (Gutzeit and Gehring, 1979; Loyd *et al.*, 1981) or by incubation of whole follicles and isolated nurse cell complexes *in vitro* (Gutzeit and Gehring, 1979). Newly synthesized proteins were then displayed by 2D gel electrophoresis. These studies show that most of the several hundred proteins visualized are synthesized throughout oogenesis, though there are a few stage-specific species, and that all the newly synthesized proteins found in the oocyte are synthesized as well in isolated nurse cells. On the other hand several nurse cell proteins are not detectable in the oocyte. Flies that are prevented from laying eggs retain their stage 14 oocytes for some days, and these oocytes, which of course lack functional nurse cells, carry out the same program of protein synthesis as do the oocyte-nurse cell complexes of earlier stages (Kuo and Garen, 1978). These observations are all consistent with a relatively simple model in which a single qualitative pattern of gene activity exists in all the active germ line cells of the polytrophic egg chamber throughout oogenesis.

Yolk is taken up by the oocyte from without, and thus only the accumulation of *nonyolk* proteins provides a useful measure of biosynthetic activity within the oocyte-nurse cell complex. The mass of total protein in the mature *Drosophila* egg is about 1.8 μg (Bownes, 1975a) of which about 22% is yolk (Warren and Mahowald, 1979; Ruddell and Jacobs-Lorena, 1983). The rate of nonyolk protein accumulation is relatively low in stages 1-9, but after this the mass of polysomes in the egg chamber increases sharply, and the rate of nonyolk protein synthesis accelerates correspondingly (Ruddell and Jacobs-Lorena, 1983). The same mass of polysomes is then present in the egg chamber from stage 10 to stage 14, and there is little further change even after fertilization (Mermod and Crippa, 1978; Ruddell and Jacobs-Lorena, 1983). Assuming that 4 or 5% of the mass of polysomes measured in egg chambers after stage 10 is mRNA (Mermod *et al.*, 1980; Galau *et al.*, 1977a), its mass would be 2.1-2.6 ng. Mermod *et al.* (1980) measured the mass of total poly(A) RNA per stage 14 oocyte as 2.8 ng, and an estimate of 3.8 ng of cytoplasmic poly(A) RNA was reported for early embryos by Anderson and Lengyel (1979). About 60-80% of the total poly(A) RNA is associated with polysomes throughout late oogenesis and into early embryogenesis (Lovett and Goldstein, 1977; Mermod *et al.*, 1980), though this can vary in mature oocytes depending on physiological conditions (Mermod and Crippa, 1978). Thus these data indicate that the amounts of total as well as of polysomal mRNA found in the *developing embryo* are already set by stage 10 of oogenesis, though the location of the mRNA shifts thereafter from nurse cells plus oocyte to mainly oocyte. This shift probably occurs primarily as a result of the injection of nurse cell cytoplasm (including polysomes) into the oocyte (see above).

Despite the high polysomal content in stage 12-14 oocytes no significant further net accumulation of egg chamber protein occurs after stage 12 (Ruddell and Jacobs-Lorena, 1983). At this point each egg chamber already contains the amount of nonyolk protein present in the mature oocyte, about 1.45 μg. Current data do not suffice to distinguish between the alternatives that protein synthesis rate in the polysomes of the late oocyte has declined by a large factor due to a precipitous decrease in translational efficiency that involves change in neither polysome size or content, or on the other hand, that there is simply a sharp increase in protein turnover rate. As noted above synthesis of a large number of diverse proteins does continue in stage 14 oocytes, though the rate is not known (see, e.g., Kuo and Garen, 1978), and when measured very shortly after fertilization the absolute rate of protein synthesis implies that a large fraction of the polysomes present are normally active. This rate is about 9.4 ng hr^{-1} embryo, as measured by Santon and Pellegrini (1981). It is clear in any case that *Drosophila* differs from species such as the sea urchin or *Xenopus* in which the quantity of polysomes increases manyfold over the period of late oogenesis, maturation and early development (cf. Chapter II).

In late *Drosophila* oocytes the significance of the nonpolysomal mRNP compartment is primarily qualitative rather than quantitative. Thus certain mRNA species are preferentially localized in the nonpolysomal compartment prior to fertilization, and these species appear in polysomes only after development begins (see, e.g., Mermod *et al.*, 1980; Fruscoloni *et al.*, 1983; see also Chapter II).

(iii) Activity of Specific Genes in the Oocyte-Nurse Cell Complex

Specific Structural Gene Products

Measurements on the accumulation of actin, tubulin, heat shock protein and histone transcripts during *Drosophila* oogenesis illustrate in particular many of the general conclusions drawn above. Some relevant data are summarized in Table 5.1. The tubulin measurements cited were obtained by Loyd *et al.* (1981), who separated stage 10B egg chambers into oocyte, nurse cell, and follicle cell fractions, and determined the proteins synthesized in

TABLE 5.1.   Accumulation Data for Three Specific mRNA Species during *Drosophila* Oogenesis

Protein	Average rate of protein accumulation per egg chamber (ng hr^{-1})	Stage	Period of net mRNA accumulation (stages of oogenesis)	Approximate final number mRNA molecules (stage 14 oocyte)
Tubulins[a] ($\alpha + \beta$)	1	9–14	<10	5×10^7
Actins[b] (2 cytoskeletal species)	0.8	9–12	7–11	4×10^7
Histones[c] (all 5 species)	—		10B-14	5×10^8

[a] Data of Loyd *et al.* (1981) obtained by radioimmune precipitation. The rate of tubulin synthesis may exceed 2 ng hr^{-1} during stage 10, and is about 1 ng hr^{-1} between stages 10 and 14. The number of molecules is calculated from the length of tubulin mRNA, ~2000 nt, and from the quantity of message, 55 pg per stage 14 oocyte, estimated by Loyd *et al.* (1981) from translational parameters to be required to support a tublin synthesis rate of ~1 ng hr^{-1}. The tubulin messages are thus a very prevalent species, accounting for about 4% of the nonmitochondrial poly(A) RNA of the egg. The α tubulin mRNAs derive mainly from two genes located at 84B3-6 (Mischke and Pardue, 1982).

[b] Data of Ruddell and Jacobs-Lorena (1984). The rate of cytoskeletal actin protein accumulation was estimated from densitometry of stained actin isolated from egg chambers of various stages by an electrophoretic procedure, using known amounts of authentic actin proteins as a standard for quantitation. mRNA accumulation data were obtained from RNA gel blots carried out with a plasmid containing a cytoskeletal actin gene which hybridizes to both actin genes expressed in the oocyte, *viz.* genes 5C and 42A (Fyrberg *et al.*, 1980). The number of actin mRNAs is calculated from the rate of actin protein accumulation listed, assuming the same translational efficiency as for tubulin, and an average mRNA length of 1850 nt (the mRNAs produced by the two cytoskeletal actin genes are reported as 1700 and 1950 nt; Ruddell and Jacobs-Lorena, 1984).

[c] Data of Ruddell and Jacobs-Lorena (1985). Ambrosio and Schedl (1985) showed that from stages 5-10A histone mRNA accumulates and the degrades cyclically in individual nurse cells, probably in synchrony with their endomitotic DNA replication cycles. Up to the end of this stage the oocyte contains <5% of the total amount of histone mRNA present in stage 14 oocytes. At stage 10B the histone genes in all the nurse cell nuclei become activated in concert and the quantity of histone mRNA thereafter increases to its final level. The stages during which histone mRNAs accumulate were estimated by blot hybridization with RNA, using a cloned probe that includes all five histone genes. Relative mRNA accumulation profiles were obtained by densitometry. The absolute quantity of histone mRNA in the egg was measured by Anderson and Lengyel (1980), using DNA excess hybridization of the same cloned genes with uniformly labeled egg RNA. The calculation shown assumes an average mRNA length of 500 nt. There are about 110 histone gene clusters, each cluster containing a gene for all of the 5 histone species in *Drosophila* (Lifton *et al.*, 1978). However, the fraction of these genes active during oogenesis is not established.

each compartment by 2D gel electrophoresis. Tubulin synthesis occurs in both nurse cells and oocytes at stage 10B. At this point the oocyte contains 0.6 ng of tubulin, and the nurse cells about 4-6 ng, while at the end of oogenesis the stage 14 oocyte contains about 18 ng of tubulin. After stage 11 the continued synthesis of tubulin evidently occurs in the oocyte on the templates provided earlier by the nurse cells. A more or less constant rate of tubulin synthesis obtains after stage 9 (Table 5.1), suggesting the maintenance of a constant amount of tubulin message. This could signify either repression of tubulin genes from stage 10 on, or a steady state quantity of metabolically labile tubulin mRNA. Loyd et al. (1981) also found that in stage 10B egg chambers cytoskeletal actins are synthesized by both oocyte and nurse cells, as well as by follicle cells. The accumulation of these proteins has been described by Ruddell and Jacobs-Lorena (1984). Two cytoskeletal actin genes are utilized during oogenesis, and from the rate of protein accumulation their mRNAs can be calculated to be almost as prevalent as the tubulin mRNAs (see Table 5.1). Net accumulation of actin protein ceases at stage 12, though like most oocyte mRNAs the actin message remains loaded on polysomes thereafter (Ruddell and Jacobs-Lorena, 1984). Messages for three heat shock proteins also originate in nurse cells (in normal nonstressed adults), and they are released into the oocyte at stage 10-11 (Zimmerman et al., 1983). There are 1-2 × 10^7 molecules of each species of heat shock mRNA, coding for the 83 kd, the 28 kd and the 26 kd heat shock proteins, in the egg at fertilization, i.e., somewhat less than the quantities of the specific mRNAs listed in Table 5.1. The histone mRNAs provide a different example, in that these messages continue to accumulate right up to stage 14 (Ruddell and Jacobs-Lorena, 1985). Up to stage 10A the nurse cell histone mRNA is apparently utilized primarily to provide histone for the replicating nurse cell DNA (Ambrosio and Schedl, 1985; see also legend to Table 5.1). Since polyploidization in the nurse cells is then complete and there is no further DNA synthesis in the oocyte-nurse cell complex, the translation products of the histone mRNA synthesized after this must be destined for use in the embryo. The source of the histone mRNA after stage 12 is mysterious, since the nurse cells are then walled off from the oocyte by the chorion, while the chromosomes of the late oocyte nucleus have been thought to be transcriptionally quiescent on the basis of their condensed structure (see, e.g., Nokkala and Puro, 1976) and autoradiographic evidence (Mahowald and Tiefert, 1970).

The import of Table 5.1 can be summarized as follows. The sets of genes examined are regulated differently in detail during oogensis, and the accumulation of their protein products follows different kinetics. Nonetheless, these examples individually support the generalization that maternal mRNAs of the *Drosophila* egg are mainly synthesized in the nurse cell nuclei and injected into the oocyte, where they are found associated with polysomes and engaged in protein synthesis during most or all of the ensuing periods of oogenesis.

Genetic Evidence for Germ Line Functions That Affect Oocyte Structure

Several germ line mutations have been described that block oogenesis after the completion of the cystocyte divisions and formation of the egg chamber. Except that the locus of action of these mutations is the nurse cells or oocyte, for the most part the primary physiological or structural defects they cause are not known, and none are yet analyzed at the molecular level. Their significance here is to remind us that there are many additional genetically controlled aspects of oogenesis besides synthesis of maternal transcripts and proteins required after fertilization.

An interesting example analyzed by Waring *et al.* (1983) concerns a recessive female sterile mutation, *fs29* (other alleles are *fs117* and *fs445*), that maps to region 12E1-12F1 on the X chromosome. The germ line function of this gene was demonstrated by transplanting homozygous mutant pole cells into normal eggs, and mutant pole cells into normal eggs. The visible effect of the homozygous mutation is a slightly abnormal chorion structure, particularly at the anterior dorsal side, where in *fs29* oocytes the respiratory appendages project at the wrong angle and the adjacent anterior region of the chorion is abnormally formed. A consequence is failure of fertilization, which normally occurs via the micropyle, an anterior chorion structure. Waring *et al.* (1983) found a sharp decrease in the amount of yolk protein taken up by *fs29* oocytes, although synthesis of yolk occurs normally in both fat body and follicle cells. *Drosophila* follicle cells are the source of about 35% of two of the three major yolk proteins (Brennan *et al.*, 1982; Bownes, 1982). The deficiency in *fs29* animals appears to lie in the mechanism by which yolk is sequestered by the oocyte, which includes an extremely specific binding to surface receptors followed by pinocytotic internalization (reviewed in Engelmann, 1979). Probably the interruption of follicle cell-oocyte contacts that produces the fatal structural abnormality in the chorion is a peripheral side effect of decreased turgidity in the anterior region of the oocyte, due to decreased yolk content. A second mutation, *fs(1)K10*, that affects chorion structure in the same region and also acts in the germ line, has been described by Wieschaus *et al.* (1978). In normal egg chambers morphogenesis of the dorsal appendages is carried out by nests of follicle cells apposed bilaterally to the anterior surface of the oocyte (King, 1970, p. 47), and the dorsoventral polarity of the oocyte is foreshadowed by the thicker follicle cell layer on the dorsal side. Mutations at the *fs(1)K10* locus dorsalize the egg chamber, in that a thick follicle cell layer occurs on the future ventral side as well, and they result in eggs with enlarged dorsal appendages (Wieschaus, 1979). In the few cases in which development is initiated, the embryos lack ventral structures and display a dorsalized phenotype (see discussion of maternal mutants of the *dorsal* class in Chapter VI). Function of the *K10⁺* gene is necessary only in the female germ line (Wieschaus *et al.*, 1978), and normal progeny can be derived from transgenic *fs1(K10)* females bearing the wild-type gene (Haenlin *et al.*, 1985).

Another female sterile mutation that according to pole cell transplantation tests acts in the germ line, though it affects follicle cell function, is *tiny* (King, 1970, pp. 174 ff.; DiMario and Hennen, 1982). Though it maps to the same region of the X chromosome, *tiny* complements *fs29* and any of its known alleles. In homozygous (or hemizygous) *tiny* egg chambers the distal follicle cell migration occurring early in oogenesis is blocked, and an abnormally thick follicular wall that prevents expansion of the oocyte is formed. This in turn produces a thickened, irregular chorion. DiMario and Hennen (1982) suggested that the primary lesion in *tiny* egg chambers is a defect in the nurse cell-oocyte surfaces over which the follicle cells are supposed to migrate. Additional mutations with effects similar to these examples are described by Perrimon and Gans (1983), Bishop and King (1984), and Konrad *et al.* (1985).

Ribosomal RNA Synthesis

Two different mechanisms by which ribosomal RNA is supplied to the oocyte have been observed in meroistic insects. In most forms the ribosomal RNA derives wholly from the nurse cells, along with other transcript species. As this is of course the bulk form of RNA in the oocyte its transfer is easily detected. In *Drosophila* and *Calliphora* egg chambers this occurs mainly when the nurse cell cytoplasm is injected into the oocyte, while in the *Hyalophora* ovariole most of the rRNA is transferred to the oocyte prior to the terminal injection (reviewed in Telfer, 1975). Early vitellogenic *Hyalophora* oocytes already contain about 1 μg of total RNA, most of which is undoubtedly ribosomal, and this is also the RNA content of the seven nurse cells taken together. During vitellogenesis the oocyte doubles its total RNA content, and this is increased to 3 μg by the terminal injection process (Pollack and Telfer, 1969). Ribosome transfer from nurse cells to oocyte was demonstrated in the polytrophic *Antheraea* egg chamber by Hughes and Berry (1970), and in the telotrophic egg chamber of *Oncopeltus* by Davenport (1976). In these experiments the ovariole was exposed to isotopic precursor, and the subsequent appearance of labeled ribosomes in the oocyte was monitored. Entry of newly synthesized rRNA could be interrupted in *Oncopeltus* by ligation of the nutritive cord and in *Antheraea* did not occur after removal of the nurse cell cap.

In many organisms the rRNA genes are amplified during oogenesis, a special mechanism required to meet the enormous demand for preformed ribosomes early in embryonic development (Chapter II). However, significant ribosomal gene amplification, i.e., relative to the remainder of the genome, has been reported not to occur in the polyploid nurse cells of a number of meroistic insect species. In *Calliphora* rRNA synthesis takes place in extrachromosomal nucleoli (Ribbert and Bier, 1969), but even here the fraction of ovariole DNA that is ribosomal is only 1.3 times that measured in diploid brain cells (Renkawitz and Kunz, 1975). In four dipteran species, *Drosophila hydei, Drosophila virilis, Sarcophaga barbata* and *Rynchosciara angelae* the ribosomal DNA is actually underreplicated by about a factor of two during nurse cell polytenization (Gambarini and Lara, 1974; Renkawitz and Kunz, 1975; Endow and Gall, 1975; Kunz *et al.*, 1982). Nor is there preferential accumulation of ribosomal DNA during oogenesis in *Drosophila melanogaster* (Mohan and Ritossa, 1970; Mermod *et al.*, 1977; Hammond and Laird, 1985); in the silk moths *Antheraea pernyi* and *Bombyx mori* (Cave, 1978); or in *Oncopeltus* (Cave, 1975), where the nurse cell nuclei of the telotrophic egg chambers are polyploid only to the extent of 32-64C. These measurements imply that the total number of ribosomal genes present per egg chamber, i.e., the product of nurse cell ploidy, rRNA genes per haploid genome, and the number of nurse cells, is sufficient to provide the oocyte with the requisite quantity of ribosomes in the time available, so that ribosomal gene amplification is unnecessary. This can be seen explicitly for the *Drosophila* oocyte, where the transcription rate and other necessary parameters can be estimated. There are about 250 rRNA genes per haploid *Drosophila* genome (Ritossa, 1976), and so assuming neither under- nor over- DNA replication, at stage 9 when nurse cell polyploidization is complete, the number of rRNA genes per egg chamber is about 3.8×10^6. At a polymerase translocation rate of 10 nt sec^{-1} (Anderson and Lengyel, 1979), and assuming a minimum polymerase spacing of about 100 nt it would require only about 30 hr for this number of genes to produce the approximately 4.5×10^{10} ribosomal RNA molecules stored in the mature egg (there are 160 ng of rRNA per egg; Anderson and Lengyel, 1979). Mermod *et al.* (1977) showed that earlier in oogenesis the rate of egg chamber rRNA synthesis is directly proportional to the degree of nurse cell polyploidization and the total number of rRNA genes. Ribosomal RNA synthesis could be a rate limiting process in *Drosophila* oogenesis. Thus in *bobbed* mutants in which various portions of the ribosomal gene cluster are deleted, the length of time required for oogenesis depends on the number of ribosomal genes remaining (Mohan, 1971). For example, in mutant egg chambers containing only 1/3rd the normal complement of rDNA, oogenesis required 206 hr compared to 75 hr in controls. It is interesting that as a result of this compensatory mechanism the mature eggs of *bobbed* females contain a normal quantity of rRNA (Ritossa, 1976).

Ribosomal DNA amplification does occur in the *oocytes* of some insect species, including water beetles such as *Dytiscus* and *Colymbetes*; the dipteran *Tipula*; and the neuropteran *Chrysoma* (reviewed in Telfer, 1975;

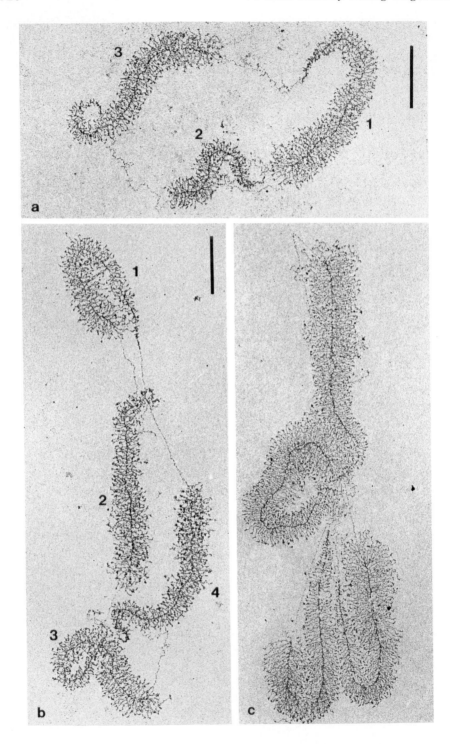

Trendelenberg *et al.*, 1977). In these examples a single "DNA body" that contains rRNA genes is found in the definitive oocyte nucleus, but not in nurse cell nuclei. Autoradiographic observations and *in situ* hybridizations with ribosomal sequence probes indicate that rDNA amplification in these bodies begins during the cystocyte divisions, and continues in the early stages of previtellogenic oogenesis. The best studied example is the water beetle *Dytiscus*. Gall and Rochaix (1974) found that the DNA body in this species contains about 3×10^6 rRNA genes, arranged in small circles, each containing one to five genes. This number is remarkably close to that present in the nurse cell chromosomes of *Drosophila* (see above). In midoogenesis the DNA body disperses and intense rRNA transcription begins. This is illustrated in the electron micrographs reproduced in Fig. 5.5. No other transcriptional activity can be detected in the oocyte nuclei. The chromosomes are present in a condensed mass, the karyosphere, and as usual in meroistic oocytes, they appear silent in autoradiograph experiments (Telfer, 1975).

The meroistic form of oogenesis is usefully considered from a logistic point of view. A great many copies of genes are put to work cooperatively in the meroistic egg chamber, all producing sequences required by the oocyte. The obvious adaptive value of this elaborate strategy is that it accelerates by a large factor the overall rate of oogenesis.

2. THE TRANSCRIPTIONAL ROLE OF OOCYTE LAMPBRUSH CHROMOSOMES

The oocytes of many animals contain spectacular meiotic prophase chromosomes that are distinguished by the presence of thousands of lateral loops. Such chromosomes were first observed by Flemming (1882) in sections taken through urodele oocytes, and they were the subject of a detailed study by Rückert (1892), carried out on isolated germinal vesicles of shark oocytes. Rückert recognized that they are paired chromosomes from which the numerous lateral loops project, and this aspect of their structure suggested to him the designation *lampbrush chromosomes*. These chromosomes are never found in the germinal vesicles of meroistic oocytes, which as we have seen are generally quiescent in regard to transcription, nor do they occur in the oocytes of many other species in which the maternal

---

Fig. 5.5. Transcriptionally active amplified rDNA genes from the oocyte nucleus of *Dytiscus*. Scales: 2μm. (a) X10,925; (b) X9,975. (a) Ring containing 3 matrix units, one of which appears truncated (unit 2); (b) and (c) rings containing 4 matrix units each. In each ring the ribosomal genes appear maximally loaded with nascent transcripts. Trendelenberg *et al.* (1976) concluded from the heterogeneity of spacer and gene lengths, as illustrated, that each ring seems to represent a different region of the nucleolar organizer. [From M. F. Trendelenberg, W. W. Franke, and U. Scheer (1977). *Differentiation* **7**, 133.]

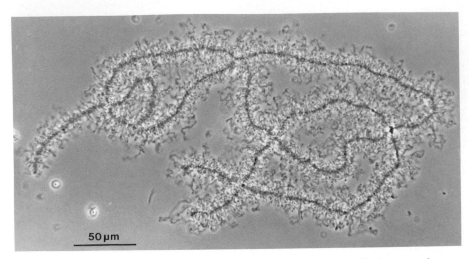

Fig. 5.6.   Unfixed lampbrush chromosomes of *Triturus viridescens*. The loops can be seen projecting laterally from the main chromosomal axis. Note the chiasmata, three of which can be clearly discerned. [From J. G. Gall (1966). *Methods Cell Physiol.* **2**, 37.]

transcripts derive solely from the germinal vesicle. Though there remain some mysteries, and their patterns of transcription are in some respects unusual, we shall conclude from molecular and cytological analyses that their basic function is probably the synthesis and maintenance of the large pool of maternal transcripts resident in the oocyte cytoplasm.

(i) Lampbrush Chromosome Structure: Transcription Units and Chromomeres

The modern era of lampbrush chromosome cytology began with the establishment of two basic structural tenets. First, lampbrush chromosomes were shown to be bivalent meiotic prophase structures in which each loop of an apposing pair contains a single DNA duplex, while the central axis contains two such duplexes. The meiotic homologues are typically united by several chiasmata. Second, the loops were perceived to be the sites of chromosomal RNA synthesis. Thus, an active locus is represented by four loops, two deriving from each chromosomal axis (reviewed in Davidson, 1976; Callan, 1982). The general appearance of a newt oocyte lampbrush chromosome in the phase contrast microscope is shown in Fig. 5.6.

The axes of lampbrush chromosomes from which the loops project were recognized classically to consist of a linear array of compacted beads, referred to as *chromomeres*. Gall (1954) showed that DNA is concentrated in the chromomeres, as these are the only structures in lampbrush chromosomes that can be visualized by the Feulgen reaction. The paired structure of the chromomeric axis was demonstrated by Callan (1955), by stretching local regions with microneedles to the point where the chromomeres would separate transversely into two distinct chromatin strands. From measurement of the kinetics of lampbrush chromosome breakage by DNase I, Gall (1963) deduced the presence of two DNA duplexes in the axis and of single duplexes in the loops. Important evidence also derived from the maps constructed for newt lampbrush chromosomes by Callan and Lloyd (1960). These studies showed that distinctive loops can be recognized at invariant locations in the chromosomes. Particular loop morphologies are the property of species, subspecies, or individuals (Callan, 1963). Heterozygotes, or hybrids between related species, generally produce heterozygous sets of lampbrush chromosomes which display the allelic alternatives characteristic of each parent, and the frequencies at which these alternatives appear are distributed in wild populations as predicted by a Hardy-Weinberg calculation. A general conclusion is thus that the loop structure is determined by the genetic locus, i.e., the DNA sequence which it contains. Lampbrush chromosome maps have now been assembled for a number of different species, including several urodeles (see, e.g., Mancino *et al.*, 1977; Callan, 1982); the anuran *Xenopus* (Müller, 1974); and some amphisbaenian reptiles (Macgregor and Klosterman, 1979).

Transcription Units of Lampbrush Chromosomes as Visualized in the Electron Microscope

Early autoradiographic studies carried out with the light microscope showed that intense RNA synthesis occurs throughout the length of most loops, though there are a few giant loops that display unusual labeling patterns (Gall and Callan, 1962; Callan, 1967; reviewed in Macgregor, 1980). The loops contain newly synthesized proteins as well as RNA (Gall, 1958; Gall and Callan, 1962; Izawa *et al.*, 1963). Furthermore, most loops appear to have a polarity, in that the loop matrix is thicker at one end than at the other (Callan and Lloyd, 1960; Callan, 1963). Electron microscopy carried out by the Miller spreading technique has now provided a convincing molecular interpretation of both the widespread transcriptional activity and the polarity observed at light microscope resolution.

As illustrated by the examples shown in Fig. 5.7, the transcription units of lampbrush chromosomes are packed densely with nascent RNA molecules. Observations on both *Xenopus* and *Triturus* lampbrush chromosomes show that the polymerase molecules by which the nascent transcripts are anchored to the DNA fibril of the loop are typically only 100-200 nt apart (Miller and Bakken, 1972; Hill, 1979). Though some variation is observed both within and among transcription units, the structural feature that immediately distinguishes lampbrush chromosomes is that almost all the transcription units functioning in lampbrush chromosomes are synthesizing

Fig. 5.7. Transcription units of amphibian oocyte lampbrush chromosomes, visualized for electron microscopy. (a) Low magnification view of a contiguous region of a *Xenopus laevis* oocyte chromosome. Three highly active transcription units can be seen (1-3) separated by two inactive "spacer" regions (*S*). These regions and the inactive axial chromatin strands (*C*) display a beaded, nucleosomal configuration. [From R. S. Hill (1979). *J. Cell Sci.* **40**, 145.] (b) Initial portion of a large transcription unit from *Triturus viridescens* lampbrush chromosome.

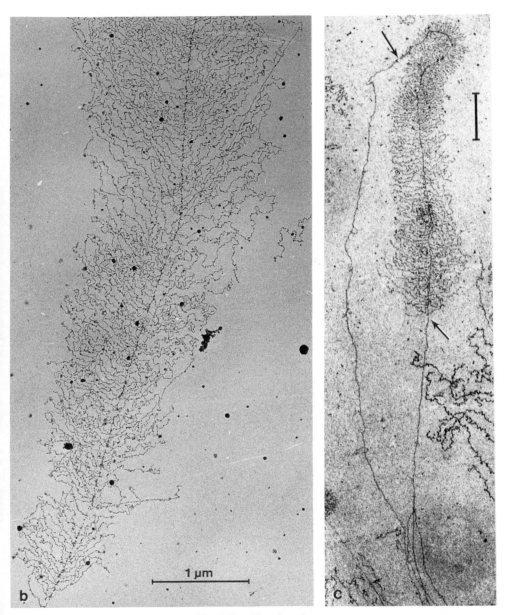

Fig. 5.7.   (Continued)

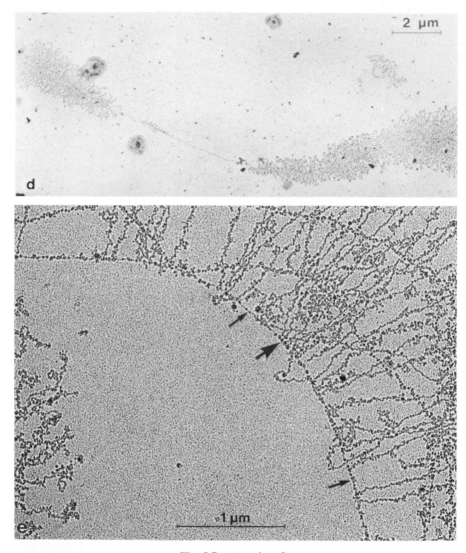

2 µm

1 µm

d

e

Fig. 5.7.   (continued)

The region shown is more than 30 kb in length. The transcription unit begins below the lower left corner and continues beyond the upper boundary of the photograph. The transcripts appear to be spaced only about 100 ntp apart. [From O. L. Miller and A. H. Bakken (1972). *Acta Endocrinol. (Stockholm), Suppl.* No. **168**, 155.] (c) A complete transcription unit at least 50 kb in length, from a lampbrush chromosome loop of *Triturus cristatus*. The transcripts again appear to originate only one to a few hundred nucleotide pairs apart. The start and termination sites of transcription are indicated by arrows, and the transcription unit is preceded and followed by long sequences of nontranscribed DNA. There are at least 100 kb of silent genomic sequence surrounding the transcription unit, and at higher magnification these regions can be

RNA at the maximum possible rate. That is, the nascent transcripts are closely packed because initiation is occurring as frequently as permitted by the polymerase translocation rate. Figure 5.7(a) displays a characteristic region in which all three of the separate transcription units evident are being intensely transcribed. By contrast, when the same methods of visualization are applied to active somatic cell types, e.g., sea urchin or *Drosophila* embryonic cells (reviewed in Chapter III), only about 10% of the transcription units are densely packed with nascent transcripts (Laird and Chooi, 1976; McKnight and Miller, 1977; Busby and Bakken, 1979), just as expected from consideration of the RNA prevalence distributions characteristic of such cells (Davidson and Britten, 1979; see also Chapter III).

The lengths of the transcription units observed in lampbrush chromosomes vary from about 2 kb to at least 70 kb of extended chromosomal DNA in *Xenopus* (Hill, 1979), and from about 10 to 100 kb in several urodeles (Miller and Bakken, 1972; Angelier and Lacroix, 1975; Scheer *et al.*, 1976a; Macgregor, 1980). The enormous size of transcription matrices such as those illustrated in Fig. 5.7(b) and (c) accounts easily for the polarized form of the prominent loops observed by classical methods, since the nascent transcripts associated with a given transcription unit increase in length toward the distal end. However, it is necessary to take into account the effects of the spreading procedure required for visualization. The detergent treatments involved in these preparations probably result in the removal of some proteins, and the surface tension to which the sample is exposed converts what might be imagined a solid ribonucleoprotein core of increasing diameter into the open, two-dimensional form illustrated. A useful calculation of Macgregor (1980) shows that average dimensions of the solid ribonucleoprotein matrices from which the typical transcription complexes visualized in the electron microscope would have derived are consistent with those actually observed in the phase microscope.

The relation between the number of loops and of transcription units is not straightforward, though many loops probably contain single transcription units. However, it is not uncommon to observe two transcription units separated by a short nontranscribed region (Angelier and Lacroix, 1975; Scheer *et al.*, 1976a). This is particularly striking when the transcription matrices are oriented in opposite directions, indicating initiation sites on opposite strands, as illustrated in Fig. 5.7(d). A given region may also contain multi-

---

seen to be packed with nucleosomes. Scale bar, 2 μm. [From U. Scheer, W. W. Franke, M. F. Trendelenburg, and H. Spring (1976a). *J. Cell Sci.* **22**, 503.] (d) A region of a loop showing parts of two transcriptional units of opposite polarity separated by a nontranscribed region several kb in length. The nucleosomal structure of the intergenic region can be observed. [From N. Angelier and J. C. Lacroix (1975). *Chromosoma* **51**, 323.] (e) Transcription unit from a *Xenopus* oocyte lampbrush chromosome. Nucleosomal chromatin structure in the internal regions between the transcribing polymerases is here evident. Small arrows denote nucleosomes, large arrow denotes polymerase. [From K. Martin, Y. N. Osheim, A. L. Beyer, and O. L. Miller (1980). *Results Probl. Cell Differ.* **11**, 373.]

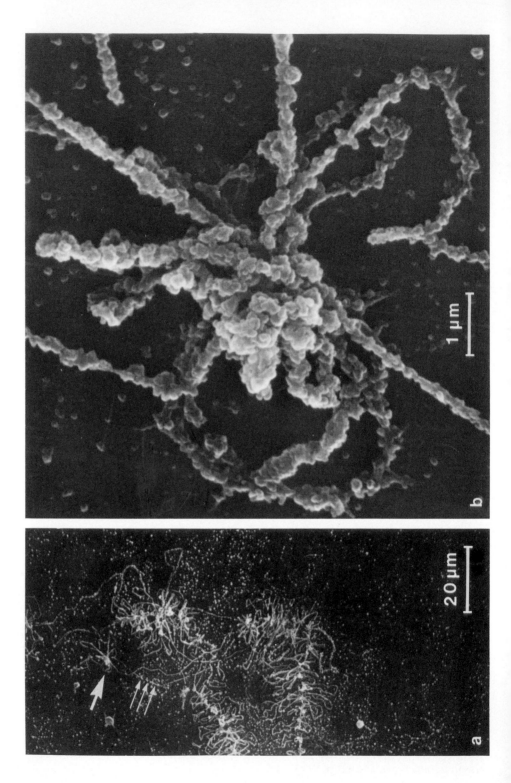

ple transcription units oriented similarly, and occasionally a series of matrices of identical size and orientation are observed that probably indicate the transcription of a tandemly repeated gene family (Scheer *et al.*, 1976a; Scheer, 1981). Careful light microscope observations of discontinuities in the matrices of giant loops have also provided indications that some loops bear multiple transcription units (reviewed in Macgregor, 1980). It follows that the number of transcription units is by some unknown factor greater than the number of loops. The number of loops, or of loops and associated chromomeres, was classically assumed to provide an approximation of the number of genetic loci, or at least of active loci. Given that the physical manifestation of an active locus is a transcription complex, this venerable correlation is clearly incorrect. Furthermore, while at electron microscope resolution the active transcription units of lampbrush chromosomes are easily defined, the distinction between the loop and chromomere domains inferred from light microscopy in most regions is not at all obvious. Closely apposed transcription units such as are illustrated in Fig. 5.7(a) and (d) no doubt derive from the same loop, while at the other extreme the long inactive axial fibrils that can also be observed [e.g., Fig. 5.7(c)] could represent the chromomeric DNA of the classical model. After spreading by the Miller procedure all of the inactive regions retain a similar beaded structure, indicative of nucleosomal conformation (Scheer *et al.*, 1976a; Scheer, 1978; Hill, 1979; Macgregor, 1980; Martin *et al.*, 1980). Nucleosomes can be perceived even *within* active transcription units, where the nascent fibrils and polymerase molecules are not so densely packed (Martin *et al.*, 1980). An example is shown in Fig. 5.7(e). An implication of the latter observation is that nucleosomal structures must be able to reform within less than a minute following the passage of a polymerase and the associated nascent transcript.

A reasonable interpretation of the structure of lampbrush chromosomes as classically observed in the light microscope is that chromomeres occur where there are long regions of inactive chromatin that *in situ* are condensed into higher order structures. These are disaggregated during the preparation of transcription spreads by the Miller procedure. Shorter inactive regions may bunch together, giving rise to a complex aggregate that would also be recognized as a chromomere in the light microscope, from which protrudes an array of multiple loops, all containing active transcription units. Hill (1979) pointed out that the alternative interpretation would require that

Fig. 5.8. Scanning electron microscopy of regions of a lampbrush chromosome from the oocyte of the newt *Pleurodeles waltlii*. Parts of chromosome bivalent 12 are shown. Small arrows seen in the low magnification view in (a) represent a characteristic double bridge, and the large arrow displays a prominent terminal chromomere. Note in (a) the appearance of *multiple* loops protruding from each condensed axial element (i.e., chromomere). (b) High magnification view of terminal chromomere marked by large arrow in (a). A group of small loops extend in all directions from the central nucleoprotein mass. [From N. Angelier, M. Paintrand, A. Lavaud, and J. P. Lechaire (1984). *Chromosoma* **89**, 243.]

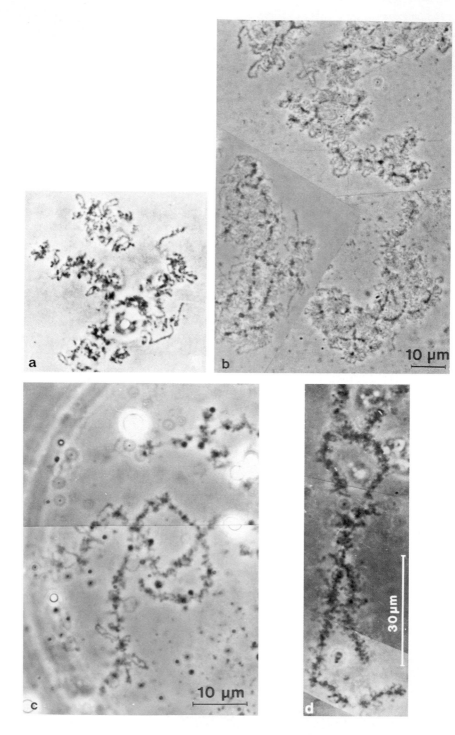

much of the length of an average loop be occupied by inactive intermatrix spacer sequences, contrary to indications from light microscope autoradiography. A scanning electron microscope study of Angelier *et al.* (1984), in which the same lampbrush chromosomes were also visualized in the phase microscope, supports the view that chromomeres may consist of associated regions of condensed nucleoprotein from which many loops extend (see Fig. 5.8). It follows that the chromomere is not an invariant structure equivalent to a single genetic locus. It is rather to be considered a compacted aggregate of transcriptionally inactive DNA, the manifestation of which depends on the spacing of the surrounding active transcription units. In different organisms among which the mode of genomic sequence organization varies, the prominence and distribution of chromomeres might also vary. Thus the differences in the number of chromomeres and the size of loops observed in comparing the lampbrush chromosomes of various species (see, e.g., Callan and Lloyd, 1960; Vlad and Macgregor, 1975; Macgregor and Klosterman, 1979) probably reflect primarily differences in the distribution and lengths of the nontranscribed DNA sequences that define the termini of the loops. This interpretation clears away the paradox that develops from the classical view that such differences would indicate fundamental variation in the number of functional genetic loci among related species.

Phylogenetic Occurrence of Lampbrush Chromosomes

For no other organisms do there exist ultrastructural or molecular observations on lampbrush chromosomes comparable to those available for amphibians. Yet these structures occur in the oocytes of many other vertebrates and in a great many invertebrates as well, and at least as perceived in the phase microscope they appear completely homologous in their cytological organization to the amphibian examples. Photomicrographs of invertebrate lampbrush chromosomes are shown in Fig. 5.9. Here are displayed lampbrush chromosomes from an orthopteran insect, *Decticus albifrons* (Kunz, 1967a); a squid, *Sepia officinalis* (Ribbert and Kunz, 1969); a snail, *Bithynia tentaculata* (Bottke, 1973); and a starfish, *Echinaster sepositus* (DeLobel, 1971). The general structural similarity of lampbrush chromosomes is illustrated in the detailed study of DeLobel (1971) on *Echinaster* lampbrush chromosomes. The map constructed for the chromosomes of this

Fig. 5.9. Lampbrush chromosomes of invertebrates. Chromosomes were isolated from living germinal vesicles in salt solutions, by methods essentially similar to those utilized for amphibian chromosomes. (a) Lampbrush chromosomes from the oocyte of an orthopteran insect, *Decticus albifrons*. [From W. Kunz (1967a). *Chromosoma* **21**, 446.] (b) Lampbrush chromosomes from the oocyte of a cephalopod mollusc, the squid *Sepia officinalis*. [From D. Ribbert and W. Kunz (1969). *Chromosoma* **28**, 93.] (c) Lampbrush chromosomes from the oocyte of a gastropod mollusc, *Bithynia tentaculata*. [From W. Bottke (1973). *Chromosoma* **42**, 175.] (d) Lampbrush chromosomes from the oocyte of an echinoderm, the starfish *Echinaster sepositus*. [From N. DeLobel (1971). *Ann. Embryol. Morphog.* **4**, 383.]

TABLE 5.2.   Occurrence of Lampbrush Chromosomes in Oocytes and Duration of the Lampbrush Stage

Taxonomic affiliation of animals in which lampbrush chromosomes have been reported	Estimated duration of lampbrush stage (where available)
Deuterostomes	
Chaetognaths[a]	
Echinoderms	
Starfish[b]	
Chordates	
Cyclostomes[c]	Several weeks in lamprey *Petromyzon*[d]
Elasmobranchs[e]	
Teleosts[e]	
Amphibians[f]	
Urodele	About 7 months in *Triturus*[g]
Anurans	*Xenopus:* minimum length 3–6 months[h]; 1.3 months[i] maximum length up to 2 yr[j]
	1–1.5 months in *Engystomops*[k]
Reptiles	
Lizards[l]	Some months[m]
Amphisbaenia[n]	
Birds[o]	3 weeks in chick[p]
Protostomes	
Molluscs	
Gastropods[q]	
Cephalopods[r]	
Insects	
Orthopterans[s]	3 months in cricket[t]

[a] Benoit (1930); [b] DeLobel (1971); [c] Okkelberg (1921); [d] Lewis and McMillan (1965); [e] Rückert (1892); Maréchal (1907); [f] Callan (1957); Duryee (1950); [g] Callan (1963); [h] Davidson (1968), Keem et al. (1979); [i] Scheer (1973); [j] Callen et al. (1980); [k] Davidson and Hough (1969); [l] Loyez (1905); [m] Loyez (1905), Boyd (1941); [n] Macgregor and Klosterman (1979); [o] D'Hollander (1904), Romanoff (1960); [p] D'Hollander (1904); [q] Davidson (1968), Bottke (1973); [r] Callan (1957), Ribbert and Kunz (1969); [s] Kunz (1967a,b); Bier et al. (1969); [t] Ribbert and Bier (1969).

organism displays the same kinds of special structures, such as giant loops of unusual conformation, that serve as landmarks in amphibian lampbrush chromosomes (see, e.g., Callan and Lloyd, 1960), and the average loop dimensions are similar to those of *Xenopus* lampbrush chromsomes.

In Table 5.2 are collected data regarding the overall distribution of lampbrush chromosomes, and where possible the duration of the lampbrush stage. It is clear that lampbrush chromosomes occur in many major groups, both deuterostome and protostome. Thus, like the process of oogenesis itself, lampbrush chromosomes are probably of very ancient evolutionary origin. Lampbrush chromosomes evidently perform some fundamental function in oogenesis, since they have been retained throughout most of metazoan evolution.

The list of organisms in which lampbrush chromosomes have been reported is of course limited by the choices made by investigators and the difficulty of observing them in some material. Nonetheless, it is now established that lampbrush chromosomes are not ubiquitous. Their absence from mouse oocytes has been clearly demonstrated by Bachvarova *et al.* (1982), and they are also lacking in at least some sea urchin species (O. L. Miller, B. A. Hamkalo, B. R. Hough-Evans, and E. H. Davidson, unpublished data). A generalization suggested by their known distribution is that lampbrush chromosomes occur in large oocytes that contain relatively huge pools of heterogeneous maternal RNA, while they are absent from small oocytes. The mouse egg, for example, contains less than 1/3000 the amount of maternal poly(A) RNA that is present in the egg of *Xenopus*, and the sea urchin egg less than 1/500 this amount. The *Xenopus* egg, in turn, is small compared to some other anuran eggs and to urodele eggs. The same relation pertains among the echinoderms. Thus the egg of the starfish *Echinaster sepositus*, which contains lampbrush chromosomes, is several hundred times the volume of the egg of the sea urchin *Strongylocentrotus purpuratus*, which does not. The correlation suggests a simple interpretation of lampbrush chromosome function. This is that these structures exist where the logistic demands of supplying the growing oocyte with maternal transcripts in the allowed time require that almost all transcription units function at maximum rate. As we have seen, widespread and intense transcription is the definitive property of oocyte lampbrush chromosomes. A logistic function for lampbrush chromosomes is also suggested by the exclusive relationship between meroistic oogenesis and the presence of lampbrush chromosomes in the oocyte nucleus. That is, lampbrush chromosomes probably perform the same function that the polytene nurse cell chromosomes do in meroistic oocytes, *viz.* the provision of maternal RNAs. This function appears to be exercised continuously over the relatively long period required for oocyte growth. Thus, as shown in Table 5.2, the lampbrush stage lasts for weeks, months, or even longer in species for which estimates are available. In some amphibian species a mechanism has evolved that almost approaches the meroistic mode of oogenesis. Here each oocyte contains multiple nuclei. Macgregor and Kezer (1970) showed that in the tailed frog *Ascaphus*, for example, the growing oocyte has eight germinal vesicles, formed initially by oogonial nuclear divisions that are not accompanied by cytokinesis. Each nucleus is endowed with a complete set of lampbrush chromosomes, and thus there are 32 copies of every active locus functioning per oocyte. Even more extreme examples have been described by del Pino and Humphries (1978) in studies of oogenesis in marsupial frogs. In these forms development proceeds directly from egg to juvenile, with a reduced or nonexistent larval tadpole stage. In several genera of marsupial frogs the previtellogenic oocyte can be seen to contain hundreds of nuclei. Each oocyte is the product of many oogonia, formed by the disappearance of cell membranes within an oogonial cyst. During the growth phase each of the multiple nuclei is endowed with lampbrush chromosomes and all are active in RNA synthesis. Later in oogenesis all but one nucleus disappears. It may be relevant that the eggs of species carrying out direct development are relatively enormous, ranging up to 9 mm in diameter. The egg of *Xenopus*, for comparison, is about 1.2 mm in diameter.

In summary, the comparative biology of occurrence, when combined with ultrastructural evidence for intense transcriptional activity in all or most of the loops, suggests that the basic function of lampbrush chromosomes is to provide the growing oocyte with the maximum possible flow of newly synthesized transcripts.

(ii) Complexity, Average Structural Characteristics, and Synthesis Rates for Lampbrush Chromosome RNAs

Size of Primary Transcripts and Association with Specific Proteins

Newly synthesized germinal vesicle RNAs have been extracted from the oocytes of several amphibian species, and the distribution of their molecular lengths determined under denaturing conditions. In one study carried out on vitellogenic oocytes of the newt *Pleurodeles poireti* transcripts as large as 10-30 kb were observed, though most RNAs recovered were smaller (Denoulet *et al.*, 1977). RNAs up to at least 60 kb in length have been extracted from *Triturus* oocytes (Sommerville and Malcolm, 1976). Scheer and Sommerville (1982) also reported measurements on extracted germinal vesicle RNAs carried out by electron microscopy. Length distributions obtained in this study extended up to about 20 kb for both *Triturus* and *Necturus* oocyte nRNAs, and to about 10 kb for *Xenopus* oocyte nRNA. However, the number average size of the molecules recovered from the isolated germinal vesicles of all three species is around that of mature mRNA, about 2 kb. The RNAs examined in these studies could have suffered strand scissions during their preparation, despite all possible precautions, and whether for this reason, or because endonucleolytic cleavages occur in the course of processing while the transcripts are still nascent, the *extracted* molecules clearly fail to match the enormous dimensions of the larger transcription units observed in the electron microscope (e.g., Fig. 5.7). As expected for a high complexity population of nuclear transcripts, lampbrush chromosome RNA has a DNA-like base composition, except for a still unexplained bias towards unusually high uridylic acid content reported for both *Xenopus* (Davidson *et al.*, 1964; Anderson and Smith, 1977) and *Triturus* oocyte nRNAs (Sommerville, 1973).

The nascent transcripts of the loop matrix are associated with proteins, which results in a significant compaction. Hill (1979) noted that even after preparation for electron microscopy the apparent nascent RNA length is less than half the length of the DNA from which it is transcribed, and in native RNP the degree of compaction is undoubtedly much higher (Sommerville and Malcolm, 1976). In high voltage electron microscopic images of thick sections through the loop matrices RNA fibrils can be seen connecting particles of about 20 nm diameter, as shown in Fig. 5.10(a). The RNP assemblages have been isolated from *Triturus* oocytes by differential centrifugation and their structure analyzed *in vitro* (Malcolm and Sommerville, 1974, 1977). Linear chains of the 20 nm ribonucleoprotein particles in different states of aggregation are shown in Fig. 5.10(b) and (c), and their disaggregation into monomers after mild RNase treatment in Fig. 5.10(d). The particles consist largely of proteins, and more than 20 distinct polypeptides have been isolated from such preparations (Malcolm and Sommerville, 1977). Antibodies prepared against these polypeptides generally react with the matrices of all loops [Fig. 5.10(e)], even though different loops often display individual morphologies, due to the various conformations into which the 20 nm particles are arranged (Malcolm and Sommerville, 1974), perhaps a function of the primary RNA sequences. However, there are certain proteins that are present only on a small number of specific loop pairs out of the whole germinal vesicle complement, as illustrated in Fig. 5.10(f) and (g) (Sommerville *et al.*, 1978b). The same proteins are included in the ribonucleoprotein complexes released into the nuclear sap from these loops when transcription has been completed [e.g., Fig. 5.10(g)]. In their physical properties and constitution the ribonucleoprotein particles that contain newly synthesized chromosomal RNAs of the amphibian oocyte directly resemble the heterogeneous nuclear RNP complexes of somatic cells (reviewed in Malcolm and Sommerville, 1977; see also Beyer *et al.*, 1977; Pederson and Munroe, 1981; Pederson, 1983). This suggests that the functions mediated by these assemblages, whether transcript packaging, transport or processing, may also be similar.

Rates of RNA Synthesis in Lampbrush Chromosomes

The lampbrush chromosome stage begins in *Xenopus* in previtellogenic, early diplotene oocytes only 50-60 μm in diameter. According to the staging criteria of Dumont (1972), these are stage 1 oocytes just beginning their growth phase. Hill and Macgregor (1980) showed that the chromosomes of even these very young oocytes contain heavily loaded transcription matrices typical of those present in midvitellogenic (stage 3) lampbrush chromosomes. Transcription is initiated even earlier, in oocytes that are only 25-40 μm in diameter. Though the nascent transcripts are at first 5-10-fold more widely separated than in the later stage 1 oocytes the average *length* of the earliest transcription units is already the same as at the maximum lampbrush stage. This study yields the important inference that a high rate of chromosomal RNA synthesis is instituted very soon after the stage 1 oocyte begins to grow, rather than only at the midlampbrush stage, as earlier assumed. Martin *et al.* (1980) showed, furthermore, that mature lampbrush chromosome transcription units persist even in fully grown (Dumont stage 6) oo-

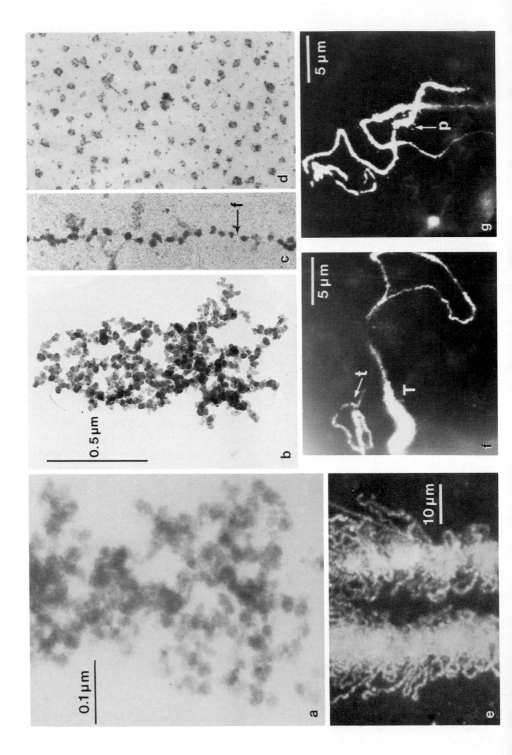

cytes. Lampbrush chromosomes are thus present almost from the beginning to the end of oogenesis, although in the light microscope the early and late oocyte lampbrush chromosomes are less easily resolved. The main difference in morphology between the transcription matrices of very early and very late oocytes, on the one hand, and of the Dumont stage 3-5 vitellogenic oocytes classically described as "maximum lampbrush stage," on the other, reflects the degree to which the nascent *transcripts* are compacted in their RNP assemblages. The lateral loops of Dumont stage 6 oocytes appear shorter, and in Dumont stage 1 and 2 oocytes the nascent transcripts are about 2-fold less extended in preparations spread for electron microscopy than in stage 3 oocytes (Hill and Macgregor, 1980; Martin *et al.*, 1980).

A valuable series of direct synthesis rate measurements carried out on stage 3 and stage 6 *Xenopus* oocytes by Anderson and Smith (1977, 1978) is summarized in Fig. 5.11 and Table 5.3. Both stage 3 and stage 6 oocytes synthesize high molecular weight (>40S or over 7 kb), unstable, nucleus-confined RNAs. The accumulation kinetics of these transcripts are illustrated in Fig. 5.11(a). Transcripts of the same size range, which turn over with a half-life of about 20-45 min are of course found in somatic nuclei as well (cf. Chapter III; reviewed in Davidson and Britten, 1973). In addition both stage 3 and stage 6 oocytes synthesize a somewhat greater amount of unstable, nucleus-confined RNA of smaller size which accounts nicely for the relatively low number average size of the nRNAs isolated by Scheer and Sommerville (1982) noted above. In vitellogenic stage 3 oocytes the unstable 4-40S RNA fraction turns over rapidly, at a rate similar to the >40S RNA, while in stage 6 oocytes the rate of turnover of the 4-40S nRNA class is somewhat lower. This results in a large increase in the steady state nRNA content in the stage 6 oocyte nucleus (Table 5.3). Anderson and Smith (1977)

Fig. 5.10. Ribonucleoprotein (RNP) of *Triturus* lampbrush chromosome loop matrices. (a) High voltage electron micrograph of portion of lateral loop showing 20 nm subparticles. [From D. B. Malcolm and J. Sommerville (1974). *Chromosoma* **48**, 137.] (b) Nuclear RNP centrifugally isolated from a homogenate of *Triturus* ovaries, after purification by sedimentation through a sucrose velocity gradient. Note the particulate structure of the partially disaggregated RNP. (c) RNP as in (b), but after treatment with 85% formamide, revealing individual 20 nm particles and RNA fibril (f). (d) Similar preparation as in (b) after treatment with 1 µg ml⁻¹ RNaseA. The individual particles are rapidly released. (b)-(d) X47,560. [(b)-(d) From D. B. Malcolm and J. Sommerville (1977). *J. Cell Sci.* **24**, 143.] (e) Immunofluorescence of lampbrush chromosome loops reacting with antibody prepared against 40-50 kd proteins isolated from preparations as in (a). All loops in the region shown fluoresce. [From J. Sommerville, D. B. Malcolm, and H. G. Callan (1978a). *Philos. Trans. R. Soc. London, Ser.* B **283**, 359.] (f) and (g) Immunofluorescence of specific chromosome loops that react with antibody prepared against a 30-35 kd chromosomal matrix protein. About 10 loops display this reactivity per haploid genome. In (f) thin (t) and thick (T) ends of the transcription matrix can be discerned, and in (g) fluorescent ribonucleoprotein particles (p) appear to have just been released from the thick terminus of one of the loops. [(f) and (g) From J. Sommerville, C. Crichton, and D. Malcolm (1978b). *Chromosoma* **66**, 99.]

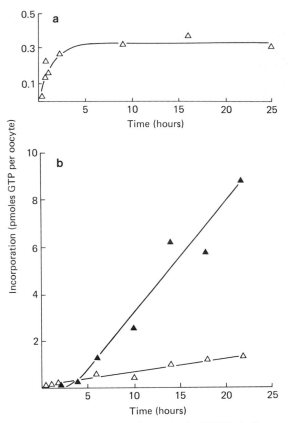

Fig. 5.11. Accumulation kinetics for newly synthesized RNAs in Dumont stage 3 *Xenopus* oocytes. Oocytes were individually dissected from the ovary, and the follicular cell layer was removed. They were injected with ³H-GTP, the precursor pool specific activity was determined, and the radioactivity incorporated in each class of RNA was measured as a function of time. Cytoplasmic fractions were prepared by manual enucleation. (a) Synthesis and turnover kinetics for >40S heterogeneous RNA. Though measured on whole oocytes this RNA fraction is confined to the nucleus. (b) Kinetics of accumulation in cytoplasmic RNA for newly synthesized 4-40S heterogeneous RNA (open triangles) and ribosomal RNA (closed triangles). [From D. M. Anderson and L. D. Smith (1978). *Dev. Biol.* **67**, 274.]

suggested as a possible explanation a decrease in processing efficiency late in oogenesis, a proposition discussed further in a following section.

Table 5.3 shows that only 11% of the total heterogeneous RNA synthesized in the stage 3 *Xenopus* germinal vesicle is destined for the cytoplasm, and in stage 6 oocytes this value has fallen to 5%. The RNA transported to the cytoplasm is detected kinetically as the *stable* 4-40S component of total newly synthesized heterogeneous RNA. Thus Anderson and Smith (1978) prepared enucleate cytoplasms by manual dissection and showed that the rate of entry of the 4-40S RNA into the cytoplasm equals its rate of synthe-

TABLE 5.3. Heterogeneous Nuclear and Cytoplasmic RNA Synthesis in *Xenopus* Oocytes[a]

Dumont stage	RNA class[b]	Synthesis rate (pg min^{-1})	Cytoplasmic RNA synthesis (%)[c]	$t_{1/2}$ (min)	Steady state RNA content per nucleus (pg)
3[d]	Unstable, nucleus confined				
	>40S	5.0		48	350
	4-40S	13.3		38	730
	Total	18.3			1080
	Stable 4-40S	2.2	12	ND[e]	
	Total synthesis	20.5			
6[f]	Unstable, nucleus confined				
	>40S	12.5		31	560
	4-40S	21.5		312	9700
	Total	32.3			10300
	Stable 4-40S	1.7	5	ND[e]	
	Total synthesis	34.0			
	Stable poly(A) RNA[g]	1.1		ND[e]	

[a] Data are for females *not* stimulated by injection of chorionic gonadotropin. Anderson and Smith (1978) showed that in stage 3 oocytes, stimulation results in a 1.5- to 2-fold increase in synthesis rate for all classes of RNA listed, but this hormone has no consistent effect on RNA synthesis in stage 6 oocytes.

[b] Heterogeneous RNAs were estimated as total newly synthesized RNA minus newly synthesized rRNA. Size estimates, and identification of RNA species, were made by gel electrophoresis, and by correction for RNA content after hybridization with ribosomal DNA.

[c] All the newly synthesized *stable* heterogeneous 4-40S RNA enters the cytoplasm (Anderson and Smith, 1977, 1978), and at least at stage 6 (Dolecki and Smith, 1979) most of the stable RNA is polyadenylated, as shown. A maximum of about 15% of this RNA could consist of mitochondrial transcript (Anderson and Smith, 1978; Dolecki and Smith, 1979; Anderson *et al.*, 1982). The remainder consists of mRNA and interspersed RNAs, as discussed in text.

[d] Data from Anderson and Smith (1978).

[e] ND: not determined. This RNA was too stable for its turnover to be measured over the 25-40 hr labeling periods utilized. Therefore the steady state content could not be measured kinetically.

[f] Data from Anderson and Smith (1977).

[g] Data from Dolecki and Smith (1979).

sis. The accumulation kinetics for the stable 4-40S RNA component are displayed in Fig. 5.11(b), compared to those for rRNA. In absolute terms the rates of heterogeneous RNA synthesis shown in Table 5.3 are enormous. For comparison the rate at which total heterogeneous nuclear RNA is synthesized in postgastrula *Xenopus* embryo cells was calculated in Chapter III as about 0.015 pg min^{-1} per (2C) nucleus. It is clear that the lampbrush stage oocyte nucleus is a thousand times more active. The difference lies mainly in the number of polymerases transcribing each functional region in lampbrush chromosomes. As calculated earlier (see Chapter III) the density of nascent heterogeneous nuclear RNA molecules is typically less than one per 10^4 nt in somatic cells, compared to one per 10^2 nt in the active lampbrush transcription units. Transcription units may also be longer in the lampbrush chromosome, as further discussed below, and the 4C oocyte genome contributes an additional factor of two to the comparison. However, the fractions of newly synthesized heterogeneous RNA exported to the cytoplasm from the lampbrush chromosome stage germinal vesicle, here 5-12%, are not very different from those found in embryonic or other somatic cells (Chapter III). It follows that during the lampbrush stage *the rate of flow of stable transcripts into the cytoplasm is also as much as 1000 times greater than in other cells*. The significance of the observation that heterogeneous RNAs are being exported at high rates to the cytoplasm throughout the midlampbrush chromosome stage is that it directly relates lampbrush chromosome transcription to the function of supplying such RNAs to the cytoplasm. Dolecki and Smith (1979) showed, furthermore, that much of this heterogeneous RNA flowing into the cytoplasm is polyadenylated. The 4-40S RNAs are at least moderately stable, since no turnover can be detected in the form of the accumulation curve during the labeling period [e.g., see Fig. 5.11(b)].

Complexity of Germinal Vesicle RNA

There have been no direct experimental estimates of the complexity of the nuclear RNA of *Xenopus* oocytes, though as reviewed in Chapter II the complexity and sequence content of the cytoplasmic RNAs stored in the mature egg are relatively well known. However, the measurements of transcription rate summarized in Table 5.3 can be utilized to calculate the approximate total length of the DNA included in the transcription matrices of the stage 3 oocyte chromosome, and from this the single copy sequence complexity may be estimated. These calculations are shown in Table 5.4 (see note *a*). The result is that around 30% of the *Xenopus* genome appears to be transcribed into newly synthesized nRNA. This fraction is not significantly different from that observed for many somatic cell nuclear RNAs (reviewed in Davidson and Britten, 1973; Chapter III). Table 5.4 also indicates that the stably accumulated cytoplasmic RNA of the *Xenopus* oocyte includes only about 4-6% of the calculated *complexity* of the nuclear RNA. This ratio falls at the lower end of the range commonly observed for somatic cells, about 5% to 20% (reviewed in Davidson and Britten, 1979). The conclusion that the RNA transcribed in lampbrush chromosomes is no more complex than the nuclear RNA of somatic cells is also consistent with several early hybridization studies which showed that only a minor fraction of the diverse genomic repetitive sequences is represented in the newly synthesized RNA of stage 3 *Xenopus* oocytes (see, e.g., Davidson *et al.*, 1966; Crippa *et al.*, 1967; Hough and Davidson, 1972; reviewed in Davidson, 1976). Speculations regarding lampbrush chromosome function that require the whole genome to

TABLE 5.4. Estimates of the Lengths of Genome Included in Lampbrush Chromosome Transcription Units and of Oocyte nRNA and mRNA Complexities

	Xenopus laevis	*Triturus cristatus*
Length of genomic sequence represented in nRNA (ntp)	9.1×10^{8}[a]	1.5×10^{9}[b]
Total genomic length transcribed (%)	30[c]	5[b]
Single copy complexity of germinal vesicle nRNA (ntp)	6.8×10^{8}[d]	$\geq 9.6 \times 10^{8}$[e]
Single copy complexity of mature egg RNA (ntp)	2.7-4.0×10^{7}[f]	2×10^{7}[g]
$\dfrac{\text{Complexity mature egg RNA}}{\text{Complexity germinal vesicle nRNA}} \times 100$	~4–6%	~2%

[a] Calculated from average polymerase translocation rate, S, measured for stage 3 *Xenopus* oocytes by Anderson and Smith (1978), or 15 nt sec^{-1}. The length of sequence transcribed is $L = (R \times 100) \div (4 \times S)$, where R is the total rate of transcription (Table 5.3), 18.3 pg min^{-1} or 5.5×10^{8} nt sec^{-1}; 100 nt is the spacing between polymerases in lampbrush chromosome transcription units (see text); and the factor 4 is required to convert from transcription of a 4C to a haploid genome. Considering uncertainties in these parameters, e.g., the 2-fold difference in the value of S for polymerase II depending on hormonal state of the animal (Anderson and Smith, 1978), and the fact that polymerase packing densities in fact vary slightly from one transcription unit to another (Hill, 1979), with 100 nt probably a minimum value, the listed number may be correct only within a factor of 2 or 3.

[b] Cytological measurements of *Triturus* lampbrush loop length, assuming a compaction ratio of 1.0 for lampbrush chromosome loops (see text), indicate that 4-5% of the genome is included in the loops (Callan, 1963; Macgregor, 1980). As reviewed in text, most of the loop length consists of transcription matrices. The genome size of *Triturus cristatus* is reported as 29 pg or 3×10^{10} ntp (quoted in Macgregor, 1980). The value shown in the Table could be a twofold underestimate, since the compaction ratio for loops could be as high as about 2 (reviewed in Macgregor, 1980).

[c] The haploid genome size of *Xenopus* is taken as 3×10^{9} nt (Dawid, 1965).

[d] The complexity of *Xenopus* germinal vesicle nRNA has not been measured directly. However, Davidson *et al.* (1973) showed that 75% of the genome is single copy sequence, and it is assumed that the sequence organization of the nRNA reflects that of the DNA, as found, e.g., for sea urchin nRNA (Smith *et al.*, 1974).

[e] Based on direct measurements of Sommerville and Malcolm (1976), carried out by reacting a single copy tracer with excess nuclear RNA extracted from purified RNP particles (see Fig. 5.10). The authors reported that about 4% of the tracer hybridized, though the reaction may not have been terminated. From renaturation of sheared *Triturus* DNA about 40% of DNA appears to be single copy (Sommerville and Malcolm, 1976). However, this may be an underestimate as well, due to repetitive sequence interspersion (Davidson *et al.*, 1973). Since transcription is asymmetric, the complexity is therefore $\geq 0.08 \times 0.4 \times 3 \times 10^{10}$ ntp or at least 9.6×10^{8} ntp of single copy sequence.

[f] Davidson and Hough (1971); Rosbash and Ford (1974).

[g] Rosbash *et al.* (1974).

be transcribed in lampbrush chromosomes (see, e.g., Callan and Lloyd, 1960; Callan, 1967; León, 1975; Perlman *et al.*, 1977) are thus not supported. Furthermore, RNAs coding for hemoglobin and vitellogenin, both expressed specifically in terminally differentiated adult tissues, have been shown to be absent from *Xenopus* egg RNA (Schaefer *et al.*, 1982), as are mRNAs for certain though not all of the heat shock proteins that can be expressed in somatic cells (Bienz, 1984). The possibility is not excluded, however, that the rapidly decaying nuclear transcripts synthesized in lampbrush chromosomes include these particular sequences. What is clear is that *a specific, though large, set of sequences is transcribed and that these in turn give rise to a specific array of stable cytoplasmic RNAs.*

Table 5.4 also presents estimates for the length of sequence represented in *Triturus* lampbrush chromosome RNA. The basis in measurement is here almost complementary to that available for *Xenopus*. While there are no synthesis rate measurements for *Triturus* oocytes, direct attempts have been made to estimate the nRNA complexity by the single copy saturation hybridization method (Sommerville and Malcolm, 1976). Furthermore, *Xenopus* lampbrush chromosomes are small, thus rendering quantitative cytological examination at the light microscope level difficult, while extensive cytological measurements have been carried out on *Triturus* lampbrush chromosomes. A visual comparison between *Xenopus* and *Triturus* lampbrush chromosomes, photographed at the same magnification, is shown in Fig. 5.12. There are about 20,000 loops in the whole chromosome set of *Triturus* (Callan, 1963), or about 5000 loops per haploid set. From direct length measurements these are estimated to contain about 5% of the genomic DNA (Gall, 1955; Macgregor, 1980). In Table 5.4 the length of DNA transcribed in *Triturus* oocytes on this basis is also listed, i.e., assuming that on the average each loop consists only of transcription matrix (see note *b*). The value obtained, i.e., 1.5×10^9 nt, is not significantly different from the measured single copy complexity, i.e., $\geq 9.6 \times 10^8$ nt, taking into account that about 40% of the *Triturus* genome is repetitive (see notes to Table 5.4).

It can be seen in Table 5.4 that the estimated lengths of genomic sequence represented in the nRNA of *Xenopus* and *Triturus* oocytes probably differ only by a factor of about 1.7, though the genome size of *Triturus* is 10 times larger than that of *Xenopus*. On the other hand as much as a six times greater *fraction* of the *Xenopus* genome is apparently being transcribed (30%) than of the *Triturus* genome (5%). This conclusion is directly contrary to the earlier assumption that because the lampbrush chromosomes of organisms of greater genome size have larger loops, an equivalent *fraction* of the genomic DNA is always being expressed. From that assumption has risen the famous mystery known as the "C-value paradox." The question posed in statements of the "C-value paradox" is why animals of similar biological organization should display manyfold differences in the length of genomic information transcribed in homologous cells, or in more extreme form, why

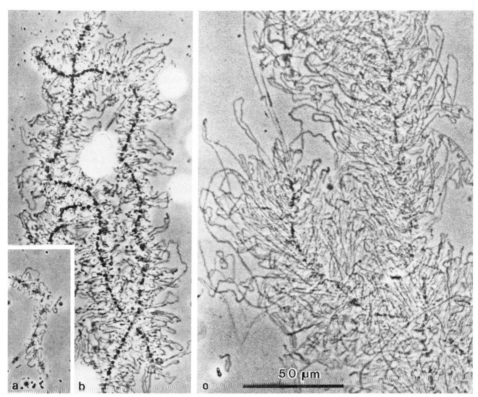

Fig. 5.12. Oocyte lampbrush chromosomes of three amphibian species differing in genome size. The three preparations were photographed through the phase microscope, all at the same magnification. (a) *Xenopus laevis.* The haploid genome size is about 3 pg (Dawid, 1965). (b) *Triturus cristatus.* The genome size is 29 pg (quoted in Macgregor, 1980). (c) *Necturus maculosus.* The genome size is 78 pg (Sexsmith, 1968). [From U. Scheer and J. Sommerville (1982). *Exp. Cell Res.* **139**, 410.]

the homologous expressed "genes" of one organism should on the average include manyfold longer DNA sequences than in a related organism. Table 5.4 indicates that with respect to amphibian oogenesis this paradox in fact does not exist, though the enormous fraction of silent DNA accumulated during evolution in urodele genomes indeed remains a mysterious phenomenon. Thus the sixfold greater *relative* activity of the *Xenopus* genome reduces the ratio of *expressed* genome sizes to 1.65, which might suggest a slightly greater length of sequence utilized in *Triturus* lampbrush chromosomes. This last factor, which as described in note *b* of Table 5.4 could be a little larger, probably indicates that *Triturus* transcription units on the average include about twice as much noncoding sequence that is confined to the nucleus as do *Xenopus* transcripts. However, Sommerville and Scheer

(1982) found no striking difference in the fraction of oocyte nRNAs from these organisms that consists of repetitive sequences.

The significance of the difference in loop lengths among species requires reexamination as well. For one thing, contrary to the premise of the "C-value paradox," Macgregor (1980) showed that *average* loop lengths in fact vary far less than proportionately with genome size. It is indeed the case that in *Xenopus* the loops average only 5-10 μm, and large loops are about 10-15 μm in length (Callan, 1963; Müller, 1974) [though the average length has also been estimated as 12.5 μm (Macgregor, 1980)]. In *Triturus* species the loops average 50 μm and the largest range up to 200 μm (Gall, 1955; Callan, 1963). In urodeles with even larger genomes than *Triturus cristatus*, e.g., *Necturus maculosus* [Fig. 5.12(c)] the average loop length is slightly greater (Macgregor, 1980). However, since average loop length is overall a complex function of sequence organization, i.e., the distance between the relatively long regions of silent DNA that are compacted in chromomeric structures, together with the spacing of genes active in oogenesis, the most reasonable conclusion is simply that transcription units are arranged differently in anuran and urodele genomes. In the latter a particularly large fraction of the genome would appear to consist of very long blocks of silent sequences, but the active regions bunched together in the loops include an amount of single copy genomic sequence that according to Table 5.4 is within a factor of two the same as is expressed in the genome of *Xenopus*.

The sequence complexity of the stable heterogeneous RNAs stored in the cytoplasm of the *Triturus* egg was measured by Rosbash *et al.* (1974) and Sommerville and Malcolm (1976). Table 5.4 shows that the values obtained are very similar to those found for *Xenopus* egg RNA (Davidson and Hough, 1971; Rosbash and Ford, 1974). Thus in *Triturus* only about 2% of the initially transcribed single copy sequence length is ultimately represented in the stored maternal RNA.

(iii) Transcription of Specific Sequences and Classes of Sequence in Lampbrush Chromosomes

Transcription of specific sequences has been visualized by *in situ* hybridization of labeled DNA probes with the nascent matrix RNA of lampbrush chromosome loops. Such hybridization is abolished by prior treatment with RNase, and the signal obtained depends directly on the dense packing of the nascent transcripts. The first application of this procedure to lampbrush chromosomes was an attempt to identify the locus of the active 5S rRNA genes in *Triturus* lampbrush chromosomes (Pukkila, 1975). The probe consisted of labeled 5S DNA that had been purified from genomic DNA by isopycnic centrifugation. As it may have contained some other sequences, the results of this particular investigation remain equivocal (Macgregor, 1980). However, the *in situ* hybridization method, when utilized with cloned probes, has provided useful information in regard to both the locus of particular genes, and of greater importance, the general properties of transcription in lampbrush chromosomes.

Transcription of Repetitive Sequences

A great variety of repetitive sequences is transcribed in amphibian oocyte nuclei, and is also included in the interspersed poly(A) RNAs that by mass form the dominant fraction of the heterogeneous maternal transcripts stored in amphibian egg cytoplasm (Anderson *et al.*, 1982; reviewed in Chapter II). Heterogeneous nuclear RNAs that have an interspersed sequence organization have also been extracted directly from the germinal vesicles of several amphibian species (Sommerville and Scheer, 1982). When renatured and spread for electron microscopy these RNAs form partially duplexed multimolecular structures similar to those observed in studies of renatured oocyte cytoplasmic poly(A) RNA (Anderson *et al.*, 1982). In addition many low molecular weight RNAs are transcribed from repeated genes. Aside from 5S rRNA these include U1 and U2 snRNAs which are synthesized in the *Xenopus* oocyte (Forbes *et al.*, 1984; Fritz *et al.*, 1984), and the 181 nt cytoplasmic species known as OAX (Wakefield *et al.*, 1983; Ackerman, 1983). There are other, yet unidentified tandemly repeated sequences transcribed in lampbrush chromosomes as well, among them the clusters of short, densely packed transcription units visualized in the electron microscope by Scheer (1981).

Molecular measurements carried out on genomic DNA show that the members of most short repetitive sequence families in amphibian genomes are widely interspersed, rather than tandemly repeated (Davidson *et al.*, 1973; Sommerville and Malcolm, 1976; Graham and Schanke, 1980). Since many *different* interspersed repeat families are actively transcribed during oogenesis (Hough and Davidson, 1972; Anderson *et al.*, 1982; Sommerville and Scheer, 1982), *in situ* hybridizations to loop matrix RNA carried out with probes representing such sequences as a class would be expected to reveal reactions at multiple loci. Just this result was obtained by Macgregor and Andrews (1977), whose probe consisted of a low C_0t DNA fraction labeled *in vitro*, and also by Sommerville and Scheer (1982) who utilized an RNA probe prepared from renatured RNase digested nRNA, labeled *in vitro* with [125]I. An unexpected observation in the study of Macgregor and Andrews (1977) was that only certain regions of some loops reacted with the repeat sequence probe. These loops are interpreted to contain more than one transcription unit, and some of these lack sufficient repetitive sequence to react detectably. An example is shown in Fig. 5.13(a). Exactly the same pattern of labeling is present in the homologous loops of chromosomes from different oocytes, from both the same and different animals. This particular observation refutes an early theory of lampbrush chromosome function,

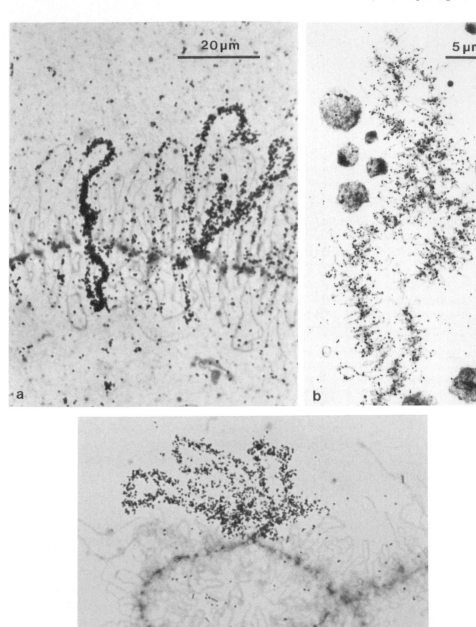

according to which the genomic DNA is constantly being spun out into the loops from the chromomeres, so that at any specific time a different set of sequences is being transcribed in given loops [Gall and Callan, 1962; molecular evidence had in any case long before rendered this idea untenable, as may be noted in the previous editions of this book (Davidson, 1968, 1976)].

As expected a widely distributed pattern of hybridization is also observed with cloned probes representing *single interspersed repeat sequence families*. An example is shown in Fig. 5.13(b), from a study of Jamrich *et al.* (1983) carried out on *Xenopus* lampbrush chromosomes. The probe represents a sequence present in about 10^3 copies per haploid genome. These appear to be so extensively interspersed that they are present in a significant fraction of the lampbrush loops visible. A similar result was reported by Kay *et al.* (1984), who found that about 100 pairs of lampbrush chromosome loops react with a different cloned repetitive sequence probe. In another experiment (Mahon and Gall, 1984) a probe representing satellite 2 of the *Notophthalmus viridescens* genome was hybridized to the lampbrush chromosomes. Tandem sequence blocks of this satellite are scattered throughout the genome. However, very little hybridization was observed with the loop matrix RNAs. Satellite 2 may thus be an example of the large, widely distributed silent DNA regions inferred above to be characteristic of urodele genomes.

Much of the repetitive sequence hybridization observed by Macgregor and Andrews (1977) in *Triturus cristatus carnifex* lampbrush chromosomes occurs in loops of the heteromorphic arms of chromosome I. These arms are invariably distinct in structure in the maternally and paternally derived genomes of any one animal and they bear several recognizable landmark loops (Callan and Lloyd, 1960). According to Giemsa staining and additional cytological criteria they appear to consist mainly of satellite DNA and other repetitive sequences (reviewed in Morgan *et al.*, 1980). A surprising result obtained by Varley *et al.* (1980a,b) is that satellite DNA is actively tran-

---

Fig. 5.13. Repetitive sequence transcripts in oocyte lampbrush chromosomes, visualized by *in situ* hybridization to loop matrix RNA. (a) Region of the longer heteromorphic arm of chromosome 1 of *Triturus cristatis carnifex*, hybridized to a probe consisting of C_0t 0.2-C_0t 50 ^{3}H-DNA labeled *in vitro* with polymerase I. Portions of several pairs of labeled loops can be seen in the autoradiograph. Note, loop that is labeled only part of the way along its length. [From H. C. Macgregor and C. Andrews (1977). *Chromosoma* **63**, 109.] (b) *In situ* hybridization between *Xenopus laevis* lampbrush chromosome loop matrix RNA and a ^{3}H-cRNA probe transcribed from a cloned, short interspersed repetitive sequence present in about 10^3 copies per genome. [From M. Jamrich, R. Warrior, R. Steele, and J. G. Gall (1983). *Proc. Natl. Acad. Sci. U.S.A.* **80**, 3364.] (c) *In situ* hybridization of a cloned satellite DNA probe to transcripts present on a cluster of several loops in the long heteromorphic arm of lampbrush chromosome I of *Triturus c. carnifex*. [From J. M. Varley, H. C. Macgregor, and H. P. Erba (1980a). Reprinted by permission from *Nature (London)* **283**, 686. Copyright © 1980 Macmillan Journals Limited.]

scribed in certain loops located on the heteromorphic arms. This is demonstrated in Fig. 5.13(c). Experiments were carried out with both satellite DNA probes and probes prepared by direct isolation of satellite sequences from genomic DNA. Satellite transcription is unexpected on the basis of the lack of expression of this type of DNA sequence in somatic cells (i.e., where investigated; see for example Flamm *et al.*, 1969); its usual presence in inactive heterochromatic regions of the chromosomes; and the extremely low complexity and tandem repetition of these sequences. Thus it would appear that in lampbrush chromosomes sequences may be transcribed that are not expressed in other cell types, though of course this result in itself provides only a suggestion to this effect, as measurements on transcription of the same satellites in somatic *Triturus* cells have not been carried out. Further exploration of satellite transcription in lampbrush chromosomes has provided a new insight into the functional characteristics of these unique structures.

Readthrough Transcription of Satellite Sequences at the Histone Gene Loci of Newt Lampbrush Chromosomes

A series of investigations on histone gene transcription in oocytes of *Notophthalmus viridescens* has provided direct evidence that initiation from histone gene promoters can result in the transcription of downstream satellite DNA sequences (Diaz *et al.*, 1981; Gall *et al.*, 1981; Stephenson *et al.*, 1981; Diaz and Gall, 1985). There are about 600-800 copies of each of the histone genes per haploid genome, organized primarily in 9 kb clusters that contain all five of the genes plus intragenic "spacer" DNA in the order H$\overline{1}$, H$\overline{3}$, H$\overline{2}$b, H$\overline{2}$a, H$\overline{4}$ (Stephenson *et al.*, 1981). The 9 kb gene clusters are separated by long stretches of satellite 1 DNA, some of which extend for more than 50 kb. This satellite consists of tandem repeats of a 222 nt sequence. Gall *et al.* (1981) showed by *in situ* hybridization to the DNA of both somatic mitotic chromosomes and oocyte lampbrush chromosomes that the histone genes are located predominantly in two major clusters, on chromosomes 2 and 6. Additional minor sites may exist as well. The number of histone genes at each major site is approximately equal, i.e., each includes 300-400 of the 9 kb clusters. *In situ* hybridization of the histone probes to lampbrush chromosome matrix RNA reveals clusters of labeled loops at the corresponding regions of these chromosomes. An example of histone probe hybridization at the chromosome 2 locus is reproduced in Fig. 5.14(a). From the size and number of the labeled loops it may be surmised that in this oocyte a significant fraction of the histone genes present on chromosome 2 are being transcribed. This may not be the case at the other histone gene locus or in different animals, since the DNA/DNA hybridizations revealed some chromomeric labeling. A significant correlation is observed between the chromosomal locations of the histone gene clusters and the presence of landmark bodies, known as "spheres," which are attached to the chromosomal axis (Callan and Lloyd, 1960, 1975; Callan, 1982). The spheres are composed of an acid protein, conceivably accumulated for the special purpose of transporting or processing histone transcripts. The sphere loci are the sites of histone probe hybridization in *Triturus cristatus* and *Triturus alpestris* as well as in *Notophthalmus,* though in each case they are found on different chromosomes (Gall *et al.,* 1981). Sphere loci also exist in *Xenopus* lampbrush chromosomes (Müller, 1974), but in this organism the location of the histone genes is not yet known.

Lampbrush chromosome loops at the *Notophthalmus* sphere loci react with cloned satellite 1 probes as well as with histone probes. This is shown, for example, in the experiment reproduced in Fig. 5.14(b). The most important observations have been obtained with asymmetric probes transcribed from M13 templates (Diaz *et al.,* 1981). These experiments demonstrate that within certain loops there are multiple transcription units, each many kilobases in length, since only sharply defined regions of these loops are labeled by the probe. Thus transcription units of opposite polarity can be seen to abut directly within the same loop. Some striking examples are shown in Fig. 5.14(c) and (d). Both strands of the satellite sequence are represented in the loop RNAs in different transcription units (Diaz *et al.,* 1981; Diaz and Gall, 1985). These findings are rationalized as shown in Fig. 5.14(e). Transcription initiates at a histone promoter within the gene cluster, and continues in the downstream direction across the remaining histone genes within the cluster and on into the flanking satellite DNA. The large size and the morphology of the transcription matrices implies that transcription fails to stop until another gene cluster is reached, or until a transcription unit initiated in the opposite direction from an adjoining cluster is encountered. Thus, in *Notophthalmus* lampbrush chromosomes, termination fails to occur at the end of the histone gene sequences. Unfortunately, nothing is known in regard to histone gene transcription in somatic cells of this newt, and thus it is not yet demonstrated that inefficient termination is a pecularity of the transcription of these genes in lampbrush chromosomes. Mature histone messages accumulate in amphibian oocyte cytoplasm during oogenesis (reviewed below) and must be derived from such readthrough transcripts. Processing of these transcripts thus must occur, probably by a strand scission at the 3′ terminus of the initial histone mRNA in each transcript, since the satellite 1 sequences are not found outside the nucleus (Diaz and Gall, 1985). The satellite transcripts together with the readthrough histone gene transcripts are presumably degraded within the germinal vesicle.

It remains to be determined whether readthrough transcription is a special feature of the histone genes or a general explanation for the enormous size of some urodele transcription units. Since there occur in lampbrush chromosome loops well defined transcription units separated by short regions of silent DNA, e.g., those shown in Fig. 5.9(a) (see also Scheer, 1981), there

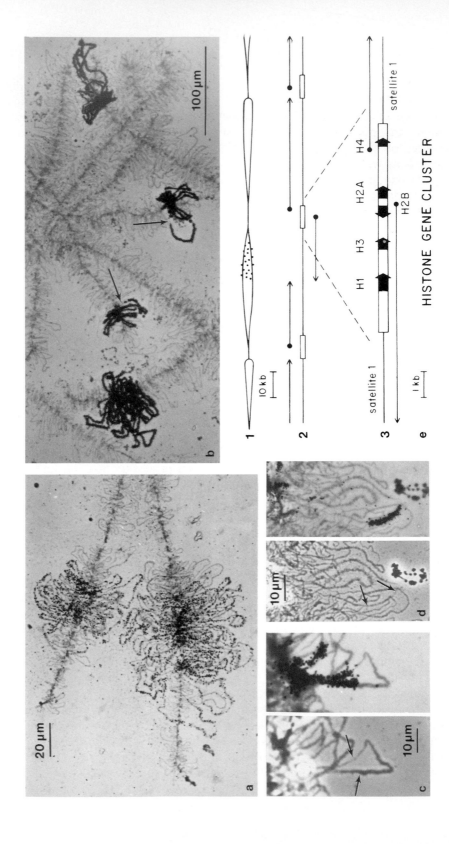

HISTONE GENE CLUSTER

probably exist some means of termination within at least some loops other than encounter with an adjacent transcription unit. Nonetheless, it is possible that readthrough is a common feature of transcription in lampbrush chromosomes, perhaps an indirect consequence of the generally high rate of initiation, i.e., relative to the rate at which termination complexes could form. Readthrough transcription initiated at a normal gene promoter could account for the satellite sequences detected in the heteromorphic arms of the *Triturus c. carnifex* lampbrush chromosomes by Varley *et al.* (1980a,b) [see Fig. 5.13(c)]. This region of the genome must contain some functional genes, since homozygosity for the heteromorphic arms is lethal in this newt species (Macgregor and Horner, 1980). Ribosomal RNA sequences are also transcribed in chromosome I, though this is not the site of the true nucleolus organizers, where the vast majority of ribosomal genes are located, and which appear silent in lampbrush chromosomes (Morgan *et al.*, 1980). The transcription of the ectopic rRNA sequences in lampbrush chromosomes is mediated by polymerase II rather than polymerase I, and therefore it is likely

---

Fig. 5.14. Readthrough transcription of histone genes in the lampbrush chromosomes of *Notophthalmus viridescens*. (a) Homologous sphere loci of lampbrush chromosome 2, after hybridization with a ^{3}H-cRNA probe containing H2a, H2b, and H3 gene sequences. [From J. G. Gall, E. C. Stephenson, H. P. Erba, M. O. Diaz, and G. Barsacchi-Pilone (1981). *Chromosoma* **84**, 159.] (b) Four sphere loci labeled by hybridization with an asymmetric ^{3}H-cRNA probe transcribed from an M13 clone of satellite 1. The arrows indicate loops in which unlabeled as well as labeled matrix appears in the same loop. (c) and (d) Detailed views of several loops at the chromosome 2 sphere loci, after hybridization with asymmetric ^{3}H-cRNA probes transcribed from M13 clones of satellite 1 DNA. Left, phase photomicrographs; right, autoradiographs of same loops. (c) An unlabeled transcription unit separates two labeled transcription units of opposite polarity. The *left arrow* indicates a junction in which the terminations of the contiguous transcription units abut, and the *right arrow* a junction in which two initiation sites abut. (d) Same as (c) except that the probe represents the opposite strand of satellite 1, and an oocyte from a different animal was utilized. A single labeled transcription unit is observed, flanked on either side by unlabeled transcription units. The *left arrow* indicates a terminus-to-terminus junction, and the *right arrow* an initiation-to-initiation junction. (e) Diagrammatic interpretation of readthrough transcription patterns in the *Notophthalmus* histone genes. Transcription initiates normally at individual histone gene promoters and continues through the adjacent satellite sequences until another transcription complex proceeding in the opposite direction, or another newly transcribed histone gene cluster, is encountered. (1) Autoradiographic signal obtained with a probe that represents one strand of the satellite 1 sequence. (2) Interpretation showing individual transcription units beneath the corresponding cytological features indicated in (1). The interspersion of the histone gene clusters (boxes) with satellite 1 (lines) is shown at large scale. (3) Detailed view of one histone gene cluster, containing all five genes oriented ($5' \rightarrow 3'$) as indicated. The labeled region shown in (1) contains readthrough transcripts initiated at the leftward oriented H2b promoter, and extends to the terminus of this transcription unit, which abuts the terminus of a rightward matrix originating in the next gene cluster. In (3) a second transcription unit oriented rightward can be seen originating in the same histone gene cluster at the H4 promoter. The H1 and H3, and H2a promoters would be utilized in other gene clusters. [(b)-(e) From M. O. Diaz, G. Barsacchi-Pilone, K. A. Mahon, and J. G. Gall (1981). *Cell* **24**, 649. Copyright by M.I.T.]

that it too is initiated at a functionally unrelated upstream promoter. It is reasonable to suppose in the absence of additional knowledge that transcription of any sequence that is found by *in situ* hybridization to be represented in lampbrush chromosome RNA could have been initiated at the normally regulated promoter of another nearby gene. Note that there is no evidence in the best studied example, the histone genes, that *transcriptional initiation* occurs at abnormal locations in lampbrush chromosomes; that is, as indicated in Fig. 5.14(e) the promoters at which initiation occurs are those of the various histone genes.

(iv) Processing of mRNA Precursors in the *Xenopus* Oocyte Nucleus

Several of the observations thus far reviewed reflect on the capacity of the amphibian oocyte nucleus to process and selectively transport mRNA precursors. For example there is the presence in the mature egg of a large quantity of nontranslatable poly(A) RNA that differs from most mature mRNA in its content of interspersed repetitive sequences (Chapter II). Further evidence reviewed below shows that in *Xenopus* oocytes RNA of this nature is actively exported to the cytoplasm during the lampbrush stage. It is clear, on the other hand, that a stringent selection of those transcripts destined for the cytoplasm does take place, since as shown in Table 5.4, the complexity of germinal vesicle RNA is 15-25-fold greater than that of cytoplasmic RNA. These observations completely exclude the possibility that the oocyte germinal vesicle randomly "leaks" heterogeneous nuclear RNA into the cytoplasm. The measurements summarized in Table 5.3 indicate in addition that the flow of newly synthesized heterogeneous RNA into the cytoplasm includes only 5% (stage 6 oocytes) to 12% (stage 3 oocytes) of the nucleotides initially polymerized into heterogeneous nuclear RNA.

Processing has been demonstrated in the *Xenopus* oocyte nucleus in several studies on the molecular fate of specific precursors, either synthesized in the germinal vesicle from injected plasmids, or injected directly. One such experiment that may be relevant to the possible disposition of the readthrough histone transcripts was carried out by Krieg and Melton (1984). A histone precursor RNA was synthesized *in vitro* from a plasmid containing a chicken H2b histone gene, and introduced into the oocyte nucleus. Generation of the native histone message from this precursor requires a correct 3' endonucleolytic processing reaction, and this was observed. Furthermore, the appearance of a correctly terminated histone mRNA was not affected by the presence of several hundred nucleotides of additional vector sequence distal to the proper termination site. Birnstiel and associates also have demonstrated the production of correctly terminated histone mRNAs after injection of cloned sea urchin histone genes into *Xenopus* oocyte nuclei. Both a highly conserved 3' terminal sequence element in the histone gene (Birchmeier *et al.*, 1982), and a small ribonucleoprotein (Galli *et al.*, 1983) appear necessary for the formation of normally terminated messages. These experiments indicate the existence of the mechanisms that would be required for production of mature histone messages from the 5' terminal gene sequences of the readthrough histone transcripts synthesized in the lampbrush chromosomes.

Direct evidence that *Xenopus* oocytes also possess the capacity to carry out RNA splicing reactions has been obtained by injecting cloned genes that include introns. In several such experiments synthesis of the protein coded by the exogenous sequences is reported to occur. Examples include the ovalbumin gene (Wickens *et al.*, 1980), in which correct processing involves the precise removal of no less than seven introns, and the SV40 T antigen, requiring precise excision of one intron (Rungger and Türler, 1978). Accurate splicing of a yeast tRNA^{tyr} precursor in the *Xenopus* oocyte nucleus has also been demonstrated (Melton *et al.*, 1980; Nishikura and De Robertis, 1981). Processing of the yeast tRNA^{tyr} precursor involves removal of a 5' leader sequence and also of extra nucleotides at the 3' end and a sequential series of base modifications, as well as the splicing reactions. All of the enzymatic machinery required for these activities is confined to the nucleus (De Robertis *et al.*, 1981).

The observations of Anderson and Smith (1977, 1978) that in late oocytes there seems to be an increased retention of nuclear RNAs and a lower intranuclear turnover rate (see Table 5.3) suggest that the efficiency of the processing reactions might be relatively low. By "efficiency" is meant the fraction of precursor molecules that are converted into mature messages. A kinetic study of the processing of SV40 transcripts in the stage 6 *Xenopus* oocyte nucleus suggests that the efficiency with which processing is carried out is indeed poor. Miller *et al.* (1982) injected SV40 DNA into the oocyte nuclei, and found that 90% of the primary transcript is degraded within the nucleus. The half-life of these SV40 primary transcripts is only 20-40 min, just as for the endogenous nRNA. However, a small fraction of the transcription products appears to be converted into correctly processed 19S poly(A) mRNA that accumulates in the cytoplasm. An even smaller fraction of the precursor is converted into 16S mRNA (Fradin *et al.*, 1984). It seems unlikely that these results are artifactual consequences of overloading the processing capacity of the oocyte with SV40 transcripts, since the amount of SV40 transcript synthesized in the study of Miller *et al.* (1982) was less than the amount of endogenous nRNA transcript, and the total amount exported to the cytoplasm was only about 10-20% of the endogenous newly synthesized poly(A) RNA flow into the cytoplasm. A further investigation of the processing of late SV40 transcripts by Wickens and Gurdon (1983) demonstrates that the requirements for transport to the cytoplasm are endonucleolytic cleavage at the site of the 3' terminus of the mature message, and polyadenylation. Both spliced and unspliced derivatives with these terminal features are found in the cytoplasm, while nonpolyadenylated transcripts that include sequences distal to the mature 3' terminus are retained in the nucleus. These results suggest that the amount of processing may be a function of the *relative rates* at which transcription, degradation, preparation of the 3′ terminus, and the splicing reactions occur. Where the rates of the latter are relatively insufficient, unprocessed transcripts will either be degraded, or if they have acquired mature 3′ termini, exported to the cytoplasm where further processing reactions cannot take place. An investigation of Green *et al.* (1983) in which human β-globin precursor RNA was injected into oocyte nuclei shows that molecules lacking the 5′ cap structure are rapidly degraded. The β-globin precursor molecules were synthesized enzymatically *in vitro*, and only if capped prior to injection were they found to be stable (similar results are reported for chicken lysozyme mRNA; Drummond *et al.*, 1985). Accurate splicing out of both β-globin precursor introns occurs in the oocyte germinal vesicle, and for this a terminal poly(A) sequence is not required. However, these splicing reactions are also inefficient, in that the majority of the globin precursor molecules are never processed (Green *et al.*, 1983). A sequence-specific form of selective nRNA processing has been demonstrated in the oocyte nucleus as well. Thus Bozzoni *et al.* (1984) showed that two of the nine introns included in primary transcripts produced from an injected ribosomal protein gene are not spliced out, while all nine introns from a different ribosomal protein gene transcript are removed. Most of the incompletely processed precursor molecules are confined to the germinal vesicle, but again some are recovered from the cytoplasmic compartment.

A role for snRNPs in at least some splicing reactions carried out in *Xenopus* oocytes has been demonstrated by Fradin *et al.* (1984). These experiments showed that injection of various antibodies against U1 RNP sharply inhibit splicing of the SV40 late transcript, resulting in an accumulation of unspliced precursor. Many of these antibodies had no effect on the splicing of the early transcript, however, suggesting that a variety of other snRNP cofactors could be required for the processing of different precursor molecules. It may be relevant that the *Xenopus* oocyte germinal vesicle *lacks* several snRNA species that are synthesized in embryos after the blastula stage. Among these are two particular U1 snRNA species found in somatic cells, though other U1 snRNAs are synthesized during oogenesis (Forbes *et al.*, 1984; Gelfand and Smith, 1986). The mature *Xenopus* egg contains about 8×10^8 molecules of the oocyte types of U1 snRNA, sufficient to provision the several thousand nuclei that have been formed by the time transcription resumes at the blastula stage (Forbes *et al.*, 1984; cf. Chapter III). It remains to be determined whether deficiencies in specific snRNAs or other characteristics of the germinal vesicle snRNA pool are in any respect responsible for the inefficiency of nRNA processing in the oocyte.

In summary, it is clear that the oocyte nucleus can carry out all known RNA processing reactions; that it does so inefficiently, however; and that if they have been polyadenylated, transcripts that are not mature messages may be exported to the cytoplasm. These conclusions provide a reasonable interpretation for the presence of the interspersed cytoplasmic poly(A) RNAs. However, this proposition must remain inferential, since the results have all been obtained with exogenously introduced sequences, many of which are foreign to the *Xenopus* oocyte. The posttranscriptional disposition of an endogenous primary transcript of the lampbrush chromosomes, read-through or otherwise, has yet to be described in similar detail.

(v) Messenger RNAs and Interspersed Heterogeneous RNAs of the Oocyte Cytoplasm: The Logistic Role of Lampbrush Chromosomes

Accumulation and Turnover of Cytoplasmic Poly(A) RNA

We have so far regarded the newly synthesized oocyte transcripts mainly from an intranuclear perspective. However, the key issue, in considering the ultimate function of lampbrush chromosomes, is the relation between the chromosomal transcription products and the cytoplasmic RNAs accumulated in the course of oogenesis. This relation is by no means an obvious one. In *Xenopus*, the only organism for which extensive measurements exist, the quantity of total poly(A) RNA present at the end of oogenesis is about 80–90 ng (Dolecki and Smith, 1979; Sagata *et al.*, 1980) and the same amount is to be found in midvitellogenesis stage 3 oocytes. Relative measurements carried out at a series of different stages of oogenesis by Rosbash and Ford (1974) and Golden *et al.* (1980) show that in fact the final quantity of oocyte poly(A) RNA has already accumulated before the end of Dumont stage 2. Data from the study of Golden *et al.* (1980) are reproduced in Fig. 5.15, where it can be seen that the net content of oocyte poly(A) RNA achieves a plateau value just before or at the beginning of vitellogenesis (see legend). Golden *et al.* (1980) also reported that the patterns of accumulation of several cloned maternal sequences follow that of the total poly(A) RNA, all achieving their final levels, as judged from RNA gel blots, by the beginning of vitellogenesis. The only exceptions observed were mitochondrial poly(A) RNAs, which continue to accumulate throughout oogenesis. The mature *Xenopus* oocyte contains about 10^7 mitochondria, and about 4 ng of mitochondrial DNA, or ~12 mitochondrial genomes per mitochondrion (Webb *et al.*, 1975; Marinos, 1985). However, mitochondrial transcripts account for only about 15% of the newly synthesized cytoplasmic poly(A) RNA of the stage 6 oocyte (Webb *et al.*, 1975; Dolecki and Smith, 1979; Anderson *et al.*, 1982). The early appearance in oogenesis of the final content of cytoplasmic poly(A) RNA seemed particularly paradoxical before the demonstration by Hill and Macgregor (1980) that even in growing stage 1 oocytes only 60 μm in diameter the chromosomes contain densely packed transcription units similar to those found in Dumont stage 3 oocytes.

A certain amount of confusion has been caused by the fact that continuous transcriptional activity takes place in the lampbrush chromosome in stages

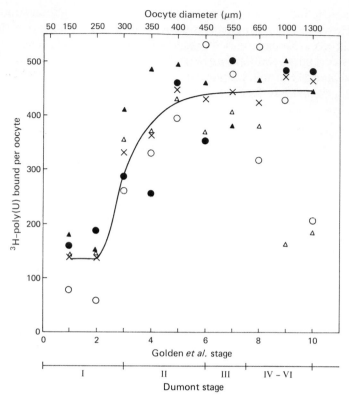

Fig. 5.15. Poly(A) content of *Xenopus* oocytes, estimated by ³H-poly(U) hybridization. Oocytes were individually staged by size and color. Top abscissa shows oocyte diameter (μm). Each symbol indicates oocytes from a different female. The numbers above the lower abscissa show stages utilized by Golden *et al.* (1980) as follows: 1, 50-150 μm previtellogenic clear oocytes; 2, 150-250 μm growing previtellogenic oocytes; 3, 250-300 μm translucent white oocytes; 4, 300-350 μm early vitellogenic oocytes; 5, 350-400 μm vitellogenic oocytes, now opaque; 6, 400-450 μm vitellogenic oocytes; 7, 450-550 μm oocytes, in which pigment disposition is beginning; 8, 550-650 μm dark brown oocytes; 9, 650-1000 μm oocytes, animal hemisphere dark, vegetal hemisphere light; 10, 1200-1300 μm banded mature oocytes. The standard Dumont (1972) stages are indicated below the abscissa. These are matched to the stages of Golden *et al.* (1980) by oocyte diameter. RNA was extracted from each pool of oocytes and an amount equivalent to one oocyte of that stage hybridized with ³H-poly(U). The line shown represents the average values (X) calculated from each set of four measurements reported by Golden *et al.* (1980) at each stage (other symbols), except for stages 9 and 10. The three very low data points reported for these stages have been discarded in calculation of these averages as experimental aberrations, since Dolecki and Smith (1979) proved by direct measurement that Dumont stages 3 and 6 oocytes contain equal amounts of poly(A) RNA. [From L. Golden, U. Schafer, and M. Rosbash (1980). *Cell* **22**, 835. Copyright by M.I.T.]

2-6 oocytes, while the net mass of the cytoplasmic poly(A) RNA remains constant. Several discussions appear in the literature in which it is concluded that because there is no further *accumulation* of cytoplasmic poly(A) RNA, newly synthesized poly(A) RNA must not be exported to the cytoplasm during most of the lampbrush chromosome stage. This is obviously a faulty deduction, however, in that it is directly contradicted by the measurements of Anderson and Smith (1977, 1978) and Dolecki and Smith (1979). As reviewed in Table 5.3 these demonstrate a continuous flow of newly synthesized heterogeneous RNA into the cytoplasm in Dumont stages 3 and 6 oocytes, most of which is polyadenylated. The poly(A) RNA accumulation function shown in Fig. 5.15 means simply that the poly(A) RNAs emerging from the germinal vesicle accumulate until a steady state content is reached, and thereafter this is maintained by the balance between new synthesis and turnover. The lampbrush chromosomes provide the synthetic activity that throughout oogenesis drives the kinetic process. The following calculations show that the maximum rate of transcription in virtually all the transcription units represented in maternal RNA is required in order for this function to be carried out.

According to the evidence reviewed in Chapter II there are on the average about 4×10^6 copies of each maternal mRNA sequence in the *Xenopus* egg, representing $1-2 \times 10^4$ different transcription units. The polymerase II translocation rate in *Xenopus* oocyte nuclei is 10-20 nt sec^{-1} (Anderson and Smith, 1978). Given that the transcribing polymerases are packed only 100 nt apart (i.e., as in lampbrush chromosomes), to synthesize 4×10^6 molecules of an average transcript on four single copy genes per nucleus would require about 80 days [($4 \times 10^6 \times 7$ sec/transcript) ÷ (4 genes/chromosome set × 8.64×10^4 sec/day)]. The *average* duration of the growth phase of stage 1 and stage 2 when net poly(A) RNA is accumulating has been estimated to be as long as 8-9 months (Callen *et al.*, 1980). However, 2-3 months is a reasonable minimal estimate for young frogs. Thus, vitellogenic oocytes appear four months after metamorphosis is complete (Kalt and Gall, 1974), while at three months after metamorphosis no oocytes have yet advanced into stage 2 (Van Dongen *et al.*, 1981; see also Scheer, 1973; Keem *et al.*, 1979; Davidson, 1976, p. 343). A complementary calculation based on the complexity of the maternal RNA was presented by Anderson *et al.* (1982). In stage 3 and stage 6 *Xenopus* oocytes about 1.7-2.2 pg min^{-1} of stable newly synthesized 4-40S RNA, most of which is polyadenylated, enters the cytoplasm (Table 5.3). Of this perhaps 10% consists of repetitive sequence transcript and ≤15% of mitochondrial transcripts (Anderson *et al.*, 1982). Thus a minimum value of 1 pg min^{-1} can be taken as the flow into the cytoplasm of newly synthesized transcripts representing single copy genomic sequence. The complexity of this RNA is about 4×10^7 ntp (Davidson and Hough, 1971; Rosbash and Ford, 1974), and we assume for the limit calculation that *all* of the maternal species are continuously being synthesized. Thus the number of copies of *each sequence* exiting from the oocyte nucleus is about 45 per minute (1 pg min^{-1}, or 1.8×10^9 nt min$^{-1} \div 4 \times 10^7$ ntp), or 11 transcripts per minute attributable to each of the four active genes per nucleus. However, this rate requires one initiation per 5.5 sec, which is almost the same as calculated independently from the measured polymerase translocation rate, if one assumes only 100 nt between transcribing polymerases (i.e., 100/15 sec = 7 sec/initiation). Therefore, virtually all of the transcription units active in synthesizing cytoplasmic poly(A) RNA must be maximally packed with transcribing polymerases. Since the mean size of *Xenopus* oocyte poly(A) RNA is about 2.2 kb (Rosbash and Ford, 1974; Anderson *et al.*, 1982), the number of such transcription units must be close to the total number included in the lampbrush chromosome loops, i.e., almost 2×10^4 ($4 \times 10^7 \div 2.2 \times 10^3$). It follows that to account for the measured flow of stable heterogeneous RNA into the oocyte cytoplasm *the large majority of the transcription units active in the oocyte nucleus must be densely packed with polymerases*. We may conclude that the transcriptional structure of lampbrush chromosomes is in fact *required* to explain both the rate at which the poly(A) RNA of the oocyte initially accumulates (stages 1-2) and the rate at which it is synthesized and exported at steady state (stages 2-6).

Given a steady state pool as large as 80 ng of poly(A) per oocyte, and a flow rate of 1 pg min^{-1} (1.44 ng day^{-1}), the half-life of the newly synthesized cytoplasmic poly(A) RNA would be about 35 days (ln 2 $\times$ 80 ng $\div$ 1.44 ng day^{-1}). Of course a fraction of the poly(A) RNA may be turning over at a higher rate. Since no detectable turnover was observed in the kinetic experiments of Anderson and Smith (1978) and Dolecki and Smith (1979) [e.g., see Fig. 5.11(b)] the half-life of such a fraction would still have to be greater than several days. In any case it is clear that many if not all of the newly synthesized cytoplasmic transcripts ultimately turn over at a rate which these calculations predict would be too low to permit detection in labeling experiments, and yet would still be sufficient to provide a kinetic steady state.

Transport of Interspersed RNAs to the Cytoplasm at the Midlampbrush Stage

The interspersed cytoplasmic RNAs considered in Chapter II were those that are stored in mature oocytes, where they constitute about 70% of the poly(A) RNA mass. This class of RNA can be experimentally defined by its ability to renature, forming multimolecular networks held together by repetitive sequence duplexes. Anderson *et al.* (1982) showed by renaturing poly(A) RNA extracted from stage 3 oocytes that interspersed RNAs have already been deposited in the cytoplasm by the midlampbrush stage. Electron micrographs of the partially duplexed multimolecular structures recovered are shown in Fig. 5.16(a)-(c). These structures included about 50% of the total poly(A) RNA mass examined. In Fig. 5.16(d)-(f) is illustrated an

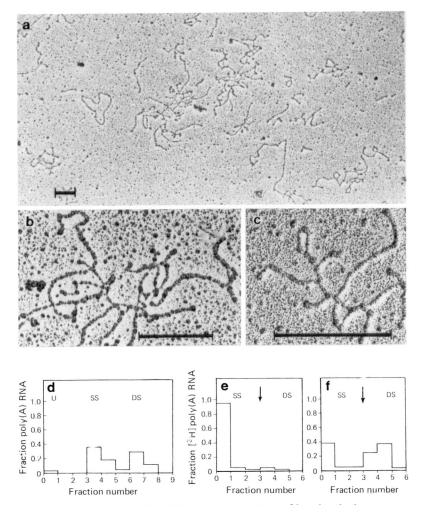

Fig. 5.16. Interspersed poly(A) RNAs in the cytoplasm of lampbrush chromosome stage *Xenopus* oocytes. (a)-(c) Electron micrographs of poly(A) RNA extracted from stage 3 oocytes, after renaturation to an RNA C_0t of 600 M sec. Size markers indicate a single-stranded RNA length of 1000 nt. Controls show that the presence of multimolecular structures is completely dependent on renaturation. (d)-(f) Ethanol-cellulose chromatography experiments demonstrating that most newly synthesized *cytoplasmic* poly(A) RNA in early stage 6 oocytes is *interspersed* poly(A) RNA. Oocytes were labeled by microinjection of ³H-GTP, and incubated *in vitro* for 48 hr. They were then manually enucleated and the cytoplasmic ³H-poly(A) RNA extracted (tracer). This was mixed with a 100-fold mass excess of cytoplasmic poly(A) RNA from stage 6 oocytes (driver). (d) On renaturation to C_0t 600 about 50% of the driver RNA elutes from a cellulose-ethanol column in the single-stranded fraction, and 50% in the double-stranded fraction. Almost no material is seen in the unbound fraction (U), where nucleotides elute. Since by electron microscopy 70% of the renatured RNA preparation consists of molecules that include duplex structures the binding efficiency of the column is about 0.7 (average 0.67). (e) Elution of the ³H-poly(A) RNA in the tracer-driver mixture, immediately after heat denatura-

experiment which demonstrates that newly synthesized interspersed poly(A) RNA is being exported to the cytoplasm in early stage 6 oocytes. The oocytes were labeled by microinjection of precursor, and after 2 days the newly synthesized poly(A) RNA was extracted from the cytoplasm. About 90% of the labeled molecules fractionated as interspersed RNA. Thus in *quantitative* terms, the major product of lampbrush chromosome transcription that is exported to the cytoplasm at least at this late stage of oogenesis is *interspersed poly(A) RNA*. These experiments confirm the view that newly synthesized lampbrush chromosome transcripts continue to enter the cytoplasm even in late oogenesis. The function and the ultimate disposition of this large class of heterogeneous maternal RNAs are still to be determined.

The Quantity of Functional mRNA in Xenopus Oocytes

During oogenesis the rate of protein synthesis per oocyte increases over 100-fold from about 0.18 ng hr^{-1} to about 22 ng hr^{-1}, as illustrated in Fig. 5.17 (data from Taylor and Smith, 1985; Richter and Smith, 1983). The quantity of ribosomal RNA per oocyte increases to about the same extent. Thus the *fraction* of ribosomes engaged in protein synthesis remains approximately constant. L. D. Smith *et al.* (1984) report this as 1.6% at stage 1 and 2.4% and 2.3% at stages 3 and 6, respectively, in good agreement with the earlier estimation of about 2% made by Woodland (1974). Since the mass of total poly(A) RNA remains constant after stage 2, an increasing fraction of the poly(A) RNA pool must be involved in the translational apparatus as oogenesis proceeds. In Table 5.5 the amount of mRNA engaged on the oocyte polysomes at stages 3 and 6 is compared to the total amount of translatable message estimated to be present in the oocyte, calculated from the total poly(A) RNA content, after deducting mitochondrial poly(A) RNA and interspersed poly(A) RNA. The latter is known to be largely nontranslatable (Richter *et al.*, 1984). Thus the mass of true maternal message in the stage 6 oocyte is estimated to be about 21 ng. To account for the quantity of polysomes present in the embryo up to the blastula stage, when mRNA begins to be synthesized anew (see Chapter III), would require about the same quantity of *bona fide* message as by these calculations is apparently present in the stage 6 oocyte, i.e., assuming that the mRNA is stable after fertilization.

---

tion. All of the ^{3}H-RNA elutes in the single stranded fraction. (f) On incubation of the mixture to driver RNA C$_0$t 154 about 60% of the ^{3}H-poly(A) RNA now elutes in the double-stranded fraction. Controls show that the self-reaction of the tracer (i.e., to C$_0$t 1.54) results in insignificant duplex formation. Thus, considering the column efficiency, almost 90% of the tracer must have reacted with repetitive sequences in the driver RNA. Therefore, the tracer consists mainly of newly synthesized interspersed poly(A) RNA molecules that were exported to the cytoplasm within the labeling period. [From D. M. Anderson, J. D. Richter, M. E. Chamberlin, D. H. Price, R. J. Britten, L. D. Smith, and E. H. Davidson (1982). *J. Mol. Biol.* **155**, 281.]

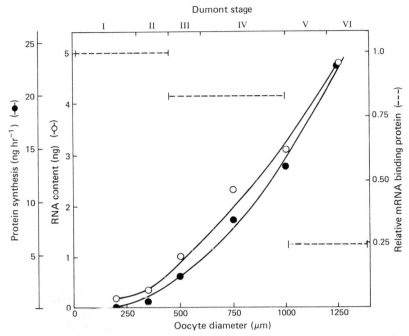

Fig. 5.17. Protein synthesis and mRNA binding proteins in *Xenopus* oocytes. Rates of protein synthesis were measured on isolated oocytes of the desired stages, injected or incubated with radioactive amino acids. The precursor pool size and specific activities were determined, and the protein synthesis rate calculated from measurements of acid precipitable radioactivity. Of the total RNA measured in oocytes of these stages, 25% is rRNA at Dumont stage 1 (Mairy and Denis, 1971); 85% is rRNA at stage 3; and 90% is rRNA at stage 6 (Scheer, 1973). Data for protein synthesis rate and total RNA content are from Taylor and Smith (1985). The relative amounts of mRNA binding proteins per oocyte were calculated from measurements of the quantity of ^{125}I-globin mRNA bound by extracts from oocytes of given stages. As described in text, a discrete set of proteins of known size is associated with translationally inactive mRNA. Data for the proteins included in classes A and B of Richter and Smith (1983) have been pooled for this Figure.

Thus during maturation the polysome content increases about 2-fold, and a further gradual increase in the fraction of polysomes occurs as additional maternal mRNAs are recruited in early embryogenesis (see Chapter II). By blastula stage about 15% of the ribosomes are engaged in polysomes (Woodland, 1974). The mass of polysomes at blastula stage is therefore about six times that in the stage 6 oocyte (15%/2.3%), while the mass of putative mRNA in the oocyte is about 5-fold in excess of that being utilized (21.4 ng/4 ng). The conclusion is that the unused *bona fide* mRNA in the stage 6 oocyte just constitutes a sufficient supply of maternal mRNA for the embryo, assuming no significant contribution from the interspersed poly(A) RNA class.

Table 5.5 also demonstrates an impressive difference between stage 3 and stage 6 oocytes. There is at least as much putative mRNA in stage 3 as in

TABLE 5.5.   Poly(A) RNA Classes in *Xenopus* Oocytes

	Stage 3 (ng)	Stage 6 (ng)
Poly(A) RNA[a]	90	90
Nontranslatable interspersed poly(A) RNA[b]	41.3	55.1
Mitochondrial poly(A) RNA[c]	9.0	13.5
Putative mRNA[d]	39.7	21.4
Polysomal mRNA[e]	0.82	4.0
(% of putative mRNA)	(2.1%)	(19.0%)

[a] Dolecki and Smith (1979); Sagata *et al.* (1980).

[b] Anderson *et al.* (1982) reported 68% of poly(A) RNA to be included in the interspersed fraction at stage 6, and 51% at stage 3. Ninety percent of the content of interspersed poly(A) RNA is deducted from the total poly(A) RNA mass, since the experiments of Richter *et al.* (1984) show this class of transcript to be 90% nontranslatable.

[c] About 15% of stage 6 oocyte poly(A) RNA is mitochondrial (Webb *et al.*, 1975; Dolecki and Smith, 1979; Anderson *et al.*, 1982), and analysis of cDNA clone libraries suggests that of the total poly(A) RNA in the early gastrula stage embryo, the bulk of which is maternal, ~10% consists of mitochondrial transcripts (Dworkin *et al.*, 1981). Direct measurements of mitochondrial RNA are not available for stage 3 oocytes. However, Webb and Smith (1977) showed that the amount of mitochondrial DNA in stage 3 is 67% of that in stage 6. Assuming that major mitochondrial poly(A) RNA transcripts increase proportionately in concentration as the oocyte grows, at stage 3 there would be about $(0.67 \times 13.5$ ng) per oocyte. However, even if the effect of mitochondrial poly(A) RNA on the stage 3 calculation is ignored completely, or alternatively is considered the same as in stage 6, the import of the calculation is unchanged.

[d] I.e., [90 ng − (interspersed poly(A) RNA) − (mitochondrial poly(A) RNA)].

[e] Calculated as 4% of the mass of polysomal RNA (L. D. Smith *et al.*, 1984).

stage 6 oocytes, and there could be almost twice as much (see notes to Table 5.5). However only about 2% of the putative stage 3 mRNA i.e., about 20 pg, is included on polysomes compared to almost 20% of the stage 6 mRNA (L. D. Smith *et al.*, 1984). Thus throughout the *early* stages of oogenesis, all but a few percent of the putative mRNA present is sequestered and stored in a translationally inactive form. Richter and Smith (1983) and Darnbrough and Ford (1981) isolated poly(A) RNA binding proteins from early oocytes and showed that the quantity of these mRNPs per oocyte declines throughout oogenesis, as illustrated in Fig. 5.17. Messages coding for one such poly(A) RNA binding protein also decrease during oogenesis (Lorenz and Richter, 1985). There are five major mRNA binding proteins that are regulated downward during oogenesis, 94, 68, 56, 55, and 50 kd in mass. Richter and Smith (1984) showed that if the mRNA binding proteins of the early oocyte are reconstituted with message *in vitro,* the resulting particles cannot be translated on injection into *Xenopus* oocytes, unlike particles similarly reconstituted with histones, or with a variety of message-associated proteins extracted from other cells. As the amount of the translational repressor declines through oogenesis, and the quantity of message engaged on polysomes increases, mRNA molecules stored early in oogenesis are probably released for translation at later stages. Thus the sequestration proteins could serve the function of controlling the rate of biosynthesis during the growth phase of oogenesis.

The mechanism responsible for the failure of the remaining 80% of the putative stage 6 oocyte mRNA to be translated during oogenesis remains unclear. The bulk of this fraction of the poly(A) RNA clearly is capable of protein synthesis when extracted and tested *in vitro,* or injected into *Xenopus* oocytes (Richter *et al.,* 1984). Perhaps within the later oocyte it remains translationally repressed by the same mRNA binding proteins as present earlier (Fig. 5.17). However, this is an unnecessary hypothesis. A sufficient alternative explanation has derived from an exhaustive series of experiments that have demonstrated a limitation in some component(s) of the stage 6 oocyte translational apparatus, *other* than the supply of accessible mRNA (Laskey *et al.,* 1977; Richter and Smith, 1981). The key observation is that in stage 6 oocytes injected mRNA can compete for translational components with endogenous mRNA, or with other injected messages. One limiting factor required for translation of membrane-bound mRNA is associated with rough endoplasmic reticulum, since coinjection of a rat liver cytoplasmic membrane fraction enhances the ability of the oocyte to translate injected messages of this class (Richter *et al.,* 1983). The translational component(s) that are limiting for cytosol messages remain unknown. Limitation of translational capacity is a specific feature of late oocytes, however. Thus injection of globin mRNA into stage 3 and stage 4 oocytes causes an *increase in total protein synthesis,* and no competitive depression of endogenous protein synthetic activity is observed, while the same exogenous mRNA injected into stage 6 oocytes competes with the endogenous messages (Taylor *et al.,* 1985). The highest rate of protein synthesis that can be induced by injection of mRNA in stage 4 oocytes approaches that of stage 6 oocytes, i.e., ~21 ng hr^{-1}. This suggests that by stage 4 the maximum translational capacity of the oocyte has been set (i.e., until it is expanded at maturation), but that this capacity is not fully utilized until late in oogenesis, perhaps because of the prevalence of repressive mRNA binding proteins in younger oocytes (Fig. 5.17).

mRNAs for Specific Proteins

Many diverse proteins have been identified in late *Xenopus* and *Rana* oocytes, stored in relatively enormous quantities for use after fertilization (reviewed in Smith and Richter, 1985). Examples include histones, of which a sufficient supply exists to assemble the chromatin of over 20,000 cells (Woodland and Adamson, 1977; Woodland et al., 1983); nucleoplasmin, which is apparently involved in the mobilization of histones for chromatin assembly (Kleinschmidt et al., 1985) and which constitutes 10% of the total nuclear protein (Krohne and Franke, 1980; Earnshaw et al., 1980); RNA polymerases, equal in amount to that contained in about 10^5 larval amphibian cells (Roeder, 1974; Hollinger and Smith, 1976); DNA ligases (Carré et al., 1981); tubulins, also resident in a huge pool amounting to about 1% of all soluble oocyte protein (Pestell, 1975); some even more prevalent actin species that constitute >8% of the total soluble protein in stage 6 oocytes (Merriam and Clark, 1978); the ribosomal proteins included in the 10^{12} ribosomes of the mature oocyte; proteins binding specifically to snRNAs (Zeller et al., 1983; De Robertis et al., 1982); and a variety of other nucleoproteins. Several of the latter are considered in the following section, in connection with 5S RNA synthesis during oogenesis, and others, e.g., those associated with cytoplasmic mRNAs (Darnbrough and Ford, 1981; Richter and Smith, 1983) have already been mentioned. DNA polymerase provides a particularly striking example, since there is no nuclear DNA synthesis during the growth phase of oogenesis. The number of molecules of $\alpha_1 + \alpha_2$ DNA polymerase increase continuously from about 10^8 in the stage 1 oocyte, to about 10^9 in the stage 3 oocyte and 2×10^{10} in the unfertilized egg (Benbow et al., 1975; Fox et al., 1980; Zierler et al., 1985). For comparison, in the hatched tadpole, a level of about 7.5×10^4 molecules of $\alpha_1 + \alpha_2$ DNA polymerase is maintained per cell by new synthesis. The egg thus stores sufficient α-polymerases to provision over 2×10^5 embryonic cells.

Though sequences represented in the maternal RNA continue to be transcribed throughout the lampbrush stage there are only a few examples that provide evidence on the origins of specific maternal mRNAs. Anderson et al. (1982) showed that transcripts labeled to high specific activity by in vitro incubation of isolated stage 6 oocyte nuclei reacted with six different cloned cDNAs derived from the maternal poly(A) RNA. However there is yet no direct demonstration, in these or any other experiments, of the entrance of particular newly synthesized mRNAs into the cytoplasm in late oocytes. The problem is technically difficult because specific activities adequate for the detection of individual mRNA species cannot be obtained in whole oocytes, even after injection of high specific activity precursor nucleotides, and thus only the overall flow rates have so far been measured (Table 5.4). Quantitative studies of protein synthesis during oogenesis have shown by two dimensional gel electrophoresis that most protein species continue to be synthesized at the same relative rates throughout, which would be consistent with later utilization of a pool of mRNAs stored from stage 1-2 (Darnbrough and Ford, 1976; Ruderman and Pardue, 1977; Harsa-King et al., 1979). Though a considerable number of stage-specific changes have been reported in these studies, they are apparently the result of alterations in the fraction of mRNA of given species that are loaded on the oocyte polysomes. Thus King and Barklis (1985) reported that the *in vitro* translation products of poly(A) RNA extracted from stages 2, 3 and 6 oocytes are essentially identical.

We turn now to two sets of identified proteins for which cloned probes have been generated, the histones, and the ribosomal proteins (rP). As summarized in Table 5.6, the *Xenopus* oocyte contains about 4×10^8 molecules of each nucleosomal core histone mRNA, and on the average 8×10^7 molecules of each of the rP mRNAs investigated, though the mRNA quantities for individual rP species differ up to fourfold with respect to one another (Baum and Wormington, 1985). The greater content of histone mRNAs is more than accounted for by the difference in the respective gene copy numbers. Thus there are about 40 genes for each histone species per haploid genome in *Xenopus* (Perry *et al.,* 1985), and 2-5 copies of each rP gene (Bozzoni *et al.*, 1982). The number of RNA polymerase mRNA molecules, estimated indirectly in Table 5.6 from the subunit synthesis rate measurements of Hollinger and Smith (1976), is severalfold lower. A similar number of molecules of the mRNA for the 70 kd heat shock protein is also reported, i.e., about 3×10^6 (Bienz, 1984). For comparison, the prevalence of the *average* poly(A) RNA species in the egg or stage 6 oocyte is about 4×10^6 molecules (see Chapter II).

Table 5.6 shows that for both the rP mRNAs and the histone mRNAs the maximum level of accumulation is achieved by the beginning of vitellogenesis at Dumont stage 2, just as for the bulk of the poly(A) RNA (Fig. 5.15). Pierandrei-Amaldi *et al.* (1982) and Baum and Wormington (1985) noted that the rP mRNAs have declined in quantity by stage 6, and Dixon and Ford (1982) also reported that synthesis of rP is much decreased in late oogenesis. On the other hand, the quantity of histone mRNA remains about the same throughout the later phases of oogenesis (Van Dongen *et al.*, 1981). This difference in behavior is directly interpretable in terms of the temporal requirements for the respective proteins. The assembly of ribosomes occurs most rapidly in midoogenesis, and is largely completed by the end of oogenesis. mRNAs coding for ribosomal proteins achieve their peak concentrations just prior to institution of the maximum rate of rRNA synthesis, which occurs at stage 3 (Scheer, 1973). Thus, by stage 2, 50% of the rP mRNA is loaded on polysomes, compared to only a few percent of total oocyte mRNA (Table 5.5), and the high level of translational utilization of the rP messages persists throughout oogenesis (Baum and Wormington, 1985). Much of the rP mRNA is degraded after fertilization, and some species of ribosomal proteins are not synthesized again until the neurula stage, though the mRNA begins to be produced several hours earlier at gastrulation (Pierandrei-Amaldi *et al.*, 1982; see Chapter III). In contrast, both the large pool of histone protein synthesized during oogenesis, and the mass of maternal histone mRNA shown in Table 5.6 are destined for use in early development.

TABLE 5.6.   Specific mRNAs in *Xenopus* Oocytes

mRNA	Maximum accumulation attained[a]	Estimated molecules mRNA per oocyte
Individual nucleosomal histones	—	~3.3×10^8 [b]
Histone H3	Early vitellogenesis[c]	4.2×10^8 [d]
Ribosomal proteins[e]	Early vitellogenesis[c,f]	~8×10^7 [g]
RNA polymerases I, IIa, IIb	—	$0.6–1.2 \times 10^7$ [h]
70 kd heat shock protein[i]	—	3×10^6

[a] Stage at which peak transcript level is first attained.

[b] Calculated from an estimate of Woodland *et al.* (1983) and Woodland and Wilt (1980b), that to account for the instantaneous rate of histone protein synthesis in eggs (~2500 pg hr^{-1}) mature oocytes must contain about 50 pg of each core histone mRNA. However, histone mRNA has a half-life of only about 3 hr in embryos (Woodland *et al.*, 1979; Woodland and Wilt, 1980b), and thus *at least* 2×50 pg of stored mRNA of each core species would be required (Woodland *et al.*, 1983), since the stored mRNA must provide for all histone synthesis up to the blastula stage.

[c] Van Dongen *et al.* (1981).

[d] Calculated from estimate of 138 pg of histone H3 mRNA per oocyte obtained by hybridization kinetics, using a cloned *Xenopus* histone DNA probe. The probe represents a rare H3 gene type not known to be expressed, and this could have resulted in an underestimate (Woodland *et al.*, 1983).

[e] Identified by cDNA clones (Bozzoni *et al.*, 1981).

[f] Pierandrei-Amaldi *et al.* (1982); Baum and Wormington (1985). Data are available for over 10 different ribosomal protein mRNAs.

[g] Calculated from estimate of Baum and Wormington (1985) that cDNA clones to ribosomal protein mRNAs react with at least 0.05% of poly(A) RNA, assuming 90 ng of poly(A) RNA per oocyte (Table 5.5) and an average mRNA length of about 10^3 nt (Pierandrei-Amaldi *et al.*, 1982).

[h] Hollinger and Smith (1976) isolated these polymerases and estimated their rates of synthesis in mature *Rana pipiens* eggs as about 0.1–0.2% of total protein synthesis. It is assumed the same would be true of *Xenopus* since there is a similar accumulation of polymerase during oogenesis in this species (Roeder, 1974). The subunit size observed was about 2×10^5 daltons (Hollinger and Smith, 1976). Thus the number of 6 kb mRNAs that would constitute 0.1–0.2% of the mRNA of the oocyte can be calculated, taking the amount of *bona fide* mRNA as 21.4 ng (Table 5.5).

[i] Bienz (1984) estimated this mRNA content from S1 nuclease protection experiments with a cloned probe. Were the four copies of this gene in the oocyte nucleus active at the rate per gene observed in *heat shocked* somatic cells, 70 days would be required to accumulate the observed pool of mRNA.

The histones begin to be synthesized at a high rate immediately upon maturation, and maternal histone mRNAs thereafter support all new histone synthesis until the blastula stage (Adamson and Woodland, 1977; Woodland, 1980; reviewed in Chapter II).

Sufficient measurements on accumulation of both histone mRNA and histone proteins during oogenesis in *Xenopus* exist so that a quantitative image of the kinetics of this process emerges. We have seen that accumulation of the maternal histone mRNA pool occurs very early in oogenesis. Assuming that the polymerases synthesizing histone transcripts are packed only 100 nt apart; that the translocation rate is 15 nt sec^{-1} (Anderson and Smith, 1978); and that there are 40 genes for each core histone (Perry *et al.*, 1985) in each of the four haploid genomes of the 4C oocyte nucleus; a period of 4–5 months would be required to synthesize the 3-4 $\times$ 10^8 histone mRNA molecules that according to Table 5.6 constitute for each core species the maternal histone mRNA pool. As discussed earlier, this is similar to the estimated time required for accumulation of the steady state content of an average poly(A) RNA coded by a a single copy gene, about 3 months. This calculation requires the histone transcription units to be in the densely packed lampbrush chromosome configuration during early oogenesis. In addition, it requires that the histone mRNA be stable during its initial accumulation phase, and thereafter be sufficiently long-lived so that it decays no faster than it can be replaced by new synthesis. Woodland and Wilt (1980b) showed that sea urchin histone mRNA injected into oocytes turns over with a half-life of only a few hours. There are at least two known factors that could contribute to the stability of *endogenous* histone mRNA in the oocyte. Unlike these injected messages the endogenous histone mRNA might be complexed with protective proteins, such as the mRNPs analyzed by Darnbrough and Ford (1981) and Richter and Smith (1983). In addition, 50-75% of the histone mRNA in the *Xenopus* oocyte is polyadenylated (reviewed in Woodland *et al.*, 1983). It has been demonstrated directly by Huez *et al.* (1978) that prior polyadenylation increases the stability of HeLa cell histone mRNA injected into *Xenopus* oocytes.

The four nucleosomal core histones are synthesized at a rate of about 50 pg hr^{-1} during oogenesis (Adamson and Woodland, 1974). This is only a very small fraction of the rate that could be supported by the mass of stored histone message, and indeed during maturation the rate increases 50-fold to about 2500 pg hr^{-1}. The final histone content is about 135 ng (Woodland and Adamson, 1977). This includes about 3-4 pmoles of each of the core histones, and 0.5 pmoles of H1 (Woodland, 1980). Van Dongen *et al.* (1983b) found that accumulation of H1a, the major H1 species, is completed during stage 3. Though some H1 synthesis continues after this (Flynn and Woodland, 1980), it must be balanced by turnover. At the rate of 50 pg hr^{-1} for the core histones, it would take about 100 days to accumulate the amounts of stored core histone proteins in the egg. Again, this is not inconsistent with the time usually required for *Xenopus* oocytes to progress from stage 2 onward as discussed earlier (Table 5.3; see also Keem *et al.*, 1979).

The stored histones are mainly accumulated within the germinal vesicle of the oocyte, where they are stored in stoichiometric complexes of several kinds with the nucleophilic proteins N$_1$, N$_2$, and nucleoplasmin (Kleinschmidt *et al.*, 1985). N$_1$ and N$_2$ are synthesized exclusively in germ line cells, including lampbrush stage oocytes, but not after fertilization in the

embryo, or in somatic cells (Wedlich *et al.*, 1985). The embryonic role of
these maternal proteins may be to bind, neutralize, and transport into the
blastomere nuclei the newly synthesized histones that are translated from
maternal mRNA during maturation and in early cleavage. Nucleoplasmin
(reviewed in Laskey *et al.*, 1985) and N_1 and N_2 (Wedlich *et al.*, 1985) are not
the only important proteins destined for use after fertilization that are stored
in the germinal vesicle. For example most of the α DNA polymerase (Fox *et
al.*, 1980), the RNA polymerase (Hollinger and Smith, 1976), the actins
(Merriam and Clark, 1978) and the snRNPs (Zeller *et al.*, 1983) are similarly
localized in the enormous nucleus of the stage 6 oocyte. Germinal vesicle
breakdown at maturation releases all of these components into the cytoplasm, where they remain until after fertilization, when they are redistributed to the newly formed nuclear compartments of the embryo.

Summary: The Functional Role of Lampbrush Chromosomes

A coherent interpretation of the data so far reviewed is as follows.
Lampbrush chromosomes are the cytological manifestation of a transcriptional apparatus in which the distinguishing—and entirely unique—feature
is that almost all the active transcription units are operating at the maximum
possible initiation rate. The basic role of lampbrush chromosomes is to
provide the transcription products loaded into the oocyte cytoplasm. This
function can be considered in two phases. Early in the lampbrush stage (in
Xenopus, Dumont stages 1 and 2) net accumulation of maternal poly(A)
RNAs occurs until the levels that will be maintained thereafter are achieved.
This has been illustrated in many different ways, e.g., by measurements of
the total poly(A) RNA content, analyses of the spectrum of proteins synthesized at different stages, and observations on several specific transcripts.
The second phase begins with vitellogenesis. Throughout this process the
oocyte is synthesizing the enormous supplies of proteins that will be required during early development, of which the best understood example is
provided by the histones. The role of the lampbrush chromosomes during
this phase (Dumont stages 3 to 6) is evidently to maintain in kinetic steady
state the required level of the maternal transcripts. This is demonstrated
directly by measurements of the flow of newly synthesized poly(A) RNA
into the cytoplasm in stage 3 and stage 6 oocytes. We have seen that when
the observed flow rates are considered in conjunction with the complexity of
the RNA, the maximum possible transcription rates would be required
merely to preserve the levels of transcripts present, even were the half-lives
of these in excess of a month. Thus it may be concluded that *lampbrush
chromosomes provide the transcriptional solution to the logistic problem of
building and maintaining very large oocytes.* The same problem is solved in
a completely different manner in meroistic oogenesis. As the following sections of this chapter illustrate there are still other solutions utilized in the

same oocytes in answer to the special logistic demands posed by the requirements for 5S rRNA and 18 and 28S rRNA.

There could exist adaptive advantages for a system of oogenesis in which the maternal constituents are continuously renewed, even though this entails (in larger eggs) the maintenance of an extremely high rate of synthesis right to the end of oogenesis. Many amphibians, for example the tropical temporary water breeder *Engystomops pustulosus*, store mature oocytes within the ovary until an appropriate environment for egg laying becomes available (Davidson and Hough, 1969), and this probably also is characteristic of *Xenopus*. Continuous molecular turnover and renewal could provide a means of escaping a rigid requirement to complete oogenesis and shed eggs on a certain schedule, thereby improving adaptive flexibility in response to uncertain environmental conditions. In any case it should not seem peculiar to assume that the transcripts of the oocyte are not absolutely stable. In all living cells that have been examined, except for special examples of dormancy such as cryobiotic systems, seeds, and spores, mRNAs are found to decay, usually stochastically, and almost always with kinetics far faster than implied by the minimum half-lives required in the foregoing considerations of amphibian oocyte transcripts.

The subject of lampbrush chromosome function has a curious intellectual history. Lampbrush chromosomes have provoked a remarkable series of colorful, but wrong interpretations, among which are the once widely accepted idea that the loops are the *products* of the chrommomeric genes; the "master-slave" theory; the proposition that new DNA sequences are progressively unwound into the loops from the chromomeres; the "C-value paradox"; and most recently, the assertion that the intense RNA synthesis in the loops is all nonproductive, in the sense that none of the newly synthesized RNA is exported to the cytoplasm. The value of these proposals has lain mainly in the efforts they have stimulated to disprove them, and none seem at this point likely avenues to further understanding. But it is scarcely the case that we have reached the end of this curious progression. Though we may now understand in logistical terms what lampbrush chromosomes do, we clearly do not understand why they are doing it. Thus the majority of the heterogeneous transcripts exported from the lampbrush stage oocyte nucleus are nontranslatable, interspersed RNAs, of unknown function. Perhaps these are useless products of readthrough transcription, exported to the cytoplasm as a side effect of some yet unknown characteristic of the post-transcriptional processing apparatus of the oocyte. In this case there remains to be discovered an unexpected feature of the mechanism by which transcripts are handled in the oocyte germinal vesicle, a mechanism so valuable that it is retained even at the expense of loading the cytoplasm with 70% nonfunctional poly(A) RNA. Or these transcripts may serve as unprocessed nRNAs destined for modification and translational use after fertilization, or perhaps they constitute some totally different class of maternal RNA required in developing systems about which we yet understand nothing. A further solution to the problem of lampbrush chromosome function, and perhaps a basic new understanding of oogenesis, could lie somewhere among the answers to these newly arisen mysteries.

(vi) The Accumulation of Ribosomal RNAs

The mechanisms by which 5S rRNA and the 18S and 28S rRNAs are accumulated are for *Xenopus* relatively well described. About 10^{12} ribosomes must be assembled per oocyte, and this logistical requirement is met by three completely different strategies. The synthesis of ribosomal proteins has already been discussed, and in this section we compare the means by which the heavy ribosomal RNAs and 5S rRNA are synthesized and accumulated. Though they can only be summarized briefly here, the molecular details of the processes by which these two classes of ribosomal genes are controlled provide interesting new insights into the mechanisms by which the genes transcribed by polymerase I and III are developmentally regulated.

5S rRNA Synthesis in Xenopus Oocytes

Low molecular weight RNAs, mainly tRNAs and 5S rRNA, are the major transcription products of Dumont stage 1 and 2 oocytes. Thomas (1974) showed that after 24 hr of labeling about 24% of the radioactive RNA in the cytoplasm of these previtellogenic oocytes is tRNA and 39% is 5S RNA, and data of Rosbash and Ford (1974) indicate that 5S RNA and tRNA account for about 70% of the mass of total RNA then present. In previtellogenic oocytes the molar ratio of newly synthesized tRNA to heavy ribosomal RNA is as high as 25, while for 5S RNA it is over 100 (Ford, 1971). Both tRNA and 5S RNA continue to be synthesized, but as the rate of synthesis of 18 and 28S rRNA accelerates after stage 2 (see below), these ratios shift dramatically. Previtellogenic oocytes are thus very unusual in the large fraction of their total accumulated RNA accounted for by stable low molecular weight species.

About 50% of the 5S RNA synthesized in previtellogenic oocytes is stored in a 7S particle, where it is complexed with a single 38.5 kd protein. The remaining 50% is found in a multimeric 42S particle, which contains in addition to the same 38.5 kd protein and 5S RNA, tRNA and two other protein molecules of ~50 kd mass (Picard and Wegnez, 1979; Denis and Mairy, 1972; Picard *et al.*, 1980; Barrett *et al.*, 1983). The latter proteins are believed to be bound to the tRNA, which is present in the aminoacyl form, and in the 42S particles of previtellogenic oocytes turn over with half-lives of 4-36 hr, depending on the tRNA species (Denis and le Maire, 1985). Both forms of storage particle are prominent only in previtellogenic oocytes. The proteins of the 42S particle are reported to be involved in transport of 5S RNA to the nucleolar site of ribosome assembly (Barrett *et al.*, 1984). This occurs as newly synthesized 18S and 28S rRNAs become available (Mairy and Denis, 1972). In the ribosome the 5S RNA is complexed with a ribosomal protein that differs in molecular weight and amino acid composition from the storage particle proteins (Picard and Wegnez, 1979). The proteins of the 7S and 42S particles are synthesized at the highest rate in stage 1 oocytes (Dixon and Ford, 1982). The same general mechanism, *viz.*, early synthesis of 5S RNA and storage in 7S and 42S RNP particles, followed during vitellogenesis by incorporation of the stored 5S RNA into ribosomes, may occur in many large vertebrate oocytes that also bear lampbrush chromosomes. Thus homologous 42S and 7S particles have been reported in previtellogenic oocytes of anurans, urodeles, and teleost fish (Mazabraud *et al.*, 1975; Picard and Wegnez, 1979).

There are two classes of 5S RNA genes in *Xenopus*, those active in oocytes and those active in both somatic cells and oocytes, termed the somatic 5S RNA genes. Oocyte-type 5S RNA differs from somatic 5S RNA at 6 out of 120 nt (Brownlee *et al.*, 1974; Ford and Brown, 1976). Only the oocyte-type of 5S RNA is stored in the 7S and 42S RNP particles. The products of the somatic 5S genes are apparently unstable in the oocyte, since though they are synthesized these transcripts fail to accumulate (Ford and Southern, 1973; Denis and Wegnez, 1977). After fertilization and up until the mid-blastula stage, both classes of 5S RNA genes are silent, as are all other nuclear genes (Chapter III). When transcription resumes some oocyte 5S RNA synthesis occurs, as does somatic 5S RNA synthesis. After the gastrula stage, however, synthesis of oocyte-type 5S RNA dwindles to an insignificant level (Peterson *et al.*, 1980; Wakefield and Gurdon, 1983; Wormington and Brown, 1983). Thus the oocyte-type genes are regulated, in that they are mainly transcribed only during oogenesis, while the somatic 5S RNA genes are expressed constitutively. The very high rate of oocyte 5S RNA synthesis in early oogenesis is accounted for by the large number of oocyte-type genes. Thus there are about 20,000 oocyte 5S RNA genes and about 400 somatic 5S RNA genes per haploid genome (Peterson *et al.*, 1980). During oogenesis each oocyte-type 5S RNA gene is responsible for the synthesis of about 10^7 5S RNA molecules.

Chromatin containing the 5S RNA genes can be transcribed with fidelity *in vitro*, by added polymerase III, as first shown by Parker and Roeder (1977). Exploitation of this advantage has led to a relatively advanced understanding of the chromatin regulatory factors involved in developmental control of these genes. A DNA sequence within the gene, extending from nucleotide +45 to +97 and known as the "internal control region," is required for the formation of a transcription complex and for initiation of transcription by polymerase III (Bogenhagen *et al.*, 1980; Sakonju *et al.*, 1980; Engelke *et al.*, 1980). The active transcription complex contains three protein species, TFIIIA, B, and C, aside from the polymerase itself (Lassar *et al.*, 1983; Schlissel and Brown, 1984). One of the most important properties of the transcription complex is that once formed it remains stable and functions for many rounds of transcription (Bogenhagen et al., 1982). Though little is so far known of the transcription factors B and C, it is established that a single molecule of TFIIIA is bound directly to the internal control region of each active 5S RNA gene. This has been shown explicitly by mapping the DNA protein contact points, and by determination of the binding stoichiometry (Sakonju and Brown, 1982; D. R. Smith et al., 1984). In the absence of TFIIIA a transcription complex cannot be assembled and the genes instead may form a stably repressed complex with histone (Bogenhagen et al., 1982; Schlissel and Brown, 1984). This is apparently the fate of the oocyte-type 5S RNA genes in the nuclei of somatic cells such as adult erythrocytes or kidney tissue culture cells, since these genes can be reactivated and assembled into active transcription complexes following treatments that remove histone H1 (Bogenhagen et al., 1982; Korn and Gurdon, 1981). Furthermore, oocyte type 5S DNA chromatin, i.e., the DNA in nucleosomal form, can be reconstituted in vitro together with histone H1 to reproduce the repressed structure. In tissue culture cells the somatic 5S RNA genes are instead included in stable transcription complexes, and they require only the addition of polymerase III for transcription (Schlissel and Brown, 1984). The oocyte 5S RNA genes are found in such complexes only during oogenesis.

A simple binary choice between TFIIIA binding or histone H1 binding may constitute a sufficient explanation for the regulation of the 5S RNA genes. In vitro experiments with TFIIIA demonstrate a 4 : 1 preference for somatic over oocyte-specific genes (Wormington et al., 1981), apparently the result of the three nucleotide difference in the oocyte and somatic gene sequences within an important area of the internal control region (e.g., D. R. Smith et al., 1984). However, when a mixture of oocyte and somatic type 5S gene derivatives is injected into fertilized Xenopus eggs a preference for the somatic 5S gene sequences of up to 200-fold is observed (Brown and Schlissel, 1985). In vivo the discrimination between the endogenous somatic and oocyte-type genes for a limiting amount of regulatory factors may indeed be the cause of the 1000-fold greater activity observed (on the average) for each of the somatic-type 5S RNA genes in somatic cells, relative to each of the oocyte-type genes (Korn and Gurdon, 1981). This type of explanation depends on the decrease in TFIIIA concentration that occurs during embryogenesis. In the fertilized egg there are about 3×10^9 molecules of TFIIIA; in the gastrula stage embryo about 9×10^4 molecules per cell; and in tadpole stage embryos and adult kidney about $1\text{-}2 \times 10^4$ per cell (Shastry et al., 1984). The relative kinetics with which there forms either an active transcription complex with TFIIIA or an inactive complex with histone may be determinant, since the somatic cell content of TFIIIA is still sufficient to bind about 1/4th of all the cellular 5S RNA genes, i.e., including the oocyte-type genes. Thus Gargiulo et al. (1984) showed that on coinjection of excess oocyte and somatic 5S RNA genes into the Xenopus oocyte nucleus, the greater affinity of the somatic genes for a limiting factor, presumably TFIIIA, results in the irreversible formation of inactive histone complexes with the oocyte type genes. This result mimics the events that occur in newly formed somatic cells, and demonstrates that in the presence of histones the binding preference of TFIIIA for the somatic 5S RNA gene control sequences is apparently sufficient to account for the activation of somatic-type 5S RNA genes and the stable repression of oocyte-type 5S RNA genes. Furthermore, Brown and Schlissel (1985) demonstrated that injection of TFIIIA itself into syncytial *Xenopus* embryos that have been prevented from undergoing cytokinesis results in a sharp increase in the fraction of endogenous 5S rRNA transcript deriving from the oocyte-type genes, i.e., after transcriptional reactivation at the blastula stage. This experiment provides a direct indication that a determining parameter controlling the selective expression of the somatic type 5S RNA genes is the *limited concentration of TFIIIA* in late embryonic and somatic cells. The observation noted above that in normal late blastula stage embryos when there are still about ten molecules of TFIIIA per 5S RNA gene there is a transient reactivation of the oocyte-type 5S RNA genes is also consistent with this interpretation.

The changing prevalence of the positive regulatory factor TFIIIA also helps to explain directly the course of 5S RNA synthesis during oogenesis. A key observation made by Pelham and Brown (1980) and Honda and Roeder (1980) is that TFIIIA is the 38.5 kd protein that is complexed with the 5S RNA in the 7S particles of previtellogenic oocytes. In other words, this protein binds to both the anticoding strand of the 5S RNA gene and to the identical sequence of the RNA product. In the presence of the large excess of TFIIIA that exists in early previtellogenic stages, when there are about 10^{12} molecules of this protein per oocyte (Shastry *et al.*, 1984), the 5S RNA genes are all actively transcribed. However, as the available TFIIIA is progressively sequestered together with the newly synthesized 5S RNA in the 7S and 42S storage particles, a decrease of 5S RNA transcription occurs. When these particles disappear later in oogenesis the total amount of TFIIIA in the oocyte has decreased by a factor of about two (Shastry *et al.*, 1984). The TFIIIA gene has been cloned, and estimates of the level of TFIIIA mRNA provided by Ginsberg *et al.* (1984). There are about 5×10^6 molecules of TFIIIA message in the early oocyte, classifying this transcript as a member of the low abundance sequence class (cf. Table 5.6). This falls to about 1×10^6 molecules by stage 4-6 of oogenesis, and to only about 1 per cell in the swimming tadpole.

In summary, the 5S RNA of the oocyte ribosomes is provided by the active transcription of about 80,000 genes in the 4C oocyte nucleus, occurring mainly during the previtellogenic period of oogenesis. The transcription of these genes is regulated by the formation of a specific transcription complex including three protein factors plus polymerase III. As the concentration of 5S RNA builds, one of these factors, TFIIIA, becomes bound instead to the 5S RNA product. Competitive interactions between a limiting amount of TFIIIA and histones are probably involved in the regulatory events that result in the extremely accurate discrimination between the oocyte and somatic types of 5S RNA genes later in development.

Ribosomal RNA Synthesis during Oogenesis

Though the same enormous number of the large rRNAs are required to be synthesized during oogenesis as of 5S rRNA, there are typically no more than a few hundred genes for these rRNAs per haploid genome. In organisms whose oocytes contain lampbrush chromosomes the ribosomal genes are amplified early in oogenesis, and are present as extrachromosomal circular DNA molecules that are packaged into nucleolar structures (Peacock, 1965; Miller, 1966; Lane, 1967; Hourcade et al., 1973; Gall and Rochaix, 1974; Buongiorno-Nardelli et al., 1976). Transcription of ribosomal DNA during oogenesis occurs in the amplified rDNA. Ribosomal DNA amplification has been observed in a number of urodele and anuran amphibians, in teleosts, and in orthopteran insects (Cave, 1973) all of which, as shown in Table 5.2, utilize lampbrush chromosomes in their oocytes. Ribosomal gene amplification is also reported in coleopteran and in a few dipteran insects, which carry out meroistic oogenesis, and to a low level in an echiuroid worm and in lamellibranch molluscs (Kidder, 1976; see reviews in Brown and Dawid, 1968; Davidson, 1968; Gall, 1969). As reviewed in the following section ribosomal gene amplification probably does not occur in the mouse, though it does in primates, and it is absent in those sea urchin oocytes that have been examined, as well as in some other invertebrates, e.g., the nematode *Panagrellus* (Noble et al., 1976).

Replication of the rDNA occurs mainly at the pachytene stage in amphibian oocytes (Gall, 1968; Pardue and Gall, 1969; Van Gansen and Schram, 1974), though the process begins in the premeiotic oogonia (Kalt and Gall, 1974). In orthopteran oocytes amplification also takes place during the pachytene stage (Cave, 1973). The original copies used for rDNA replication are of chromosomal origin (Brown and Blackler, 1972) and usually but probably not always the amplified rDNA of each original circle derives from a single genomic rRNA gene. This inference is supported by the observations that spacer sequence heterogeneity, which is observed frequently in adjacent genomic rDNA repeat units (Wellauer et al., 1976a) is only occasionally seen within the tandem repeats of a given *amplified* rDNA molecule (Wellauer et al., 1976b; Scheer et al., 1977). Most of the amplified rDNA is synthesized by some form of rolling circle replication of preexistent extrachromosomal rDNA circles (Hourcade et al., 1973, 1974; Bird et al., 1973; Rochaix et al., 1974; Buongiorno-Nardelli et al., 1976).

In *Xenopus* oocytes the extrachromosomal rDNA formed by the amplification process is packaged in a large number of extrachromosomal nucleoli, which can be observed in the germinal vesicle sap throughout oogenesis. On germinal vesicle breakdown the extrachromosomal rDNA is released to the cytoplasm, where it remains during early embryogenesis. By the gastrula stage it has disappeared (Busby and Reeder, 1982). Thiébaud (1979) showed that the number of extrachromosomal nucleoli per oocyte varies from 500 to 2500 in oocytes of the same female. The maximum number is attained at stage 4, and after this some nucleoli apparently fuse, resulting in a lower value in later oocytes. The number of rRNA genes per nucleolus also varies, ranging from 500-11,000. On the other hand, the *total amount of amplified rDNA per nucleus is always the same*, about 30-35 pg (Perkowska *et al.*, 1968; Thiébaud, 1979) or approximately 2.5×10^6 rRNA genes. Since there are about 450 rRNA genes per haploid genome (Brown and Dawid, 1968), this represents for the oocyte about a 1400-fold increase in gene number. However, even after amplification a long period is required for the accumulation of the large rRNAs, and rRNA synthesis is likely among the rate-limiting processes in the growth of the amphibian oocyte, just as in the growth of the *Drosophila* oocyte (see above).

In the earliest previtellogenic oocytes there is little rRNA synthesis, though extrachromosomal replication of the rDNA has been completed. Scheer *et al.* (1976b) showed that in the newt *Triton alpestris* previtellogenic oocytes (equivalent to Dumont stage 1 in *Xenopus*) synthesize rRNA at only about 0.01-0.5% of the rate measured in vitellogenic oocytes. This relatively low synthetic rate is correlated with sparse packing of transcripts in the extrachromosomal nucleolar genes, and with the presence of many totally inactive gene regions. In contrast, 90-95% of the nucleolar ribosomal genes are being transcribed in midvitellogenic oocytes, and the transcripts visible on these are tightly packed, with as many as 130 transcripts per gene region. From the known length of the amphibian 40S rRNA precursor, the spacing between polymerases in the extrachromosomal nucleolar genes is calculated to be only about 100 ntp (Miller and Beatty, 1969a,b,c; Scheer *et al.*, 1976b). Figure 5.18(a) displays active transcription units in the nucleolar rDNA of midvitellogenic *Triturus* oocytes (Miller and Beatty, 1969a). Such structures have been observed in the oocytes of several other urodeles (see, e.g., Scheer *et al.*, 1976b; Angelier *et al.*, 1979), and in *Xenopus* oocytes (Miller and Beatty, 1969b), where transcription of nucleolar genes occurs at high rates after Dumont stage 2 (Scheer, 1973).

In Table 5.7 several estimates of the rate of rRNA synthesis for stage 3 *Xenopus* oocytes are shown. The rate apparently varies greatly depending on the state of gonadotropic hormonal stimulation of the female. As ribosomes are a major constituent of the cytoplasm, hormonal determination of their rate of accretion, and also of the rate of yolk uptake (Keem *et al.*, 1979; Wallace and Misulovin, 1978) appear to provide direct physiological controls over the kinetics of oocyte growth. The difference in synthesis rate between stimulated and nonstimulated females is due to the frequency of initiation in

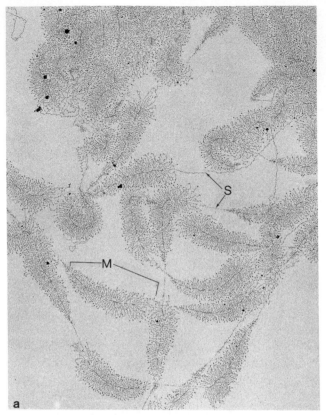

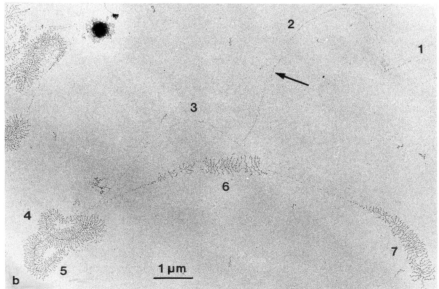

TABLE 5.7. Estimates of rRNA Synthesis Rates in Stage 3 *Xenopus* Oocytes

Measurement (hormonal stimulation, S or NS)	Rate (pg min^{-1})
Synthesis kinetics (NS)	7[a]
Synthesis kinetics (S)	24.3[a]
Net increase in rRNA content (S)	76[b]
Expectation for maximum activity of all genes (S)	93[c]

S, stimulated; NS, nonstimulated.

[a] Data of Anderson and Smith (1978) for rate of synthesis of 40S rRNA precursor in stage 3 oocytes, converted to pg min^{-1}. These authors showed that the translocation rate for polymerase I is the same, about 15 nt sec^{-1}, with or without chorionic gonadotropin stimulation.

[b] Calculated from data of Scheer (1973) who showed that in a female maximally stimulated by partial oviectomy, in which oogenesis took place at the highest rate so far reported (cf. Table 5.2), 3.8 μg of rRNA was synthesized in 38 days, with the maximum rates of rRNA flow into the cytoplasm at stages 3-5. The rate of precursor synthesis at this time, 2.3×10^9 nt sec^{-1}, is equivalent to 76 pg min^{-1}.

[c] Calculated on the basis that there are 2.5×10^6 rRNA genes active; in all of which polymerases are advancing at 15 nt sec^{-1} (Anderson and Smith, 1978); and that the transcribing polymerases are packed only 100 nt apart. The weight of the 40S precursor molecule is 2.5×10^6 daltons.

the active extrachromosomal nucleolar genes, and in the fraction of these genes that are transcribed at all, since hormone treatment does not affect the fundamental polymerase translocation rate (Anderson and Smith, 1978). Table 5.7 shows that under optimal stimulation the rate of rRNA production in midvitellogenic oocytes is close to that expected were all of the 2.5×10^6 extrachromosomal nucleolar genes functioning at the maximum rate. Thus

Fig. 5.18. Ribosomal transcription units of urodele oocytes. (a) Portion of nucleolar core isolated from *Triturus viridescens* oocyte showing rRNA transcription units. Matrix-covered axis segments (M) are separated by matrix-free axis segments (S). Matrix units are ~2.5 μm long and the matrix-free segments range from 1/3rd to 10 times or more the length of the matrix unit. [From O. L. Miller and B. R. Beatty (1969a). Reproduced from *J. Cell Physiol.* **74**, Suppl. 1, 225 by copyright permission of the Rockefeller University Press.] (b) Active, partially active, and inactive ribosomal RNA transcription units in mature oocytes of the newt *Triton alpestris*. A region of a single amplified rDNA molecule is seen. The arrow indicates an axial rDNA chromatin fibril that contains three inactive rRNA genes (1-3) in nucleosomal conformation, in series with four active genes (4-7). These display various degrees of polymerase packing. [From U. Scheer, M. F. Trendelenburg, and W. W. Franke (1976b). *J. Cell Biol.* **69**, 465.]

over 3×10^5 rRNA molecules are being synthesized per second, compared
to 10-100 sec^{-1} in somatic cells. The instantaneous rate of rRNA synthesis
exceeds the rate of total heterogeneous RNA synthesis in the lampbrush
chromosomes, and is many times the rate of stable heterogeneous RNA flow
into the cytoplasm (Table 5.5). The newly formed rRNA molecules exit into
the cytoplasm through "pore complexes," which occupy 25% of the area
of the nuclear membrane. From the number of these structures Scheer
(1973) calculated that 1-2 molecules of rRNA pass through each pore per
minute.

Net rRNA accumulation continues at least under some conditions in stage
6 *Xenopus* oocytes (Taylor and Smith, 1985). Under conditions of hormonal
stimulation synthesis of 40S ribosomal precursor occurs in these oocytes at
an average rate about 50% of that in stage 3 oocytes (Anderson and Smith,
1977, 1978). However, this rate varies greatly from female to female, controlled by unknown physiological factors. The 40S ribosomal precursor synthesized in stage 6 oocytes seems not to be processed with the normal
efficiency (Anderson and Smith, 1977). On germinal vesicle breakdown the
precursors are released into the cytoplasm where they persist into early
cleavage (Busby and Reeder, 1982; Gelfand and Smith, 1983). In species
other than *Xenopus* more sharply decreased levels of rRNA synthesis are
observed towards the end of oogenesis. For example, in postvitellogenic
mature oocytes of *Triton alpestris* the nucleolar genes contain only about
15% as many transcripts per ribosomal gene region as do growing oocytes,
and the rate of synthesis is about 13% of that observed in midoogenesis
(Scheer *et al.*, 1976b). In mature oocytes of hibernating *Rana pipiens* at least
70% of the extrachromosomal ribosomal gene units appear completely devoid of nascent transcripts when examined in the electron microscope (Trendelenburg and McKinnell, 1979). In summary, the rate of rRNA synthesis on
the extrachromosomal nucleolar genes increases from a low initial level to a
maximum at midvitellogenesis and then falls again at maturity, to an extent
that varies according to species. These regulatory changes reflect large differences in the frequency of chain initiation and in the fraction of the rRNA
genes utilized. It has been noticed that in either very young or mature oocytes when not all the extrachromosomal genes are functioning, adjacent
transcription units may display large differences in activity (Scheer *et al.*,
1976b; Trendelenburg and McKinnell, 1979). An example is shown in Fig.
5.18(b) from the study of Scheer *et al.* (1976b). The rate of initiation of *each
gene* in the tandem nucleolar arrays thus appears to be regulated independently during oogenesis.

Transcription of the ribosomal genes is controlled by two upstream regions, *viz.* a proximal promoter, and a set of regulatory elements located
distally in the intergenic spacer. The promoter located immediately at the 5'
end of the gene has been mapped by observing the effects on transcription of
various 5' deletions. These experiments were carried out either by injecting

synthetic deletion mutants into the *Xenopus* oocyte nucleus, or by transcribing them in a cell-free polymerase I system that consists essentially of an
oocyte nuclear homogenate (Moss, 1982; Sollner-Webb *et al.*, 1983). Transcripts of exogeneous origin were distinguished by utilization of mutant
Xenopus laevis gene constructs in the *Xenopus borealis* transcription systems. The proximal promoter is found to extend from nt -142 (i.e., 142 nt
upstream of the start of the transcript) to nt $+6$. The 13 nt sequence from -7
to $+6$ is the major requirement for accurate transcription in injected oocyte
nuclei, and this sequence element is conserved among the rRNA genes of
several *Xenopus* species (Bach *et al.*, 1981). This region is evidently directly
utilized for polymerase I initiation. At -27 is a T_6 sequence that occurs in an
analogous position to the "TATA box" of genes transcribed by polymerase
II. The remainder of the promoter region is apparently required at least
under conditions of limited initiation in order to obtain maximal transcription
rates. Evidence for this conclusion is shown in Fig. 5.19(a) and (b), and the
sequence of the promoter region is given in Fig. 5.19(c). There is some
evidence that portions of the promoter region extending out to -142 serve as
the binding site for a transcription factor (Reeder *et al.*, 1983). An interesting
indirect light on the mechanism by which these genes are activated derives
from an analysis of the effects of topological constraint on rDNA transcription (Pruitt and Reeder, 1984a). A supercoiled configuration was found to be
required for continued expression of injected plasmids, but not for transcription of the endogenous rRNA genes in the oocyte nucleus. Thus on injection
of restriction endonucleases, both classes of gene are digested, but while this
abolishes transcription from the injected plasmids it does not effect endogenous rRNA synthesis. The torsional strain to which the supercoiled plasmids
are subjected may thus mimic the effects of the assembled transcription
complex on the endogenous rDNA.

A novel aspect of regulation in the tandem rRNA gene sets is the role of
sequences located between about 200 and 2500 nt upstream in the nontranscribed spacer. A diagram of this region is shown in Fig. 5.19(d). The
importance of these spacer sequences for transcription of the rRNA genes
has been shown in several ways. Moss (1983) found that in oocyte nuclear
injection experiments deletion of a portion of the spacer region reduces the
amount of transcription 20-fold compared to a coinjected wild-type gene,
and a similar conclusion was drawn by Busby and Reeder (1983) from experiments in which deletion mutants lacking far upstream sequences were injected into *Xenopus* eggs, so as to observe the effect on developmental
regulation in the embryo. After the midblastula transition the injected genes
are activated, and deletion of the spacer sequences reduces their transcription in the embryo 5-10-fold. Additional experiments (Reeder *et al.*, 1983)
show that when different rDNA plasmids are coinjected into the oocyte
nucleus in sufficient quantity to induce competition, those plasmids containing the more extensive spacer sequences always produce the most tran-

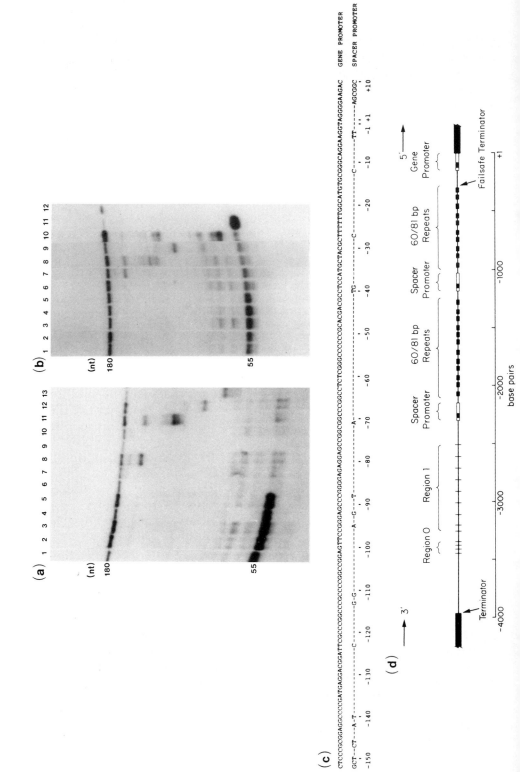

script. The key spacer elements responsible for these effects are the repetitive sequences identified in Fig. 5.19(d) as the 60/81 base pair repeat. Both forms of this repeat include a 42 ntp sequence also found in the proximal promoter [Fig. 5.19(c)] at position −73 to −114. The 42 ntp elements have the properties of enhancer sequences. Thus they endow a plasmid carrying them with the ability to compete successfully with a second plasmid that lacks them, and their effect occurs exclusively on genes to which they are in *cis* relation (Labhart and Reeder, 1984). Furthermore, they operate equally well in either orientation, or several kb away from the gene. This interpretation also explains an old mystery of *Xenopus* species hybridization experiments. When *X. laevis* and *X. borealis* are crossed, the *X. laevis* rRNA genes are always utilized predominantly in the hybrid embryos, irrespective of the direction of the cross (Honjo and Reeder, 1973; Cassidy and Blackler, 1974). *X. laevis* rRNA spacers turn out to contain more than 20 copies of the 42 ntp enhancer elements while the *X. borealis* genes contain only four. Thus "nucleolar dominance" can be recreated experimentally by injecting genes from these two *Xenopus* species into oocyte nuclei of either, under conditions where competition will occur (Reeder and Roan, 1985). The intergenic spacer regions also contain several nearly perfect duplications of the complete proximal promoter sequence [see Fig. 5.19(c) and (d)]. These upstream promoters are capable of initiating transcription themselves under

---

Fig. 5.19. Transcriptional control regions of *Xenopus laevis* ribosomal RNA genes. (a) and (b) S1 nuclease analysis of the transcription products of a series of 5′ deletion mutants. The mutants were constructed by progressive Bal31 exonuclease digestion of a cloned fragment of a *Xenopus laevis* rRNA gene, and were ligated into plasmid pBR322 with synthetic BamHI linkers. They were incubated with extracts of *Xenopus borealis* oocyte nuclei and the RNA was isolated and hybridized with a 180 nt end-labeled probe that includes 55 nt of the transcribed sequence. (a) Lane 1, control (nondeleted); lane 2, deletion to −245; lane 3, to −210; lane 4, to −166; lane 5, to −158; lane 6, to −127; lane 7, to −115; lane 8, to −95; lane 9, to −85; lane 10, to −75; land 11, to −65; lane 12, to −45; lane 13, to −27. (b) A second experiment using a more active extract in which the low level of transcription mediated by the inner domain (−7 to +6; see text) can be observed. Lane 1, control (nondeleted); lane 2, deleted to −245; lane 3, to −210; lane 4, to −166; lane 5, to −158; lane 6, to −127; lane 7, to −95; lane 8, to −65; lane 9, to −45; lane 10, to −7; lane 11, control, with *X. laevis* RNA; lane 12, control, with *X. borealis* RNA. [From B. Sollner-Webb, J. A. K. Wilkinson, J. Roan, and R. H. Reeder (1983). *Cell* **35**, 199. Copyright by M.I.T.](c). Nucleotide sequences of proximal promoter (top) and of a spacer region promoter (bottom), from −150 to +10. Homology between the spacer and gene promoters is indicated by dashes. [From S. C. Pruitt and R. H. Reeder (1984b). *Mol. Cell. Biol.* **4**, 2851.] (d). Diagram of the spacer regulatory elements. The transcribed gene sequences are the heavy bars shown at either end. Promoters are white bars which include the 42 ntp region, indicated by the thin black bar within. The 42 ntp sequence also accurs 22 times in the spacer within the 60/81 nt repeats. The spacer promoters are marked by BamHI sites. The repetitive sequences indicated in regions 0 and 1 are not homologous with the remainder of the spacer. The terminator on the left stops transcription of the preceding gene. The "failsafe" terminator on the right blocks transcription beginning from the spacer promoters. [From S. J. Busby and R. H. Reeder (1983). *Cell* **34**, 989. Copyright by M.I.T.]

certain circumstances, both experimentally and naturally occurring (Scheer *et al.*, 1977; Boseley *et al.*, 1979; Morgan *et al.*, 1983; Pruitt and Reeder, 1984b), and they are thus to be regarded as potentially functional. However, they are followed by a "failsafe" terminator at position -243 that prevents readthrough from them into the gene proper (Moss, 1983). The role played by the duplicated upstream promoters is not yet apparent, though it is known that the requirements for their function are subtly different than for the proximal promoter (Morgan *et al.*, 1983; Pruitt and Reeder, 1984b). They could serve in some way as ancillary elements necessary for the function of the contiguous enhancer sequences.

These detailed researches into the regulation of the genes that synthesize the rRNAs may provide a preview of the immense molecular complexity of the transcriptional apparatus that functions during oogenesis. Though major in amount, the 5S and 40S rRNAs are but two of the $1\text{-}2 \times 10^4$ diverse RNA species that are being transcribed in the lampbrush chromosome stage nucleus. A lesson that we have encountered several times in this chapter, as well as earlier, is that the detailed mode of regulation is particular to each gene, or functionally related set of genes. It is likely that we can anticipate the discovery of many new regulatory structures of diverse properties when equivalent studies are carried out on individual single copy transcription units of the lampbrush chromosomes.

3. NONMEROISTIC OOGENESIS WITHOUT LAMPBRUSH CHROMOSOMES

In the oogenesis of many species neither nurse cells nor lampbrush chromosomes are utilized. Nurse cells and lampbrush chromosomes are best interpreted as alternative devices for the supply of a large variety of transcripts at elevated rates, as we have seen. The requirement for high synthesis rates is not an invariant feature of oogenesis, however. Such requirements do not exist in animals in which the final quantities of maternal RNAs are relatively low, and the allowed time for completion of oogenesis is sufficient. In such animals the structure of the oocyte transcriptional apparatus as well as the kinetics of RNA synthesis are very different from those we considered earlier. Yet in qualitative terms the end result, *viz.* the transcript pool of the mature oocyte, has the same distinctive characteristics as in the species discussed earlier. Thus where investigated, oocytes that develop without either nurse cells or lampbrush chromosomes also contain sufficient ribosomes and stored maternal mRNAs to support protein synthesis after fertilization, and at least in sea urchins, interspersed maternal poly(A) RNAs as well. The characteristics of transcription in oocytes that do not utilize lampbrush chromosomes provide interesting comparisons, which illuminate those aspects of oocyte nuclear function that are intrinsic to the process of creating an egg, irrespective of any particular *logistic* strategy.

(i) RNA Synthesis in *Urechis* and Sea Urchin Oocytes

The echiuroids represent an unsegmented lower protostome grade of organization, while the echinoderms constitute a major branch of deuterostome evolution. Yet the strategies utilized in these most unrelated creatures for the preparation of the maternal transcript pools are strikingly homologous. Although the evidence is still incomplete it appears that the oocytes of neither form possess lampbrush chromosomes. In the light microscope the diplotene meiotic prophase chromosomes of growing *Urechis* oocytes appear to assume a diffuse configuration (Das, 1976) that is difficult to resolve. However, it does not resemble a true lampbrush chromosome structure. Electron microscope observations on chromatin spread from *Urechis* oocytes have not been reported. Lampbrush chromosomes were thought to be present in sea urchin oocytes on the basis of light microscopy (Jörgenssen, 1913; Davidson, 1968). However, despite determined attempts to visualize them, typical lampbrush chromosome transcription matrices could not be demonstrated in electron microscope preparations of *Strongylocentrotus purpuratus* oocyte chromatin. Except for densely packed ribosomal gene matrices, only sparsely distributed nascent transcripts were observed (O. L. Miller, B. A. Hamkalo, B. R. Hough-Evans, and E. H. Davidson, unpublished data). Since the prevalence of maximally packed transcription matrices provides a definitive ultrastructural characteristic for true oocyte lampbrush chromosomes, it can be concluded that such structures are absent from the growing oocytes of at least this sea urchin species.

Both the *Urechis* egg and the sea urchin egg are relatively small, and they contain amounts of rRNA that are proportional to their respective volumes. These and some other relevant data are listed in Table 5.8. The amount of poly(A) RNA or of heterogeneous RNA stored in the *Urechis* egg is not reported, though from measurements both of hybridization kinetics (Davis, 1975) and of the content of poly(A) tracts (Davis and Davis, 1978) it would appear to be less than 1% of the total RNA. However, Rosenthal and Wilt (1986) found that most maternal mRNAs in fully-grown *Urechis* oocytes lack poly(A) tracts, though after fertilization they are rapidly adenylated. In the egg of the sea urchin *Strongylocentrotus purpuratus* there is about 30 pg of mRNA, about half of which is polyadenylated, plus approximately twice this amount of interspersed poly(A) RNA that is not translatable (see Chapter II). As shown in Table 5.8 the *average* prevalence of each RNA species of the heterogeneous sequence class is only 1-2 × 10^3 molecules per egg. The data in Table 5.8 suffice to illustrate the quantitative contrast between the *Strongylocentrotus* and *Urechis* eggs on the one hand, and on the other, eggs

TABLE 5.8.   Maternal RNAs in the Eggs of *Urechis caupo* and
Strongylocentrotus purpuratus

	Urechis	Strongylocentrotus
Mature oocyte		
diameter (μm)	125	80
rRNA content (ng)	8.0[a]	2.4[b]
Maternal RNA		
complexity (nt $\times$ 10^{-6})	31–47[c]	37 $\pm$ 4[d]
Average molecules		
of each sequence	—	~1.6 $\times$ $10^{3\ d,e}$
Length of time		
estimated for the growth		
phase of oogenesis	$\geq$4 months[f]	$\geq$8 months[g]

[a] Davis and Wilt (1972); Das (1976); Gould-Somero (1975). About 90% of the total RNA is ribosomal.

[b] Goustin and Wilt (1981).

[c] Calculated by Davidson (1976) from data of Davis (1975).

[d] Galau *et al.* (1976); Hough-Evans *et al.* (1977).

[e] Note that this refers to the low abundance sequences that dominate the complexity measurement. More prevalent sequences exist as well in the egg RNA (Flytzanis *et al.*, 1982), though most RNA species are prevent at $\leq$10[4] molecules per egg. See Chapter II.

[f] Das (1976) calculated a minimum period of 135 days for oogenesis using the rate of ribosome production.

[g] Inferred from data of Gonor (1973a,b), and Holland and Giese (1965), as described in text. The period indicated includes the previtellogenic growth phase ($\geq$6 months) plus the vitellogenic growth phase (2–3 months).

such as that of *Xenopus*. The volume of the *Xenopus* egg is about 3000 and 1000 times greater than the volumes of the *Strongylocentrotus* and *Urechis* eggs, respectively; the rRNA content is about 1500 and 500 times greater, respectively; and the prevalence per egg of an average complex class maternal poly(A) RNA species in *Xenopus* is again about 1000 times the corresponding value shown in Table 5.8 for the sea urchin egg (reviewed in Chapter II). Yet the complexities of the maternal RNA in sea urchin, *Urechis*, and *Xenopus* eggs, as well as the length of time required to complete the growth phase of oogenesis, are all very similar. A difference of about three orders of magnitude therefore exists in the quantitative demands exerted on the transcriptional apparatus in *Xenopus* compared to sea urchin or *Urechis* oocytes.

In echiuroid worms and several other groups of marine organisms, oogenesis is of the solitary type, in which follicle and nurse cells are absent, and the oocytes develop autonomously in the coelomic cavity. In the ease of labeling, accessibility of oocytes of diverse stage, and the possibility of removing and then reimplanting growing oocytes in the coelom, these forms offer some unique advantages for the study of gene activity during oogenesis, though as yet they have been relatively little exploited (see reviews in Gould-Somero, 1975; Das, 1976; Fischer, 1984; Fischer and Pfannenstiel, 1984). For example, immature oocytes of the polychaete annelid *Platynereis dumerilii* can be transferred between the coeloms of animals of different genotype and mature eggs recovered that are able to undergo embryological development. Fischer (1977) utilized this method to demonstrate maternal inheritance of larval eye color, which is evidently the consequence of the autonomous expression during oogenesis of genes affecting the pigmentation pathways. Measurements carried out by Lee and Whiteley (1984) on growing oocytes of another polychaete annelid, *Schizobranchia insignis*, show that the rate of RNA synthesis during oogenesis decreases sharply as the oocyte grows, and that the egg RNA appears to include high complexity components similar to those observed in other forms (cf. Table 2.1). In addition, a variety of molecular measurements that are directly relevant to the transcriptional processes carried out by relatively small solitary oocytes have been reported for *Urechis*.

Patterns of Transcription during Oogenesis in Urechis

The course of RNA synthesis in *Urechis* oocytes differs in several ways from that in amphibian and other oocytes that bear lampbrush chromosomes. As in the amphibians the ribosomal RNA genes of *Urechis* oocytes are amplified, but only to the moderate extent of about six times the number of rRNA genes included in the 4C chromosomal complement (Dawid and Brown, 1970). The excess rDNA is located in a single very large nucleolus present in the germinal vesicle throughout oogenesis (Das, 1976). In *Urechis* oocytes rRNA is the major transcript species synthesized at *all* stages of oogenesis (Davis and Wilt, 1972; Das, 1976). The rate of rRNA synthesis actually increases 4-5-fold in late vitellogenic oocytes, and rRNA continues to be produced, though at a lower rate, even in mature unfertilized eggs (Gould-Somero, 1975; Das, 1976). There is no evidence for preferential accumulation of 4S and 5S RNAs in previtellogenic oocytes, as in amphibian and teleost oogenesis. These low molecular weight species are instead synthesized continuously in *Urechis* oocytes at all stages (Davis and Wilt, 1972). In *Urechis* about 8% of the growing ribosome pool is utilized for protein synthesis throughout the growth phase of oogenesis. Thus as the oocyte develops it engages an increasing mass of mRNA. Heterodisperse RNA that is detectable after 2 hr of labeling also continues to be synthesized at all stages of oogenesis, and poly(A) tracts, presumably associated with the heterogeneous RNA molecules, accumulate (Davis and Wilt, 1972; Davis and Davis, 1978). Some specific examples have been reported by Rosenthal and Wilt (1986), in a study of the representation in total RNA of a series of transcripts identified by cDNA clones. Certain of these transcripts accumulate continuously throughout the period of oocyte growth, though the quantity of others, presumably those required only for translation during oogenesis, has decreased sharply by the final stage. At least for many of those species that remain polyadenylated throughout, this provides another contrast with *Xenopus*, where as we have seen, the final net quantity of poly(A) RNA is attained very early in oogenesis (Fig. 5.15). In quantitative terms, the rate of accumulation of poly(A) RNA in the cytoplasm of *Urechis* oocytes is probably less than 1% of the steady state flow into the cytoplasm of newly synthesized poly(A) RNA in midlampbrush chromosome stage *Xenopus* oocytes.

Transcription in Sea Urchin Oocytes

Most sea urchins have an annual reproductive cycle (reviewed in Boolootian, 1966; Piatagorsky, 1975), and from observations on oocyte size and number during this cycle it is possible to estimate the approximate lengths of the various phases of oogenesis. Oogonial multiplication occurs anew each year. Intertidal populations of *Strongylocentrotus purpuratus* spawn intermittently from late November to May or June, and in these animals oogonial multiplication takes place primarily between April and November, when the latest oogonial DNA synthesis is observed in each cycle (Holland and Giese, 1965; Gonor, 1973a,b). The number of previtellogenic oocytes increases during the winter months to a maximum value that is attained in March, and remains constant until the following September, when the first group of oocytes enters the vitellogenic growth phase. Between March and September the newly formed oocytes grow slowly in volume (Gonor, 1973b), from about 10-15 μm diameter to about 30 μm. Their rate of growth then increases about 2-fold, and mature oocytes 80 μm in diameter are found about 2-3 months later. Oocytes in all stages of the growth process can be observed in the ovary at any one time during the winter months. It may be concluded that the previtellogenic growth phase requires *at least* six months (March to September), and the more rapid vitellogenic growth phase about 2-3 months (September to November-December). Oocytes that begin vitellogenesis later in the spawning period must spend an even longer time in previtellogenic stages, though the length of the previtellogenic *growth* phase may depend largely on environmental conditions such as nutrient supply (see, e.g., Gonor, 1973a). The previtellogenic oocytes are located in the walls of the ovarian tubules, where they exist in close contact with other cells, though no specific follicular structures are evident (reviewed in Piatagorsky, 1975). As the oocytes increase in size they lose their associations with the epithelial wall, and the mature and late vitellogenic oocytes accumulate in the lumen of the ovarian tubules. Both meiotic reduction divisions are completed in this location, i.e., prior to release from the ovary.

RNA synthesis has not been studied in any detail in *previtellogenic* oocytes, and it is known only that they synthesize transcripts of all classes, including low molecular weight RNAs, 18 and 28S rRNAs, and heterogeneous RNAs (Sconzo *et al.*, 1972; earlier studies reviewed in Piatagorsky, 1975). Hough-Evans *et al.* (1979) showed that the nuclear RNA of previtellogenic oocytes has a complexity of at least 1.6×10^8 nt, i.e., close to 30% of the single copy genome. Note that this is about the same as the complexity of embryo and adult sea urchin nuclear RNAs (Hough *et al.*, 1975; Kleene and Humphreys, 1977; Wold *et al.*, 1978). Furthermore, the particular single copy sequence set represented in the previtellogenic oocyte nuclear RNA is within the limits of experimental detection wholly represented in gastrula stage embryo nuclear RNAs, as illustrated in Fig. 5.20 (Hough-Evans *et al.*, 1979). The significance of this experiment is that it shows that *most if not all of the transcription units operative in midembryogenesis have already been activated at the previtellogenic stage of oogenesis.*

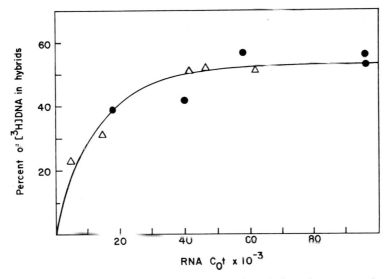

Fig. 5.20. Hybridization with sea urchin gastrula and previtellogenic oocyte nuclear RNAs of single copy ^{3}H-DNA recovered from duplexes with previtellogenic oocyte nuclear RNA. A pseudo-first-order function was used to fit data from the reaction of this ^{3}H-DNA with previtellogenic oocyte nuclear RNA (●), and with gastrula nuclear RNA (△), with the assumptions that there is a single kinetic component and that the ordinate intercept is zero. ^{3}H-DNA molecules representing the previtellogenic oocyte nRNA sequence set were concentrated 4–5-fold by prereaction with this RNA. Thus the termination value for the curve shown is $53 \pm 2\%$, compared to about 13% for the starting single copy tracer preparation. The termination value and the rate constant of $7 \pm 1.4 \times 10^{-5} \ M^{-1} \ sec^{-1}$ were obtained by least-squares analysis. [From B. R. Hough-Evans, S. G. Ernst, R. J. Britten, and E. H. Davidson (1979). *Dev. Biol.* **69**, 258.]

In the sea urchin stable heterogeneous maternal transcripts accumulate *sequentially* during oogenesis. The cytoplasmic fraction of previtellogenic *S. purpuratus* oocytes thus contains only about 45% of the ultimate maternal RNA sequence set (Hough-Evans *et al.*, 1979). The prevalence of typical maternal sequences is about 1-2 × 10³ molecules per egg (Table 2.1), and even for sequences present at ten times this concentration less than an hour would be required for their transcription from single copy genes and accumulation, were these genes being initated at the maximum possible rate. This calculation assumes the polymerase translocation rate measured for sea urchin embryo nuclei, about 9 nt sec⁻¹ (Aronson and Chen, 1977). However, as discussed earlier, few of the transcription units in the oocyte seem to be heavily loaded with nascent transcripts. The actual time course of cytoplasmic heterogeneous RNA accumulation during sea urchin oogenesis has yet to be measured.

Diverse RNA species are also synthesized in *vitellogenic* oocytes, including both heterogeneous and ribosomal RNAs. Ruderman and Schmidt (1981) showed that at this stage the majority of precursor incorporated per oocyte in a 2 hr labeling period is located in mitochondrial rather than nuclear transcripts, just in unfertilized eggs and early embryos (Chapter III). Messenger RNAs for the histones were identified among the newly synthesized transcription products of the germinal vesicle, and a large fraction of the newly synthesized histone mRNA is rapidly assembled into polysomes and translated (Ruderman and Schmidt, 1981). There are about 10⁶ copies of each core nucleosomal histone mRNA of the *early* or α-subtypes per egg (Mauron *et al.*, 1982; reviewed in Chapter II). However, Angerer *et al.* (1984) showed by *in situ* hybridization with early histone gene probes that the α-histone mRNAs do not accumulate in vitellogenic oocytes, but appear only after maturation, when they are synthesized and stored within the egg pronucleus. Were all 400 histone genes coding for each early histone species in the haploid pronucleus as active as are these genes in cleavage stage embryos (see Chapter III), the accumulation of ~10⁶ mRNA molecules of each histone species would require less than two days. The histone mRNAs synthesized in vitellogenic oocytes are thus likely to be of the cleavage stage (CS) type. As reviewed in Chapter II, CS histones are stored in the unfertilized egg in sufficient quantity to accommodate the DNA of a number of embryonic division cycles. In contrast to the α-histone messages, the total poly(A) RNA, and several other specific mRNAs, are distributed throughout the cytoplasm of the mature egg, and total poly(A) RNA is accumulated both in the germinal vesicle and the cytoplasm during the vitellogenic phase of oogenesis (Venezky *et al.*, 1981; Angerer *et al.*, 1984). It is likely, by analogy with the case in *Xenopus*, that the α-histone mRNAs are sequestered in the pronuclear compartment by some specific protein to which they are bound. Messenger RNA for actins and tubulins are also prevalent in growing oocytes, and these proteins are actively synthesized during oogenesis. However, at maturation these proteins cease to be translated at a high rate

(Grainger *et al.,* 1986), and as we have seen (cf. Chapters II and IV), only a small amount of maternal mRNA coding for these proteins is stored in the unfertilized egg.

A different picture emerges in considering rRNA accumulation during sea urchin oogenesis. Measurements of absolute rRNA synthesis rate were carried out in *Tripneustes gratilla* oocytes by Griffith *et al.* (1981). The average rate of synthesis per oocyte is about 1.1×10^5 molecules of rRNA precursor per hour. The mature egg contains about 4×10^8 ribosomes, and thus 4 to 5 months would be required for accumulation. Since all stages of oocyte were included in these experiments, it is possible that previtellogenic oocytes synthesize rRNA less actively, and that an even longer time is necessary, e.g., the whole of the oocyte growth period. In echinoids rDNA amplification does not occur (Vincent *et al.,* 1969; Griffith *et al.,* 1981). The 4C *T. gratilla* oocyte genome contains about 200 rRNA genes (Griffith *et al.,* 1981), and in order to account for the measured rate of synthesis initiation must occur on each rRNA gene about every 6.5 sec. These rRNA synthesis measurements lead to the conclusion that essentially all of the rRNA genes are operating at close to the maximum possible rate throughout oogenesis. Thus if the polymerases were packed only 100 nt apart, a translocation rate of about 15 nt sec^{-1} would be required. Aronson and Chen (1977) determined rates of 13 nt sec^{-1} for both polymerase I and II in feeding *S. purpuratus* larvae, and 6-9 nt sec^{-1} for embryos of this species, which are cultured at a temperature several degrees lower than are *T. gratilla* embryos. It follows that for sea urchins, as for other diverse creatures considered in this chapter, e.g., *Drosophila* and *Xenopus,* the rate of accumulation of rRNA may be among the parameters that determine the overall pace of oogenesis.

Seventy percent of the poly(A) RNA stored in sea urchin eggs consists of nontranslatable, interspersed transcripts, just as in amphibian eggs. The structural characteristics of the interspersed RNAs that can be recovered from mature sea urchin eggs, and their disposition in the embryo, are reviewed in the discussion of maternal RNAs in Chapter II. From their complexity (Costantini *et al.,* 1980) it is clear that the interspersed RNAs are the stored products of a great many discrete transcription units operative during oogenesis. Posakony *et al.* (1983) estimated that the prevalences of a set of individual interspersed poly(A) RNAs fall in the range 10^3 to 10^4 molecules per egg, and accurate molecular titration data for two cloned examples (Lev *et al.,* 1980; Calzone *et al.,* 1986b) indicate a few hundred to about a thousand copies per egg. Thus the abundance of these sequences is about 2-3 orders of magnitude lower than for the interspersed poly(A) RNAs of *Xenopus* eggs, just as are the respective abundances in these two systems of histone mRNAs, rRNAs, total poly(A) RNA, or average complex class transcripts. Yet both the structure and the amount of interspersed transcripts relative to total egg poly(A) RNA is the same in the eggs of these two species. This surprising correspondence clearly does not favor the specific explanation that interspersed poly(A) RNAs are incidentally released to the cytoplasm because processing in the germinal vesicle cannot keep pace with the exceptionally high rate of transcript production, e.g., as in lampbrush chromosomes. That is, the rate of heterogeneous RNA transcription in sea urchin oocytes is probably quite low. Perhaps the flow of interspersed poly(A) RNAs into the cytoplasm is a tangential consequence of a basic inefficiency of processing in the oocyte nucleus that is yet unexplained, though very general in occurrence. Or perhaps their storage in the cytoplasm in the course of two such different forms of oogenesis implies that like other maternal transcript species, these also perform a functional role after fertilization.

(ii) Transcription and the Accumulation of Maternal RNAs during Oogenesis in the Mouse

Overall Rate of Heterogeneous RNA Synthesis and the Absence of Lampbrush Chromosomes

Oogonial multiplication, premeiotic DNA synthesis, and the initial stages of the meiotic prophase occur in eutherian mammals during fetal life. In the neonatal mouse the primary oocytes are arrested at early diplotene in a nongrowing state, termed the *dictyate* phase. Beginning a few days after birth and throughout the reproductive life of the female small groups of oocytes periodically withdraw from the dictyate pool, and enter the growth phase of oogenesis. Dictyate mouse oocytes may thus persist in a quiescent though viable state for anywhere from a few weeks to 18 months or more, and in the human where a similar course of events obtains, for over 40 years. Once activated the process of oocyte growth in the mouse requires only two weeks, during which the diameter of the oocyte increases from about 12 μm to 85 μm, a greater than 300-fold change in volume. The maternal RNAs found in the mature egg are synthesized and accumulated during this period, and many cytological and biochemical changes that lie outside the scope of this discussion occur as well (see review in Wassarman, 1983). Within several days after growth is completed the follicular structure surrounding the oocyte undergoes an enormous expansion, and the meiotic prophase terminates at ovulation with the completion of the first reduction division.

Midgrowth phase oocytes denuded of follicle cells can be cultured in the presence of a feeder layer of somatic follicular cells or in excised follicles, under conditions in which growth continues at about 70% of the normal rate (Eppig, 1977; Bachvarova *et al.*, 1980). The synthesis and turnover rates of RNAs synthesized in cultured oocytes that had been labeled for various periods of time were measured by Bachvarova (1981). Some results of this study are summarized in Table 5.9. As in *Xenopus* lampbrush chromosome oocytes, by far the major fraction of the newly synthesized heterogeneous RNA is unstable, displaying a typical nuclear RNA half-life estimated at about 20 min. Table 5.9 shows that less than 2% of the heterogeneous RNA synthesized is destined to become stable transcription products (0.0075 pg

TABLE 5.9.   RNA Synthesis Rates in Growing
Mouse Oocytes

RNA Class	Synthesis rate (pg min^{-1})
Stable RNAs	
18 + 28S rRNA	0.015[a]
Heterogeneous RNA (≤36S)	0.0075[a]
Poly(A) RNA	0.005[b]
Unstable RNA	
Heterogeneous RNA ($t_{1/2}$ = 20 min)	0.43[a]
Total heterogeneous RNA	0.44[c]

[a] From data of Bachvarova (1981), after correcting for a revised estimate of rRNA content. Bachvarova (1981) assumed 400 pg of rRNA per egg, but subsequent measurements of Kaplan *et al.* (1982) provide a more accurate value of 300 pg. The synthesis rate data were obtained from the amount of radioactive precursor incorporated as a function of time into RNAs of various classes as resolved on electrophoretic gels. It is assumed that the same precursor pool feeds both rRNA and heterogeneous RNA synthesis. Since newly synthesized rRNA is completely stable during oogenesis (Jahn *et al.*, 1976; Brower *et al.*, 1981), the rate of synthesis of all classes of RNA could be estimated from the net rate of rRNA mass accumulation and the rate of label incorporation into rRNA.
[b] De Leon *et al.* (1983). Two-thirds of the newly synthesized stable heterogeneous RNA is polyadenylated.
[c] Sum of stable and unstable heterogeneous RNA.

min^{-1} ÷ 0.44 pg min^{-1}). This value is similar to those estimated from the similar comparisons shown in Table 5.4 for amphibian oocytes. The low ratio of stable RNA transcript production to total heterogeneous RNA transcription is thus clearly not dependent on the overall *rate* of chromosomal RNA synthesis, but is rather an intrinsic aspect of transcription and processing in the oocyte nucleus. The major result shown in Table 5.9 is for our purposes the *low absolute rate* of heterogeneous RNA synthesis. The comparable rate was estimated in Table 5.3 as 18.3 pg min^{-1} in the lampbrush chromosome stage *Xenopus* oocyte, which is over forty times the rate measured for the mouse oocyte. Yet these organisms have about the same haploid genome size, approximately 3 pg.

A light and electron microscope investigation of the cytological state of mouse and Rhesus monkey oocyte chromosomes during the growth phase was carried out by Bachvarova *et al.* (1982). At light microscope magnification the oocyte chromosomes appear fuzzy and diffuse, with central proteinaceous cores surrounded by loose bundles of fibrils. In basic organization these structures bear no homology with the lampbrush chromosomes of amphibian oocytes. The key observation as illustrated in Fig. 5.21 is that when examined in the electron microscope the transcription units of growing mouse oocyte nuclei are generally populated only sparsely with nascent RNA molecules. As discussed earlier, this provides direct evidence that true lampbrush chromosomes are absent where the rate of heterogeneous RNA transcription is relatively low. Lampbrush chromosomes are to be considered the cytological manifestation of a condition in which the maximal transcriptional initiation rate is established in essentially all active regions of the oocyte genome.

The complexity of the stored maternal RNA has not been measured for any mammalian egg. If it is assumed to be approximately that of sea urchin and *Xenopus* maternal poly(A) RNAs, and thus that something over 10^4 diverse transcription units are represented, the rate of synthesis of 0.005 pg min^{-1} per nucleus shown in Table 5.9 for stable poly(A) RNA (or of 0.0075 pg min^{-1} per nucleus for total stable heterogenous RNA) implies that on the average no more than one or a few nascent transcripts would be found on the active genes of the growing oocyte. Thus 0.005 pg min^{-1} would represent a production of about 5000 molecules min^{-1} per nucleus, assuming a mean length of 2000 nt for the poly(A) RNA (Bachvarova and De Leon, 1980; Clegg and Pikó, 1983a), distributed among at least 40,000 transcription units in the 4C oocyte nucleus. It follows that initiation would occur on each productive transcription unit, on the average, only once every eight minutes, and thus for a 37°C polymerase translocation rate on the order of 20-50 nt sec^{-1}, the nascent transcripts would be spaced $1-2.5 \times 10^4$ nt apart. This result predicts exactly the low transcript frequency illustrated in Fig. 5.21(a) and (b), though of course were processing very inefficient, or if there are transcription units that produce RNAs *wholly* confined to the nucleus, the transcript density in some regions could be markedly higher.

Accumulation of Stable Maternal RNAs

The general pattern of RNA synthesis appears not to change during the growth phase of oogenesis. A number of investigations have shown that at all stages mouse oocytes synthesize 18 and 28S rRNAs, 5S rRNA, tRNAs and poly(A) RNAs, and that throughout, the ratio of these various components in the newly synthesized RNA remains constant (Jahn *et al.*, 1976; Bachvarova and De Leon, 1980; Brower *et al.*, 1981; Sternlicht and Schultz, 1981; De Leon *et al.*, 1983; Boreen *et al.*, 1983). An early growth stage characterized primarily by synthesis of tRNA and 5S RNA thus does not exist in the mouse (Boreen *et al.*, 1983), which in this respect resembles the

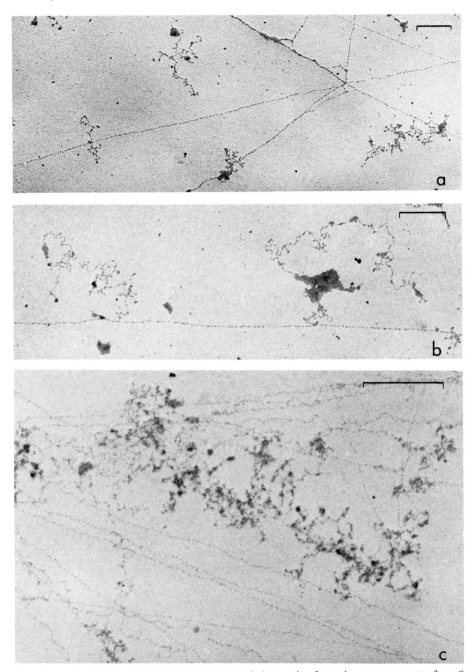

Fig. 5.21. Transcription units found in spread chromatin of growing mouse oocytes from 8 or 12 day old mice. (a) Three single nascent RNP fibrils and one free RNP molecule are visible. (b) Two fibrils on one transcription unit. (c) Multiple fibrils on one transcription unit. Bars represent 0.5 μm. [From R. Bachvarova, J. P. Burns, I. Spiegelman, J. Choy, and R. S. K. Chaganti (1982). *Chromosoma* **86**, 181.]

sea urchin and *Urechis* examples, and differs from those vertebrates that utilize lampbrush chromosomes. By day 14-15 the oocyte has accumulated its final maternal complement of rRNA, about 300 pg, as well as of stable poly(A) RNA, though its volume is only 50-60% of that of a mature 21 day oocyte (Sternlicht and Schultz, 1981; Kaplan *et al.*, 1982). Though it continues to grow and to accumulate protein for the following 7 days, 14-15 day oocytes are already functionally mature, in the sense that they have attained competence to resume the meiotic division process (Sorenson and Wassarman, 1976). RNA synthesis, but not accumulation, also continues in late oocytes, and becomes undetectable only at germinal vesicle breakdown and meiotic maturation (reviewed in Wassarman, 1983).

For the first 9 days of the growth phase the rate of rRNA synthesis is about 2/3 of that shown in Table 5.9, or about 0.01 pg min^{-1}, while the rate of 0.015 pg min^{-1} obtains between days 9 and 14-15 (Kaplan *et al.*, 1982). Ribosomal gene amplification has not been reported in the mouse, though a fourfold rDNA amplification has been noted both in diplotene baboon oocytes (Wolgemuth *et al.*, 1980) and in human oocytes (Wolgemuth *et al.*, 1979). However, ribosomal gene amplification is not required to account for the measured rate of rRNA accumulation in mouse oocytes. Thus Kaplan *et al.* (1982) calculated that the rate of 0.015 pg min^{-1} represents the synthesis of about one rRNA molecule per 15 sec for each of the 1200 genomic rRNA genes present in the 4C oocyte nucleus. This rate is below that measured in somatic mouse tissue culture cells (Brandhorst and McConkey, 1974).

The mouse egg is unusual in the large quantity of maternal poly(A) RNA that it contains, relative to total RNA. Bachvarova and De Leon (1980) labeled oocytes at all stages of the growth period by injection of radioactive precursor into the bursal sac, and on recovery of matured eggs from 4 to 19 days later found that about 8.3% of the mass of labeled RNA retained is poly(A) RNA. The length of the interval between injection of precursor and collection of ova did not affect this result. Calculations based on the mass of poly(A) per egg (Levey *et al.*, 1978; Clegg and Pikó, 1983a) indicate the quantity of stored poly(A) RNA is about 25 pg, or 5.5% of the total RNA. This represents about 2.4×10^7 poly(A) RNA molecules of average length (Clegg and Pikó, 1983a). Since the bulk of the RNA accumulating during growth is rRNA, which is completely stable, the close agreement between the calculated mass of poly(A) RNA and the fraction of total egg RNA labeled during oogenesis that is poly(A) RNA shows that the poly(A) RNA synthesized in the oocyte is largely stable as well. This has also been demonstrated directly in pulse-chase experiments carried out on cultured follicles by Brower *et al.* (1981). As it accumulates the poly(A) RNA is deposited in the cytoplasm of the growing oocytes, according to *in situ* hybridizations carried out with ^{3}H-poly(U) (Sternlicht and Schultz, 1981).

About half the poly(A) RNA accumulated in the course of oogenesis disappears during the maturation period, though it is not clear whether in general it is degraded or is only deadenylated (De Leon *et al.*, 1983). The oocyte actin messages, however, have been shown to be deadenylated rather than degraded during maturation though total stable RNA content does decline during this period by about 20% (Kaplan *et al.*, 1985). In any case, the fraction of *newly synthesized* oocyte RNA that is poly(A) RNA is about twice the fraction of mature, unfertilized egg RNA that is poly(A) RNA. Thus about 18-20% of precursor incorporated in the relatively stable transcripts of *growing oocytes* is found in poly(A) RNA (Brower *et al.*, 1981; De Leon *et al.*, 1983). The newly synthesized poly(A) RNA of the oocyte distributes into two functionally distinct compartments. About 23% is assembled onto polysomes and utilized directly for oocyte protein synthesis, and the half-life of this fraction is about 6 days, according to measurements carried out on cultured oocytes in contact with follicle cells or in whole follicles (De Leon *et al.*, 1983). The remainder is stored, and it displays no detectable turnover. The pool of poly(A) RNA present in the oocyte at termination of growth is thus the sum of the steady state quantity of polysomal poly(A) RNA, and the quantity of nontranslated poly(A) RNA that has been synthesized and accumulated during oogenesis. Given that the rate of synthesis of total poly(A) RNA is 0.005 pg min^{-1} (Table 5.9), the steady state amount of polysomal poly(A) RNA would be about 14 pg, and the final quantity of stored poly(A) RNA about 78 pg, for a total of 92 pg, a fifth of all the RNA in the oocyte. This value may be compared with the equivalent fraction in the mature *Xenopus* oocyte, which is about 2.2% (90 ng poly(A) RNA ÷ 4000 ng total RNA; see above).

Protein Synthesis during Oogenesis in the Mouse

The absolute rate of protein synthesis increases proportionately with oocyte diameter from shortly after the beginning of the growth phase to its termination (Schultz *et al.*, 1979b). Protein synthesis rate measurements were obtained on cultured oocytes, and are based on determination of the methionine pool specific activity. The maximum rate of synthesis is about 42 pg hr^{-1} in full grown oocytes of about 80 μm diameter, which is close to forty times the rate observed in 10-15 μm oocytes. Of the total newly synthesized protein only 1-2% represents proteins encoded by the 10^5 mitochondrial genomes present per egg (Cascio and Wassarman, 1981). De Leon *et al.* (1983) showed that throughout the period of oocyte growth about 35% of the ribosomes are engaged in polysomal structures. The remainder may be aggregated in large assemblages utilized for storage of ribosomes in an inactive form, as they can be sedimented from homogenates at very low centrifugal forces (Brower and Schultz, 1982). Assuming a ribosomal translocation rate of 3 codons sec^{-1} per ribosome (values for 37°C vertebrate systems range from about 2-8 codons sec^{-1} per ribosome; e.g., Kafatos, 1972), the rate of protein synthesis measured in fully-grown oocytes is close to that expected, providing that 35% of the ribosomes are included in the translational apparatus (i.e., 2.6×10^7 ribosomes). However, this rate of translation is 4-8-fold *lower* than expected for the steady state quantity of *polysomal* poly(A) RNA, i.e., 14 pg (see above). The major cause would seem to be a low rate of translational initiation, since the ratio of the mass of mRNA to the mass of polysomal rRNA is about four times greater than the 4% value obtained for fully-loaded polysomes (i.e., 14 pg mRNA ÷ 0.35 × 300 pg rRNA = 0.13). The rate of ribosomal translocation may also be depressed about 2-fold. A caveat pointed out by Kaplan *et al.* (1982) is that the protein synthesis rate measurements of Schultz *et al.* (1979b) were carried out under culture conditions inadequate to support net growth, and this could conceivably have adversely affected the translational activity of the oocytes. In any case, assuming the rates measured, it can be calculated that if all the protein synthesized during oogenesis were stably accumulated, this would account for only about 60% of the 25 ng of protein in the fully-grown oocyte (Schultz *et al.*, 1979b). This is in fact an overestimate, since only about 60% of the newly synthesized protein is stable (Kaplan *et al.*, 1982). The oocyte thus probably absorbs the balance of its protein from the blood, by endocytosis, just as yolk is taken up in lower vertebrates (see Wassarman, 1983, for review and discussion of this possibility).

Qualitative analysis by two-dimensional gel electrophoresis has indicated that most of the ~400 species resolved are synthesized throughout oogenesis, and that the newly synthesized species are the same as the species detected by sensitive staining procedures (Schultz *et al.*, 1979b; Wassarman, 1983). However, there are a few specific proteins whose rate of synthesis increases sharply in fully-grown oocytes (Schultz *et al.*, 1979b; Kaplan *et al.*, 1982). At meiotic maturation, as reviewed in Chapter II, the overall pattern of protein synthesis changes noticeably, and these changes are programmed at a posttranscriptional level since there is little new mRNA synthesis at this time. The stored maternal mRNA accumulated during oogenesis does not appear to be utilized until maturation, when a qualitative pattern of protein synthesis is established that persists until well after fertilization (Chapter II).

Synthesis of a number of known protein species has been studied in mouse oocytes. In Table 5.10 are presented synthesis rate data for several examples. Though the estimates for the number of functioning mRNA molecules are of limited accuracy (see note *b*), they suffice to indicate the minute accumulations of mRNA in the mouse oocyte compared, for example to the specific message contents in *Xenopus* oocytes (see Table 5.6). These estimates do not include *stored* maternal mRNAs coding for the same protein species (see, e.g., the measurements shown for total stored *vs.* translated actin message). However, in all the cases shown, the rate of synthesis *falls* rather than increases during maturation, when the maternal mRNA begins to be utilized. The histones and ribosomal proteins, which were also considered in respect to protein synthesis in *Xenopus* oocytes, again provide interesting contrasts. In the mouse, the 60 pg of core histones accumulated

TABLE 5.10. Synthesis Rates for Some Specific Proteins Synthesized in Fully Grown Mouse Oocytes

Protein	Total synthesis (%)	Synthesis rate (pg hr^{-1})[a]	Quantity in egg (pg)	Approximate number of active molecules mRNA per oocyte[b]
α and β tubulin[c]	1.5	0.60	250–300	2×10^5
β and γ actin[d]	0.3	0.10	—	3.8×10^4
			—	(3.8×10^5)
Lactate dehydrogenase[e]	1.7	0.7	240	2×10^5
All (~70) ribo- somal proteins[f]	~1.5	~0.6	200	2×10^5
4 nucleosomal core histones[g]	0.4	0.17	60	5×10^4

[a] Calculated from the rate of total protein synthesis in full grown oocytes, 42 pg hr^{-1} (Schultz *et al.*, 1979b; see text).

[b] Calculated assuming that the percent of protein synthesis is also the percent of polysomal mRNA molecules. This calculation is only approximate, since the rate of translation for different mRNAs may vary severalfold. Furthermore, the number of mRNA molecules will be overestimated when the mRNAs are unusually large (relative to the mean of ~2000 nt; Clegg and Pikó, 1983a), and underestimated if the mRNAs are small. The amount of mRNA is taken as 14 pg, the steady state quantity of polysomal poly(A) RNA according to the data of De Leon *et al.* (1983) (see text). The total number of functioning mRNA molecules is thus (14 $\times$ 10^{-12} g $\times$ 6 $\times$ 10^{23}) $\div$ (2000 $\times$ 330 g), or 1.3 $\times$ 10^7.

[c] Schultz *et al.* (1979b). A slightly larger fraction of total protein synthesis is tubulin synthesis in midgrowth oocytes. The final quantity of tubulin is 1000 times less than in *Xenopus* eggs, and about 2 times the amount found in sea urchin eggs (Wassarman, 1983).

[d] Wassarman (1983) estimated that actin synthesis accounts for only 0.3% of the total. However, a direct measurement of actin mRNA by filter hybridization in the mature egg indicates about 430 fg, or 3.8 $\times$ 10^5 molecules of actin message (shown in parentheses; Giebelhaus *et al.*, 1985). The tenfold discrepancy between this value and that derived from the relative translation rate suggests either that actin is translated far less efficiently in the oocyte than in other proteins, or perhaps more likely, that there is a large pool of stored actin message not participating in translation at all. Moor and Osborn (1983) reported that in sheep oocytes actin synthesis accounts for 10% of total protein synthesis.

[e] Cascio and Wassarman (1982).

[f] LaMarca and Wassarman (1979). Estimate extrapolated from studies of 12 ribosomal proteins, assuming that these are representative of all 70 species. The molar rates of synthesis of the various species vary about 4-fold.

[g] Extrapolated from measurements of Wassarman and Mrozak (1981) on histone H4, assuming equimolar synthesis of H2a, H2b, H3, and H4. Values shown are the calculated sum of all four species. The measured rate of synthesis for H4 was 0.043 pg hr^{-1}.

during oogenesis would suffice for the chromatin of only about 10 cells. As will be recalled, mature *Xenopus* oocytes contain histone protein sufficient for over 10^4 cells. The leisurely pace of cell division in the mouse imposes no urgency on new histone synthesis, and whereas in *Xenopus* there is a 50-fold increase in the rate of histone synthesis during maturation, in the mouse the rate actually declines about 40% during maturation (Wassarman and Mrozak, 1981). Thus it is unlikely that there is a very large store of maternal histone mRNA, such as is found, e.g., in both sea urchin and *Xenopus* eggs. An interesting aspect of the data included in Table 5.10 for ribosomal proteins is that many of these are encoded by mRNAs that are distinctly of the rare message class. For example, LaMarca and Wassarman (1979) showed that synthesis of several individual ribosomal proteins represents less than 10^{-4} of total protein synthesis, and on the basis of the calculation in Table 5.10 there would be no more than about a thousand molecules of the messages for such individual species functioning in the oocyte. The quantity of each functional ribosomal protein mRNA species in the mouse oocyte is thus equivalent to ≤ 10 molecules per typical somatic cell (i.e., normalizing 1.3×10^7 polysomal poly(A) RNA molecules in the full-grown mouse oocyte to the amount of poly(A) RNA in a somatic cell, $\sim 10^5$ molecules). The molar rates of synthesis of the individual ribosomal proteins differ by as much as a factor of four (LaMarca and Wassarman, 1979). However, this imbalance is compensated by a differential accumulation of newly synthesized molecules of the various species in the germinal vesicle, so that within this compartment an equimolar ratio is established (LaMarca and Wassarman, 1984). The total amount of newly synthesized protein that is distributed to the germinal vesicle is about 0.9 pg hr^{-1} in the full-grown mouse oocyte, and of this histone accounts for about 10% and the ribosomal proteins about 25% (LaMarca and Wassarman, 1984). After maturation ribosomal proteins continue to be produced in the absence of any new rRNA synthesis. It is not clear whether the maternal messages utilized for this synthesis are the same as were already functioning in the polysomes prior to maturation, or alternatively, are recruited from the previously inactive stored poly(A) RNA pool.

Three secreted proteins, the major constituents of the zona pellucida, account for at least 10% of the newly synthesized proteins in growing oocytes (Wassarman, 1983). The zona pellucida is the thick glycoprotein coat that surrounds the oocyte, and its constituents are synthesized exclusively during the growth phase of oogenesis. Denuded oocytes actively incorporate amino acids or labeled fucose into these proteins. An *O*-linked oligosaccharide component of these glycoprotein species, designated ZP3 ($\sim$83 kd) apparently serves as the specific sperm receptor molecule (Wassarman and Bleil, 1982; Florman and Wassarman, 1985). A second protein, ZP2 ($\sim$120 kd), which constitutes over 50% of the total zona pellucida protein, is involved in the hardening of the egg coat on fertilization. This process is mediated by proteolytic enzymes released from cortical granules, just as in the sea urchin. The messages for these proteins must be relatively very prevalent, and the genes for ZP1-3 are likely to be transcribed much more intensely than the genes coding for typical maternal poly(A) RNAs required after fertilization.

A complementary example has been described by Cascio and Wassarman (1982). This is a set of proteins designated the fertilization proteins. FP1-4

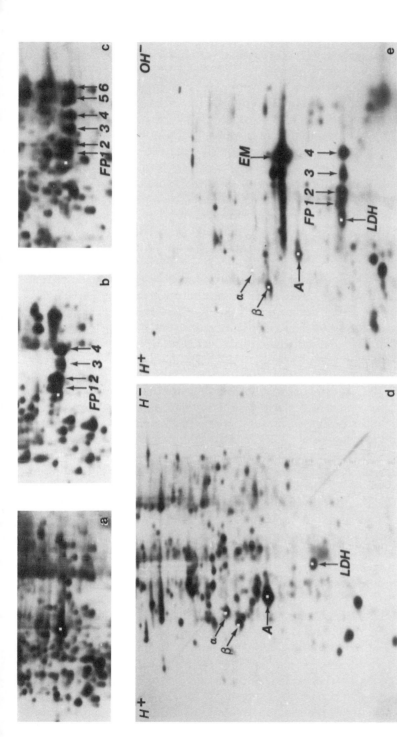

Fig. 5.22. Fertilization proteins (FP1-6) synthesized *in vivo* and in cell-free extracts from mouse oocytes, eggs, and embryos. (a) The 35 kd region of a two-dimensional electrophoretic gel displaying proteins synthesized by ³H-methionine labeled, full-grown oocytes; (b) 1-cell embryos; (c) 2-cell embryos. None of these proteins (arrows) are synthesized in 8-cell embryos. Position of lactate dehydrogenase (LDH) is marked by the white spot in (a)-(c). (d) Proteins synthesized *in vivo* in growing (40 μm) oocytes cultured *in vitro*; (e) similarly analyzed proteins synthesized in a cell-free rabbit reticulocyte translation system programmed with RNA extracted from 40 μm oocytes. Positions of LDH, α and β tubulin, and actin (A) are indicated. EM, products of endogenous reticulocyte message. [From S. M. Cascio and P. M. Wassarman (1982). *Dev. Biol.* **89**, 397.]

are synthesized at very low levels, each accounting for about 0.05% of total protein synthesis in growing oocytes, and FP5 and FP6 are not detectable at all. However, after fertilization, FP1-6 become major species, together representing 3-5% of the total protein synthesis in one and two cell embryos. The dramatic change in their rate of synthesis which occurs as a result of maturation and fertilization is shown in Fig. 5.22(a)-(c). The fertilization proteins are among those coded primarily by the stored, *inactive* poly(A) RNA of the oocyte (Braude *et al.*, 1979; Cascio and Wassarman, 1982). This is demonstrated in the cell-free translation experiment shown in Fig. 5.22(d) and (e). The partition between polysomal and stored poly(A) RNA in the oocyte thus encompasses some sharp *qualitative* delineations, illustrated respectively by the zona pellucida proteins and the fertilization proteins.

We have now traversed much of the known landscape, in respect to transcriptional activity in oogenesis. The three major exemplars, *Drosophila*, *Xenopus*, and the mouse, utilize their germ line genomes in very different ways in the preparation of their oocytes. In each case the strategies they employ can be understood in logistic terms, and on finer scale examination, this is seen to be true as well for a variety of individual genes and sets of genes. From this survey emerges an image of the various devices by which the maternal transcript pools are created during oogenesis.

The major conceptual issues that lie ahead divide into two general domains. These can be described loosely as the mechanisms that effect gene transcription during oogenesis; and the molecular nature and source during oogenesis of the regulatory information stored in mature eggs. Among many others, problems subsumed in the first of these categories include definition of the transcriptional components responsible for the imposition in lampbrush chromosomes of maximum initiation rates in all or most functional loci; the structural determinants that distinguish loop from chromomeric regions of the genome; the posttranscriptional fate of readthrough transcripts; and the nature of those processes that take place specifically in the germ line, that endow nurse cell and oocyte genomes alone with the capacity to express the particular sets of genes represented in the maternal RNA. Exploration of the second of these domains has hardly begun. It is evident that the egg contains molecules that in some manner determine the patterns of gene expression in the early embryo. As documented earlier, for instance, genes normally functional in the oocyte but quiescent in somatic cells can be reactivated when injected into oocyte cytoplasm. More generally, the set of genes that is expressed during oogenesis is largely utilized in the early embryo nuclei as well, at least in those organisms investigated. Thus the descendants of the previously quiescent paternal genome seem to be "instructed" by their new environment to behave in the same specific manner as did the oocyte (or nurse cell) genomes. In many examples, reviewed in the following chapter, regulatory information that ultimately affects gene activity becomes spatially localized in the egg cytoplasm. Whether the topological distribution of such information occurs before, during, or after maturation and fertilization, the responsible molecules are probably synthesized along with other stored maternal components during oogenesis. Thus, within the oocyte transcript populations there must be included molecules whose function is primarily regulatory. The nature of such transcripts, and of the regulatory genes that they represent, are questions that are of fundamental interest. Various speculations on this subject have included the ideas that maternal mRNAs which encode regulatory triggers are localized asymmetrically; that maternal mRNAs direct the synthesis of many sequence-specific gene regulatory proteins which are ultimately targeted to the blastomere nuclei; or that transcripts synthesized and stored during oogenesis find their way to the blastomere nuclei, where their interactions with newly synthesized embryonic transcripts might affect processing in the embryo (see, e.g., Davidson and Britten, 1971, 1979; Costantini *et al.*, 1980; Posakony *et al.*, 1983).

The subject of the following chapter is the process of regional differentiation in the initial stages of embryogenesis. To understand this process it will ultimately be necessary to define the causal role of molecules synthesized during oogenesis in generating the first asymmetric patterns of genome function in the embryo.

VI

Cytoplasmic Localization

Localization as used here is the specification of cell fate according to the sector of egg cytoplasm inherited by an embryonic cell lineage. The localization phenomenon is particularly interesting to contemporary students of development because it suggests that molecules triggering specific programs of development are sequestered in the egg. As the early blastomeres divide up the egg cytoplasm, they appear, in some organisms, to inherit "instructions" for various kinds of cell differentiation, including specific patterns of macromolecular synthesis. Among the main problems posed by the localization phenomenon are the molecular nature of the morphogenetic agents stored in the cytoplasm; the level(s) of control at which these agents might operate; their subcellular location, and the cytoskeletal structures to which they are anchored; and the means by which they are distributed to given regions of the egg or early embryo.

In one form or another these problems have been studied ever since the founding of the field of cellular developmental biology. The classic review of the extensive and sometimes incisive experimentation carried out on localization in the late 19th and early 20th centuries remains E. B. Wilson's monumental treatise, "The Cell in Development and Heredity" (1925). Many of Wilson's clarifications and insights into the localization phenomenon are still relevant. The modern form of this area of developmental biology was shaped directly by classical experimentation, and by the often vigorous conceptual discussions that the subject aroused. After 1925 there ensued a relative decline in the level of interest in localization. Though what could be termed classical experimental methods continued to be applied, the field no longer occupied the fashionable, intellectually central position that it had earlier. Its partial eclipse was probably due mainly to successful competition from new approaches, in particular the rise of cellular and developmental genetics, the attractiveness of embryonic induction systems, and the growing accessability of metabolic, and then molecular measurements that could be carried out on embryos. Nonetheless, during the intervening decades many important advances occurred, resulting in a major improvement in the veracity of the data relevant to the phenomenon of localization, if not in our understanding of its mechanisms.

There has now occurred a significant revival of interest in this fascinating aspect of early development. The following review is written from the vantage point of the contemporary developmental molecular biologist, for whom the fundamental objective is to discern the nature of the cytoplasmic control of genomic processes that is implied by the localization phenomenon. The initial problem, however, is to convince ourselves that cytoplasmic localization of morphogens in eggs is a demonstrable reality, and not an illusion, and to determine how widespread is this developmental mechanism in the animal kingdom. These problems were also faced by classical biologists. Thus this chapter begins with a brief review of several convincing studies of localization carried out around the turn of the century, and a summary of the major conclusions, as well as points of contention, deriving from the classical focus on this subject. While the intellectual history of "the localization problem," as Wilson termed it, is a fascinating subject in its own right, the intent is rather to provide the basic foundation on which current efforts at experimental analysis of localization still squarely rest.

1. CLASSICAL DEFINITIONS OF THE PHENOMENON OF CYTOPLASMIC LOCALIZATION IN EMBRYOS

(i) Cell Lineage and Cell Fate

The observations that led initially to the concept of cytoplasmic localization were the early studies of embryonic cell lineage. The description by C. O. Whitman (1878) of cleavage and cell fate in the leech *Clepsine* was the first complete study of this kind, and greatly influenced the field. In this organism, as in many others shortly thereafter the subject of cell lineage analysis, the number of individual blastomeres at the cleavage and blastula stages is relatively small, and they can be distinguished by size, shape, and position, so that the developmental destiny of each cell and its lineal descendants can be followed. Whitman showed that the mesodermal and neural "germ bands" of the leech embryo are budded forth from bilateral stem cells, the teloblasts (annelid teloblasts had first been noticed by Kowalevsky, 1871). Whitman's most important finding was that the ventral nerve cords of *Clepsine* develop from bilateral sets of four individual neuroblast stem cells, themselves deriving from early bilateral blastomeres. The mesoderm also forms from individual bilateral stem cells, the mesoblasts. Several of Whitman's figures and other contemporary illustrations that display annelid teloblasts are reproduced in Fig. 6.1. Though the discussion in Whitman's 1878

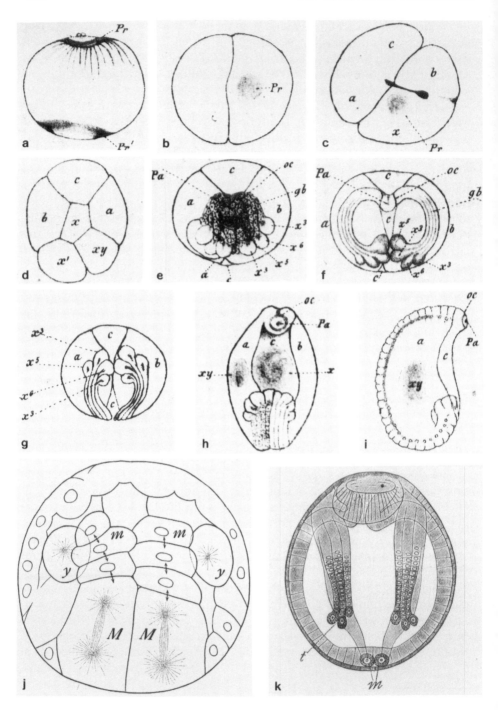

publication is largely focused on the evolutionary affinities implied by his observations, their significance for the cell biology of development was widely recognized. This was the demonstration that highly differentiated structures, such as nerve cords, can be derived lineally from *specific individual blastomeres that arise only in certain regions of the egg, by a precise pattern of cleavage, and that normally inherit certain segments of the egg cytoplasm.*

In the following decades elegant and in some cases extremely detailed cell lineage investigations were carried out on animals belonging to diverse phylogenetic groups, including numerous molluscs and annelids, the nematode *Ascaris*, and various ascidians, flatworms, and ctenophores. References to many such studies can be found in Wilson (1925), Richards (1931), Costello (1956), and Reverberi (1971a). In many cases the fates of *every* cell of the embryo were determined. Sufficient data were obtained for the annelid-mollusc group to support comparative analyses of the developmental values of homologous blastomeres by Wilson (1898); by Child (1900) in his study of cell lineage in the polychaete *Arenicola*; and by Conklin (1897), in his classic description of the cell lineage of the mollusc *Crepidula*. Many of the investigators who carried out cell lineage studies were deeply interested in the evolutionary implications of cellular homologies in the developmental pro-

---

Fig. 6.1. Mesoblast and neuroblast stem cells in the cell lineage of annelids, as observed by 19th century microscopists. (a)-(i) Observations of C. O. Whitman on *Clepsine parasitica* and *Clepsine complanata* (1878). (a) Uncleaved fertilized egg (Pr, Pr': "polar rings," discernable in the uncleaved egg cytoplasm). Pr marks the future oral pole of the embryo; the Pr-Pr' axis is the future longitudinal axis of the animal. (b) 1st cleavage. Pr is confined to one of the two initial blastomeres. (c) 2nd cleavage. Pr is confined to cell x. This is the progenitor of the future mesoblasts, i.e., the mesoderm stem cells. a,b,c, are the other three macromeres. (d) Later cleavage viewed from the vegetal pole. Macromere x of (c) has given rise to the two mesoblasts, now called x and xy, and to x', the first neuroblast stem cell. (e) Gastrula, viewed from the oral pole where the mouth will be located. (Pa, pharyngeal orifice). The germ bands (gb) can now be observed. Whitman showed that each germ band derives from one mesoblast (x or xy), now buried within the embryo, and four neuroblast stem cells on each side (x³, x⁵, x⁶, and x³ on right). The germ band extends in an anterior direction as columns of cells are progressively budded off from these stem cells. (f) The four neuroblast columns on each germ band extend forward around the embryo (oc, pharyngeal clefts). (g) Embryo viewed from posterior end, showing neuroblast bands beginning to fuse. (h) Fusion complete, shortly before hatching. Embryo is now one day old. (i) Same, viewed from side. Ultimately there are 33 ganglia formed from the 8 rows of neuroblasts, and 33 segments. [From C. O. Whitman (1878). *Q. J. Microsc. Sci.* **18**, 215.] (j) Mesoblast stem cells (M) in a polychaete annelid, *Aricia foetida*, giving rise to the mesoblast bands (m); (y), cells deriving from other blastomeres. [From E. B. Wilson (1898). *Ann. N.Y. Acad. Sci* **11**, 1.] (k) Germ bands in an oligochaete annelid, *Allolophora foetida*. Gastrula viewed from the ventral side. (m) represents mesoblast stem cells, from which the bilateral mesoblast columns can be seen extending in an anterior direction. Three teloblasts (t), i.e., neuroblast stem cells, can be observed on each of the germ bands. These are giving rise to the columns of neuroblasts from which segmental ganglia will arise. [From E. B. Wilson (1896b). "The Cell in Development and Inheritance," p. 274. Macmillan, New York.]

cess. Thus a by-product of these studies was the construction of an imposing morphological argument for the phylogenetic relationships of various protostome phyla, particularly polyclad worms, annelids, and molluscs. For us, as for contemporaries focused primarily on the problems of developmental mechanism, the main significance of the cell lineage determinations lies in a different direction. Conklin's description of the lineage of *Crepidula* in many ways illustrates the major theoretical impact of these studies. Conklin followed the fate of individual blastomere lineages beyond the 100-cell stage, from their initial appearance to their participation in differentiated structures, and thus defined exactly the embryonic cellular origins of all the organ systems and structures of the advanced veliger larva (Fig. 6.2). The overall result was to confirm that every blastomere has a precise developmental fate, leading to a specific, differentiated multicellular structure. Discussing his own experiments on molluscan eggs, Wilson (1904a) wrote:

> The entire cleavage pattern represents ... a *mosaic-work* of self-differentiating cells ... [italics added]

Fig. 6.2. (See color plate following p. 256.) Cell lineage in the gastropod mollusc *Crepidula fornicata*. The nomenclature system is as follows. The first four large blastomeres (macromeres) A, B, C, D define the quandrants into which the embryo is initially divided. 1a, 1b, 1c, 1d form the first "quartet" of small cells (micromeres) given off by the yolk-laden macromeres A–D, respectively. Thereafter the cells descended from 1a are $1a^1$ and $1a^2$, $1a^{1.1}$, $1a^{1.2}$ and $1a^{2.1}$, $1a^{2.2}$, etc; and the quartets of small cells given off at the next cleavages from the macromeres A–D are 2a–d and 3a–d. The subsequent lineage is defined wholly in terms of these cells, i.e., by naming their descendants. (a) Fourth cleavage embryo seen from above. The 1st and 2nd quartets are visible. The 1st quartet cells ultimately give rise to the larval sensory organs and nervous system, and to a portion of the larval velum. The second (and third) quartet cells form lineages that contribute to mouth, stomodaeum, velum, shell gland, foot, and other external structures. (b) 29-cell embryo, same view. The mesoblast stem cell 4d or ME is colored (cf. cell x of Fig. 6.1). This cell marks the definitive posterior pole of the embryo. It appears precociously, since in the A, B, and C quadrants only the third quartet cells have so far appeared. (c) Side view of egg at about same stage as in (b). (d) Side view of 52-cell embryo showing division of 4d (ME) into ME^1 and ME^2. This division marks the imposition of bilateral symmetry. (e) The A, B, and C macromeres have divided to form new yolky macromeres, 4A, 4B, 4C. These macromeres do not form mesoderm, as this is an exclusive property of the D-quadrant product ME. The first progeny of ME^1 and ME^2 are m^1 and m^2. Bilateral entodermal stem cells E^1 and E^2 are now observed as well, colored blue. These cells give rise to the larval intestine and stomach. (f) 86-cell stage. 1st quartet cells forming the apical "cross" are outlined. Transverse arms of the cross give rise to part of the velum. (g) Gastrula, from top. The apical cross structure has shifted toward anterior. Extensive mesoblast columns can be seen. (h) Late gastrula, showing V_1 and V_2, rows of velar cells; St, stomodaeum; Ap, apical pole of embryo; AnC, anal cells (cells $1b^{122221}$ and $1b^{22222}$ are slightly mislabeled in figure, according to Conklin). (i) Veliger larva, nervous system indicated in red. The organ systems and structures shown in the larvae are AC, apical cell plate; V, velar ridge; T, tentacle; Pre Ov and Post Ov, pre oral and post oral velum; ExK, external kidney; Op, operculum; Int, intestine (in blue); P, foot; Ot, otocyst; nervous system (in red): CG, cerebral ganglion; CC, cerebral commissure. [From E. G. Conklin (1897). *J. Morphol.* **13**, 1.]

It was evident, however, that the specific nature of the "self-differentiation" undergone by each blastomere lineage depends in some way on the cytoplasmic substances that it has inherited from the egg. The cleavage pattern serves to distribute these substances to the appropriate cells, or as Wilson, continuing his discussion put it:

> the *germ regions* prelocalized in the unsegmented egg are ... accurately marked off by the subsequent lines of cleavage ...

Contemporary writers therefore stressed that the determinants in mosaic developmental processes are the maternal cytoplasmic substances, rather than being a particular property of particular embryonic blastomeres. Thus Child (1900) concluded from his comparative annelid cell lineage analysis:

> A comparison of the process of formation of ... the prototroch in different animals seems to show how absolutely distinct cell-formation and *differentiation* are. The prototroch may consist of eight or of sixteen cells, of twenty-five, of thirty, or of many ... Whether the material (i.e., the egg cytoplasmic substances) is contained within one cell or many is a mere incident so far as the result is concerned. [italics added]

(ii) Localization of Morphogenetic Potential in the Egg of *Cynthia* (*Styela*) *partita* and in Some Other Eggs That Display Determinate Cleavage

The cases of localization that most impressed classical experimentalists were those in which areas of future cell fate could be visibly mapped out on the uncleaved egg cytoplasm. A spectacular example is Conklin's (1905a) study of development in the ascidian *Cynthia* (*Styela*), which greatly expanded and confirmed earlier experiments on ascidian embryos by Chabry (1887). In the egg of *Styela* pigmented areas of cytoplasm corresponding to a morphogenetic fate map for the cells of the early embryo can be distinguished. Some of Conklin's elegant hand-drawn figures are reproduced in Fig. 6.3. These display the relationship between the various pigmented regions of the egg cytoplasm and the tissues ultimately formed from these regions. Five kinds of cytoplasm were noted by Conklin: a dark yellow granular cytoplasm eventually included in the tail muscles of the larva; a light yellow material later segmented into the coelomic mesoderm of the larva; a light gray substance inherited by notochord and neural plate cells; an opaque gray material segregated into the endoderm cell lineage; and a transparent cytoplasm later present only in ectodermal cells. In Fig. 6.3(m) and (n) the dark yellow granular cytoplasm referred to by Conklin as the "yellow crescent" is shown in recent photomicrographs of eggs that had been extracted with detergent (Jeffery and Meier, 1983). As described in the legend, the persistence of the yellow crescent in these eggs indicates that the pigment granules that Conklin saw are embedded in the cytoskeletal matrix of the cytoplasm.

Conklin focused on the striking correspondence between the sets of blastomeres that give rise to specific groups of differentiated larval cells, and the distribution of the five types of egg cytoplasm that he could visually distinguish throughout cleavage. Summarizing his conclusions, Conklin (1905b) wrote:

> As early as the close of the first cleavage ... these strikingly different ooplasmic substances ... are differentiated for particular ends, and ... they give rise to organs of a particular kind. These materials are, therefore, "organ forming substances," and the areas of the egg in which they are localized are "organ forming regions."

The definitive distribution of the "organ forming substances" only appears after fertilization. It is not a primordial property of the egg. The process by which this distribution develops was observed carefully by Conklin and some of his illustrations are reproduced in Fig. 6.4. One axis of polarity is already present in the unfertilized egg. The establishment of the anterior-posterior axis depends on the acentric movement of the male pronucleus.

Fig. 6.3. (See color plate following p. 256.) Eggs and early embryos of *Cynthia (Styela) partita* as drawn by Conklin (1905a), and as photographed after detergent extraction (Jeffery and Meier, 1983). (a) Right side view of living fertilized egg showing the formation of the crescent (*cr*) from the yellow hemisphere (*y.h.*); in (a)-(c) the future dorsal pole is below. The yellow crescent marks the posterior end. Above the yellow crescent is an area of clear protoplasm (*c.p.*). (b) First cleavage of an egg, viewed from the posterior region and showing the form taken by the yellow crescent during the division, the area of clear protoplasm, and the polar bodies (*p.b.*). (c) Left side view of egg of same stage as (b) showing the lateral limits of the yellow crescent, the clear protoplasm in the upper (future ventral) hemisphere, and the yolk (*yk.*) in the lower. The anterior portion of the lower hemisphere is composed of light gray material; this is the gray crescent and is ultimately included in chorda and neural plate cells. (d) Four-cell stage seen from the vegetal pole [*v.p.*, which corresponds to the future dorsal pole in (a)-(c)]; the yellow crescent covers about half of the posterior blastomeres. (*n*), nucleus. (e) 8-cell stage viewed from the right side showing a small amount of yellow protoplasm around all the nuclei. Note the yellow crescent. *A*, anterior; *P*, posterior; *D*, dorsal. (f) 22-cell stage from the vegetal pole; the embryo now contains 4 mesoderm cells (yellow); 10 endoderm, chorda, and neural plate cells (gray); and 8 ectoderm cells (clear). (g) Same stage viewed from the posterior region. (h) 44-cell stage, posterior view, showing separation of another mesenchyme cell (*m'ch.*) from a muscle stem cell (*ms.*). (i) 74-cell stage, dorsal view, showing division of 4 chorda (*ch.*) and 4 neural-plate (*n.p.*) cells; there are 10 mesenchyme and 6 muscle cells, besides 10 endoderm cells. (j) 116-cell stage showing the beginning of gastrulation, and also, the neural plate, chorda, muscle, and mesenchyme cells. (k) Late gastrula; the yellow cells in the midline are mesenchyme cells, the others, presumptive muscle cells. (l) Young tadpole seen from dorsal side, neural groove open in front and closed behind, small-celled mesenchyme in front of large muscle cells. [(a)-(l) From E. G. Conklin (1905a). *J. Acad. Nat. Sci. Philadelphia* **13**, 1.] (m) Photomicrograph of fertilized *Styela* egg at about same stage as in (a). The egg has been extracted with a buffer containing the detergent Triton X-100. This treatment removes over 90% of the lipid and carbohydrate, and about 80% of the protein, leaving behind cytoskeletal elements to which the yellow pigment granules visualized by Conklin are bound. (n) Detergent-extracted two-cell stage viewed from the posterior pole, and displaying the yellow crescent in the vegetal regions of the blastomeres [cf. (b) and (c) above]. [(m) and (n) From W. R. Jeffery and S. Meier (1983). *Dev. Biol.* **96**, 125.]

The fusion nucleus comes to lie near the future posterior pole, and there the yellow crescent cytoplasm later incorporated in the embryonic muscle cells is localized.

Conklin regarded the ascidian embryo as a "mosaic-work," and as had the other observers cited above, he also argued that "the mosaic is one of organ forming substances rather than of cleavage cells." His concept of ascidian development was supported by an extensive series of experiments in which he determined the developmental potentialities of individual blastomeres and sets of blastomeres in embryos in which the other blastomeres had been killed. In general the surviving blastomeres were found to establish cell lineages that differentiate in their respective normal directions, though the overall organization of the embryo is of course affected. The drawings reproduced in Fig. 6.5 show that each embryo fraction contains the potentiality of forming certain presumptive tissue types, such as notochord, mesoderm, neural plate, gut, or ectoderm, even though the isolates were often not cultured sufficiently long to reveal completely their developmental capabilities [see, however, Fig. 6.5(e)]. In recent years Conklin's experiments have been repeated and then expanded in great detail, by Reverberi, Ortolani and others of their colleagues (reviewed in Reverberi, 1971b; Whittaker, 1979a). The ascidian cell lineage shown in Fig. 6.5(j) includes some corrections of Conklin's original assignments, mainly obtained by the method of horserad-ish peroxidase injection into individual blastomeres of 8-cell embryos of *Ciona intestinalis*, *Ascidia abodori*, and *Halocynthia roretzi* (Nishida and Satoh, 1983), and into blastomeres of 16- and 32-cell embyos of *Halocynthia* (Nishida and Satoh, 1985; see also Zalokar and Sardet (1984) for similar assignments obtained for some lineages of *Phallusia mammillata* embryos by a different method). This lineage describes diagrammatically the ultimate capacities of the various early blastomeres *in situ*, and in general *in partial embryos as well*, and thus provides a developmental linkage between Conklin's "organ forming substances" and embryonic cell differentiation. How-

---

Fig. 6.4. (See color plate following p. 256.) Figures of the living eggs of *Cynthia* (*Styela*) *partita* maturation and fertilization period only. (a) Unfertilized egg before the breakdown of the germinal vesicle (*g.v.*), showing central mass of gray yolk (*yk.*), peripheral layer (*p.l.*) of yellow cytoplasm, test cells (*t.c.*), and chorion (*cn.*). (b) Similar egg during the disappearance of the nuclear membrane, showing the spreading of the clear cytoplasm of the germinal vesicle at the animal pole. (c) Another egg about 5 min after fertilization, showing the streaming of the peripheral protoplasm to the lower pole, thus exposing the gray yolk (*yk.*) of the upper hemisphere. (d) Later stage in the collection of the yellow cytoplasm. Clear cytoplasm lies beneath and extends a short distance beyond the edge of the yellow cap. In (e), viewed from the vegetal pole, the area of yellow cytoplasm is smaller, and the sperm nucleus, (♂ *n.*) is a small clear area. (f) Shows the same egg about 20 min later. (g) Shows the spreading of this yellow cytoplasm until it covers nearly the whole of the lower hemisphere [yellow hemisphere (*y.h.*)]; at the same time the sperm nucleus and aster move toward one side and the crescent (*cr.*) begins to form at this side. [From E. G. Conklin (1905a). *J. Acad. Nat. Sci. Philadelphia* **13**, 1.]

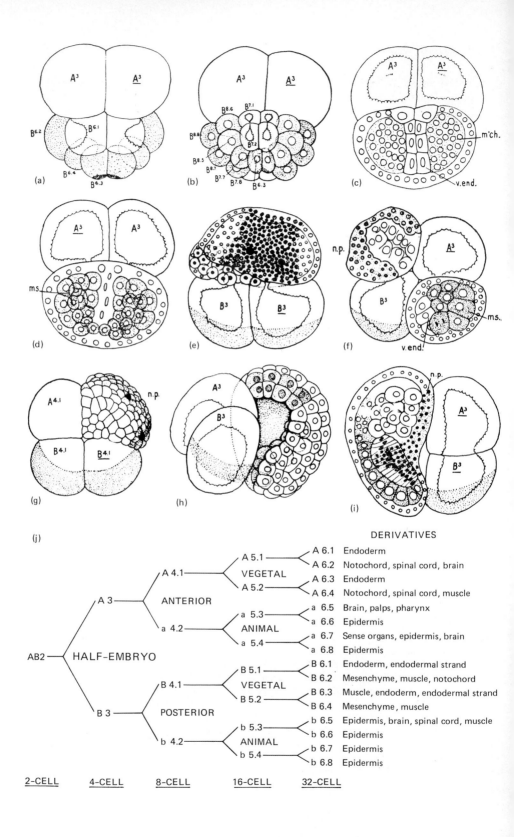

2-CELL	4-CELL	8-CELL	16-CELL	32-CELL	DERIVATIVES
			A 5.1	A 6.1	Endoderm
		A 4.1		A 6.2	Notochord, spinal cord, brain
		VEGETAL	A 5.2	A 6.3	Endoderm
	A 3	ANTERIOR		A 6.4	Notochord, spinal cord, muscle
			a 5.3	a 6.5	Brain, palps, pharynx
		a 4.2		a 6.6	Epidermis
		ANIMAL	a 5.4	a 6.7	Sense organs, epidermis, brain
AB2	HALF-EMBRYO			a 6.8	Epidermis
			B 5.1	B 6.1	Endoderm, endodermal strand
		B 4.1		B 6.2	Mesenchyme, muscle, notochord
		VEGETAL	B 5.2	B 6.3	Muscle, endoderm, endodermal strand
	B 3			B 6.4	Mesenchyme, muscle
		POSTERIOR	b 5.3	b 6.5	Epidermis, brain, spinal cord, muscle
		b 4.2		b 6.6	Epidermis
		ANIMAL	b 5.4	b 6.7	Epidermis
				b 6.8	Epidermis

ever, the relation between embryonic cell fate and cell lineages is in some areas not as simple as implied by Conklin's (1905a) proposal that *segregation of morphogenetic agents* has been completed by the 64-cell stage, and thus that the progeny of each 64-cell blastomere contribute to only one type of tissue. The lineage chart shown in Fig. 6.5(j) indicates that some functionally asymmetric divisions must occur later than the 64-cell stage, since certain 32-cell blastomeres contribute to at least three different types of tissue. Examples of clonal determinations by the 64-cell stage, and also of later determination, are to be found in the muscle cell lineages, as discussed later in this chapter.

The concept that early embryos are a "mosaic-work" of determined blastomeres was widely challenged. Beginning in the late 1880's, blastomeres were isolated and cultured from many different embryos, in order to determine whether their division products indeed develop *in vitro* as do their

---

Fig. 6.5. Fate of partial embryos of *Cynthia (Steyla) partita*. (a) 32-cell stage, after anterior blastomeres have been killed; dorsal view [cf. Fig. 6.3(f-h)]. The cleavage of the posterior half is altogether normal. The anterior blastomeres (A^3) were killed by spurting the embryo from a pipette at the 4-cell stage, fixed 1 hr later. (b) Cleaving posterior half, 76-cell stage [cf. Fig. 6.3(i)]. Spurted in the 4-cell stage, fixed 2 hr later. Two rows of yellow crescent cells are present, the inner being mesenchyme, the outer muscle cells; the anterior pair of mesenchyme cells ($B^{8.6}$) are larger than normal. There are two pairs of caudal endoderm cells ($B^{7.1}$ and $B^{7.2}$). A pair of clear ventral ectoderm cells is visible in the midline behind $B^{6.3}$. (c) An embryo, spurted in the 4-cell stage, fixed at 4 hr, deep focus, showing the double row of ventral endoderm cells (v. end.) in the midline, and on each side of this a mass of mesenchyme cells (m'ch). (d) Ventral view of posterior half-embryo of the same stage as the preceding, showing the muscle (ms,) and mesenchyme cells beneath the ectoderm and on each side of the strand of ventral endoderm. (e) Anterior half-embryo, dorsal view. Spurted in the 4-cell stage, fixed 22 hr later. The granular yellow crescent material is plainly visible in the injured cells. Sense spots are present, but the neural plate never forms a tube. The chorda cells lie in a heap at the left side. There is no trace of muscle substance or of a tail in this anterior half embryo. Normal larvae of this stage are undergoing metamorphosis. (f) Left anterior and right posterior (diagonal) quarter embryo, dorsal view; spurted in the 4-cell stage, fixed 5 hr later. The anterior quarter shows thickened ectoderm cells, probably neural plate (n.p.) around the endoderm cells; in the posterior quarter are 8 muscle and 3 caudal endoderm cells. (g) Right anterior dorsal eighth embryo, 14 hr after injury, showing endoderm, chorda, and neural plate cells with sense spots. (h) Right half-gastrula of about 200-cell stage; spurted in the 4-cell stage and fixed 3 hr later. The neural plate, chorda, and mesoderm cells are present only on the right side *and in their normal positions and numbers*. (i) Living left half-embryo, dorsal view, showing the endoderm cells forming exogastrulae and the yellow crescent cells at the surface. [(a)-(i) From E. G. Conklin (1905b). *J. Exp. Zool.* **2**, 145.] (j) Cell lineage diagram of ascidian embryonic development. [Constructed from data of Conklin and Ortolani by J. R. Whittaker (1979a). *In* "Determinants of Spatial Organization" (S. Subtelney and I. R. Konigsberg, eds.), p. 29. Academic Press, New York; revised in accordance with Crowther and Whittaker (1984), and with results of horseradish peroxidase cell lineage studies reported for *Halocynthia roretzi* by Nishida and Satoh (1985). *Dev. Biol.* **110**, 440.] Different species may have slightly different cell lineages, though fates of 8-cell blastomere progeny are identical for three species (Nishida and Satoh, 1983; see text).

normal lineages in the whole embryo. Diverse results were obtained with different organisms, and a ferocious controversy over the basic nature of embryological development soon arose. The conflicting theories that followed are reviewed briefly in the next section, while in this we consider further examples of the evidence supporting one side of this controversy, *viz.*, those isolated blastomere experiments that provided powerful evidence for early fixation of morphogenetic potential in localized regions of the embryo.

A method discovered by Herbst (1900) much facilitated this line of research. This was the disaggregation of marine embryos in low Ca^{2+} sea water, which in many cases yields viable single blastomeres or clusters of blastomeres. These could be cultured in isolation, rather than in contact with the physical remains of killed sister blastomeres such as are present in Conklin's *Styela* experiments [e.g., Fig. 6.5(a-i)]. One of the most successful examples of this kind of experiment is illustrated in Fig. 6.6, which is compiled from Wilson's (1904a) report on experiments with embryos of the mollusc *Patella coerulea*. In this study Wilson isolated a number of different presumptive cell types in low calcium seawater and compared their subsequent development to that expected if they had remained in the context of the whole embryo. The isolated blastomeres were found to follow exactly their normal developmental fate. Thus Fig. 6.6 shows that primary tro-

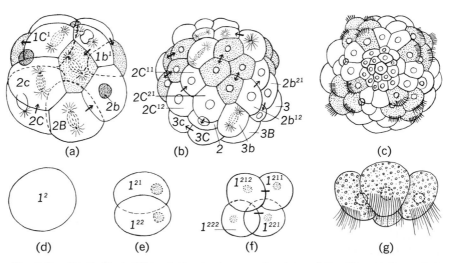

Fig. 6.6. Trochoblast differentiation in the normal embryo of *Patella coerulea* and after isolation. The lineage of each cell is defined by its name (cf. Fig. 6.2). (a)-(c) Normal development of *Patella*. (a) 16-cell stage, from the side (primary trochoblasts shaded); (b) 48-cell stage; (c) "ctenophore" stage, about 10 hr, from upper pole, primary trochoblasts ciliated. (d)-(g) Isolated primary trochoblasts cultured *in vitro*: (d) primary trochoblast; (e) result of first division; (f) after second division; (g) product of (f). [After E. B. Wilson (1904a). *J. Exp. Zool.* **1**, 197.]

choblasts isolated from *Patella* embryos carry out the correct number of cell divisions and later become ciliated on a normal schedule. Wilson (1904a) concluded:

> The history of these cells gives indubitable evidence that they possess within themselves all the factors that determine the form and rhythm of cleavage, and the characteristic and complex differentiation that they undergo, wholly independently of their relation to the remainder of the embryo.

Consistent results were obtained with other isolated cell types and with partial *Patella* embryos. Among these were isolated one-sixteenth embryo macromeres, which produced endodermal gut rudiments, and isolated apical progenitors, which differentiated *in vitro* into apical sensory and ectodermal cells.

Isolated blastomeres of annelid eggs, as well as of several additional species of mollusc, were likewise found to cleave *in vitro* according to normal patterns, and where early differentiation could be observed, to develop according to expectation. Several examples are presented in following sections of this chapter, and reviews can be found in Wilson (1925), Morgan (1927), and Reverberi (1971a). The morphogenetic fates of the cell lineages descendant from early embryonic blastomeres were studied in marine annelids, such as the polychaetes *Sabellaria* (Hatt, 1932) and *Nereis* (Costello, 1945; see review in Reverberi, 1971c), and an illustrative series of experiments was carried out by Penners (1926; reviewed in Morgan, 1927) on the oligochaete annelid *Tubifex rivulorum*. This animal develops essentially as described for *Clepsine* by Whitman (Fig. 6.1). Thus a neuroblast germ band and mesodermal germ band both derive from the D quadrant of the embryo (blastomere nomenclature here is as described in the legend to Fig. 6.2). At the 4-cell stage of the *Tubifex* embryo the D macromere is the largest, and with respect to both size and rate of division, its products remain distinct from those of the A, B, and C quadrants. Penners found that each of these blastomeres would continue its normal course of development even if all the others were killed *in situ* by UV microbeam irradiation. Thus if A, B, and C are killed, the D macromere nonethless adheres to its unique cleavage pattern. It gives rise to an ectodermal neuroblast stem cell, 2d (equivalent to the neuroblast stem cells shown in Fig. 6.1), and to a primary mesodermal stem cell, 4d (equivalent to the mesoblast stem cells in Fig. 6.1). After this, columnar ectodermal and mesodermal germ bands are produced by anterior budding. Penners showed that if the 4d cell is individually killed, the neuroblast germ bands still form, but the embryo lacks coelomic mesoderm. Recent investigation has demonstrated that the individual neuroblast stem cells of annelids each give rise to certain specific neurons. In the leech, for instance, ablation of a given germ band neuroblast results ultimately in the absence of one member of each of the three bilateral segmental pairs of larval neurons that contain serotonin, and ablation of a different neuroblast causes the absence of one member of each pair of segmental body wall neurons that contain dopamine (Blair, 1983).

(iii) The Early History of the Idea of Cytoplasmic Localization

The initial discovery of determinate, asymmetric cell lineages was among the antecedents of the proposal that morphogenetic substances are localized in the cytoplasm of the very early embryo. There was also a parallel theoretical development, the concept of "precocious segregation" of maternal molecular determinants. The origins of this idea lie in the nineteenth century controversy between preformationist and epigenetic theories of development. The participants in the final stages of this controversy included many eminent biologists engaged personally in research on localization. Huxley (1878), Hertwig (1894), Bourne (1894), and Whitman (1895a) all published discussions of the implications of localization in which the main issue considered was whether the embryo actually increases in biological complexity or, on the other hand, merely reveals progressively to the observer an organizational complexity that is already resident in the unfertilized egg. A major accomplishment of mid-19th century biology had been to establish the apparently epigenetic nature of early development. The first investigations directly supporting epigenetic interpretations were those of Wolff (1759, 1768), which showed that embryonic chick blood vessels and gut develop from undifferentiated tissues; Pander's (1817) description of epigenetic development in the chick from primitive germ layers; and the studies of von Baer (1828, 1837), as a result of which the germ layer theory was generalized to other animals. Von Baer showed that skin develops epigenetically from ectoderm, muscular and skeletal systems from mesoderm, etc. Kowalevsky (1867) suggested that the germ layers of many animal phyla are also formed epigenetically. These observations were regarded as crucial refutations of the traditional preformationist theories previously current, such as those of Bonnet (1762; see translation and discussion in Whitman, 1895b). Bonnet had argued that little change in complexity actually occurs in embryological development, and that "organized bodies preexist from the beginning." According to Bonnet's rather crude ovist theory of preformation (1745), a complete embryo is patterned in every egg, and each such embryo contains an ovary, with eggs that bear embryos within, etc.

The matter seemingly settled by scientific investigation, it was thus a striking event when in the 1870's a novel and sophisticated new form of preformationism was proposed, by Wilhelm His and others. His was the teacher of Johann F. Miescher, the discoverer of DNA, and was a proponent of the view that satisfactory explanations of biological phenomena can only be obtained at the molecular level. In 1874, several years before the publication of Whitman's *Clepsine* study, His suggested that the epigenetic character of early chick development is only apparent, the underlying phenomenon being the "coalescence of preformed germs":

> It is clear, on the one hand, that every point in the embryonic region of the blastoderm must represent a later organ or part of an organ, and on the other hand, that every organ developed from the blastoderm has its preformed germ in a definitely localized region. *The material of the germ is already present, but is not yet ... directly recognizable.* [His (1874); translated by Wilson (1896b); italics added]

A farsighted statement of this school was made by Lankester (1877), who further extended the "molecular preformation" hypothesis:

> Though the substance of an [egg] cell may appear homogeneous under the most powerful microscope, excepting for the fine granular matter suspended in it, it is quite possible, indeed certain, that it may contain, *already formed and individualized,* various kinds of physiological molecules. The visible process of segregation is only the sequel of a differentiation already established. ... Thus, since the fertilised egg already contained hereditarily acquired molecules, ... invisible though differentiated, there would be a possibility that these ... molecules should part company, *not* after the egg-cell had broken up into many cells as a morula, but at the very first step in the multiplication of the egg-cell ... We should not be able to recognize these molecules by sight; the two cleavage-cells would present an identical appearance, and yet the segregation ... had already taken place. This hypothesis may be called that of PRECOCIOUS SEGREGATION. [Lankester's italics and emphasis]

The close relationship between these ideas and the subsequent concept of cytoplasmic localization is obvious. However, these passages were written a decade before it was first realized that the hereditary determinants are confined to the chromosomes of the cell nucleus. To obtain from them the idea of cytoplasmic localization as we now think of it requires the conceptual separation of localized *cytoplasmic* determinants from the *genomic* determinants shared equally among all the blastomere nuclei.

The next major step in the foundation of the localization theory was taken by Wilhelm Roux (1888). Roux proposed that two kinds of processes leading to embryonic cell differentiation occur in development, *self-differentiation* and *correlative differentiation.* The first of these describes the determinate behavior of blastomeres and their cell lineages in mosaic forms of embryogenesis, and predicts that when isolated, embryonic blastomeres will continue to display their normal potentialities. Correlative differentiation was invoked by Roux to explain the regulative behavior of partial embryos. Through their powers of correlative differentiation embryonic cells could secondarily modify their normally determined fate. Roux suggested that correlative differentiation is mediated by means of interactions between cells.

Unfortunately, Roux (1883) and Weismann (1885, 1892) initially attempted to explain blastomere self-differentiation as the consequence of the qualitative partitioning during cleavage of *nuclear* genetic determinants, a proposal that was almost immediately discredited. Experiments of Driesch on sea urchin eggs, Wilson on *Nereis* eggs, and Yatsu on nemertine eggs, among others, showed that treatments which *redistribute* nuclei destined for given blastomeres to other blastomeres nonetheless often result in normal development (reviewed in Wilson, 1925, pp. 1057–1062). The most convincing experiments were carried out by the transient application of pressure to cleaving eggs, e.g., by exposing them to the weight of a cover slip. The result is to inhibit normally occurring transverse cleavages, and to induce extra rounds of vertical cleavage. Three-dimensional development resumes on release of the pressure, and nuclei normally assigned to given blastomeres are now found to be located in different blastomeres. For example, in Wilson's *Nereis* experiments (1896a) the nuclei that would in normal embryos populate the entoblasts (i.e., the endodermal stem cells of Fig. 6.2) were shown to be able to direct the development of first quartet micromeres, and vice versa. In other words these nuclei must contain genes for endodermal as well as ectodermal differentiation. Generalizing, it follows that any cleavage stage nucleus contains all the zygote genes, and therefore that the differences in cell fate displayed by given lineages must be due to segregation of *cytoplasmic* factors.

The promulgation of Roux's mosaic theory of development—ignoring the nuclear segregation portion—initiated a most interesting intellectual divergence which to some extent is still with us. While tests of the developmental capacities of isolated blastomeres in many cases directly supported the mosaic view of embryogenesis, some regulative capacity was often noted even in blastomeres from eggs that were described overall as "mosaic-works." Most importantly, it was discovered that in some organisms isolated blastomeres could give rise to complete rather than partial larvae. The latter result was obtained with at least some of the early cleavage blastomeres of species of sea urchins, hydroid coelenterates, cephalochordates, nemertean worms, amphibians, and fish [for a complete summary of the classical partial embryo experiments see Morgan (1927), chapters 16 and 17]. Proponents of various forms of correlative differentiation regarded the amazing regulatory behavior revealed in these partial embryo experiments as a decisive blow to the mosaic theory that the embryo is the sum of its self-differentiating blastomeres. The basic difference in attitude thus related to the significance of specific substances in specific cells. Wilson and Conklin, for example, considered there to be convincing evidence that the fate of a given blastomere or set of blastomeres may be *determined* by its specific cytoplasmic contents. In contrast, Driesch, who first discovered the totipotency of sea urchin blastomeres, wrote:

> The relative position of a blastomere in the whole determines in general what develops from it; if its position be changed, it gives rise to something different; ... *its prospective value is a function of its position.* [Translated by Wilson (1896b)].

Similarly, for Child (1900):

> The appearance of definite protoblasts [founder cells] in cleavage does not necessarily imply that they contain a specific material necessary for the formation of the organ in question. Protoblasts are ... to be regarded as centers of distribution of the material of the egg ... The material separated as a result of precocious segregation may, I believe, be perfectly indifferent material except as regards position. ...

Whitman, who in a widely quoted essay (1893) entitled "The inadequacy of the cell theory of development" took much the same view, was nonetheless bothered by its quality of vagueness:

> If the formative processes cannot be referred to cell division [i.e., cleavage] ... can they be referred to cellular interaction? That would only be offering a misleading name for what we cannot explain. ...

The subsequent history of these two fundamental classes of developmental theory is familiar to the modern reader. The focus on the overall organizational properties of the embryo, and on the role of *position* can be regarded as ancestral to those 20th century schools of embryology oriented toward analysis of organizing centers, morphogenetic fields, and gradients, and that more recently has given rise to the concepts of positional information. The opposing *cellular* focus was soon bolstered by the rediscovery of Mendelism and the general acceptance of the presence in blastomere nuclei of genomic determinants for all the specific traits manifested by differentiated cells. An event of considerable significance at the time was the demonstration of Boveri (1902) that blastomere gene action is specifically required for early development (reviewed in Davidson, 1985). This study concerned the causes of developmental arrest in dispermic sea urchin eggs. Boveri showed that the tetrapolar and tripolar mitoses that occur in these eggs usually result in aneuploidy, so that particular chromosomes are missing or present in greater than diploid number in some blastomeres. He demonstrated that the developmental potential of isolated blastomeres containing partial genomes is invariably incomplete, while in contrast, as Driesch had shown, complete plutei could be obtained from isolated blastomeres of normal embryos. Boveri concluded:

> Only a precise combination of chromosomes, probably only the totality of those which are contained within each pronucleus, represent the entire nature of the form of the organism, ... [in] the embryonic development of which unfold the qualities of the nuclei ...

The classical *cellular* theory of localization, and Boveri's essentially modern assertion of the role of the genome in differentiation, are the traditions that led to the persistent effort to understand embryo cell function, which today is largely molecular in nature. It is an interesting historical note that in the modern pursuit of the mechanism by which initial specification of embryo cell function occurs we once again find ourselves face to face with "the localization problem."

(iv) Summary of Classical Conclusions

It was apparent to perspicacious classical writers such as Roux, Conklin, and Wilson that the dichotomy sketched above is to some extent polemic rather than real. The main lesson to be drawn was that *both* "correlative," or as we would say, *inductive* differentiation, and self-differentiation, are involved in the early development of most creatures. Some organisms rely more on one, and some more on the other, a matter to which we return often in the following. Cytoplasmic localization of substances that determine blastomere fate occurs in embryos that also display regulative powers in isolated blastomere experiments, as well as in embryos that do not. For example the eggs of amphibians and echinoderms display striking regulatory powers, in that in both groups partial embryos and fused embryos can form whole larvae. This was shown for amphibian eggs by Spemann (1903), McClendon (1910), Mangold (1920), and Ruud (1925), and has been demonstrated in many ways since. Driesch (1891) found that a single sea urchin blastomere possesses the capability of forming a complete embryo, and subsequently, that the first two blastomeres and the individual blastomeres of the 4-cell stage sea urchin embryo also possess this capability. The sea urchin blastomere recombination experiments carried out by Hörstadius (1939) and discussed later in this chapter are among the most impressive demonstrations of regulative ability. These experiments proved that the morphogenetic fate of given tiers of blastomeres can be completely changed depending on which other blastomeres are present. Yet there is also very clear evidence for cytoplasmic localization in sea urchin eggs, and modern descriptions of the events leading to the establishment of the dorsal-ventral axis of polarity in the amphibian egg provide an excellent example of the establishment of a definitive localization pattern involving the distribution of maternal cytoplasmic components.

Wilson (1925) pointed out with great clarity that blastomere totipotency and cytoplasmic localization are not necessarily exclusive. In Fig. 6.7 is reproduced a heuristic diagram which illustrates this point. Here localized areas of morphogenetic potential are mapped out for the eggs of an echinoderm, a gastropod or annelid, an ascidian, and a hydromedusa. The diagram shows how mesodermal and other cytoplasmic determinants are partitioned among the blastomeres. The developmental potency of the early blastomeres of each type of egg can be understood in terms of the relation between the planes of cleavage and the distribution of the morphogenetically significant regions in the egg. For example, if the mesodermal determinants are asymmetrically distributed with respect to the cleavage planes, all of the blastomeres cannot be totipotent [Fig. 6.7(d-f)].

To quote Wilson (1925):

> Totipotence on the part of the early blastomeres is dependent primarily on a symmetrical or merely quantitative distribution of the protoplasmic stuffs of the cleavage. In the

hydromedusa [Fig. 6.7(j-l)] the original grouping of these materials is, broadly speaking, concentric about the center of the egg, and all of the radial cleavages accordingly are quantitative. ... Since the first five cleavages are of this type, complete dwarfs may be produced from any of the blastomeres up to the 16-cell stage ... when the first qualitative divisions begin by the delamination-cleavages parallel to the surface. In the sea urchin the ooplasmic stuffs are polarized, displaying a symmetrical horizontal stratification at right angles to the axis of the egg. Since the first two cleavages pass exactly through the axis and cut all the strata symmetrically [Fig. 6.7(a-c)] the first four or two blastomeres receive equal allotments of these strata in their normal proportions and hence remain totipotent...we shold expect the third cleavage to be qualitative; this is borne out by both observation and experiment.

Similarly, as shown in Fig. 6.7(d-f), in many annelids, molluscs, and some other creatures undergoing spiral cleavage, the mesoderm determinants initially located in the polar region of the egg are distributed only to the D blastomere. Consequently this cell alone retains the capacity to develop into a qualitatively complete dwarf embryo. In the ascidian egg [Fig. 6.7(g-i)] only the posterior blastomeres A and D retain the mesodermal determinants, but other essential determinants not figured (e.g., those for neural ectoderm) are missing from these cells (see Fig. 6.3 and 6.5). As Conklin showed in *Styela*, no one blastomere gives rise to a complete embryo in this form (Fig. 6.5). The potentialities of blastomeres isolated from embryos of the types shown in Fig. 6.7 would depend in part on these cytoplasmic localization patterns. Because of the relation between these patterns and the cleavage planes, the sea urchin and hydromedusa embryo might be characterized as "regulative", while the annelid would be characterized as "mosaic," if the criterion were simply the developmental potentialities of isolated blastomeres. Yet the development of all four of these embryos depends at least in part on localized cytoplasmic determinants.

In the remainder of this section is presented a summary of other specific conclusions that derived from the half century of research on cytoplasmic localization prior to 1925. These points provide a convenient and relevant framework for organizing the following review of modern research.

(a) Localization may result in the determination of embryo cell lineages. This was first recognized in the studies of annelid and molluscan cleavage, where individual cell lineages are easily identified as the progeny of particular founder cells (cf. Fig. 6.1). Extensive investigations such as those of Conklin on *Styela* (Figs. 6.3, 6.4 and 6.5) generalized to large regions of the embryo the relations between the fixed regions of the egg cytoplasm inherited by given blastomeres, and the ultimate differentiations displayed by their lineages.

(b) Localization is progressive and occurs by means of redistribution of cytoplasmic components during maturation or after fertilization. Conklin's classic illustration of cytoplasmic movements in the *Styela* egg is reproduced in Fig. 6.4. Various observations to the same effect were made on sea urchin, annelid, molluscan, and nemertean eggs, as well as many others. Sum-

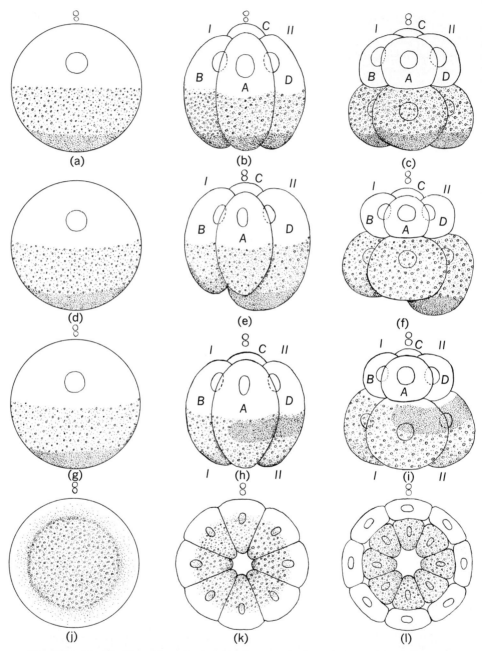

Fig. 6.7. Localization of morphogenetic substances in the early blastomeres of four types of egg. Diagrams of the primary stratification in the eggs of the sea urchin (a)-(c) and the annelid or gastropod (d)-(f). The first two cleavage planes are designated as I or II. The upper or white zone of each egg is ectoblastic, the middle or granular one the entoblastic, the lower or stippled one the mesoblastic. In (a)-(c) all the zones are equally divided; in (d)-(f) only the two upper

marizing his observations on the developmental origins of the patterns of localization in *Cerebratulus* eggs, Wilson (1903) wrote:

> The specification of the germinal regions of the egg is a progressive process ... the ontogeny assumes more and more of the character of a mosaic work as it goes forward.

There are nonetheless examples of *pre*localized morphogenetic determinants in eggs, that are present from the beginning. A commonly cited example is the animal-vegetal axis, which in many organisms, e.g., *Clepsine* (Whitman, 1878) or the sea urchin (Boveri, 1901a), is indeed a primordial property of the egg.

(c) The localized components are maternal in origin. The most important conclusion for us is of course implicit in the foregoing. It is that *there are morphogenetic determinants resident in the unfertilized egg cytoplasm.* This fact is separable from the fascinating problem of how these determinants achieve their definitive topological localization after development begins. Wilson removed various regions of molluscan egg cytoplasm prior to cleavage and discovered that the results resemble those obtained when the blastomeres normally inheriting this cytoplasm are ablated (Wilson, 1904a; reviewed in a following section of this chapter). He concluded:

> The cleavage mosaic is a mosaic of specifically differentiated cytoplasmic materials, in which are somehow involved corresponding morphogenetic factors ... [Wilhelm] His' principle of organ forming egg substances ... is thus shown to have a true causal significance ...

(d) The localized determinants are often attached to a "persistent framework" in the egg. The phrase is Conklin's, and refers to what he called the "spongioplasm," and we today term the cytoskeleton. Conklin (1917) centrifuged *Crepidula* eggs, and showed that while all the microscopically visible inclusions such as pigment granules and oil droplets can be stratified, the polarity of the egg and its developmental capacities remain unaffected. Similar results were obtained by Lillie (1909) with *Chaetopterus* eggs and by Morgan and Spooner (1909) and Spooner (1911) with sea urchin eggs. Conklin also proposed that cytoplasmic streaming and intracellular movements in eggs are caused by contraction of the cytoskeleton, which he prophetically imagined to be an elastic and contractile structure.

(e) Localization is ultimately the product of nuclear function in oogenesis. This insight is due to Driesch, Boveri, and Wilson, and it produced a convenient resolution to the allegation that the ideas leading to the theory of

---

zones are thus divided, the lower one being segregated entirely to the D quadrant. (g)-(i) Primary stratification in the ascidian egg, and (j)-(l) in the hydromedusan egg. In the ascidian egg the lower mesoblastic stratum is equally divided between A and D quadrants. In the hydromedusan egg this stratum is absent, and the remaining two are equally distributed up to the time of delamination (l). [From E. B. Wilson (1925). "The Cell in Development and Heredity," 3rd ed., pp. 1072-1076. Macmillan, New York.]

cytoplasmic localization represent a return to preformationism. Driesch pointed out [translation by Wilson (1896b)] that:

> Cytoplasmic organization, while affording the immediate conditions for development, is itself the result ... of the nuclear substance which represents the totality of heritable potence.

Thus the only real preformation is that of the genome itself, "which in the course of development finds expression in a process of cytoplasmic epigenesis" (Wilson, 1925). This statement, as it applies to localization, describes an essentially modern view. Further than this, however, classical writers did not go. The morphogenetic determinants present in the egg cytoplasm were always referred to simply as "organ forming substances." Their nature, function, and mode of interaction remained mysterious, as to a large extent they are yet. The great contributions of the intensive classical studies on cytoplasmic localization were to demonstrate experimentally the reality of this mechanism of early development; to identify for us well described and accessible embryonic systems for its further analysis; and ultimately to clarify an enormous melange of conflicting and often cloudy ideas. The resolution summarized here illuminates localization as a manifestation of developmentally significant interactions of blastomere nuclei with spatially defined elements of the egg cytoplasm.

2. DETERMINATION OF SPECIFIC CELL LINEAGES BY LOCALIZED COMPONENTS OF THE EGG CYTOPLASM

Three examples of embryonic cell lineages that are apparently determined by processes of cytoplasmic localization are considered in this section. One difference between the classical studies just reviewed and the modern investigations we now review is the use of specific ultrastructural and biochemical markers as indices of differentiated cell lineage function. The first of these examples is the determination of the ascidian muscle cell lineage, for which current evidence on the role of localization is probably the most complete; the second is the determination of the gut cell lineage in a nematode; and the third concerns determination of germ cell lineages in certain nematodes, insects, and anuran amphibians.

(i) Egg Cytoplasmic Determinants for the Larval Muscle Cell Lineage of Ascidian Embryos

The majority of the muscle cells of the ascidian larva descend from the bilateral B4.1 stem cells, which inherit the yellow crescent cytoplasm of the egg (Whittaker, 1979a). In most ascidian groups so far investigated there are 18 tail muscle cells on either side of the completed larva (though in *Halocyn-*

thia there are 21). As shown in Fig. 6.8(a), on each side 14 muscle cells are descended from the B4.1 progenitor cells; two arise from the A4.1 cells and two (five in *Halocynthia*) from the b4.2 cells. However, as discussed below the mechanism of muscle cell specification in the B4.1 derivatives appears to differ from that in the b4.2 and A4.1 lineages. At 4th cleavage the yellow cytoplasm of the B4.1 cells is divided between the B5.1 cells and the B5.2 cells, and with the appearance of the B7.4, B7.5, and B7.8 cells two divisions later three pure muscle lineage clones are established on each side. A specific acetylcholinesterase (ACE) that can be detected histochemically has been used as a marker of muscle differentiation [Fig. 6.8(b)]. Under normal conditions this enzyme does not appear until the neurula stage, when there are more than a hundred cells over all. By this time several rounds of division have elapsed in the muscle lineage since the initial formation of the exclusively specified myogenic progenitor cells of the B4.1 lineage [Fig. 6.8(a)], a phenomenon similar to those discussed in Chapter IV, in relation to lineage-specific differentiation in sea urchin and *Caenorhabditis* embryos.

When B4.1 cells are isolated and cultured their progeny synthesize detectable ACE at about the same time as do the presumptive muscle cells of intact control embryos (Whittaker *et al.*, 1977). Results of one such experiment are shown in Fig. 6.8(c). The appearance of ACE is correlated with general cytotypic muscle differentiation that can be visualized at the ultrastructural level. For example, myofibrillar bundles can be observed in the progeny of isolated B4.1 cells that have been cultured *in vitro*, as shown in Fig. 6.8(d) (Crowther and Whittaker, 1983). The electron micrograph reproduced in Fig. 6.8(e) demonstrates that the progeny of isolated B5.2 cells also retain the capacity to construct myofibrillar bundles (Crowther and Whittaker, 1983). In the classic sense of Roux and Wilson the differentiation of the embryonic tail muscle lineage descendant from the B4.1 blastomeres appears to be a true *self-differentiation*.

The ultimate appearance of muscle ACE is blocked by treatment with actinomycin (Whittaker, 1979a), as is the appearance of myosin heavy chains (Meedel, 1983), suggesting the requirement for newly synthesized transcripts. Meedel and Whittaker (1983) and Perry and Melton (1983) showed directly that in normal embryos ACE mRNA accumulates sharply after the midgastrula stage. In these studies embryonic RNAs extracted from ascidian embryos of various stages were injected into *Xenopus* oocytes. The presence of ACE messages was identified by immunoprecipitation of the translation products (Meedel and Whittaker, 1983), and by the appearance of ACE enzymatic activity in the injected oocytes (Perry and Melton, 1983). No ACE mRNA is detected prior to gastrulation. Muscle differentiation in this embryo is thus a lineage-specific process that requires activation of specific genes for myogenesis, an event that occurs only some hours following the segregation of the B4.1 muscle cell lineages from other prospective cell types at the 64-cell stage [Fig. 6.8(a)].

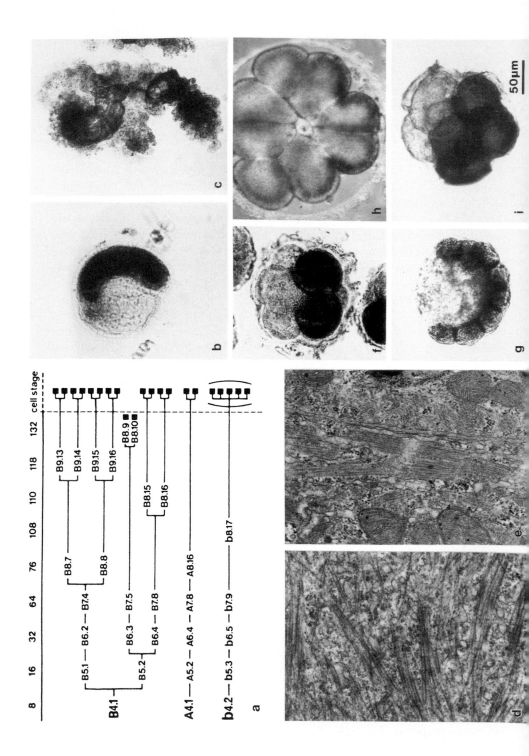

50µm

An illuminating observation made by Whittaker (1973, 1983) is that ACE appears in the myogenic B4.1 quadrant cells on schedule, and in approximately normal quantity, even when cleavage is blocked with cytochalasin. This drug interferes with cytokinesis, possibly by affecting microfilament assembly, but does not interrupt nuclear division. The treated blastomeres thus become syncytial, multinucleate structures, and the number of presumptive muscle cells is frozen from the moment of cytochalasin addition. If cytochalasin is added at the 4-cell stage, when the muscle lineage cytoplasm, or myoplasm, is confined to the two posterior B3 cells, these two cells synthesize ACE after some hours of culture; if it is added at the 8-cell stage only the two B4.1 cells eventually display ACE, etc. Examples are shown in Fig. 6.8(f)-(g). Cytochalasin-treated blastomeres produce myofibrillar structures as well as ACE (Crowther and Whittaker, 1983). Another specific differentiation that appears in the muscle cell lineage of cytochalasin-treated ascidian embryos is the organization of Na^+ and Ca^{2+} membrane channels (Takahashi and Yoshii, 1981). All of these results demonstrate that muscle determination occurs specifically *in whatever cells contain the pigmented*

Fig. 6.8. Determination of the embryonic muscle cell lineage in ascidian embryos. (a) Lineages of muscle cells in the ascidian embryo of *Halocynthia roretzi*, from the 8-cell stage until the terminal differentiation stage, derived from horseradish peroxidase cell lineage studies of Nishida and Satoh (1983, 1985), and from cell lineage data of Zalokar and Sardet (1984). Broken line indicates stage up to which cleavage of individual cells can be observed by light microscopy. In *Ciona intestinalis* and *Ascidia ahodori* the number of muscle cells descendant from b4.2 is two, while in *Halocynthia roretzi* it is five (parenthesis). Only those daughter cells of each cleavage that are ancestral to muscle cells are shown. [From H. Nishida and N. Satoh (1985). *Dev. Biol.* **110**, 440.] (b) Histochemical localization of acetylcholinesterase in normal tailbud embryo of *Ciona intestinalis*. The tail is in a folded position due to the presence of the chorion. (c) Acetylcholinesterase present in the progeny of four isolated B4.1 blastomeres that were cultured for the period required by control embryos [e.g., (b)] to attain tailbud stage (7 hr at 22°C). Note the presence of cells not displaying the enzyme, probably belonging to the endodermal lineage also descending from B4.1 (see text). [(b) and (c) From J. R. Whittaker, G. Ortolani, and N. Farinella-Ferruzza (1977). *Dev. Biol.* **55**, 196.] (d) Transmission EM displaying myofibrillar structures in the progeny of an isolated *C. intestinalis* B4.1 cell that had been cultured for 20 hr. Glycogen granules can also be seen. X20,175. (e) Myofibrillar structures near the border of two cells developing from an isolated B5.2 cell pair from *C. intestinalis*. X20,175. [(d) and (e) Unpublished, according to R. J. Crowther and J. R. Whittaker (1983). *Dev. Biol.* **96**, 1.] (f) Acetylcholinesterase in embryos of *C. intestinalis* that had been treated with cytochalasin B at the eight-cell stage and cultured for 15 hr. The B4.1 cell pair displays the enzyme. (g) Acetylcholinesterase in embryo of *Ciona* treated with cytochalasin at about the 64-cell stage. The eight cells displaying the enzyme are probably the B8.7, B8.8, B7.5 and B7.8 cells [cf. (a)]. [(f) and (g) From J. R. Whittaker (1973). *Proc. Natl. Acad. Sci. U.S.A.* **70**, 2096.] (h) *Styela plicata* embryo compressed at 3rd cleavage. The yellow myoplasm appears as a dark granular material in the *four lower* blastomeres. (i) ACE in *four* blastomeres of an embryo compressed as in (h), then treated with cytochalasin and cultured for 16 hr. Compare the uncompressed cytochalasin-treated 8-cell embryo in (f). [(h)-(i) From J. R. Whittaker (1980). *J. Embryol. Exp. Morphol.* **55**, 343.]

myoplasm at the time the drug is added, and that for this molecular and ultrastructural differentiation the cellularity of the lineage is unimportant. Cytochalasin treatment also provides a vastly simplified preparation for analysis of the effect of experimental alteration of the cleavage pattern.

Nuclear DNA synthesis is in some way required to trigger the transcriptional events that lead to the appearance of ACE in cytochalasin-treated blastomeres. Aphidicolin, an inhibitor of DNA polymerase, was shown by Satoh and Ikegami (1981a) to block 98% of DNA synthesis in the embryos of an ascidian, and if applied prior to the 64-cell stage this treatment also prevents subsequent appearance of ACE. An interesting experiment performed by Satoh and Ikegami (1981b) indicates in addition that in *Halocynthia* embryos whose cleavage is arrested by cytochalasin a set number of DNA replication cycles seems to be required for ACE expression. Thus if aphidicolin and cytochalasin are applied together at the 64-cell stage ACE is never synthesized, while if they are applied at the gastrula stage, the majority of the presumptive muscle cells then present continue to differentiate. Further experiments of Mita-Miyazawa *et al.* (1985) on isolated B5.1 cells of *Ciona* show that if these cells are allowed to undergo only two further divisions in culture they do not generate ACE, while if they are allowed to complete three divisions they produce ACE [cf. Fig. 6.8(a)]. The third division is the first that is not asymmetric with respect to daughter cell fate, in that both daughters (i.e., B8.7 and B8.8) are muscle cell progenitors. Thus muscle cell differentiation is activated following one division beyond the initial establishment of a clonal muscle cell lineage element. One possibility is that the attainment of a certain nucleus to cytoplasm, or DNA to cytoplasm, ratio could be required to trigger differentiation, so that when DNA replication is blocked as early as the 64-cell stage this ratio is never attained, whether in cleavage-arrested or normal embryos.

The studies reviewed so far, like those of Conklin (1905a), provide excellent correlative evidence for the determinative role of the yellow crescent cytoplasm, but they stop short of demonstrating that this cytoplasm can actually *cause* muscle cell differentiation. This would require introduction or redistribution of this cytoplasm into a cell not normally receiving it, and demonstration that ectopic muscle differentiation results. Whittaker (1980, 1982) reported several experiments of this kind. A variant of the classical pressure plate method was used in one of these studies to alter the 3rd cleavage, with the result that the myoplasm is distributed to four blastomeres rather than two [Fig. 6.8(h)]. On release, the embryos were treated with cytochalasin and three or four blastomeres (i.e., all those receiving the myoplasm) were now found to synthesize ACE rather than the two usually observed when cytochalasin-treated control 8-cell embryos are cultured. An example of this result is shown in Fig. 6.8(i). This experiment also suffices to demonstrate that localization of muscle-forming potential is a cytoplasmic rather than nuclear phenomenon, since the particular nuclei included in the ectopic myoblasts would normally be located in ectoderm cells. In similar experiments Deno and Satoh (1984) transferred cytoplasm from B4.1 cells into A4.1 cells by microinjection, and then treated the embryos with cytochalasin. A small fraction of the experimental embryos displayed three rather than two blastomeres that subsequently developed ACE (i.e., the two B4.1 plus the A4.1 cell), a phenomenon that never occurs in the cytochalasin-treated controls. The implication of these studies (Whittaker 1980, 1982; Deno and Satoh, 1984) is that the myoplasm includes morphogenetic factors that can act in an "instructive" way, since nuclei exposed to it may be induced to express a molecular program associated with myogenesis.

The lineage diagram shown in Fig. 6.8(a) indicates that A4.1 and b4.2 also serve as muscle progenitor cells in normal embryos. Thus experiments carried out on cleavage stage *Ciona* embryos show that when the B4.1 cell pair is removed, 90% of the resulting partial embryos nonetheless develop ACE at the caudal tip of the tail, and this is associated with myofibril formation (Deno *et al.*, 1984). Furthermore isolated b4.2 and A4.1 cells may also give rise to progeny that produce ACE, though the frequency with which this occurs is only 1-10%, depending on species, compared to 80-100% for the progeny of isolated B4.1 cells (Deno *et al.*, 1984, 1985). The difference between the latter results and those obtained with otherwise whole embryos lacking the B4.1 cells suggests that in the b4.2 and A4.1 lineages ACE expression, and muscle differentiation, depend on intercellular interactions. Additional evidence is that in *cytochalasin-treated embryos* only cells of the B4.1 lineages produce ACE (Deno *et al.*, 1985). Thus the morphogenesis of the normal multicellular enivirons might be required for the expression of muscle characters in b4.2 and A4.1 lineages, while the B4.1 lineages differentiate *autonomously*. This interesting contrast illustrates the diverse developmental pathways that may be required in the construction of a single embryonic structure, the tail muscle. The B4.1 muscle progeny are clonally self-differentiating lineages, apparently specified by means of their inheritance of maternal cytoplasmic morphogens, while the A4.1 and b4.2 muscle progeny are probably products of what Roux would have termed "correlative differentiation," and we would today describe as inductive differentiation.

Though this discussion has been focused on muscle cell lineage determination, evidence exists as well for localized cytoplasmic factors that induce gut cell and brain melanocyte determination. Alkaline phosphatase serves as a histochemical marker for gut cell differentiation (Whittaker, 1977); melanotic pigment formation as a marker for the differentiation of brain melanocytes; ciliary membranes for differentiation of a "pressure organ"; extracellular test material as marker of epidermal differentiation; and vacuolar matrix formation as a marker for notochord differentiation (Whittaker, 1979b; Crowther and Whittaker, 1984). In cytochalasin-treated 4-cell embryos all of these types of differentiation occur according to expectation from the cell lineage fate map [Fig. 6.5(j)]. Furthermore, all are sensitive to actinomycin D treatment (Crowther and Whittaker, 1984). Thus it seems clear that in ascidian embryos localized determinants exist for several different cell types, and the conclusions that can be drawn for the self-differentiating myogenic lineage may obtain for many others as well. Crowther and Whittaker (1986) showed that cytochalasin treated zygotes of *Ciona intestinalis* and *Ascidia ceratodes*, which develop as uncleaved single cell syncytia, generate ultrastructural cytoplasmic structures specific to muscle, notochord, neural, and epidermal cell types as well as ACE activity. These embryos possess an approximately normal number of nuclei at the time histospecific markers become evident, and the development of the various differentiated cytological markers is not exclusive. That is, different regions of the same multinucleate single cell produce markers of at least two and sometimes four distinct states of differentiation. This result suggests that the maternal morphogenetic factors act positively and autonomously. It is inconsistent with alternatives such as the possibilities that such factors function by means of potentiating intercellular inductive interactions; or that they function as repressors of regionally *non*expressed genes rather than as activators of regionally expressed genes, since in this case multiple forms of differentiation would not be expected to occur so frequently within the same single cell.

(ii) The Gut Cell Lineage of *Caenorhabditis elegans*

As reviewed in Chapter IV, development in the nematode *Caenorhabditis elegans* is a highly determinate process. The timing and orientation of each embryonic cell division is programmed genetically, and under normal circumstances the ultimate fate of each cell is rigidly fixed. The complete lineage, from egg to hatched larva is shown in Fig. 4.1 (Sulston *et al.*, 1983; see pocket at back of book). The simplified diagram in Fig. 6.9 illustrates the origin in the initial cleavages of the six embryonic founder cells (Deppe *et al.*, 1978; Wood *et al.*, 1983; Sulston *et al.*, 1983). The first cleavage produces an anterior cell, AB, and a posterior cell, P1, and the latter then undergoes several unequal divisions, each resulting in a germ line precursor cell (P1-P4) plus a primordial somatic cell. The endoderm derives from the somatic founder cell EMS, the descendants of which give rise to endoderm, various mesoderm structures, and some neurons. The endoderm lineage is definitively segregated out at the first division of EMS, with the formation of E (endoderm) and MS (mesoderm-stomadaeum) founder cells. Since these cells all arise from topologically specific regions of the egg, and since the destiny of the lineage descendant from them is fixed, a possible interpretation of the initial lineage assignments is that each is determined by the localized cytoplasmic elements that it inherits. In this section we consider

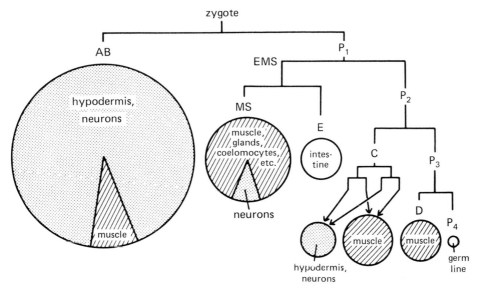

Fig. 6.9. Founder cells of the *Caenorhabditis elegans* embryo, and diagrammatic summary of cell types derived from them. Areas of circles and sectors are proportional to number of descendent cells in the hatched larva. Cell types usually regarded as ectodermal are stippled, and those usually regarded as mesodermal are striped. The AB, MSt and C founder cells are multipotential in terms of these germ layer designations. [From J. E. Sulston, E. Schierenberg, J. G. White, and J. N. Thomson (1983). *Dev. Biol.* **100**, 64.]

evidence related to this proposition in respect to the gut cell lineage descendant from the E blastomere, and in the following, the homologous arguments that can be applied to the germ line lineage deriving from the P4 stem cell.

As reviewed in Chapter IV two markers that specficially indicate gut cell differentiation are the appearance of fluorescent particles that contain tryptophan catabolites, called *rhabditin granules* (Chitwood and Chitwood, 1974), and a specific esterase that can be identified by histochemical staining (McGhee, 1986). These appear in the gut cells a few cycles prior to the terminal divisions of their lineage. Laufer *et al.* (1980) and Cowan and McIntosh (1985) used the appearance of rhabditin granules to determine the cellular location of gut cell determinants in embryos whose cleavage had been blocked with cytochalasin and colchicine, and similar experiments monitoring the appearance of esterase activity in cytochalasin treated eggs were reported by Edgar and McGhee (1986). The drugs were introduced after bursting the impermeable chorion by application of pressure, or shear forces. The results of these cleavage arrest experiments are essentially similar to those of the ascidian cytochalasin studies just reviewed. When cytokinesis is arrested at the 2-cell stage, rhabditin granules and esterase appear after a number of hours in the P1 cell but not in AB; when it is arrested at the

4-cell stage the fluorescent bodies and esterase are observed in EMS, but in no other blastomere; and in arrested 8-cell embryos only the E cell develops these granules or gut esterase. By the time the differentiation reflected by the presence of the cytotypic rhabditin granules has occurred, the arrested blastomeres have accumulated an amount of DNA equivalent to many genomes. It is evident that gut differentiation does not require the appearance of the individual gut cell precursors normally produced during subsequent cleavage of the E lineage, and probably does not depend on intercellular interactions among them, or with other cells. Whether the genomic multiplication is necessary (see above) is not known. However, unlike the case in ascidian embryos, cytochalasin-blocked zygotes are unable to produce rhabditin granules, though they do produce hypodermal markers (Cowan and McIntosh, 1985), nor do blocked one-cell embryos produce the gut esterase (Edgar and McGhee, 1986). In any case, the capacity to synthesize rhabditin granules and gut esterase is shown to be an exclusive and intrinsic property of the E blastomere, and the EMS and P1 blastomeres from which it derives. This is confirmed in studies on the development of burst partial embryos in which given blastomeres have lysed (Laufer et al., 1980). Rhabditin granules are observed only in the E cell derivatives, and not, for example, in isolated AB, P1 or P2 cell partial embryos.

These results are consistent with a determinant role for cytoplasmic factors that are segregated into the E cell during cleavage. However, demonstration of such a role would require that gut differentiation occur in cells normally destined to develop into other structures as a consequence of the experimental introduction of E or EMS cytoplasm. One test of this possibility was carried out by fusing AB.a blastomeres (i.e., the anterior daughter of the AB cell) to the contiguous EMS cell by puncture of the membranes separating them with a laser microbeam (Wood et al., 1983). Following the next cleavage, either the MS cell or the E cell was destroyed by UV irradiation. In 90% of cases when the E cell was irradiated no rhabditin granules formed, though they appeared in the remaining 10% of the experimental embryos, while most control embryos in which the E cell was preserved developed these granules. However, when cleavage was arrested with cytochalasin after the AB.a-EMS fusion, it could be seen that only the E cell developed granules, never the AB.a cell. An alternative experimental design has yielded more positive results. At the 2-cell stage the P1 blastomere was punctured by laser microbeam and the nucleus extruded (Wood et al., 1984). In such embryos only the nucleated AB cell divides further and no rhabditin granules are ever formed (see Fig. 6.9). However, if nuclei of the progeny of the AB cell are exposed to the enucleate P1 cytoplasm, by laser puncture of the intervening membranes, rhabditin granules form amongst the AB cell progeny in 50% of cases. These blastomere fusion experiments thus provide preliminary support for the causal specification of the gut cell lineage by maternal cytoplasmic factors segregated into the P1, EMS, and then the E blastomeres. The perfect precision, and the predictability of the developmental fate of each blastomere indeed suggest this solution, just as did the same features of many other mosaic embryos to classical writers. However, there is no evidence in this or the foregoing ascidian examples, as to the mode of action, or the directness with which such maternal factors may function. At one extreme, they could immediately bind to and activate gut-specific genes, and at the other, they could serve merely as the initial triggers for a cascade of subsequent regulatory events, an ultimate consequence of which is the activation of these genes.

(iii) Localization of Germ Cell Determinants

In most species that have been examined the definitive germ cell lineage can be observed after midembryogenesis, and thereafter be traced directly to the gametogenic cells of the larval or juvenile gonads. Primordial stem cells which serve as the exclusive progenitors of the germ line are known to arise in specific embryonic locations at the outset of development in some organisms, but in others, where germ line specification occurs later, the origins of the germ cell lineage remain controversial or obscure. A unique, localized region of the egg from which the germ line arises can probably be excluded in the case of mammals, for example, since any of the first four or even eight blastomeres can give rise to a complete and fertile embryo. Grafting experiments suggest that in the mouse determination of germ line stem cells occurs in the primitive ectoderm, but not until about 6-7 days of development, as the cells of this region remain totipotent until this time (see Eddy *et al.*, 1981, for review). It is interesting that very early germ cell determination from localized regions of the egg occurs in phylogenetically distant groups, e.g., nematodes, insects, and anuran amphibians. Thus specification of the germ line by means of a cytoplasmic localization process may be an extremely ancient mechanism, since it is found at least occasionally in both deuterostomes and protostomes (see Fig. 6.10).

Determination of the Germ Cell Lineage in Nematodes

Cytoplasmic localization of determinants for the primordial germ cells of the parasitic nematode *Ascaris* was inferred by Boveri and his associates from experiments carried out around the turn of the century. In *Ascaris*, as in *Caenorhabditis*, the definitive germ line stem cell is the product of an early series of unequal cleavages. A special feature of *Ascaris* development not observed in *Caenorhabditis* is the phenomenon of *chromosome diminution*. At the first cleavage one daughter cell retains the complete chromosome complement while the other eliminates a large fraction of the genome in deeply staining pycnotic granules. The stem cell not undergoing chromosome diminution retains the capacity to give rise to the germ cell lineage,

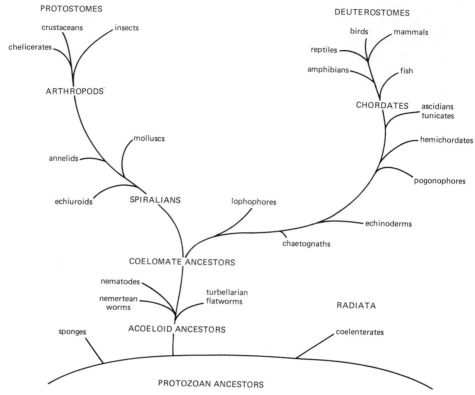

Fig. 6.10. Phylogenetic tree of the metazoa, essentially in accordance with the ideas of Hyman (1940).

while at each of the next several cleavages it produces one daughter which becomes a somatic cell. After fifth cleavage there are 31 somatic cells and one definitive germ line stem cell, the only blastomere that retains all of the zygotic DNA. This process is illustrated in Fig. 6.11, after Boveri (1899, 1910). Boveri (1909, 1910) and Boveri and Hogue (1909) showed that retention of the complete genome, and hence the initial step in determination of the germ line, depends on a special sector of polar egg cytoplasm. In eggs undergoing abnormal cleavage induced by polyspermy or centrifugation, nuclei other than the normal germ line stem cell nucleus that are surrounded by the polar cytoplasm now behave like normal germ line nuclei, in that they are spared chromatin loss at mitosis, while all nuclei distributed into the remainder of the cytoplasm undergo diminution. As many as 3 prospective germ line stem cells may form compared to the normal one, depending on the relation of the altered cleavage planes to the polar cytoplasm, as shown in Fig. 6.11. An additional experiment carried out on *Ascaris* eggs by

Stevens (1909) showed that UV irradiation of the polar egg cytoplasm results in failure of any primordial germ cells to develop. Centrifugation experiments on *Ascaris* eggs were performed by Guerrier (1967), with results basically in agreement with Boveri's conception of the organization of this egg. Boveri's studies do not address completely the issue of germ line determination, since they do not test the ultimate function of this lineage, *viz.*, the provision of fertile gametes. However, they demonstrate that the key first step of this process is indeed specifically induced by the polar egg cytoplasm.

The relation between the embryonic germ line stem cells and the definitive germ cells ultimately located in the gonads has been traced clearly in *C. elegans*, by using fluorescent antibodies that react specifically with granular inclusions of the germ cell cytoplasm (Strome and Wood, 1982, 1983). These inclusions, which have been called P granules, are dispersed throughout the cytoplasm at the time of fertilization, but, as shown in Fig. 6.12(a)-(d), they coalesce in the posterior polar region of the egg during pronuclear migration, and are thence segregated to the P1 cell. This process is probably mediated by cytoskeletal microfilaments, as it can be blocked by cytochalasin D treatment (Strome and Wood, 1983). As cleavage progresses the P granules are segregated successively to P2, P3, and ultimately P4 cells [Fig. 6.12(e)-(f)]. The P granules revealed by the fluorescent antibody are later present exclusively in the germ line progenitor cells Z2 and Z3, the products of P4 [Fig. 6.12(g)-(h)], and in differentiated larvae they exist only in the gonadal germ cells [Fig. 6.12(i)-(j)]. Electron-dense bodies that are probably identical with the P granules have also been visualized in the polar cytoplasm of *C. elegans* eggs, and in the embryonic germ line blastomeres (Wolf *et al.*, 1983). In the uncleaved fertilized egg these structures are located in the peripheral regions of the polar cytoplasm, and by midcleavage they have assumed a perinuclear location in the primordial germ cells. The immunofluorescent P granules are similarly disposed, and they remain perinuclear throughout oogenesis, when they again disperse into the distal cytoplasm of the egg (Wood *et al.*, 1983). It is not clear whether the anterior-posterior axis of the *C. elegans* egg is determined as a result of events following fertilization or is a preformation that is set up during oogenesis (see, e.g., Albertson, 1984).

Maternal mutations are known in *C. elegans* that alter the early cleavage pattern so that two cells form in place of P1 (Schierenberg *et al.*, 1980; Wood *et al.*, 1980). Wood *et al.* (1983) showed that in one such mutant the germ line specific P granules are distributed to the two cells that share the cytoplasm normally destined only for the P2 blastomere, a result that seems exactly to parallel that observed by Boveri in centrifuged and dispermic *Ascaris* (Fig. 6.11).

Distinctive cytoplasmic inclusions that are similar in distribution and ultrastructure to the *C. elegans* P granules and are confined to the germ cell lineage are observed in many forms. They were noticed by classical cytolo-

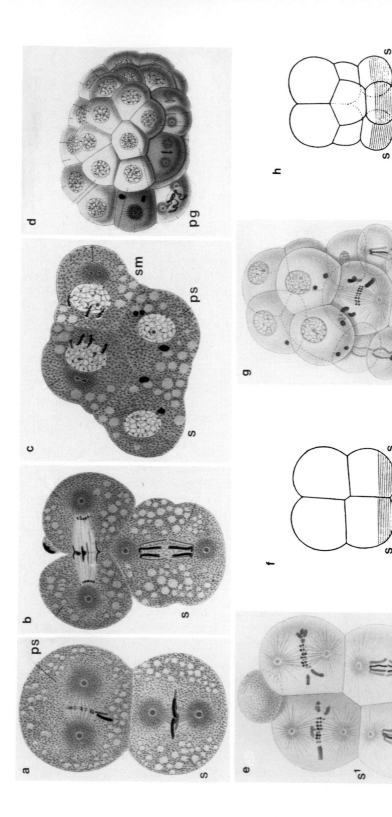

gists in the germ cells of crustaceans (Amma, 1911), chaetognaths, and insects (reviewed in Hegner, 1914). These structures have now been recognized at the ultrastructural level in germ cells of many other animals, among which are annelids, dipteran insects, ascidians, teleosts, anuran amphibians, and various mammals, in which they are usually referred to as *nuage* (reviewed in Kerr and Dixon, 1974; Eddy, 1975; Eddy *et al.*, 1981; Mahowald, 1977; Mahowald *et al.*, 1979a; Smith *et al.*, 1983). In animals where germ cell determination is precocious, and the germ cell lineage is set aside early in cleavage, the germinal granules are typically found in the region of egg cytoplasm that is to be included in the primordial germ cells, as are the *C. elegans* P granules. In animals such as the mouse, where germ cell determination does not occur until later in embryogenesis, they are not evident until the definitive germ cells can be identified (Eddy *et al.*, 1981). Classical writers such as Hegner (1914) thought that the germinal granules of the polar egg cytoplasm, often referred to as *polar granules*, might themselves function as localized germ cell determinants. However, as the following review of germ cell lineage specification in insects and amphibians shows, it is not yet possible to distinguish between this view, which attributes causal significance to the presence of polar granules, and an alternative. This is that these inclusions may function (in some unknown way) in later stages of germ cell differentiation, but not in the *determination* of the germ cell lineage. In organisms where germ line segregation is at least initiated by means of localized elements of the egg cytoplasm, as in *Ascaris,* such granules could be preformed during oogenesis and distributed to that section of cytoplasm which is to be inherited by the germ line stem cell, much as other maternal components of subsequent use during development appear to be localized

---

Fig. 6.11. Chromosomal diminution and determination of primordial germ cells in *Ascaris megalocephala.* Normal development (a)-(d). ps, primordial somatic cell, yet to undergo diminution; sm, somatic cell; s, germ line stem cell. (a) Second cleavage in progress. Chromosome diminution is occurring in the primordial somatic cell. (b) Later in second cleavage, elimination-chromatin at equator of upper spindle (T stage). (c) 4-cell stage showing eliminated chromatin in upper two cells. From the beginning the germ line stem cell possesses a distinct cytological appearance, and in the early cleavages the germ cell mitotic spindles are oriented perpendicularly to those of the somatic cells. (d) About 32 cells, fourth diminution in progress, primordial germ cell (pg) in prophase. [(a)-(d) From T. Boveri (1899). *In* "Festschrift f. C. von Kupffer," p. 383. Fischer, Jena.] (e) Abnormal cleavage in centrifuged egg displaying two prospective germ line stem cells (S^1 and S^2) rather than the single stem cell normally present at this stage [cf. (b) and (c)]. (f) Diagram of same cleavage stage as in (e) showing the ectopic location of the presumptive germinal cytoplasmic determinants. (g) Abnormal later cleavage of a dispermic egg. Three prospective germ line stem cells are recognized by the absence of chromosome diminution and the presence of the same four large chromosomes as seen in the zygote fusion nucleus. (h) Diagram of 8-cell stage of a dispermic egg such as will give rise to the abnormal form shown in (g), with presumptive germinal determinants located in the three lower cells. [(e)-(h) From T. Boveri (1910). *In* "Festschrift f. R. Hertwigs," Vol III, p. 133. Fischer, Jena.]

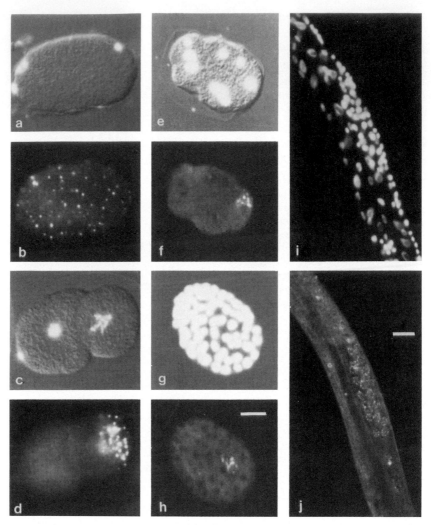

Fig. 6.12. P granules of germ line cells in *Caenorhabditis elegans*. P granules (see text) are displayed by immunofluorescent staining. A monoclonal antibody that reacts with a 40,000 d P granule protein is used in (b) and (d). The antibody used in (f), (h) and (j) was obtained from the serum of a rabbit that possibly had acquired a nematode infection at some time prior to bleeding. (a) Uncleaved fertilized egg following second meiotic division, viewed with Nomarski optics and stained with DAPI to reveal DNA of sperm nucleus at right, and DNA of polar bodies and egg pronucleus at left. (b) Fluorescent antibody staining of same embryo shows P granules scattered throughout cytoplasm. (c) Two-cell embryo stained with DAPI. The P1 blastomere is at right. (d) Same embryo stained with fluorescent anti-P granule antibody. [From S. Strome and W. B. Wood (1983). *Cell* **35**, 15. Copyright by M.I.T.] (e) Six-cell embryo stained with DAPI; the P2 cell is at the right hand pole. (f) Same embryo displaying immunofluorescence of P granules in this cell. (g) DAPI stained embryo of >100 cells, shortly after division of P4 into the primordial germ cells Z2 and Z3 (see Fig. 4.1). (h) Immunofluorescnece of P granules in Z2 and Z3 cells of the same embryo. (i) DAPI stained stage midregion of a late stage 1 larva. (j) Immunofluorescence of P granules in proliferating nest of germ cells. Bars = 10 μm. [From S. Strome and W. B. Wood (1982). *Proc. Natl. Acad. Sci. U.S.A.* **79**, 1558.]

(e.g., tubulin, yolk, mitochondria, etc.). At present it indeed seems reasonable to regard germinal granules as cytotypic markers for the germ cell lineage, as illustrated so clearly for *C. elegans* in the experiments reproduced in Fig. 6.12, but it remains to be determined whether they participate in the initial specification of that lineage.

Determination of Germ Cell Lineage in the Eggs of Drosophila and Some Other Insects

Primordial germ cell differentiation was studied in chrysomelid beetles by Hegner (1911, 1914). Germ cell formation can be said to initiate when nuclei arrive at the polar region of the oblong egg, and enter a special region of cytoplasm which appears to function as the germ cell determinant. Hegner succeeded in selectively destroying the polar cytoplasm with a hot needle before the peripheral movement of the nuclei had occurred. The injury induced by the needle is rapidly walled off by the forming blastoderm, and normal development of a differentiated gastrula and ultimately a hatching insect takes place. However, primordial germ cells are absent in embryos descended from the cauterized eggs, and adults developing from them are sterile. No genomic material is directly affected, since no nuclei are in the vicinity of the polar cytoplasm at the time of cauterization. Hegner's experiments show that when the polar cytoplasm is destroyed, the capacity of the embryo to elicit germ cell differentiation is lost.

The primordial germ cells in *Drosophila* eggs arise at the posterior pole of the egg in much the same manner. Huettner (1923) described this process in detail and concluded:

> The deciding factor which determines whether a nucleus shall become somatic or germinal appears to be the posterior polar plasm, which I regard as a differentiated ooplasm. Any nucleus of the developing egg may be differentiated into a polar nucleus if it comes accidentally into the region of the posterior polar plasm.

The *pole cells* are easily distinguished, as shown in Fig. 6.13. They are the first cells formed in the embryo and they lie outside of the blastoderm wall. Geigy (1931) showed that UV irradiation of vegetal pole cytoplasm prior to the migration of the cleavage nuclei into this region results in otherwise normal, but agametic animals. Many subsequent experiments demonstrated that the same cleavage nuclei that when surrounded by polar cytoplasm give rise to pole cells, instead become somatic blastoderm nuclei if the polar region of the egg is destroyed, so that they are surrounded by other cytoplasm (reviewed in Counce, 1973). Observations of similar import have been made with a large number of other insect species. For example, Brown and Kalthoff (1983) and von Brunn and Kalthoff (1983) demonstrated that UV irradiation of a sharply defined region of the polar cytoplasm in *Smittia* eggs that had not yet initiated nuclear division interferes with the migration of nuclei into the polar cytoplasm, prevents subsequent pole cell formation,

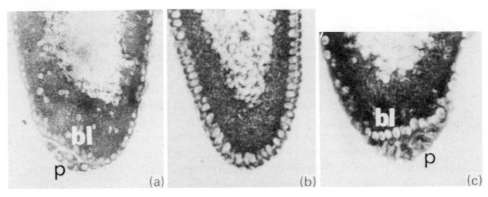

Fig. 6.13.   Pole cells of *Drosophila* eggs and rescue of UV irradiated eggs by injection of polar egg cytoplasm. Longitudinal sections of the posterior regions of eggs fixed at the blastoderm stage. (a) Normal control egg. Complete blastoderm and pole cells (P) are evident. X300. (b) Egg was irradiated at cleavage stage. Blastoderm has formed over the entire egg, but no pole cells are found. X300. (c) Eggs were irradiated at cleavage stage and subsequently injected with polar cytoplasm. Blastoderm (bl) and pole cells (P) similar to those of normal eggs can be seen. Damage on right side is an artifact of sectioning. X450. [From M. Okada, I. A. Kleinman, and H. A. Schneiderman (1974b). *Dev. Biol.* **37**, 43.]

and also affects the survival of those pole cells that do form. This case is of particular interest because the action spectrum for UV inhibition of pole cell formation closely approximates the UV absorption spectrum of RNA, except for a slightly higher relative efficiency at wavelengths >260 nm. In addition all the effects of UV irradiation on the pole cells are photoreversible, which is usually taken to suggest a nucleic acid target (Jäckle and Kalthoff, 1980). Togashi and Okada (1983) reported an action spectrum for UV inhibition of *Drosophila* pole cell formation, with resultant adult sterility, that peaks at 280 nm and is also high at 254 nm. In both this and the *Smittia* study the sensitive target areas were shown to be confined to a thin polar region of the cortical cytoplasm. The UV rays could not have reached the zygote nucleus in the *Smittia* experiment, nor in the case of *Drosophila* any of the nuclei, which at the stage when the irradiation was performed are located deep in the interior of the egg.

The sterilizing effect of UV irradiation offers an experimental opportunity to demonstrate that the polar egg cytoplasm is specifically required for the formation of functional primordial germ cells. Thus Okada *et al.* (1974b), Warn (1975), and Ueda and Okada (1982) showed that injection of unfractionated posterior pole cytoplasm, but not anterior pole cytoplasm, rescues eggs that would otherwise have been sterilized by UV irradiation. The injected eggs are able to produce pole cells, as shown in Fig. 6.13(c), and frequently developed into adults that contain fertile gametes, in contrast to the agametic adults obtained from the irradiated controls. It is interesting that the cytoplasmic substances required for rescue of pole cell formation can be physically separated from those necessary to restore adult fertility. Thus while unfractionated posterior pole cytoplasm contains both activities, Ueda and Okada (1982) found that a 28,000 × g centrifugal pellet fraction promotes pole cell formation only. Experiments of Okada and Togashi (1985) indicate that the component required for pole cell formation is an mRNA that can be extracted from unfertilized eggs. However, though injection of UV irradiated eggs with this RNA fraction restores the formation of pole cells of apparently normal structure, fertile flies are never obtained.

Convincing evidence that *causal pole cell determinants* are localized in the posterior region of the *Drosophila* egg is provided by the cytoplasmic injection experiments that were carried out by Illmensee and Mahowald (1974, 1976). In these studies posterior pole cytoplasm was injected into the anterior or ventral regions of recipient eggs. Pole cells displaying cytological structures unique to the germ cell lineage were thereby induced to form at the ectopic locations where the cytoplasm had been injected. Among the ultrastructural features used to identify pole cells are intranuclear organelles that consist of an electron dense cortex and in some species an electron lucid core, called *nuclear bodies* (reviewed in Mahowald, 1977). Illmensee *et al.* (1976) also found that posterior pole cytoplasm from stage 13-14 ovarian oocytes induces the formation of ectopic pole cells when injected into the anterior ends of recipient eggs, but cytoplasm extracted from younger oocytes (stages 10-12) does not display this activity. The polar cytoplasm of these oocytes contains some morphologically identifiable polar granules, but evidently lacks some necessary constituents. The special significance of this demonstration is that the localized determinants for pole cell formation are shown to be present prior to fertilization and ovulation. They are thus in the classical sense a *prelocalization*, of a morphogenetic substance that is synthesized and anchored to the polar region of the egg during ovarian oogenesis. An elegant feature of the cytoplasmic transfer experiments of Illmensee *et al.* was the use of donor and recipient strains bearing different genetic marking. For example the anterior pole cell induction experiment was carried out in flies of *mwh e* (*multiple wing hair, ebony*) genotype, and to test the ability of the ectopic pole cells that were formed to serve as germ line progenitors these were transplanted to the posterior regions of *y w sn* (*yellow, white, singed*) hosts. A very small percent of the resulting adults displayed germ line mosaicism, as expected if the induced pole cells gave rise to functional gametes (Illmensee and Mahowald, 1974). A similar result was reported for pole cells induced to form ectopically in ventral regions of the embryo (Illmensee and Mahowald, 1976).

This review briefly summarizes much evidence that the posterior polar cytoplasm can specifically induce *pole cell formation*, and that it contains substances *necessary* for germ cell formation. However, additional demonstrations of the ability of ectopically induced pole cells to give rise to functional gametes is clearly required to secure the proposal that the polar cytoplasm actually *determines the germ cell lineage*. The demonstration of Ueda and Okada (1982) that injection of certain cytoplasmic fractions into UV irradiated eggs restores pole cell formation but not adult fertility shows that this is not simply a semantic issue. A difficulty is that in *Drosophila* not all pole cells give rise to germ cells, though the ultimate fate of those that do not find their way to the larval gonad remains controversial (Mahowald *et al.*, 1979a).

The germinal or polar granules of *Drosophila* eggs have been characterized in great detail. At oviposition these granules consist of dense, membrane-free particles surrounded by clouds of ribosomes (Mahowald, 1968). In various species of *Drosophila* these particles fuse or fragment during early embryogenesis, and the ribosomes and polyribosomes initially associated with them are no longer observed after the pole cell stage of embryogenesis (Mahowald, 1971a). By the time the germ cells are located in the larval gonad, the polar granules have apparently given rise to characteristic fibrillar structures applied to the nuclear membrane (Mahowald, 1971b). This progression of forms is similar to what is observed in both amphibian and mammalian oogonia. The fibrillar structures remain throughout the oogonial stage, but are absent in oocytes. Polar granules can again be observed in *Drosophila* oocytes during vitellogenesis. Though they are already localized in stage 10-12 oocytes, as noted above the polar cytoplasm of these oocytes is not yet competent to induce pole cell formation (Illmensee *et al.*, 1976). This observation at least distinguishes the cytological entities recognized as ovarian polar granules from the functional properties of the mature posterior pole cytoplasm. The *Drosophila* polar granules have been partially purified from pole cells (Waring *et al.*, 1978), and there is evidence from two-dimensional analyses of their proteins for a single major species of about 95 kilodaltons. This protein is synthesized during oogenesis, prior to stage 10 (Mahowald *et al.*, 1979a).

The effects of several maternal mutations that result in sterility have recently been interpreted as consistent with a mechanistic role for the polar granules in germ cell determination. Among the examples of interest are a *grandchildless* (*gs*) mutation of *Drosophila subobscura* which in homozygotes prevents pole cell formation and thus results in sterility. Ultrastructural examination (Mahowald *et al.*, 1979b) shows that among other things this genetic lesion affects the deposition of polar granules during oogenesis. Some *gs* eggs lack any polar granules, while others possess small numbers of polar granules that are often incorrectly localized. Unfortunately, this is not the only cytological lesion in *gs* eggs, as both anterior and posterior poles appear abnormally organized, and during early cleavage, nuclei never descend into the posterior pole cytoplasm. A *gs* mutation of *D. melanogaster*, *gs(1)N26*, also prevents nuclear migration into the posterior cytoplasm but has no effect on polar granule disposition (Niki, 1984). On the other hand, various alleles of another *D. melanogaster grandchildless* mutant, *tudor*, provide a good correlation between the amount of polar granule material and the ability to produce pole cells (Mahowald, 1983; Boswell and Mahowald, 1985). Again, however, this is not the unique effect of the mutation, some alleles of which also disturb embryonic segmentation. Perhaps this mutation interferes with a (cytoskeletal?) process by which several different determinants are localized during early development, and the abnormal polar granule deposition in mutant eggs could be merely a visible index of the disturbance, rather than the immediate cause of sterility. A number of other recessive maternal mutations of the *grandchildless* phenotype are also known, but most of these also produce somatic embryonic defects (reviewed in Konrad *et al.*, 1985). With respect to the polar granules *per se*, neither modern genetic nor cytological evidence provides a counter to the skepticism expressed by Huettner (1923):

> It is the opinion of most observers at the present time that the polar granules exert no influence on the differentiation of the polar cells ... I assume that they are by products of the posterior polar plasm which have nothing to do with the *causal differentiation* of the germ cells. [italics added]

Determination of the Germ Cell Lineage in Anuran Amphibians

It has been known for many years that in frogs the germ cell lineage originates early in cleavage from blastomeres forming at the vegetal pole of the egg. In normal development the vegetal pole cytoplasm or "germ plasm" is distributed to all of the first four blastomeres, and each has the capacity to generate a functional germ cell lineage. The primordial germ cells reside in the endoderm during early development, where they undergo several divisions, and after gastrulation they migrate to the germinal ridges where they are located by stage 25 in *Rana pipiens* tadpoles, or by stage 44-47 in *Xenopus laevis* tadpoles (Whitington and Dixon, 1975; Smith and Williams, 1979). At this point they are in a state of mitotic arrest. Although if undisturbed these cells subsequently give rise to the functional germ cells of the organism, they are not yet irreversibly determined to do so, even by this late stage. Thus Wylie *et al.* (1985) showed that fluorescein-labeled primordial germ cells from stage 45 *Xenopus* tadpoles give rise to a variety of different ectodermal, endodermal, and mesodermal cell types if implanted into the blastocoel of a late blastula stage embryo.

Bounoure (1937) and Bounoure *et al.* (1954) showed that UV irradiation of the vegetal region of the cytoplasm of frog eggs results in a sharp decrease in the number of primordial germ cells, and ultimtely, in some animals, in agametic gonads and sterility. These classical experiments suggested that peripheral components of this cytoplasmic region, which is incorporated in the initial precursors of the germ cell lineage, might be required for specification of this lineage. The significance of vegetal pole cytoplasm for germ line differentiation was confirmed in many subsequent experiments in which this region of the cytoplasm of uncleaved eggs or of 2- or 4-cell embryos was severely damaged by ultraviolet irradiation or by microsurgical means (reviewed in Blackler, 1970; Smith and Williams, 1979).

The number of primordial germ cells observed in the early tadpole (about 35 in normal stage 47 *Xenopus* tadpoles) has traditionally been used to assay the effect of exposing eggs and early embryos to UV irradiation. Thus the germinal ridges of stage 47 tadpoles derived from vegetally irradiated *Xenopus* eggs, zygotes, or 2-cell embryos were often found to be devoid of germ cells. Irradiation at the 8-cell stage, however, has no specific effect on the number of tadpole primordial germ cells (e.g., Ijiri, 1977). Similar results were obtained earlier by Smith (1966) in *Rana pipiens*, and control experiments showed that parallel irradiation of the animal pole cytoplasm produces no visible defects, and certainly none in the germ cell line. Furthermore, Smith (1966) demonstrated that the damage to cytoplasmic targets resulting from ultraviolet irradiation of *Rana* eggs can be compensated by injection into the vegetal pole of cytoplasm from the vegetal pole of an unirradiated egg. In a significant fraction of cases the recipient eggs developed into stage 25 tadpoles in which the germinal ridges bore primordial germ cells, while controls receiving no vegetal cytoplasm or cytoplasm from the animal pole lacked germ cells. Rescue of UV irradiated *Rana chensinensis* and *Xenopus* eggs can also be accomplished by injection of vegetal pole cytoplasm (Wakahara, 1977, 1978). A crude centrifugal fraction of vegetal but not animal pole cytoplasm from nonirradiated eggs was shown to be effective in restoring germ cells to the tadpole germinal ridge. In addition, injection of these fractions into nonirradiated eggs is reported to induce the formation of up to twice the normal number of primordial germ cells (Wakahara, 1978).

The surprising observation has recently been made in several laboratories (Züst and Dixon, 1977; Smith and Williams, 1979; Subtelny, 1980; Williams and Smith, 1984) that primordial germ cells eventually reappear in the germinal ridges of tadpoles raised from some vegetally irradiated eggs, albeit on a much delayed schedule. Thus while the germinal ridges of stage 25 *Rana pipiens* tadpoles grown from UV irradiated eggs lack primordial germ cells, two weeks later these are present, and they ultimately give rise to differentiated gametes. Ikenishi and Kotani (1979) showed that in *Xenopus* embryos raised from vegetally irradiated eggs, fewer primordial germ cell divisions occur, and after the tailbud stage many of the primordial germ cells initially formed disappear. Furthermore, the surviving germ cells remain in the central endoderm at tailbud stage, while in control embryos they have already migrated to the lateral and dorsal regions. These observations occasion a somewhat different interpretation than classically proposed of the effect of cytoplasmic UV irradiation. Since surgical removal or disturbance of the vegetal pole cytoplasm does eventually result in sterility or partial sterility

(Buehr and Blackler, 1970; Gipouloux, 1971; Sakata and Kotani, 1985) there may indeed be a cytoplasmic determinant for the anuran germ cell lineage localized at the vegetal pole of the egg. However the *UV sensitive component* seems to control only the replicative or migratory behavior of the primordial germ cells rather than their fundamental specification. We are reminded of the parallel observations in *Drosophila* where, as we have seen, the determination of the germ cell lineage is a process that has several components, and pole cell formation can be separated experimentally from the specification of a functional germ cell lineage.

The role of polar granules in the anuran germ cell lineage specification process is also indistinct. These bodies, also known as the "germinal plasm," are initially found in the outer regions of the vegetal pole cytoplasm, and in ultrastructure they closely resemble the polar granules of insect eggs. Their appearance in transmission electron micrographs is shown in Fig. 6.14. As illustrated in Fig. 6.14(a)-(b), they are present in unfertilized eggs, having been localized just beneath the vegetal pole cortex during maturation (Smith and Williams, 1975, 1979). They may be anchored to the egg cortex, as if the eggs are inverted, they do not redistribute gravitationally as do the yolk platelets (Wakahara *et al.*, 1984). During cleavage these organelles are incorporated in the primordial germ cell precursors, and they are still present, with associated ribosomes, in these cells at the blastula and gastrula stages [Fig. 6.14(c)-(d)]. By the time the primordial germ cells are localized in the germinal ridges they no longer contain identifiable polar granules. Instead a fibrous component is found applied to the nuclear membrane. Though transition stages have not been convincingly described, these could derive from the earlier polar granules, an interpretation that is supported by the analogous progression of form undergone by insect polar granules. In anuran amphibians the fibrillar perinuclear structures persist through most of oogenesis, though they are absent in mature oocytes. Typical polar granules reappear at about the time of germinal vesicle breakdown. Several correlative items of evidence, in addition to the natural history of these germ cell specific organelles, have been adduced in support of the idea that polar granules are involved in germ cell specification. Thus loss of sensitivity to UV irradiation by the 8-cell stage occurs just when the polar granules move inward from the cortex of the egg so that they no longer lie within the shallow penetration range of the ultraviolet irradiation (Tanabe and Kotani, 1974). Furthermore, when *Xenopus* eggs are centrifuged so as to move the polar granules inward precociously, irradiation at the 2-cell stage fails to affect the number of primordial germ cells in the germinal ridges of stage 47 tadpoles. This result is significant, since only large, dense particles would be likely to have been affected by the low centrifugation forces applied in this experiment, 150 g for 60 seconds. On the other hand there is no visible effect of UV irradiation on polar granule ultrastructure (Smith and Williams, 1979), and even if these particles were the essential UV targets, their function in

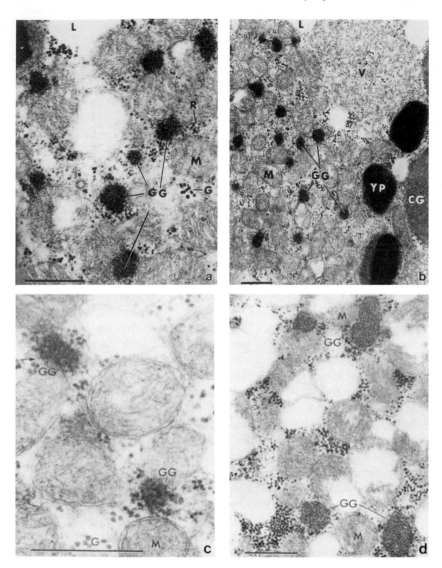

Fig. 6.14. Ultrastructure of region containing polar granules in anuran eggs. GG, polar or germinal granules; M, mitochondria; MY, mitochondria which contain yolk; R, ribosomes; and G, glycogen; YP, yolk platelet; L, lipid; CG, cortical granule. Scale lines equal 0.5 μm. (a) Unfertilized egg of *Rana pipiens*. Numerous polar granules are present together with mitochondria. (b) Maturing oocyte, enucleated, 23 hr after stimulation with progesterone. This observation shows that germinal granule deposition does not require coincident nuclear function. [(a) and (b) From L. D. Smith and M. Williams (1979). *In* "Maternal Effects in Development" (D. R. Newth and M. E. Balls, eds.), p. 167. Cambridge Univ. Press, London and New York.] (c) 4-cell embryo, vegetal pole region; arrow points to fibril connecting polar granule and ribosomes. (d) 16-cell embryo. [(c) and (d) From M. A. Williams and L. D. Smith (1971). *Dev. Biol.* **25**, 568.]

germ cell differentiation could be confined to effects on replication and migration (see above).

Polar granules are not evident in the vegetal pole cytoplasm of cleavage stage urodele eggs, and germ cells in these eggs apparently arise from dispersed regions of the marginal zone of the animal ectodermal cap rather than from the vegetal endoderm, as in anurans (Sutasurya and Nieuwkoop, 1979; Michael, 1984). Though they may develop from specific cells of the marginal zone, the differentiation of primordial germ cells in urodeles probably requires inductive influences from adjacent caudal mesoderm. The earliest reported appearance of structures resembling polar granules in urodele eggs is at gastrula (Smith *et al.*, 1983), though the definitive primordial germ cells of urodele larvae ultimately resemble those of anurans in their ultrastructural characteristics. Thus the germ cell lineages in urodeles would appear to arise by a different route than in anurans (reviewed in Smith *et al.*, 1983), though in both groups the primordial germ cells ultimately possess the same cytoplasmic organelles. The least artificial view may again be that amphibian polar granules are cytoplasmic structures required later in germ cell differentiation, and that they appear together with whatever localized germ cell determinants may exist in the egg cytoplasm only when the germ cell lineage is set aside at the very beginning of cleavage, as in the anurans.

3. LOCALIZATION OF DORSOVENTRAL DETERMINANTS IN THE CYTOPLASM OF MOLLUSCAN AND ANNELID EGGS

In the examples we have so far considered localization results in, or may result in, the specification of those cell lineages inheriting given elements of egg cytoplasm. We now turn to the more complex, and also more general case, in which there are in addition secondary inductive effects of localization. Axial determination in molluscan eggs involves the specification of a series of cell lineages that produce axial structures, e.g., the apical and neuronal tissues of the larva. While the initial axial specifications occur in some cases by means of cytoplasmic localization, the later ones are characteristically inductive.

A characteristic feature of early cleavage in many molluscan and annelid forms is the transient extrusion of a lobe of vegetal pole cytoplasm, which following cytokinesis is resorbed into one of the embryonic macromeres. In the typical case the first cleavage *polar lobe* is resorbed into the CD macromere, the second cleavage polar lobe into the D macromere, the third cleavage polar lobe into the 1D macromere, and so forth. Photographs of this process are shown for the first two cleavages of the egg of the gastropod *Ilyanassa obsoleta* in Fig. 6.15. As will be recalled it is the D quadrant of molluscan and annelid eggs that gives rise to the mesentoblasts and is responsible for the dorsoventral organization of the embryo (see Fig. 6.2). The

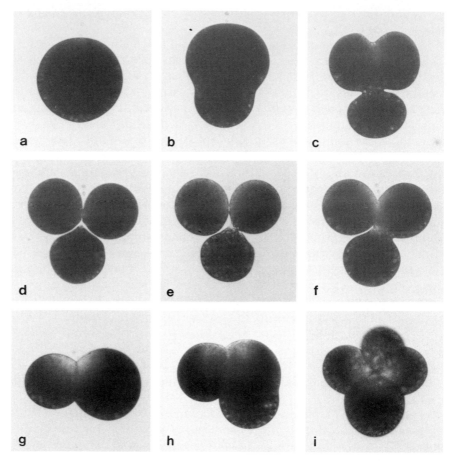

Fig. 6.15. Formation and resorption of the first and second cleavage polar lobes in eggs of *Ilyanassa obsoleta*. (a) Mature egg, following fertilization and extrusion of polar bodies. (b)-(d) Progressive stages in extrusion of the first polar lobe as first cleavage progresses. The lobe can be distinguished as the more refractile of the three spheres at the "trefoil" stage, (d). At this point the lobe is attached to the future CD blastomere by a thin cytoplasmic stalk. (e)-(g) Resorption of the polar lobe into the CD blastomere. In (g) first cleavage is complete and the CD blastomere is the larger one. (h) Extrusion of the second polar lobe as second cleavage begins. (i) Four-cell stage. The second polar lobe has been resorbed by the D macromere which is consequently the largest of the four embryonic blastomeres. In (a)-(h) the animal pole is toward the top; in (i) it is facing the observer. [From A. C. Clement (1976). *Am. Zool.* **16**, 447.]

polar lobe is often attached to the remainder of the embryo by only a thin strand of cytoplasm, as illustrated in Fig. 6.15(d) and (e). Its removal is easily accomplished without immediate injury to the embryo, which continues to divide approximately on schedule. However, the ultimate developmental consequences of the removal of this anucleate element of egg cytoplasm are both dramatic and specific. In this section we review the experimental effort that has been made over the last 90 years to establish a causal relation between the polar lobe cytoplasm and the establishment of the dorsoventral axis of the embryo, and to understand the nature of the localized determinants that it apparently contains.

(i) Morphogenetic Significance of the Polar Lobe Cytoplasm

The developmental consequences of polar lobe removal were first investigated by Crampton (1896), who was then a student of E. B. Wilson. Crampton found that deletion of the first cleavage polar lobe of *Ilyanassa* eggs causes the asymmetric cleavage pattern and the delayed division schedule that normally mark the D quadrant to disappear. Cleavage of embryos from which the polar lobe had been removed is radially rather than bilaterally symmetric, and Crampton was unable to recognize in these embryos the 4d stem cell from which the mesentoblasts derive. Wilson (1904b) followed up these preliminary results with an extraordinary study of the effect of polar lobe removal on embryogenesis in the scaphopod mollusc *Dentalium*. These experiments demonstrated that deletion of the first cleavage polar lobe irreversibly blocks the appearance of all the special D quadrant cell lineages. He reported that the mesentoblasts and the mesodermal germ bands never form, and that other larval structures are missing after lobe removal, such as the apical tuft, and the larval shell and foot. Wilson (1904b) compared the capacities of isolated blastomeres with those manifested by lobeless embryos, and showed that the latter develop in exactly the same way (or fail to develop in exactly the same way) as do embryos deriving from AB blastomeres or from single A, B, and C blastomeres, none of which produce the mesodermal primordium. From these observations it was possible to infer the existence in the polar lobe of cytoplasmic determinants for the D quadrant cell lineages that build the dorsoventral structures of the embryo.

Polar lobe removal is now known to result in the same characteristic complex of developmental defects in a number of molluscan and annelid species (see reviews in Cather, 1971; Dohmen and Verdonk, 1979; Dohmen, 1983). Among the organisms in which this phenomenon has been studied are the gastropod molluscs *Bithynia* (Cather and Verdonk, 1974) and *Ilyanassa* (Clement, 1952); the lamellibranch mollusc *Mytilus* (Rattenburg and Berg, 1954); the annelid *Sabellaria* (Hatt, 1932; Novikoff, 1938), as well as *Dentalium*, the subject of recent investigations by Verdonk (1968), Van Dongen (1976), Van Dongen and Geilenkirchen (1975), and Cather and Verdonk (1979), that have confirmed and extended the 1904 conclusions of Wilson. This is not an invariable phenomenon in eggs bearing polar lobes, however. Thus removal of the small polar lobe of the polychaete annelid *Chaetopterus* has only minor developmental effects, and instead vegetal region cytoplasm included in the polar lobe is functionally analogous to the polar lobe cytoplasm in the other species mentioned (Henry, 1986).

In Fig. 6.16 the later cleavage and cell lineage of *Ilyanassa* embryos are diagrammed, with special emphasis on the morphogenetically significant D quadrant (Clement, 1952). Normal development is shown in Fig. 6.16(a)-(g), and the radially symmetric cleavage of lobeless embryos, in which the D

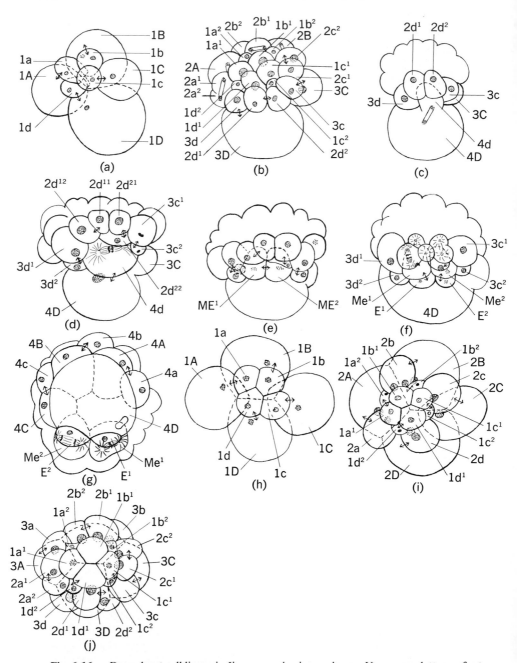

Fig. 6.16. D quadrant cell linege in *Ilyanassa obsoleta* embryos. Upper case letters refer to macromeres and lower case to micromeres. Preceding numerals indicate the macromere division at which the micromeres are given off, and superscripts indicate subsequent micromere division products. E[1] and E[2], primary entoblasts; Me[1] and Me[2], primary mesentoblasts. (a) and

quadrant cells cannot be distinguished, is illustrated for comparison in Fig. 6.16(h)-(j). The structures of the advanced larvae descendant from lobeless embryos are compared to those of normal *Ilyanassa* larvae in the drawings of Fig. 6.17, from observations of Atkinson (1971). Lobeless larvae fail to develop heart, intestine, statocyst, operculum, velum, external shell, eyes, and foot. On the other hand, lobeless larvae possess active muscle, nerve ganglia and nerve endings, stomach, some velar tissue with cilia, digestive gland, mantle gland tissue, and pigment cells. Removal of the first cleavage polar lobe cytoplasm thus does not block all cell differentiation, only *certain* differentiation, though ultimately it produces a cascade of delayed organizational defects.

The catastrophic though specific effects of first polar lobe removal illustrated in these figures suggests that diverse morphogenetic determinants are located in the vegetal pole cytoplasm of the CD blastomere by the 2-cell stage. The most straightforward presumption would be that during later cleavage these determinants are physically transferred to the stem cells from which lobe dependent lineages arise, e.g., 4d. The latter cell is the direct ancestor of both the primary mesentoblasts (Me^1 and Me^2) and the primary entoblasts (E^1 and E^2; see Fig. 6.16). Thus only embryos in which 4d is present produce heart and intestine at the veliger larval stage. In *Ilyanassa* deletion of 1d produces no observable effect, but deletion of 2d causes absence or impairment of the shell, while 3d forms the primordium for the left half of the larval foot (Clement, 1976). The inference that morphogenetic determinants for these tissues are transferred directly into the d quadrant micromeres during cleavage was tested in *Ilyanassa* by Clement (1962) and in *Dentalium* by Cather and Verdonk (1979), by determining the effect of ablation of the D macromere at successive cleavages. Deletion of the D macromere after the 1d micromere has been given off tests whether the morphogenetic factors of the polar lobe still reside in the macromere or have been shunted into the micromere, and similarly for D macromere deletion after 2d, 3d, or 4d have arisen. Clement (1962) found that in *Ilyanassa* removal of the 4D macromere after the formation of 4d has no effect on later differentiation, and the resulting embryo is normal except for its small size. This experiment also helps to eliminate the possibilities that the effects of polar lobe removal are due to general injuries, starvation for nutrients, or to disruption of an embryo-wide "morphogenetic gradient." On the other hand

---

(b) Normal cleavage showing first, second, and third quartet micromeres, from camera lucida drawings of stained whole mount preparations. (c)-(f) Normal cleavage showing the early derivatives of the first mesentoblast progenitor cell, 4d. (g) The egg has been oriented so that the vegetal pole is toward the observer. Further division of Me^1 and Me^2 will produce the primordial mesoderm columns. (h)-(j) Symmetrical cleavage after removal of the polar lobe at the trefoil stage. The D quadrant is named arbitrarily, as it cannot be differentiated by inspection from the B quadrant. [From A. C. Clement (1952). *J. Exp. Zool.* **121**, 593.]

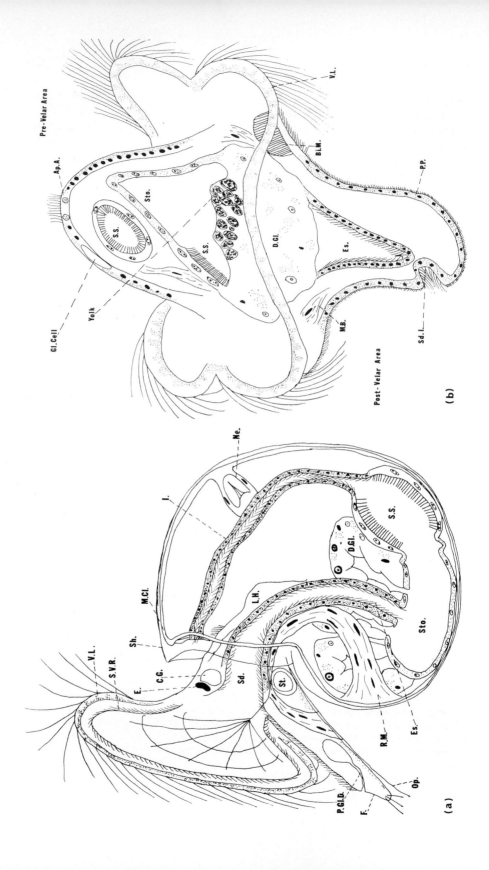

Pre-Velar Area

V.L.

B.M.

P.P.

Ap.A.

Sto.

S.S.

S.S.

D.Gl.

Es.

Gl.Cell

Yolk

M.B.

Sd.l.

Post-Velar Area

(b)

Ne.

I.

M.Cl.

S.S.

D.Gl.

L.H.

Sh.

V.L.

S.V.R.

C.G.

Sto.

E.

Sd.

St.

R.M.

Es.

P.Gl.

Op.

F.

(a)

removal of the D macromere before the 2d micromere has been formed results in as severe an inhibition of morphogenesis as removal of the first cleavage polar lobe or of the whole D quadrant at the 4-cell stage. If the 3D macromere is deleted, i.e., after 1d, 2d, and 3d have been formed, the resulting larvae display velum, eyes, foot, and some shell, all of which are polar lobe dependent structures (see Fig. 6.17), but they still lack the mesodermal primordium. The morphogenetically significant polar lobe contents therefore appear to be shunted into 3d and 4d micromeres. Essentially similar results are reported for *Dentalium*. Here again removal of the D macromere when only 1d has been produced is equivalent in its effects to deletion of the polar lobe, while consequences of decreasing severity occur if the macromere is ablated after 2d, or both 2d and 3d micromeres have appeared. *Dentalium vulgare* embryos from which 4D is removed also develop into normal larvae (Cather and Verdonk, 1979). The general conclusion from both sets of experiments is that the morphogenetic elements localized in the polar lobe are progressively distributed during cleavage to the specific micromeres where they will function.

Localization of determinants for the ciliary apical organ of the larva provides a particularly clear case. In annelid and molluscan larvae this organ consists of a plate of differentiated cells, several of which bear long cilia, overlying a plexus of neurosecretory cells that is later associated with the cerebral commissure (Lacalli, 1981). *Dentalium* embryos from which the first polar lobe has been removed form no apical tuft, but differentiation of the apical organ proceeds normally in embryos from which the second polar lobe has been deleted (Cather and Verdonk, 1979). The apical organ forms from progeny of the 1c and 1d micromeres (Van Dongen and Geilenkirchen, 1974). Microsurgical experiments indicate that the apical organ determinants are localized in the upper region of the first polar lobe and they are later shunted into the upper regions of the C and D blastomeres. Thus the whole lower end of the lobe and 50-60% of the volume of cytoplasm that it contains can be ablated without loss of apical tuft formation (Geilenkirchen *et al.*, 1970; Render and Guerrier, 1984). Determination of the cell lineage that produces these tuft cilia, and probably of the neuronal components of the apical organ as well, occurs after resorption of the lobe into the CD macro-

---

Fig. 6.17. Normal and lobeless *Ilyanassa obsoleta* larvae. (a) Composite reconstruction of normal 8- to 9-day veliger larva of *Ilyanassa*, with left velar lobe omitted. (b) Reconstruction of a lobeless larva based on whole mount and sectional material. Ap.A., apical area; Bi.M., birefringent mass; C.G., cerebral ganglion; D.Gl., digestive gland; E., eye; Es., esophagus; F., foot; Gl. Cell, ectodermal gland cell; I., intestine; L.H., larval heart; M.B., muscle block; M.C., mantle collar; Ne., nephridium; Op., operculum; P.P., posterior protusion; P.Gl.D., pedal gland duct; R.M., retractor muscle; S.S., style sac; Sd., stomodaeum; Sd.I., stomodeal-like invagination; Sh., shell; St., statocyst; Sto., stomach; S.V.R., second velar row; V.L., velar lobe. [From J. W. Atkinson (1971). *J. Morphol.* **133**, 339.]

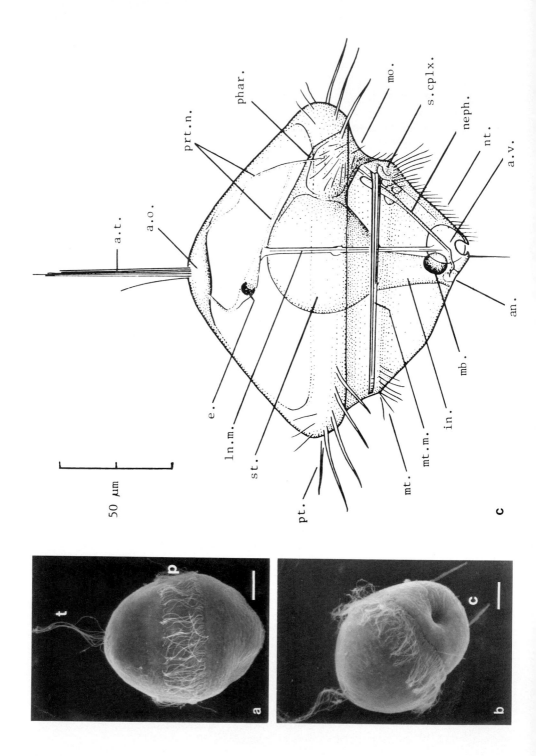

50 μm

a.t.
a.o.
prt.n.
phar.
mo.
s.cplx.
neph.
nt.
a.v.
an.
mb.
in.
mt.m.
mt.
pt.
st.
ln.m.
e.

t
p
c

a
b
c

mere, evidently as a consequence of the inclusion of elements of the upper lobe cortex or cytoplasm in the 1c and 1d micromeres.

Cather (1973) demonstrated that in *Ilyanassa* the polar lobe cytoplasm *represses* the ability of micromere derivatives other than those normally destined for the apical plate to produce cilia, and this aspect of apical organ differentiation has also been analyzed in the polychaete annelid *Sabellaria*. In the latter organism the first cleavage polar lobe bears the determinants for the apical tuft and for the posttrochal chaetae (Hatt, 1932; Novikoff, 1938; Render, 1983) (see Fig. 6.18). As in *Dentalium*, *Sabellaria* embryos from which the second cleavage polar lobe has been removed produce apical tufts. Render (1983) found that any combination of isolated macromeres that includes the C macromere, i.e., ABC, BC, or C alone, will form an apical tuft. Cultured D macromeres never give rise to an apical tuft, but, if the *second* cleavage polar lobe is removed and the C and D macromeres are then cultured, *both* are now able to produce apical tufts. It follows that the second cleavage polar lobe contains a repressor of apical tuft formation, while the first contains only the determinants that induce apical organ differentiation. The elements that induce apical tuft formation are left in the CD macromere after resorption of the first polar lobe, whence they are distributed to both the C and D macromeres, and eventually to the apical precursor micromeres. The elements that repress apical tuft formation are included in the smaller second polar lobe, and thus are confined to the D macromere after resorption, so that only the C quadrant micromeres are able to produce apical tufts. In accord with this interpretation Render (1983) showed that suppression of second polar lobe formation by treatment with a dilute SDS solution prevents apical tuft formation in both C and D isolates, since the repressive cytoplasmic factors are now included in both macromeres. For formation of posttrochal chaetae, on the other hand, the D cell must be included in the isolate. Determinants for the posttrochal chaetae also remain in the second polar lobe, and are returned to the D macromere alone.

Fig. 6.18. Trochophore larvae of polychaete annelids. (a) and (b) Scanning electron micrographs of one and two day old larvae of *Sabellaria cementarium*. Bar represents 10 μm. Apical tuft (t), the equatorial band of cilia or prototroch (p), posttrochal chaetae (c). [From J. A. Render (1983). *Wilhelm Roux's Arch. Dev. Biol.* **192**, 120.] (c) Schematic diagram of the structure of a 2 day old trochophore larva of *Spirobranchus polycerus*. Apical tuft (a.t.), apical organ (a.o.), prototroch (pt.), metatroch (mt.), stomach (st.), pharynx (phar.), the single eye (e), protonephridium (neph.) (only one of pair shown), pretrochal nerves (prt.n.), mouth (mo.), anus (an.), suboral complex (s.cplx.), neurotroch (nt.), anal vesicle (a.v.), mesoblast (mb.), intestine (in.), metatrochal muscle (mt.m.), and longitudinal muscle (ln.m.). [From T. C. Lacalli (1984). *Philos. Trans. R. Soc. London, Ser. B* **306**, 79.]

(ii) Polar Lobe Dependent Inductive Effects

Deletion of the polar lobe results in developmental failures that involve cell lineages other than those descendant directly from the D quadrant, and this implies *inductive interactions* between cells receiving polar lobe elements and other cells of the embryo. In *Dentalium*, for example, development of eyes and tentacles does not occur in lobeless embryos, even though these do not descend directly from D quadrant cells (Van Dongen, 1976; Cather and Verdonk, 1979). Induction was shown by Clement (1967) to be required for eye development in *Ilyanassa*. In this species neither the eye, the foot, nor the shell is formed directly from descendants of 3d, and yet as we have seen, Clement's (1962) experiments show that their formation occurs only if the 3d micromere has been given off at the time of D macromere deletion. The *Ilyanassa* shell gland is normally formed mainly from the second quartet micromeres 2d and 2c. Cather (1967) showed that any experimentally formed combination of micromeres and macromeres, or even the 4d micromere and an isolated polar lobe, will interact inductively to produce shell. However in normal embryos this induction is evidently mediated specifically by the 2D macromere, for ablation of 2D even after the formation of 2d blocks shell formation. In normal embryos, furthermore, this inductive D quadrant stimulus is coupled with a *repression* of the capacity demonstrated for shell formation by isolated A, B, and C quadrant cells.

This brief review shows that the polar lobe cytoplasm influences development in several ways: (a) It affects cleavage patterns and thus the initial axial specification of the embryo; (b) It is required for generation of the 4d cell and the mesentoblasts, which give rise to the mesodermal primordium and the entoblasts, and thereby to the larval and adult organs directly derived from the entoblasts and the mesoderm; (c) It is indirectly responsible for several structures arising by inductive interaction between D quadrant cells and derivatives of the A, B, and C quadrants; and (d) It inductively represses certain differentiations in cells other than the appropriate ones. The polar lobe cytoplasm is thus involved *both* in self-differentiation and in differentiation occurring as a result of inductive interactions between cells that are already dissimilar.

(iii) Determination of the Dorsoventral Axis in Equally and Unequally Cleaving Molluscan Eggs

In molluscan eggs that do not produce polar lobes and in which the first four blastomeres are of equal size the D quadrant is the source of the mesentoblast and entoblast cell lineages, just as in unequally cleaving, polar lobe forming species. Comparative analysis of the process of dorsoventral axial fixation has proved illuminating, particularly since both developmental forms are to be found within a single taxonomic class, the gastropods. One possibility is that the dorsoventral axis of the embryo is actually predetermined in eggs of both types, and that polar lobe formation is merely a prominent manifestation of this preformation, or an element of the mechanism by which it is translated into a three dimensional embryo.

A decisive experiment carried out by Guerrier *et al.* (1978) on eggs of *Dentalium vulgare* disproves this interpretation. Cytochalasin was applied after the beginning of first cleavage to suppress polar lobe formation, and then washed out. Two equal sized blastomeres are formed by the treated eggs, both of which proceed to elaborate a second polar lobe, as shown in Fig. 6.19(a). After second cleavage the embryo can be said to consist of two C and two D macromeres, where D is the macromere into which the lobe is subsequently resorbed, and C is its sister macromere. There are no A or B type macromeres. Development of these embryos results in a duplication of the normal D quadrant cleavage pattern [Fig. 6.19(b) and (c)]. They give rise to larvae containing double the normal number of polar lobe dependent structures, e.g., two dorsal shell glands [Fig. 6.19(d)], four statocysts, and four mantle cavities, but lack any A or B quadrant derivatives such as the stomodaeum. This experiment shows that the *disposition of the polar lobe cytoplasm is sufficient to determine the dorsoventral axis*, since two such axes form if this cytoplasm is distributed to both macromeres at first cleavage. Choice of the blastomere into which the first cleavage polar lobe flows in the normal egg may be a matter of chance, i.e., the exact position of the cleavage plane with reference to the thin cytoplasmic stalk by which the lobe is connected at the trefoil stage (cf. Fig. 6.15). The future dorsal end of the egg is in any case clearly *not irreversibly predetermined*. On the other hand the *vegetal* location of the first polar lobe is a genuine preformation, since the annular constriction by which it arises always occurs in a plane perpendicular to the animal-vegetal axis. Cytological evidence reviewed below shows that the animal-vegetal axis is already clearly evident prior to fertilization.

In equally cleaving gastropod eggs the macromeres remain for some time both morphologically and morphogenetically equivalent. It has been shown for three different species that either of the first two blastomeres can give rise to complete larvae (reviewed in Guerrier and van den Biggelaar, 1979), a result that contrasts directly with the restricted potential of the AB blastomere in unequally cleaving molluscan eggs (Wilson, 1904a; see also Fig. 6.6). In equally cleaving gastropod eggs the macromere that gives rise to the primary mesentoblast, 4d, cannot be distinguished until after 5th cleavage. This macromere, 3D, is then identified by the central position that it assumes in the embryo, its nonsynchronous cleavage, and the asymmetric disposition of its progeny. Deletion experiments have shown that prior to 5th cleavage all of the four macromeres retain the capacity to serve as the "D macromere," and to produce the primary mesentoblasts (van den Biggelaar and

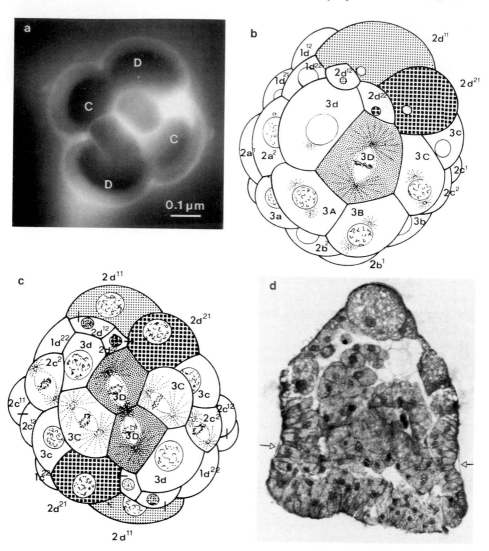

Fig. 6.19. Development of *Dentalium vulgare* embryos after suppression of first polar lobe formation with cytochalasin B. (a) Photomicrograph of four-cell stage embryo containing two opposing cells, labeled D, each in process of fusing with one of the two polar lobes. The sister blastomeres are labeled C. (b) Diagram of normal 32-64-cell stage embryo with individual blastomeres identified. (c) Diagram of 64-cell embryo developed from cytochalasin treated embryo such as shown in (a); note duplication of both C and D quadrant micromeres and absence of A and B quadrant micromeres. (d) Larva of embryo that had been treated with cytochalasin B, with apical organ at the top and two shell glands with shell anlagen marked by arrows. [From P. Guerrier, J. A. M. van den Biggelaar, C. A. M. Van Dongen, and N. H. Verdonk (1978). *Dev. Biol.* **63**, 233.]

Guerrier, 1979). In normal embryos the 3D macromere is usually, but not necessarily, one of the two "vegetal cross furrow" macromeres shown in Fig. 6.20 (a). The event that specifies which will become the D macromere, and hence that determines the dorsoventral axis, is apparently an inductive interaction with the first quartet micromeres across the blastocoelic cavity, as illustrated in Fig. 6.20. D quadrant function is assumed by that macromere first achieving the greatest number of contacts with micromeres during the interval between 4th and 5th cleavage. This has been established for *Lymnaea stagnalis* (Arnolds *et al.*, 1983) and *Patella vulgata* (van den Biggelaar and Guerrier, 1979) embryos by deletion of first quartet micromeres. Removal of a sufficient number of these micromeres precludes the differentiation of any D quadrant macromere, and in partial deletions the macromere that assumes the role of 3D can be determined by the position of the remaining micromeres. If the intercellular interaction between macromeres and micromeres is prevented by cytochalasin treatment, the micromere cap remains completely or partially radially symmetric (Martindale *et al.*, 1985). These experiments show furthermore that in *Lymnaea palustris* all four quadrants of the micromere cap remain undetermined and capable of forming dorsal, lateral, or ventral head structures until after the crucial cell contacts occur, i.e., during the pause in cleavage following the 24-cell stage. It follows from these and the other results reviewed that the dorsoventral axis of neither the micromere nor macromere blastomere tiers is primordially established, both arising inductively by means of the interactions between them. A variety of other cell contacts form elsewhere in the embryo during the same interval in which these micromere-macromere interactions occur. For example, in *Patella* embryos functional intercellular coupling is also established at this point, as indicated by the diffusion of iontophoretically injected lucifer yellow dye from blastomere to blastomere. However, contacts of this kind appear not to be required for dorsoventral determination, since they are absent from the 3D-micromere junctions, and conversely, they still occur in embryos from which the first quartet micromeres have been deleted (Dorresteijn *et al.*, 1983).

Establishment of the dorsoventral axis in equally cleaving gastropod eggs is thus accomplished by an *intercellular* interaction, while the same end is achieved in unequally cleaving gastropod eggs through an *intracellular* mechanism. The contrast is nicely illustrated by the effect of first quartet micromere deletion in embryos that form polar lobes. Though severe defects in larval head formation result, the whole first quartet can be removed from *Bithynia tentaculata* (Van Dam and Verdonk, 1982) or *Ilyanassa* (Clement, 1967) embryos without affecting determination of the dorsoventral axis and of the specific cell lineages responsible for its construction. The role of the polar lobe is illuminated by the comparison. It is the device by which determinative cytoplasmic agents or structures initially resident in the egg cytoplasm are spatially disposed to the appropriate blastomeres. This can be

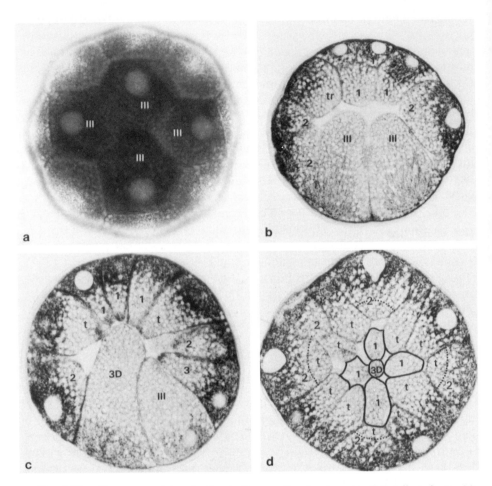

Fig. 6.20. Macromere determination in the equally cleaving egg of *Patella vulgata*. (a) External vegetal pole view of 32-cell embryo. The third quartet has been given off and the four equipotential macromeres are marked III. The upper and lower macromeres form the "vegetal cross furrow." (b) Median section through 32-cell embryo. The vegetal pole is at bottom. Micromeres of the first quartet are marked 1, and those of the second quartet are marked 2. Trochoblast cell, tr. The two cross furrow macromeres are visible, extending toward the micromeres at the animal pole. (c) Median section long after 5th cleavage. Cell contact has been achieved between the cross furrow macromere (3D), that will give rise to the primary mesentoblasts and several micromeres of the first quartet, 1, including primary trochoblasts, t. Cells of the second quartet, 2, and the 3rd quartet, 3, are visible in the section. (d) Cross section of embryo of similar stage as in (c), displaying the apex of 3D, surrounded by 1st quartet micromeres. [From J. A. M. van den Biggelaar, A. W. C. Dorresteijn, S. W. de Laat, and J. G. Bluemink (1981). *In* "International Cell Biology 1980–1981" (H. G. Schweiger, ed.), p. 526. Springer-Verlag, Berlin and New York.]

seen in detail in the examples of the apical tuft determinants, and the determinants for mesentoblast formation, reviewed above. Whereas both unequally and equally cleaving gastropod embryos rely extensively on inductive interactions for organogenesis, the equally cleaving embryos use this mechanism for dorsoventral axial determination as well, while in polar lobe forming eggs axial determination involves the inbuilt location of the constrictions by which the polar lobes are formed, and the relative location and timing of the successive cleavage planes. It is interesting that in both mechanisms chance appears to play a decisive role, in equally cleaving eggs in determining which macromere first achieves sufficient micromere contacts, and in unequally cleaving eggs in determining which macromere resorbs the first cleavage polar lobe cytoplasm.

(iv) What Are the Functional, Structural, and Molecular Constituents of the Polar Lobe Cytoplasm?

Molecular Measurements

An obvious hypothesis that at least in principle has the advantage of testability is that in the polar lobe are sequestered a particular set of regulatory maternal mRNAs. These could code for proteins that directly or indirectly elicit the differential patterns of gene activity required of the cell lineages that are determined through the action of polar lobe cytoplasm. Such lineages sooner or later express different sets of genes than do other embryonic cell lineages, and the basic function of polar lobe constituents must be to set in train events that eventually impose the appropriate states of genomic activity in the nuclei exposed to them. As might be expected, the absence of certain differentiated tissues in advanced larvae grown from lobeless eggs has gross molecular consequences. About the same amount of ribosomal RNA is present in lobeless and normal *Ilyanassa* embryos at three days of development, but after this the lobeless embryo accumulates significantly less total RNA (Collier, 1975). Similarly DNA content increases faster in normal embryos than in lobeless embryos after four days. Several alkaline phosphatase and esterase isozyme forms that normally appear in advanced larvae are also lacking in larvae derived from lobeless embryos (Freeman, 1971).

The presence of maternal mRNA in polar lobes was established initially by Clement and Tyler (1967), who showed that *Ilyanassa* polar lobes continue to synthesize protein for at least 24 hr after isolation at the trefoil stage. The spectrum of proteins synthesized in the isolated polar lobes was analyzed by 2-dimensional gel electrophoresis, by Brandhorst and Newrock (1981), Collier and McCarthy (1981), and Collier (1981). Out of the several hundred species visualized there are no newly synthesized polar lobe proteins that are not also synthesized after equivalent incubations in lobeless or normal cleavage stage embryos. Though this result provides no support for the proposal that certain species of maternal mRNA are sequestered exclusively in the polar lobe, nor is it excluded. Proteins whose pI's fall outside of the range of the isoelectric focussing gradient, i.e., any whose pI is >8 or <5, and proteins coded by rare messages, would not be included in these analyses. The latter class of course includes the large majority of the diverse species of maternal mRNA present, if measurements carried out on other embryonic material can be considered a guide (cf. Chapter II). An interesting observation of Brandhorst and Newrock (1981) and Collier and McCarthy (1981) is that several proteins are synthesized in isolated polar lobes after 25 hr of incubation *in vitro* that are not synthesized at 4 hr, and that these same changes occur in normal and in lobeless embryos, and after actinomycin treatment of these embryos (by 25 hr the normal *Ilyanassa* embryo is at the mesentoblast stage, 29 cells). This set of molecular changes thus originates at the translational (or posttranslational) level, and the polar lobe evidently includes the cytoplasmic elements necessary for their occurrence.

The prevalence of a few other newly synthesized protein species increases sharply during the development of normal *Ilyanassa* embryos, but not in isolated polar lobes (Brandhorst and Newrock, 1981). Their appearance could be the result of embryonic transcriptional activity, as also implied by the effect of actinomycin, which alters the pattern of protein synthesis during the first 24 hr (Newrock and Raff, 1975; Collier and McCarthy, 1981). However the same changes occur in lobeless embryos as in normal embryos (Brandhorst and Newrock, 1981) and thus there is no evidence that the regulation of these protein species is dependent on polar lobe constituents. The sole observation that assigns to the polar lobe any direct effect on embryonic transcriptional activity is a quantitative one. In mesentoblast or gastrula stages the absolute rate of total RNA synthesis in lobeless embryos is only about 55-60% of that in normal embryos (Davidson *et al.*, 1965; Collier, 1977). The difference may be significant, since cell division in lobeless and normal embryos occurs at about the same rate and they possess the same number of nuclei throughout the period studied (Davidson *et al.*, 1965; Cather, 1971). During these stages the newly synthesized RNA measured by labeling is mainly of heterogeneous, nonribosomal nature, and is probably a combination of nRNA and mRNA (Newrock and Raff, 1975; Koser and Collier, 1976).

Cytological Observations

Numerous ultrastructural studies have been carried out on the polar lobe cytoplasm (reviewed in Dohmen and Verdonk, 1979). Unusual membrane-bound organelles are present in the cytoplasm of the large polar lobes of *Ilyanassa* and *Dentalium*, as well as mitochondria and other undefined particulate inclusions. However, none of these structures are required for polar lobe function, and they cannot contain the morphogenetic determinants of the lobe, since they can be redistributed to other parts of the egg by low speed centrifugation without disturbing normal development. This was shown for *Ilyanassa* by Clement (1968) and for *Dentalium* by Verdonk (1968). On the other hand a special organelle termed the vegetal body that by the same test is functionally important has been demonstrated in the small polar lobe of the gastropod *Bithynia*, and similar structures are present in other species that produce small polar lobes. The vegetal body consists of a dense aggregation of small vesicles, as shown in Fig. 6.21(a) and (b). Though the *Bithynia* polar lobe occupies <1% of the egg volume, its morphogenetic determinants are functionally similar to those of the *Dentalium* and *Ilyanassa* polar lobes. Thus, lobeless *Bithynia* embryos fail to form mesentoblasts and lack eyes, foot, intestine, organized shell, operculum, etc., though they successfully develop digestive gland, ganglia, and some muscle (Cather and Verdonk, 1974). In *Bithynia*, however, the developmental capacity of the C macromere is the same as that of the D macromere, and lobe-dependent structures form if either the C or D macromeres are present (Verdonk and Cather, 1973; Cather and Verdonk, 1974). The progressive distribution of the vegetal body vesicles parallels that of the lobe morphogenetic determinants. The vegetal body is shunted into the CD cell at first cleavage, whereupon the vesicles disperse, and are inherited by both the C and D macromeres. The staining reactions of the vegetal body suggest that its vesicles contain RNA (Cather and Verdonk, 1974), but this remains to be established by biochemical measurement. A direct functional relationship between the vegetal body and the polar lobe determinants was demonstrated by Van Dam *et al.* (1982). Centrifugation at speeds of about 1400 × g dislocates the vegetal body away from the polar region of the uncleaved egg in about 50% of cases, and these eggs form polar lobes that lack the vegetal body [Fig. 6.21(c)]. Van Dam *et al.* (1982) found that about 50% of centrifuged eggs from which the first polar lobes had been removed developed into *normal embryos*, a result that is never observed after deletion of the lobe from uncentrifuged eggs. Inclusion of the vegetal body in one of the two blastomeres, in the absence of the polar lobe structure, thus suffices for normal development, and it follows that the vegetal body is the site of the lobe determinants. Seen in this light the function of the polar lobe appears essentially to be the distribution of the special package of vesicles to one or the other of the first two blastomeres.

In eggs with large polar lobes organelles similar to the vegetal body are not present, and as we have seen centrifugal forces that totally redistribute the particulate cytoplasmic elements fail to dislodge the morphogenetic determinants. Thus it is likely that these determinants are anchored in the cortical structures of large polar lobes, and that they exist initially as regional differentiations of the lobe plasma membrane and cortex. For example, the apical tuft determinants in the first cleavage *Dentalium* polar lobe are apparently

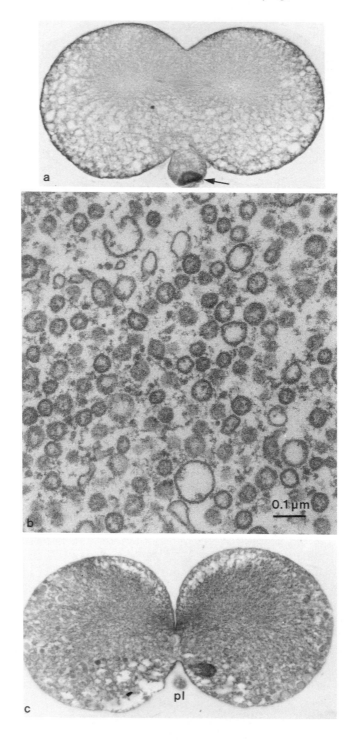

fixed in the upper region of the lobe. The extrusion and resorption of the lobe can be considered as mechanisms for the spatial disposition of the cytoplasmic determinants to the appropriate blastomeres. These movements are accompanied by a complex of cytoskeletal changes, as described by Conrad (1973).

Scanning electron micrographs have provided dramatic instances of regional cortical differentiations at the polar lobe. Examples are shown in Fig. 6.22(a)-(e), from a study of Dohmen and van der Mey (1977). The vegetal pole is seen to be clearly marked by distinct surface features prior to the onset of cleavage. Furthermore, as shown in Fig. 6.22(f) at least in the egg of *Crepidula* the plane of first cleavage is also predetermined, since the longitudinal orientations of the polar microvillar patch bears a fixed relation to the future plane of cleavage, which is retained during the subsequent cleavages (Dohmen, 1983).

The initial specification of the vegetal pole in polar lobe bearing eggs may occur far back in oogenesis. Animal-vegetal differences in distribution of microvilli and other cytological features can be observed in the uncleaved egg of the gastropod *Nassarius reticulata* (Speksnijder *et al.*, 1985), and could derive originally from the polarized morphology of the oocyte-follicle cell complex. Thus during the growth phases of oogenesis the area of contact is confined to the base of the oocyte, and this becomes the future vegetal pole (Dohmen and van der Mey, 1977). An interesting inference is that the animal-vegetal polarity of the egg is imprinted by intercellular surface interactions occurring in the ovary. Prelocalization of vegetal pole structures extends as well to internal structures such as the vegetal body, which is already localized at the future vegetal pole in growing *Bithynia* oocytes (Van Dam *et al.*, 1982). Maternal components are also responsible for the asymmetric "spiral" cleavage of molluscan eggs, as shown by genetic studies on the maternal inheritance of left and right handed body forms in the gastropod *Lymnaea* (Boycott and Diver, 1923; Sturtevant, 1923; Boycott *et al.*, 1930; Freeman and Lundelius, 1982). The orientation of the adult body form in gastropods is determined by the direction in which the micromeres are given off during early cleavage (see the discussion of spiral cleavage in Wilson,

---

Fig. 6.21. Vegetal bodies of the polar lobe in the egg of *Bithynia tentaculata*. (a) Light micrograph of an egg at first cleavage, stained for RNA with methyl green pyronin. The vegetal body is indicated by the arrow. (b) Electron micrograph of the vegetal body. The vesicles are bound with double membranes and after fixation in OsO_4 and staining with lead-uranylacetate many appear to be filled with an electron dense substance. [(a)-(b) From M. R. Dohmen and N. H. Verdonk (1974). *J. Embryol. Exp. Morphol.* **31**, 423.] (c) Ectopic vegetal body in a centrifuged egg, viewed by light microscopy as in (a), except stained with iron haemotoxylin and eosin. "pl" indicates a tangential section of the edge of the polar lobe, which now lacks the vegetal body. [(c) From W. I. Van Dam, M. R. Dohmen, and N. H. Verdonk (1982). *Wilhelm Roux's Arch Dev. Biol.*, **191**, 371.]

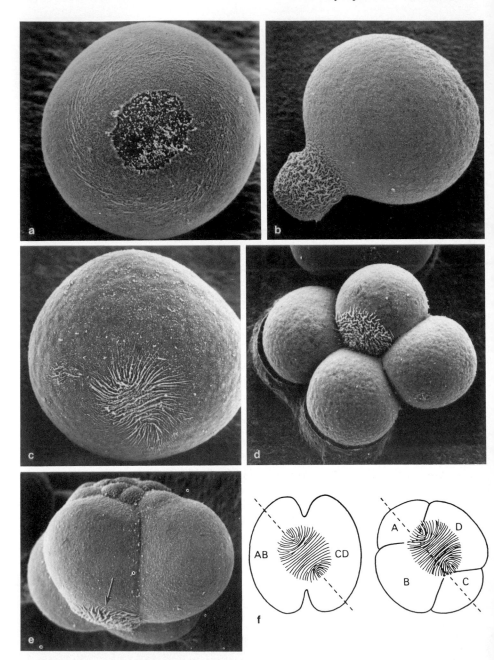

Fig. 6.22. Surface structures of the vegetal pole in gastropod eggs. (a)-(c) Scanning electron micrographs. (a) Vegetal pole of egg of *Nassarius reticulata*, 1st meiotic division. Vegetal pole is marked by a circular interruption in the external carbohydrate coat, containing mi-

1925). Freeman and Lundelius (1982) showed that cytoplasm from right handed or dextral eggs injected into genetically sinistral eggs can transfer the property of dextrality. The direction of cleavage is thus not determined by the orientation of the ovarian oocyte with respect to the follicle cells. The cytoplasmic elements that confer this property are synthesized during oogenesis, and provide an excellent additional example of the preformation of components specifying spatial organization in the molluscan embryo.

The studies of polar lobe function reviewed in this section provide an observer of early development with a comprehensive and detailed description of the biological aspects of a cytoplasmic localization process. It seems inescapable that a characteristic set of dorsoventral determinants is carried in the polar lobe cytoplasm, and that these are responsible for specification of many of the early cell lineages of the embryo. The spatial distribution of these determinants is an epigenetic, developmental process mediated by the disposition of the polar lobe constituents. However, the structures or substances that later exert morphogenetic effects evidently include or consist largely of preformed maternal components. Their origins in oogenesis, their nature, and their molecular mode of action, which directly or indirectly ultimately affects the embryo genomes, remain unknown.

4. THE ONTOGENY OF LOCALIZATION

The distinction was drawn earlier between spatial assignments in the egg that are prelocalizations, and those that are constructed ontogenically following fertilization. Of course, as Driesch and Wilson pointed out, prelocalized features of the mature egg are the product of prior ontogenic processes that have taken place during oogenesis. The animal-vegetal axes of ascidian and some molluscan eggs, for example, are established during oogenesis, while the anterior-posterior or dorsoventral axes that provide the bilateral

---

crovilli, and surrounded by a spiral array of folds. The ring of folds marks the location of the constriction that will give rise to the first polar lobe. About X325. (b) Polar lobe of uncleaved *Buccinum undatum* egg. The lobe is covered with branched surface ridges absent from rest of egg. About X180. (c) Uncleaved egg of *Crepidula fornicata* after first meiotic division. Oriented folds are present at the vegetal pole. About 400X. (d) *Crepidula* embryo at 4-cell stage. Surface folds are confined to D macromere, in region where 2nd polar lobe has been resorbed. About X210. (e) *Crepidula* embryo, 24-cell stage. Micromeres are visible at the animal pole. The surface folds marked by arrow are visible on 3D macromere. About X225. [(a)-(e) From M. R. Dohmen and J. C. A. van der Mey (1977). *Dev. Biol.* **61**, 104.] (f) Diagram showing relation of bilaterally symmetrical pattern of folds at the vegetal pole of *Crepidula* egg to the 1st and 2nd cleavage planes. The 1st cleavage plane is oriented at about 50° from the axis of symmetry marked by the dashed lines. [From M. R. Dohmen (1983). *In* "Time, Space, and Pattern in Embryonic Development" (W. R. Jeffery and R. A. Raff, eds.), p. 197. Alan R. Liss, New York.]

organization of the embryo are established by the developmental redistribution of maternal cytoplasmic elements. While common in marine eggs, this scheme of development is by no means universal. In some animals both axes are prelocalized in the unfertilized egg and in others neither. The examples considered in the following provide some insight into the possible nature of both prelocalized and ontogenically localized determinants, and the processes by which they may be deposited in given regions of the egg.

(i) Prelocalization in the Cortex of Dipteran Eggs

By the cellular blastoderm stage the cortical region of the *Drosophila* egg is already a mosaic of determined cell types. Thus when allowed to develop in intracoelomic culture, cells from anterior and posterior regions of the blastoderm give rise only to the expected anterior or posterior adult structures (Chan and Gehring, 1971). In addition, Schubiger (1976) demonstrated that partial embryos formed by ligation at the blastoderm stage can produce all of the anterior, thoracic, or caudal adult parts that would normally derive from them.

Blastoderm fate maps were based initially on histological reconstructions of the development of organ anlagen (Poulson, 1950), and then on analyses of genetic mosaics induced during the first few divisions of the zygote nuclei, in which the developmental disposition of clones of cells bearing visible morphological markers can be followed. By the latter method have been determined the approximate location and number of the primordial cells for adult cuticular structures, as well as for some larval organs (Garcia-Bellido and Merriam, 1969; Hotta and Benzer, 1973; Kankel and Hall, 1976; reviewed in Janning, 1978). A more direct approach to mapping the morphogenetic values of the blastoderm cells is microablation, by pricking the egg (see, e.g., Howland and Child, 1935), local cautery (see, e.g., Bownes, 1975b), or UV irradiation. An UV laser microbeam was utilized by Lohs-Schardin *et al.* to map the primordia for the thoracic and abdominal epidermis of the first instar larva (1979a) and for adult structures (1979b). Cephalic larval structures were mapped by microsurgical blastoderm cell deletions by Underwood *et al.* (1980). In the experiments of Lohs-Schardin *et al.* (1979a) the UV microbeam was 20 μm in diameter, and affected a patch of about 15 cells (the embryo is about 500 μm $\times$ 175 μm, and at the stage of irradiation in these experiments it contains about 5000 cortical cells). Specific defects in the larval epidermis resulted from the localized irradiation, and these experiments demonstrate that about one-third of the blastoderm cells comprise the progenitors of the entire larval epidermis. Extensive fate map data have also been obtained by injection of horseradish peroxidase into individual blastoderm cells and other cytological observations (Hartenstein *et al.*, 1985; Hartenstein and Campos-Ortega, 1985; Technau and Campos-Ortega, 1985). In Fig. 4.17(a) is reproduced a detailed morphogenetic map derived from these studies that displays the positions of anlagen for all major structures of the late embryo on the two-dimensional surface of the completed blastoderm. The overall result of these studies is to confirm that long before the appearance of visible manifestations of differentiation (except for the precociously appearing pole cells) the blastoderm has acquired a complex pattern of specification, in which are encoded both embryonic axes as well as many specific regional assignments. The following evidence shows that in contrast to any of the eggs so far discussed, *both embryonic axes are prelocalized* in the unfertilized dipteran egg cytoplasm, just as are the determinants for pole cell formation.

Origin of the Axial Determinants in Drosophila: *Morphological and Genetic Evidence*

The bilateral rather than radial symmetry of the unfertilized dipteran egg is evident from its external morphology. Thus the future dorsal side is less convex; the future anterior pole of the embryo is marked by the micropyle, the opening in the chorion through which the sperm enters; the two chorionic filaments are located bilaterally at the future dorsal posterior region; and the germinal cytoplasm is already localized at the future posterior pole, as discussed earlier.

A number of maternal mutations are known that disturb primary axial determination in the embryo. There are so far identified 10 different maternally functioning genes, mutations in which alter the prospective fates of ventral and lateral cells of the blastoderm to that of the prospective dorsal epidermis (Anderson and Nüsslein-Volhard, 1984) A well known mutation of this type is *dorsal* (*dl*) (Nüsslein-Volhard, 1979; Nüsslein-Volhard *et al.*, 1980; the effects of this mutation on the pattern of neuroblast specification in the embryo were discussed earlier in Chapter IV). There are two *dorsal* phenotypes. Embryos from homozygous *dl/dl* females express the "recessive phenotype," which is characterized by a lack of ventral, mesodermal, and neural structures. Though at the blastoderm stage the embryo appears normal, invagination of the presumptive mesoderm fails, and the embryo becomes a yolk-filled epidermal tube that lacks any internal organs. A "dominant phenotype" is temperature sensitive. The eggs of *dl^D/+* females develop normally at 22°C but arrest at 29°C, due to the absence of muscles and portions of the ventral epidermis, tissues arising normally from the ventral region of the blastoderm. Nüsslein-Volhard *et al.* (1980) concluded that ventral cells of these embryos carry out assignments normally relegated to the more lateral cells, in particular the formation of epidermis instead of mesodermal primordia. These and other observations show that embryonic specification of ventral cell types requires the maternal products of the wild-type *dorsal* gene, and also of the other maternally acting genes that when mutated produce *dorsal* phenotypes. The *dorsal* gene has been cloned (Steward *et al.*, 1984) and found to produce a single 2.8 kb poly(A) RNA species, which is present in eggs up to the blastoderm stage, after which it disappears. Studies with germ line mosaics demonstrate that the activity of the *dorsal*[+] gene is required only in germ line cells (Schupbach and Wieschaus, 1986a), and *in situ* hybridization using a *dl* probe shows that the gene is active in the nurse cell nuclei at stages 5-11 of oogenesis (Steward *et al.*, 1985). After stage 11 *dl*[+] RNAs are transferred to the oocyte, as are the products of many other genes (cf. Chapter V).

Of the ten maternally acting loci known to be required for specification of the normal dorsoventral egg axis, the effects of eight, *viz., dorsal, tube, pipe, snake, easter, Toll*[rec]*, spätzle,* and *pelle,* have been shown to be at least partially rescued by injection of wild-type egg cytoplasm, and all except *dorsal* can also be rescued to some extent by injection of poly(A) RNA extracted from very early wild-type embryos (Santamaria and Nüsslein-Volhard, 1983; Anderson and Nüsslein-Volhard, 1984). The rescuing activity for *pelle* mutants disappears from the poly(A) RNA fraction at an earlier stage of cleavage than from the cytoplasm, suggesting that translation of the maternal mRNA is by this stage complete (Müller-Holtkamp *et al.*, 1985), while for certain other of these mutants, the rescuing activity is instead preferentially located in the poly(A) RNA fraction of the cytoplasm (Anderson and Nüsslein-Volhard, 1984). Thus translation of *dorsal* mRNA may occur particularly early in development (and/or in oogenesis). The *dorsal* cytoplasmic component required for rescue is by the blastoderm stage spatially localized, since cytoplasm removed from the ventral regions of donor embryos is the most effective in rescue experiments (Santamaria and Nüsslein-Volhard, 1983). The mutant effects of the *Toll* locus that can be rescued by RNA injection are those caused by recessive mutations, which produce phenotypes similar to those of the other dorsalizing maternal loci. Dominant *Toll* mutations (*Toll*[D]) have the opposite effects, resulting in the appearance of patches of ventral denticles over the entire dorsoventral circumference, and the absence rather than overextension of dorsal structures (Anderson *et al.*, 1985a). Injection of wild-type cytoplasm into *Toll*[−] mutants restores their ability to produce dorsal structures, and the site of injection defines the ventral region of the structures formed (Anderson *et al.*, 1985b). Thus, whereas lack of *Toll* function results in *dorsalization,* as does lack of function of the other nine loci (all recessive), these observations are consistent with the observation that the dominant *Toll*[D] gene results in *ventralization,* due to production of excess *Toll* product. Double mutants combining *Toll*[D] and recessive mutant alleles of *gastrulation, defective, snake, easter,* etc., lack *both* ventral and dorsal pattern elements but develop lateral structures, while the *Toll*[D] and *dorsal* combination is totally dorsalized (Anderson *et al.*, 1985a). In any case the demonstration that dorsoventral polarity can be restored to mutant embryos by injection of wild-type egg cytoplasm or poly(A) RNA shows explicitly that *maternal transcripts synthesized in the* *meroistic egg chamber contain genetic information required for normal axial specification in the embryo.* The complete morphological rescue of a homozygus *snake* embryo by injected poly(A) RNA is shown in Fig. 6.23 (Anderson and Nüsslein-Volhard, 1984).

A large number of maternally acting genes are also required specifically for establishment of the anterior-posterior morphogenetic organization of the *Drosophila* egg (Gans *et al.*, 1975; Nüsslein-Volhard, 1979; Schupbach and Weischaus, 1986b). Expression of mutant alleles of seven such genes in cells of the female germ line carried in mosaic mothers has been shown in each case to suffice for complete production of the mutant embryonic pheno-

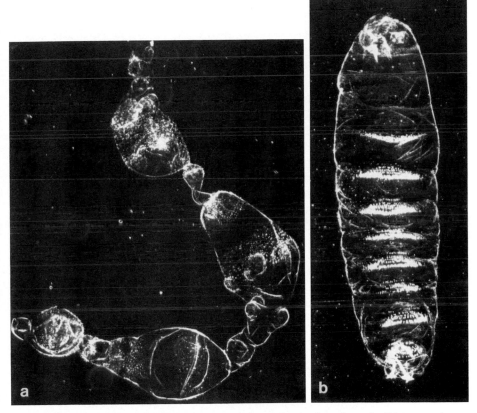

Fig. 6.23. Rescue of a dorsalizing mutation, *snake*, by injection of wild-type embryo poly(A) RNA. (a) Cuticle of *snake* embryo. The embryo consists of a tube of dorsal epidermis. (b) Cuticle of embryo of same genotype after injection of poly(A) RNA extracted from 0-2 hr (cleavage stage) wild-type embryos. Ventral denticle belts of normal dimensions can be observed, and complete dorsoventral structural differentiation has taken place (cf. Fig. 4.16 and 4.19). [From K. V. Anderson and C. Nüsslein-Volhard (1984). Reprinted by permission from *Nature (London)* **311**, 223. Copyright © 1984 Macmillan Journals Limited.]

(Schupbach and Weischaus, 1986a). The developmental abnormalities caused by these seven recessive maternal lethals fall into three morphological classes (Schupbach and Weischaus, 1986b). Mutations in *torso* and *trunk* result in failure to develop structures of the extreme anterior and posterior ends of the embryo, and in addition they cause development of structures normally arising more towards the midregion to occur in the vicinity of the anterior and posterior poles. Thus, for example, *torso* embryos typically terminate at the posterior end with cuticular patterns characteristic of abdominal segment A7. Yet, these mutations do not prevent cellularization of the blastoderm, and their effects apparently result from alterations of blastoderm cell assignments. These effects are large scale, disturbing the global fate map of the egg. The same conclusion may be drawn for other classes of morphological maternal mutation that affect anterior-posterior specification in the embryo. Thus maternal mutations at *valois, vasa, staufen,* and *tudor* loci that prevent the formation of polar granules and pole cells also cause deletions of abdominal segments, and in the case of *staufen*, head structures; and mutations at the *exuperantia* locus cause replacement of head structures by an inverted posterior end. It is not clear, however, whether loci such as these are *directly* responsible for the spatial patterns in which maternal morphogens might be distributed. Thus, for example, there is some evidence that another gene that when mutated may produce a *torso*-like phenotype, *viz.*, *lethal (1) pole hole*, may be involved primarily in some common cellular function such as proliferation or cell formation (Perrimon *et al.*, 1985), and similarly, many *valois* eggs display severe defects in cellularization of the blastoderm (Schupbach and Weischaus, 1986b).

Among maternally acting mutations that affect establishment of the anterior-posterior axis are genetic defects that interfere with determination of *either* anterior or posterior embryonic structures, which implies (but does not in itself prove) the existence of separate maternal factors necessary for cephalic and for caudal cell specification. A well known example is provided by *bicaudal (bic)*, a recessive maternal point mutation discovered by Bull (1966) and studied by Nüsslein-Volhard (1977). Two other loci, which are semidominant, *BicC* and *BicD*, produce similar defects (Mohler and Wieschaus, 1985). Morphological defects produced by mutations at any of these three loci consist either of deletions of various amounts of anterior structure, with otherwise normal retention of embryo polarity, or of replacement of the deleted anterior structures with duplicated posterior structures in reverse orientation. Possibly this mirror image symmetry is the result of a secondary reorganization along the anterior-posterior axis induced in embryos suffering the most extensive deletions of anterior segments (Mohler and Wieschaus, 1985). Determination of pole cells is not affected in *bicaudal* embryos, and these remain situated normally at one end. The *bicaudal* mutations affect both the specification of particular cell lineages, for example causing malpighian tubules to form from cells that in normal embryos would give rise to cephalic structures, and the general pattern, for example the number of posterior segments in the embryo. The variation in the phenotypes of double abdomen embryos is interesting, in that the most anterior sturctures formed may be located anywhere from the third to the sixth abdominal segment (Mohler and Wieschaus, 1985), suggesting that the *bicaudal* mutations affect the distribution of a morphogen that extends over large regions of the egg, rather than being sharply localized to one area of the cytoplasmic cortex, for example. The temperature sensitivity of the *BicD* lesions have been used to demonstrate that, as expected, function of the gene is required only during oogenesis, and it is interesting that oogenesis in females homozygous for *BicC* mutations is blocked at stage 8 (Mohler and Wieschaus, 1985).

It seems likely that the distribution of products of maternally acting genes that effect anterior-posterior embryonic axis specification would reflect the polarity of the egg. We have already noted evidence for dorsoventral localization of maternal products of genes affecting that axis. Mlodzik *et al.* (1985) isolated a gene by its sequence homology with a *Ubx* homeobox (cf. Chapter IV), and though the morphological significance of this gene, named *caudal*, is unknown, the distribution of its maternal transcripts provides a possible example. These transcripts are produced in the nurse cells during oogenesis, and at fertilization are present in the egg. *In situ* hybridization shows *caudal* RNAs to be more or less evenly distributed during cleavage, but at the syncytial blastoderm stage when they are located in the cortex, they have largely disappeared from the anterior end of the egg up to about 20% of total egg length. The concentration of these transcripts increases monotonically for the next 50% of egg length, and they are present at their maximum level for the posterior 30% of egg length. At cellular blastoderm stage their distribution sharpens, possibly by means of zygotic transcription, and they accumulate in a single sharp band not far from the posterior pole, where are located precursors of the proctodaeum and regions of abdominal segments 9 and 10 [cf. Fig. 4.17(a)]. Such observations focus attention on the *origins of the primordial polarity of the egg*, which provides an initially oriented matrix for the establishment of differential concentrations of maternal gene products.

An interesting insight into the processes by which polarity might initially be imposed derives from morphological observations on another maternal mutation, *dicephalic*, which produces an abnormality converse to that arising from *bicaudal* mutations (Lohs-Shardin, 1982). As the name implies, *dicephalic* embryos of extreme phenotype develop head furrows at both poles, and their cuticular structures suggest a mirror image duplication of the cephalic segments, while differentiation of posterior structures is suppressed. In contrast to *bicaudal* and *dorsal*, the *dicephalic* mutation visibly affects the construction of the nurse cell-oocyte complex. In *dicephalic* ovarioles the 15 nurse cells form two clusters positioned at opposite ends of the oocyte, in place of the normal asymmetric arrangement in which the nurse cells are all at the future anterior pole. The chorion of *dicephalic* eggs contains two micropyles, one at each end. The implication is that the anterior determinants are localized near the sites of the nurse cell oocyte junctions, or perhaps are loaded into the oocyte by the nurse cells. Thus the spatial organization of the ovariole may be translated into the spatial organization of the egg, and the preformation of the anterior-posterior axis may descend from intercellular interactions that took place during oogenesis. We have encountered a mechanism of this nature earlier, *viz.* the determination of the vegetal pole of the *Nassarius* oocyte at the unique site of its contact with follicle cells.

Distribution of Anterior and Posterior Determinants

The existence in the dipteran egg of anterior and posterior determinants is required as well by the results of a variety of micromanipulation experiments (reviewed in Sander, 1976, 1981; Kalthoff, 1983). Experimental evidence has also been reported for a "middle" determinant in the egg of a homopteran, the leaf hopper *Euscelis* (Vogel, 1983). Among the interpretations that have often been proposed for the insect egg is that the structures formed along the anterior-posterior axis are specified by gradients of *diffusible* morphogens originating at the two polar regions. However, a series of results that are not easily consistent with this venerable assumption have recently been obtained with *Euscelis* eggs (Vogel, 1978, 1982a,b, 1983). For example these eggs produce complete germ bands displaying the normal anterior-posterior segmental differentiation after severe physical deformations that should have completely destroyed the postulated gradients (Vogel, 1982a). Another study contrary to the predictions of gradient theory shows that middle egg fragments isolated during early cleavage are unable to regulate so as to produce complete embryos, but regain the ability to do so if injected with posterior pole cytoplasm, providing they also include some anterior cytoplasm (Vogel, 1982b). A consistent alternative interpretation requires cortically localized, anterior, posterior, and perhaps middle determinants, and implies that in the *Euscelis* egg only short range, not embryo wide, interactions occur, possibly between nearby sets of blastoderm cells.

A similar situation perhaps obtains in *Drosophila*. Anterior egg fragments isolated by ligation during the syncytial cleavage stage form head structures, and posterior fragments give rise to abdominal structures, but midregion segments are not formed by either half (Herth and Sander, 1973; Schubiger, 1976; Schubiger *et al.*, 1977; see also Wieschaus and Gehring, 1976). This result contrasts with the complete determination patterns present by the cellular blastoderm stage, and it might imply a requirement for early developmental interactions between anterior and posterior regions in the determination of the middle structures of the embryo. Schubiger *et al.* (1977) supported this inference by demonstrating that if the cellular barrier that comes to separate the anterior from the posterior compartments of ligated eggs is punctured, a complete set of differentiated middle segments will now be formed. Prior to the blastoderm stage the state of anterior and posterior regional specification is exclusively a feature of the cytoplasm, since nuclei from all regions of the preblastoderm embryo remain totipotent in nuclear transfer experiments (Illmensee, 1972; Okada *et al.*, 1974c; reviewed by Kauffman, 1981). However, it is important to keep in mind that combinatorial sets of autonomous genomic functions in the nuclei of the early blastoderm cells are required for the establishment of their states of specification (cf. discussion of metamerization in Chapter IV). That is, acquisition of blastoderm cell fate requires more than a passive response to a distribution of "instructive" prelocalized morphogens.

The distribution of anterior and posterior determinants has been studied by other methods in the egg of the chironomid midge *Smittia* (reviewed in Kalthoff, 1979, 1983). UV irradiation of the cortical cytoplasm of unfertilized *Smittia* eggs, or of early syncytial cleavage stages, results in a high frequency of "double abdomen" embryos (Kalthoff, 1971a). The head, thorax, and anterior abdominal segments are replaced in these embryos by posterior abdominal segments, in mirror image symmetry to the normal posterior end, just as in *bicaudal Drosophila* mutants. Puncturing the anterior pole with a needle also produces the double abdomen phenotype (Schmidt *et al.*, 1975). Double cephalon embryos [Fig. 6.24(c)] occur after posterior UV irradiation as well, and this abnormality is very efficiently induced in chironomid eggs by centrifugation (Yajima, 1960; Rau and Kalthoff, 1980). Centrifugation, which visibly stratifies the cytoplasmic contents of the egg, may also result in double abdomens, or in normal embryos the polarity of which has been completely inverted with respect to the asymmetric egg shell, as shown in Fig. 6.24. Kalthoff *et al.* (1982) and Kalthoff (1983) interpreted the combined effects of UV irradiation and centrifugation to indicate that mutually repressive cytoplasmic determinants for anterior and posterior differentiation are present throughout the *Smittia* egg, rather than being confined to their respective polar regions, but that the anterior determinants "predominate" at one end, while the reverse is true at the other. The outcome of experimental operations would depend on the relative regional levels of these two determinants (or sets of determinants) that result. Whatever the literal accuracy of this explanation, further exploration of the UV sensitivity displayed by the anterior determinants in *Smittia* eggs has resulted in some interesting evidence regarding their molecular nature and location.

Characterization of the Prelocalized Anterior Determinants in Smittia

In the unfertilized egg the UV sensitive elements of the anterior determinants are symmetrically distributed within a polar cone of cytoplasm about 20 μm in diameter (Kalthoff, 1971b, 1973; Ripley and Kalthoff, 1983). Since

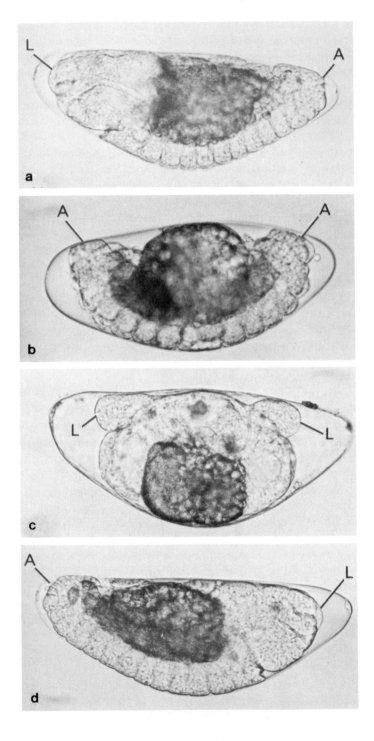

little of the incident UV energy penetrates beyond 5-10 μm of the surface, *these elements must lie in or near the cortex of the egg*. After 5-7 hr, when in *Smittia* the nuclei begin their migration toward the egg surface, the UV sensitive targets disperse over the entire anterior half of the egg. Double abdomen embryos can then be produced by a low flux of widespread anterior irradiation, a treatment that is not effective earlier. The action spectrum of double abdomen induction by UV light displays a major peak at 285 nm, and a minor peak at about 257 nm (Kalthoff, 1973, 1979). An interpretation consistent with this result is that the anterior determinants include a ribonucleoprotein target. Additional evidence that also suggests a functional RNA component includes the observation of Kandler-Singer and Kalthoff (1976) that anterior determination can be blocked by RNase admitted through a small puncture made in the anterior end of the egg. Up to 40% of eggs so treated display the double abdomen syndrome. Punctures made in other regions of the egg, denatured RNase, or other enzymes do not have this effect. Furthermore, the UV induction of double abdomen is photoreversible. As much as 60% increase in the frequency of normal embryos occurs if the irradiated eggs are subsequently exposed to long wavelength UV or blue visible light (Kalthoff, 1971a, 1979). UV irradiation sufficient to induce double abdomen embryos causes extensive formation of pyrimidine dimers in the ribosomal RNA of these embryos, and photoreactivation treatments that rescue UV irradiated embryos also stimulate the repair of most of these dimers. The same would presumably be true of other RNA species within the UV penetration zone. In any case the conclusion that localized maternal RNAs are required for axial specification in *Smittia* is consistent with that drawn earlier for *Drosophila* from the demonstration that maternal mutants such as *snake* can be rescued by injected wild-type RNA (cf. Fig. 6.23).

Potentially useful protein markers for both anterior and posterior determination have recently been identified in *Smittia* eggs (Jäckle and Kalthoff, 1981). An *anterior indicator protein* of about 35 kd is synthesized in the anterior but not posterior blastoderm cells of normal embryos. This protein is produced in cells from both halves of the blastoderm in double cephalon embryos, and again exclusively in the anterior end after photoreversal of irradiated eggs. A *posterior indicator protein* of about 50 kd displays a complementary distribution. This protein is synthesized in both ends of the

---

Fig. 6.24. Normal and abnormal segment patterns in *Smittia* embryos developing from centrifuged eggs. Eggs were centrifuged for 20 min at 6700 × g at the early intravitelline cleavage stage. These embryos were photographed 53-78 hr after deposition, at the shortened germ band stage, ventral side down and anterior pole of egg shell to the left. A, abdomen; L, cephalic structures. (a) Normal embryo; (b) double abdomen; (c) double cephalon; (d) normal embryo inverted with respect to polarity of shell. [From K. G. Rau and K. Kalthoff (1980). Reprinted by permission from *Nature (London)* **287**, 635. Copyright © Macmillan Journals Limited.]

blastoderm in UV induced double abdomen embryos, and its appearance is restricted to the posterior end in irradiated embryos rescued by photoreversion. Synthesis of both indicator proteins occurs long in advance of any visible anterior-posterior differentiation. As the regional synthesis of these proteins is predicted perfectly by experimental interventions that affect the prelocalized cytoplasmic determinants, it follows that molecular regulation of these blastoderm cell functions occurs during the earliest stages of the axial specification process.

(ii) Disposition of Maternal Cytoplasmic Determinants during Early Cleavage

The Cleavage Process and the Ontogenic Establishment of Localization Patterns in Ctenophore Eggs

At the opposite extreme from the eggs of dipterans and other insects in which both embryonic axes are prelocalized are the eggs of ctenophores and hydrozoans, in which until development begins neither axis is specified (Freeman, 1977, 1981). Cytoplasmic determinants are nonetheless present in ctenophore eggs, as was shown convincingly by classical experimentalists. The potentialities of isolated ctenophore blastomeres were investigated by Chun (1880), and by Driesch and Morgan (1896), Fischel (1898, 1903), and Yatsu (1912). These early experiments as well as more recent blastomere isolation studies were reviewed by Reverberi (1971a). At the 8-cell stage the embryo contains two pairs of external cells denoted "E" cells, located on either side of the four inner cells, which are denoted "M" cells [see Fig. 6.25(a)]. The derivatives of E cells and M cells have different morphogenetic fates which are expressed accurately in isolated blastomere experiments (Freeman and Reynolds, 1973). In normal embryos the e micromeres given off by the E macromeres produce fused rows of large swimming cilia, or comb plates, and the micromeres derived from M macromeres form other structures, including mouth, apical organ, and most spectacularly, photocytes. These cells contain a calcium activated photoprotein which enables them to produce flashes of light [Fig. 6.25(b) and (c)]. The embryo gives rise to a bilaterally symmetrical lava organized as indicated in the diagram of Fig. 6.25(d).

Freeman (1977) showed that in three species of ctenophore the embryonic oral-aboral axis is determined by the site of the first cleavage furrow. Though in normal eggs this usually coincides with the location of the polar bodies, which provide a stable orientation marker on the egg surface, that is not invariably the case, and in centrifuged eggs it is often not the case. The cleavage furrow apparently forms near the zygote nucleus, which is easily shifted by low centrifugal forces. Normal development ensues in centrifuged eggs whatever the plane of first cleavage, and the consequent orientation of the oral-aboral axis, with respect to the polar bodies. The future oral pole develops wherever the cleavage initiates, and the plane of this cleavage is the future sagittal plane of the larva [Fig. 6.25(a) and (d)]. The second cleavage occurs at right angles to the first and its plane becomes the future tentacular plane of the larva. Thus neither axis is prelocalized in the egg, and the first and second cleavages are the causal events in the orientation of both.

The comb plate and photocyte determinants are localized at different times, and in the unfertilized egg are resident in different components of the cytoplasm. The comb plate determinants are associated with the cortex (Freeman and Reynolds, 1973). Segregation of these factors occurs between the 2- and 8-cell stages, when they become localized in the M and E macromeres respectively [see Fig. 6.25(b)]. Experiments of Freeman (1976a,b, 1977) imply that localization in this egg occurs as a *consequence* of cleavage. Prior to the 2-cell stage the morphogenetic potential for comb plate formation is not spatially restricted, and it begins to be localized to the aboral pole after first cleavage. If the cleavage schedule is altered by transient exposure to cytochalasin or 2,4-dinitrophenol, or the sequence of cleavage planes is disturbed by application of pressure, the result is altered localization of the ability of the embryonic cells to form comb plate cilia or photocytes when cultured. The nature of the localization pattern that develops in the treated eggs was found to depend on the character of the cleavage abnormality that had been induced. Furthermore, cleavage triggers a timing mechanism, and normal localization requires that the correct plane of cytokinesis occur at the appropriate time. This is shown in the experiment diagrammed in Fig. 6.25(e). The peripheral cytokinesis that in normal embryos gives rise to the E macromeres at third cleavage can be induced to occur at second cleavage in two different ways. If second cleavage is reversibly blocked for one cycle and the inhibitor is later removed, the planes of the following cytokinesis are located similarly to those of third cleavage in normal embryos, though the treated embryo has only four cells. Such embryos display normal localization in the E macromeres of the potential to give rise to comb plates, since these macromeres are segregated out at the appropriate time after first cleavage. E macromeres can also be made to form one cycle too early by causing second cleavage to take place under pressure. In these embryos, though they are of identical form with those produced by cleavage delay, comb plate forming potential is found to be present in both E and M macromere derivatives. It can be concluded that in ctenophore embryos the localization patterns are spatially dependent on the orientation of the cleavages, but the ontogenic process by which these patterns are established is paced by a developmental clock that specifies the window within which the localizing function of each cleavage may operate.

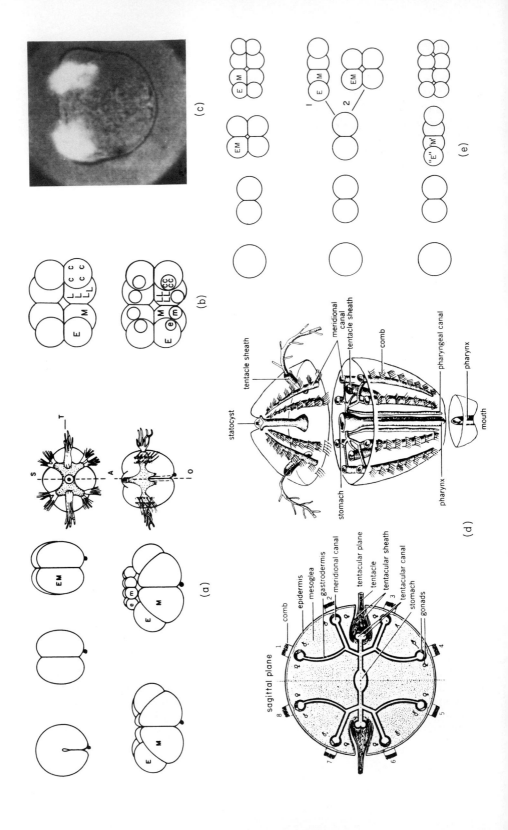

(a)

(b)

(c)

(d) sagittal plane

comb
epidermis
mesoglea
gastrodermis
meridional canal
tentacular plane
tentacle
tentacular sheath
tentacular canal
stomach
gonads

statocyst
tentacle sheath
meridional canal
tentacle sheath
comb
pharyngeal canal
pharynx
mouth
pharynx
stomach

(e)

Ontogeny of Localization in Nemertean Eggs and the Role of Asters

The eggs of nemertean worms of the genus *Cerebratulus* were also favorite subjects of classical research on localization. Wilson (1903), Zeleny (1904), and Yatsu (1904, 1910) showed that establishment of the definitive localization patterns in the *Cerebratulus* egg begins with the meiotic divisions set in train by fertilization, and is not complete until the 8-cell stage. The morphogenetic factors that specify the apical tuft and the gut, among other structures, have by then segregated to their respective progenitor blastomeres. Thereafter the blastomeres behave in an irreversibly determined fashion (Hörstadius, 1937a). However, in the unfertilized egg the morphogenetic factors are globally distributed, although the animal-vegetal axis is clearly predetermined. Thus, for example, Yatsu (1910) showed that normal larvae may be obtained either from fertilized animal or vegetal half eggs. A developmental map for the apical tuft and gut determinants of the *Cerebratulus lacteus* egg was constructed by Freeman (1978), by determining the morphogenetic potential of egg fragments isolated at various stages after fertilization. As diagrammed in Fig. 6.26, these factors independently move

---

Fig. 6.25. Cleavage and localization in the ctenophore egg. (a) Normal development. First cleavage and the 2-, 4-, 8- and 16-cell stages of embryogenesis are shown, viewed from the side. A mark placed on the site where the first cleavage furrow originates is traced through the different cleavage stages where it resides in the future oral region. M and E denote macromeres and m and e the micromeres present at the 16-cell stage. Right, advanced cydippid larva (see (d)), to indicate tentacular plane (T); sagittal plane (S); A, aboral; O, oral surface. [From G. Freeman (1979). *In* "Determinants of Spatial Organization" (S. Subtelny and I. R. Konigsberg, eds.), p. 53. Academic Press, New York.] (b) The localizations of developmental potential for photocyte and comb plate cilia differentiation at the 8- and 16-cell stages, viewed from the aboral pole. L indicates the localization of developmental potential which specifies photocyte differentiation. C indicates the localization of developmental potential which specifies comb plates. Note that at the 16-cell stage the potential to form photocytes is still associated with the M macromeres. [From G. Freeman (1976a). *Dev. Biol.* **49**, 143.] (c) Side view of embryo at developmental stage when light production is first detected. The embryo was viewed with black illumination and stimulated by a weak electric pulse. The embryo is 170 µm in diameter. The light-producing regions are in the aboral portion of the embryo. Photographed after image intensification. [From G. Freeman and G. T. Reynolds (1973). *Dev. Biol.* **31**, 61.] (d) Diagrams of cydippid larvae. Left shows cross section and right a side view sectioned in stomach area. The oral axis is vertical in the right drawing. The photocytes are located along the canals leading to the comb plates. [From A. Kaestner (1967). *In* "Invertebrate Zoology" (H. W. Levi and L. R. Levi, transl.), Vol. 1, p. 145. Wiley (Interscience), New York.] (e) Diagram of cleavage in normal and abnormal ctenophore embryos viewed from the aboral pole. Top, uncleaved egg, 2-, 4-, and 8-cell stages of a normal embryo. Middle, embryo in which the second cleavage has been reversibly inhibited. (1) shows the most common configuration that occurs after cleavage is initiated following the block; this configuration is similar to the configuration normally generated by the third cleavage. (2) shows the blastomere configuration generated in the remaining cases; it is similar to the normal 4-cell stage. Bottom, embryo in which the second cleavage occurred under compression. [From G. Freeman (1979). *In* "Determinants of Spatial Organization" (S. Subtelny and I. R. Konigsberg, eds.), p. 53. Academic Press, New York.]

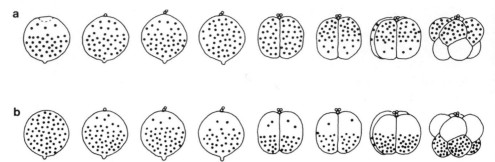

Fig. 6.26.   Progressive localization in the eggs of *Cerebratulus lacteus*. Diagrammatic representation of the distribution of factors specifying apical tuft formation (a) and gut formation (b) from the unfertilized egg stage to the 8-cell stage. The period when the meiotic maturation divisions occur is indicated by the appearance of the polar bodies. [From G. Freeman (1978). *J. Exp. Zool.* **206**, 81.]

in opposite directions during the meiotic maturation stages. They achieve their final distribution just before the transverse third cleavage. In this egg, in contrast to the molluscan examples examined earlier, cytochalasin treatment does not interfere with the progressive mechanism of localization. Nor does alteration of the cleavage planes by pressure. On the other hand a drug that suppresses aster formation, ethyl carbamate, arrests the localization process for as long as it is present. When the drug is washed out new asters form, and localization resumes. Freeman (1978) also showed that removal of the meiotic asters with a microneedle has the same effect as ethyl carbamate, while precocious induction of asters with hypertonic sea water accelerates the localization of the gut and apical tuft determinants. Though it is clear from these results that aster formation is required for localization in the *Cerebratulus* egg, the function of these organelles in regard to localization remains unknown.

There are clearly many diverse mechanisms by which maternal cytoplasmic determinants are spatially distributed. These include intercellular interactions during oogenesis, the disposition of maternal cytoplasm in polar lobes, and the mediation of localization by cytokinesis in early cleavage, or by global cytoplasmic movements triggered at maturation or fertilization. An underlying element of homology in some of these various processes may be the engagement of the cytoskeletal apparatus of the egg, as long ago argued by Conklin (1905a). The crucial role of asters in the localization processes of *Cerebratulus* eggs may be interpretable in these terms. In the following we consider some recent evidence on the participation of cytoskeletal structure in the ontogeny of localization patterns.

(iii) Localization and the Cytoskeleton: Evidence from Ascidian and Annelid Eggs

We have touched on a number of cases in which cytoplasmic *movements* occurring after fertilization are involved in localization, and yet the determinants appear to be *anchored* in place, in that they are resistant to centrifugal forces that stratify the egg cytoplasm (see, e.g., review in Morgan, 1927; Clement, 1968; Verdonk, 1968). The cytoskeleton of the egg provides both the possibility of contractile movement and the opportunity of localized anchorage for particles, organelles, and macromolecules. Current evidence specifically implicates the cortical cytoskeleton of ascidian eggs, both in the initial bipolar contraction of the surface elements of the cytoplasm that occurs immediately after fertilization (Sawada and Osanai, 1981, 1985), and in the localization of the posterior muscle cell determinants. As will be recalled (see Fig. 6.3 and 6.4) the yellow myoplasm that marks the positions of these determinants is localized by means of cytoplasmic movements that occur prior to cleavage. Conklin (1931) showed that low speed centrifugation fails to interfere with the localization patterns of ascidian eggs, and indirect evidence for the association of the muscle determinants with the cytoskeleton was shown in Fig. 6.3(n). As described in the legend to that figure the pigmented myoplasm remains in place after detergent extraction that removes three-fourths of the egg protein and almost all of the bulk RNA and lipid. In Fig. 6.27(a) is reproduced a scanning electron micrograph taken by Jeffery and Meier (1983) that displays at the ultrastructural level the engagement of the myoplasm pigment granules with the cortical cytoskeleton. The pigment granules are seen beneath a mesh-like structure termed the plasma membrane lamina, which is composed primarily of membrane proteins, and is supported by actin microfilaments (Jeffery and Meier, 1984; Sawada and Osanai, 1985). Beneath this lies a complex filamentous lattice [Fig. 6.27(b) and (c)], that may consist largely of the intermediate filament proteins desmin and vimentin, and to which the pigment granules are attached (Jeffery and Meier, 1983). After fertilization the plasma membrane lamina, which initially covers much of the egg, contracts over the surface carrying its inclusions, and thus forms the localized yellow crescent (cf. Conklin's drawings in Fig. 6.4). The actin microfilaments of the cortical cytoskeletal complex could contribute to the structural integrity of the cortical myoplasm domain, and also provide components of the contractile engine that drives the spatial segregation of this domain (Jeffery and Meier, 1984). By analogy with what is known of echinoderm eggs (see below) cytoskeletal myosin could be present, and might provide the motive force. A dramatic scanning electron micrograph in which this contraction process has been caught is shown in Fig. 6.27(d). Jeffery (1982) showed that in eggs of the ascidian *Boltenia villosa* the location of the myoplasm cresent can be specified by

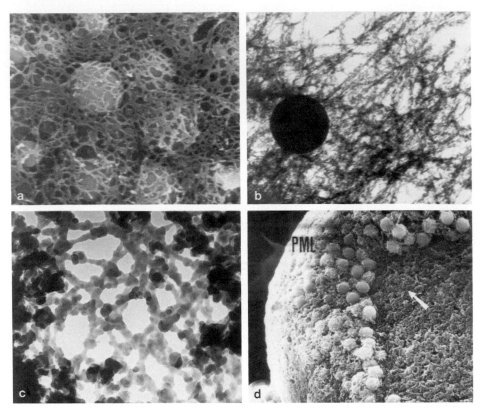

Fig. 6.27. Cortical cytoskeletal structure of ascidian egg myoplasm. (a) Scanning EM of the surface of a fertilized *Styela plicata* egg that has been extracted with Triton X-100. The surface coat surrounding the egg is thus removed, revealing the plasma membrane lamina, overlying the spherical pigment granules. X7,400. (b) Transmission EM of of egg extracted as in (a), and showing a pigment granule embedded in cytoskeletal filaments. X19,500. (c) Same filamentous structure. X118,400. (d) Scanning EM of surface of a fertilized egg of *Boltenia villosa,* after extraction as above, fixed while in process of cytoplasmic segregation. The cortical domain containing the pigment granules is elevated above the surface of the contiguous cytoskeleton. The arrow indicates the probable direction of the contraction. Plasma membrane lamina, PML. [From W. R. Jeffery and S. Meier (1983). *Dev. Biol.* **96**, 125.]

presentation of a Ca^{2+} ionophore, which apparently acts as the focal point for the cortical contraction, just as might the penetrating sperm in the untreated egg.

A potentially interesting aspect of cytoskeletal localization in ascidian eggs is that a major fraction of the maternal poly(A) RNA is retained after detergent extraction, though most of the ribosomal RNA is not (Jeffery and

Meier, 1983; Jeffery, 1984). The possibility thus exists that within spatially localized cytoskeletal domains could be sequestered *specific* maternal poly(A) RNA sequences. However, a two-dimensional examination of the translation products of mRNA extracted from isolated yellow crescents failed to reveal any such mRNA localization (Jeffery, 1985a). Among the mRNAs present in the yellow crescent myoplasm are some coding for actin (Jeffery *et al.,* 1983; Jeffery, 1984), but it is important to note that these maternal actin messages code for cytoskeletal rather than the muscle actin isoform. Since only the most prevalent mRNAs of the yellow crescent have been examined, perhaps the main significance is the observation that there are some mRNAs tightly associated with it. Cytoskeletal components other than the actin microfilaments are responsible for binding these mRNAs since the cortical actin lattice can be disrupted without causing their release (Jeffery, 1984). The possibility remains to be excluded that lower prevalence maternal mRNAs could be specifically in the myoplasm.

Maternal poly(A) RNA is also anchored to large cytoplasmic structures in the egg of *Chaetopterus pergamentaceus.* A typical annelid localization pattern is established in this egg by means of a dramatic series of cytoplasmic rearrangements occurring during the meiotic maturation stages that immediately precede first cleavage (Lillie, 1906; Jeffery and Wilson, 1983). Lillie (1906) showed that regional differentiation occurrs in *Chaetopterus* eggs even if cytokinesis is repressed by treatment with KCl, just as it does in cytochalasin treated ascidian zygotes (see above). Cytoplasmic reorganization in both KCl activated and fertilized *Chaetopterus* eggs is blocked by colchicine, which affects microtubule organization, but not by cytochalasin (Brachet and Donini-Denis, 1978; Eckberg, 1981). Stratification of the internal cytoplasm by centrifugation also fails to disturb the localization patterns of *Chaetopterus* eggs, and centrifuged eggs develop normally (Lillie, 1909; Morgan, 1910). *In situ* hybridization with ^{3}H-poly(U) demonstrates that the egg cortex, where the key morphogenetic movements of the cytoplasm take place, is also the site of maternal poly(A) RNA accumulation, as can be seen in Fig. 6.28(a) (Jeffery and Wilson, 1983). The cortical localization of the maternal poly(A) RNA persists in first cleavage, and extends to the polar lobe formed at that time [Fig. 6.28(b) and (c)]. After low speed centrifugation most of the maternal poly(A) RNA is sedimented to the centrifugal pole of the egg [Fig. 6.28(d)]. This observation shows that correct localization does not require the normal cortical disposition of the bulk of the maternal poly(A) RNA. Nonetheless, the result is interesting because it indicates that in this egg heterogeneous RNAs are associated with particulate elements, which for certain sequences could as well be anchored in structural components of the cytoskeleton, just as are the yellow pigment granules and the maternal mRNAs of the ascidian myoplasm.

We have encountered in this review several kinds of evidence suggesting

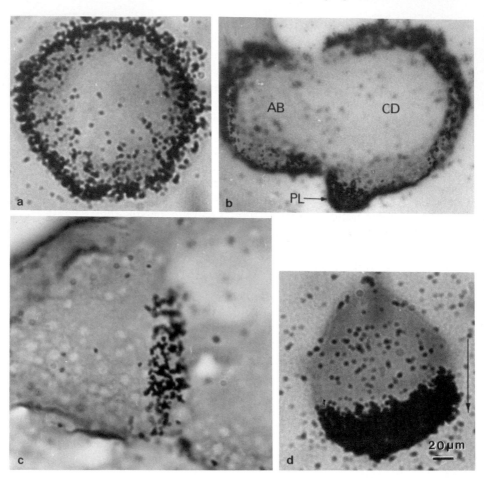

Fig. 6.28. Localization of poly(A) RNA in eggs of *Chaetopterus pergamentaceus*, established by *in situ* hybridization of ^{3}H-poly(U). Autoradiographs of hybridized sections are shown. (a) Mature egg 10 min after germinal vesicle breakdown. The grains are primarily located in the cortical cytoplasm. X350. (b) Trefoil stage embryo. Grains are concentrated in the cortical cytoplasm of the animal hemisphere and in the polar lobe (PL). X350. (c) Section along animal-vegetal axis of cleaving embryo, showing accumulation of grains over cortical granules in the vegetal region of the cleavage furrow. X780. (a)-(c) From W. R. Jeffery and L. J. Wilson (1983). *J. Embryol. Exp. Morphol.* **75**, 225. (d) Mature *Chaetopterus* egg centrifuged at 500 × g for 5 min. The arrow indicates the direction of centrifugal force. The grains are concentrated at the centrifugal end of the egg. [(d) From W. R. Jeffery (1985b). *Dev. Biol.* **110**, 217.]

that among the morphogenetic determinants of the egg cytoplasm are maternal RNAs, which may be prelocalized in the egg, or progressively localized after fertilization. Examples for which this has been proposed include the UV-sensitive targets involved in germ cell determination in anuran amphibians and dipteran insects; the anterior determinants of the *Smittia* egg; the early embryo poly(A) RNAs that on injection may result in rescue of maternal axial specification mutants in *Drosophila*; and the polar lobe determinants of the *Bithynia* vegetal body. Several additional instances are reviewed by Jeffery (1983). A theory of localization based on regionally confined maternal regulatory RNAs was originally proposed by Davidson and Britten (1971). The attractiveness of the proposal that the egg stores regulatory RNAs that are or will be localized is based on two features. First, there is the argument in principle that regulatory information transcribed during oogenesis is ultimately required to induce new, differential patterns of genomic function in embryonic cell lineages, information that could reside either in maternal mRNAs coding for regulatory polypeptides, or in regulatory RNA sequences functioning in some other manner. Second, there are observations reviewed in this section that at least in some eggs maternal poly(A) RNAs are associated with regionally localized cytoskeletal elements. This provides in principle the outline of a mechanism for the regional distribution of morphogens. To proceed further will be a difficult task, as it would require experimental demonstration that specific maternal RNA or protein determinants are both necessary and sufficient for cell lineage specification in the embryo.

5. SPATIAL ASSIGNMENTS, REGULATION, AND LOCALIZATION IN THE EGGS OF SEA URCHINS, AMPHIBIANS, AND MAMMALS

Cytoplasmic factors of maternal origin that ultimately affect the specification of embryonic cells appear to function in the early development of all the creatures so far considered, though the means by which such factors are spatially localized are many and various. The comparative arguments for localization as a general and basic mechanism for early development are thus strong, as classical writers such as Wilson (1925) also stressed. Nonetheless, as for our predecessors, the problem of explaining regulative behavior remains difficult. The localization theory has seemed to some observers incomplete, in that it does not obviously predict that subelements of the embryo, when isolated or placed in combination with other subelements, will in some cases produce structures other than those they are normally destined to form, including complete, normally constructed larvae. Regulative behavior has been regarded as a major challenge ever since the first reproducible observations of this kind were described by Driesch (1891). Several different factors are now understood to contribute to regulative developmental processes in experimentally perturbed embryos, none of which in any way exclude specification, in the *normal* course of events, by means of cytoplasmic localization. Thus, Wilson pointed out (see Fig. 6.7) that even on the strict assumption of cell lineage determination by localized morphogenetic agents, certain blastomeres would be expected to behave in a *totipotent* manner if isolated and cultured, while others should give rise to partial embryos, according only to the geometry of the localization patterns with respect to the planes of cleavage. We have also seen that localization patterns are in some organisms established ontogenically, at a relatively leisurely pace, so that the morphogenetic potential displayed by a given subembryonic fragment depends on the stage at which it is tested. In addition, a strong relation exists between the extent to which cell lineages are determined through inductive intercellular interactions during early embryogenesis, and the capacity for regulative behavior that subelements of the embryo will manifest. As an example already considered, in equally cleaving gastropod eggs in which the dorsoventral axis is determined *inductively,* any of the four macromeres can give rise to this axis, while in unequally cleaving gastropod eggs only the D macromere retains the necessary maternal cytoplasmic factors. Another variable factor is that the structural elements responsible for the spatial disposition of localized determinants are sometimes apparently very delicate, so that in early stages localization is *physically labile*, and is easily reversed by experimental manipulation. The eggs of certain deuterostome groups, including sea urchins, amphibians and mammals, provide illuminating examples both of lability and of early inductive interactions between blastomeres. Recent experimental analyses of these features illuminate both the role of localization in early development, and the mechanisms underlying regulative developmental behavior.

(i) Localization and Regulation in the Sea Urchin Egg

Axial Determination

Localization of determinants along the animal-vegetal axis of the unfertilized sea urchin egg was demonstrated by Boveri (1901a,b), in a study carried out on eggs of *Paracentrotus lividis*. In this species a fortuitous subequatorial band of pigment granules arranged in a plane orthogonal to the animal-vegetal axis indicates the future orientation of the embryo. These granules also indicate the location of cytoplasm that is later included in the cells of the archenteron. By shaking the eggs into pieces and fertilizing the fragments, Boveri showed that only fragments containing the pigmented cytoplasm are able to gastrulate and to carry out gut formation. Hörstadius (1928) confirmed that the pigment layer indicates the eventual axial orientation of the embryo, and showed that animal half egg fragments develop in a manner similar to the animal halves of 16-cell cleavage-stage embryos, while the partial morphogenesis carried out by vegetal egg fragments corresponds to that observed in 8- or 16-cell vegetal half embryos. Hörstadius (1937b) extended these findings to *Arbacia punctulata*, orienting the eggs individually soon after fertilization, and cutting them in half at the pronuclear fusion stage. The animal halves cleaved equally, producing no micromeres, and gave rise to spherical structures with enlarged apical tufts. The vegetal halves, however, formed micromeres and gastrulated, with archenteron and skeleton formation ensuing. It follows that cytoplasmic factors required for micromere, skeleton, and archenteron formation are localized in the vegetal region of the egg cytoplasm even before fertilization. The orientation and distribution of these materials remain unchanged as the egg cytoplasm begins to be divided up among the blastomeres. Like lobeless *Dentalium* and *Ilyanassa* eggs, the animal egg fragments in these experiments possess complete genomes, but fail to differentiate several important cell lineages.

Boveri (1901a) noticed two additional external features that identify the prelocalized animal pole in banded *Paracentrotus* eggs. These are the minute channel known as the *jelly canal*, and the position of the polar bodies. Schroeder (1980a,b) showed that the jelly canal, visualized by immersion of unfertilized eggs or oocytes in ink, can be used to define the location of the animal pole in unfertilized sea urchin eggs that lack an oriented band of pigment granules, as illustrated in Fig. 6.29. On elevation of the fertilization membrane the ink mark left at the site of the canal remains visible, and as can be seen in Fig. 6.29(e) this mark occurs *opposite* the micromeres at the 16-cell stage. The animal-vegetal prelocalization is thus a general characteristic of sea urchin eggs, and is not confined to those with banded pigment layers. Contrary to Boveri and other authors, both classical and modern (see, e.g., Hörstadius, 1973), the jelly canal is not a micropyle. Sperm can enter and fertilize the sea urchin egg at any point on its surface with respect to the animal pole (Hörstadius, 1928; Schroeder, 1980b; Schatten, 1982), as shown in Fig. 6.29(c)-(d). Irrespective of the point of sperm entry, the orthogonal first and second planes of cleavage intersect along the preformed animal-vegetal axis. Fig. 6.29(a) demonstrates that the polar orientation of the egg is a prelocalization that extends back into oogenesis, since the jelly canal is present even in immature oocytes. A subequatorial band of pigment granules similar to that present in *unfertilized Paracentrotus* eggs [Fig. 6.29(f)] appears *during early cleavage* in *Arbacia* eggs [Fig. 6.29(g)]. Observations of Schroeder (1980a) suggest that the pigment band forms by means of an upward contraction of the cortical cytoplasm that carries the pigment granules away from the vegetal pole, as diagrammed in Fig. 6.29(h). The implication is that a similar event has already occurred in *Paracentrotus* eggs by the time they become available for fertilization, and in this species the pigment granules have been shown to be structurally associated with the cortical cytoskeleton in both unfertilized and fertilized eggs (Sardet and

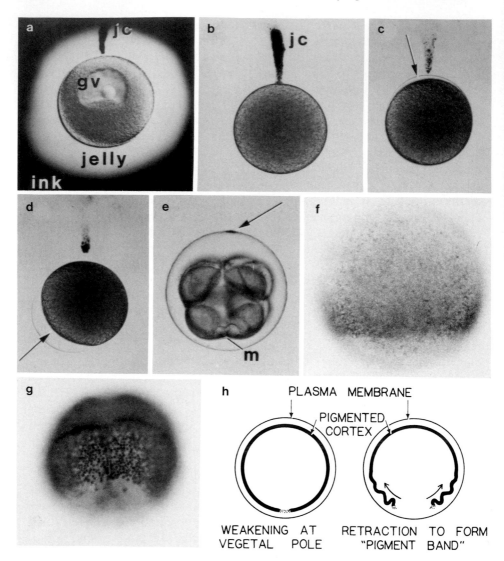

Fig. 6.29. Manifestations of animal-vegetal axial prelocalization in sea urchin eggs. (a)-(e) Jelly canal of *Strongylocentrotus dröbachiensis* oocytes and eggs visualized by immersion in ink. (a) A full-sized primary oocyte surrounded by ink. The ink-filled jelly canal (jc) indicates the animal pole, toward which the asymmetric germinal vesicle (gv) is displaced. (b) Unfertilized egg without polar bodies after immersion in ink and washing. The jelly canal (jc) locates the animal pole which is otherwise not identifiable. (c) An egg about 45 sec after fertilization. The sperm entry point (arrow) is indicated by the center of the lifting vitelline layer, near the jelly canal. (d) Another egg soon after fertilization. The sperm entry site (arrow) is distant from the jelly canal. (e) A 16-cell embryo whose jelly coat is no longer present. The fertilization envelope retains a small spot of ink (arrow) from the original jelly canal, overlying the animal pole of the embryo, and opposite the vegetal pole where the micromeres (m) are located. X200. [(a)-(e) From T. E. Schroeder (1980b). *Exp. Cell Res.* **128**, 490.] (f)-(h) Pigment granules. (f) Unfertilized egg of *Paracentrotus lividus* seen in bright field, animal pole at top. Granules are concen-

Chang, 1985). A prelocalized animal-vegetal polarity is also manifested in various cytoskeletal features of starfish oocytes, including the distribution of actin-filled external spikes and of certain acidic vacuoles, both emeshed in the contractile cortex; the distribution of pigment granules; and the position of the premeiotic aster (Schroeder, 1985).

Cytological studies of meiotic activation in starfish oocytes and of fertilization in sea urchin eggs that lie outside the scope of this discussion have revealed a complex cytoskeletal architecture in the echinoderm egg (see, e.g., Schatten and Schatten, 1981; Coffe *et al.*, 1982; Schroeder and Stricker, 1983; Otto and Schroeder, 1984a; Cline *et al.*, 1984; Schroeder, 1985; reviewed in Schatten, 1982). Networks of microtubules and microfilaments are present in the cortex, and cortical myosin, actin, and fascin have all been identified in the starfish oocyte (Mabuchi, 1976; Otto and Schroeder, 1984a,b). The imperviousness of the axial organization to stratifying centrifugal force exerted in any direction with respect to the egg axis (Morgan and Spooner, 1909; Harvey, 1956; Hörstadius, 1973), and the observations interpreted in Fig. 6.29(h), imply that the oriented elements important for animal-vegetal prelocalization reside within, or are anchored to, a regionally differentiated, preformed, cortical cytoskeleton. Mabuchi and Okuno (1977) showed that in the starfish egg cytokinesis, which involves a cytoplasmic contraction that is normal rather than tangential to the surface, is inhibited by microinjected anti-myosin antibodies. A reasonable interpretation is that the tangential cortical contractile processes occurring during localization in echinoderm, ascidian, amphibian, and many other eggs are also driven by contractile actin-myosin complexes of the cytoskeleton, and that these structures are associated with the cortex as a whole by actin microfilaments that in many eggs can be dissociated by cytochalasin treatment.

As reviewed in Chapter IV the oral-aboral axis of the sea urchin embryo is not evident at a gross level until the gastrula stage, when the ectodermal wall on that side thickens and shortens, the archenteron develops an asymmetric tilt to the oral side, the oral rods of the skeleton appear, and the secondary mesenchyme cells contact the inner wall of the blastocoel. In undisturbed

---

trated in the pigment band, but are also present in the entire animal hemisphere at lower density. They are absent from the vegetal region. X440. (g) Pigment granules in *Arbacia lixula*, seen at 16-cell stage from the side. Focus is at embryo surface to display granules, and micromeres, which lie within the vegetal clear zone, are not visible. X600. (h) Model for the mechanism by which cortical pigment moves to form the vegetal clear zone and "pigment band." Cortical granules move coherently, in conjunction with the matrix in which the granules lie. In this model the plasma membrane is not involved. A weakening of the cortical matrix at the vegetal pole (stippling) is postulated, and the remaining cortex retracts toward the equator. The area evacuated by pigment cortex becomes the vegetal clear zone; the compacted cortex in the subequatorial zone becomes the "pigment ring" with its abrupt vegetal margin and gradient of pigment concentration which decreases toward the equator. [(f)-(h) From T. E. Schroeder (1980a). *Dev. Biol.* **79**, 428.]

embryos the oral-aboral axis is apparently specified as early as the 8-cell stage of cleavage, however (cf. Fig. 4.5), although even at late cleavage stages meridional half embryos and embryos in which the animal half has been rotationally reoriented with respect to the vegetal half are all able to reproduce new oral-aboral axes and thence to develop normally (reviewed in Hörstadius, 1973).

Localization and the Determination of the Micromere Cell Lineage

The micromeres are already definitively specified as the precursors of the skeletogenic mesenchyme cells when they first appear at the 16-cell stage. Molecular and biological aspects of skeletogenic mesenchymne cell differentiation are reviewed in Chapter IV. As discussed there, under appropriate *in vitro* culture conditions isolated micromeres will divide the same number of times as do the micromeres and their progeny in the embryo, and then on schedule produce spicules. A number of macromolecular markers have been discovered that characterize this cell lineage, including antigens absent from all other cell types, specific proteins that can be visualized in 2-dimensional gels, and transcripts deriving from genes expressed only in mesenchyme cells. However, though irreversibly determined from the initial formation of their lineages (see Fig. 4.5; see pocket in back of book), the progeny of the micromeres express most of these markers only after completing their clonal multiplications, and in most cases during or after the ingression into the blastocoel of their terminal lineage products, the skeletogenic mesenchyme cells.

The *initial specification of the micromere cell lineage* apparently occurs by means of cytoplasmic localization. Only vegetal pole cytoplasm can induce formation of a cell line that gives rise to mesenchyme cells and endows the embryo with spicules. This has been shown by culturing vegetal as opposed to animal half eggs, and various isolated blastomeres and combinations of blastomeres (Boveri, 1901b; reviewed in Hörstadius, 1939). Experiments of Maruyama *et al.* (1985) show that cytoplasmic factors required for the specification of primary skeletogenic mesenchyme cells are already localized in the vegetal regions of the egg before fertilization, as shown by fertilization and culture of half-eggs oriented by means of the jelly canal (cf. Fig. 6.29). The vegetal halves of equatorially bisected eggs, and all meriodional half-eggs give rise to skeletogenic primary mesenchyme, while animal half-eggs do not. The extreme polar region of the egg cytoplasm that is normally sequestered to the micromeres by the most vegetal of the transverse planes of fourth cleavage, is not the only sector of egg cytoplasm that contains determinants capable of eliciting micromere formation. Hörstadius (1939) reported that if cultured in isolation the tier of vegetal blastomeres lying just above the micromeres will give rise to an imperfect gastrula-like embryo that constructs spicules, and spicules can also be induced to form from secondary mesenchyme cells by first removing the primary (i.e., skeletogenic) mesenchyme cells (Fukushi, 1962; see Chapter IV for discussion of origins and functions of primary and secondary mesenchyme cells and their descendant lineages). However, the secondary mesenchyme cells are ultimately also descendants of the original vegetal blastomeres (see Fig. 4.5). It has been conventional to regard micromere formation as an example of a localization process paced by a developmental "clock," on the basis of experiments in which cleavage was transiently blocked by various treatments, or micromeres were induced to form when there are only eight rather than 16 cells in the embryo. Micromere formation was observed in these experiments only after 4th cleavage had occurred in control embryos (Rustad, 1960; Dan and Ikeda, 1971). However, this proposition is less convincing than it once was, since Kitajima and Okazaki (1980) demonstrated that normally differentiating micromeres can in fact be induced at 3rd cleavage by treatment with a quinoline derivative, i.e., well before the occurrence of 4th cleavage in normal embryos. A general conclusion that can be extracted from such experiments is that micromere determination occurs independently of any exact prior pattern of cleavage. Thus, for example, Langelan and Whiteley (1985) showed in *Dendraster excentricus* that skeletogenic mesenchyme formation is not prevented in embryos treated with dilute SDS solutions, in which the 16-cell cleavage stage consists of two horizontal tiers of eight cells each (cf. normal 16-cell embryo in Fig. 4.5, inset). Nor does equalization of 4th cleavage by SDS treatment preclude spicule formation in *Strongylocentrotus purpuratus* or *Lytechinus pictus* embryos (Langelan and Whiteley, 1985) On the other hand, in *Hemicentrotus pulcherrimus*, treatment with SDS that prevents the unequal horizontal 4th cleavage by which the micromeres are segregated also prevents skeleton formation (Tanaka, 1976; Dan, 1979). The mode of action of SDS is not straightforward, however, as it has general effects on embryonic morphogenesis in sea urchins (reviewed in Langelan and Whiteley, 1985), and it also disrupts the normal timing of the cleavage divisions (Filosa *et al.*, 1985; see also Fig. 4.5). Thus results demonstrating the *independence* of skeletogenic mesenchyme specification from the particular pathway of 4th cleavage normally observed would seem the more significant.

Among the proposals that have been advanced to explain the role of the maternal vegetal cytoplasm in micromere determination is the now familiar idea that the micromeres inherit a special set of maternal RNA species segregated in the polar region of the egg. As an initial test, both the protein synthesis patterns and the RNA sequence content of 16- to 32-cell embryo micromeres have been analyzed with respect to the remainder of the embryo. Tufaro and Brandhorst (1979) and Harkey and Whiteley (1983) found all of the clearly discernible protein species that can be resolved in two-dimensional gels to be synthesized alike by micromeres, mesomeres, and macromeres. Similarly, Ernst *et al.* (1980) showed that within the limits of resolution of the RNA excess hybridization method, just the same set of polysomal mRNA sequences is present in micromeres as in the other embyro blastomeres. While these studies provide no support for the RNA segregation hypothesis, nor do they exclude it, since the two-dimensional protein synthesis analyses concern only those proteins produced by the moderately prevalent and abundant messages (see Chapter III), and the complexity comparison is incapable of resolving small differences, e.g., anything less than about 10% of the 10^4 or so diverse transcripts in the polysomal RNA.

There are several fundamental molecular as well as cytological properties that are already known to be special to the micromeres. Schroeder (1982) pointed out that from the moment of their formation micromeres must contain less plasma membrane, and less polymerized actin, fascin, and pigment granules (cf. Dan *et al.*, 1983). Several specific surface protein species, probably of maternal origin, are also localized in the micromeres of *Arbacia* and *Strongylocentrotus* embryos (De Simone and Spiegel, 1985). Micromeres have a nucleus/cytoplasm ratio several times greater than do the other blastomeres, and consequently the ratio of newly synthesized message to maternal message could be much higher than elsewhere in the 4th or 5th cleavage stage embryo. Among other effects this results in a significant increase in the ratio of histone to nonhistone protein synthesis (Senger and Gross, 1978). The relatively lower amount of (maternal) cytoplasm suggests that genomic activity might determine biosynthetic patterns in micromeres at an earlier time than in the other blastomeres. At the nuclear level two additional differences have been noticed. Synthesis of high complexity nuclear RNA is delayed in micromeres until well after 5th cleavage, while it is active in other embryo nuclei before this (Ernst *et al.*, 1980). Finally, an interesting observation that emerges from the comparative RNA complexity studies carried out on total RNA by both Rodgers and Gross (1978) and Ernst *et al.* (1980) is that micromeres *lack* a significant component of the maternal RNA. This is the fraction not represented on 16-cell embryo polysomes, and it probably consists mainly of interspersed maternal transcripts (see Chapter II). Whatever the specific significance of this result, a conclusion that can be drawn is that the micromere nuclei reside in a *distinct cytoplasmic domain*, since maternal cytoplasmic RNA sequences present elsewhere are missing from the micromeres. The key feature implied is the interaction of elements localized specifically in the vegetal cytoplasm with these nuclei, rather than any other special characteristics, or the time of appearance, of the micromeres *per se*. Whether by virtue of the absence of maternal repressors or the presence of special activators, or both, the properties of this vegetal cytoplasmic domain perhaps provide the specific conditions that induce the genome level alterations ultimately resulting in micromere lineage specification.

Liability and Regulation

It is clear from the foregoing, and the cell lineage evidence reviewed in Chapter IV, that axial differentiation in the sea urchin embryo, including the determination of archenteron and skeletogenic primary mesenchyme, is specified by means of classic processes of cytoplasmic localization. Yet, many of the blastomere assignments in the sea urchin embryo turn out to be subject to spectacular revisions. A renowned series of experiments by Hörstadius (1928, 1935; reviewed in 1939, 1973) show that except for the micromere lineage, the fate of which is irreversibly committed from the beginning, the roles of the progeny of all the other tiers of blastomeres in the 16-cell embryo can be altered by placing them in chimeric combinations with other blastomeres. Several examples are diagrammed in Fig. 6.30, in which the cleavage stage embryo is portrayed according to Hörstadius in terms of tiers of equipotential cells, rather than as a set of lineages as shown in Fig. 4.5. Hörstadius' experiments show that the behavior of the animal pole cap of cells depends on which other blastomere tiers are included in the chimera. In the combination shown in Fig. 6.30(e), for example, the archenteron, which normally forms from the progeny of the vegetal tier of blastomeres proximal to the micromeres [veg_2 in Fig. 6.30(a)], differentiates instead from the equatorial vegetal tier (veg_1). Similarly, both archenteron and skeletal structures derive from veg_2 cells in the absence of micromeres, as shown in the combination represented in Fig. 6.30(d). The traditional interpretation of these results invokes gradients of morphogens originating at the animal and vegetal poles (see, e.g., Hörstadius, 1973), but there is as yet no evidence that uniquely requires such gradients. However, the results illustrated in Fig. 6.30 clearly indicate that decisive *inductive interactions* occur between adjacent blastomere tiers. The inductive potency of isolated micromeres has been demonstrated directly by implanting them into the other blastomere tiers, and into meridional half or quarter embryos (Hörstadius, 1935). Micromere implantation inhibits apical tuft formation by animal pole isolates, turns presumptive ectoderm into endoderm, and in some cases creates new embryonic axes. Similar activities can be elicited from the veg_2 tier, just as these cells also share with micromeres the innate (i.e., localized) capacity to give rise to mesenchyme and skeleton. Many additional constructions and embryonic recombinations have been described (reviewed in Hörstadius, 1973), with the general result that the fate of given blastomeres is always found to be affected by the apposition of different neighboring cells than adjoin them in normal embryos. The regulative behavior manifested in these experiments can be reconciled with the demonstrated role of localization in the undisturbed sea urchin embryo in the following way. Localized maternal cytoplasmic determinants specify *certain* cells in the normal embryo, in particular (though probably not exclusively) the micromeres and the archenteron precursors near the vegetal pole. These cells then determine induc-

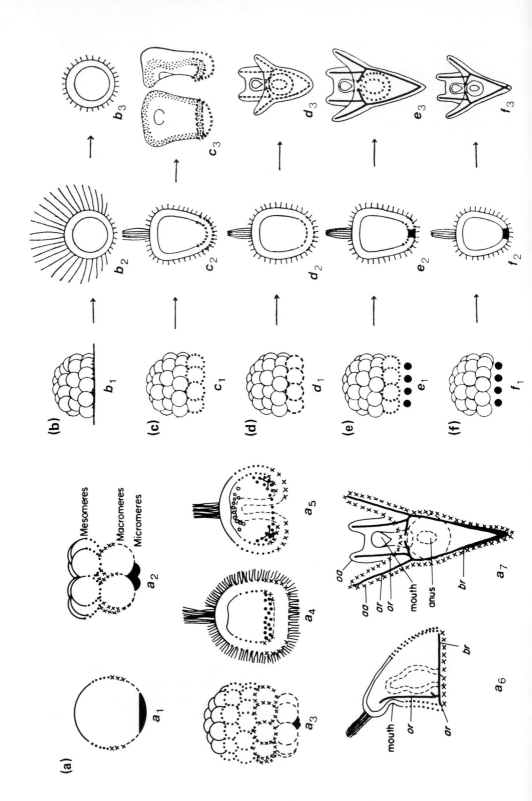

(a)

a_1

a_2 — Mesomeres, Macromeres, Micromeres

a_3

a_4

a_5

a_6 — mouth, or, ar, br

a_7 — oa, oa, ar, or, mouth, anus, br

(b) b_1 b_2 b_3

(c) c_1 c_2 c_3

(d) d_1 d_2 d_3

(e) e_1 e_2 e_3

(f) f_1 f_2 f_3

tively the fates of neighboring blastomeres, which interact in turn with their neighbors. This mechanism *requires plasticity* in the midcleavage stage blastomeres until they have engaged in the appropriate intercellular contacts, and the induction has been completed. It is this plasticity that is prominently displayed in the chimeric recombination experiments.

An independent though related demonstration of the lability of the natural blastomere assignments is found in the effects of a large number of chemicals that have "animalizing" or "vegetalizing" activity. Agents of this kind delay, repress, and in some cases cause substitutions, of one differential pattern of expression for another in the cells of the developing embryo (see, e.g., Kominami, 1984; see also Chapter IV for effects on particular genes expressed in given cell types). When applied during early cleavage they may cause whole echinoderm embryos to develop as do isolated animal or vegetal half embryos, respectively (Herbst, 1893; reviewed in Runnström and Immers, 1971; Lallier, 1975). The implication, consistent with the experimental demonstrations of Hörstadius, is that many of the blastomeres retain potentialities other than those they normally express, and for some time these blastomeres are only reversibly specified, as required for a developmental system that depends to a large extent on induction. In contrast, there is no such requirement for lability or plasticity in largely self-differentiating embryonic systems, e.g., the ascidian or nematode embryos. This argument may account in part for the oft observed correlation between developmental plasticity and regulative capacity on the one hand, and on the other the irreversibility of the self-differentiations carried out by blastomeres that are specified by localized cytoplasmic determinants, rather than by inductive interactions.

It is to be noted that the mechanisms described for the sea urchins considered in this section cannot be generalized to other echinoderms, or even to sea urchins outside of the subclass Euechinoidea (i.e., "modern" sea urchins). For example, the more "primitive" cidaroid sea urchins form no *primary* mesenchyme cells, i.e., defining these as cells of a skeleton-forming lineage that ingress prior to archenteron formation. Neither micromeres nor primary mesenchyme cells are found in embryos of the echinoderm classes

---

Fig. 6.30. Developmental potential of sea urchin blastomere tiers, and their fate in chimeric combinations. (a) Normal development, embryonic blastomere tiers as in a_3, as follows: an_1, continuous lines; an_2, dotted; veg_1, crosses; veg_2, broken lines; micromeres black. Cytoplasmic elements included in these tiers shown in uncleaved egg, a_1; 16-cell stage, a_2; 64-cell stage, a_3; mesenchyme blastula, a_4; gastrula, a_5; prism, oral side to left, a_6; pluteus, a_7. *aa*, anal arm; *ar*, anal rod; *br* body rod; *oa*, oral arm; *or*, oral rod; *stom*, stomodaeum; *vtr*, ventral transverse rod. (b)-(f) Development of chimeras, showing the inductive influence of veg_1, veg_2 and micromeres on animal blastomeres (mesomeres and their descendants). The chimeras were constructed with 16-cell stage mesomere caps. Note that here an_1 and an_2 derivatives are both indicated as continuous lines. [From S. Hörstadius (1935). *Pubbl. Staz. Zool. Napoli* **14**, 251; (1939). *Biol. Rev. Cambridge Philos. Soc.* **14**, 132.]

Crinoidea, Holothuroidea, or Asteroidea (Schroeder, 1981). In at least one starfish, furthermore, all 1/8th blastomeres may give rise to complete dwarf larvae (Dan-Sohkawa and Satoh, 1978), suggesting that the role of prelocalized determinants might be much reduced in this species. The earliest developmental mechanisms are not necessary evolutionarily conservative even within the taxon, a view that is suggested as well by the diverse means of primary embryonic axis determination demonstrated amongst the gastropod molluscs.

(ii) Localization in Amphibian Eggs

Amphibian eggs have been studied since ancient times, and from the 19th century on this familiar material has been the subject of experimental attempts to understand the relation between the visible structure of the egg and the axial coordinants of the embryo. The animal-vegetal axis is a preformed feature, which evidently originates far back in oogenesis. Thus, in immature amphibian oocytes the germinal vesicle is always located toward the future animal pole, and during vitellogenesis the large yolk platelets are concentrated mainly toward the vegetal pole (reviewed in Nieuwkoop, 1977). In species whose eggs are colored the pigment granules are usually confined to the animal hemisphere, probably an adaptation that provides protection from both above and below, since due to the dense vegetal yolk platelets the eggs normally float animal pole upward. Certain macromolecular constituents also seem to be distributed along the animal-vegetal axis even in oocytes and in uncleaved zygotes. For instance, Capco and Jeffery (1982) found that the poly(A) RNA in advanced vitellogenic *Xenopus* oocytes is concentrated along the vegetal pole cortex and in a perinuclear region on the vegetal side of the germinal vesicle. Shortly after fertilization poly(A) RNA concentrations are higher toward the animal pole (Phillips, 1982). Out of several hundred protein species resolvable on two-dimensional gels a few are particularly localized in the vegetal cytoplasm of matured *Ambystoma* oocytes (Jäckle and Eagleson, 1980), and a variety of protein species have been reported to be distributed in a gradient along the animal-vegetal axis in newly fertilized *Xenopus* eggs, though most are not (Moen and Namenwirth, 1977; Jäckle and Eagleson, 1980; Dreyer *et al.*, 1982). A detailed study was reported by King and Barklis (1985), in which poly(A) RNA was isolated from animal, middle and vegetal thirds of eggs sectioned perpendicular to the animal-vegetal axis and translated in a cell-free system. The proteins synthesized were displayed by two-dimensional electrophoresis, and by this means five mRNA species of the approximately 600 whose products were resolved are found to be confined to the vegetal region of the unfertilized egg, while others are preferentially (i.e., 10 to 100-fold) concentrated there. One species was identified that is significantly more prevalent in the animal region. Several clearly localized maternal mRNAs have also been recovered from a cDNA clone library constructed from unfertilized *Xenopus* egg poly(A) RNA, including one clone representing a transcript localized at the vegetal pole, and three representing animal pole messages (Rabagliati *et al.*, 1985). All of these transcripts belong to the low abundance class of maternal messages of which there are $\geq 10^4$ diverse species stored in the egg (cf. Chapter II). Though primordially imposed during oogenesis, the animal-vegetal axis of amphibian eggs can be at least partially reversed by causing the eggs to develop in an inverted orientation (Chung and Malacinski, 1983). Eggs treated in this manner arrest after gastrulation, but transplantation experiments reveal that reversal of some axial specifications has taken place, so that, for example, neural structures now derive from explanted cells of the original vegetal hemisphere, and similar results with respect to production of an epidermis-specific molecular marker were cited in Chapter IV.

Unlike the animal-vegetal axis, the dorsoventral axis of the amphibian egg is an ontogenic construction, that is specified only after fertilization. In appropriately pigmented eggs the location of the future dorsal pole is forecast by the appearance of a zone of cortical cytoplasm that is depleted of pigment granules, traditionally known as the *grey crescent*. Classical observers, beginning with Newport (1854), noted that the plane of first cleavage usually becomes the sagittal plane of the embryo, and that in most (though not all) cases this plane bisects the grey crescent. The latter normally forms *opposite* to the point of sperm entrance, as first shown experimentally by Roux (1887, 1888, 1903), and further studied by Schultze (1900), Morgan and Boring (1903), Brachet (1904), and many other investigators (reviewed in Morgan, 1927, Chapter XI). Since the sperm can enter, or will fertilize if experimentally deposited at any spot on the animal hemisphere, the egg is initially radially symmetric, and before fertilization all equatorial loci thus possess the capacity to participate in the formation of either ventral or dorsal structures.

At gastrulation the lower edge of the grey crescent becomes the dorsal lip of the blastopore, which is the site of formation of the dorsal structures of the neurula. The inductive process by which these structures arise was revealed in a series of studies begun by Spemann and associates (Spemann and Mangold, 1924; early experiments reviewed in Spemann, 1938; Holtfreter and Hamburger, 1956). Dorsal structures such as notochord, somites and neural tube form as a result of a sequential series of inductive interactions (reviewed in Chapter IV; see Fig. 4.13). Attention was initially focused on the induction of dorsal axial structures during gastrular invagination, carried out by cells of the "Spemann organizer", which is located above the dorsal lip of the blastopore. So powerful is the effect of the organizer and so plastic the condition of the cells prior to invagination that organizer cells transplanted to the prospective ventral side of the early gastrula induce there a secondary dorsal axis.

A consistent and until recently widely accepted interpretation was that the cells of the Spemann organizer are initially specified by maternal cytoplasmic factors localized after fertilization in the grey crescent area of the egg (see, e.g., Wilson, 1925; Spemann, 1938; Pasteels, 1946; Davidson, 1976; Brachet, 1977). This theory found direct support in experiments of Curtis (1962), who reported that prior to the 8-cell stage transplantation of a small piece of cortical grey crescent egg cytoplasm to the prospective ventral side of the embryo suffices to produce a second dorsal axis, just as does ventral transplantation of cells of the dorsal organizer region at the gastrula stage. Unfortunately these results appear to have been the consequence of experimental artefacts (Kirschner et al., 1980; Gerhart et al., 1981), and a wholly different understanding of the role of localization in the specification of dorsoventral axis of amphibian eggs has emerged from recent cell lineage, transplantation and cytological observations.

Origin of the Dorsoventral Axis

The fate map of the pregastrular amphibian embryo shows that the cells located in the dorsal marginal zone ultimately give rise to mesodermal structures of the dorsal axis. It is some of these cells which in the Spemann experiments display the capacity to induce dorsal structures. The problem we now address is whether the initial determination of these cells as mesoderm progenitors with dorsal inductive capabilities in fact occurs as a self-differentiation, such as might follow from their inheritance of cytoplasmic factors localized on the dorsal side. Explantation experiments carried out in *Triturus* embryos (Nakamura and Takasaki, 1970) and in *Xenopus* embryos (Nakamura et al., 1970b) suggest that the dorsal marginal cells of the animal blastomere cap have become fully determined by the early blastula stage. Thus, if these cells are excised after this stage and cultured, they produce various mesodermal structures, including myotomes, notochord, mesothelium, etc., but if isolated earlier they give rise only to ciliated epithelium. From these and similar studies (Koebke, 1977; reviewed in Gimlich and Gerhart, 1984), it can be concluded that in *Xenopus* determination of dorsal mesoderm precursors takes place between the 64-cell and the 512-cell stage. In *Rana* and *Ambystoma* as well, *the capacity to induce a second axis*, i.e., behave as a Spemann organizer, appears to have been fixed in dorsal marginal cells in the course of the blastula stage (Malacinski et al., 1980). However, as these tests challenge the state of determination of the dorsal organizer precursor cells, they do not reveal directly the mode of their initial specification.

It is now evident that both the self-differentiation of early dorsal equatorial blastomeres, and the inductive differentiation of cells in this region by vegetal pole blastomeres on the dorsal side lying *below* the original gray crescent are involved. Inductive determination of mesoderm in animal cap cells by vegetal region cells was demonstrated in the urodele *Ambystoma* by

Nieuwkoop (1969), and in *Xenopus* by Sudarwati and Nieuwkoop (1971), in experiments in which the normal axial mesodermal precursors were excised at the blastula stage, and the remaining animal hemisphere ectoderm, when combined with the vegetal hemisphere, was shown to be able to recreate the axial mesoderm anew. The dorsoventral orientation of the induced mesoderm is controlled by that of the vegetal hemisphere in these recombinants, and the induced dorsal mesoderm is then capable of behaving as a Spemann organizer (reviewed in Nieuwkoop, 1977). Gimlich and Gerhart (1984) demonstrated explicitly that the ability to induce dorsal mesoderm is originally resident in a few *vegetal* blastomeres at the 64-cell stage. In their experiments the original dorsoventral axial orientation of the egg was destroyed by UV irradiation of the vegetal hemisphere, and the irradiated embryos were shown to be rescued by transplantation of one to three blastomeres from the quadrant underlying the prospective dorsal marginal region. Injection of fluorescent label demonstrates that the transplanted vegetal cells are themselves ancestral to no mesodermal derivatives, and their progeny are later found incorporated only in the gut. On the other hand, similar experiments, in which two dorsal equatorial blastomeres drawn from the third cell tier of a normal 32-cell embryo are transplanted into equivalent positions in UV-irradiated hosts, show that these blastomeres possess the capacity to rescue dorsal axis formation by *self-differentiation* (Gimlich, 1986). Thus injection of fluorescent lineage tracers demonstrates that the progeny of these transplanted cells contribute directly to the notochordal and other mesodermal structures of the organizer region. In the same way, the evidence reviewed in Chapter IV shows that there are cells inheriting vegetal subequatorial cytoplasm on the dorsal side that have the capacity to self-differentiate in culture as somitic muscle, while the animal cap blastomeres that in normal embryos also give rise to mesoderm are specified inductively. As summarized in Fig. 6.31, the inductive activity of the vegetal blastomeres is shown by the experiments of Gimlich and Gerhart (1984) to be sufficient to account for the specification of the future organizer region. Nonetheless, the evidence reviewed clearly implies that there exist maternal cytoplasmic factors that specify the cells inheriting them to give rise to dorsal mesoderm, including that of the Spemann organizer, and that these factors are localized in the dorsal equatorial blastomeres of the early cleavage state embryo. The inductive mechanisms also operating no doubt assist in imposing the appropriate spatial patterns of dorsal specification (cf. Chapter IV). We have seen that the vegetal pole cytoplasm is also the location of determinants for the germ cells, at least in anuran amphibians. Though as described in Chapter IV, the animal hemisphere blastomeres display innate tendencies to differentiate as epidermis, they seem in general to remain plastic and undetermined into the early blastula stage, pending their reception of appropriate signals from self-differentiating blastomeres below. In the sea urchin embryo as well the early vegetal pole blastomeres appear to be more highly determined *ab initio*, and the fates of the animal cap blastomeres to be inductively determined, partic-

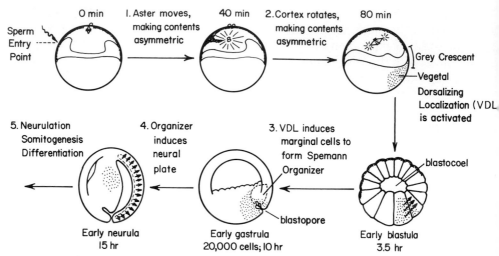

Fig. 6.31. Schematic diagram of a possible sequence of events in inductive dorsal-ventral axis formation in anurans. As discussed in text, self-differentiation is also involved in formation of the dorsal mesoderm (Gimlich, 1986). At upper left the radially symmetric unfertilized egg is shown at the time of fertilization, with the sperm entering in the animal hemisphere randomly at one point. The internal contours represent the three layers of cytoplasm visible: an upper layer containing the germinal vesicle contents; a middle layer of small yolk platelets embedded in cytoplasm; and a lower layer of large, densely-packed yolk platelets (Klag and Ubbels, 1975). In the period until 0.45 (45% of the time interval to first cleavage), the sperm microtubule organizing center forms a large aster and moves to the center of the egg, coordinating the migration of the sperm and egg pronuclei, which complete DNA synthesis and meet at approximately 0.45. Aster formation and movement from one side of the egg may cause a slight regional difference in the animal hemisphere cortex. In the period 0.45-0.75, the cortex shifts relative to the deep contents of the egg, by a rotation toward the site of sperm entry, as revealed by the emergence of the gray crescent. Deep contents adhering to the cortex are drawn up on the gray crescent side. This reorganization could result in the activation of maternal dorsal determinants in the vegetal hemisphere, producing the "vegetal dorsalizing localization," (VDL). This region is inherited by vegetal cells arising in the next several cleavages. By the 512-cell stage the vegetal cells have induced adjacent animal hemisphere cells to become prospective chordal mesoderm, including the prospective Spemann organizer. During gastrulation, the prospective dorsal mesoderm migrates up the blastocoel wall and in turn induces the neural plate, which in subsequent steps will form definitive dorsal organs of the nervous system. [From J. Gerhart, S. Black, and S. Scharf (1983). *In* "Modern Cell Biology. Spatial Organization of Eukaryotic Cells" (R. McIntosh, ed.), Vol. 2, p. 483. Alan R. Liss, New York.]

ularly with respect to the second axis. In considering the mechanism underlying this mode of development it is important to note that in the amphibian embryo the initial inductive interactions required for the determination of the dorsal marginal cells of the animal hemisphere takes place *in the complete absence of genomic transcription.* As reviewed earlier (see Chapter III), transcription does not resume in amphibian eggs until the midblastula stage.

The Mechanism of Localization

We now return to the initial process by which the dorsal determinants are localized on the grey crescent side of the egg. As indicated in Fig. 6.31 the primary dorsoventral asymmetry is caused by the movement of the sperm nucleus, and a displacement of internal cytoplasmic constituents to one side that may be due to formation of the large sperm aster (Ubbels *et al.*, 1983). This is followed, in normal development, by a strong cortical contraction toward the future ventral side, occurring in the plane of the future dorsoventral axis of symmetry, that results in the rotation of the cortex with respect to the deeper cytoplasmic contents. Cortical microtubules and microfilaments are evidently involved in these cytoplasmic movements (see, e.g., Elinson, 1983) though the contractile process itself is not sensitive to injected cytochalasin (Merriam and Sauterer, 1983). Direct evidence for the involvement of myosin in this contraction has been obtained by Christensen *et al.* (1984). The consequence of the rotational contraction is the appearance of the grey crescent at the future dorsal side. This is a zone of cortical cytoplasm that has been depleted of pigment granules (Klag and Ubbels, 1975; Gerhart *et al.*, 1983). As in the ascidian egg (cf. Fig. 6.27) the pigment granules are engaged in a network of cytoskeletal actin microfilaments (Franke *et al.*, 1976; Merriam and Sauterer, 1983). The same contractile movements result in elevation of vegetal region yolk along the future ventral side, and in an asymmetric distribution of poly(A) RNA, which appears to be related to the redistribution of the yolk platelet mass (Phillips, 1985). However, there is no evidence that the internal redistribution of yolk is of functional significance (Neff *et al.*, 1984).

The mechanism by which the dorsal determinants are distributed is sensitive to application of pressure, to transient exposure to cold (1°C for 4 min) (Scharf and Gerhart, 1983), and to UV irradiation. These affects of UV irradiation were investigated extensively by Grant and Wacaster (1972), Malacinski *et al.* (1974, 1977), and Chung and Malacinski (1975), following earlier studies (see, e.g., Baldwin, 1915; Grant, 1969). Malacinski *et al.* (1974) and Chung and Malacinski (1975) found the UV-sensitive neuralizing factors localized mainly in the future dorsal region of the vegetal egg cortex in both urodele and anuran eggs. UV irradiation has a purely cytoplasmic effect, since nuclei from irradiated eggs are able to support development when injected into enucleated eggs as well as do control nuclei, and the irradiation penetrates only 5-10µm or less into the egg cortex (Grant and Wacaster, 1972). However, ultimately UV irradiation clearly affects subsequent differential gene function in the embryonic cells, and thus alters the distribution of protein products that are confined to dorsal or ventral tissues (Smith and Knowland, 1984). Cold, hydrostatic pressure, and UV irradiation all prevent axial determination if applied to the zygote between about 0.4 and 0.8 of the interval between fertilization and first cleavage (cf. Fig. 6.31).

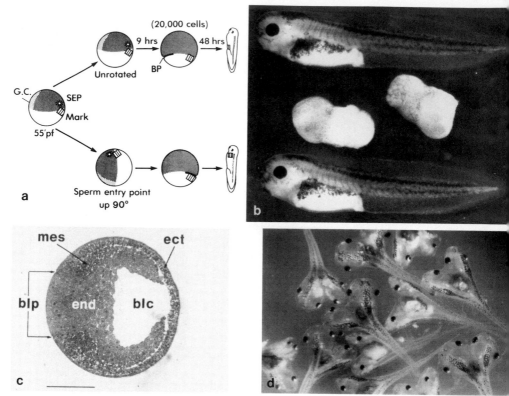

Fig. 6.32. Relocation, impairment, and rescue of axial determination in *Xenopus* eggs. (a) Schematic diagram of 90° rotation experiments in *Xenopus*. Fertilized eggs are dejellied and marked below the sperm entrance point (SEP). The sperm entrance mark is denoted by a small circle with rays, the dye mark by a square striped patch, the animal hemisphere by stippling, and the grey crescent by lighter stippling. In the control series (top line) the unrotated egg initially forms its blastopore lip, 180° opposite the mark, which denotes the site of sperm entrance. In eggs rotated sperm entrance side up (grey crescent down) for the same period, the opposite result is obtained: the blastopore forms on the same side as the mark (sperm entrance side). [From M. Kirschner, J. C. Gerhart, K. Hara, and G. A. Ubbels (1980). *In* "The Cell Surface: Mediator of Developmental Processes" (S. Subtelny and N. K. Wessells, eds.), p. 187. Academic Press, New York.] (b) Rescue of UV-irradiated eggs by gravity-driven reorganization of the egg contents. Eggs were irradiated heavily with UV on the vegetal hemisphere at 0.4 of the interval between fertilization and first cleavage, to block dorsal development, as shown by the ventral embryos in the center of the figure. A tadpole from an unirradiated control egg is shown in the upper portion of the figure. Irradiated eggs were turned 90° off axis for at least 30 min, and then returned to the normal orientation and allowed to develop. An apparently normal tadpole developed, as shown below. Equivalent rescue is obtained for cold or pressure-treated eggs by rotation. [From S. R. Scharf and J. C. Gerhart (1980). *Dev. Biol.* **79**, 181.] (c) Morphology of totally axis-deficient embryo induced by vegetal UV irradiation. A median section of the radially symmetric "gastrula" is shown: blp, blastopore; mes, mesenchyme cells; ect, ectoderm; end, endoderm; blc, blastocoel. Bar length, 0.5 mm. [From S. R. Scharf and J. C. Gerhart (1983). *Dev. Biol.* **99**, 75.] (d) Double axial embryos produced by centrifugation of eggs. Fertilized, dejellied eggs were embedded in gelatin and centrifuged at 30 × g for 4 min with centrifugal

Thus the subsequent differentiation of axial dorsal structures is suppressed, to a variable extent, by these treatments, and when most effective they result in embryos that consist only of an ectodermal tube containing endoderm and some mesenchyme cells (Scharf and Gerhart, 1983). An example is shown in Fig. 6.32(c). All three treatments probably affect the cytoskeleton of the egg, and interfere with the redisposition of cytoplasmic constituents occurring during the period of sensitivity that precedes first cleavage. Microtubules are stabilized by D_2O, and Scharf and Gerhart (1983) found that if immersed in D_2O, cold treated eggs display much better axial development. The effect of UV irradiation on the cytoskeleton is not known. However, to some extent D_2O also rescues axial formation in irradiated eggs. Unlike some examples considered earlier, the action spectrum for the UV inhibition of axial differentiaton in amphibian eggs peaks at 280 nm, indicating a protein rather than a nucleic acid target (Youn and Malacinski, 1980). An additional argument that the UV effect could be due to damage to processes mediated by microtubules is that UV irradiation also blocks grey crescent formation as well (Manes and Elinson, 1980), and this is known to be sensitive to treatment with colchicine as well (Manes *et al.*, 1978).

The localization pattern established in the sensitive interval before first cleavage is remarkably labile. It has been known for a century that the dorsoventral axis can be shifted merely by rotating the egg and holding it in a turned position after grey crescent formation (Born, 1885). Schultze (1894) showed that double axes form in anuran eggs that are inverted just before and during first cleavage. This observation was extended to the 2-cell stage by Penners and Schleip (1928a,b), and was analyzed as a consequence of the gravitationally driven movements of the heavy internal yolk mass with reference to fixed cortical components by Pasteels (1938, 1939; see review in Brachet, 1977). Ancel and Vintemberger (1948) showed that in parthenogenically activated *Rana* eggs, which lack the aster normally induced by the sperm, the location of the grey crescent and the future dorsal axis can be perfectly controlled by a brief rotation. This suggests that gravitational rearrangment of the inner cytoplasmic constituents may *substitute* for the cytoskeletally mediated rearrangements that occur in normally fertilized eggs. To examine this inference, Gerhart *et al.* (1981) developed a procedure by which *Xenopus* eggs could be tipped at any given angle off vertical, and

force directed 90° to the animal-vegetal axis. The dish of eggs was then removed from the rotor and placed flat in the incubator so that gravity would move the egg contents back to their original positions. Maximum twinning was found for eggs centrifuged at 0.4 in the interval from fertilization to first cleavage. 100% twins were obtained. Note that the double axes join in some cases at the posterior trunk or tail level. All twins share ventral structures. [From J. Gerhart, S. Black, and S. Scharf (1983). *In* "Modern Cell Biology. Spatial Organization of Eukaryotic Cells" (R. McIntosh, ed.), Vol. 2, p. 483. Alan R. Liss, New York.]

maintained in the altered position as long as desired. This was accomplished by immersing the eggs in Ficoll, thus dehydrating the perivitelline space and destroying its natural function as a lubricating bearing within which the egg freely rotates. Regardless of the original point of sperm entry or the position of the grey crescent, a 90° rotation suffices to determine a new dorsoventral axis. Thus as diagrammed in Fig. 6.32(a) the blastopore can be made to form at, rather than across from, the point of sperm entry. The eggs are most sensitive to gravitational rearrangement before grey crescent formation, but by application of low centrifugal force, the dorsoventral axis can also be shifted later in the precleavage interval. Centrifugation at two successive times in different orientations, or centrifugation followed by gravitational reorganization, induces the formation of doubly axiated larvae, such as are illustrated in Fig. 6.32(d) (Gerhart *et al.*, 1983). This result is consistent with the view mentioned above, that potential dorsal determinants are initially distributed radially around the egg, and that in normal development they become active on the future dorsal side following contact with, or additions from, the relocated deep cytoplasm. The major translocation occurring in normal eggs is a rotation in the plane of the future dorsoventral axis, that results in a 30° displacement of the subcortical cytoplasm relative to the egg surface (Vincent *et al.*, 1986).

The destructive effects of cold or pressure treatment, and of UV irradiation can be overcome merely by tipping the egg (Scharf and Gerhart, 1980, 1983). Thus, as shown in Fig. 6.32(b) eggs irradiated so as to completely inhibit dorsal development are rescued by being held for 30 min 90° off axis. It follows that in tipped eggs the gravitationally driven reorganization substitutes for the cytoskeletally driven reorganization that occurs in unirradiated fertilized eggs. In both cases a necessary interaction along the future dorsoventral axis must occur between the deep cytoplasm and the cortical components. There is no evidence that such interactions are involved in localization of germ cell determinants at the vegetal pole of the anuran egg, and rotation does not rescue primordial germ cell formation from the effects of UV irradiation as it does axial determination (Thomas *et al.*, 1983). Nor are germ cell determinants redistributed in inverted eggs (Neff *et al.*, 1984).

As for the eggs of several other species that we have considered, there is evidence that amphibian eggs also contain localized maternal morphogens that appear to promote certain pathways of self-differentiation in the blastomeres that inherit them. Thus, for example, blastomeres of the subequatorial zone tend to produce differentiated muscle cells; blastomeres of the vegetal pole give rise to endodermal cell types; the progeny of cultured blastomeres of the animal pole region differentiate as epidermis (see Chapter IV for review). The studies considered here show that dorsoventral axial determination depends on *structural rearrangements*, mediated in normally developing eggs by cytoskeletal cortical movements. There is no evidence that the cortical rearrangements which specify axial orientation are to be equated with the direct localization of specific maternal morphogens. Thus a series of centrifugation experiments of Black and Gerhart (1984) show that exertion of gravitational force at various angles with respect to the animal-vegetal axis may provoke the appearance of an embryonic dorsal pole at unexpected locations, the immediate cytoskeletal structure of which should not have been perturbed asymmetrically in the treated eggs. For example, eggs centrifuged down along the animal-vegetal axis produce dorsal poles in equatorial regions located randomly with respect to the sperm entry point. Rather than directly redisposing a specific maternal "dorsal morphogen," treatments (and normal processes) that affect the cortical cytoskeleton of the egg should perhaps be thought of as altering a spatially oriented *matrix,* in which such morphogens may at some later time regionally associate. Ultimately there indeed is localized in the vegetal blastomeres of the 32-cell embryo the capacity to induce the overlying cells to give rise to the "organizer" tissue on which later axial development depends (Gimlich and Gerhart, 1984).

(iii) Cell Lineage Determination in Mammalian Eggs

Mammalian embryos differ from all others considered in this review in both their slow absolute rate of early development and in the relative temporal order in which various developmental events occur. Thus first cleavage in the mouse embryo occurs only toward the end of the second day after ovulation; there are only 10-30 cells on day 3; and 100 cells are not attained until day 4 (Ellem and Gwatkin, 1968; Olds *et al.,* 1973). The rabbit egg, which is about 8 times the volume of the mouse egg, cleaves somewhat more rapidly, and at two days it contains 16 cells; at 3 days 128 cells; and at 4 days over 1000 cells (Daniel, 1964). A major distinction of mammalian embryos is the early developmental stage at which they switch from the use of maternal mRNA to mRNA synthesized in the embryo nuclei. As reviewed in Chapter II, in the mouse embryo, which has been studied the most intensively, maternal mRNA has largely disappeared as early as the two-cell stage. The synthesis of new proteins from embryo transcripts becomes a prominent feature at this time. Furthermore, the early mammalian embryo develops from the beginning in a nutrient environment, so that relative to other embryos, net growth, including rRNA synthesis, starts very early (see Chapter III). The storage of relatively huge amounts of maternal cytoplasmic constituents such as ribosomes and yolk is for the same reason unnecessary.

Most of the differentiated cell lineages giving rise to the tissues and organs of the fetus appear after implantation of the blastocyst in the uterine wall, which in the mouse occurs at day 6. The first overt differentiation of cell types in the embryo can be distinguished at blastocyst formation late in the

5th cleavage stage, with the separation of trophectoderm cells from the inner cell mass (ICM). The trophectoderm is a functionally differentiated cell lineage, as discussed in detail below, which after implantation gives rise only to extraembryonic membranes. The ICM includes the progenitors of all cell lineages participating in the construction of the embryo proper, as well as of some additional extraembryonic structures. The first specific cell layer to derive from the ICM appears with the delamination of the primitive endoderm. This takes place late on the fourth day of development in the mouse (Nadijcka and Hillman, 1974). The slow pace of determination within the ICM has been demonstrated by transplanting genetically marked primitive endoderm and other ICM cells into dissimilar host blastocysts. Gardner (1982, 1984) demonstrated by this means that primitive endoderm cells are already definitively committed since single cells of this layer from 5 day embryos participate only in the development of endodermal tissues in the postimplantation fetus. However, they give rise both to the visceral yolk sac endoderm, and to the parietal endoderm sheath. Even in postimplantation (7 day) embryos the endoderm cells remain plastic with respect to their ability to be included in either parietal or visceral structures. This decision apparently depends on contacts with other cell layers (Gardner, 1982, 1984). Individual cells other than those of the primitive endoderm of the 5th day ICM (usually described by the misnomer "primitive ectoderm") may give rise to all other types of fetal tissues, including the definitive embryonic gut, and mesodermal and neural structures (Gardner and Rossant, 1979; Hogan and Tilly, 1977; Beddington, 1983; Lawson *et al.*, 1986). We have already seen that germ cell determination in the mouse embryo also occurs in the "primitive ectoderm," and not until after the 7th day of development. Probably many of the primitive ectoderm cells have the capacity to give rise to germ cells (Gardner *et al.*, 1985). It is most unlikely that the late determination of cell lineages within the ICM is in any way influenced by localized maternal cytoplasmic components, though it is difficult to exclude this categorically, in view of the extreme plasticity observed in chimeric recombinations of mammalian blastomeres (see below). Determination in the postimplantation mammalian embryo instead is probably mediated by intracellular interactions, but little is yet known of the detailed mechanisms (see review in Beddington, 1983).

Discussion of the role of cytoplasmic localization in mammalian embryogenesis has thus focused on the separation of trophectoderm from ICM cells at the early blastocyst stage, a process which has been studied in unusually great detail. The caveat should be kept in mind, however, that the separation of a determined layer of *extraembryonic* cells from the totally uncommitted embryonic progenitor cells of the ICM is not necessarily a phenomenon that is analogous to the determination of muscle or germ line or skeleton-forming cell lineages *within* the embryo.

Blastomere Totipotency

A fundamental observation is that the cells of the early mammalian embryo appear completely totipotent when tested by culture of isolated blastomeres or of partial embryos, and in chimeric blastomere recombination experiments. Individual blastomeres from 2-, 4-, and 8-cell rodent and rabbit embryos can develop into complete blastocysts (Nicholas and Hall, 1942; Seidel, 1960; Tarkowski, 1959a,b; Mulnard, 1965; Tarkowski and Wróblewska, 1967; Kelly, 1975, 1977). Blastomeres from 8-cell mouse embryos fused with late morulae are able to participate in normal development (Stern and Wilson, 1972), and fused embryos up to the late morula stage also develop normally (Tarkowski, 1961; 1963; Mintz, 1962, 1965; Gardner, 1968). Furthermore, although cells from the inside of the morula normally give rise to ICM and those from the outside normally give rise to the trophectoderm, inside cells isolated from 32-cell morulae or from the ICM's of blastocysts remain totipotent (Handyside, 1978; Hogan and Tilly, 1978a,b; Spindle, 1978; Rossant and Lis, 1979). Not until the late blastocyst stage do the ICM cells lose the ability to regenerate trophectoderm. The outside cells of 16-cell morulae also retain the capacity to contribute to the ICM (Rossant and Vijh, 1980). Along the same lines Ziomek *et al.* (1982a) showed that experimentally produced aggregates of pure inside cells, pure outside cells, or of mixed blastomere populations from 16-cell morulae will all give rise to blastocysts that implant and develop normally. An example of an aggregate of outside 16-cell stage blastomeres is shown in Fig. 6.33(a), and a blastocyst and a fetus deriving from such aggregates are shown in Figs. 6.33(b) and (c). In normal embryos cells ancestral to both trophectoderm and ICM are present as late as the 32-cell stage (Balakier and Pedersen, 1982). Marked clones of cells descended from single outer blastomeres that had been injected with horseradish peroxidase at the 16-cell morula stage are illustrated in Fig. 6.33(d). These clones can be seen to include progeny in both trophectoderm and ICM. After the 32-cell stage in normal embryos there is no further allocation of new cells to the ICM, and this process seems mainly to be complete by the 16-cell stage (Smith and McLaren, 1977; Copp, 1978; Graham and Lehtonen, 1979; reviewed in Balakier and Pedersen, 1982; Ziomek *et al.*, 1982b; Johnson *et al.*, 1984; Cruz and Pedersen, 1985). Soon after its formation the mouse blastocyst contains about 14 prospective ICM cells, and there are about 20 prospective trophectoderm cells (Johnson *et al.*, 1984).

The totipotency of the blastomeres of rodent embryos at least through cleavage and morula stages implies a large reliance on inductive intercellular processes during normal early development. The plasticity manifested by these cells under the artificial conditions of the experiments reviewed also indicates the difficulty of determining whether cytoplasmic localization

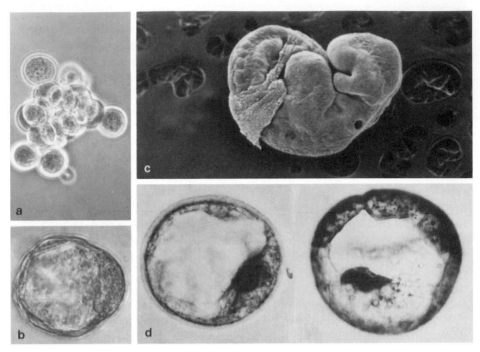

Fig. 6.33. Totipotency of mouse embryo blastomeres at the 16-cell stage. (a) Aggregate of 16 "outer" blastomeres from 16-cell morulae that had been disaggregated in a Ca^{2+}-free medium, and induced to adhere by addition of phytohemaglutinin. (b) Blastocyst formed from aggregate such as in (a), after 29 hr of culture. (c) Postimplantation fetus formed from a similar aggregate. [(a)-(c) From C. A. Ziomek, M. H. Johnson, and A. H. Handyside (1982a). *J. Exp. Zool.* **221**, 345.] (d) Two blastocysts cultured for 20-24 hr after injection of horseradish peroxidase into single outer cells at the 16-cell stage. The peroxidase activity is visualized as a brown reaction product formed from diaminobenzidine in the presence of H_2O_2, to which the fixed preparation had been exposed. In both embryos the stain is visible both in the ICM, particularly in the region from which the endoderm will delaminate, and in the polar trophectoderm. [(d) From H. Balakier and R. A. Pedersen (1982). *Dev. Biol.* **90**, 352.]

might be involved in the differentiation of the trophectoderm in undisturbed embryos. Thus even if particular spatial relations predicting cell fate were to exist in the normal embryo, these assignments might be very easily reversed, and hence not easily detected.

Origin of the Trophectoderm Cell Lineage in Normal and Experimentally Perturbed Embryos

By the 6th cleavage there are about 45 trophectoderm cells, and these can be functionally distinguished from the ICM cells by a number of cytological, physiological, and molecular criteria (reviewed in Gardner and Papaioannou, 1975). The trophectoderm cells are linked to one another by characteristic junctional complexes, and they are distinguished in ultrastructure from ICM cells by a number of both surface and integral cytoplasmic features (Nadijcka and Hillman, 1974; Johnson and Ziomek, 1982). They perform the special function of pumping fluid into the interior of the blastocoel. On presentation to the uterine wall they elicit an implantation response, and in their absence implantation cannot occur (Rossant and Lis, 1979). Ultimately they give rise to the extraembryonic ectoplacental cone, to a population of endopolyploid giant cells that divide no further, and probably to chorionic ectoderm (Johnson and Rossant, 1981). Two-dimensional analyses of the proteins synthesized by the trophectoderm cells of preimplantation embryos have revealed a number of species not included in the set of proteins synthesized by ICM cells at the same stages (Van Blerkom *et al.*, 1976; Handyside and Johnson, 1978; Dewey *et al.*, 1978; Brûlet *et al.*, 1980). After implantation some of these proteins continue to be synthesized in the extraembryonic layers descended from the trophectoderm (Johnson and Rossant, 1981). A monoclonal antibody against intermediate filament proteins of trophectoderm cells has been described (Brûlet *et al.*, 1980), and a cDNA clone representing the message for this trophoblast-specific protein has also been recovered (Brûlet and Jacob, 1982).

Since cells on the outside of the morula usually give rise to the trophectoderm, it was proposed that *outside position* is the causal factor in the determination of this cell lineage, and conversely, that *inside position* leads to the alternative developmental pathway, participation in the ICM (Mintz, 1965; Tarkowski and Wróblewska, 1967). This idea was supported by experiments of Hillman *et al.* (1972), in which the fate of marked cells inserted in chimeric blastomere combinations was shown to depend in the expected way on their positions. Though as we have seen each blastomere of the 4-cell embryo is totipotent, Hillman *et al.* (1972) showed that if placed on the outside of a 4- to 16-cell embryo the progeny of the test blastomere is often (though not invariably) found only in the trophectoderm, and after implantation, in the yolk sac. If surrounded by other blastomeres, on the other hand, its progeny is usually recovered in the ICM and in the postimplantation fetus. Blastomeres from the inner regions of the 16-cell embryo can also be steered toward either trophectoderm or ICM differentiation by being placed on the outside or inside, respectively, of arrays of 15 other blastomeres of the same age (Ziomek and Johnson, 1982). In these last experiments the test blastomere and its clonal progeny were labeled with a fluorescent stain. However, Ziomek and Johnson observed that by this stage test outer blastomeres display a predisposition to give rise to trophectoderm, though not an irreversible one. Ultimately the ICM is covered by a trophectoderm layer, and this may be the event that represses further delamination of trophectoblasts, by permanently eliminating an "outside" environment (Fleming *et al.*, 1984). Several other kinds of experiments contribute to the impression that external or internal position exerts a powerful influence. Thus for example alkaline phosphatase, which is an inner cell mass marker, is reported to appear in chimeric aggregates only when these include blastomeres totally surrounded by other blastomeres (Izquierdo and Ortiz, 1975). Stern (1973) flattened mouse embryos under a glass plate and showed that in these embryos trophoblast cells appear even in internal regions where the blastomeres abut the glass interface rather than other cells.

These observations may be interpreted in two distinct ways. The external cells could become committed to the trophectoderm pathway as a direct result of their position on the outside of the embryo, as proposed earlier, or alternatively, in undisturbed embryos their natural process of commitment could be mediated by their *inheritance* of cytoplasm localized radially at the periphery of the embryo. The latter interpretation is allowed because observations on intact embryos have shown that in the course of cleavage, morula and blastocyst formation, relatively little radial or migratory displacement of cells occurs. Thus cells marked with fluorescent or enzymatic stains, or by injection of oil droplets, give rise to progeny that remain clonally clustered (Wilson *et al.*, 1972; Garner and McLaren, 1974; Balakier and Pedersen, 1982; Ziomek and Johnson, 1982; reviewed in Johnson and Pratt, 1983). A detailed study of the cell lineage of the cleavage stage mouse embryo and of the geometry of early intercellular contacts has been carried out by Graham and his colleagues (Kelly *et al.*, 1978; Graham and Deussen, 1978; reviewed in Graham and Lehtonen, 1979). At third cleavage one of each pair of daughter cells normally lies deep within the embryo while the other occupies a peripheral position. By injecting oil drops into inside and outside blastomeres Graham and Deussen (1978) showed that as expected the inner cells of these pairs contribute preferentially to the ICM, while oil drops injected into the outside cells are normally recovered in trophectoderm cells. The important point is that this radial correlation with blastomere fate reflects an earlier intracellular cytoplasmic distribution. Thus oil droplets injected in the peripheral cytoplasm of 2-cell stage blastomeres also tend to be distributed to later trophectoderm cells, while those injected centrally ultimately appear in both trophectoderm and ICM (Wilson *et al.*, 1972). Thus in normal development there appears to be a relatively fixed radial relation between the peripheral cytoplasm at early cleavage, the peripheral cell layer of the 8-cell embryo, and the outer trophectoderm progenitors of the 16- and 32-cell morula. There is unfortunately insufficient evidence to indicate whether relevant regional localization preexists in the uncleaved mammalian egg or the fertilized ovum.

A closer focus on the events occurring at the 8-cell stage reveals a radial intracellular *polarization* of the blastomere cytology, which is mediated by intercellular contact. At 4th cleavage the peripheral cytoplasmic regions are physically segregated to the outer cells of the 16-cell morula. The early 8-cell stage blastomere is spherical, and both surface and internal features are arranged in a homogeneous way. By the end of the 8-cell stage a major cytoplasmic reorganization has occurred (reviewed in Johnson and Pratt, 1983; Johnson *et al.*, 1984). The nuclei move toward the basal, inner surface of the cell, where contact with an adjoining cell is established. At the apical or external poles endocytotic vesicles accumulate. These have been displayed in cytological preparations by enzymatic staining of ingested horseradish peroxidase (Reeve, 1981). Cytoplasmic actin also accumulates at the polar regions, as visualized with fluorescent antibodies (Johnson and Maro, 1984). The most extensively studied changes occur near the surfaces. At the basal side gap junctions and tight junctions form; arrays of microfilaments and microtubules are erected parallel to the lateral walls of the blastomere; and the apical surfaces are distinguished by localized arrays of short microvilli. These can be easily detected at light microscope magnifications by the density of binding sites for various fluorescent ligands that they provide. The overall consequence of this polarization is a change in form of the embryo from a loose aggregate of spherical cells to a tightly compacted, radially organized embryo, in which intercell contacts have been maximized within, while the outer surface has become covered with microvilli. When 16-cell morulae are disaggregated into couplets of sister blastomeres and examined, the *outer cells* display the same polar cytological features as do the *outer regions* of the 8-cell blastomeres. The inner cells, which are derived from the basal and lateral portions of the 8-cell blastomeres, remain nonpolar (Johnson and Ziomek, 1981a; Johnson and Pratt, 1983). Outer cell features, such as the external microvillar surface, are also retained by the outside cells of the 32-cell morula and the trophectoderm cells of the early blastocyst (Johnson and Ziomek, 1982). The fate of the progeny of individual polar cells at the 8-cell stage depends on the plane of cleavage. If this bisects the pole, two new polar cells are formed, both of which become trophectoderm, while if it is normal to the polar axis a trophectoderm and an ICM progenitor cell are formed (Johnson and Pratt, 1983).

These observations imply directly that the peripheral cytoplasm of the polarized 8-cell stage blastomere acts as a determinant for trophectoderm differentiation, while the cells inheriting the internal cytoplasm are destined to become ICM cells, i.e., unless they are placed in circumstances where they are induced to polarize anew. The early determination of trophectoderm and ICM precursors at once affects their biosynthetic activities. Thus Handyside and Johnson (1978) showed that marker polypeptides synthesized only in the trophectoderm cells of the definitive blastocyst are not synthesized in inner cells of the 25-30 cell morula, while polypeptides characteristic of the later ICM are.

Polarization is an inductive process which requires intercellular contact. Furthermore, continued cell contact is needed to maintain the states of polarization initially imposed. When cultured *in vitro*, polar 16-cell blastomeres display trophectoderm-like behavior, and they wrap around nonpolar cells, while apolar cells cultured together tend to polarize, and to produce

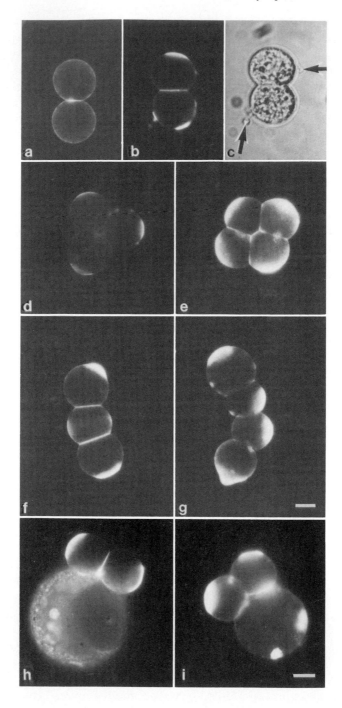

aggregates including both external trophectoderm-like cells, and inner, ICM-like cells (Johnson and Ziomek, 1983). Polarization occurs *in vitro* when a blastomere from an 8-cell embryo that has not yet undergone compaction is brought into asymmetric contact with other blastomeres (Ziomek and Johnson, 1980; reviewed in Johnson and Pratt, 1983; Johnson *et al.*, 1984). Polarization is induced in the 8-cell stage blastomere by contact with the surface of a 2-cell, 4-cell, 8-cell or 16-cell stage blastomere but not with an uncleaved zygote or an unfertilized egg (Johnson and Ziomek, 1981b). As we have seen, the 2-cell stage, when this inducing ability first appears, is also the point at which the biosynthetic processes of the embryo become dependent on its own transcription. Induction of polarity does not require the formation of intercellular junctional complexes, since in normal embryos it precedes the formation of these structures; since 8-cell blastomeres polarize when placed in contact with 2-cell blastomeres, but form no junctional complexes; and since ionic media, antisera, and drugs that block junction formation do not prevent polarization (Pratt *et al.*, 1982; Johnson *et al.*, 1984). Perhaps the most convincing demonstration that contacts with competent inducing cells account for polarization is that the axis of polarity can be controlled by the geometry of the contacts. The apical pole always forms at a point opposite the contact, and in the case of multiple contacts, at a point equidistant from them (Johnson and Ziomek, 1981b). Examples are shown in Fig. 6.34. Within the intact 8-cell embryo, this property would naturally result in a radially organized embryo in which the apical faces of the blastomeres are oriented outwards, since the greatest number of intercellular contacts is in the inside (Graham and Lehtonen, 1979; Ziomek and Johnson, 1980).

   The ability of the early cells of the rodent embryo to undergo polarization as a result of cell contact lies at the root of their totipotent behavior in

---

Fig. 6.34. Induction of polarity by intercellular contact in 8-cell stage mouse embryo blastocysts. Embryos were disaggregated and individual blastomeres or combinations of blastomeres were treated with phytohemaglutinin to promote adherence and cultured. After various periods of culture they were stained with fluorescein-labeled concanavalin A, a ligand which binds largely at the apical microvillar patch forming on polarization, and photographed in the fluorescence microscope. (a) Pair of 8-cell stage blastomeres 2 hr after reaggregation. No polarization is visible. (b) Similar pair after 7 hr, displaying axial polarization. (c) Same pair seen in a phase micrograph, in which midbodies (arrows) can be observed. These are remnants of the previous cytokinesis, and from their positions it is clear that the polarization seen in (b) is a result of the *new* plane of contact. (d) Triplet. (e) Quartet. (f) Linear triplet; note equatorial fluorescence in center cell. (g) Linear quartet. The position of the pole appears to respect all contacts in these aggregates. (h) Two sister blastomeres of an 8-cell stage embryo aggregated with an unfertilized egg and cultured for 9 hr. The original polarization is retained, and the new contact has no effect. (i) Two sister 8-cell stage blastomeres aggregated with a blastomere isolated from a 2-cell embryo. The position of the apical pole is determined by the inducing 2-cell blastomere. Bars, 10 μm. [From M. H. Johnson and C. A. Ziomek (1981b). Reproduced from *J. Cell Biol.* **91**, 303 by copyright permission of The Rockefeller University Press.]

isolated blastomere and chimeric recombination experiments (see discussions in Johnson and Pratt, 1983; Johnson *et al.*, 1984). When isolated, the blastomeres divide, and then induce polarization in each other, with the eventual segregation of cells that inherit the apical cytoplasm of the new external poles and give rise to trophectoderm precursors. In chimeric aggregates cells placed on the inside do not polarize, because they are symmetrically surrounded by contacts on all sides. However, on the outside they polarize anew, whatever their original orientation, and thus give rise to trophectoderm cells. Since outside cells tend to form extended contacts with underlying cells and to wrap around them, their dimensions expand in the plane normal to the axis of polarity. The plane of cytokinesis in these cells thus is usually radial, and two new polarized outside cells are generally formed at each division in the undisturbed morula or in complex aggregates. However, isolated outside cells may appear totipotent since the plane of cleavage now frequently bisects the axis of polarity, producing both new outside and a new, nonpolar inside cell. In short, the experiments that display the totipotency of mouse embryo blastomeres perturb these cells in two fundamental ways. They cause repolarization, and they result in reorientation of cleavage planes. Neither effect is inconsistent with the view that in *undisturbed embryos* the determination of the trophectoderm cell lineage begins with the segregation of radially localized cytoplasm into the outer cells, though their specification clearly remains reversible at least through the 32-cell stage (Balakier and Pedersen, 1982; Johnson and Ziomek, 1983).

Conclusions

Some aspects of the determination of the trophectoderm and ICM cell lineages recall processes in other organisms that we have considered in this Chapter. For example, the role of inductive intercellular contacts in early cell lineage determination is thoroughly documented in the processes by which the dorsoventral axis is determined in molluscan eggs, and inductive interaction among blastomeres is obviously important in amphibian and sea urchin embryos as well. Cytoplasmic localization, as we have seen, is usually, though certainly not always, an ontogenic process occurring after fertilization, rather than a preformation, and sometimes it proceeds rather gradually with respect to developmental stage. The invertebrate examples that could be most analogous to the mammalian case in this respect are the *Cerebratulus* egg, in which segregation of embryonic cell lineage determinants occurs during cleavage and is not accomplished until the 8-cell stage, and the ctenophore egg, in which the embryonic axes are oriented only as a result of cleavage. As we have seen, if there are radial localizations in mammalian eggs, they must be extremely labile when subjected to experimental interference, but there are other eggs, e.g., amphibian eggs, in which such lability is known. A process of inductive polarization analogous to that occurring in the mouse embryo may take place in the Hörstadius sea urchin blastomere chimeras in which, for example, missing micromeres are replaced by new mesenchymal progenitors. Many other such parallels are evident, and the possibility that trophectoderm specification is mediated by cytoplasmic determinants ultimately localized in outside cells of the morula would seem partially to bridge the apparent distinctions between mammalian and other embryos. Nonetheless, from the more general vantage point of this review it is the basic *differences* between mammalian embryos and the other embryos considered that seem most significant.

The early mammalian embryo remains spherically organized well into cleavage, in that there is no animal-vegetal axis of symmetry. There is as yet no evidence for developmental polarity in the zygote or ovum, and even were the trophectoderm cell lineage determined by peripheral cytoplasmic elements at the compaction and early morula stages, there is still no hint of localized determinants for any cell lineage *within the embryo proper* even at these stages. The first ICM cell type to become differentiated is the embryonic endoderm of the late blastocyst, which gives rise to extraembryonic tissues, and most embryonic cell types proper become determined only well after implantation. The most fundamental difference is that transcriptional control of embryonic biosynthesis occurs long in advance of even the earliest events in trophectoderm determination, in the mouse at the 2-cell stage, and that due to the extremely slow pace of morphological development maternal transcripts have already by then largely disappeared.

The very different mode of embryogenesis in mammals suggests a general interpretation. Localization, as it functions in the development of most animal organisms, is perhaps best viewed as an adaptive mechanism. It is the means by which cell lineage specification can be organized in advance of the point at which embryo genome action controls the events of development. In dipteran insects the initial specification of cell fate has taken place by the blastoderm stage, and in amphibian embryos prior to the midblastula transition, in the latter example in the virtual absence of embryo nuclear transcription. These eggs, and more generally the eggs of all animals that develop in the external environment or within isolating barriers, are endowed with sufficient logistic supplies, cytoplasmic capabilities, and biological instructions to provide for rapid development of a larva that is sufficiently complex to feed itself. At the outset this often requires a huge cytoplasm to nucleus ratio, which precludes an early influence of embryo transcripts on many of the general biosynthetic activities of the embryo, irrespective of whether transcription begins early in development. A developmental system in which there is no adverse selective consequence of slow early development, and in which embryonic nutrition is accomplished maternally, is not subject to these constraints. Thus inductive mechanisms of intercellular interaction, which probably dominate later morphogenesis in all organisms, are utilized almost exclusively in mammalian embryos right from the beginning. It follows that localization is not to be viewed as an invariant, special feature of development from eggs, which if one looks hard enough might be observed even in mammals. Instead it can be considered primarily as an *adaptive mechanism* invented early in the evolution of the metazoa, which has taken on a great many forms, and which, with the exception of warm-blooded viviparous animals, is always required by the unprotected conditions in which early development occurs. Viviparous development occurs in a range of creatures other than mammals, e.g., in pelagic tunicates, and in certain spiders, amphibians, ophiuroids, etc. These unusual examples may offer the opportunity of testing the prediction that relative to free living embryonic forms of the same taxa, developmental pathways will have altered in the direction of slower pace, less reliance on initial lineage assignments by means of cytoplasmic localization, and an earlier and more important role for induction.

For the developmental molecular biologist there is another and more immediate significance to the localization phenomenon. The ultimate result of localization is specification of particular patterns of gene activity in given cell lineages. Whether by direct or indirect mechanisms, indifferent cell nuclei are induced to express specific sets of transcription units by the cytoplasm to which they are distributed. The egg cytoplasm of many nonmammalian organisms thus appears to contain gene regulatory agents, or it can give rise to internal environments that specifically regulate nuclear genes. The cytoskeletal apparatus of the egg cortex in some cases evidently provides a two-dimensional distribution of these regulatory molecules or structures in the cortex of the egg. The localization phenomenon, as it occurs in lower chordates, and in many of the invertebrate forms that have been investigated, thus offers an unusual opportunity for resolution of the molecular basis of cell lineage specification in early development.

APPENDIX I

Sequence Complexity and the Quantitative Analysis of Renaturation and Hybridization Experiments Carried Out in Solution

1. DEFINITION OF SEQUENCE COMPLEXITY

The basic concept underlying the following treatment is that of nucleic acid *sequence complexity*. The complexity of a population of nucleic acid molecules is the total length of diverse sequence represented. Suppose, for example, that an RNA population consists of 100 molecules of sequence "a," 10 molecules of sequence "b," and one molecule each of sequences "c," "d," and "e." The complexity is the sum of the diverse sequences present, i.e., (a + b + c + d + e). Complexity is usually given in terms of nucleotides (for RNA) or nucleotide pairs (for DNA), but daltons or any other mass units can also be applied. If each of the species ("a" through "e") in our imaginary nucleic acid population were 10^3 nt in length, the complexity would be 5×10^3 nt. The term *representation* is often used for the frequency with which given transcripts occur in an RNA population. In

525

this example the representation of sequence "a" is 100 times that of sequence "e."

Repetitive as well as nonrepetitive sequences in animal DNA are transcribed. A difficult issue may arise as to the meaning of complexity when the sequences in question are the typical moderately repetitive sequences found in all animal DNAs (Britten and Kohne, 1968a). It is now well known that these repetitive sequences are usually not perfect replicates (Britten and Kohne, 1968a; Davidson and Britten, 1973; Klein *et al.*, 1978; Scheller *et al.*, 1981b). Thus when such sequences are renatured, the duplexes formed include mismatched bases. The genome of the sea urchin *Strongylocentrotus purpuratus* provides a good example. Most of the moderately repetitive sequences in this DNA range in frequency of occurrence from about 100 to about 3000 times per haploid genome. About three-fourths of these repetitive sequences are only a few hundred nucleotides long, with a mean length around 300 nt (Graham *et al.*, 1974; Britten *et al.*, 1976). Klein *et al.* (1978) showed that duplexes formed by renaturating individual (i.e., cloned) repetitive sequences with their genomic homologues melt anywhere from only a few degrees to >15°C lower than do equally long native DNA duplexes. Since 1% base pair mismatch gives rise to approximately 1°C decrease in duplex thermal stability (reviewed by Britten *et al.*, 1974; Wetmur, 1976), it follows that a range of primary sequence differences exists among the homologous repetitive elements. The amount of divergence is an important characteristic of individual repetitive sequence families. For example, Klein *et al.* (1978) found that of 18 cloned *S. purpuratus* repetitive sequence families examined, three displayed <4°C intrafamilial divergence, 13 showed <10°C intrafamilial divergence, and five were more highly divergent. A useful definition of sequence complexity for a set of repetitive sequence families, the members of which though not identical are sufficiently homologous to permit renaturation, is simply the sum of the complexities of any one member from each of the individual repetitive sequence families. Thus, suppose a genome contains three repetitive DNA sequence families, each composed of 10 slightly divergent sequences, $a_1 \ldots a_{10}$, $b_1 \ldots b_{10}$, and $c_1 \ldots c_{10}$. The complexity of the repetitive DNA would be calculated as $(a + b + c)$, which under the renaturation conditions used is the same as $(a_3 + b_7 + c_9)$ or $(a_1 + b_1 + c_1)$, etc.

In Chapters II and III are cited several hybridization experiments which show that some fraction of the repetitive DNA sequence is represented in RNA. This, of course, means some fraction of the repetitive DNA sequence *families*, since any RNA or DNA capable of reacting with one sequence of a repetitive sequence family can also react with all others. For example, an RNA transcribed only from sequence a_3 would hybridize with 33% of the total repetitive DNA in the example given above. The complexity of the RNA would be stated as 33% of the repetitive DNA complexity.

2. BASIC RELATIONS BETWEEN RENATURATION RATE CONSTANT AND SEQUENCE COMPLEXITY

We now consider the estimation of sequence complexity by measurement of renaturation kinetics. The main object of the following discussion is to review briefly the relations needed for analyses of relevant RNA-DNA hybridization and DNA-DNA renaturation experiments. For derivations of some of these relations and detailed physicochemical data on nucleic acid renaturation the reader is referred elsewhere: The technical foundations of this area are to be found in papers by Britten and Kohne (1967, 1968a,b) and by Wetmur and Davidson (1968). A useful review incorporating much detailed information is that of Wetmur (1976), and special treatments of renaturation and hybridization kinetics for various particular circumstances are presented in papers by Britten *et al.* (1974), Smith *et al.* (1975), Davidson *et al.* (1975), Britten and Davidson (1976), Galau *et al.* (1977c,d), and Van Ness and Hahn (1982), among others. The latter authors have provided particularly useful revisions of the effect of salt concentration on the rate of RNA-DNA hybrid formation, the effect of mixing on small reactions, and the difference in behavior of cDNA and sheared genomic DNA tracers.

The rate-limiting step in renaturation is the bimolecular reaction of single-stranded regions bearing complementary nucleotide sequences. The process by which a fruitful collision of strand pairs occurs, the region of complementarity is recognized, and base pair formation begins is termed *nucleation*. Under most conditions, if not all, the continuation of base-pair formation to the end of the complementary region is very fast compared to the rate of nucleation. When assayed in certain ways, the kinetics of DNA renaturation appear to be approximately second order, and this provides convincing evidence that nucleation is rate limiting. Second-order renaturation kinetics were clearly demonstrated in a series of studies on prokaryote DNAs in which most sequences appear only once per genome. Britten and Kohne (1968a) observed second-order renaturation kinetics in experiments in which hydroxyapatite chromatography was used to follow the course of the reaction. Hydroxyapatite, a calcium phosphate complex, binds double-stranded DNA at certain phosphate buffer concentrations, while releasing single-stranded DNA fragments. In addition, Britten and Kohne (1968a) and Wetmur and Davidson (1968) demonstrated second-order kinetics for at least the initial portion of the reaction by measuring the decrease in the optical absorbance of DNA during renaturation. It should be noted that earlier workers (see, e.g., Marmur *et al.*, 1963) had also suggested second-order kinetics for the process of DNA renaturation.

Equation (1) defines a second-order reaction. Here C is the concentration of nucleotides remaining single stranded (conveniently expressed in moles nucleotide liter^{-1}) at time t (in seconds), and k is the observed second-order rate constant:

$$\frac{dC}{dt} = -kC^2 \tag{1}$$

The units of k are thus $M^{-1}\text{sec}^{-1}$. A very useful form of the solution to equation (1) is

$$\frac{C}{C_0} = \frac{1}{1 + kC_0 t} \tag{2}$$

where C_0 is the total DNA concentration or the concentration of single-stranded nucleotides at the initiation of the reaction (Britten and Kohne, 1967, 1968a). As mentioned above C/C_0 can be measured directly by optical hypochromicity or hydroxyapatite chromatography as well as by several other methods. It is the usual practice to extract the rate constant k from the data by least-squares methods, as illustrated in many examples in the text.

For a given DNA the observed rate constant k is found to vary sharply with monovalent cation concentration. For example, the value of k is about five times higher at $0.6\ M$ Na$^+$ than at $0.18\ M$ Na$^+$. Tables for conversion of renaturation rate constants to their equivalent values under "standard conditions" (i.e., $0.18\ M$ Na$^+$, 60°C) are to be found in Britten *et al.* (1974). In this book all renaturation rates and related data are cited after conversion to their values under standard conditions.

The value of the observed rate constant also depends significantly on the DNA fragment length. Wetmur and Davidson (1968) showed that the rate of the reaction varies directly with the square root of the fragment length. Arguments have been made (Wetmur and Davidson, 1968; Wetmur, 1971) that this length dependence means that the nucleation process is inhibited by limitations on the freedom of the incident nucleic acid strand to penetrate the region of solution within which the elements of another strand are likely to be found. This is known as the "excluded volume effect." The dimensions of the excluded volume are a function of the length and the flexibility of the nucleic acid chain under the particular environmental conditions applied. Detailed interpretation of the excluded volume phenomenon remains a subject for further research.

We now consider the renaturation of single copy DNAs of differing complexity. Since the nucleation event is rate limiting, the rate of the reaction for each DNA depends directly on the concentration of each sequence in the mixture. The *sequence concentration* determines the frequency with which fragments bearing a given sequence encounter other fragments that include elements of complementary sequence. Therefore, for a given total DNA concentration, the greater the complexity the slower the reaction, since the concentration of each sequence is lower. The useful principle emerging from this logic is that the rate of renaturation is inversely proportional to the complexity of the renaturing nucleic acid. This provides a powerful tool for measuring nucleic acid sequence complexity.

The relation between sequence complexity and sequence concentration can be seen in the following formalisms (Britten and Kohne, 1967; Britten, 1969): Consider a genome which contains only single copy sequence, and is G nucleotides in length. G is thus the complexity as well as the genome size. Any particular sequence, "i," occurs once per genome. Therefore, the concentration of any given nucleotide in any specific sequence in the mixture is

$$C_i = \frac{C_0}{G} \tag{3}$$

This concentration determines the rate of duplex formation when the DNA is allowed to renature. Equation (2) states that for any given nucleotide in the sequence the fraction present on fragments remaining single-stranded (SS) at time t is

$$\frac{C_{i(SS)}}{C_i} = \frac{1}{1 + k_i C_i t} = \frac{1}{1 + k_i (C_0/G)t} \tag{4}$$

It follows that the *observed rate constant* of renaturation for any sequence (or set of sequences) is inversely proportional to G, the genome size, i.e.,

$$k = \frac{k_i}{G} \tag{5}$$

Here k_i is the basic nucleation rate, which depends on salt concentration, temperature, microscopic viscosity, etc. (see review in Wetmur, 1976, for a summary of these effects). The effect of fragment length is included in the observed rate constant k in two ways. As noted above it affects the nucleation rate, but the fragment length also determines the yield in base-paired nucleotides resulting from each fruitful collision (Britten and Davidson, 1976). For any given set of conditions, k_i in equation (5) can be evaluated by measuring the rate of renaturation of a single copy DNA from a genome of known size. For example, under standard conditions (0.18 M Na$^+$, 60°C) 450 nt fragments of *E. coli* DNA react with an observed second-order rate constant of 0.25 M^{-1}sec^{-1} (where M = moles liter^{-1}), measured by hydroxyapatite assay (Britten and Kohne, 1967). Since the *E. coli* genome contains about 4.2×10^6 ntp, the value of k_i for these conditions is 1.05×10^6 ntp M^{-1} sec^{-1}. A frequently used relation by which an unknown genome size may be estimated is derived from equation (5). If G_1 and k_1 are the genome size and observed renaturation rate of a known DNA (e.g., *E. coli* DNA), and G_2 and k_2 are the equivalent parameters for an unknown DNA,

$$G_2 = \frac{k_1}{k_2} G_1 \tag{6}$$

In the examples we have so far considered the complexity of the DNA equals the genome size, but this is of course a special case. The complexity of any purified DNA fraction in which all sequences are present in equal concentration can be calculated from equation (6).

Experimental verification of the relations shown in equations (5) and (6) is presented in Fig. AI.1. This graph is reproduced from a paper of Laird (1971) and demonstrates the inverse proportionality of genome size and the observed second-order rate constant. The data shown cover four orders of

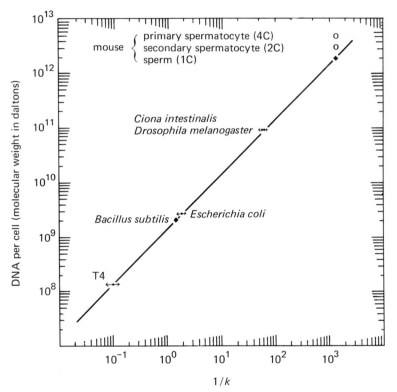

Fig. AI.1. Correlation between genome size and $1/k$ predicted by equation (5). k is the observed rate constant for renaturation of total DNA (prokaryotic DNA) or of single copy sequence components in whole genomic DNA (eukaryotic DNA) fit to the data by assuming second order kinetics [see equation (2) and text]. Original sources of the genome size and renaturation rate data are given by Laird (1971). Observations were made by the hydroxyapatite procedure. Horizontal lines about each point represent the 95% confidence limits ($\pm 2\sigma$) for the rate constant determinations. The DNA had been sheared by passage through a needle valve at 12,000 lb in^{-2}, and was thought to be about 400 nt. However, the absolute values of the renaturation rates obtained suggest that the fragments may actually have been somewhat longer. Values of haploid DNA content per cell are plotted for *Ciona* and *Drosophila*, and for the mouse, where sperm was used for the genome size determination. [From C. D. Laird (1971). *Chromosoma* **32**, 378.]

magnitude, and other observations concerning naturally occurring DNAs extend the proportionality down to complexities of a few nucleotides. Such low complexities are found in some satellite DNAs, the renaturation kinetics of which were studied by many investigators, including Waring and Britten (1966), Britten and Kohne (1967), Flamm *et al.* (1969), Hutton and Wetmur (1973), Brutlag and Peacock (1975), and Cordeiro-Stone and Lee (1976). We note that the proportionality shown in Fig. AI.1 between renaturation rate and genome size is not specifically dependent on the functional form of the reaction kinetics. As discussed briefly below, the renaturation of randomly sheared DNA is in fact not exactly a second-order process, though for most purposes the differences are slight. However, renaturation *is* dependent kinetically on the rate of occurrence of successful nucleations with which the pairing of complementary fragments begins. This basic fact is implicit in the relation symbolized in equation (5).

3. RENATURATION OF REPETITIVE SEQUENCES

The concept of sequence concentration used in equations (3) and (4) also provides the basis for understanding the renaturation of repetitive sequences. Since the rate of the reaction depends on sequence concentration, a repetitive sequence present many times per genome will react at a rate which is proportionately higher than a sequence present once per genome. Suppose a particular class of repetitive sequences occurs at a frequency of F copies per genome. If all sequences of this repetition class are taken together, they occupy a fraction α of the genome, the total size of which we again term G. Following the definition for repetitive sequence complexity given above, the complexity of this set of repetitive sequences, N, is thus

$$N = \frac{\alpha G}{F} \tag{7}$$

This relation of course reduces to $N = G$ for the case where the genome is entirely nonrepetitive. When the DNA is sheared, the sequence concentration, C_r, for a given family of repetitive sequences is [see equation (3)]

$$C_r = \frac{FC_0}{G} \tag{8}$$

where C_0 is again the total concentration of nucleotides. Therefore, as in equation (4),

$$\frac{C_{r(SS)}}{C_r} = \frac{1}{1 + (k_i F/G)C_0 t} \tag{9}$$

From this expression, the observed rate constant for reaction of the repetitive sequences can be seen to be proportional to F, the repetition frequency, and inversely proportional to G, the genome size. It follows from equations (9) and (4) that the value of F for a given class of repetitive sequences can be directly measured simply by renaturing whole, sheared DNA. That is,

$$F = \frac{k_r}{k_{sc}} \tag{10}$$

where k_r is the observed rate at which the repetitive sequences in the whole DNA renature, and k_{sc} is the rate at which the single copy sequences renature in the same experiment. Equation (10) states that a repetitive DNA fraction renatures faster than a single copy sequence in the same genome in proportion to the number of copies comprising each repetitive sequence family. It will be noted that in a DNA which includes both repetitive and nonrepetitive sequences, k_{sc} is still inversely proportional to the genome size, as shown in Fig. AI.1 for mouse and *Drosophila* DNAs.

Figure AI.2 illustrates in detail the renaturation kinetics of a typical animal DNA, that of the sea urchin *S. purpuratus*. The ordinate shows the fraction of the DNA fragments present in duplex-containing structures, as measured by binding to hydroxyapatite under temperature conditions equivalent to those at which the renaturation was permitted to occur. The abscissa of Fig. AI.2 is calibrated on a log scale in units of concentration × time for the reacting DNA solutions; that is, the "C_0t" term of equation (2). This mode of presenting renaturation data is known as a "C_0t plot," and was introduced by Britten and Kohne (1968a). The units of C_0 are customarily moles nucleotide liter^{-1} sec. Figure AI.2 shows that the renaturation of sea urchin DNA occurs over a range of more than six decades in C_0t. Evaluation of equation (2) shows that a single kinetic component can occupy only two decades between 10% and 90% of the reaction. Thus, a number of individual kinetic components evidently contribute to the overall curve shown. Equation (10) provides a means of resolving these into different frequency classes. The most slowly renaturing fraction consists of fragments which contain recognizable lengths only of nonrepetitive sequence (curve a). The dotted lines (curves b, c, and d) show the faster kinetic components resolved from the overall curve. These represent the reaction of sets of fragments bearing repetitive sequences. Moving from right to left, each component of the reaction represents the renaturation of increasingly repetitive sequences. It is important to realize that in an experiment such as that shown in Fig. AI.2, the DNA fragments renature and become bindable by hydroxyapatite at rates appropriate for the most highly repetitive sequence which they contain. Thus, because repetitive and nonrepetitive sequences are interspersed, many fragments contain elements of both these sequence classes. These will be bound to hydroxyapatite after incubation to C_0t's permitting the reaction of the repetitive sequences only. Therefore the quantity of interspersed repetitive sequence is overestimated and the quantity of nonrepetitive sequence underestimated in hydroxyapatite asays of renaturation. The magni-

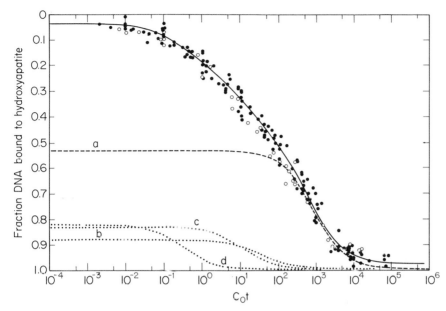

Fig. AI.2. Renaturation kinetics of 450 nt sea urchin DNA fragments. Ordinate shows fraction of fragments bound to hydroxyapatite after incubation to various C_0t's. C_0t (DNA concentration × time, in moles nucleotide liter^{-1}sec) is plotted on a log scale on the abscissa. Closed circles show data obtained at 60°C, and open circles show data obtained at 50°C. Reactions at 60°C were carried out in 0.12 *M* phosphate buffer (0.18 *M* Na$^+$) or are plotted at the C_0t values calculated to be equivalent to 0.18 *M* Na$^+$ on the basis of the acceleration in rate expected for Na$^+$ concentrations above 0.18 *M* (Wetmur and Davidson, 1968; Britten *et al.*, 1974). Reactions at 50°C were run in 0.14 *M* phosphate buffer. A computer operating with a nonlinear least-squares program has been used to fit the renaturation data to equation (11). The dashed line (a) shows the renaturation in whole DNA of fragments bearing only single copy sequence. The rate constant for this reaction is 1.25 × 10^{-3} *M*$^{-1}$sec^{-1}, and this component includes 47% of the DNA. The dotted lines (b-d) show various repetitive sequence kinetic components. These components probably all represent averages of many individual sequence families differing in repetitiveness, and several alternative kinetic components can be resolved with little change in the root mean square error. Furthermore, each fragment renatures at the rate characteristic of the most highly repetitive sequence of recognizable length which it contains, and less repetitive sequences may also be present on it. For curve b in this particular solution the rate constant is 3 × 10^{-2} *M*$^{-1}$sec^{-1}; the average repetition of each sequence would be about 20-30, and the fraction of fragments bearing these sequences is 12%. For curve c the rate constant is 1 × 10^{-1} *M*$^{-1}$sec^{-1}; the average repetition of each sequence is about 80, and the fraction of fragments included is 17%. For curve d the rate is 3.3 *M*$^{-1}$sec^{-1}; the average repetition of each sequence is about 2600, and the fraction of fragments included is 16%. [From D. E. Graham, B. R. Neufeld, E. H. Davidson, and R. J. Britten (1974). *Cell* **1**, 127; and additional unpublished data from author's laboratory.]

tude of this effect depends on the fragment length and on the spacing of the interspersed repetitive and nonrepetitive sequences in the genome (Davidson *et al.*, 1973).

The kinetic analysis shown in Fig. AI.2 is based on the principle that the overall renaturation is the sum of a series of second-order components, each behaving according to equation (2). The calculation is thus carried out as shown in the following expression:

$$\frac{C}{C_0} = \beta + \frac{\alpha_1}{1 + k_1 C_0 t} + \frac{\alpha_2}{1 + k_2 C_0 t} + \frac{\alpha_3}{1 + k_3 C_0 t} + \cdots \qquad (11)$$

Here β is the fraction of DNA remaining unreassociated at the termination of the reaction and each kinetic component, representing a fraction α_j of the genome, renatures with the observed rate constant k_j. The various k's depend on the repetition frequency of the sequence components they represent, as in equation (10). Customarily the parameters of equation (11) are derived from renaturation data by nonlinear least-squares or equivalent procedures with the aid of a computer. The kinetic components calculated for the data shown in Fig. AI.2 are illustrated by the dotted lines, and the values of k_j and α_j in equation (11) are listed in the legend to the figure.

It is important to note that kinetic components such as those portrayed in Fig. AI.2 are only numerical averages. For example there are repetitive sequences in the sea urchin genome which are present over a range from 10^1 to 10^3 copies per haploid complement of DNA, and for which the *average* representation is 10^2-fold repetition. This has been demonstrated explicitly by analysis of the repetition frequency of a series of individual cloned repetitive sequences (Klein *et al.*, 1978). A histogram representing the frequency distribution actually observed is shown in Fig. AI.3.

An isolated kinetic fraction of DNA renatures more rapidly than it does in the presence of all the other sequences in the DNA, since its sequence concentration increases as it is purified. The observed (or calculated) rate constant for the renaturation of a purified frequency component is called "k_{pure}" where, as above, α is the fraction of the genome occupied by this component, and k is its observed renaturation rate constant in whole DNA.

$$k_{\text{pure}} = \frac{k}{\alpha} \qquad (12)$$

The complexity of the repetitive component can either be calculated directly from k_{pure} by means of equations (5) or (6) or estimated from its repetition frequency and quantity by means of equation (7).

An important feature included in Fig. AI.2 is that data obtained at 50°C (open circles) are only slightly different from those obtained at 60°C (closed circles). This is a fairly typical, though not universal result. It cannot be assumed that the renaturation of a previously unstudied DNA will be insensitive to such criterion changes. As noted above, moderately repetitive se-

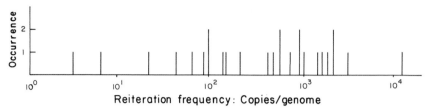

Reiteration frequency: Copies/genome

Fig. AI.3. Histogram showing frequency of occurrence of repetitive sequence families of different sizes in the genome of the sea urchin *Strongylocentrotus purpuratus*. The repetitive sequence family is experimentally defined as that set of genomic sequences reacting with a given cloned repeat probe, under specified conditions of renaturation. Data are based on analysis of the genomic repetition frequency of 26 clones randomly selected from a clone library made by partially renaturing the DNA, and cloning the S1 nuclease resistant repetitive duplexes in a plasmid vector. Low frequency repeat sequences (<20 copies per haploid genome) could be underrepresented, since the DNA was initially renatured to a C_0t of 40 M sec. Repetition frequencies were determined by the kinetics of the terminally labeled cloned DNA inserts renatured with sheared genomic DNA, in the presence of appropriate internal kinetic standards. [Data from W. H. Klein, T. L. Thomas, C. Lai, R. H. Scheller, R. J. Britten, and E. H. Davidson (1978). *Cell* **14**, 889.]

quences in animal DNA are typically divergent, but the stability of most repetitive duplexes formed by renaturation is high enough so as not to affect strongly the rate of renaturation under usual conditions. The reason for this is that the rate of renaturation remains close to the optimum over the whole range from 15-30°C below the melting temperature of the renatured strand pairs (Bonner *et al.*, 1973; Wetmur, 1976). Only as the incubation temperature approaches the melting temperature does the rate drop sharply. Most of the duplexes scored in the experiments shown in Fig. AI.2 melt between 73°C and 83°C, and thus, as expected, renaturation at 50°C is found to be equivalent to renaturation at 60°C. Of course, if the temperature were increased above 60°C, some effect on the renaturation kinetics would be expected.

4. KINETICS OF DISAPPEARANCE OF SINGLE-STRAND NUCLEOTIDES AND RNA HYBRIDIZATION IN DNA EXCESS

It was noted earlier that the renaturation of randomly sheared DNA fragments is not an ideal second-order reaction, and this fact has some interesting and significant consequences. Random shearing produces fragments which begin at as many different sites as there are nucleotide pairs in the genome. Therefore, when any two fragments of DNA bearing complementary sequences react, single-stranded tails will almost always remain after the complementary regions have become paired. For the fragment length distributions produced by the usual random shearing procedures, the mean length of duplex resulting from a single nucleation event is about 55% of the mean single-strand length (Smith *et al.*, 1975). The result is that, except at the beginning and end of a renaturation reaction, the fraction of *nucleotides* remaining single-stranded is always significantly greater than the fraction of *fragments* remaining completely single-stranded. It is the latter, of course, which is measured by the hydroxyapatite binding assay. Morrow (1974) and Smith *et al.* (1975) showed that the fraction of nucleotides remaining single-stranded, S, is approximated by the expression

$$\frac{S}{C_0} = \left(\frac{1}{1 + kC_0t}\right)^n \tag{13}$$

Here C_0 is the total DNA concentration and t the time, as above. The observed rate constant k also has exactly the same meaning as in equation (2). It is found empirically that the best value of n in equation (13) is about 0.45. The form of equation (13) is clearly non-second-order, i.e., unless the exponent $n = 1.0$, in which case equation (13) reduces to equation (2). If shearing occurs at specific rather than random locations, as, for example, when restriction enzymes are used to cut a single copy DNA, the value of n in equation (13) does in fact equal 1. Thus Morrow (1974) showed that the disappearance of single-stranded nucleotides in SV40 DNA treated with restriction endonucleases follows perfect second-order kinetics.

In practice S is usually measured by the use of single-strand-specific nucleases or by optical methods. Most commonly used is the single-strand-specific S1 nuclease derived from *Aspergillus*. In Fig. AI.4 the renaturation kinetics of randomly sheared 700 nt *E. coli* DNA are shown. The two curves represent the kinetics of the reaction as assayed by hydroxyapatite binding and by S1 nuclease resistance. The hydroxyapatite data (solid circles) are fit according to equation (2), and the S1 nuclease data according to equation (13). Figure AI.4 demonstrates that at any given C_0t value less of the DNA is resistant to S1 nuclease than binds to hydroxyapatite. The explanation for the exact form of the S1 nuclease kinetics illustrated here is complex. This subject was further developed by Smith *et al.* (1975) and Britten and Davidson (1976) but lies outside of the scope of the present discussion. It should be stressed that when renaturation of randomly sheared DNA is assayed by hydroxyapatite binding, the actual deviations from perfect second-order kinetics are so small as to be nearly undetectable, as can be seen in Fig. AI.4. Furthermore, when the S1 nuclease kinetics are fit with the form shown in equation (13), the basic relations between the observed rate constant, the complexity, and the repetition frequency remain unaffected. Thus, as mentioned previously, equations (5), (6), and (10) do not depend on the exact order of the reaction kinetics or the means of assay, so long as appropriate methods are used to evaluate the observed rate constant, k.

For our present purpose the main importance of the S1 nuclease kinetic

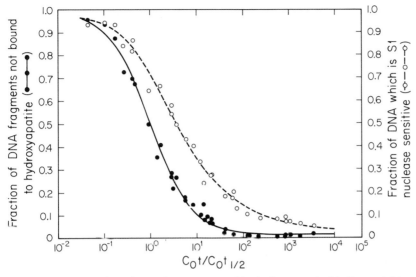

Fig. AI.4. Renaturation of *E. coli* DNA assayed by hydroxyapatite binding and S1 nuclease resistance. The DNA was randomly sheared to about 700 nt. DNA duplex formation was measured by binding to hydroxyapatite at 60°C in 0.12 *M* phosphate buffer (●), and by resistance to digestion with single-strand-specific S1 nuclease (○). Two sets of data were pooled by normalizing both to a hydroxyapatite $C_0t_{1/2}$ of 1.0 (i.e., $C_0t/C_0t_{1/2}$ for each data set). The solid line represents the least-squares solution of the renaturation kinetics as assayed by hydroxyapatite binding, according to equation (2) in text. The dashed line represents the least-squares solution for the S1 nuclease kinetics, according to equation (13) in text. The normalized rate constant k is 1.0 for both curves and the best value for n in equation (13) for the S1 nuclease curve is 0.453. [From M. J. Smith, R. J. Britten, and E. H. Davidson (1975). *Proc. Natl. Acad. Sci. U.S.A.* **72**, 4805.]

determination illustrated in Fig. AI.4 is that it provides a basis for analysis of a useful class of hybridization and renaturation reactions. These are reactions in which a labeled RNA or DNA present in trace quantities is reacted with excess DNA. Such reactions are required in order to determine whether an RNA is transcribed from repetitive or single copy sequences, or both. In an experiment of this type the excess DNA is termed the "driver DNA." The tracer is capable of reacting with all of the single-stranded DNA sequence present in the driver at any given time [i.e., the DNA whose concentration is denoted by S in equation (13)] and not solely with those fragments which remain completely single-stranded at that time [i.e., the DNA whose concentration is denoted by C in equations (1) and (2)]. Therefore equation (13) must be used to determine the concentration of reactive driver DNA sequence. Furthermore, it often happens that the rate of reaction of the tracer nucleic acid with single-stranded driver DNA sequence differs from that of the driver fragments with each other. For example, the tracer fragment length may be shorter, or the tracer may be RNA. For reasons which remain obscure, in DNA-driven hybridization reactions the rate of RNA-DNA duplex formation is significantly lower than that of DNA-DNA duplex formation even when fragment lengths are carefully taken into account (Melli *et al.*, 1971; Galau *et al.*, 1977c). The following expression has been derived for the purpose of analyzing such reactions (Davidson *et al.*, 1975). Suppose the observed rate constant for tracer DNA duplex formation is called h; S is the concentration of single-stranded nucleotides in driver DNA as above; and U is the fraction of the tracer fragments remaining unreacted at time t. Then

$$\frac{dU}{dt} = -hUS \tag{14}$$

Substituting equation (13) for S and solving, we have

$$\frac{U}{U_0} = \exp \frac{h[1 - (1 + kC_0t)^{1-n}]}{k(1 - n)} \tag{15}$$

Note that k is the same observed rate constant as in equations (2) and (13), here applied to the driver DNA.

5. RNA EXCESS HYBRIDIZATION KINETICS

RNA excess hybridization reactions are required for measurement of the sequence complexity of RNA populations and for comparisons of the sets of sequences present in various RNA populations. These reactions are usually carried out with labeled DNA, and the RNA excess is such that the concentration of RNA at the start of the experiment, R_0, remains essentially unchanged throughout. The reaction is thus pseudo-first-order in form. We let D represent the unreacted DNA tracer concentration at time t, R_0 the starting (and final) RNA concentration, and k_h the observed rate constant for RNA-DNA duplex formation. The rate of change in the concentration of unreacted DNA tracer is then

$$\frac{dD}{dt} = -k_hR_0D \tag{16}$$

Thus,

$$\frac{D}{D_0} = e^{-k_hR_0t} \tag{17}$$

Here D/D_0 is the fraction of DNA tracer remaining unreacted.

For each kinetic component of the form given by equation (17) the pseudo-first-order hybridization rate constant, k_h, is inversely proportional to the RNA complexity. The *complexity* of the class of RNA reacting with these kinetics can be measured independently, however, simply by determining the fraction of the total DNA tracer hybridized at the termination of the reaction. Thus if the RNA complexity is known, the concentration in the total RNA of those species driving the reaction can be calculated. For this calculation, we require the rate constant expected for a pure RNA of the measured complexity. This is obtained by use of a proportionality constant relating RNA complexity and the pseudo-first-order rate constant k_h [see equation (5)].

Figure AI.5 displays a set of measurements which provides the necessary information (Galau *et al.*, 1977c). Here two pseudo-first-order reactions are shown, which follow closely the kinetics described by equation (17). The open circles represent the reaction of ϕX174 ^{3}H-DNA tracer driven by excess ϕX RNA. The RNA had been transcribed enzymatically from the (-) strand of ϕX174 DNA. The complexity of the ϕX174 genome (and of the

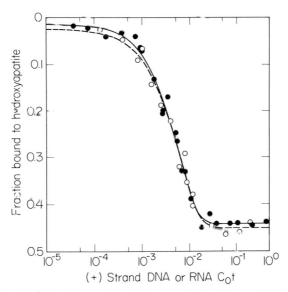

Fig. AI.5. Pseudo-first-order reactions driven by ϕX174 RNA and DNA. Trace quantities of 300 nt ϕX174 replicative form (RF) ^{3}H-DNA were reacted with excess ϕX174 (+) strand RNA or DNA in 0.12 M phosphate buffer at 60°C. The (+) strand DNA was obtained from mature phage and the RNA was transcribed *in vitro* using *E. coli* polymerase. The fraction of ^{3}H-DNA in duplex was assayed by binding to hydroxyapatite in the same phosphate buffer at 60°C. Data were fit to the pseudo-first-order function shown in equation (17) by least-squares methods. The maximum possible reaction is 50%, since both driver nucleic acids are complementary to only one strand of the ^{3}H-DNA tracer. (●—●) Reaction of ^{3}H-DNA with excess 300 nt (+) strand DNA. The best fit rate constant is 196 $M^{-1}sec^{-1}$. (○--○) Reaction of ^{3}H-DNA with excess 360 nt long (+) strand RNA. The best fit rate constant is 169 $M^{-1}sec^{-1}$. [Modified from G. A. Galau, R. J. Britten, and E. H. Davidson (1977c). *Proc. Natl. Acad. Sci. U.S.A.* **74**, 1020.]

RNA transcript) is about 5400 nucleotides, and k_h is evaluated from the data in Fig. AI.5 at about 169 $M^{-1}\text{sec}^{-1}$. The expected rate constant for the reaction of any other RNA of known complexity can be calculated by application of these data to equation (6), i.e., for these conditions and fragment lengths.

In the reaction displayed by the closed circles in Fig. AI.5 (+) strand ϕX174 DNA is used to drive the reaction of the same tracer. This reaction is directly analogous to the one just considered, in that like the (+) strand RNA, the driver DNA cannot react with itself. The reaction is therefore again pseudo-first-order, as demonstrated by the kinetics shown in Fig. AI.5. The significance of this reaction is that its rate is almost indistinguishable from that of the RNA-driven reaction. The best fit value of k_h for this reaction is 196 $M^{-1}\text{sec}^{-1}$, close to that expected on the basis of the second-order reaction kinetics. The experiment shows that in RNA excess reactions the basic rate of formation of RNA-DNA duplex is almost the same (probably within 20%) as the basic rate of formation of DNA-DNA duplex.

Once the complexity of an RNA is known and the expected hybridization rate is calculated, the fraction of the total RNA serving as driver can be easily obtained. The driver fraction which we shall call Df, is simply the ratio of the expected rate, $k_{h_{exp}}$, calculated for a pure RNA of the measured complexity, to the observed rate actually fit to the data points on the basis of the total RNA in the reaction mixture, $k_{h_{obs}}$.

$$Df = \frac{k_{h_{obs}}}{k_{h_{exp}}} \tag{18}$$

For example, suppose a hundredfold excess of ribosomal RNA were added to the ϕX174 RNA reaction shown in Fig. AI.5, and the RNA C_0t were calculated on the basis of total RNA concentration. The rate constant would then be 1.7 $M^{-1}\text{sec}^{-1}$ rather than 170 $M^{-1}\text{sec}^{-1}$, and equation (18) would show that the reacting RNA species constitutes 1% of the total RNA. Equation (18) thus provides approximate information as to the quantity of the hybridizing RNAs in the preparation.

APPENDIX II

Measurement of Transcript Prevalence by Single-Stranded Probe Excess Titration

Titration by solution hybridization reactions is the method of choice for measurement of the absolute number of transcript molecules of a species for which a cloned probe is available. The reactions are carried out with a single-stranded, labeled probe, *that is present in excess*. This procedure differs fundamentally from kinetic methods, in that the prevalence determination depends on the amount of probe hybridized at termination in each sample rather than on the observed rate of the reaction. The RNA preparation that includes the molecular species of interest is mixed at various mass ratios with the excess tracer, and reacted to completion in solution. In practice, over ten times the half reaction point is usually easy to obtain, since the reactions are pseudo-first-order [Appendix I, equation (17)] in respect to the cloned excess tracer, which is always of low complexity. Thus, all the samples have essentially the same rapid reaction kinetics. This removes most of the uncertainties and difficulties associated with prevalence determination by kinetic methods, in which it is often difficult to be certain that the hybridization reactions are actually completed, a problem that is particularly acute for rare RNA species that require very extensive hybridization for their reaction (see, e.g., Van Ness and Hahn, 1982).

In the most frequently encountered situation a given sequence is represented asymmetrically in the RNA population. For this case the amount of hybridization obtained in each sample is proportional to the RNA/probe ratio, as shown in equation (1) (Wallace *et al.*, 1977; Scheller *et al.*, 1978). Here P is the fraction of the strand-separated, labeled probe hybridized at kinetic termination, in reactions carried out at RNA/probe mass ratios of R,

541

and α is the *fraction* of the RNA that is complementary to the probe sequence:

$$P = \alpha R \tag{1}$$

Hybridization is preferably measured by a nuclease protection method, so that the number of nucleotides actually paired is determined, but if appropriate length corrections are made hydroxyapatite binding values can also be used to establish the value of P in equation (1) (see, e.g., Lev *et al.*, 1980). To obtain R it is necessary to have an accurate knowledge of the probe specific activity.

In practice, where equation (1) can be applied, α is usually obtained by linear least-squares estimate of the slope of the line given by equation (2):

$$T = \alpha' S \tag{2}$$

Here T is the number of counts per minute (cpm) of tracer hybridized, rather than the fraction of cpm hybridized (i.e., $P = T/T_0$) and S is the mass of RNA in the reaction, rather than the ratio of RNA/probe (i.e., $R = S/T_0$), as in equation (1). This gives α' in terms of cpm hybridized ng^{-1} RNA, which is converted to α, the mass fraction of RNA included in the specific sequence, by application of the probe specific activity.

In certain circumstances the probe sequence will be represented by complementary RNA molecules that can react with each other as well as with the excess, strand-separated probe. This situation is encountered, for example, when the probe is a typical interspersed repetitive sequence. Scheller *et al.* (1978) and Costantini *et al.* (1978) showed that such repeat sequences are usually represented by transcripts complementary to both strands of the canonical genomic repeat element. This is due to contributions from many different repeat sequence elements oriented in both directions in respect to the transcription unit of which they are a part, rather than to symmetrical transcription of any given repeat sequence element (Posakony *et al.*, 1983). For this case a nonlinear form obtains:

$$P = \frac{1}{1 + \dfrac{1}{\alpha} R} \tag{3}$$

To obtain α the titration data (i.e., values of P, R) may be easily fit to the form of equation (3) by use of a nonlinear least-squares program such as discussed in Appendix I. Examples of titration measurements utilizing this form can be seen in text Fig. 2.4(a)-(b). However, the initial portion of the curve described by equation (3), i.e., the function P for very low RNA/probe ratios, is given by equation (1), providing a simple method in which a linear least-squares analysis will suffice.

The limiting factor in application of the titration method is availability of the single-stranded probe. Such probes must be uniformly labeled in internal positions if a nuclease assay procedure is to be utilized. Almost any cloned DNA fragment can be strand-separated and then labeled *in vitro* with radioactive iodine (see, e.g., Lev *et al.*, 1980), but this process is often time consuming and difficult. At present the preferred methods are preparation of a labeled single-strand DNA probe using an M13 vector (Hu and Messing, 1982), or of a labeled single-stranded RNA probe, using a vector based on the Sp6 or T7 phage promoters (Kassavetes *et al.*, 1982; Butler and Chamberlin, 1982; Melton *et al.*, 1984). The sensitivity of titration assays using probes synthesized enzymatically *in vitro* is very high, due to the high specific activities that can be obtained. Thus it is possible to measure specific RNA concentrations down to $<10^{-1}$ molecules per cell. RNA-RNA hybrids are somewhat easier to assay than are RNA-DNA hybrids because of the complete resistance of RNA-RNA duplexes to RNase at monovalent cation concentrations >0.24 M (see, e.g., Lee *et al.*, 1986a).

The number of molecules, say per cell, N_c, where Q_c is the content of total RNA per cell, L is the length of the transcript hybridized by the probe, and 350 is average weight in daltons of a ribonucleotide, is then given by:

$$N_c = \alpha Q_c \left(\frac{6 \times 10^{23}}{350L} \right) \tag{4}$$

APPENDIX III

Measurement of Rates of Synthesis or Entry into the Cytoplasm, and of Transcript Turnover Rate, from the Kinetics of Incorporation of Labeled Precursor

The ideal method would be one which does not perturb cellular metabolism, in contrast to procedures that involve blocking transcription with drugs, or the "chase" of labeled precursor with relatively huge quantities of unlabeled precursor. Effective chase is impossible in many embryos in any case, due to the very large size of the stored pools, their low turnover rate, and their relative inelasticity. Most synthesis and turnover rate constants cited in this book have been obtained from measurements of the kinetics of incorporation of precursor into a specific newly synthesized species of transcript, or class of transcripts, and of the precursor pool specific activity determined over the same time course. This method involves minimal physiological insult to the embryos, except for the possibility of radiation damage on long exposure, and except when severe measures are required to permeabilize the embryos, as, e.g., the immersion of *Drosophila* embryos in octane (Limbourg and Zalokar, 1973).

1. BEHAVIOR OF EMBRYO NUCLEOSIDE TRIPHOSPHATE POOLS

A possible difficulty that requires consideration in any determination of absolute synthesis rate is whether the chemical precursor pool is *compartmentalized*, so that the pool specific activity measured is not the specific activity of the actual immediate RNA precursor. In the best studied example, the sea urchin embryo, it has been shown that compartmentalization of RNA precursor pools does not occur (see text, Chapter III).

Methods of determining pool specific activity lie outside the scope of this discussion, and the reader is referred to the variety of procedures described in the individual studies cited in Chapter III. Several investigations directed specifically at the size and specific activity of sea urchin embryo nucleoside triphosphate pools have been carried out (Kijima and Wilt, 1969; Aronson and Wilt, 1969; Wilt *et al.*, 1969; Wilt, 1970; Emerson and Humphreys, 1970; Brandhorst and Humphreys, 1971, 1972; Wu and Wilt, 1974; Galau *et al.*, 1977a; Cabrera *et al.*, 1984). The S-adenosylmethionine pool of these embryos, from which RNA cap and internal methyl groups are derived, has also been studied (Nemer, 1979). Some evidence also exists for the RNA precursor pools of embryos of other species. The behavior of nucleoside triphosphate pools in permeabilized *Drosophila* embryos was described by Anderson and Lengyel (1979, 1981). Most modern synthesis rate studies carried out on *Xenopus* embryos have not been performed on whole embryos due to their relative impermeability, but rather on dissociated cells (references in Chapter III). The behavior of mouse embryo pools after addition of exogenous precursor and under chase conditions was described by Clegg and Pikó (1977). In the following brief discussion the sea urchin embryo pools serve as examples, as these are the most extensively characterized.

When low concentrations of labeled nucleoside are added to a culture of sea urchin embryos the precursor is rapidly taken up and the specific activity of the internal nucleoside triphosphate pool soon attains a maximum value. The specific activities of some pools, e.g., the ATP and UTP pools of cleavage and blastula stage embryos, then remain constant for many hours. This behavior is shown in Fig. AIII.1(a)-(c). In the more general case, however, the pool specific activity cannot be assumed constant throughout the experiment. Thus in blastula and gastrula stage sea urchin embryos, the specific activities of the GTP pool and of the S-adenosylmethionine pool change rapidly after the exogenous precursor is added to the medium (i.e., ^{3}H-guanosine or ^{3}H-methylmethionine, respectively). These pools are quickly labeled as the precursor is absorbed from the sea water but after reaching a peak their specific activities decline sharply, as the pools are flushed with unlabeled molecules derived from the degradation of nonradioactive RNA that was synthesized prior to addition of the precursor. Eventually the pool specific activities settle to a steady state level, but this may require a long

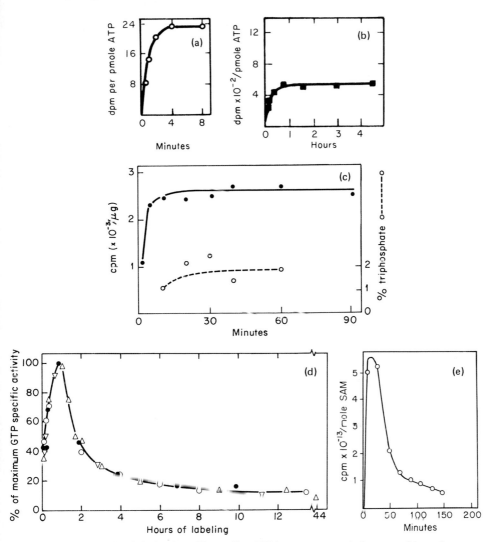

Fig. AIII.1. Kinetic behavior of immediate RNA precursor pools in sea urchin embryos. (a-d) Nucleoside triphosphate pools. (a) Rapid approach to saturation of ATP pool specific activity (in dpm pmole^{-1} ATP), in an experiment in which blastula stage *Lytechinus pictus* embryos were labeled with ^{3}H-adenosine (4×10^{-7} M; 28 Ci mmole^{-1}) in 10% suspension. At the indicated times they were diluted and the specific activity of the acid-soluble ATP was measured. (b) ATP pool specific activity (dpm $\times$ 10^2 pmole^{-1} ATP) in a 2.5% suspension of blastula stage embryos continuously exposed to the same precursor as in (a), at 1×10^{-7} M. [(a) and (b) Reprinted with permission from B. P. Brandhorst and T. Humphreys (1971). *Biochemistry* **10**, 877. Copyright by the American Chemical Society.] (c) Specific activity (left ordinate) of nucleotides in a 1% suspension of *S. purpuratus* blastula labeled with ^{3}H-uridine (4×10^{-8} M; 26.6 Ci mmole^{-1}). Specific activity values are given as cpm $\times$ 10^{-3} μg^{-1} of total acid-soluble nucleosides and nucleotides. About 2% of the total labeled nucleotides are UTP (right ordinate). [(c) From F. H. Wilt, A. I. Aronson, and J. Wartiovaara (1969). *In* "Problems in Biology: RNA in

time relative to the synthesis and turnover kinetics of the RNA species of interest. Examples of this form of pool specific activity behavior are shown in Fig. AIII.1(d) for the GTP pool and in (e) for the S-adenosylmethionine pool. The condition resulting in this form of pool specific activity time function is a pool that is small relative to the amount of synthesis that it feeds, and that thus turns over rapidly. This phenomenon is more likely to be encountered in late embryos when there are more nuclear sites of synthesis, than early in development when there are few nuclei and for many precursors a very large maternal pool. Thus Kijima and Wilt (1969) observed that the specific activity of the acid-soluble pool in the presence of about 4×10^{-7} M exogenous ^{3}H-guanosine declines only slightly from its initial maximum value in cleavage-stage embryos, while at later stages the pool behaves as illustrated in Fig. AIII.1(d). If the exogenous nucleoside concentration is greatly increased, e.g., to 10^{-3} M, the pools can be expanded, though only by a small factor (Grainger and Wilt, 1976). More exogenous nucleoside can then be incorporated in the pool, thus facilitating labeling with radioactive or density isotopes. Expansions on the order of 2-fold were measured by Grainger and Wilt (1976) for the ATP pool at these high concentrations of exogenous precursor.

2. EXTRACTION OF SYNTHESIS RATE, CYTOPLASMIC ENTRY RATE, AND DECAY RATE CONSTANTS FROM THE KINETICS OF PRECURSOR INCORPORATION

The mathematical approach utilized here is applied both to measurements of RNA synthesis kinetics and to measurements of the entry of newly synthesized message into the cytoplasm. In this book constants describing the rate of synthesis of nuclear RNA or the rate of entry into the cytoplasm of messenger RNA are denoted k_s. The units in which k_s is expressed per cell are usually molecules min^{-1}, and for the decay rate constant, k_d, the units

---

Development" (E. W. Hanly, ed.), p. 331. Univ. of Utah Press, Salt Lake City; A. I. Aronson and F. H. Wilt (1969). *Proc. Natl. Acad. Sci. U.S.A.* **62**, 186.] (d) Specific activity of GTP in late blastulae and early gastrulae of *S. purpuratus* with ^{3}H-guanosine (5×10^{-7} M; 0.5-2.6 Ci mmole^{-1}). Several different experiments are indicated by the various symbols, and for ease of presentation these are normalized to the maximum specific activity value, i.e., at 50 min. Data are then expressed as percent of maximum specific activity (ordinate). The actual specific activity obtained was directly correlated with the external specific activity. For example, when this was 2 Ci mmole^{-1}, the 50 min (peak) GTP specific activity was about 2×10^{11} dpm mmole^{-1} GTP, or about 4% of the specific activity of the medium. [(d) From G. A. Galau, E. D. Lipson, R. J. Britten, and E. H. Davidson (1977a). *Cell* **10**, 415. Copyright by M.I.T.] (e) S-adenosylmethionine pool specific activity in gastrula stage *S. purpuratus* embryos labeled in 10% suspension with methyl ^{3}H-methionine (80 Ci mole^{-1}) at 6×10^{-8} M. [(e) From M. Nemer (1979). *Dev. Biol.* **68**, 643.]

are min^{-1}. In some cases, the entry rate constant is expressed in mass rather than molecular units, i.e., as pg min^{-1}. When expressed in mass terms, the rate constant is conventionally denoted k'_s.

For the simplest case (Kafatos and Gelinas, 1974; Davidson, 1976) it is assumed that the population of RNA molecules considered is synthesized at an unchanging rate throughout the experiment; that these molecules turn over rapidly in relation to the time scale of the observations; that when labeled precursor is introduced, the precursor pool equilibrates rapidly; and that the specific activity of the pool then remains constant throughout the labeling period [i.e., as shown in Fig. AIII.1(a)-(c)]. The kinetics with which a labeled precursor accumulates in newly synthesized, unstable RNA can be conveniently analyzed as follows. Let the specific activity of the precursor be σ (cpm pg^{-1}; dpm molecule^{-1}; or cpm pmole^{-1}; etc.), and the amount of radioactivity accumulated in the RNA at time t, be $R(t)$ (dpm). The rate of appearance of labeled RNA is thus:

$$\frac{dR(t)}{dt} = \sigma k_s - k_d R(t) \tag{1}$$

Here the mass units of k_s must be compatible with those chosen for s. For example if σ is expressed as dpm pg^{-1}, k'_s expressed as pg min^{-1}. The *mass* of the newly synthesized RNA at any point in time, $C(t)$ (here in pg), would then be R/σ, and thus the rate of synthesis of the RNA is

$$\frac{dC(t)}{dt} = k'_s - k_d C(t) \tag{2}$$

The assumption in equations (1) and (2) is that decay of the RNA population is stochastic, and thus first order in form, as shown by numerous experimental studies (see text). That is, if in equation (2) all new synthesis were stopped so that $k_s = 0$,

$$C(t) = C_0 e^{-k_d t} \tag{3}$$

where C_0 is the amount of newly synthesized RNA at the moment the experiment is begun. Thus to obtain the half-life, $t_{1/2}$

$$t_{1/2} = \frac{\ln 2}{k_d} \tag{4}$$

The number of molecules of newly synthesized (i.e., labeled) RNA accumulates at t min, assuming no labeled RNA present at the initiation of the experiment, is from equation (2):

$$C(t) = \frac{k_s}{k_d} (1 - e^{-k_d t}) \tag{5}$$

Equation (5) describes a saturation curve [e.g., see Fig. 3.4(a)], the *initial slope* of which gives the synthesis rate constant k_s, and the rate of curvature of which is determined by the decay rate constant, so that the position of the half-maximum quantity of newly synthesized RNA on the time scale (i.e., abscissa) is given by equation (4). The saturation or equilibrium value is the *steady state* quantity of newly synthesized RNA, as can be seen by considering equation (5) at $t = \infty$. The steady state level of newly synthesized RNA, C_∞ (i.e., in molecules or pg, depending on the units of k_s) is thus:

$$C_\infty = \frac{k_s}{k_d} \tag{6}$$

The following treatment permits kinetic analysis of incorporation experiments in the most general case in which the precursor pool specific activity varies during the course of the experiment. $R(t)$ again represents the quantity of radioactive precursor incorporated in the RNA at any time t, as in equation (1), and the analogous statement is thus:

$$\frac{dR(t)}{dt} = k_s S(t) - k_d R(t) \tag{7}$$

where $S(t)$ is the value of the precursor pool specific activity *at any given time* (i.e., cpm pg^{-1}, cpm pmole^{-1}, or cpm molecule^{-1}, as above). Equation (7) is utilized in this book mainly for analysis of entry of newly synthesized mRNA into the cytoplasm, where an empirically determined lag, L, precedes the appearance of labeled molecules (i.e., the intranuclear synthesis, processing, and transport time). Thus, the form of the solution to equation (7) that is most useful for our purposes is

$$R(t) = k_s \int_L^t S(t')\exp[-k_d(t - t')_{dt'}] \tag{8}$$

Here k_s is the rate of flow of newly synthesized mRNA into the cytoplasm, and t' is a variable of integration. Equation (8) was utilized initially by Galau *et al.* (1977a). In practice a nonlinear least squares analysis is required to extract the best values of k_s and k_d from data that consist of experimental measurements of $R(t)$ and $S(t)$, obtained from samples of the same batch of embryos. The analysis is done with the aid of a computer using an algorithm in which varying values of k_s and k_d are applied, and the solution is obtained by numerical integration, assuming the form given by equation (8) (see Cabrera *et al.*, 1984, for description of this calculation). Alternatively a coupled set of equations describing nuclear precursor and processing kinetics can be used to provide the lag (L) in the appearance of labeled molecules in the cytoplasm (see, e.g., Nemer *et al.*, 1979). Or, where $S(t)$ is approximated by a known mathematical function, equation (8) can be solved in closed form (see, e.g., Anderson and Lengyel, 1981).

Once the best values of k_s and k_d have been extracted they can be utilized to generate the mass accumulation function for the newly synthesized RNA by use of equation (2). Data fit to the form of equation (8), and the resulting mass accumulation curves, are illustrated in text Fig. 3.6.

It is sometimes required to calculate the accumulation of a transcript in an embryo where the synthesis rate per cell can be assumed constant, but where the number of nuclear sites of synthesis is increasing during development as a result of mitosis. For stable sequences not displaying measurable turnover, the rate of transcript accumulation per embryo would be given as

$$\frac{dC(t)}{dt} = k_s N_0 e^{k_g t} \tag{9}$$

Here k_s is the synthesis rate per cell; C is the amount of newly synthesized transcript per embryo; N_0 is the number of synthetically active cells present at the beginning of the experiment; and k_g is a growth constant selected to fit the rate of increase in the number of synthetically active cells per embryo. Where the transcript is not stable the accumulation of transcripts per embryo at any point is simply $C_\infty N(t)$ where $N(t)$ is the number of cells at that time, and C_∞ is the steady state per cell obtained as in equation (6).

References

Ackerman, E. J. (1983). Molecular cloning and sequencing of OAX DNA: An abundant gene family transcribed and activated in *Xenopus* oocytes. *EMBO J.* **2**, 1417.

Adamson, E. D., and Ayers, S. E. (1979). The localization and synthesis of some collagen types in developing mouse embryos. *Cell* **16**, 953.

Adamson, E. D., and Woodland, H. R. (1974). Histone synthesis in early amphibian development: Histone and DNA synthesis are not coordinated. *J. Mol. Biol.* **88**, 263.

Adamson, E. D., and Woodland, H. R. (1977). Changes in the rate of histone synthesis during oocyte maturation and very early development of *Xenopus laevis*. *Dev. Biol.* **57**, 136.

Akam, M. (1984). A common segment in genes for segments of *Drosophila*. *Nature (London)* **308**, 402.

Akam, M. E., and Martinez-Arias, A. (1985). The distribution of *Ultrabithorax* transcripts in *Drosophila* embryos. *EMBO J.* **4**, 1689.

Akam, M., Moore, H., and Cox, A. (1984). *Ultrabithorax* mutations map to distant sites within the bithorax complex of *Drosophila*. *Nature (London)* **309**, 635.

Akasaka, K., Amemiya, S., and Terayama, H. (1980). Scanning electron microscopical study of the inside of sea urchin embryos (*Pseudocentrotus depressus*). Effects of aryl β-xyloside, tunicamycin, and deprivation of sulfate ions. *Exp. Cell Res.* **129**, 1.

Akhurst, R. J., Calzone, F. C., Lee, J. J., Britten, R. J., and Davidson, E. H. (1986). Structure and organization of the CyIII actin gene subfamily of the sea urchin, *Strongylocentrotus purpuratus*. Submitted for publication.

Al-Atia, G. R., Fruscoloni, P., and Jacobs-Lorena, M. (1985). Translational regulation of mRNAs for ribosomal proteins during early *Drosophila* development. *Biochemistry* **24**, 5798.

Albertson, D. G. (1984). Formation of the first cleavage spindle in nematode embryos. *Dev. Biol.* **101**, 61.

Alexandraki D., and Ruderman, J. V. (1981). Sequence heterogeneity, multiplicity, and genomic organization of α- and β-tubulin genes in sea urchins. *Mol. Cell. Biol.* **1**, 1125.

Alexandraki, D., and Ruderman, J. V. (1985a). Multiple polymorphic α- and β-tubulin mRNAs are present in sea urchin eggs. *Proc. Natl. Acad. Sci. U.S.A.* **82**, 134.

Alexandraki, D., and Ruderman, J. V. (1985b). Expression of α- and β-tubulin genes during development of sea urchin embryos. *Dev. Biol.* **109**, 436.

Ambrosio, L., and Schedl, P. (1985). Two discrete modes of histone gene expression during oogenesis in *Drosophila melanogaster*. *Dev. Biol.* **111**, 220.

Amemiya, S. (1986). Fibril network as a possible substratum for the migration of primary mesenchyme cells in sea urchin embryo. Submitted for publication.

Amemiya, S., Akasaka, K., and Terayama, H. (1982). Scanning electron microscopy of gastrulation in a sea urchin (*Anthocidaris crassispina*). *J. Embryol. Exp. Morphol.* **67**, 27.

Amma, K. (1911). Über die Differenzierung der Keimbahnzellen bei den Copepoden. *Arch. Zellforsch.* **6**, 497.

Ancel, P., and Vintemberger, P. (1948). Reserches sur le déterminisme de la symétrie bilatérale dans l'oeuf des Amphibiens. *Biol. Bull. Fr. Belg., Suppl.* **31**, 1.

Anderson, D. M., and Smith, L. D. (1977). Synthesis of heterogeneous nuclear RNA in full-grown oocytes of *Xenopus laevis* (Daudin). *Cell* **11**, 663.

553

Anderson, D. M., and Smith, L. D. (1978). Patterns of synthesis and accumulation of heterogeneous RNA in lampbrush stage oocytes of *Xenopus laevis* (Daudin). *Dev. Biol.* **67**, 274.

Anderson, D. M., Galau, G. A., Britten, R. J., and Davidson, E. H. (1976). Sequence complexity of the RNA accumulated in oocytes of *Arbacia punctulata*. *Dev. Biol.* **51**, 138.

Anderson, D. M., Scheller, R. H., Posakony, J. W., McAllister, L. B., Trabert, S. W., Beall, C., Britten, R. J., and Davidson, E. H. (1981). Repetitive sequences of the sea urchin genome. Distribution of members of specific repetitive families. *J. Mol. Biol.* **145**, 5.

Anderson, D. M., Richter, J. D., Chamberlin, M. E., Price, D. H., Britten, R. J., Smith, L. D., and Davidson, E. H. (1982). Sequence organization of the poly(A) RNA synthesized and accumulated in lampbrush chromosome stage *Xenopus laevis* oocytes. *J. Mol. Biol.* **155**, 281.

Anderson, J. N. (1984). The effect of steroid hormones on gene transcription. *In* "Biological Regulation and Development: Hormone Action" (R. F. Goldberger and K. R. Yamamoto, eds.), Vol. 3B, p. 169. Plenum, New York.

Anderson, K. V., and Lengyel, J. A. (1979). Rates of synthesis of major classes of RNA in *Drosophila* embryos. *Dev. Biol.* **70**, 217.

Anderson, K. V., and Lengyel, J. A. (1980). Changing rates of histone mRNA synthesis and turnover in *Drosophila* embryos. *Cell* **21**, 717.

Anderson, K. V., and Lengyel, J. A. (1981). Changing rates of DNA and RNA synthesis in *Drosophila* embryos. *Dev. Biol.* **82**, 127.

Anderson, K. V., and Lengyel, J. A. (1984). Histone gene expression in *Drosophila* development: Multiple levels of gene regulation. *In* "Histone Genes, Structure, Organization, and Regulation" (G. S. Stein, J. L. Stein, and W. F. Marzluff, eds.), p. 135. Wiley, New York.

Anderson, K. V., and Nüsslein-Volhard, C. (1984). Information for the dorsal–ventral pattern of the *Drosophila* embryo is stored as maternal mRNA. *Nature (London)* **311**, 223.

Anderson, K. V., Jürgens, G., and Nüsslein-Volhard, C. (1985a). Establishment of dorsal–ventral polarity in the *Drosophila* embryo: Genetic studies on the role of the *Toll* gene product. *Cell* **42**, 779.

Anderson, K. V., Bokla, L., and Nüsslein-Volhard, C. (1985b). Establishment of dorsal-ventral polarity in the *Drosophila* embryo: The induction of polarity by the *Toll* gene product. *Cell* **42**, 791.

Angelier, N., and Lacroix, J. C. (1975). Complexes de transcription d'origines nucléolaire et chromosomique d'ovocytes de *Pleurodeles waltii* et *P. poireti* (Amphibiens, Urodèles). *Chromosoma* **51**, 323.

Angelier, N., Hemon, D., and Bouteille, M. (1979). Mechanisms of transcription in nucleoli of amphibian oocytes as visualized by high-resolution autoradiography. *J. Cell Biol.* **80**, 277.

Angelier, N., Paintrand, M., Lavaud, A., and Lechaire, J. P. (1984). Scanning electron microscopy of amphibian lampbrush chromosomes. *Chromosoma* **89**, 243.

Angerer, L. M., and Angerer, R. C. (1981). Detection of poly(A)$^+$ RNA in sea urchin eggs and embryos by quantitative *in situ* hybridization. *Nucleic Acids Res.* **9**, 2819.

Angerer, L. M., DeLeon, D. V., Angerer, R. C., Showman, R. M., Wells, D. E., and Raff, R. A. (1984). Delayed accumulation of maternal histone mRNA during sea urchin oogenesis. *Dev. Biol.* **101**, 477.

Angerer, L., DeLeon, D., Cox, K., Maxson, R., Kedes, L., Kaumeyer, J., Weinberg, E., and Angerer, R. (1985). Simultaneous expression of early and late histone messenger RNAs in individual cells during development of the sea urchin embryo. *Dev. Biol.* **112**, 157.

Angerer, R. C., and Davidson, E. H. (1984). Molecular indices of cell lineage specification in sea urchin embryos. *Science* **226**, 1153.

Anstrom, J. A., Chin, J. E., Leaf, D. S., Parks, A. L., and Raff, R. A. (1986). Immunolocalization and characterization of a sea urchin primary mesenchyme cell lineage specific surface antigen, msp130. In preparation.

Arceci, R. J., and Gross, P. R. (1977). Noncoincidence of histone and DNA synthesis in cleavage cycles of early development. *Proc. Natl. Acad. Sci. U.S.A.* **74**, 5016.

Arceci, R. J., and Gross, P. R. (1980). Histone gene expression: Progeny of isolated early blastomeres in culture make the same change as in the embryo. *Science* **209**, 607.

Arceci, R. J., Senger, D. R., and Gross, P. R. (1976). The programmed switch in lysine-rich histone synthesis at gastrulation. *Cell* **9**, 171.

Arnolds, W. J. A., van den Biggelaar, J. A. M., and Verdonk, N. H. (1983). Spatial aspects of cell interactions involved in the determination of dorsoventral polarity in equally cleaving gastropods and regulative abilities of their embryos, as studied by micromere deletions in *Lymnaea* and *Patella*. *Wilhelm Roux's Arch. Dev. Biol.* **192**, 75.

Aronson, A. I. (1972). Degradation products and a unique endonuclease in heterogeneous nuclear RNA in sea urchin embryos. *Nature (London) New Biol.* **235**, 40.

Aronson, A. I., and Chen, K. (1977). Rates of RNA chain growth in developing sea urchin embryos. *Dev. Biol.* **59**, 39.

Aronson, A. I., and Wilt, F. H. (1969). Properties of nuclear RNA in sea urchin embryos. *Proc. Natl. Acad. Sci. U.S.A.* **62**, 186.

Aronson, A. I., Wilt, F. H., and Wartiovaara, J. (1972). Characterization of pulse-labeled nuclear RNA in sea urchin embryos. *Exp. Cell Res.* **72**, 309.

Artavanas-Tsakonis, S., Muskavitch, M. A. T., and Yedvobnick, B. (1983). Molecular cloning of *Notch*, a locus affecting neurogenesis in *Drosophila melanogaster*. *Proc. Natl. Acad. Sci. U.S.A.* **80**, 1977.

Artavanas-Tsakonis, S., Grimwade, B. G., Harrison, R. G., Markopoulou, K., Muskavitch, M. A. T., Schlesinger-Bryant, R., Wharton, K., and Yedvobnick, B. (1984). The *Notch* locus of *Drosophila melanogaster*: A molecular analysis. *Dev. Genet.* **4**, 233.

Arthur, C. G., Weide, C. M., Vincent, W. S., and Goldstein, E. S. (1979). mRNA sequence diversity during early embryogenesis in *Drosophila melanogaster*. *Exp. Cell Res.* **121**, 87.

Atkinson, J. W. (1971). Organogenesis in normal and lobeless embryos of the marine proso-branch gastropod *Ilyanassa obsoleta*. *J. Morphol.* **133**, 339.

Auclair, W., and Meismer, D. M. (1965). Cilia development and associated protein synthesis in the sea urchin embryo. *Biol. Bull. (Woods Hole, Mass.)* **129**, 397.

Auclair, W., and Siegel, B. W. (1966). Cilia regeneration in the sea urchin embryo: Evidence for a pool of ciliary proteins. *Science* **154**, 913.

Bach, R., Allet, B., and Crippa, M. (1981). Sequence organization of the spacer in the ribosomal genes of *X. clivii* and *X. borealis*. *Nucleic Acids Res.* **9**, 5311.

Bachvarova, R. (1981). Synthesis, turnover, and stability of heterogeneous RNA in growing mouse oocytes. *Dev. Biol.* **86**, 384.

Bachvarova, R., and Davidson, E. H. (1966). Nuclear activation at the onset of amphibian gastrulation. *J. Exp. Zool.* **163**, 285.

Bachvarova, R., and De Leon, V. (1980). Polyadenylated RNA of mouse ova and loss of maternal RNA in early development. *Dev. Biol.* **74**, 1.

Bachvarova, R., Davidson, E. H., Allfrey, V. G., and Mirsky, A. E. (1966). Activation of RNA synthesis associated with gastrulation. *Proc. Natl. Acad. Sci. U.S.A.* **55**, 358.

Bachvarova, R., Baran, M. M., and Tejblum, A. (1980). Development of naked growing mouse oocytes *in vitro*. *J. Exp. Zool.* **211**, 159.

Bachvarova, R., Burns, J. P., Spiegelman, I., Choy, J., and Chaganti, R. S. K. (1982). Morphology and transcriptional activity of mouse oocyte chromosomes. *Chromosoma* **86**, 181.

Bachvarova, R., De Leon, V., Johnson, A., Kaplan, G., and Paynton, B. V. (1985). Changes in total RNA, polyadenylated RNA, and actin mRNA during meiotic maturation of mouse oocytes. *Dev. Biol.* **108**, 325.

Bäckström, S., and Gustafson, T. (1953). Lithium sensitivity in the sea urchin in relation to the stage of development. *Ark. Zool.* **6**, 185.

Baker, B. S. (1973). The maternal and zygotic control of development by *cinnamon*, a new mutant in *Drosophila melanogaster*. *Dev. Biol.* **33**, 429.

Balakier, H., and Pedersen, R. A. (1982). Allocation of cells to inner cell mass and trophecto-derm lineages in preimplantation mouse embryos. *Dev. Biol.* **90**, 352.

Baldwin, W. M. (1915). The action of ultra-violet rays upon the frog's eggs. I. The artificial production of spina bofida. *Anat. Rec.* **9**, 365.

Ballantine, J. E. M., and Woodland, H. R. (1985). Polyadenylation of histone mRNA in *Xenopus* oocytes and embryos. *FEBS Lett.* **180**, 224.

Ballantine, J. E. M., Woodland, H. R., and Sturgess, E. A. (1979). Changes in protein synthesis during the development of *Xenopus laevis*. *J. Embryol. Exp. Morphol.* **51**, 137.

Ballinger, D. G., Bray, S. J., and Hunt, T. (1984). Studies of the kinetics and ionic requirements for the phosphorylation of ribosomal protein S6 after fertilization of *Arbacia punctulata* eggs. *Dev. Biol.* **101**, 192.

Baltus, E., and Hanocq-Quertier, J. (1985). Heat-shock response in *Xenopus* oocytes during meiotic maturation and activation. *Cell Differ.* **16**, 161.

Banerji, J., Olson, L., and Schaffner, W. (1983). A lymphocyte-specific cellular enhancer is located downstream of the joining region in immunoglobulin heavy chain genes. *Cell* **33**, 729.

Bantock, C. R. (1970). Experiments on chromosome elimination in the gall midge, *Mayetiola destructor*. *J. Embryol. Exp. Morphol.* **24**, 257.

Barrett, P., Kloetzel, P.-M., and Sommerville, J. (1983). Specific interaction of proteins with 5S RNA and tRNA in the 42S storage particle of *Xenopus* oocytes. *Biochim. Biophys. Acta* **740**, 347.

Barrett, P., Johnson, R. M., and Sommerville, J. (1984). Immunological identity of proteins that bind stored 5S RNA in *Xenopus* oocytes. *Exp. Cell Res.* **153**, 299.

Baum, E. Z., and Wormington, W. M. (1985). Coordinate expression of ribosomal protein genes during *Xenopus* development. *Dev. Biol.*, **111**, 488.

Beachy, P. A., Helfand, S. L., and Hogness, D. S. (1985). Segmental distribution of bithorax complex proteins during *Drosophila* development. *Nature (London)* **313**, 545.

Beall, C. J., and Hirsh, J. (1984). High levels of intron-containing RNAs are associated with expression of the *Drosophila* dopa decarboxylase gene. *Mol. Cell. Biol.* **4**, 1669.

Beckendorf, S. K., and Kafatos, F. C. (1976). Differentiation in the salivary glands of *Drosophila melanogaster*: Characterization of the glue proteins and their developmental appearance. *Cell* **9**, 365.

Bédard, P.-A., and Brandhorst, B. P. (1983). Patterns of protein synthesis and metabolism during sea urchin embryogenesis. *Dev. Biol.* **96**, 74.

Bédard, P.-A., and Brandhorst, B. P. (1986a). Cytoplasmic distributions of translatable messenger RNA species and the regulation of patterns of protein synthesis during sea urchin embryogenesis. *Dev. Biol.* **115**, 261.

Bédard, P.-A., and Brandhorst, B. P. (1986b). Translational activation of maternal mRNA encoding the heat shock protein hsp90 during sea urchin embryogenesis. *Dev. Biol.* **117**. In press.

Beddington, R. (1983). The origin of the foetal tissues during gastrulation in the rodent. *In* "Development in Mammals" (M. H. Johnson, ed.), Vol. 5, p. 1. Elsevier, Amsterdam.

Beermann, W., and Clever, U. (1964). Chromosome puffs. *Sci. Am.* **210**, 50.

Belayew, A., and Tilghman, S. M. (1982). Genetic analysis of α-fetoprotein synthesis in mice. *Mol. Cell. Biol.* **2**, 1427.

Bell, E., and Reeder, R. (1967). The effect of fertilization on protein synthesis in the egg of the surf clam, *Spisula solidissima*. *Biochim. Biophys. Acta* **142**, 500.

Benbow, R. M., Pestell, R. Q. W., and Ford, C. C. (1975). Appearance of DNA polymerase activities during early development of *Xenopus laevis*. *Dev. Biol.* **43**, 159.

Bender, W., Akam, M., Karch, F., Beachy, P. A., Peifer, M., Spierer, P., Lewis, E. B., and Hogness, D. S. (1983). Molecular genetics of the bithorax complex in *Drosophila melanogaster*. *Science* **221**, 23.

Benoit, J. (1930). Contribution à l'étude de la lignée germinale chez le poulet. Destruction précoce des genocytes primaires par les rayons ultra-violet. *C. R. Seances Soc. Biol. Ses. Fil.* **104**, 1329.

Benson, S. C., Benson, N. C., and Wilt, F. (1986a). The organic matrix of the skeletal spicule of sea urchin embryos. *J. Cell Biol.* **102**, 1878.

Benson, S. C., Sucov, H. M., Stephens, L., Davidson, E. H., and Wilt, F. (1986b). A lineage specific gene encoding a major matrix protein of the sea urchin embryo spicule. I. Authentication of the cloned gene and developmental expression. Submitted for publication.

Berg, W. E., and Mertes, D. H. (1970). Rates of synthesis and degradation of protein in the sea urchin embryo. *Exp. Cell. Res.* **60**, 218.

Berger, E. R. (1967). A study on the size, shape, and distribution of mitochondria and yolk granules in the sea urchin egg. *Abstr., Annu. Meet. Am. Soc. Cell Biol. 7th* **35**, 11a.

Bernstein, S. I., Mogami, K., Donady, J. J., and Emerson, C. P. (1983). *Drosophila* muscle myosin heavy chain encoded by a single gene in a cluster of muscle mutations. *Nature (London)* **302**, 393.

Beyer, A. L., Christensen, M. E., Walker, B. W., and LeStourgeon, W. M. (1977). Identification and characterization of the packaging proteins of core 40S hnRNP particles. *Cell* **11**, 127.

Bibring, T., and Baxandall, J. (1977). Tubulin synthesis in sea urchin embryos: Almost all tubulin of the first cleavage mitotic apparatus derives from the unfertilized egg. *Dev. Biol.* **55**, 191.

Bibring, T, and Baxandall, J. (1981). Tubulin synthesis in sea urchin embryos. II. Ciliary A tubulin derives from the unfertilized egg. *Dev. Biol.* **83**, 122.

Bienz, M. (1984). Developmental control of the heat shock response in *Xenopus*. *Proc. Natl. Acad. Sci. U.S.A.* **81**, 3138.

Bienz, M., and Gurdon, J. B. (1982). The heat shock response in *Xenopus* oocytes is controlled at the translational level. *Cell* **29**, 811.

Bier, K. H. (1960). Der Karyotyp von *Calliphora erythrocephala* Meigen unter besonderer Berücksichtigung der Nährzellkernchromosomen im gebündelten und gepaarten Zustand. *Chromosoma* **11**, 335.

Bier, K. (1963). Synthese, interzellulärer Transport, und Abbau von Ribonukleinsäure in Ovar der Stubenfliege *Musca domestica*. *J. Cell Biol.* **16**, 436.

Bier, K. (1967). Oogenese, das Wachstum von Riesenzellen. *Naturwissenschaften* **54**, 189.

Bier, K., Kunz, W., and Ribbert, D. (1969). Insect oogenesis with and without lampbrush chromosomes. *Chromosomes Today* **2**, 107.

Birchmeier, C., Grosschedl, R., and Birnstiel, M. L. (1982). Generation of authentic 3' termini of an H2a mRNA *in vivo* is dependent on a short inverted DNA repeat and on spacer sequences. *Cell* **28**, 739.

Birchmeier, C., Folk, W., and Birnstiel, M. L. (1983). The terminal RNA stem-loop structure and 80 bp of spacer DNA are required for the formation of 3' termini of sea urchin H2a mRNA. *Cell* **35**, 433.

Birchmeier, C., Schümperli, D., Sconzo, G., and Birnstiel, M. L. (1984). 3' Editing of mRNAs: Sequence requirements and involvement of a 60 nucleotide RNA in maturation of histone mRNA precursors. *Proc. Natl. Acad. Sci. U.S.A.* **81**, 1057.

Bird, A. P., Rochaix, J.-D., and Bakken, A. H. (1973). The mechanism of gene amplification in *Xenopus laevis* oocytes. *In* "Molecular Cytogenetics" (B. A. Hamkalo and J. Papaconstantinou, eds.), p. 49. Plenum, New York.

Bird, A., Taggart, M., and Macleod, D. (1981). Loss of rDNA methylation accompanies the onset of ribosomal gene activity in early development of *X. laevis*. *Cell* **26**, 381.

Birnstiel, M. L., Busslinger, M., and Strub, K. (1985). Transcription termination and 3' processing: The end is in site! *Cell* **41**, 349.

Bishop, D. L., and King, R. C. (1984). An ultrastructural study of ovarian development in the *otu*[7] mutant of *Drosophila melanogaster*. *J. Cell. Sci.* **67**, 87.

Black, S. D., and Gerhart, J. C. (1984). Experimental control of the site of embryonic axis formation in *Xenopus laevis* eggs centrifuged before first cleavage. *Dev. Biol.* **108**, 310.

Blackler, A. W. (1970). The integrity of the reproductive cell line in the Amphibia. *Curr. Top. Dev. Biol.* **5**, 71.

Blair, S. S. (1983). Blastomere ablation and the developmental origin of identified monoamine-containing neurons in the leech. *Dev. Biol.* **95**, 65.

Blau, H. M., Pavlath, G. K., Hardeman, E. C., Chiu, C.-P., Silberstein, L., Webster, S. G., Miller, S. C., and Webster, C. (1985). Plasticity of the differentiated state. *Science* **230**, 758.

Blumenthal, A. B., Kriegstein, H. J., and Hogness, D. S. (1974). The units of DNA replication in *Drosophila melanogaster* chromosomes. *Cold Spring Harbor Symp. Quant. Biol.* **38**, 205.

Bogenhagen, D. F., Sakonju, S., and Brown, D. D. (1980). A control region in the center of the 5S RNA gene directs specific initiation of transcription: II. The 3' border of the region. *Cell* **19**, 27.

Bogenhagen, D. F., Wormington, W. M., and Brown, D. D. (1982). Stable transcription complexes of *Xenopus* 5S RNA genes: A means to maintain the differentiated state. *Cell* **28**, 413.

Boivin, A., Vendrely, R., and Vendrely, C. (1948). Biochimie de l'heredite. -L'acide desoxyribonucleique du noyan cellulaire, dépositaire des caractéres héréditaires; arguments d'ordre analytique. *C. R. Hebd. Seances Acad. Sci.* **226**, 1061.

Bonner, T. I., Brenner, D. J., Neufeld, B. R., and Britten, R. J. (1973). Reduction in the rate of DNA reassociation by sequence divergence. *J. Mol. Biol.* **81**, 123.

Bonner, W. M. (1975a). Protein migration into nuclei. I. Frog oocyte nuclei accumulate microinjected histones, allow entry to small proteins, and exclude larger proteins. *J. Cell Biol.* **64**, 421.

Bonner, W. M. (1975b). Protein migration into nuclei. II. Frog oocyte nuclei accumulate a class of microinjection oocyte nuclear proteins and exclude a class of oocyte cytoplasmic proteins. *J. Cell Biol.* **64**, 431.

Bonnet, C. (1745). "Traité d'Insectologie," 2 vols. Chez Durand, Paris. (Last Ed., 1779.)

Bonnet, C. (1762). "Considérations sur les Corps Organisés," 2 vols. Chez Marc-Michel Rey, Amsterdam. (Reprinted 1768, 1776.)

Boolootian, R. A. (1966). "Physiology of Echinodermata." Wiley (Interscience), New York.

Boreen, S. M., Gizang, E., and Schultz, R. M. (1983). Biochemical studies of mammalian oogensis: Synthesis of 5S and 4S RNA during growth of the mouse oocyte. *Gamete Res.* **8**, 379.

Born, G. (1885). Über den Einfluss der Schwere auf das Froschei. *Arch. Mikrosk. Anat.* **24**, 475.

Borst, P., Frasch, A. C. C., Bernards, A., Van der Ploeg, L. H. T., Hoeijmakers, J. H. J., Arnberg, A. C., and Cross, G. A. M. (1981). DNA rearrangements involving the genes for variant antigens in *Trypanosoma brucei*. *Cold Spring Harbor Symp. Quant. Biol.* **45**, 935.

Boseley, P., Moss, T., Mächler, M., Portmann, R., and Birnstiel, M. (1979). Sequence organization of the spacer DNA in a ribosomal gene unit of *Xenopus laevis*. *Cell* **17**, 19.

Bossy, B., Hall, L. M. C., and Spierer, P. (1984). Genetic activity along 315 kb of the *Drosophila* chromosome. *EMBO J.* **3**, 2537.

Boswell, R. E., and Mahowald, A. P. (1985). *tudor*, a gene required for assembly of the germ plasm in *Drosophila melanogaster*. *Cell* **43**, 97.

Bottke, W. (1973). Lampenbürstenchromosomen und Amphinukleolen in Oocytenkernen der Schnecke *Bithynia tentaculata* L. *Chromosoma* **42**, 175.

Bounoure, L. (1937). Les suites de l'irradiation du déterminant germinal, chez la Grenouille rousse. Par les rayons ultra-violets: Résultats histologiques. *C. R. Seances Soc. Biol. Ses. Fil.* **125**, 898.

Bounoure, L., Aubry, R., and Huck, M. L. (1954). Nouvelles recherches experimentales sur les origines de la lignée reproductive chez la Grenouille rousse. *J. Embryol. Exp. Morphol.* **2**, 245.

Bourne, G. C. (1894). Epigenesis or evolution. *Sci. Prog. (Oxford)* **1**, 8.

Boveri, T. (1899). Die Entwickelung von *Ascaris megalocephala* mit besonderer Rücksicht auf die Kernverhältnisse. *In* "Festschrift f. C. von Kupffer," p. 383. Fischer, Jena.

Boveri, T. (1901a). Die Polarität von Oocyte, Ei und Larve des *Strongylocentrotus lividus. Zool. Jahrb., Abt. Anat. Ontog. Tiere* **14**, 630.

Boveri, T. (1901b). Über die Polarität des Seeigel-Eies. *Verh. Phys. Med. Ges. Würzburg* **34**, 145.

Boveri, T. (1902). Über mehrpolige Mitosen als Mittel zur Analyse des Zellkerns. *Verh. Phys. Med. Ges. Würzburg* **35**, 67. [Parts of this paper have been translated by B. R. Voeller (1968). "The Chromosome Theory of Inheritance," p. 85. Appleton, New York.]

Boveri, T. (1904). "Ergebnisse über die Konstitution der Chromatischen Substanz des Zellkerns." Fischer, Jena.

Boveri, T. (1909). Die Blastomerenkerne von *Ascaris megalocephala* und die Theorie der Chromosomenindividualität. *Arch. Zellforsch.* **3**, 181.

Boveri, T. (1910). Die Potenzen der *Ascaris*—Blastomeren bei abgeänderter Furchung. Zugleich ein Beitrag zur Frage qualitativ-ungleicher Chromosomen-Teilung. *In* "Festschrift f. R. Hertwigs," Vol. III, p. 133. Fischer, Jena.

Boveri, T. (1918). Zwei Fehlerquellen bei Merogonieversuchen und die Entwicklungsfähigkeit merogonischer partiellmerogonischer Seeigelbastarde. *Arch. Entwicklungsmech. Org.* **44**, 417.

Boveri, T., and Hogue, M. J. (1909). Über die Möglichkeit, *Ascaris*-Eier zur Teilung in zwei gleichwertige Blastomeren zu veranlassen. *Sitz. Ber. Phys. Med. Ges. Würzburg.*

Bownes, M. (1975a). A photographic study of development in the living embryo of *Drosophila melanogaster. J. Embryol. Exp. Morphol.* **33**, 789.

Bownes, M. (1975b). Adult deficiencies and duplications of head and thoracic structures resulting from microcautery of blastoderm stage *Drosophila* embryos. *J. Embryol. Exp. Morphol.* **34**, 33.

Bownes, M. (1982). Ovarian yolk-protein synthesis in *Drosophila melanogaster. J. Insect Physiol.* **28**, 953.

Boycott, A. E., and Diver, C. (1923). On the inheritance of sinistrality in *Limnaea peregra. Proc. R. Soc. London, Ser. B* **95**, 207.

Boycott, A. E., Diver, C., Garstang, S. T., and Turner, F. M. (1930). The inheritance of sinistrality in *Limnaea peregra* (Mollusca, Pulmonata). *Philos. Trans. R. Soc., London, Ser. B* **219**, 51.

Boyd, M. M. M. (1941). The structure of the ovary and the formation of the mortal corpus luteum in *Hoplodactylus maculatus*, Gray. *Q. J. Microsc. Sci.* **82**, 337.

Bozzoni, I., Beccari, E., Luo, Z. X., and Amaldi, F. (1981). *Xenopus laevis* ribosomal protein genes: Isolation of recombinant cDNA clones and study of the genomic organization. *Nucleic Acids Res.* **9**, 1069.

Bozzoni, I., Tognoni, A., Pierandrei-Amaldi, P., Beccari, E., Buongiorno-Nardelli, M., and Amaldi, F. (1982). Isolation and structural analysis of ribosomal protein genes in *Xenopus laevis*. Homology between sequences present in the gene and in several different messenger RNAs. *J. Mol. Biol.* **161**, 353.

Bozzoni, I., Fragapane, P., Annesi, F., Pierandrei-Amaldi, P., Amaldi, F., and Beccari, E. (1984). Expression of two *Xenopus laevis* ribosomal protein genes in injected frog oocytes. A specific splicing block interferes with L1 RNA maturation. *J. Mol. Biol.* **180**, 987.

Brachet, A. (1904). Recherches experimentales sur l'oeuf de *Rana fusca. Arch. Biol.* **21**, 103.

Brachet, J. (1933). Recherches sur la synthèse de l'acide thymonucléique pendant le développement de l'oeuf d'Oursin. *Arch. Biol.* **44**, 519.

Brachet, J. (1949). L'hypothèse des plasmagènes dans le développement et la différenciation. *Colloq. Int. C. N. R. S.* **8**, 145.

Brachet, J. (1977). An old enigma: The gray crescent of amphibian eggs. *Curr. Top. Dev. Biol.* **11**, 133.

Brachet, J., and Denis, H. (1963). Effects of actinomycin D on morphogenesis. *Nature (London)* **198**, 205.

Brachet, J., and Donini-Denis, S. (1978). Studies on maturation and differentiation without cleavage in *Chaetopterus variopedatus*. *Differentiation* **11**, 19.

Brachet, J., Decroly, M., Ficq, A., and Quertier, J. (1963a). Ribonucleic acid metabolism in unfertilized and fertilized sea urchin eggs. *Biochim. Biophys. Acta* **72**, 660.

Brachet, J., Ficq, A., and Tencer, R. (1963b). Amino acid incorporation into proteins of nucleate and anucleate fragments of sea urchin eggs: Effect of parthenogenetic activation. *Exp. Cell Res.* **32**, 168.

Brandhorst, B. P. (1976). Two-dimensional gel patterns of protein synthesis before and after fertilization of sea urchin eggs. *Dev. Biol.* **52**, 310.

Brandhorst, B. P. (1980). Simultaneous synthesis, translation and storage of mRNA including histone mRNA in sea urchin eggs. *Dev. Biol.* **79**, 139.

Brandhorst, B. P., and Bannet, M. (1978). Terminal completion of poly(A) synthesis in sea urchin embryos. *Dev. Biol.* **63**, 421.

Brandhorst, B. P., and Humphreys, T. (1971). Synthesis and decay rates of major classes of deoxyribonucleic acid-like ribonucleic acid in sea urchin embryos. *Biochemistry* **10**, 877.

Brandhorst, B. P., and Humphreys, T. (1972). Stabilities of nuclear and messenger RNA molecules in sea urchin embryos. *J. Cell Biol.* **53**, 474.

Brandhorst, B. P., and McConkey, E. H. (1974). Stability of nuclear RNA in mammalian cells. *J. Mol. Biol.* **85**, 451.

Brandhorst, B. P., and Newrock, K. M. (1981). Posttranscriptional regulation of protein synthesis in *Ilyanassa* embryos and isolated polar lobes. *Dev. Biol.* **83**, 250.

Brandhorst, B. P., Verma, D. P. S., and Fromson, D. (1979). Polyadenylated and nonpolyadenylated messenger RNA fractions from sea urchin embryos code for the same abundant proteins. *Dev. Biol.* **71**, 128.

Brandis, J. W., and Raff, R. A. (1978). Translation of oogenetic mRNA in sea urchin eggs and early embryos. Demonstration of a change in translational efficiency following fertilization. *Dev. Biol.* **67**, 99.

Brandis, J. W., and Raff, R. A. (1979). Elevation of protein synthesis is a complex response to fertilization. *Nature (London)* **278**, 467.

Braude, P., Pelham, H., Flach, T., and Lobatto, R. (1979). Posttranscriptional control in the early mouse embryo. *Nature (London)* **282**, 102.

Braude, P. R., Johnson, M. H., Bolton, V. N., and Pratt, H. P. M. (1983). Cleavage and differentiation of the early embryo. *In* "*In Vitro* Fertilization and Embryo Transfer" (P. G. Crosignani and B. L. Rubin, eds.), p. 211. Academic Press, London.

Bravo, R., and Knowland, J. (1979). Classes of proteins synthesized in oocytes, eggs, embryos, and differentiated tissues of *Xenopus laevis*. *Differentiation* **13**, 101.

Brennan, M. D., Weiner, A. J., Goralski, T. J., and Mahowald, A. P. (1982). The follicle cells are a major site of vitellogenin synthesis in *Drosophila melanogaster*. *Dev. Biol.* **89**, 225.

Breuer, M. E., and Pavan, C. (1955). Behavior of polytene chromosomes of *Rhynchosciara angelae* at different stages of larval development. *Chromosoma* **7**, 371.

Brickell, P. M., Latchman, D. S., Murphy, D., Willison, K., and Rigby, P. W. J. (1983). Activation of a *Qa/Tla* class I major histocompatibility antigen gene is a general feature of oncogenesis in the mouse. *Nature (London)* **306**, 756.

Briggs, R. (1972). Further studies on the maternal effect of the *o* gene in the Mexican axolotl. *J. Exp. Zool.* **181**, 271.

Briggs, R., and Cassens, G. (1966). Accumulation in the oocyte nucleus of a gene product essential for embryonic development beyond gastulation. *Proc. Natl. Acad. Sci. U.S.A.* **55**, 1103.

Briggs, R., and King, T. J. (1952). Transplantation of living nuclei from blastula cells into enucleated frogs' eggs. *Proc. Natl. Acad. Sci. U.S.A.* **38**, 455.

Briggs, R., and King, T. J. (1957). Changes in the nuclei of differentiating endoderm cells as revealed by nuclear transplantation. *J. Morphol.* **100**, 269.

Briggs, R., Green, E. U., and King, T. J. (1951). An investigation of the capacity for cleavage and differentiation in *Rana pipiens* eggs lacking 'functional' chromosomes. *J. Exp. Zool.* **116**, 455.

Brinster, R. L., Chen, H. Y., Trumbauer, M. E., and Avarbock, M. R. (1980). Translation of globin messenger RNA by the mouse ovum. *Nature (London)* **283**, 499.

Britten, R. J. (1969). The arithmetic of nucleic acid reassociation. *Year Book Carnegie Inst. Washington* **67**, 332.

Britten, R. J., and Davidson, E. H. (1969). Gene regulation for higher cells: A theory. *Science* **165**, 349.

Britten, R. J., and Davidson, E. H. (1971). Repetitive and nonrepetitive DNA sequences and a speculation on the origins of evolutionary novelty. *Q. Rev. Biol.* **46**, 111.

Britten, R. J., and Davidson, E. H. (1976). Studies on nucleic acid reassociation kinetics: Empirical equations describing DNA reassociation. *Proc. Nat. Acad. Sci. U.S.A.* **73**, 415.

Britten, R. J., and Kohne, D. E. (1967). Nucleotide sequence repetition in DNA. *Year Book Carnegie Inst. Washington* **65**, 78.

Britten, R. J., and Kohne, D. E. (1968a). Repeated sequences in DNA. *Science* **161**, 529.

Britten, R. J., and Kohne, D. E. (1968b). Repeated nucleotide sequences. *Year Book Carnegie Inst. Washington* **66**, 73.

Britten, R. J., Graham, D. E., and Neufeld, B. R. (1974). Analysis of repeating DNA sequences by reassociation. *In* "Nucleic Acids and Proteins Synthesis," Part E (L. Grossman and K. Moldave, eds.), Methods in Enzymology, Vol. 29, p. 363. Academic Press, New York.

Britten, R. J., Graham, D. E., Eden, F. C., Painchaud, D. M., and Davidson, E. H. (1976). Evolutionary divergence and length of repetitive sequences in sea urchin DNA. *J. Mol. Evol.* **9**, 1.

Brock, H. W., and Reeves, R. (1978). An investigation of *de novo* protein synthesis in the South African clawed frog, *Xenopus laevis. Dev. Biol.* **66**, 128.

Brogan, K. L., and Goldstein, E. S. (1985). mRNA usage during *Drosophila melanogaster* embryonic development. Analysis of nine cloned DNA segments. *Exp. Cell Res.* **158**, 95.

Browder, L. W., ed. (1985). "Developmental Biology. A Comprehensive Survey: Oogenesis," Vol. 1. Plenum, New York.

Browder, L. W., and Williamson, J. H. (1976). The effects of *cinnamon* on xanthine dehydrogenase, aldehyde oxidase, and pyridoxal oxidase activity during development in *Drosophila melanogaster. Dev. Biol.* **53**, 241.

Brower, D. L. (1985). The sequential compartmentalization of *Drosophila* segments revisited. *Cell* **41**, 361.

Brower, P. T., and Schultz, R. M. (1982). Biochemical studies of mammalian oogenesis: Possible existence of a ribosomal and poly(A)-containing RNA-protein supramolecular complex in mouse oocytes. *J. Exp. Zool.* **220**, 251.

Brower, P. T., Gizang, E., Boreen, S. M., and Schultz, R. M. (1981). Biochemical studies of mammalian oogenesis: Synthesis and stability of various classes of RNA during growth of the mouse oocyte *in vitro. Dev. Biol.* **86**, 373.

Brown, D. D., and Blackler, A. W. (1972). Gene amplification proceeds by a chromosome copy mechanism. *J. Mol. Biol.* **63**, 75.

Brown, D. D., and Dawid, I. B. (1968). Specific gene amplification in oocytes. *Science* **160**, 272.

Brown, D. D., and Gurdon, J. B. (1964). Absence of ribosomal RNA synthesis in the anucleolate mutant of *Xenopus laevis. Proc. Natl. Acad. Sci. U.S.A.* **51**, 139.

Brown, D. D., and Gurdon, J. B. (1966). Size distribution and stability of DNA-like RNA synthesized during development of anucleolate embryos of *Xenopus laevis. J. Mol. Biol.* **19**, 399.

Brown, D. D., and Littna, E. (1966a). Synthesis and accumulation of low molecular weight RNA during embryogenesis of *Xenopus laevis. J. Mol. Biol.* **20**, 95.

Brown, D. D., and Littna, E. (1966b). Synthesis and accumulation of DNA-like RNA during embryogenesis of *Xenopus laevis. J. Mol. Biol.* **20**, 81.

Brown, D. D., and Schlissel, M. S. (1985). A positive transcription factor controls the differential expression of two 5S RNA genes. *Cell* **42**, 759.

Brown, D. D., and Weber, C. S. (1968). Gene linkage by RNA-DNA hybridization. I. Unique DNA sequences homologous to 4S RNA, 5S RNA and ribosomal RNA. *J. Mol. Biol.* **34**, 661.

Brown, D. T., Morris, G. F., Chodchoy, N., Sprecher, C., and Marzluff, W. F. (1985). Structure of the sea urchin U1 RNA repeat. *Nucleic Acids Res.* **13**, 537.

Brown, E. E., and Schubiger, G. (1981). Segmentation of the central nervous system in ligated embryos of *Drosophila melanogaster. Wilhelm Roux's Arch. Dev. Biol.* **190**, 62.

Brown, E. H., and King, R. C. (1964). Studies on the events resulting in the formation of an egg chamber in *Drosophila melanogaster. Growth* **28**, 41.

Brown, J. E., and Weiss, M. C. (1975). Activation of production of mouse liver enzymes in rat hepatoma-mouse lymphoid cell hybrids. *Cell* **6**, 481.

Brown, P. C., Johnston, R. N., and Schimke, R. T. (1983). Approaches to the study of mechanisms of selective gene amplification in cultured mammalian cells. *In* "Gene Structure and Regulation in Development" (S. Subtelny and F. C. Kafatos, eds.), p. 197. Alan R. Liss, New York.

Brown, P. M., and Kalthoff, K. (1983). Inhibition by ultraviolet light of pole cell formation in *Smittia* sp (Chironomidae, Diptera): Action spectrum and photoreversibility. *Dev. Biol.* **97**, 113.

Brownlee, G. G., Cartwright, E. M., and Brown, D. D. (1974). Sequence studies on the 5S DNA of *Xenopus laevis. J. Mol. Biol.* **89**, 703.

Brûlet, P., and Jacob, F. (1982). Molecular cloning of a cDNA sequence encoding a trophectoderm-specific marker during mouse blastocyst formation. *Proc. Natl. Acad. Sci. U.S.A.* **79**, 2328.

Brûlet, P., Babinet, C., Kemler, R., and Jacob, F. (1980). Monoclonal antibodies against trophectoderm-specific markers during mouse blastocyst formation. *Proc. Natl. Acad. Sci. U.S.A.* **77**, 4113.

Brun, R. B. (1978). Developmental capacities of *Xenopus* eggs, provided with erythrocyte or erythroblast nuclei from adults. *Dev. Biol.* **65**, 271.

Bruskin, A. M., Tyner, A. L., Wells, D. E., Showman, R. M., and Klein, W. H. (1981). Accumulation in embryogenesis of five mRNAs enriched in the ectoderm of the sea urchin pluteus. *Dev. Biol.* **87**, 308.

Bruskin, A. M., Bedard, P.-A., Tyner, A. L., Showman, R. M., Brandhorst, B. P., and Klein, W. H. (1982). A family of proteins accumulating in ectoderm of sea urchin embryos specified by two related cDNA clones. *Dev. Biol.* **91**, 317.

Brutlag, D. L., and Peacock, W. J. (1975). Sequences of highly repeated DNA in *Drosophila melanogaster. In* "The Eukaryote Chromosome" (W. J. Peacock and R. D. Brock, eds.), p. 35. Aust. Natl. Univ. Press, Canberra.

Bryan, P. N., Olah, J., and Birnstiel, M. L. (1983). Major changes in the 5' and 3' chromatin structure of sea urchin histone genes accompany their activation and inactivation in development. *Cell* **33**, 843.

Buehr, M. L., and Blackler, A. W. (1970). Sterility and partial sterility in the South African clawed toad following the pricking of the egg. *J. Embryol. Exp. Morphol.* **23**, 375.

Bull, A. L. (1966). Bicaudal, a genetic factor which affects the polarity of the embryo in *Drosophila melanogaster. J. Exp. Zool.* **161**, 221.

Büning, J. (1980). RNA synthesis during early embryogenesis of mass cultured highly synchronized coleopteran embryos. *Wilhelm Roux's Arch. Dev. Biol.* **188**, 215.

Buongiorno-Nardelli, M., Amaldi, F., and Lava-Sanchez, P. A. (1976). Electron microscope analysis of amplifying ribosomal DNA from *Xenopus laevis. Exp. Cell Res.* **98**, 95.

Burch, J. B. E., and Weintraub, H. (1983). Temporal order of chromatin structural changes associated with activation of the major chicken vitellogenin gene. *Cell* **33**, 65.

Burke, R. D. (1980). Morphogenesis of the digestive tract of the pluteus larva of *Strongylocentrotus purpuratus*: Shaping and bending. *Int. J. Invert. Reprod.* **2**, 13.

Burke, R. D. (1983). Development of the larval nervous system of the sand dollar, *Dendraster excentricus*. *Cell Tissue Res.* **229**, 145.

Burns, R. G. (1973a). Kinetics of the regeneration of sea urchin cilia. *J. Cell Sci.* **13**, 55.

Burns, R. (1973b). The regeneration of cilia of *Arbacia punctulata* blastulae. *Biol. Bull. (Woods Hole, Mass.)* **153**, 417.

Burnside, B. (1971). Microtubules and microfilaments in newt neurulation. *Dev. Biol.* **26**, 416.

Busby, S., and Bakken, A. (1979). A quantitative electron microscopic analysis of transcription in sea urchin embryos. *Chromosoma* **71**, 249.

Busby, S., and Bakken, A. H. (1980). Transcription in developing sea urchins: Electron microscopic analysis of cleavage, gastrula, and prism stages. *Chromosoma* **79**, 85.

Busby, S. J., and Reeder, R. H. (1982). Fate of amplified nucleoli in *Xenopus laevis* embryos. *Dev. Biol.* **91**, 458.

Busby, S. J., and Reeder, R. H. (1983). Spacer sequences regulate transcription of ribosomal gene plasmids injected into *Xenopus* embryos. *Cell* **34**, 989.

Busslinger, M., and Barberis, A. (1985). Synthesis of sperm and late histone cDNAs of the sea urchin with a primer complementary to the conserved 3' terminal palindrome: Evidence for tissue-specific and more general histone gene variants. *Proc. Natl. Acad. Sci. U.S.A.* **82**, 5676.

Busslinger, M., Portmann, R., Irminger, J. C., and Birnstiel, M. L. (1980). Ubiquitous and gene-specific regulatory 5' sequences in a sea urchin histone DNA clone coding for histone protein variants. *Nucleic Acids Res.* **8**, 957.

Busslinger, M., Hurst, J., and Flavell, R. A. (1983). DNA methylation and the regulation of globin gene expression. *Cell* **34**, 197.

Butler, E. T., and Chamberlin, M. J. (1982). Bacteriophage SP6-specific RNA polymerase. I. Isolation and characterization of the enzyme. *J. Biol. Chem.* **257**, 5772.

Cabrera, C. V., Jacobs, H. T., Posakony, J. W., Grula, J. W., Roberts, J. W., Britten, R. J., and Davidson, E. H. (1983). Transcripts of three mitochondrial genes in the RNA of sea urchin eggs and embryos. *Dev. Biol.* **97**, 500.

Cabrera, C. V., Lee, J. J., Ellison, J. W., Britten, R. J., and Davidson, E. H. (1984). Regulation of cytoplasmic mRNA prevalence in sea urchin embryos: Rates of appearance and turnover for specific sequences. *J. Mol. Biol.* **174**, 85.

Calabretta, B., Robberson, D. L., Maizel, A. L., and Saunders, G. F. (1981). mRNA in human cells contains sequences complementary to the *Alu* family of repeated DNA. *Proc. Natl. Acad. Sci. U.S.A.* **78**, 6003.

Caldwell, D. C., and Emerson, C. P. (1985). The role of cap methylation in the translational activation of stored maternal histone mRNA in sea urchin embryos. *Cell* **42**, 691.

Callan, H. G. (1955). Recent work on the structure of cell nuclei. *Union Int. Sci. Biol., Ser. B* No. 21, 89.

Callan, H. G. (1957). The lampbrush chromosomes of *Sepia officianalis* L., *Anilocra physiodes* L. and *Scyllium catulus* Cuv. and their structural relationship to the lampbrush chromosomes of Amphibia. *Pubbl. Staz. Zool. Napoli* **29**, 329.

Callan, H. G. (1963). The nature of lampbrush chromosomes. *Int. Rev. Cytol.* **15**, 1.

Callan, H. G. (1967). The organization of genetic units in chromosomes. *J. Cell Sci.* **2**, 1.

Callan, H. G. (1974). DNA replication in the chromosomes of eukaryotes. *Cold Spring Harbor Symp. Quant. Biol.* **38**, 195.

Callan, H. G. (1982). Lampbrush chromosomes. *Proc. R. Soc. London, Ser. B* **214**, 417.

Callan, H. G., and Lloyd, L. (1960). Lampbrush chromosomes of crested newts *Triturus cristatus* (Laurenti). *Philos. Trans. R. Soc. London, Ser. B* **243**, 135.

Callan, H. G., and Lloyd, L. (1975). Working maps of the lampbrush chromosomes of amphibia. *In* "Handbook of Genetics" (R. C. King, ed.), Vol. 4, p. 57. Plenum, New York.

Callen, J. C., Dennebouy, N., and Mounolou, J. C. (1980). Kinetic analysis of entire oogenesis in *Xenopus laevis*. *Dev., Growth Differ.* **22**, 831.

Calzone, F. J., Lee, J. J., Le, N., Britten, R. J., and Davidson, E. H. (1986a). A novel family of maternal gene transcripts present in the egg and embryo of *Strongylocentrotus purpuratus*. I. The structure and expression of an interspersed maternal transcript. In preparation.

Calzone, F. J., Lee, J. J., Le, N., Britten, R. J., and Davidson, E. H. (1986b). A novel family of maternal gene transcripts present in the egg and embryo of *Strongylocentrotus purpuratus*. II. The developmental expression and subcellular localization of a family of untranslatable repetitive gene transcripts. In preparation.

Cameron, R. A., and Hinegardner, R. T. (1971). Initiation of metamorphosis in laboratory cultured sea urchins. *Biol. Bull. (Woods Hole, Mass.)* **146**, 335.

Cameron, R. A., and Holland, N. D. (1985). Demonstration of a granular layer and the fate of the hyaline layer during development of a sea urchin (*Lytechinus variegatus*). *Cell Tissue Res.* **239**, 455.

Campos-Ortega, J. A. (1983). Topological specificity of phenotype expression of neurogenic mutations in *Drosophila*. *Wilhelm Roux's Arch. Dev. Biol.* **192**, 317.

Campos-Ortega, J. A., and Hartenstein, V. (1985). Development of the nervous system. *In* "Comprehensive Insect Physiology, Biochemistry, and Pharmacology. Nervous System: Structure and Motor Function" (G. A. Kerkut and L. I. Gilbert, eds.), Vol. 5, p. 49. Pergamon, Oxford.

Canaani, D., and Berg, P. (1982). Regulated expression of human interferon β_1 gene after transduction into cultured mouse and rabbit cells. *Proc. Natl. Acad. Sci. U.S.A.* **79**, 5166.

Capco, D. G., and Jeffery, W. R. (1979). Origin and spatial distribution of maternal messenger RNA during oogenesis of an insect *Oncopeltus fasciatus*. *J. Cell Sci.* **39**, 63.

Capco, D. G., and Jeffery, W. R. (1982). Transient localizations of messenger RNA in *Xenopus laevis* oocytes. *Dev. Biol.* **89**, 1.

Cape, M., and Decroly, M. (1969). Mesure de la capacite 'template' des acides ribonucleiques des oeufs de *Xenopus laevis* au cours du developpement. *Biochim. Biophys. Acta* **174**, 99.

Card, C. O., Morris, G. F., Brown, D. T., and Marzluff, W. F. (1982). Sea urchin small nuclear RNA genes are organized in distinct tandemly repeating units. *Nucleic Acids Res.* **10**, 7677.

Carpenter, C. D., Bruskin, A. M., Hardin, P. E., Keast, M. J., Anstrom, J., Tyner, A. L., Brandhorst, B. P., and Klein, W. H. (1984). Novel proteins belonging to the troponin C superfamily are encoded by a set of mRNAs in sea urchin embryos. *Cell* **36**, 663.

Carpenter, G., and Cohen, S. (1979). Epidermal growth factor. *Annu. Rev. Biochem.* **48**, 193.

Carrasco, A. E., McGinnis, W., Gehring, W. J., and De Robertis, E. M. (1984). Cloning of an *X. laevis* gene expressed during early embryogenesis coding for a peptide region homologous to *Drosophila* homeotic genes. *Cell* **37**, 409.

Carré, D., Signoret, J., Lefresne, J., and David, J. C. (1981). Enzymes involved in DNA replication in the axolotl. I. Analysis of the forms and activities of DNA polymerase and DNA ligase during development. *Dev. Biol.* **87**, 114.

Carroll, A. G., and Ozaki, H. (1979). Changes in the histones of the sea urchin *Strongylocentrotus purpuratus* at fertilization. *Exp. Cell Res.* **119**, 307.

Carroll, E. J., Acevedo-Duncan, M., Justice, R. W., and Santiago, L. (1986). Structure, assembly, and function of the surface envelope (fertilization envelope) from eggs of the sea urchin, *Strongylocentrotus purpuratus*. *In* "Molecular and Cellular Biology of Fertilization" (J. L. Hendrick, ed.). Alan R. Liss, New York. In press.

Carroll, S. B., and Scott, M. P. (1985). Localization of the *fushi tarazu* protein during *Drosophila* embryogenesis. *Cell* **43**, 47.

Carroll, S. B., and Scott, M. P. (1986). Zygotically active genes that affect the spatial expression of the *fushi tarazu* segmentation gene during early *Drosophila* embryogenesis. *Cell* **45**, 113.

Carson, D. D., and Lennarz, W. J. (1979). Inhibition of polyisoprenoid and glycoprotein biosynthesis causes abnormal embryonic development. *Proc. Natl. Acad. Sci. U.S.A.* **76**, 5709.

Carson, D. D., Farach, M. C., Earles, D. S., Decker, G. L., and Lennarz, W. J. (1985). A monoclonal antibody inhibits calcium accumulation and skeleton formation in cultured embryonic cells of the sea urchin. *Cell* **41**, 639.

Cascio, S. M., and Wassarman, P. M. (1981). Program of early development in the mammal: Synthesis of mitochondrial proteins during oogenesis and early embryogenesis in the mouse. *Dev. Biol.* **83**, 166.

Cascio, S. M., and Wassarman, P. M. (1982). Program of early development in the mammal: Posttranscriptional control of a class of proteins synthesized by mouse oocytes and early embryos. *Dev. Biol.* **89**, 397.

Cassada, R., Isnenghi, E., Culotti, M., and von Ehrenstein, G. (1981). Genetic analysis of temperature-sensitive embryogenesis mutants in *Caenorhabditis elegans*. *Dev. Biol.* **84**, 193.

Cassidy, D. M., and Blackler, A. W. (1974). Repression of nucleolar organizer activity in an interspecific hybrid of the genus *Xenopus*. *Dev. Biol.* **41**, 84.

Cassidy, J. D., and King, R. C. (1972). Ovarian development in *Habrobracon juglandis* (Ashmead) (Hymenoptera: Braconidae). I. The origin and differentiation of the oocyte-nurse cell complex. *Biol. Bull. (Woods Hole, Mass.)* **143**, 483.

Cather, J. N. (1967). Cellular interactions in the development of the shell gland of the gastropod, *Ilyanassa*. *J. Exp. Zool.* **166**, 205.

Cather, J. N. (1971). Cellular interactions in the regulation of development in annelids and molluscs. *Adv. Morphog.* **9**, 67.

Cather, J. N. (1973). Regulation of apical cilia development by the polar lobe of *Ilyanassa* (Gastropoda: Nassariidae). *Malacologia* **12**, 213.

Cather, J. N., and Verdonk, N. H. (1974). The development of *Bithynia tentaculata* (Prosobranchia, Gastropoda) after removal of the polar lobe. *J. Embryol. Exp. Morphol.* **31**, 415.

Cather, J. N., and Verdonk, N. H. (1979). Development of *Dentalium* following removal of D-quadrant blastomeres at successive cleavage stages. *Wilhelm Roux's Arch. Dev. Biol.* **187**, 355.

Cattanach, B. M., and Kirk, M. (1985). Differential activity of maternally and paternally derived chromosome regions in mice. *Nature (London)* **315**, 496.

Cave, M. D. (1973). Synthesis and characterization of amplified DNA in oocytes of the house cricket, *Acheta domesticus* (Orthoptera: Gryllidae). *Chromosoma* **42**, 1.

Cave, M. D. (1975). Absence of ribosomal DNA amplification in the meroistic (telotrophic) ovary of the large milkweed bug, *Oncopeltus fasciatus* (Dallas) (Hemiptera: Lygaeidae). *J. Cell Biol.* **66**, 461.

Cave, M. D. (1978). Absence of amplification of ribosomal DNA in the polytrophic meroistic ovary of the giant silkworm moth, *Antheraea pernyi* (Lepidoptera: Saturniidae). *Wilhelm Roux's Arch. Dev. Biol.* **184**, 135.

Chabry, L. (1887). Contribution à l'embryologie normale et tératologique des Ascidies simples. *J. Anat. Physiol., Paris* **23**, 167.

Chada, K., Magram, J., Raphael, K., Radice, G., Lacy, E., and Costantini, F. (1985). Specific expression of a foreign β-globin gene in erythroid cells of transgenic mice. *Nature (London)* **314**, 377.

Chalfie, M., Horvitz, H. R., and Sulston, J. E. (1981). Mutations that lead to reiterations in the cell lineages of *C. elegans*. *Cell* **24**, 59.

Chamberlain, J. P., and Metz, C. B. (1972). Mitochondrial RNA synthesis in sea urchin embryos. *J. Mol. Biol.* **64**, 593.

Chan, L.-N., and Gehring, W. (1971). Determination of blastoderm cells in *Drosophila melanogaster*. *Proc. Natl. Acad. Sci. U.S.A.* **68**, 2217.

Chandler, V. L., Maler, B. A., and Yamamoto, K. R. (1983). DNA sequence bound specifically by glucocorticoid receptor *in vitro* render a heterologous promoter hormone receptor *in vivo*. *Cell* **33**, 489.

Chao, M. V., Mellon, P., Charnay, P., Maniatis, T., and Axel, R. (1983). The regulated expression of β-globin genes introduced into mouse erythroleukemia cells. *Cell* **32**, 483.

Charnay, P., Treisman, R., Mellon, P., Chao, M., Axel, R., and Maniatis, T. (1984). Differences in human α- and β-globin gene expression in mouse erythroleukemia cells: The role of intragenic sequences. *Cell* **38**, 251.

Chase, J. W. (1970). Formation of mitochondria during embryogenesis of *Xenopus laevis. Year Book Carnegie Inst. Washington* **68**, 517.

Chase, J. W., and Dawid, I. B. (1972). Biogenesis of mitochondria during *Xenopus laevis* development. *Dev. Biol.* **27**, 504.

Chen, H. Y., Brinster, R. L., and Merz, E. A. (1980). Changes in protein synthesis following fertilization of the mouse ovum. *J. Exp. Zool.* **212**, 355.

Chetsanga, C. J., Poccia, D. L., Hill, R. J., and Doty, P. (1970). Stage-specific RNA transcription in developing sea urchins and their chromatins. *Cold Spring Harbor Symp. Quant. Biol.* **35**, 629.

Chien, Y.-H., and Dawid, I. B. (1984). Isolation and characterization of calmodulin genes from *Xenopus laevis. Mol. Cell. Biol.* **4**, 507.

Chien, Y.-H., Gascoigne, N. R. J., Kavaler, J., Lee, N. E., and Davis, M. M. (1984). Somatic recombination in a murine T-cell receptor gene. *Nature (London)* **309**, 322.

Child, C. M. (1900). The early development of *Arenicola* and *Sternapsis. Arch. Entwicklungsmech. Org.* **9**, 587.

Child, F. M., and Apter, M. N. (1969). Experimental inhibition of ciliogenesis and ciliary regeneration in *Arbacia* embryos. *Biol. Bull. (Woods Hole, Mass.)* **137**, 394.

Childs, G., Maxson, R., and Kedes, L. H. (1979). Histone gene expression during sea urchin embryogenesis: Isolation and characterization of early and late messenger RNAs of *Strongylocentrotus purpuratus* by gene-specific hybridization and template activity. *Dev. Biol.* **73**, 153.

Childs, G., Nocente-McGrath, C., Lieber, T., Holt, C., and Knowles, J. A. (1982). Sea urchin (*Lytechinus pictus*) late-stage histone H3 and H4 genes: Characterization and mapping of a clustered but nontandemly linked multigene family. *Cell* **31**, 383.

Chitwood, B. G., and Chitwood, M. B. (1974). "Introduction to Nematology." University Park Press, Baltimore, Maryland.

Chiu, C.-P., and Blau, H. M. (1985). 5-Azacytidine permits gene activation in a previously noninducible cell type. *Cell* **40**, 417.

Chomyn, A., Mariottini, P., Cleeter, M. W. J., Ragan, C. I., Matsuno-Yagi, A., Hatefi, Y., Doolittle, R. F., and Attardi, G. (1985). Six unidentified reading frames of human mitochondrial DNA encode components of the respiratory-chain NADH dehydrogenase. *Nature (London)* **314**, 592.

Christensen, K., Sauterer, R., and Merriam, R. W. (1984). Role of soluble myosin in cortical contractions of *Xenopus* eggs. *Nature (London)* **310**, 150.

Chun, C. (1880). "Die Ctenophoren des Golfes von Neapel. Fauna und Flora des Golfes von Neapel," Monogr. I. Engelmann, Leipzig.

Chung, H. M., and Malacinski, G. M. (1975). Repair of ultraviolet irradiation damage to a cytoplasmic component required for neural induction in the amphibian egg. *Proc. Natl. Acad. Sci. U.S.A.* **72**, 1235.

Chung, H. M., and Malacinski, G. M. (1983). Reversal of developmental competence in inverted amphibian eggs. *J. Embryol. Exp. Morphol.* **73**, 207.

Citkowitz, E. (1971). The hyaline layer: Its isolation and role in echinoderm development. *Dev. Biol.* **24**, 348.

Clark, A. J., and Kidder, G. M. (1977). Polyadenylic acid in *Ilyanassa*: Estimates of the number and mean length of poly(A) tracts in embryonic and larval stages. *Differentiation* **8**, 113.

Clarke, J. D. W., Hayes, B. P., Hunt, S. P., and Roberts, A. (1984). Sensory physiology, anatomy, and immunohistochemistry of Rohon-Beard neurones in embryos of *Xenopus laevis. J. Physiol. (London)* **348**, 511.

Clegg, J. S. (1967). Metabolic studies of cryptobiosis in encysted embryos of *Artemia salina*. *Comp. Biochem. Physiol.* **20**, 801.

Clegg, J. S., and Golub, A. L. (1969). Protein synthesis in *Artemia salina* embryos. II. Resumption of RNA and protein synthesis upon cessation of dormacy in the encysted gastrula. *Dev. Biol.* **19**, 178.

Clegg, K. B., and Pikó, L. (1977). Size and specific activity of the UTP pool and overall rates of RNA synthesis in early mouse embryos. *Dev. Biol.* **58**, 76.

Clegg, K. B., and Pikó, L. (1983a). Poly(A) length, cytoplasmic adenylation and synthesis of poly(A)$^+$ RNA in early mouse embryos. *Dev. Biol.* **95**, 331.

Clegg, K. B., and Pikó, L. (1983b). Quantitative aspects of RNA synthesis and polyadenylation in 1-cell and 2-cell mouse embryos. *J. Embryol. Exp. Morphol.* **74**, 169.

Cleine, J. H., and Slack, J. M. W. (1985). Normal fates and states of specification of different regions in the axolotl gastrula. *J. Embryol. Exp. Morphol.* **86**, 247.

Clement, A. C. (1952). Experimental studies on germinal localization in *Ilyanassa*. I. The role of the polar lobe in determination of the cleavage pattern and its influence in later development. *J. Exp. Zool.* **121**, 593.

Clement, A. C. (1962). Development of *Ilyanassa* following removal of the D macromere at successive cleavage stages. *J. Exp. Zool.* **149**, 193.

Clement, A. C. (1967). The embryonic value of the micromeres in *Ilyanassa obsoleta* as determined by deletion experiments. I. The first quartet cells. *J. Exp. Zool.* **166**, 77.

Clement, A. C. (1968). Development of the vegetal half of the *Ilyanassa* egg after removal of most of the yolk by centrifugal force, compared with the development of animal halves of similar visible composition. *Dev. Biol.* **17**, 165.

Clement, A. C. (1976). Cell determination and organogenesis in molluscan development: A reappraisal based on deletion experiments in *Ilyanassa*. *Am. Zool.* **16**, 447.

Clement, A. C., and Tyler, A. (1967). Protein-synthesizing activity of the anucleate polar lobe of the mud snail *Ilyanassa obsoleta*. *Science* **158**, 1457.

Cline, C., Schatten, H., Balczon, R., and Schatten, G. (1984). Actin-mediated surface motility during sea urchin fertilization. *Cell Motil.* **3**, 513.

Coffe, G., Rola, F. H., Soyer, M. O., and Pudles, J. (1982). Parthenogenetic activation of sea urchin egg induces a cyclical variation of the cytoplasmic resistance to hexylene glycol-Triton X 100 treatment. *Exp. Cell Res.* **137**, 63.

Cohen, L. H., Newrock, K. M., and Zweidler, A. (1975). Stage-specific switches in histone synthesis during embryogenesis of the sea urchin. *Science* **190**, 994.

Cohen, R. S., and Meselson, M. (1985). Separate regulatory elements for the heat-inducible and ovarian expression of the *Drosophila* hsp26 gene. *Cell* **43**, 737.

Cohn, R. H., and Kedes, L. H. (1979a). Nonallelic histone gene clusters of individual sea urchins (*Lytechinus pictus*): Polarity and gene organization. *Cell* **18**, 843.

Cohn, R. H., and Kedes, L. H. (1979b). Nonallelic histone gene clusters of individual sea urchins (*Lytechinus pictus*): Mapping of homologies in coding and spacer DNA. *Cell* **18**, 855.

Colberg-Poley, A. M., Voss, S. D., Chowdhury, K., Stewart, C. L., Wagner, E. F., and Gruss, P. (1985). Clustered homeoboxes are differentially expressed during murine development. *Cell* **43**, 39.

Colin, A. M., and Hille, M. B. (1986). Injected mRNA does not increase protein synthesis in unfertilized, fertilized, or ammonia-activated sea urchin eggs. *Dev. Biol.* **115**, 184.

Collier, J. R. (1975). Nucleic and synthesis in the normal and lobeless embryo of *Ilyanassa obsoleta*. *Exp. Cell Res.* **95**, 254.

Collier, J. R. (1976). Nucleic acid chemistry of the *Ilyanassa* embryo. *Am. Zool.* **16**, 483.

Collier, J. R. (1977). Rates of RNA synthesis in the normal and lobeless embryo of *Ilyanassa obsoleta*. *Exp. Cell Res.* **106**, 390.

Collier, J. R. (1981). Protein synthesis in the polar lobe and lobeless egg of *Ilyanassa obsoleta*. *Biol. Bull. (Woods Hole, Mass.)* **160**, 366.

Collier, J. R. (1983). The biochemistry of molluscan development. *In* "The Mollusca" (N. H. Verdonk, J. A. M. van den Biggelaar, and A. S. Tompa, eds.), Vol. 3, p. 253. Academic Press, New York.

Collier, J. R., and McCarthy, M. E. (1981). Regulation of polypeptide synthesis during early embryogenesis of *Ilyanassa obsoleta*. *Differentiation* **19**, 31.

Collier, J. R., and Schwartz, R. (1969). Protein synthesis during *Ilyanassa* embryogenesis. *Exp. Cell. Res.* **54**, 403.

Colot, H. V., and Rosbash, M. (1982). Behavior of individual maternal pA⁺ RNAs during embryogenesis of *Xenopus laevis*. *Dev. Biol.* **94**, 79.

Coluzzi, M., and Kitzmüller, J. B. (1975). Anopheline mosquitoes. *In* "Handbook of Genetics" (R. C. King, ed.), Vol. 3, p. 285. Plenum, New York.

Comb, D. G., Katz, S., Branda, R., and Pinzino, C. J. (1965). Characterization of RNA species synthesized during early development of sea urchins. *J. Mol. Biol.* **4**, 195.

Compere, S. J., and Palmiter, R. D. (1981). DNA methylation controls the inducibility of the mouse metallothionein-I gene in lymphoid cells. *Cell* **25**, 233.

Conklin, E. G. (1897). The embryology of *Crepidula*. A contribution to the cell lineage and early development of some marine gasteropods. *J. Morphol.* **13**, 1.

Conklin, E. G. (1905a). The organization and cell lineage of the ascidian egg. *J. Acad. Natl. Sci. Philadelphia* **13**, 1.

Conklin, E. G. (1905b). Mosaic development in ascidian eggs. *J. Exp. Zool.* **2**, 145.

Conklin, E. G. (1917). Effects of centrifugal force on the structure and development of the eggs of *Crepidula*. *J. Exp. Zool.* **22**, 311.

Conklin, E. G. (1931). The development of centrifuged eggs of ascidians. *J. Exp. Zool.* **60**, 119.

Conklin, K. F., and Groudine, M. (1984). Chromatin structure and gene expression. *In* "DNA Methylation and Biological Significances" (A. Razin, H. Cedar, and A. Riggs, eds.), p. 293. Springer-Verlag, Berlin and New York.

Conrad, G. W. (1973). Control of polar lobe formation in fertilized eggs of *Ilyanassa obsoleta* Stimpson. *Am. Zool.* **13**, 961.

Cooke, J. (1985a). The system specifying body position in the early development of *Xenopus*, and its response to early perturbations. *J. Embryol. Exp. Morphol.* **89**, Suppl., 69.

Cooke, J. (1985b). Early specification for body position in mes-endodermal regions of an amphibian embryo. *Cell. Differ.* **17**, 1.

Cooke, J., and Webber, J. A. (1985a). Dynamics of the controls of body pattern in the development of *Xenopus laevis*. I. Timing and pattern in the development of dorsoanterior and of posterior blastomere pairs, isolated at the 4-cell stage. *J. Embryol. Exp. Morphol.* **88**, 85.

Cooke, J., and Webber, J. A. (1985b). Dynamics of the control of body pattern in the development of *Xenopus laevis*. II. Timing and pattern in the development of single blastomeres (presumptive lateral halves) isolated at the 2-cell stage. *J. Embryol. Exp. Morphol.* **88**, 113.

Cooper, A. D., and Crain, W. R. (1982). Complete nucleotide sequence of a sea urchin actin gene. *Nucleic Acids Res.* **10**, 4081.

Copp, A. J. (1978). Interaction between inner cell mass and trophectoderm of the mouse blastocyst. I. A study of cellular proliferation. *J. Embryol. Exp. Morphol.* **48**, 109.

Corces, V., Pellicer, A., Axel, R., and Meselson, M. (1981). Integration, transcription, and control of a *Drosophila* heat shock gene in mouse cells. *Proc. Natl. Acad. Sci. U.S.A.* **78**, 7038.

Cordeiro-Stone, M., and Lee, C. S. (1976). Studies on the satellite DNAs of *Drosophila nasutoides*: Their buoyant densities, melting temperatures, reassociation rates, and localizations in polytene chromosomes. *J. Mol. Biol.* **104**, 1.

Costantini, F. D., Scheller, R. H., Britten, R. J., and Davidson, E. H. (1978). Repetitive sequence transcripts in the mature sea urchin oocyte. *Cell* **15**, 173.

Costantini, F. D., Britten, R. J., and Davidson, E. H. (1980). Message sequences and short repetitive sequences arc interspersed in sea urchin egg poly(A)$^+$ RNAs. *Nature (London)* **287**, 111.

Costello, D. P. (1945). Experimental studies of germinal localization in *Nereis*. I. The development of isolated blastomeres. *J. Exp. Zool.* **100**, 19.

Costello, D. P. (1956). Cleavage, blastulation and gastrulation. *In* "Analysis of Development" (B. H. Willier, P. A. Weiss, and V. Hamburger, eds.), p. 213. Saunders, Philadelphia, Pennsylvania.

Cotton, R. W., Manes, C., and Hamkalo, B. A. (1980). Electron microscopic analysis of RNA transcription in preimplantation rabbit embryos. *Chromosoma* **79**, 169.

Counce, S. J. (1973). The causal analysis of insect embryogenesis. *In* "Developmental Systems: Insects" (S. J. Counce and C. H. Waddington, eds.), Vol. 2, p. 1. Academic Press, New York.

Cowan, A. E., and McIntosh, J. R. (1985). Mapping the distribution of differentiation potential for intestine, muscle, and hypodermis during early development in *Caenorhabditis elegans*. *Cell* **41**, 923.

Cox, G. N., and Hirsh, D. (1985). Stage-specific patterns of collagen gene expression during development of *Caenorhabditis elegans*. *Mol. Cell. Biol.* **5**, 363.

Cox, K. H., Angerer, L. M., Lee, J. J., Davidson, E. H., and Angerer, R. C. (1986). Cell lineage-specific programs of expression of multiple actin genes during sea urchin embryogenesis. *J. Mol. Biol.* **188**, 159.

Craig, S. P. (1970). Synthesis of RNA in nonnucleate fragments of sea urchin eggs. *J. Mol. Biol.* **47**, 615.

Craig, S. P., and Piatigorsky, J. (1971). Protein synthesis and development in the absence of cytoplasmic RNA synthesis in nonnucleate egg fragments and embryos of sea urchins: Effect of ethidium bromide. *Dev. Biol.* **24**, 214.

Crain, W. R., Eden, F. C., Pearson, W. R., Davidson, E. H., and Britten, R. J. (1976). Absence of short period interspersion of repetitive and nonrepetitive sequences in the DNA of *Drosophila melanogaster*. *Chromosoma* **56**, 309.

Crain, W. R., Durica, D. S., and Van Doren, L. (1981). Actin gene expression in developing sea urchin embryos. *Mol. Cell. Biol.* **1**, 711.

Crampton, H. E. (1896). Experimental studies on gasteropod development. *Arch. Entwicklungsmech. Org.* **3**, 1.

Crick, F. H. C., and Lawrence, P. A. (1975). Compartments and polyclones in insect development. *Science* **189**, 340.

Crippa, M., Davidson, E. H., and Mirsky, A. E. (1967). Persistence in early amphibian embryos of informational RNAs from the lampbrush chromosome stage of oögenesis. *Proc. Natl. Acad. Sci. U.S.A.* **57**, 885.

Crosby, M. A., and Meyerowitz, E. (1986). *Drosophila* glue gene *Sgs-3*: Sequences required for puffing and transcriptional regulation. *Dev. Biol.* In press.

Crowley, T. E., Mathers, P. H., and Meyerowitz, E. M. (1984). A *trans*-acting regulatory product necessary for expression of the *Drosophila melanogaster* 68C glue gene cluster. *Cell* **39**, 149.

Crowther, R. J., and Whittaker, J. R. (1983). Developmental autonomy of muscle fine structure in muscle lineage cells of ascidian embryos. *Dev. Biol.* **96**, 1.

Crowther, R. J., and Whittaker, J. R. (1984). Differentiation of histospecific ultrastructural features in cells of cleavage-arrested early ascidian embryos. *Wilhelm Roux's Arch. Dev. Biol.* **194**, 87.

Crowther, R. J., and Whittaker, J. R. (1986). Differentiation without cleavage: Multiple cytospecific ultrastructural expressions in the same one-celled ascidian zygotes. *Dev. Biol.* In press.

Cruz, Y. P., and Pedersen, R. A. (1985). Cell fate in the polar trophectoderm of mouse blastocysts as studied by microinjection of cell lineage tracers. *Dev. Biol.* **112**, 73.

Cullen, B., Emigholz, K., and Monahan, J. (1980). The transient appearance of specific proteins in one-cell mouse embryos. *Dev. Biol.* **76**, 215.

Curtis, A. S. G. (1962). Morphogenetic interactions before gastrulation in the amphibian *Xenopus laevis*—the cortical field. *J. Embryol. Exp. Morphol.* **10**, 410.

Czihak, G. (1963). Entwicklungsphysiologische Untersuchungen an Echiniden (Verteilung und Bedeutung der Cytochromoxydase). *Wilhelm Roux's Arch. Dev. Biol. Entwicklungsmech. Org.* **154**, 272.

Czihak, G. (1971). Echinoids. *In* "Experimental Embryology of Marine and Fresh-Water Invertebrates" (G. Reverberi, ed.), p. 363. North-Holland Publ., Amsterdam.

Czihak, G., ed. (1975). "The Sea Urchin Embryo. Biochemistry and Morphogenesis." Springer-Verlag, Berlin and New York.

Dabauvalle, M.-C., and Franke, W. W. (1982). Karyophilic proteins: Polypeptides synthesized *in vitro* accumulate in the nucleus on microinjection into the cytoplasm of amphibian oocytes. *Proc. Natl. Acad. Sci. U.S.A.* **79**, 5302.

Dan, K. (1979). Studies on unequal cleavage in sea urchins. I. Migration of the nuclei to the vegetal pole. *Dev., Growth Differ.* **21**, 527.

Dan, K., and Ikeda, M. (1971). On the system controlling the time of micromere formation in sea urchin embryos. *Dev., Growth Differ.* **13**, 285.

Dan, K., Tanaka, S., Yamazaki, K., and Kato, Y. (1980). Cell cycle study up to the time of hatching in the embryos of the sea urchin, *Hemicentrotus pulcherrimus*. *Dev., Growth Differ.* **22**, 589.

Dan, K., Endo, S., and Uemura, I. (1983). Studies on unequal cleavage in sea urchins. II. Surface differentiation and the direction of nuclear migration. *Dev., Growth Differ.* **25**, 227.

Daniel, J. C., Jr. (1964). Early growth of rabbit trophoblast. *Am. Nat.* **98**, 85.

Danilchik, M. V., and Hille, M. B. (1981). Sea urchin egg and embryo ribosomes: Differences in translational activity in a cell-free system. *Dev. Biol.* **84**, 291.

Dan-Sohkawa, M., and Satoh, N. (1978). Studies on dwarf larvae developed from isolated blastomeres of the starfish, *Asterina pectinifera*. *J. Embryol. Exp. Morphol.* **46**, 171.

Darlington, G. J., Bernhard, H. P., and Ruddle, F. H. (1974). Human serum albumin phenotype activation in mouse hepatoma-human leukocyte cell hybrids. *Science* **185**, 859.

Darnbrough, C., and Ford, P. J. (1976). Cell-free translation of messenger RNA from oocytes of *Xenopus laevis*. *Dev. Biol.* **50**, 285.

Darnbrough, C. and Ford, P. J. (1981). Identification in *Xenopus laevis* of ovary specific proteins which are bound to messenger RNA. *Eur. J. Biochem.* **113**, 415.

Darribére, T., Boucher, D., Lacroix, J.-C., and Boucaut, J.-C. (1984). Fibronectin synthesis during oogenesis and early development of the amphibian *Pleurodeles waltlii*. *Cell. Differ.* **14**, 171.

Das, N. K. (1976). Cytochemical and biochemical analysis of development of *Urechis* oocytes. *Am. Zool.* **16**, 345.

Davenport, D. (1976). Transport of ribosomal RNA into the oocytes of the milkweed bug, *Oncopeltus fasciatus*. *J. Insect Physiol.* **22**, 925.

Davidson, E. H. (1968). "Gene Activity in Early Development." Academic Press, New York.

Davidson, E. H. (1976). "Gene Activity in Early Development, 2nd ed." Academic Press, New York.

Davidson, E. H. (1985). Genome function in sea urchin embryos: Fundamental insights of Th. Boveri reflected in recent molecular discoveries. *In* "A History of Embryology" (T. J. Horder, J. A. Witkowski, and C. C. Wylie, eds.), p. 397. Cambridge Univ. Press, London and New York.

Davidson, E. H., and Britten, R. J. (1971). Note on the control of gene expression during development. *J. Theor. Biol.* **32**, 123.

Davidson, E. H., and Britten, R. J. (1973). Organization, transcription and regulation in the animal genome. *Q. Rev. Biol.* **48**, 565.

Davidson, E. H., and Britten, R. J. (1979). Regulation of gene expression: Possible role of repetitive sequences. *Science* **204**, 1052.

Davidson, E. H., and Hough, B. R. (1969). Synchronous oogenesis in *Engystomops pustulosus*, a neotropic anuran suitable for laboratory studies; localization in the embryo of RNA synthesized at the lampbrush stage. *J. Exp. Zool.* **172**, 25.

Davidson, E. H., and Hough, B. R. (1971). Genetic information in oocyte RNA. *J. Mol. Biol.* **56**, 491.

Davidson, E. H., Allfrey, V. G., and Mirsky, A. E. (1964). On the RNA synthesized during the lampbrush phase of amphibian oogenesis. *Proc. Natl. Acad. Sci. U.S.A.* **52**, 501.

Davidson, E. H., Haslett, G. W., Finney, R. J., Allfrey, V. G., and Mirsky, A. E. (1965). Evidence for prelocalization of cytoplasmic factors affecting gene activation in early embryogenesis. *Proc. Natl. Acad. Sci. U.S.A.* **54**, 696.

Davidson, E. H., Crippa, M., Kramer, F. R., and Mirsky, A. E. (1966). Genomic function during the lampbrush chromosome stage of amphibian oogenesis. *Proc. Natl. Acad. Sci. U.S.A.* **56**, 856.

Davidson, E. H., Hough, B. R., Amenson, C. S., and Britten, R. J. (1973). General interspersion of repetitive with non-repetitive sequence elements in the DNA of *Xenopus*. *J. Mol. Biol.* **77**, 1.

Davidson, E. H., Hough, B. R., Klein, W. H., and Britten, R. J. (1975). Structural genes adjacent to interspersed repetitive DNA sequences. *Cell* **4**, 217.

Davidson, E. H., Hough-Evans, B. R., and Britten, R. J. (1982). Molecular biology of the sea urchin embryo. *Science* **217**, 17.

Davidson, E. H., Flytzanis, C. N., Lee, J. J., Robinson, J. J., Rose, S. J., and Sucov, H. M. (1985). Lineage-specific gene expression in the sea urchin embryo. *Cold Spring Harbor Symp. Quant. Biol.* **50**, 321.

Davis, F. C., Jr. (1975). Unique sequence DNA transcripts present in mature oocytes of *Urechis caupo*. *Biochim. Biophys. Acta* **390**, 33.

Davis, F. C. (1982). Differential utilization of ribosomes for protein synthesis during oogenesis and early embryogenesis of *Urechis caupo* (Echiura). *Differentiation* **22**, 170.

Davis, F. C., and Davis, R. W. (1978). Polyadenylation of RNA in immature oocytes and early cleavage of *Urechis caupo*. *Dev. Biol.* **66**, 86.

Davis, F. C., Jr., and Wilt, F. H. (1972). RNA synthesis during oogenesis in the echiuroid worm *Urechis caupo*. *Dev. Biol.* **27**, 1.

Dawid, I. B. (1965). Deoxyribonucleic acid in amphibian eggs. *J. Mol. Biol.* **12**, 581.

Dawid, I. B., and Brown, D. D. (1970). The mitochondrial and ribosomal DNA components of oocytes of *Urechis caupo*. *Dev. Biol.* **22**, 1.

Dean, D. C., Knoll, B. J., Riser, M. E., and O'Malley, B. W. (1983). A 5'-flanking sequence essential for progesterone regulation of an ovalbumin fusion gene. *Nature (London)* **305**, 551.

Dean, D. C., Gope, R., Knoll, B. J., Riser, M. E., and O'Malley, B. W. (1984). A similar 5' flanking region is required for estrogen and progesterone induction of ovalbumin gene expression. *J. Biol. Chem.* **259**, 9967.

de Cicco, D. V., and Spradling, A. C. (1984). Localization of a *cis*-acting element responsible for the developmentally regulated amplification of *Drosophila* chorion genes. *Cell* **38**, 45.

Deisseroth, A., and Hendrick, D. (1978). Human α-globin gene expression following chromosomal dependent gene transfer into mouse erythroleukemia cells. *Cell* **15**, 55.

DeLeon, D. V., Cox, K. H., Angerer, L. M., and Angerer, R. C. (1983). Most early-variant histone mRNA is contained in the pronucleus of sea urchin eggs. *Dev. Biol.* **100**, 197.

De Leon, V., Johnson, A., and Bachvarova, R. (1983). Half-lives and relative amounts of stored and polysomal ribosomes and poly(A)$^+$ RNA in mouse oocytes. *Dev. Biol.* **98**, 400.

DeLobel, N. (1971). Étude descriptive des chromosomes en écouvillon chez *Echinaster sepositus* (Échinoderme, Astéride). *Ann. Embryol. Morphog.* **4**, 383.

del Pino, E. M., and Elinson, R. P. (1983). A novel development pattern for frogs: Gastrulation produces an embryonic disk. *Nature (London)* **306**, 589.

del Pino, E. M., and Humphries, A. A., Jr. (1978). Multiple nuclei during early oogenesis in *Flectonotus pygmaeus* and other marsupial frogs. *Biol. Bull. (Woods Hole, Mass.)* **154**, 198.

Denich, K. T. R., Schierenberg, E., Isnenghi, E., and Cassada, R. (1984). Cell-lineage and developmental defects of temperature-sensitive embryonic arrest mutants of the nematode *Caenorhabditis elegans*. *Wilhelm Roux's Arch. Dev. Biol.* **193**, 164.

Denis, H. (1966). Gene expression in amphibian development. II. Release of the genetic information in growing embryos. *J. Mol. Biol.* **22**, 285.

Denis, H., and le Maire, M. (1985). Biochemical research on oogenesis. Aminoacyl tRNA turns over in the 42S particles of *Xenopus laevis* oocytes, but its ester bond is protected against hydrolysis. *Eur. J. Biochem.* **149**, 549.

Denis, H., and Mairy, M. (1972). Recherches biochimiques sur l'oogenèse. 2. Distribution intracellulaire du RNA dans les petits oocytes de *Xenopus laevis*. *Eur. J. Biochem.* **25**, 524.

Denis, H., and Wegnez, M. (1977). Biochemical research on oogenesis. Oocytes of *Xenopus laevis* synthesize but do not accumulate 5S RNA of somatic type. *Dev. Biol.* **58**, 212.

Denny, P. C., and Tyler, A. (1964). Activation of protein biosynthesis in non-nucleate fragments of sea urchin eggs. *Biochem. Biophys. Res Commun.* **14**, 245.

Deno, T., and Satoh, N. (1984). Studies on the cytoplasmic determinant for muscle cell differentiation in ascidian embryos: An attempt at transplantation of the myoplasm. *Dev., Growth Differ.* **26**, 43.

Deno, T., Nishida, H., and Satoh, N. (1984). Antonomous muscle cell differentiation in partial ascidian embryos according to the newly verified cell lineages. *Dev. Biol.* **104**, 322.

Deno, T., Nishida, H., and Satoh, N. (1985). Histospecific acetylcholinesterase development in quarter ascidian embryos derived from each blastomere pair of the eight-cell stage. *Biol. Bull. (Woods Hole, Mass.)* **168**, 239.

Denoulet, P., Muller, J.-P., and Lacroix, J.-C. (1977). RNA metabolism in amphibian oocytes. I. Chromosomal pre-messenger RNA synthesis. *Biol. Cell.* **28**, 101.

De Petrocellis, B., and Rossi, M. (1976). Enzymes of DNA biosynthesis in developing sea urchins. Changes in ribonucleotide reductase, thymidine, and thymidylate kinase activities. *Dev. Biol.* **48**, 250.

Deppe, U., Schierenberg, E., Cole, T., Krieg, C., Schmitt, D., Yoder, B., and von Ehrenstein, G. (1978). Cell lineages of the embryo of the nematode *Caenorhabditis elegans*. *Proc. Natl. Acad. Sci. U.S.A.* **75**, 376.

De Robertis, E. M. (1983). Nucleocytoplasmic segregation of proteins and RNAs. *Cell* **32**, 1021.

De Robertis, E. M., and Gurdon, J. B. (1977). Gene activation in somatic nuclei after injection into amphibian oocytes. *Proc. Natl. Acad. Sci. U.S.A.* **74**, 2470.

De Robertis, E. M., Black, P., and Nishikura, K. (1981). Intranuclear location of the tRNA splicing enzymes. *Cell* **23**, 89.

De Robertis, E. M., Lienhard, S., and Parisot, R. F. (1982). Intracellular transport of microinjected 5S and small nuclear RNAs. *Nature (London)* **295**, 572.

De Simone, D. W., and Spiegel, M. (1985). Micromere-specific cell surface proteins of 16-cell stage sea urchin embryos. *Exp. Cell Res.* **156**, 7.

Desplan, C., Theis, J., and O'Farrell, P. H. (1985). The *Drosophila* developmental gene, *engrailed*, encodes a sequence-specific DNA binding activity. *Nature (London)* **318**, 630.

Devlin, R. (1976). Mitochondrial poly(A) RNA synthesis during early sea urchin development. *Dev. Biol.* **50**, 443.

De Vries, H. (1889). "Intracelluläre Pangenesis." Fischer, Jena.

Dewey, M. H., Filler, R., and Mintz, B. (1978). Protein patterns of developmentally totipotent mouse teratocarcinoma cells and normal embryo cells. *Dev. Biol.* **65**, 171.

D'Hollander, F. (1904). Rechereches sur l'oogénèse et sur la structure et la signification du noyau vitellin de Balbiani chez les oizeaux. *Arch. Anat. Microsc. Morphol. Exp.* **7**, 117.

Diaz, M. O., and Gall, J. G. (1985). Giant readthrough transcription units at the histone loci on lampbrush chromosomes of the newt *Notophthalmus*. *Chromosoma* **92**, 243.

Diaz, M. O., Barsacchi-Pilone, G., Mahon, K. A., and Gall, J. G. (1981). Transcripts from both strands of a satellite DNA occur on lampbrush chromosome loops of the newt *Notophthalmus*. *Cell* **24**, 649.

Diaz-Benjumea, F. J., Gonzalez Gaitan, M. A., and Garcia-Bellido, A. (1986). Genetic and developmental bases of the vein pattern in *Drosophila* wings. In preparation.

DiBerardino, M. A. (1979). Nuclear and chromosomal behavior in amphibian nuclear transplants. *Int. Rev. Cytol., Suppl.* No. 9, 129.

DiBerardino, M. A., and Hoffner, N. (1971). Development and chromosomal constitution of nuclear-transplants derived from male germ cells. *J. Exp. Zool.* **176**, 61.

DiBerardino, M. A., and Hoffner, N. J. (1983). Gene reactivation in erythrocytes: Nuclear transplantation in oocytes and eggs of *Rana*. *Science* **219**, 862.

DiBerardino, M. A., and King, T. J. (1967). Development and cellular differentiation of neural nuclear-transplants of known karyotype. *Dev. Biol.* **15**, 102.

DiBerardino, M. A., Hoffner, N. J., and Etkin, L. D. (1984). Activation of dormant genes in specialized cells. *Science* **224**, 946.

Dickinson, W. J., and Sullivan, D. T. (1975). "Gene-Enzyme Systems in *Drosophila*," Chap. 3. Springer, New York.

Dickinson, W. J., and Weisbrod, E. (1976). Gene regulation in *Drosphila*: Independent expression of closely linked, related structural loci. *Biochem. Genet.* **14**, 709.

DiMario, P. J., and Hennen, S. (1982). Analysis of the *Drosophila* female sterile mutant, *tiny*, by means of pole cell transplantation. *J. Exp. Zool.* **221**, 219.

DiNardo, S., Kuner, J. M., Theis, J., and O'Farrell, P. H. (1985). Development of embryonic pattern in *D. melanogaster* as revealed by accumulation of the nuclear *engrailed* protein. *Cell* **43**, 59.

Dingwall, C., Sharnick, S. V., and Laskey, R. A. (1982). A polypeptide domain that specifies migration of nucleoplasmin into the nucleus. *Cell* **30**, 449.

Di Nocera, P. P., and Dawid, I. B. (1983). Transient expression of genes intoduced into cultured cells of *Drosophila*. *Proc. Natl. Acad. Sci. U.S.A.* **80**, 7095.

Dixon, L. K., and Ford, P. J. (1982). Regulation of protein synthesis and accumulation during oogenesis in *Xenopus laevis*. *Dev. Biol.* **93**, 478.

Doe, C. Q., and Goodman, C. S. (1985a). Early events in insect neurogenesis. II. The role of cell interactions and cell lineage in the determination of neuronal precursor cells. *Dev. Biol.* **111**, 206.

Doe, C. Q., and Goodman, C. S. (1985b). Early events in insect neurogenesis. I. Development and segmental differences in the pattern of neuronal precursor cells. *Dev. Biol.* **111**, 193.

Dohmen, M. R. (1983). The polar lobe in eggs of molluscs and annelids: Structure, composition, and function. In "Time, Space, and Pattern in Embryonic Development" (W. R. Jeffery and R. A. Raff, eds.), p. 197. Alan R. Liss, New York.

Dohmen, M. R., and van der Mey, J. C. A. (1977). Local surface differentiations of the vegetal role of the eggs of *Nassarius reticulatus, Buccinum undatum*, and *Crepidula fornicata* (Gastropoda, Prosobranchia). *Dev. Biol.* **61**, 104.

Dohmen, M. R., and Verdonk, N. H. (1974). The structure of a morphogenetic cytoplasm, present in the polar lobe of *Bithynia tentaculata* (Gastropoda, Prosobranchia). *J. Embryol. Exp. Morphol.* **31**, 423.

Dohmen, M. R., and Verdonk, N. H. (1979). The ultrastructure and role of the polar lobe in development of molluscs. In "Determinants of Spatial Organization" (S. Subtelny and I. R. Konigsberg, eds.), p. 3. Academic Press, New York.

Dolecki, G. J., and Smith, L. D. (1979). Poly(A)⁺ RNA metabolism during oogenesis in *Xenopus laevis*. *Dev. Biol.* **69**, 217.

Dolecki, G. J., Duncan, R. F., and Humphreys, T. (1977). Complete turnover of poly(A) on maternal mRNA of sea urchin embryos. *Cell* **11**, 339.

Dolecki, G. J., Wannakrairoj, S., Lum, R., Wang, G., Riley, H. D., Carlos, R., Wang, A., and Humphreys, T. (1986). Stage-specific expression of a homeobox containing gene in the nonsegmented sea urchin embryo. *EMBO J.* **5**, 925.

Donahue, T. F., Farabaugh, P. J., and Fink, G. R. (1982). The nucleotide sequence of the HIS4 region of yeast. *Gene* **18**, 47.

Dorresteijn, A. W. C., Wagemaker, H. A., de Laat, S. W., and van den Biggelaar, J. A. M. (1983). Dye-coupling between blastomeres in early embryos of *Patella vulgata* (Mollusca, Gastropoda): Its relevance for cell determination. *Wilhelm Roux's Arch. Dev. Biol.* **192**, 262.

Dreyer, C., and Hausen, P. (1983). Two-dimensional gel analysis of the fate of oocyte nuclear proteins in the development of *Xenopus laevis*. *Dev. Biol.* **100**, 412.

Dreyer, C., Scholz, E., and Hausen, P. (1982). The fate of oocyte nuclear proteins during early development of *Xenopus laevis*. *Wilhelm Roux's Arch. Dev. Biol.* **191**, 228.

Driesch, H. (1891). Entwicklungsmechanische Studien. I. Der Werth der beiden ersten Furchungszellen in der Echinodermenentwicklung. Experimentelle Erzeugung von Theil-und Doppelbildungen. II. Über die Beziehungen des Lichtes zur ersten Etappe der thierischen Formbildung. *Z. Wiss. Zool.* **53**, 160.

Driesch, H. (1892). Entwickelungsmechanisches. *Anat. Anz.* **7**, 584.

Driesch, H. (1894). "Analytische Theorie der organischen Entwicklung." Engelmann, Leipzig.

Driesch, H., and Morgan, T. H. (1896). Zur Analysis der ersten Entwickelungsstadien des Ctenophoreneies. I. Von der Entwickelung einzelner Ctenophorenblastomeren. *Arch. Entwicklungsmech. Org.* **2**, 204.

Droin, A., and Fischberg, M. (1984). Two recessive mutations with maternal effect upon colour and cleavage of *Xenopus l. laevis* eggs. *Wilhelm Roux's Arch. Dev. Biol.* **193**, 86.

Drummond, D. R., Armstrong, J., and Colman, A. (1985). The effect of capping and poly-adenylation on the stability, movement, and translation of synthetic messenger RNAs in *Xenopus* oocytes. *Nucleic Acids Res.* **13**, 7375.

Dubroff, L. M. (1977). Oligouridylate stretches in heterogeneous nuclear RNA. *Proc. Natl. Acad. Sci. U.S.A.* **74**, 2217.

Dubroff, L. M. (1980). Developmental changes in the molecular weight of heterogeneous nuclear RNA. *Biochim. Biophys. Acta* **608**, 378.

Dubroff, L. M., and Nemer, M. (1975). Molecular classes of heterogeneous nuclear RNA in sea urchin embryos. *J. Mol. Biol.* **95**, 455.

Dubroff, L. M., and Nemer, M. (1976). Developmental shifts in the synthesis of heterogeneous nuclear RNA classes in the sea urchin embryo. *Nature (London)* **260**, 120.

Dudler, R., and Travers, A. A. (1984). Upstream elements necessary for optimal function of the hsp 70 promoter in transformed cells. *Cell* **38**, 391.

Dumont, J. N. (1972). Oogenesis in *Xenopus laevis* (Daudin). 1. Stages of oocyte development in laboratory maintained animals. *J. Morphol.* **136**, 153.

Duncan, I. M. (1982). Polycomblike: A gene that appears to be required for the normal expression of the bithorax and Antennapedia gene complexes of *Drosophila melanogaster*. *Genetics* **102**, 49.

Duncan, I., and Lewis, E. B. (1982). Genetic control of body segment differentiation in *Drosophila*. *In* "Developmental Order: Its Origins and Regulation" (S. Subtelny and P. B. Green, eds.), p. 533. Alan R. Liss, New York.

Duncan, R. F. (1978). Maternal messenger RNA in sea urchin eggs and embryos. Ph.D. Thesis, Univ. of Hawaii, Honolulu.

Duncan, R., and Humphreys, T. (1981). Most sea urchin maternal mRNA sequences in every abundance class appear in both polyadenylated and nonpolyadenylated molecules. *Dev. Biol.* **88**, 201.

Dura, J. M. (1981). Stage-dependent synthesis of heat shock induced proteins in early embryos of *Drosophila melanogaster*. *Mol. Gen. Genet.* **184**, 381.

Dura, J. M., Brock, H. W., and Santamaria, P. (1985). *polyhomeotic*: A gene of *Drosophila melanogaster* required for correct expression of segmental identity. *Mol. Gen. Genet.* **198**, 213.

Durica, D. S., and Crain, W. R. (1982). Analyis of actin synthesis in early sea urchin development. *Dev. Biol.* **92**, 428.

Duryee, W. (1950). Chromosomal physiology in relation to nuclear structure. *Ann. N.Y. Acad. Sci.* **50**, 920.

Dworkin, M. B., and Dawid, I. B. (1980a). Construction of a cloned library of expressed embryonic gene sequences from *Xenopus laevis. Dev. Biol.* **76**, 435.

Dworkin, M. B., and Dawid, I. B. (1980b). Use of a cloned library for the study of abundant poly(A)⁺ RNA during *Xenopus laevis* development. *Dev. Biol.* **76**, 449.

Dworkin, M. B., and Hershey, J. W. B. (1981). Cellular fibers and subcellular distributions of abundant polyadenylate-containing ribonucleic acid species during early development in the frog *Xenopus laevis. Mol. Cell. Biol.* **1**, 983.

Dworkin, M. B., and Infante, A. A. (1976). Relationship between the mRNA of polysomes and free ribonucleoprotein particles in the early sea urchin embryo. *Dev. Bio.* **53**, 73.

Dworkin, M. B., Kay, B. K., Hershey, J. W. B., and Dawid, I. B. (1981). Mitochondrial RNAs are abundant in the poly(A)⁺ RNA population of early frog embryos. *Dev. Biol.* **86**, 502.

Dworkin, M. B., Shrutkowski, A., Baumgarten, M., and Dworkin-Rastl, E. (1984). The accumulation of prominent tadpole mRNAs occurs at the beginning of neurulation in *Xenopus laevis* embryos. *Dev. Biol.* **106**, 289.

Dworkin, M. B., Shrutkowski, A., and Dworkin-Rastl, E. (1985). Mobilization of specific maternal RNA species into polysomes after fertilization in *Xenopus laevis. Proc. Natl. Acad. Sci. U.S.A.* **82**, 7636.

Dworkin-Rastl, E., Shrutkowski, A., and Dworkin, M. B. (1984). Multiple ubiquitin mRNAs during *Xenopus laevis* development contain tandem repeats of the 76 amino acid coding sequence. *Cell* **39**, 321.

Dziadek, M., and Adamson, E. (1978). Localization and synthesis of alphafoetoprotein in postimplantation mouse embryos. *J. Embryol. Exp. Morphol.* **43**, 289.

Earnshaw, W. C., Honda, B. M., Laskey, R. A., and Thomas, J. O. (1980). Assembly of nucleosomes: The reaction involving *X. laevis* nucleoplasmin. *Cell* **21**, 373.

Ebert, K. M., and Brinster, R. L. (1983). Rabbit α-globin messenger RNA translation by the mouse ovum. *J. Embryol. Exp. Morphol.* **74**, 159.

Ebert, K. M., Paynton, B. V., McKnight, G. S., and Brinster, R. L. (1984). Translation and stability of ovalbumin messenger RNA injected into growing oocytes and fertilized ova of mice. *J. Embryol. Exp. Morphol.* **84**, 91.

Eckberg, W. R. (1981). The effects of cytoskeleton inhibitors on cytoplasmic localization in *Chaetopterus pergamentaceus. Differentiation* **19**, 55.

Ecker, R. E., and Smith, L. D. (1968). Protein synthesis in amphibian oocytes and early embryos. *Dev. Biol.* **18**, 232.

Eddy, E. M. (1975). Germ plasm and the differentiaton of the germ cell line. *Int. Rev. Cytol.* **43**, 229.

Eddy, E. M., Clark, J. M., Gong, D., and Fenderson, B. A. (1981). Origin and migration of primordial germ cells in mammals. *Gamete Res.* **4**, 333.

Edgar, L. G., and McGhee, J. D. (1986). Embryonic expression of a gut-specific esterase in *Caenorhabditis elegans. Dev. Biol.* **114**, 109.

Edwards, M. K., and Wood, W. B. (1983). Location of specific messenger RNAs in *Caenorhabditis elegans* by cytological hybridization. *Dev. Biol.* **97**, 375.

Egrie, J. C., and Wilt, F. H. (1979). Changes in poly(adenylic acid) polymerase activity during sea urchin embryogenesis. *Biochemistry* **18**, 269.

Elinson, R. P. (1983). Cytoplasmic phases in the first cell cycle of the activated frog egg. *Dev. Biol.* **100**, 440.

Ellem, K. A. O., and Gwatkin, R. B. L. (1968). Patterns of nucleic acid synthesis in the early mouse embryo. *Dev. Biol.* **18**, 311.

Emanuelsson, H. (1985). Autoradiographic analysis of RNA synthesis in the oocyte-nurse cell complex of the polychaete *Ophryotrocha labronica*. *J. Embryol. Exp. Morphol.* **88**, 249.

Emerson, B. M., Lewis, C. D., and Felsenfeld, G. (1985). Interaction of specific nuclear factors with the nuclease-hypersensitive region of the chicken adult β-globin gene: Nature of the binding domain. *Cell* **41**, 21.

Emerson, C. P., and Humphreys, T. (1970). Regulation of DNA-like RNA and the apparent activation of ribosomal RNA synthesis in sea urchin embryos: Quantitative measurements of newly synthesized RNA. *Dev. Biol.* **23**, 86.

Enders, A. C., Given, R. L., and Schlafke, S. (1978). Differentiation and migration of endoderm in the rat and mouse at implantation. *Anat. Rec.* **190**, 65.

Endo, Y. (1966). Fertilization, cleavage and early development. *In* "Contempory Biology: Development and Differentiation" (Isemura *et al.*, eds.), Vol. 4, pp. 1-61. Iwanami Shoten, Tokyo. (In Jpn.)

Endow, S. A., and Gall, J. G. (1975). Differential replication of satellite DNA in polyploid tissues of *Drosophila virilis*. *Chromosoma* **50**, 175.

Engelke, D. R., Ng, S.-Y., Shastry, B. S., and Roeder, R. G. (1980). Specific interaction of a purified transcription factor with an internal control region of 5S RNA genes. *Cell* **19**, 717.

Engelmann, F. (1979). Insect vitellogenin: Identification, biosynthesis, and role in vitellogenesis. *Adv. Insect Physiol.* **14**, 49.

Epel, D. (1982). The physiology and chemistry of calcium during the fertilization of eggs. *In* "Calcium and Cell Function" (W. Y. Cheung, ed.), p. 355. Academic Press, New York.

Episkopou, V., Murphy, A. J. M., and Efstratiadis, A. (1984). Cell-specific expression of a selectable hybrid gene. *Proc. Natl. Acad. Sci. U.S.A.* **81**, 4657.

Eppig, J. J. (1977). Mouse oocyte development *in vitro* with various culture systems. *Dev. Biol.* **60**, 371.

Ernst, S. G., Britten, R. J., and Davidson, E. H. (1979). Distinct single copy sequence sets in sea urchin nuclear RNAs. *Proc. Natl. Acad. Sci. U.S.A.* **76**, 2209.

Ernst, S. G., Hough-Evans, B. R., Britten, R. J., and Davidson, E. H. (1980). Limited complexity of the RNA in micromeres of sixteen-cell sea urchin embryos. *Dev. Biol.* **79**, 119.

Etkin, L. D. (1976). Regulation of lactate dehydrogenase (LDH) and alcohol dehydrogenase (ADH) synthesis in liver nuclei, following their transfer into oocytes. *Dev. Biol.* **52**, 201.

Etkin, L. D. (1982). Analysis of the mechanisms involved in gene regulation and cell differentiation by microinjection of purified genes and somatic cell nuclei into amphibian oocytes and eggs. *Differentiation* **21**, 149.

Etkin, L. D., and DiBerardino, M. A. (1983). Expression of nuclei and purified genes microinjected into oocytes and eggs. *In* "Eukaryotic Genes, Their Structure, Activity and Regulation" (N. Maclean, S. P. Gregory, and R. A. Flavell, eds.), p. 127. Butterworth, London.

Etkin, L. D., and Maxson, R. E. (1980). The synthesis of authentic sea urchin transcriptional and translational products by sea urchin histone genes injected into *Xenopus laevis* oocytes. *Dev. Biol.* **75**, 13.

Etkin, L. D., Pearman, B., Roberts, M., and Bektesh, S. L. (1984). Replication, integration and expression of exogenous DNA injected into fertilized eggs of *Xenopus laevis*. *Differentiation* **26**, 194.

Ettensohn, C. A. (1984). Primary invagination of the vegetal plate during sea urchin gastrulation. *Am. Zool.* **24**, 571.

Ettensohn, C. A. (1985). Gastrulation in the sea urchin embryo is accompanied by the rearrangement of invaginating epithelial cells. *Dev. Biol.* **112**, 383.

Evans, T., Rosenthal, E. T., Youngblom, J., Distel, D., and Hunt, T. (1983). Cyclin: A protein specified by maternal mRNA in sea urchin eggs that is destroyed at each cleavage division. *Cell* **33**, 389.

Falkenthal, S., Parker, V. P., and Davidson, N. (1985). Developmental variations in the splicing pattern of transcripts from the *Drosophila* gene encoding myosin alkali chain result in different carboxyl-terminal amino acid sequences. *Proc. Natl. Acad. Sci. U.S.A.* **82**, 449.

Fankhauser, G. (1934). Cytological studies on egg fragments of the salamander *Triton*. IV. The cleavage of egg fragments without the egg nucleus. *J. Exp. Zool.* **67**, 349.

Faust, M., Millward, S., Duchastel, A., and Fromson, D. (1976). Methylated constituents of poly(A)$^-$ and poly(A)$^+$ polyribosomal RNA of sea urchin embryos. *Cell* **9**, 597.

Fedoroff, N., Wellauer, P. K., and Wall, R. (1977). Intermolecular duplexes in heterogeneous nuclear RNA from HeLa cells. *Cell* **10**, 597.

Feldherr, C. M., Kallenbach, E., and Schultz, N. (1984). Movement of a karyophilic protein through the nuclear pores of oocytes. *J. Cell Biol.* **99**, 2216.

Felicetti, L., Metafora, S., Gambino, R., and Di Matteo, G. (1972). Characterization and activity of the elongation factors T1 and T2 in the unfertilized egg and in the early development of sea urchin. *Cell Differ.* **1**, 265.

Felsenfeld, G., and McGhee, J. (1982). Methylation and gene control. *Nature (London)* **296**, 602.

Files, J. G., Carr, S., and Hirsh, D. (1983). Actin gene family of *Caenorhabditis elegans*. *J. Mol. Biol.* **164**, 355.

Filosa, S., Andreuccetti, P., Parisi, E., and Monroy, A. (1985). Effect on inhibition of micromere segregation on the mitotic pattern in the sea urchin embryo. *Dev., Growth Differ.* **27**, 29.

Fink, R. D., and McClay, D. R. (1985). Three cell recognition changes accompany the ingression of sea urchin primary mesenchyme cells. *Dev. Biol.* **107**, 66.

Firtel, R. A., and Monroy, A. (1970). Polysomes and RNA synthesis during early development of the surf clam *Spisula solidissima*. *Dev. Biol.* **21**, 87.

Fischel, A. (1898). Experimentelle Untersuchungen am Ctenophorene. II. Von der Künstlichen Erzeugung (halber) Doppel-und Missbildungen. III. Regulation der Entwickelung. IV. Über den Entwickelungsgang und die Organisations-stufe des Ctenophoreneies. *Arch. Entwicklungsmech. Org.* **7**, 557.

Fischel, A. (1903). Entwickelung und Organ-differenzirung. *Arch. Entwicklungsmech. Org.* **15**, 670.

Fischer, A. (1977). Autonomy for a specific gene product in oocytes: Experimental evidence in the polychaetous annelid, *Platynereis dumerilii*. *Dev. Biol.* **55**, 46.

Fischer, A. (1984). Control of oocyte differentiation in nereids (Annelida, Polychaeta)—facts and ideas. *Fortschr. Zool.* **29**, 227.

Fischer, A., and Pfannenstiel, H.-D., eds. (1984). "Polychaete Reproduction. Progress in Comparative Reproductive Biology." Fischer Verlag, Stuttgart.

Fjose, A., McGinnis, W. J., and Gehring, W. J. (1985). Isolation of a homeobox-containing gene from the *engrailed* region of *Drosophila* and the spatial distribution of its transcripts. *Nature (London)* **313**, 284.

Flamm, W. G., Walker, P. M. B., and McCallum, M. (1969). Renaturation and isolation of single strands from the nuclear DNA of the guinea pig. *J. Mol. Biol.* **42**, 441.

Fleming, T. P., Warren, P. D., Chisholm, J. C., and Johnson, M. H. (1984). Trophectodermal processes regulate the expression of totipotency within the inner cell mass of the mouse expanding blastocyst. *J. Embryol. Exp. Morphol.* **84**, 63.

Flemming, W. (1882). "Zellsubstanz, Kern und Zelltheilung." F. C. W. Vogel, Leipzig.

Florman, H. M., and Wassarman, P. M. (1985). *O*-linked oligosaccharides of mouse egg ZP3 account for its sperm receptor activity. *Cell* **41**, 313.

Floyd, E. E., Gong, Z., Brandhorst, B. P., and Klein, W. H. (1986). Calmodulin gene expression during sea urchin development: Persistence of a prevalent maternal protein. *Dev. Biol.* **113**, 501.

Flynn, J. M., and Woodland, H. R. (1980). The synthesis of histone H1 during amphibian development. *Dev. Biol.* **75**, 222.

Flytzanis, C. N., Brandhorst, B. R., Britten, R. J., and Davidson, E. H. (1982). Developmental patterns of cytoplasmic transcript prevalence in sea urchin embryos. *Dev. Biol.* **91**, 27.

Flytzanis, C. N., McMahon, A. P., Hough-Evans, B. R., Katula, K. S., Britten, R. J., and Davidson, E. H. (1985). Persistence and integration of cloned DNA in postembryonic sea urchins. *Dev. Biol.* **108**, 431.

Flytzanis, C. N., Britten, R. J., and Davidson, E. H. (1986). Mechanism of ontogenic activation of a gene expressed in a specific cell lineage of the early sea urchin embryo. *Proc. Natl. Acad. Sci. U.S.A.* In press.

Forbes, D. J., Kornberg, T. B., and Kirschner, M. W. (1983a). Small nuclear RNA transcription and ribonucleoprotein assembly in early *Xenopus* development. *J. Cell Biol.* **97**, 62.

Forbes, D. J., Kirschner, M. W., and Newport, J. W. (1983b). Spontaneous formation of nucleus-like structures around bacteriophage DNA microinjected into *Xenopus* eggs. *Cell* **34**, 13.

Forbes, D. J., Kirschner, M. W., Caput, D., Dahlberg, J. E., and Lund, E. (1984). Differential expression of multiple U1 small nuclear RNAs in oocytes and embryos of *Xenopus laevis*. *Cell* **38**, 681.

Ford, P. J. (1971). Non-coordinated accumulation and synthesis of 5S ribonucleic acid by ovaries of *Xenopus laevis*. *Nature (London)* **233**, 561.

Ford, P. J., and Brown, R. D. (1976). Sequences of 5S ribosomal RNA from *Xenopus mulleri* and the evolution of 5S gene-coding sequences. *Cell* **8**, 485.

Ford, P. J., and Southern, E. M. (1973). Different sequences for 5S RNA in kidney cells and ovaries of *Xenopus laevis*. *Nature (London), New Biol.* **241**, 7.

Fowlkes, D. M., Mullis, N. T., Comeau, C. M., and Crabtree, G. R. (1984). Potential basis for regulation of the coordinately expressed fibrinogen genes: Homology in the 5' flanking regions. *Proc. Natl. Acad. Sci. U.S.A.* **81**, 2313.

Fox, A. M., Breux, C. B., and Benbow, R. M. (1980). Intracellular localization of DNA polymerase activities within larger oocytes of the frog, *Xenopus laevis*. *Dev. Biol.* **80**, 79.

Fradin, A., Jove, R., Hemenway, C., Keiser, H. D., Manley, J. L., and Prives, C. (1984). Splicing pathways of SV40 mRNAs in *X. laevis* oocytes differ in their requirements for snRNPs. *Cell* **37**, 927.

Franke, W. W., Rathke, P. C., Seib, E., Trendelenburg, M. F., Osborn, M., and Weber, K. (1976). Distribution and mode of arrangement of the filamentous structures and actin in the cortex of amphibian oocytes. *Cytobiologie* **14**, 111.

Franks, R. R., and Davis, F. C. (1983). Regulation of histone synthesis during early *Urechis caupo* (Echiura) development. *Dev. Biol.* **98**, 101.

Frederiksen, S., Hellung-Larsen, P., and Engberg, J. (1973). Small molecular weight RNA components in sea urchin embryos. *Exp. Cell Res.* **78**, 287.

Freeman, G. (1976a). The role of cleavage in the localization of developmental potential in the ctenophore *Mnemiopsis leidyi*. *Dev. Biol.* **49**, 143.

Freeman, G. (1976b). The effects of altering the position of cleavage planes on the process of localization of developmental potential in ctenophores. *Dev. Biol.* **51**, 332.

Freeman, G. (1977). The establishment of the oral-aboral axis in the ctenophore embryo. *J. Embryol. Exp. Morphol.* **42**, 237.

Freeman, G. (1978). The role of asters in the localization of the factors that specify the apical tuft and the gut of the nemertine *Cerebratulus lacteus*. *J. Exp. Zool.* **206**, 81.

Freeman, G. (1979). The multiple roles which cell division can play in the localization of developmental potential. *In* "Determinants of Spatial Organization" (S. Subtelny and I. R. Konigsberg, eds.), p. 53. Academic Press, New York.

Freeman, G. (1981). The cleavage initiation site establishes the posterior role of the hydrozoan embryo. *Wilhelm Roux's Arch. Dev. Biol.* **190**, 123.

Freeman, G., and Lundelius, J. W. (1982). The developmental genetics of dextrality and sinistrality in the gastropod *Lymnaea peregra*. *Wilhelm Roux's Arch. Dev. Biol.* **191**, 69.

Freeman, G., and Reynolds, G. T. (1973). The development of bioluminescence in the ctenophore *Mnemiopsis leidyi*. *Dev. Biol.* **31**, 61.

Freeman, S. B. (1971). A comparison of certain isozyme patterns in lobeless and normal embryos of the snail, *Ilyanassa obsoleta*. *J. Embryol. Exp. Morphol.* **26**, 339.

Fregien, N., Dolecki, G. J., Mandel, M., and Humphreys, T. (1983). Molecular cloning of five individual stage- and tissue-specific mRNA sequences from sea urchin pluteus embryos. *Mol. Cell. Biol.* **3**, 1021.

Frey, A., Sander K., and Gutzeit, H. (1984). The spatial arrangement of germ line cells in ovarian follicles of the mutant *dicephalic* in *Drosophila melanogaster*. *Wilhelm Roux's Arch. Dev. Biol.* **193**, 388.

Fritz, A., Parisot, R., Newmeyer, D., and De Robertis, E. M. (1984). Small nuclear U-ribonucleoproteins in *Xenopus laevis* development. Uncoupled accumulation of the protein and RNA components. *J. Mol. Biol.* **178**, 273.

Fromson, D., and Duchastel, A. (1975). Poly(A)-containing polyribosomal RNA in sea urchin embryos: Changes in proportion during development. *Biochim. Biophys. Acta* **378**, 394.

Fromson, D., and Verma, D. P. S. (1976). Translation of nonpolyadenylated messenger RNA of sea urchin embryos. *Proc. Natl. Acad. Sci. U.S.A.* **73**, 148.

Fruscoloni, P., Al-Atia, G. R., and Jacobs-Lorena, M. (1983). Translational regulation of a specific gene during oogenesis and embryogenesis of *Drosophila*. *Proc. Natl. Acad. Sci. U.S.A.* **80**, 3359.

Fry, B. J., and Gross, P. R. (1970). Patterns and rates of protein synthesis in sea urchin embryos. II. The calculation of absolute rates. *Dev. Biol.* **21**, 125.

Fukushi, T. (1962). The fates of isolated blastoderm cells of sea urchin blastulae and gastrulae inserted into the blastocoel. *Bull. Mar. Biol. Stn. Asamushi, Tohoku Univ.* **11**, 21.

Fyrberg, E. A., Kindle, K. L., Davidson, N., and Sodja, A. (1980). The actin genes of *Drosophila*: A dispersed multigene family. *Cell* **19**, 365.

Fyrberg, E. A., Mahaffey, J. W., Bond, B. J., and Davidson, N. (1983). Transcripts of the six *Drosophila* actin genes accumulate in a stage- and tissue-specific manner. *Cell* **33**, 115.

Gabrielli, F., and Baglioni, C. (1977). Regulation of maternal mRNA translation in developing embryos of the surf clam *Spisula solidissima*. *Nature (London)* **269**, 529.

Galau, G. A., Britten, R. J., and Davidson, E. H. (1974). A measurement of the sequence complexity of polysomal messenger RNA in sea urchin embryos. *Cell* **2**, 9.

Galau, G. A., Klein, W. H., Davis, M. M., Wold, B. J., Britten, R. J., and Davidson, E. H. (1976). Structural gene sets active in embryos and adult tissues of the sea urchin. *Cell* **7**, 487.

Galau, G. A., Lipson, E. D., Britten, R. J., and Davidson, E. H. (1977a). Synthesis and turnover of polysomal mRNAs in sea urchin embryos. *Cell* **10**, 415.

Galau, G. A., Klein, W. H., Britten, R. J., and Davidson, E. H. (1977b). Significance of rare mRNA sequences in liver. *Arch. Biochem. Biophys.* **179**, 584.

Galau, G. A., Britten, R. J., and Davidson, E. H. (1977c). Studies on nucleic acid reassociation kinetics: Rate of hybridization of excess RNA with DNA, compared to the rate of DNA renaturation. *Proc. Natl. Acad. Sci. U.S.A.* **74**, 1020.

Galau, G. A., Smith, M. J., Britten, R. J., and Davidson, E. H. (1977d). Studies on nucleic acid reassociation kinetics: Retarded rate of hybridization of RNA with excess DNA. *Proc. Natl. Acad. Sci. U.S.A.* **74**, 2306.

Gall, J. G. (1954). Lampbrush chromosomes from oocyte nuclei of the newt. *J. Morphol.* **94**, 283.

Gall, J. G. (1955). On the submicroscopic structure of chromosomes. *Brookhaven Symp. Biol.* **8**, 17.

Gall, J. (1958). Chromosomal differentiation. *In* "The Chemical Basis of Development" (W. D. McElroy and B. Glass, eds.), p. 103. Johns Hopkins Press, Baltimore.

Gall, J. (1963). Kinetics of deoxyribonuclease action on chromosomes. *Nature (London)* **198**, 36.

Gall, J. G. (1966). Techniques for the study of lampbrush chromosomes. *Methods Cell Physiol.* **2**, 37.

Gall, J. G. (1968). Differential synthesis of the genes for ribosomal RNA during amphibian oogenesis. *Proc. Natl. Acad. Sci. U.S.A.* **60**, 553.

Gall, J. G. (1969). The genes for ribosomal RNA during oögenesis. *Genetics,* **61**, Suppl., 1.

Gall, J. G., and Callan, H. G. (1962). ^{3}H-uridine incorporation in lampbrush chromosomes. *Proc. Natl. Acad. Sci. U.S.A.* **49**, 544.

Gall, J. G., and Rochaix, J.-D. (1974). The amplified ribosomal DNA of dytiscid beetles. *Proc. Natl. Acad. Sci. U.S.A.* **71**, 1819.

Gall, J. G., Stephenson, E. C., Erba, H. P., Diaz, M. O., and Barsacchi-Pilone, G. (1981). Histone genes are located at the sphere loci of newt lampbrush chromosomes. *Chromosoma* **84**, 159.

Galli, G., Hofstetter, H., Stunnenberg, H. G., and Birnstiel, M. L. (1983). Biochemical complementation with RNA in the *Xenopus* oocyte: A small RNA is required for the generation of 3' histone mRNA termini. *Cell* **34**, 823.

Gambarini, A. G., and Lara, F. J. S. (1974). Under-replication of ribosomal cistrons in polytene chromosomes of *Rhyncosciara*. *J. Cell Biol.* **62**, 215.

Gans, M., Audit, C., and Masson, M. (1975). Isolation and characterization of sex-linked female sterile mutants in *Drosophila melanogaster*. *Genetics* **81**, 683.

Garber, R. L., Kuroiwa, A., and Gehring, W. J. (1983). Genomic and cDNA clones of the homeotic locus *Antennapedia* in *Drosophila*. *EMBO J.* **2**, 2027.

Garcia-Bellido, A. (1977). Homeotic and atavic mutations in insects. *Am. Zool.* **17**, 613.

Garcia-Bellido, A., and Merriam, J. R. (1969). Cell lineage of the imaginal discs in *Drosophila* gynandromorphs. *J. Exp. Zool.* **170**, 61.

Garcia-Bellido, A., and Moscoso del Prado, J. (1979). Genetic analysis of maternal information in *Drosophila*. *Nature (London)* **278**, 346.

Garcia-Bellido, A., and Ripoll, P. (1978). The number of genes in *Drosophila melanogaster*. *Nature (London)* **273**, 399.

Garcia-Bellido, A., and Robbins, L. G. (1983). Viability of female germ line cells homozygous for zygotic lethals in *Drosophila melanogaster*. *Genetics* **103**, 235.

Garcia-Bellido, A., Ripoll, P., and Morata, G. (1973). Developmental compartmentalization of the wing disk of *Drosophila*. *Nature (London)* New Biol. **245**, 251.

Garcia-Bellido, A., Lawrence, P. A., and Morata, G. (1979). Compartments in animal development. *Sci. Am.* **241**, 102.

Garcia-Bellido, A., Moscoso del Prado, J., and Botas, J. (1983). The effect of aneuploidy on embryonic development in *Drosophila melanogaster*. *Mol. Gen. Genet.* **192**, 253.

Gardner, R. L. (1968). Mouse chimaeras obtained by the injection of cells into the blastocyst. *Nature (London)* **220**, 596.

Gardner, R. L. (1982). Investigation of cell lineage and differentiation in the extraembryonic endoderm of the mouse embryo. *J. Embryol. Exp. Morphol.* **68**, 175.

Gardner, R. L. (1984). An *in situ* cell marker for clonal analysis of development of the extraembryonic endoderm in the mouse. *J. Embryol. Exp. Morphol.* **80**, 251.

Gardner, R. L., and Papaioannou, V. E. (1975). Differentiation in the trophectoderm and inner cell mass. *In* "The Early Development of Mammals" (M. Balls and A. E. Wild, eds.), p. 107. Cambridge Univ. Press, London and New York.

Gardner, R. L., and Rossant, J. (1979). Investigation of the fate of 4-5 day *post coitum* mouse inner cell mass cells by blastocyst injection. *J. Embryol. Exp. Morphol.* **52**, 141.

Gardner, R. L., Lyon, M. F., Evans, E. P., and Burtenshaw, M. D. (1985). Clonal analysis of X-chromosome inactivation and the origin of the germ line in the mouse embryo. *J. Embryol. Exp. Morphol.* **88**, 349.

Garen, A., and Gehring, W. (1972). Repair of the lethal developmental defect in *deep orange* embryos of *Drosophila* by injection of normal egg cytoplasm. *Proc. Natl. Acad. Sci. U.S.A.* **69**, 2982.

Garfinkel, M. D., Pruitt, R. E., and Meyerowitz, E. M. (1983). DNA sequences, gene regulation and modular protein evolution in the *Drosophila* 68C glue gene cluster. *J. Mol. Biol.* **168**, 765.

Gargiulo, G., Razvi, F., and Worcel, A. (1984). Assembly of transcriptionally active chromatin in *Xenopus* oocytes requires specific DNA binding factors. *Cell* **38**, 511.

Garner, W., and McLaren, A. (1974). Cell distribution in chimaeric mouse embryos before implantation. *J. Embryol. Exp. Morphol.* **32**, 495.

Gasaryan, K. G., Hung, N. M., Neyfakh, A. A., and Ivanenkov, V. V. (1979). Nuclear transplantation in teleost *Misgurnus fossilis* L. *Nature (London)* **280**, 585.

Gautsch, J. W., and Wilson, M. C. (1983). Delayed *de novo* methylation in teratocarcinoma suggests additional tissue-specific mechanisms for controlling gene expression. *Nature (London)* **301**, 32.

Gehring, W. J. (1984). Homeotic genes and the control of cell determination. *In* "Molecular Biology of Development" (E. H. Davidson and R. A. Firtel, eds.), p. 3. Alan R. Liss, New York.

Geigy, R. (1931). Action de l'ultra-violet sur le pole germinal dans l'oeuf de *Drosophila melanogaster* (castration et mutabilité). *Rev. Suisse Zool.* **38**, 187.

Geilenkirchen, W. L. M., Verdonk, N. H., and Timmermans, L. P. M. (1970). Experimental studies on morphogenetic factors localized in the first and the second polar lobe of *Dentalium* eggs. *J. Embryol. Exp. Morphol.* **23**, 237.

Gelfand, R. A., and Smith, L. D. (1983). RNA stabilization and continued RNA processing following nuclear dissolution in maturing *Xenopus laevis* oocytes. *Dev. Biol.* **99**, 427.

Gelfand, R. A., and Smith, L. D. (1986). Oocyte patterns of small nuclear RNA synthesis in *Xenopus laevis*. Submitted for publication.

Gerber-Huber, S., May, F. E. B., Westley, B. R., Felber, B. K., Hosbach, H. A., Andres, A.-C., and Ryffel, G. U. (1983). In contrast to other *Xenopus* genes the estrogen-inducible vitellogenin genes are expressed when totally methylated. *Cell* **33**, 43.

Gergen, J. P., and Weischaus, E. F. (1985). The localized requirements for a gene affecting segmentation in *Drosophila*: Analysis of larvae mosaic for *runt*. *Dev. Biol.* **109**, 321.

Gergen, J. P., and Wieschaus, E. F. (1986a). Localized requirements for gene activity in segmentation of *Drosophila* embryos: Analysis of *armadillo, fused, giant,* and *unpaired* mutations in mosaic embryos. *Wilhelm Roux's Arch. Dev. Biol.* **195**, 49.

Gergen, J. P., and Wieschaus, E. (1986b). Dosage requirements for *runt* in the segmentation of *Drosophila* embryos. *Cell* **45**, 289.

Gergen, J. P., Coulter, D., and Wieschaus, E. (1986). Segmental pattern and blastoderm cell identities. *In* "Gametogenesis and the Early Embryo" (J. Gall, ed.). Alan R. Liss, New York. In press.

Gerhart, J. G. (1980). Mechanisms regulating pattern formation in the amphibian egg and early embryo. *In* "Biological Regulation and Development" (R. F. Goldberger, ed.), Vol. 2, p. 133. Plenum, New York.

Gerhart, J., Ubbels, G., Black, S., Hara, K., and Kirschner, M. (1981). A reinvestigation of the role of the grey crescent in axis formation in *Xenopus laevis*. *Nature (London)* **292**, 511.

Gerhart, J., Black, S., and Scharf, S. (1983). Cellular and pancellular organization of the amphibian embryo. *In* "Modern Cell Biology. Spatial Organization of Eukaryotic Cells" (R. McIntosh, ed.), Vol. 2, p. 483. Alan R. Liss, New York.

Gerhart, J., Wu, M., and Kirschner, M. (1984). Cell cycle dynamics of an M-phase-specific cytoplasmic factor in *Xenopus laevis* oocytes and eggs. *J. Cell Biol.* **98**, 1247.

Geuskens, M. (1969). Mise en evidence au microscope électronique de polysomes actifs dans des lobes polaires isolés d'*Ilyanassa*. *Exp. Cell Res.* **54**, 263.

Geyer-Duszyńska, I. (1966). Genetic factors in oogenesis and spermatogenesis in *Cecidomyidae*. *Chromosomes Today* **1**, 174.

Gibbons, J. R., Tilney, L. G., and Porter, K. R. (1969). Microtubules in the formation and development of the primary mesenchyme in *Arbacia punctulata*. *J. Cell Biol.* **41**, 201.

Gibson, A. W., and Burke, R. D. (1985). The origin of pigment cells in embryos of the sea urchin *Strongylocentrotus purpuratus*. *Dev. Biol.* **107**, 414.

Giebelhaus, D. H., Heikkala, J. J., and Schultz, G. A. (1983). Changes in the quantity of histone and actin messenger RNA during the development of preimplantation mouse embryos. *Dev. Biol.* **98**, 148.

Giebelhaus, D. H., Weitlauf, H. M., and Schultz, G. A. (1985). Actin mRNA content in normal and delayed implanting mouse embryos. *Dev. Biol.* **107**, 407.

Gillies, S. D., Morrison, S. L., Oi, V. T., and Tonegawa, S. (1983). A tissue-specific transcription enhancer element is located in the major intron of a rearranged immunoglobulin heavy chain gene. *Cell* **33**, 717.

Gimlich, R. L. (1986). Acquisition of developmental autonomy in the equatorial region of the *Xenopus* embryo. *Dev. Biol.* **115**, 340.

Gimlich, R. L., and Cooke, J. (1983). Cell lineage and the induction of second nervous systems in amphibian development. *Nature (London)* **306**, 471.

Gimlich, R. L., and Gerhart, J. C. (1984). Early cellular interactions promote embryonic axis formation in *Xenopus laevis*. *Dev. Biol.* **104**, 117.

Ginsberg, A. M., King, B. O., and Roeder, R. G. (1984). *Xenopus* 5S gene transcription factor, TFIIIA: Characterization of a cDNA clone and measurement of RNA levels throughout development. *Cell* **39**, 479.

Gipouloux, J.-D. (1971). Effets de l'extrusion totale ou partielle du cytoplasme germinal au cours des premiers stades de la segmentation sur la fertilité des larves d'amphibiens anoures. *C. R. Hebd. Seances Acad. Sci.* **273**, 2627.

Giudice, G., Sconzo, G., Albanese, I., Ortolani, G., and Cammarata, M. (1974). Cytoplasmic giant RNA in sea urchin embryos. I. Proof that it is not derived from artifactual nuclear leakage. *Cell Differ.* **3**, 287.

Glassman, E. (1965). Genetic regulation of xanthine dehydrogenase in *Drosophila melanogaster*. *Fed. Proc.* **24**, 1243.

Glišin, V. R., Glišin, M. V., and Doty, P. (1966). The nature of messenger RNA in the early stages of sea urchin development. *Proc. Natl. Acad. Sci. U.S.A.* **56**, 285.

Goldberg, D. A., Posakony, J. W., and Maniatis, T. (1984). Correct developmental expression of a cloned alcohol dehydrogenase gene transduced into the *Drosophila* germ line. *Cell* **34**, 59.

Goldberg, R. B., Galau, G. A., Britten, R. J., and Davidson, E. H. (1973). Nonrepetitive DNA sequence representation in sea urchin embryo messenger RNA. *Proc. Natl. Acad. Sci. U.S.A.* **70**, 3516.

Goldberg, R. B., Crain, W. R., Ruderman, J. V., Moore, G. P., Barnett, T. R., Higgins, R. C., Gelfand, R. A., Galau, G. A., Britten, R. J., and Davidson, E. H. (1975). DNA sequence organization in the genomes of five marine invertebrates. *Chromosome* **51**, 225.

Golden, L., Schafer, U., and Rosbash, M. (1980). Accumulation of individual pA⁺ RNAs during oogenesis of *Xenopus laevis*. *Cell* **22**, 835.

Goldstein, E. S. (1978). Translated and sequestered untranslated message sequences in *Drosophila* oocytes and embryos. *Dev. Biol.* **63**, 59.

Golub, A., and Clegg, J. S. (1968). Protein synthesis in *Artemia salina* embryos. I. Studies on polyribosomes. *Dev. Biol.* **17**, 644.

Gonor, J. J. (1973a). Reproductive cycles in Oregon populations of the echinoid, *Strongylocentrotus purpuratus* (Stimpson). I. Annual gonad growth and ovarian gametogenic cycles. *J. Exp. Mar. Biol. Ecol.* **12**, 45.

Gonor, J. J. (1973b). Reproductive cycles in Oregon populations of the echinoid, *Strongylocentrotus purpuratus* (Stimpson). II. Seasonal changes in oocyte growth and in abundance of gametogenic stages in the ovary. *J. Exp. Mar. Biol. Ecol.* **12**, 65.

Gossett, L. A., Hecht, R. M., and Epstein, H. F. (1982). Muscle differentiation in normal and cleavage-arrested mutant embryos of *Caenorhabditis elegans*. *Cell* **30**, 193.

Gould, M. C. (1969). A comparison of RNA and protein synthesis in fertilized and unfertilized eggs of *Urechis caupo*. *Dev. Biol.* **19**, 482.

Gould-Somero, M. C. (1975). Echiura. *In* "Reproduction of Marine Invertebrates. Annelids and Echiurans" (A. C. Giese and J. S. Pearse, eds.), Vol. 3, p. 277. Academic Press, New York.

Goustin, A. S. (1981). Two temporal phases for the control of histone gene activity in cleaving sea urchin embryos (*S. purpuratus*). *Dev. Biol.* **87**, 163.

Goustin, A. S., and Wilt, F. H. (1981). Protein synthesis, polyribosomes, and peptide elongation in early development of *Strongylocentrotus purpuratus*. *Dev. Biol.* **82**, 32.

Goustin, A. S., and Wilt, F. H. (1982). Direct measurement of histone peptide elongation rate in cleaving sea urchin embryos. *Biochim. Biophys. Acta* **699**, 22.

Graham, C. F., and Deussen, Z. A. (1978). Features of cell lineage in preimplantation mouse development. *J. Embryol. Exp. Morphol.* **48**, 53.

Graham, C. F., and Lehtonen, E. (1979). Formation and consequences of cell patterns in preimplantation mouse development. *J. Embryol. Exp. Morphol.* **49**, 277.

Graham, C. F., Arms, K., and Gurdon, J. B. (1966). The induction of DNA synthesis by frog egg cytoplasm. *Dev. Biol.* **14**, 349.

Graham, D. E., and Schanke, K. E. (1980). Interspersion of repetitive with repetitive sequences in an amphibian, *Rana berlandieri*. *Nucleic Acids Res.* **8**, 3875.

Graham, D. E., Neufeld, B. R., Davidson, E. H., and Britten, R. J. (1974). Interspersion of repetitive and nonrepetitive DNA sequences in the sea urchin genome. *Cell* **1**, 127.

Grainger, J. L., Winkler, M. M., Shen, S. S., and Steinhardt, R. A. (1979). Intracellular pH controls protein synthesis rate in the sea urchin egg and early embryo. *Dev. Biol.* **68**, 396.

Grainger, J. L., von Brunn, A., and Winkler, M. M. (1986). Transient synthesis of a specific set of proteins during the rapid cleavage phase of sea urchin development. *Dev. Biol.* **114**, 403.

Grainger, R. M., and Wilt, F. H. (1976). Incorporation of ^{13}C, ^{15}N-labeled nucleosides and measurement of RNA synthesis and turnover in sea urchin embryos. *J. Mol. Biol.* **104**, 589.

Grant, P. (1969). Nucleocortical interactions during amphibian development. *In* "Biology of Amphibian Tumors" (M. Mizell, ed.), p. 43. Springer-Verlag, Berlin and New York.

Grant, P., and Wacaster, J. F. (1972). The amphibian gray crescent region—a site of developmental information? *Dev. Biol.* **28**, 454.

Grant, S. R., Farach, M. C., Decker, G. L., Woodward, H. D., Farach, H. A., and Lennarz, W. J. (1985). Developmental expression of cell surface (glyco)proteins involved in gastrulation and spicule formation in sea urchin embryos. *Cold Spring Harbor Symp. Quant. Biol.* **50**, 91.

Graves, R. A., Marzluff, W. F., Giebelhaus, D. H., and Schultz, G. A. (1985). Quantitative and qualitative changes in histone gene expression during early mouse embryo development. *Proc. Natl. Acad. Sci. U.S.A.* **82**, 5685.

Green, M. R., Maniatis, T., and Melton, D. A. (1983). Human β-globin pre-mRNA synthesized *in vitro* is accurately spliced in *Xenopus* oocyte nuclei. *Cell* **32**, 681.

Greenwald, I. (1985). *lin-12*, a nematode homeotic gene, is homologous to a set of mammalian proteins that includes epidermal growth factor. *Cell* **43**, 583.

Greenwald, I. S., Sternberg, P. W., and Horvath, H. R. (1983). The *lin-12* locus specifies cell fates in *Caenorhabditis elegans*. *Cell* **34**, 435.

Grell, R. F., and Chandley, A. C. (1965). Evidence bearing on the coincidence of exchange and DNA replication in the oocyte of *Drosophila melanogaster*. *Proc. Natl. Acad. Sci. U.S.A.* **53**, 1340.

Grell, R. F., and Day, J. W. (1974). Intergenic recombination, DNA replication and synaptonemal complex formation in the *Drosophila* oocyte. *In* "Mechanisms in Recombination" (R. F. Grell, ed.), p. 327. Plenum, New York.

Grell, R. F., and Generoso, E. E. (1982). A temporal study at the ultrastructural level of the developing pro-oocyte of *Drosophila melanogaster*. *Chromosoma* **87**, 49.

Griffith, J. K., and Humphreys, T. D. (1979). Ribosomal ribonucleic acid synthesis and processing in embryos of the Hawaiian sea urchin *Tripneustes gratilla*. *Biochemistry* **18**, 2178.

Griffith, J. K., Griffith, B. B., and Humphreys, T. (1981). Regulation of ribosomal RNA synthesis in sea urchin embryos and oocytes. *Dev. Biol.* **87**, 220.

Grimwade, B. G., Muskavitch, M. A. T., Welshons, W. J., Yedvobnick, B., and Artavanis-

Tsakonas, S. (1985). The molecular genetics of the *Notch* locus in *Drosophila melanogaster*. *Dev. Biol.* **107**, 503.

Gross, K. W., Jacobs-Lorena, M., Baglioni, C., and Gross, P. R. (1973). Cell-free translation of maternal messenger RNA from sea urchin eggs. *Proc. Natl. Acad. Sci. U.S.A.* **70**, 2614.

Gross, K., Probst, E., Schaffner, W., and Birnstiel, M. L. (1976). Molecular analysis of the histone gene cluster of *Psammechinus miliaris*: I. Fractionation and identification of five individual histone mRNA. *Cell* **8**, 455.

Gross, P. R. (1967). The control of protein synthesis in embryonic development and differentiation. *Curr. Top. Dev. Biol.* **2**, 1.

Gross, P. R., and Cousineau, G. H. (1963). Effects of actinomycin on macromolecule synthesis and early development in sea urchin eggs. *Biochem. Biophys. Res. Commun.* 10, 321.

Gross, P. R., and Cousineau, G. H. (1964). Macromolecule synthesis and the influence of actinomycin on early development. *Exp. Cell Res.* **33**, 368.

Gross, P. R., Malkin, L. I., and Moyer, W. A. (1964). Templates for the first proteins of embryonic development. *Proc. Natl. Acad. Sci. U.S.A.* **51**, 407.

Gross, P. R., Kraemer, K., and Malkin, L. I. (1965). Base composition of RNA synthesized during cleavage of the sea urchin embryo. *Biochem. Biophys. Res. Commun.* **18**, 569.

Grosschedl, R., and Baltimore, D. (1985). Cell-type specificity of immunoglobulin gene expression is regulated by at least three DNA sequence elements. *Cell* **41**, 885.

Grosschedl, R., and Birnstiel, M. L. (1980a). Identification of regulatory sequences in the prelude sequences of an H2a histone gene by the study of specific deletion mutants *in vivo*. *Proc. Natl. Acad. Sci. U.S.A.* **77**, 1432.

Grosschedl, R., and Birnstiel, M. L. (1980b). Spacer DNA sequences upstream of the T-A-T-A-A-A-T-A sequence are essential for promotion of H2a histone gene transcription *in vivo*. *Proc. Natl. Acad. Sci. U.S.A.* **77**, 7102.

Grosschedl, R., Mächler, M., Rohrer, U., and Birnstiel, M. L. (1983). A functional component of the sea urchin H2a gene modulator contains an extended sequence homology to a viral enhancer. *Nucleic Acids Res.* **11**, 8123.

Grosschedl, R., Weaver, D., Baltimore, D., and Costantini, F. (1984). Introduction of a μ immunoglobulin gene into the mouse germ line: Specific expression in lymphoid cells and synthesis of functional antibody. *Cell* **38**, 647.

Groudine, M., and Conkin, K. F. (1985). Chromatin structure and *de novo* methylation of sperm DNA: Implications for activation of the paternal genome. *Science* **228**, 1061.

Groudine, M., and Weintraub, H. (1975). Rous sarcoma virus activates embryonic globin genes in chicken fibroblasts. *Proc. Natl. Acad. Sci. U.S.A.* **72**, 4464.

Groudine, M., and Weintraub, H. (1981). Activation of globin genes during chicken development. *Cell* **24**, 393.

Groudine, M., and Weintraub, H. (1982). Propagation of globin DNAase I-hypersensitive sites in absence of factors required for induction: A possible mechanism for determination. *Cell* **30**, 131.

Groudine, M., Holtzer, H., Scherrer, K., and Therwath, A. (1974). Lineage-dependent transcription of globin genes. *Cell* **3**, 243.

Groudine, M., Eisenman, R., and Weintraub, H. (1981). Chromatin structure of endogenous retroviral genes and activation by an inhibitor of DNA methylation. *Nature (London)* **292**, 311.

Grunstein, M. (1978). Hatching in the sea urchin *Lytechinus pictus* is accompanied by a shift in histone H4 gene activity. *Proc. Natl. Acad. Sci. U.S.A.* **75**, 4135.

Grunstein, M., and Schedl, P. (1976). Isolation and sequence analysis of sea urchin (*Lytechinus pictus*) histone H4 messenger RNA. *J. Mol. Biol.* **104**, 323.

Guerrier, P. (1967). Les facteurs de polarisation dans les premiers stades du développement chez *Parascaris equorum*. *J. Embryol. Exp. Morphol.* **18**, 121.

Guerrier, P., and Freyssinet, G. (1974). Protein synthesis during embryogenesis of *Sabellaria alveolata* L. (polychaete annelid). *Exp. Cell Res.* **87**, 290.

Guerrier, P., and van den Biggelaar, J. A. M. (1979). Intracellular activation and cell interactions in so-called mosaic embryos. *In* "Cell Lineage, Stem Cells and Cell Determination" (N. Le Douarin, ed.), p. 29. Elsevier/North-Holland, Amsterdam.

Guerrier, P., van den Biggelaar, J. A. M., Van Dongen, C. A. M., and Verdonk, N. H. (1978). Significance of the polar lobe for the determination of dorsoventral polarity in *Dentalium vulgare* (da Costa). *Dev. Biol.* **63**, 233.

Gurdon, J. B. (1962). Adult frogs derived from the nuclei of single somatic cells. *Dev. Biol.* **4**, 256.

Gurdon, J. B. (1963). Nuclear transplantation in amphibia and the importance of stable nuclear changes in promoting cellular differentiation. *Q. Rev. Biol.* **38**, 54.

Gurdon, J. B. (1967). Control of gene activity during the early development of *Xenopus laevis*. *In* "Heritage from Mendel" (A. Brink, ed.), p. 203. Univ. of Wisconsin Press, Madison.

Gurdon, J. B. (1976). Injected nuclei in frog oocytes: Fate, enlargement, and chromatin dispersal. *J. Embryol. Exp. Morphol.* **36**, 523.

Gurdon, J. B., and Uchlinger, V. (1966). "Fertile" intestinal nuclei. *Nature (London)* **210**, 1240.

Gurdon, J. B., and Woodland, H. R. (1969). The influence of the cytoplasm on the nucleus during cell differentiation, with special reference to RNA synthesis during amphibian cleavage. *Proc. R. Soc. London, Ser. B* **173**, 99.

Gurdon, J. B., Lingrel, J. B., and Marbaix, G. (1973). Message stability in injected frog oocytes: Long life of mammalian α and β globin messages. *J. Mol. Biol.* **80**, 539.

Gurdon, J. B., Woodland, H. R., and Lingrel, J. B. (1974). The translation of mammalian globin mRNA injected into fertilized eggs of *Xenopus laevis*. I. Message stability in development. *Dev. Biol.* **39**, 125.

Gurdon, J. B., Laskey, R. A., and Reeves, O. R. (1975). The developmental capacity of nuclei transplanted from keratinized skin cells of adult frogs. *J. Embryol. Exp. Morphol.* **34**, 93.

Gurdon, J. B., Partington, G. A., and De Robertis, E. M. (1976). Injected nuclei in frog oocytes: RNA synthesis and protein exchange. *J. Embryol. Exp. Morphol.* **36**, 541.

Gurdon, J. B., Brennan, S., Fairman, S., and Mohun, T. J. (1984). Transcription of muscle-specific actin genes in early *Xenopus* development: Nuclear transplantation and cell dissociation. *Cell* **38**, 691.

Gurdon, J. B., Mohun, T. J., Fairman, S., and Brennan, S. (1985a). All components required for the eventual activation of muscle-specific actin genes are localized in the subequatorial region of an uncleaved amphibian egg. *Proc. Natl. Acad. Sci. U.S.A.* **82**, 139.

Gurdon, J. B., Fairman, S., Mohun, T. J., and Brennan, S. (1985b). Activation of muscle-specific actin genes in *Xenopus* development by an induction between animal and vegetal cells of a blastula. *Cell* **41**, 913.

Gustafson, T., and Wolpert, L. (1963). The cellular basis of morphogenesis and sea urchin development. *Int. Rev. Cytol.* **15**, 139.

Gustafson, T., and Wolpert, L. (1967). Cellular movements and contact in sea urchin morphogenesis. *Biol. Rev. Cambridge Philos. Soc.* **42**, 442.

Gutzeit, H. O., and Gehring, W. J. (1979). Localized synthesis of specific proteins during oogenesis and early embryogenesis in *Drosophila melanogaster*. *Wilhelm Roux's Arch. Dev. Biol.* **187**, 151.

Gutzeit, H., and Koppa, R. (1982). Time-lapse film analysis of cytoplasmic streaming during late oogenesis of *Drosophila*. *J. Embryol. Exp. Morphol.* **67**, 101.

Hadorn, E. (1966). Problems of determination and transdetermination. *Brookhaven Symp. Biol.* No. 18, 148.

Haenlin, M., Steller, H., Pirrotta, V., and Mohier, E. (1985). A 43 kilobase cosmid P transposon rescues the *fs(1)K10* morphogenetic locus and three adjacent *Drosophila* developmental mutants. *Cell* **40**, 827.

Hafen, E., Kuroiwa, A., and Gehring, W. J. (1984a). Spatial distribution of transcripts from the segmentation gene *fushi tarazu* during *Drosophila* embryonic development. *Cell* **37**, 833.

Hafen, E., Levine, M., and Gehring, W. J. (1984b). Regulation of *Antennapedia* transcript distribution by the bithorax complex in *Drosophila*. *Nature (London)* **307**, 287.

Hahn, W. E., Van Ness, J., and Maxwell, I. H. (1978). Complex population of mRNA sequences in large polyadenylated nuclear RNA molecules. *Proc. Natl. Acad. Sci. U.S.A.* **75**, 5544.

Hahn, W. E., Van Ness, J., and Chaudhari, N. (1982). Overview of the molecular genetics of mouse brain. *In* "Molecular Genetic Neuroscience" (F. O. Schmitt, S. J. Bird, and F. E. Bloom, eds.), p. 323. Raven, New York.

Hallberg, R. L., and Brown, D. D. (1969). Coordinated synthesis of some ribosomal proteins and ribosomal RNA in embryos of *Xenopus laevis*. *J. Mol. Biol.* **46**, 393.

Halsell, S., Ito, M., and Maxson, R. (1986). Regulated expression of sea urchin early and late histone genes in adult tissues. Submitted for publication.

Hamaguchi, Y., Toriyama, M., Sakai, H., and Hiramoto, Y. (1985). Distribution of fluorescently labeled tubulin injected into sand dollar eggs from fertilization through cleavage. *J. Cell Biol.* **100**, 1262.

Hammond, M. P., and Laird, C. D. (1985). Chromosome structure and DNA replication in nurse and follicle cells of *Drosophila melanogaster*. *Chromosoma* **91**, 267.

Handyside, A. H. (1978). Time of commitment of inside cells isolated from preimplantation mouse embryos. *J. Embryol. Exp. Morphol.* **45**, 37.

Handyside, A. H., and Johnson, M. H. (1978). Temporal and spatial patterns of the synthesis of tissue-specific polypeptides in the preimplantation mouse embryo. *J. Embryol. Exp. Morphol.* **44**, 191.

Hara, K. (1977). The cleavage pattern of the axolotl egg studied by cinematography and cell counting. *Wilhelm Roux's Arch. Dev. Biol.* **181**, 73.

Harbers, K., Schnieke, A., Stuhlmann, H., Jähner, D., and Jaenisch, R. (1981). DNA methylation and gene expression: Endogenous retroviral genome becomes infectious after molecular cloning. *Proc. Natl. Acad. Sci. U.S.A.* **78**, 7609.

Hardin, S. H., Carpenter, C. D., Hardin, P. E., Bruskin, A. M., and Klein, W. H. (1985). Structure of the Spec1 gene encoding a major calcium-binding protein in the embryonic ectoderm of the sea urchin, *Stronglylocentrotus purpuratus*. *J. Mol. Biol.* **186**, 243.

Harkey, M. A., and Whiteley, A. H. (1980). Isolation, culture, and differentiation of echinoid primary mesenchyme cells. *Wilhelm Roux's Arch. Dev. Biol.* **189**, 111.

Harkey, M. A., and Whiteley, A. H. (1983). The program of protein synthesis during the development of the micromere-primary mesenchyme cell line in the sea urchin. *Dev. Biol.* **100**, 12.

Harland, R. M., and Laskey, R. A. (1980). Regulated replication of DNA microinjected into eggs of *Xenopus laevis*. *Cell* **21**, 761.

Harper, M. I., Fosten, M., and Monk, M. (1982). Preferential paternal X inactivation in extraembryonic tissues of early mouse embryos. *J. Embryol. Exp. Morphol.* **67**, 127.

Harrison, M. F., and Wilt, F. H. (1982). The program of H1 histone synthesis in *S. purpuratus* embryos and the control of its timing. *J. Exp. Zool.* **223**, 245.

Harsa-King, M. L., Bender, A., and Lodish, H. F. (1979). Stage specific changes in protein synthesis during *Xenopus* oogenesis. *In* "Eucaryotic Gene Regulation" (R. Axel, T. Maniatis, and C. F. Fox, eds.), p. 239. Academic Press, New York.

Hart, C. P., Awgulewitsch, A., Fainsod, A., McGinnis, W., and Ruddle, F. H. (1985). Homeobox gene complex on mouse chromosome 11: Molecular cloning, expression in embryogenesis, and homology to a human homeobox locus. *Cell* **43**, 9.

Hartenstein, V., and Campos-Ortega, J. A. (1984). Early neurogenesis in wild-type *Drosophila melanogaster*. *Wilhelm Roux's Arch. Dev. Biol.* **193**, 308.

Hartenstein, V., and Campos-Ortega, J. A. (1985). Fate-mapping in wild-type *Drosophila melanogaster*. I. The spatiotemporal pattern of embryonic cell divisions. *Wilhelm Roux's Arch. Dev. Biol.* **194**, 181.

Hartenstein, V., Technau, G. M., and Campos-Ortega, J. A. (1985). Fate-mapping in wild-type

Drosophila melanogaster. III. A fate map of the blastoderm. *Wilhelm Roux's Arch. Dev. Biol.* **194**, 213.

Hartmann, J. F., Ziegler, M. M., and Comb, D. G. (1971). Sea urchin embryogenesis. I. RNA synthesis by cytoplasmic and nuclear genes during development. *Dev. Biol.* **25**, 209.

Harvey, E. B. (1936). Parthenogenetic merogony or cleavage without nuclei in *Arbacia punctulata. Biol. Bull. (Woods Hole, Mass.)* **71**, 101.

Harvey, E. B. (1940). A comparison of the development of nucleate and nonnucleate eggs of *Arbacia punctulata. Biol. Bull. (Woods Hole, Mass.)* **79**, 166.

Harvey, E. B. (1956). "The American *Arbacia* and Other Sea Urchins." Princeton Univ. Press, Princeton, New Jersey.

Hatt, P. (1932). Essais experimentaux sur les localisations germinales dans l'oeuf d'un annelide (*Sabellaria alveolata* L.). *Arch. Anat. Microsc. Morphol. Exp.* **28**, 81.

Hauser, C. A., Joyner, A. L., Klein, R. D., Learned, T. K., Martin, G. R., and Tjian, R. (1985). Expression of homologous homeobox-containing genes in differentiated human teratocarcinoma cells and mouse embryos. *Cell* **43**, 19.

Hay, E. D. (1968). Dedifferentiation and metaplasia in vertebrate and invertebrate regeneration. *In* "The Stability of the Differentiated State" (H. Ursprung, ed.), p. 85. Springer-Verlag, Berlin and New York.

Hayes, P. H., Sato, T., and Denell, R. E. (1984). Homeosis in *Drosophila*: The *Ultrabithorax* larval syndrome. *Proc. Natl. Acad. Sci. U.S.A.* **81**, 545.

Hazelrigg, T., Levis, R., and Rubin, G. M. (1984). Transformation of *white* locus DNA in *Drosophila*: Dosage compensation, *zeste* interaction, and position effects. *Cell* **36**, 469.

Heasman, J., Wylie, C. C., Hausen, P., and Smith, J. C. (1984). Fates and states of determination of single vegetal pole blastomeres of *X. laevis. Cell* **37**, 185.

Heasman, J., Snape, A., Smith, J., and Wylie, C. C. (1985). Single cell analysis of commitment in early embryogenesis. *J. Embryol. Exp. Morphol.* **89**, Suppl., 297.

Hedgecock, E., Sulston, J., and Thomson, N. (1983). Mutations affecting programmed cell deaths in the nematode *Caenorhabditis elegans. Science* **220**, 1277.

Hedrick, S. M., Nielsen, E. A., Kavaler, J., Cohen, D. I., and Davis, M. M. (1984). Sequence relationships between putative T-cell receptor polypeptides and immunoglobulins. *Nature (London)* **308**, 153.

Hegner, R. W. (1911). Experiments with chrysomelid beetles. III. The effects of killing parts of the eggs of *Leptinotarsa decemlineata. Biol. Bull. (Woods Hole, Mass.)* **20**, 237.

Hegner, R. W. (1914). "The Germ-Cell Cycle in Animals." Macmillan, New York.

Heifetz, A., and Lennarz, W. J. (1979). Biosynthesis of *N*-glycosidically linked glycoproteins during gastrulation of sea urchin embryos. *J. Biol. Chem.* **254**, 6119.

Heikkila, J. J., Miller, J. G. O., Schultz, G. A., Kloc, M., and Browder, L. W. (1985a). Heat shock gene expression during early animal development. *In* "Changes in Eukaryotic Gene Expression in Response to Environmental Stress" (B. G. Atkinson and D. B. Walden, eds.), p. 135. Academic Press, Orlando, Florida.

Heikkila, J. J., Kloc, M., Bury, J., Schultz, G. A., and Browder, L. W. (1985b). Acquisition of the heat shock response and thermotolerance during early development of *Xenopus laevis. Dev. Biol.* **107**, 483.

Hennen, S. (1970). Influence of spermine and reduced temperature on the ability of transplanted nuclei to promote normal development in eggs of *Rana pipiens. Proc. Natl. Acad. Sci. U.S.A.* **66**, 630.

Henry, J. J. (1986). The role of unequal cleavage and the polar lobe in the segregation of developmental potential during first cleavage in the embryo of *Chaetopterus variopedatus. Wilhelm Roux's Arch. Dev. Biol.* **195**, 103.

Hentschel, C. C., and Birnstiel, M. L. (1981). The organization and expression of histone gene families. *Cell* **25**, 301.

Herbst, C. A. (1893). Experimentelle Untersuchungen über den Einfluss der veränderten chemischen Zusammensetzung des umgebenden Mediums auf die Entwicklung der

Thiere. II. Th. Weiteres über die morphologische Wirkung der Lithiumsalze und ihre theoretische Bedeutung. *Mitt. Z. Stn. Neapel.* **11**, 136.

Herbst, C. A. (1900). Über das Auseinandergehen von Furchungs-und Gewebeszellen in kalkfreiem medium. *Arch. Entwicklungsmech. Org.* **9**, 424.

Herlands, L., Allfrey, V. G., and Poccia, D. (1982). Translational regulation of histone synthesis in the sea urchin *Strongylocentrotus purpuratus*. *J. Cell Biol.* **94**, 219.

Herth, W., and Sander, K. (1973). Mode and timing of body pattern formation (regionalization) in the early embryonic development of cyclorrhaphic dipterans (*Protophormia, Drosophila*). *Wilhelm Roux's Arch. Dev. Biol.* **172**, 1.

Hertwig, O. (1885). The problem of fertilization and isotropy of the egg, a theory of inheritance. (Engl. transl.) *In* Voeller, B. R., ed. (1968). "The Chromosome Theory of Inheritance: Classic Papers in Development and Heredity." Appleton, New York.

Hertwig, O. (1894). Präformation oder Epigenese? Grundzüge einer Entwicklungstheorie der Organismen. *Z. Streitfragen Biol. (Jena)* **1**, 143.

Hieter, P. A., Hendricks, M. B., Hemminki, K., and Weinberg, E. S. (1979). Histone gene switch in the sea urchin embryo. Identification of late embryonic histone messenger ribonucleic acids and the control of their synthesis. *Biochemistry* **13**, 2707.

Hill, R. S. (1979). A quantitative electron-microscope analysis of chromatin from *Xenopus laevis* lampbrush chromosomes. *J. Cell Sci.* **40**, 145.

Hill, R. S., and Macgregor, H. C. (1980). The development of lampbrush chromosome-type transcription in the early diplotene oocytes of *Xenopus laevis*: An electron-microscope analysis. *J. Cell Sci.* **44**, 87.

Hille, M. B., and Albers, A. A. (1979). Efficiency of protein synthesis after fertilization of sea urchin eggs. *Nature (London)* **278**, 469.

Hille, M. B., Bechtold, M. A., Hall, D. C., and Yablonka-Reuveni, Z. (1980). Initiation of protein synthesis in sea urchin eggs and embryos. *Am. Zool.* **20**, 838.

Hille, M. B., Hall, D. C., Yablonka-Reuveni, Z., Danilchik, M. V., and Moon, R. T. (1981). Translational control in sea urchin eggs and embryos: Initiation is rate limiting in blastula stage embryos. *Dev. Biol.* **86**, 241.

Hillman, N., Sherman, M. I., and Graham, C. (1972). The effect of spatial arrangement on cell determination during mouse development. *J. Embryol. Exp. Morphol.* **28**, 263.

Hinegardner, R. T. (1967). Echinoderms. *In* "Methods in Developmental Biology" (F. H. Wilt and N. K. Wessells, eds.), p. 139. Crowell, New York.

Hinegardner, R. (1974). Cellular DNA content of the Echinodermata. *Comp. Biochem. Physiol. B* **49B**, 219.

Hiromi, Y., Kuroiwa, A., and Gehring, W. J. (1985). Control elements of the *Drosophila* segmentation gene *fushi tarazu*. *Cell* **43**, 603.

Hirose, G., and Jacobson, M. (1979). Clonal organization of the central nervous system of the frog. I. Clones stemming from individual blastomeres of the 16-cell and earlier stages. *Dev. Biol.* **71**, 191.

Hirsh, J., and Davidson, N. (1981). Isolation and characterization of the dopa decarboxylase gene of *Drosophila melanogaster*. *Mol. Cell. Biol.* **1**, 475.

His, W. (1874). "Unsere Körperform und das Physiologische Problem ihrer Entstehung." F. C. W. Vogel, Leipzig.

Hochman, B. (1973). Analysis of a whole chromosome in *Drosophila*. *Cold Spring Harbor Symp. Quant. Biol.* **38**, 581.

Hodgkin, J., Horvitz, H. R., and Brenner, S. (1979). Nondisjunction mutants of the nematode *Caenorhabditis elegans*. *Genetics* **91**, 67.

Hogan, B., and Gross, P. R. (1972). Nuclear RNA synthesis in sea urchin embryos. *Exp. Cell Res.* **72**, 101.

Hogan, B., and Tilly, R. (1977). *In vitro* culture and differentiation of normal mouse blastocysts. *Nature (London)* **265**, 626.

Hogan, B., and Tilly, R. (1978a). *In vitro* development of inner cell masses isolated immunosurgically from mouse blastocysts. I. Inner cell masses from 3.5 day p.c. blastocysts incubated for 24 h before immunosurgery. *J. Embryol. Exp. Morphol.* **45**, 93.

Hogan, B., and Tilly, R. (1978b). *In vitro* development of inner cell masses isolated immuno-surgically from mouse blastocysts. II. Inner cell masses for 3.5 to 4.0-day p.c. blastocysts. *J. Embryol. Exp. Morphol.* **45**, 107.

Hogan, B. L. M., Taylor, A., and Cooper, A. R. (1982). Murine parietal endoderm cells synthesize heparan sulphate and 170 K and 145 K sulphated glycoproteins as components of Reichert's membrane. *Dev. Biol.* **90**, 210.

Hogness, D. S., Lipshitz, H. D., Beachy, P. A., Peattie, D. A., Saint, R. A., Goldschmidt-Clermont, M., Harte, P. J., Gavis, E. R., and Helfand, S. L. (1985). Regulation and products of the *Ubx* domain of the bithorax complex. *Cold Spring Harbor Symp. Quant. Biol.* **50**, 181.

Holland, N. D., and Giese, A. C. (1965). An autoradiographic investigation of the gonads of the purple sea urchin *(Strongylocentrotus purpuratus)*. *Biol. Bull. (Woods Hole, Mass.)* **128**, 241.

Hollinger, T. G., and Smith, L. D. (1976). Conservation of RNA polymerase during maturation of the *Rana pipiens* oocyte. *Dev. Biol.* **51**, 86.

Holtfreter, J., and Hamburger, V. (1956). Embryogenesis: Progressive differentiation. Amphibians. *In* "Analysis of Development" (B. H. Willier, P. A. Weiss, and V. Hamburger, eds.), p. 230. Saunders, Philadelphia, Pennsylvania.

Honda, B. M., and Roeder, R. G. (1980). Association of a 5S gene transcription factor with 5S RNA and altered levels of the factor during cell differentiation. *Cell* **22**, 119.

Honjo, T. (1983). Immunoglobulin genes. *Annu. Rev. Immunol.* **1**, 499.

Honjo, T., and Reeder, R. H. (1973). Preferential transcription of *Xenopus laevis* ribosomal RNA in interspecies hybrids between *Xenopus laevis* and *Xenopus mulleri*. *J. Mol. Biol.* **80**, 217.

Hood, L., Davis, M., Early, P., Calame, K., Kim, S., Crews, S., and Huang, H. (1981). Two types of DNA rearrangements in immunoglobulin genes. *Cold Spring Harbor Symp. Quant. Biol.* **45**, 887.

Hood, L., Steinmetz, M., Goodenow, R., Eakle, K., Fisher, D., Kobori, J., Malissen, B., Malissen, M., McMillan, M., McNicholas, J., Örn, A., Pecht, M., Sher, B. T., Smith, L., Stroynowski, I., Sun, H., Winoto, A., and Zuniga, M. (1982). Genes of the major histocompatibility complex. *Cold Spring Harbor Symp. Quant. Biol.* **47**, 1051.

Hörstadius, S. (1928). Über die Determination des Keimes bei Echinodermen. *Acta Zool. (Stockholm)* **9**, 1.

Hörstadius, S. (1935). Über die Determination in Verlaufe der Eiachse bei Seeigeln. *Pubbl. Staz. Zool. Napoli* **14**, 251.

Hörstadius, S. (1937a). Experiments on determination in the early development of *Cerebratulus lacteus*. *Biol. Bull. (Woods Hole, Mass.)* **73**, 317.

Hörstadius, S. (1937b). Investigations as to the localization of the micromere, the skeleton, and the entoderm-forming material in the unfertilized egg of *Arbacia punctulata*. *Biol. Bull. (Woods Hole, Mass.)* **73**, 295.

Hörstadius, S. (1939). The mechanics of sea urchin development, studied by operative methods. *Biol. Rev. Cambridge Philos. Soc.* **14**, 132.

Hörstadius, S. (1973). "Experimental Embryology of Echinoderms." Oxford Univ. Press (Clarendon), London and New York.

Horvitz, H. R., Sternberg, P. W., Greenwald, I. S., Fixsen, W., and Ellis, H. M. (1983). Mutations that affect neural cell lineages and cell fates during the development of the nematode *Caenorhabditis elegans*. *Cold Spring Harbor Symp. Quant. Biol.* **48**, 453.

Hotta, Y., and Benzer, S. (1973). Mapping of behavior in *Drosophila* mosaics. *In* "Genetic Mechanism of Development" (F. H. Ruddle, ed.), p. 129. Academic Press, New York.

Hough, B. R., and Davidson, E. H. (1972). Studies on the repetitive sequence transcripts of *Xenopus* oocytes. *J. Mol. Biol.* **70**, 491.

Hough, B. R., Yancey, P. H., and Davidson, E. H. (1973). Persistence of maternal RNA in *Engystomops* embryos. *J. Exp. Zool.* **185**, 357.

Hough, B. R., Smith, M. J., Britten, R. J., and Davidson, E. H. (1975). Sequence complexity of heterogeneous nuclear RNA in sea urchin embryos. *Cell* **5**, 291.

Hough-Evans, B. R., Wold, B. J., Ernst, S. G., Britten, R. J., and Davidson, E. H. (1977). Appearance and persistence of maternal RNA sequences in sea urchin development. *Dev. Biol.* **60**, 258.

Hough-Evans, B. R., Ernst, S. G., Britten, R. J., and Davidson, E. H. (1979). RNA complexity in developing sea urchin oocytes. *Dev. Biol.* **69**, 258.

Hough-Evans, B. R., Jacobs-Lorena, M., Cummings, M. R., Britten, R. J., and Davidson, E. H. (1980). Complexity of RNA in eggs of *Drosophila melanogaster* and *Musca domestica. Genetics* **95**, 81.

Houk, M. S., and Epel, D. (1974). Protein synthesis during hormonally induced meiotic maturation and fertilization in starfish oocytes. *Dev. Biol.* **40**, 298.

Hourcade, D., Dressler, D., and Wolfson, J. (1973). The amplification of ribosomal RNA genes involves a rolling circle intermediate. *Proc. Natl. Acad. Sci. U.S.A.* **70**, 2926.

Hourcade, D., Dressler, D., and Wolfson, J. (1974). The nucleolus and the rolling circle. *Cold Spring Harbor Symp. Quant. Biol.* **38**, 537.

Howard, K., and Ingham, P. (1986). Regulatory interactions between the segmentation genes *fushi tarazu, hairy,* and *engrailed* in the *Drosophila* blastoderm. *Cell* **44**, 949.

Howell, A. M., Cool, D., Hewitt, J., Ydenberg, B., Smith, M. J., and Honda, B. M. (1986). The organization and unusual expression of histone genes in the sea star *Pisaster ochraceus.* Submitted for publication.

Howland, R. B., and Child, G. P. (1935). Experimental studies on development in *Drosophila melanogaster.* I. Removal of protoplasmic materials during late cleavage and early embryonic stages. *J. Exp. Zool.* **70**, 415.

Howlett, S. K., and Bolton, V. N. (1985). Sequence and regulation of morphological and molecular events during the first cell cycle of mouse embryogenesis. *J. Embryol. Exp. Morphol.* **87**, 175.

Howlett, S., Miller, J., and Schultz, G. (1983). Induction of heat shock proteins in early embryos of *Arbacia punctulata. Biol. Bull. (Woods Hole, Mass.)* **165**, 500.

Hu, N.-t., and Messing, J. (1982). The making of strand-specific M13 probes. *Gene* **17**, 271.

Huettner, A. F. (1923). The origin of the germ cells in *Drosophila melanogaster. J. Morphol.* **37**, 385.

Huez, G., Marbaix, G., Gallwitz, D., Weinberg, E., Devos, R., Hubert, E., and Cleuter, Y. (1978). Functional stabilization of HeLa cell histone messenger RNAs injected into *Xenopus* oocytes by 3'-OH polyadenylation. *Nature (London)* **271**, 572.

Hughes, M., and Berry, S. J. (1970). The synthesis and secretion of ribosomes by nurse cells of *Antheraea polyphemus. Dev. Biol.* **23**, 651.

Hultin, T. (1952). Incorporation of N^{15}-labeled glycine and alanine into the proteins of developing sea urchin eggs. *Exp. Cell Res.* **3**, 494.

Hultin, T. (1961a). Activation of ribosomes in sea urchin eggs in response to fertilization. *Exp. Cell Res.* **25**, 405.

Hultin, T. (1961b). The effect of puromycin on protein metabolism and cell division in fertilized sea urchin eggs. *Experientia* **17**, 410.

Hultin, T., and Morris, J. E. (1968). The ribosomes of encysted embryos of *Artemia salina* during cryptobiosis and resumption of development. *Dev. Biol.* **17**, 143.

Humphrey, R. R. (1966). A recessive factor (*o,* for ova deficient) determining a complex of abnormalities in the Mexican axolotl (*Ambystoma mexicanum*). *Dev. Biol.* **13**, 57.

Humphreys, T. (1969). Efficiency of translation of messenger-RNA before and after fertilization in sea urchins. *Dev. Biol.* **20**, 435.

Humphreys, T. (1971). Measurements of messenger RNA entering polysomes upon fertilization of sea urchin eggs. *Dev. Biol.* **26**, 201.

Humphreys, T. (1973). RNA and protein synthesis during early animal embryogenesis. *In* "Developmental Regulation: Aspects of Cell Differentiation" (S. J. Coward, ed.), p. 1. Academic Press, New York.

Hunt, J. A. (1974). Rate of synthesis and half-life of globin messenger ribonucleic acid. Rate of synthesis of globin messenger ribonucleic acid calculated from data of cell haemoglobin content. *Biochem. J.* **138**, 499.

Hutton, J. R., and Wetmur, J. G. (1973). Length dependence of the kinetic complexity of mouse satellite DNA. *Biochem. Biophys. Res. Commun.* **52**, 1148.

Huxley, T. H. (1878). Evolution in biology. *In* "Encyclopedia Brittanica," 9th ed., p. 187. Scribner's, New York.

Hylander, B. L., and Summers, R. G. (1982). An ultrastructural immunocytochemical localization of hyalin in the sea urchin egg. *Dev. Biol.* **93**, 368.

Hyman, L. H. (1940). "The Invertebrates: Protozoa through Ctenophora," Vol. I. McGraw-Hill, New York.

Hyman, L. H. (1955). "The Invertebrates: Echinodermata. The Coelomate Bilateria," Vol. IV. McGraw-Hill, New York.

Hynes, R. O., and Gross, P. R. (1972). Informational RNA sequences in early sea urchin embryos. *Biochim. Biophys. Acta* **259**, 104.

Ijiri, K.-I. (1977). Existence of ultraviolet-labile germ cell determinants in unfertilized eggs of *Xenopus laevis* and its sensitivity. *Dev. Biol.* **55**, 206.

Ikenishi, K., and Kotani, M. (1979). Ultraviolet effects on presumptive primordial germ cells (pPCGs) in *Xenopus laevis* after the cleavage stage. *Dev. Biol.* **69**, 237.

Illmensee, K. (1972). Developmental potencies of nuclei from cleavage, preblastoderm, and syncytial blastoderm transplanted into unfertilized eggs of *Drosophila melanogaster.* *Wilhelm Roux's Arch. Entwicklungsmech. Org.* **170**, 267.

Illmensee, K. (1976). Nuclear and cytoplasmic transplantation in *Drosophila. In* "Insect Development" (P. A. Lawrence, ed.), p. 77. Blackwell, Oxford.

Illmensee, K., and Mahowald, A. P. (1974). Transplantation of posterior polar plasm in *Drosophila.* Induction of germ cells at the anterior pole of the egg. *Proc. Natl. Acad. Sci. U.S.A.* **71**, 1016.

Illmensee, K., and Mahowald, A. P. (1976). The autonomous function of germ plasm in a somatic region of the *Drosphila* egg. *Exp. Cell Res.* **97**, 127.

Illmensee, K., Mahowald, A. P., and Loomis, M. R. (1976). The ontogeny of germ plasm during oogenesis in *Drosophila. Dev. Biol.* **49**, 40.

Imoh, H. (1978). Re-examination of histone changes during development of newt embryos *Exp. Cell Res.* **113**, 23.

Infante, A. A., and Heilmann, L. J. (1981). Distribution of messenger ribonucleic acid in polysomes and nonpolysomal particles of sea urchin embryos: Translational control of actin synthesis. *Biochemistry* **20**, 1.

Infante, A. A., and Nemer, M. (1967). Accumulation of newly synthesized RNA templates in a unique class of polyribosomes during embryogenesis. *Proc. Natl. Acad. Sci. U.S.A.* **58**, 681.

Ingham, P., Martinez-Arias, A. Lawrence, P. A., and Howard, K. (1985a). Expression of *engrailed* in the parasegment of *Drosophila. Nature (London)* **317**, 634.

Ingham, P. W., Howard, K. R., and Ish-Horowicz, D. (1985b). Transcription pattern of the *Drosophila* segmentation gene *hairy. Nature (London)* **318**, 439.

Innis, M. A., and Craig, S. P. (1978). Mitochondrial regulation in sea urchins. II. Formation of polyribosomes within the mitochondria of 4-8 cell stage embryos of the sea urchin. *Exp. Cell Res.* **111**, 223.

Ishimoda-Takagi, T., Chino, I., and Sato, H. (1984). Evidence for the involvement of muscle tropomyosin in the contractile elements of the coelom-esophagus complex of sea urchin embryos. *Dev. Biol.* **105**, 365.

Isnenghi, E., Cassada, R., Smith, K., Denich, K., Radnia, K., and von Ehrenstein, G. (1983). Maternal effects and temperature-sensitive period of mutations affecting embryogenesis in *Caenorhabditis elegans. Dev. Biol.* **98**, 465.

Ito, M., Lyons, G., and Maxson, R. (1986). Accumulation of late histone mRNA during sea urchin embryogenesis is regulated at levels of transcription and mRNA stability. In preparation.

Ivarie, R., Schacter, B., and O'Farrell, P. (1983). The level of expression of the rat growth hormone gene in liver tumor cells is at least eight orders of magnitude less than that in anterior pituitary cells. *Mol. Cell. Biol.* **3**, 1460.

Iwata, M., and Nakano, E. (1985). Fibronectin-binding acid polysaccharide in the sea urchin embryo. *Wilhelm Roux's Arch. Dev. Biol.* **194**, 377.

Izawa, M., Allfrey, V. G., and Mirsky, A. E. (1963). Composition of the nucleus and chromosomes in the lampbrush stage of the newt oocyte. *Proc. Natl. Acad. Sci. U.S.A.* **50**, 811.

Izquierdo, L., and Ortiz, M. E. (1975). Differentiation in the mouse morulae. *Wilhelm Roux's Arch. Dev. Biol.* **177**, 67.

Izquierdo, M., and Bishop, J. O. (1979). An analysis of cytoplasmic RNA populations in *Drosophila melanogaster*, Oregon R. *Biochem. Genet.* **17**, 473.

Jäckle, H. (1979). Degradation of maternal poly(A)-containing RNA during early embryogenesis of an insect (*Smittia* spec., Chironomidae, Diptera). *Wilhelm Roux's Arch. Dev. Biol.* **187**, 179.

Jäckle, H. (1980). Actin messenger in maternal RNP particles from an insect embryo (*Smittia* spec., Chironomidae, Diptera). *Wilhelm Roux's Arch. Dev. Biol.* **188**, 225.

Jäckle, H., and Eagleson, G. W. (1980). Spatial distribution of abundant proteins in oocytes and fertilized eggs of the Mexican axolotl (*Ambystoma mexicanum*). *Dev. Biol.* **75**, 492.

Jäckle, H., and Kalthoff, K. (1979). RNA and protein synthesis in developing embryos of *Smittia* spec. (Chironomidae, Diptera). *Wilhelm Roux's Arch. Dev. Biol.* **187**, 283.

Jäckle, H., and Kalthoff, K. (1980). Photoreversible UV-inactivation of messenger RNA in an insect embryo (*Smittia* spec., Chironomidae, Diptera). *Photochem. Photobiol.* **32**, 749.

Jäckle, H., and Kalthoff, K. (1981). Proteins foretelling head or abdomen development in the embryo of *Smittia* spec. (Chironomidae, Diptera). *Dev. Biol.* **85**, 287.

Jacob, E. (1980). Characterization of cloned cDNA sequences derived from *Xenopus laevis* poly(A)$^+$ oocyte RNA. *Nucleic Acids Res.* **8**, 1319.

Jacobs, H. T., Posakony, J. W., Grula, J. W., Roberts, J. W., Xin, J.-H., Britten, R. J., and Davidson, E. H. (1983). Mitochondrial DNA sequences in the nuclear genome of *Strongylocentrotus purpuratus*. *J. Mol. Biol.* **165**, 609.

Jacobson, M. (1980). Clones and compartments in the vertebrate central nervous system. *Trends Neurosci.* **3**, 3.

Jacobson, M. (1981a). Rohon–Beard neuron origin from blastomeres of the 16-cell frog embryo. *J. Neurosci.* **1**, 918.

Jacobson, M. (1981b). Rohon–Beard neurons arise from a substitute ancestral cell after removal of the cell from which they normally arise in the 16-cell frog embryo. *J. Neurosci.* **1**, 923.

Jacobson, M. (1983). Clonal organization of the central nervous system of the frog. III. Clones stemming from individual blastomeres of the 128-, 256-, and 512-cell stages. *J. Neurosci.* **3**, 1019.

Jacobson, M. (1985a). Clonal analysis of the vertebrate CNS. *Trends Neurosci.* **8**, 151.

Jacobson, M. (1985b). Clonal analysis and cell lineages of the vertebrate central nervous system. *Annu. Rev. Neurosci.* **8**, 71.

Jacobson, M., and Hirose, G. (1981). Clonal organization of the central nervous system of the frog. II. Clones stemming from individual blastomeres of the 32- and 64-cell stages. *J. Neurosci.* **1**, 271.

Jacobson, M., and Klein, S. L. (1985). Analysis of clonal restriction of cell mingling in *Xenopus*. *Philos. Trans. R. Soc. London, Ser. B* **312**, 57.

Jacobson, M., and Moody, S. A. (1984). Quantitative lineage analysis of the frog's nervous system. I. Lineages of Rohon–Beard neurons and primary motoneurons. *J. Neurosci.* **4**, 1361.

Jacobson, M., and Rutishauser, U. (1986). Induction of neural cell adhesion molecule (NCAM) in *Xenopus* embryos. *Dev. Biol.* **116**. In press.

Jahn, C. L., Baran, M. M., and Bachvarova, R. (1976). Stability of RNA synthesized by the mouse oocyte during its major growth phase. *J. Exp. Zool.* **197**, 161.

Jähner, D., Stuhlmann, H., Stewart, C. L., Harbers, K., Löhler, J., Simon, I., and Jaenisch, R.

(1982). *De novo* methylation and expression of retroviral genomes during mouse embryogenesis. *Nature (London)* **298**, 623.

Jamrich, M., Warrior, R., Steele, R., and Gall, J. G. (1983). Transcription of repetitive sequences on *Xenopus* lampbrush chromosomes. *Proc. Natl. Acad. Sci. U.S.A.* **80**, 3364.

Jamrich, M., Sargent, T. D., and Dawid, I. B. (1985). Altered morphogenesis and its effects on gene activity in *Xenopus laevis* embryos. *Cold Spring Harbor Symp. Quant. Biol.* **50**, 31.

Janning, W. (1978). Gynandromorph fate maps in *Drosophila*. *In* "Genetic Mosaics and Cell Differentiation" (W. J. Gehring, ed.), p. 1. Springer-Verlag, Berlin and New York.

Jarry, B. P., and Falk, D. R. (1974). Functional diversity within the *rudimentary* locus of *Drosophila melanogaster*. *Mol. Gen. Genet.* **135**, 113.

Jeffery, W. R. (1982). Calcium ionophore polarizes ooplasmic segregation in ascidian eggs. *Science* **216**, 545.

Jeffery, W. R. (1983). Maternal RNA and the embryonic localization problem. *In* "Control of Embryonic Gene Expression" (M. Siddiqui, ed.), p. 73. CRC Press, Boca Raton, Florida.

Jeffery, W. R. (1984). Spatial distribution of messenger RNA in the cytoskeletal framework of ascidian eggs. *Dev. Biol.* **103**, 482.

Jeffery, W. R. (1985a). Identification of proteins and mRNAs in isolated yellow crescents of ascidian eggs. *J. Embryol. Exp. Morphol.* **89**, 275. •

Jeffery, W. R. (1985b). The spatial distribution of maternal mRNA is determined by a cortical cytoskeletal domain in *Chaetopterus* eggs. *Dev. Biol.* **110**, 217.

Jeffery, W. R., and Meier, S. (1983). A yellow crescent cytoskeletal domain in ascidian eggs and its role in early development. *Dev. Biol.* **96**, 125.

Jeffery, W. R., and Meier, S. (1984). Ooplasmic segregation of the myoplasmic actin network in stratified ascidian eggs. *Wilhelm Roux's Arch. Dev. Biol.* **193**, 257.

Jeffery, W. R., and Wilson, L. J. (1983). Localization of messenger RNA in the cortex of *Chaetopterus* eggs and early embryos. *J. Embryol. Exp. Morphol.* **75**, 225.

Jeffery, W. R., Tomlinson, C. R., and Brodeur, R. D. (1983). Localization of actin messenger RNA during early ascidian development. *Dev. Biol.* **99**, 408.

Jiménez, F., and Campos-Ortega, J. A. (1982). Maternal effects of zygotic mutants affecting early neurogenesis in *Drosophila*. *Wilhelm Roux's Arch. Dev. Biol.* **191**, 191.

Johnson, D. R. (1975). Further observations on the hairpin-tail (T^{hp}) mutation in the mouse. *Genet. Res.* **24**, 207.

Johnson, J. H., and King, R. C. (1972). Studies on *fes*, a mutation affecting cystocyte cytokinesis, in *Drosophila melanogaster*. *Biol. Bull. (Woods Hole, Mass.)* **143**, 525.

Johnson, K. E. (1976). Ruffling and locomotion in *Rana pipiens* gastrula cells. *Exp. Cell Res.* **101**, 71.

Johnson, M. H. (1981). The molecular and cellular basis of preimplantation mouse development. *Biol. Rev. Cambridge Philos. Soc.* **56**, 463.

Johnson, M. H., and Maro, B. (1984). The distribution of cytoplasmic actin in mouse 8-cell blastomeres. *J. Embryol. Exp. Morphol.* **82**, 97.

Johnson, M. H., and Pratt, H. P. M. (1983). Cytoplasmic localizations and cell interactions in the formation of the mouse blastocyst. *In* "Time, Space, and Pattern in Embryonic Development" (W. R. Jeffery and R. A. Raff, eds.), p. 287. Alan R. Liss, New York.

Johnson, M. H., and Rossant, J. (1981). Molecular studies on cells of the trophectodermal lineage of the postimplantation mouse embryo. *J. Embryol. Exp. Morphol.* **61**, 103.

Johnson, M. H., and Ziomek, C. A. (1981a). The foundation of two distinct cell lineages within the mouse morula. *Cell* **24**, 71.

Johnson, M. H., and Ziomek, C. A. (1981b). Induction of polarity in mouse 8-cell blastomeres: Specificity, geometry, and stability. *J. Cell Biol.* **91**, 303.

Johnson, M. H., and Ziomek, C. A. (1982). Cell subpopulations in the late morula and early blastocyst of the mouse. *Dev. Biol.* **91**, 431.

Johnson, M. H., and Ziomek, C. A. (1983). Cell interactions influence the fate of mouse blastomeres undergoing the transition from the 16- to the 32-cell stage. *Dev. Biol.* **95**, 211.

Johnson, M. H., Ziomek, C. A., Reeve, W. J. D., Pratt, H. P. M., Goodall, H., and Handyside,

A. H. (1984). The mosaic organization of the preimplantation mouse embryo. *In* "Ultrastructure of Reproduction" (J. Van Blerkom and P. M. Motta, eds.), p. 205. Nijhoff, The Hague.

Johnston, R. N., Beverley, S. M., and Schimke, R. T. (1983). Rapid spontaneous dihydrofolate reductase gene amplification shown by fluorescence-activated cell sorting. *Proc. Natl. Acad. Sci. U.S.A.* **80**, 3711.

Jonas, E., Sargent, T. D., and Dawid, I. B. (1985). Epidermal keratin gene expressed in embryos of *Xenopus laevis. Proc. Natl. Acad. Sci. U.S.A.* **82**, 5413.

Jones, C. W., and Kafatos, F. C. (1980). Structure, organization and evolution of developmentally regulated chorion genes in a silkmoth. *Cell* **22**, 855.

Jones, E. A., and Woodland, H. R. (1986a). Development of the ectoderm in *Xenopus*: Tissue specification and the role of cell association and division. Cell **44**, 345.

Jones, E. A., and Woodland, H. R. (1986b). The development of animal cap cells in *Xenopus*, the commitment to neuroepidermis or mesoderm. Submitted for publication.

Jones, R. E. (1978). "The Vertebrate Ovary. Comparative Biology and Evolution." Plenum, New York.

Jörgenssen, M. (1913). Die Ei-und Nährzellen von *Pisciola. Arch. Zellforsch.* **10**, 127.

Judd, B. H., Shen, M. W., and Kaufman, T. C. (1972). The anatomy and function of a segment of the X chromosome of *Drosophila melanogaster. Genetics* **71**, 139.

Jürgens, G. (1985). A group of genes controlling the spatial expression of the bithorax complex in *Drosophila. Nature (London)* **316**, 153.

Jürgens, G., Wieschaus, E., Nüsslein-Volhard, C., and Kluding, H. (1984). Mutations affecting the pattern of the larval cuticle in *Drosophila melanogaster.* II. Zygotic loci on the third chromosome. *Wilhelm Roux's Arch. Dev. Biol.* **193**, 283.

Kaestner, A. (1967). "Invertebrate Zoology" (H. W. Levi and L. R. Levi transl.), Vol. 1, p. 145. Wiley (Interscience), New York.

Kafatos, F. C. (1972). The cocoonase zymogan cells of silk moths: A model of terminal cell differentiation for specific protein synthesis. *Curr. Top. Dev. Biol.* **7**, 125.

Kafatos, F. C. (1983). Structure, evolution, and developmental expression of the chorion multigene families in silkmoths and *Drosophila. In* "Gene Structure and Regulation in Development" (S. Subtelny and F. C. Kafatos, eds.), p. 33. Alan R. Liss, New York.

Kafatos, F. C., and Gelinas, R. (1974). mRNA stability and the control of specific protein synthesis in highly differentiated cells. *In* "MTP International Review of Science—Biochemistry of Differentiation and Development" (J. Paul, ed.), Vol. 9, p. 223. MTP Press, Oxford.

Kalfayan, L., and Wensink, P. (1982). Developmental regulation of *Drosophila* α-tubular genes. *Cell* **29**, 91.

Kalt, M. R., and Gall, J. G. (1974). Observations on early germ cell development and premeiotic ribosomal DNA amplification in *Xenopus laevis. J. Cell Biol.* **62**, 460.

Kalthoff, K. (1971a). Photoreversion of UV induction of the malformation "double abdomen" in the egg of *Smittia* spec. (Diptera, Chironomidae). *Dev. Biol.* **25**, 119.

Kalthoff, K. (1971b). Position of targets and period of competence for UV-induction of the malformation "double abdomen" in the egg of *Smittia* spec. (Diptera, Chironomidae). *Wilhelm Roux's Arch. Entwicklungsmech. Org.* **168**, 63.

Kalthoff, K. (1973). Action spectra for UV induction and photoreversal of a switch in the developmental program of the egg of an insect (*Smittia*). *Photochem. Photobiol.* **18**, 355.

Kalthoff, K. (1979). Analysis of a morphogenetic determinant in an insect embryo (*Smittia Spec., Chironomidae, Diptera*). *In* "Determinants of Spatial Organization" (S. Subtelny and I. R. Konigsberg, eds.), p. 97. Academic Press, New York.

Kalthoff, K. (1983). Cytoplasmic determinants in dipteran eggs. *In* "Time, Space, and Pattern in Embryonic Development" (W. R. Jeffery and R. A. Raff, eds.), p. 313. Alan R. Liss, New York.

Kalthoff, K., Rau, K.-G., and Edmond, J. C. (1982). Modifying effects of ultraviolet irradiation on the development of abnormal body patterns in centrifuged insect embryos (*Smittia* sp., Chironomidae, Diptera). *Dev. Biol.* **91**, 413.

Kambysellis, M. P. (1974). Ultrastructure of the chorion in very closely related *Drosophila* species endemic to Hawaii. *Syst. Zool.* **23**, 507.

Kandler-Singer, I., and Kalthoff, K. (1976). RNase sensitivity of an anterior morphogenetic determinant in an insect egg (*Smittia* sp., Chironomidae, Diptera). *Proc. Natl. Acad. Sci. U.S.A.* **73**, 3739.

Kane, R. E. (1973). Hyalin release during normal sea urchin development and its replacement after removal at fertilization. *Exp. Cell Res.* **81**, 301.

Kankel, D. R., and Hall, J. C. (1976). Fate mapping of nervous system and other internal tissues in genetic mosaics of *Drosophila melanogaster*. *Dev. Biol.* **48**, 1.

Kaplan, G., Abreu, S. L., and Bachvarova, R. (1982). rRNA accumulation and protein synthetic patterns in growing mouse oocytes. *J. Exp. Zool.* **220**, 361.

Kaplan, G., Jelinek, W. R., and Bachvarova, R. (1985). Repetitive sequence transcripts and U1 RNA in mouse oocytes and eggs. *Dev. Biol.* **109**, 15.

Karch, F., Weiffenbach, B., Peifer, M., Bender, W., Duncan, I., Celniker, S., Crosby, M., and Lewis, E. B. (1985). The abdominal region of the bithorax complex. *Cell* **43**, 81.

Karin, M., Haslinger, A., Holtgreve, H., Richards, R. I., Krauter, P., Westphal, H. M., and Beato, M. (1984). Characterization of DNA sequences through which cadmium and glucocorticoid hormones induce methallothionein-II$_A$ gene. *Nature (London)* **308**, 513.

Karp, G. C., and Solursh, M. (1974). Acid mucopolysaccharide metabolism, the cell surface, and primary mesenchyme cell activity in the sea urchin embryo. *Dev. Biol.* **41**, 110.

Karp, G. C., and Whiteley, A. H. (1973). DNA-RNA hybridization studies of gene activity during the development of the gastropod, *Acmaea scutum*. *Exp. Cell Res.* **78**, 236.

Karr, T. L., and Alberts, B. M. (1986). Organization of the cytoskeleton in early *Drosophila* embryos. *J. Cell Biol.* **102**, 1494.

Karr, T. L., Ali, Z., Drees, B., and Kornberg, T. (1985). The *engrailed* locus of *D. melanogaster* provides an essential zygotic function in precellular embryos. *Cell* **43**, 591.

Kassavetes, G. A., Butler, E. T., Roulland, D., and Chamberlin, M. J. (1982). Bacteriophage SP6-specific RNA polymerase II. Mapping of SP6 DNA and selective *in vitro* transcription. *J. Bol. Chem.* **257**, 5779.

Katow, H., and Solursh, M. (1980). Ultrastructure of primary mesenchyme cell ingression in the sea urchin *Lytechinus pictus*. *J. Exp. Zool.* **213**, 231.

Katow, H., and Solursh, M. (1981). Ultrastructural and time lapse studies of primary mesenchyme cell behavior in normal and sulfate-deprived sea urchin embryos. *Exp. Cell Res.* **136**, 233.

Katow, H., and Solursh, M. (1982). *In situ* distribution of concanavalin A-binding sites in mesenchyme blastulae and early gastrulae of the sea urchin *Lytechinus pictus*. *Exp. Cell Res.* **139**, 171.

Katow, H., Yamada, K. M., and Solursh, M. (1982). Occurrence of fibronectin on the primary mesenchyme cell surface during migration in the sea urchin embryo. *Differentiation* **22**, 120.

Kauffman, S. A. (1981). Pattern formation in the *Drosophila* embryo. *Philos. Trans. R. Soc. London, Ser. B* **295**, 567.

Kaufman, T. C., Lewis, R. A., and Wakimoto, B. T. (1980). Cytogenetic analysis of chromosome 3 in *Drosophila melanogaster*: The homeotic gene complex in polytene chromosome interval 84A-B. *Genetics* **94**, 115.

Kaumeyer, J. F., Jenkins, N. A., and Raff, R. A. (1978). Messenger ribonucleoprotein particles in unfertilized sea urchin eggs. *Dev. Biol.* **63**, 266.

Kay, B. K., Jamrich, M., and Dawid, I. B. (1984). Transcription of a long, interspersed, highly repeated DNA element in *Xenopus laevis*. *Dev. Biol.* **105**, 518.

Kaye, P. L., and Church, R. B. (1983). Uncoordinated synthesis of histones and DNA by mouse eggs and preimplantation embryos. *J. Exp. Zool.* **226**, 231.

Kaye, P. L., and Wales, R. G. (1981). Histone synthesis in preimplantation mouse embryos. *J. Exp. Zool.* **216**, 453.

Kedes, L. H., and Gross, P. R. (1969). Synthesis and function of messenger RNA during early embryonic development. *J. Mol. Biol.* **42**, 559.

Kedes, L. H., Gross, P. R., Cognetti, G., and Hunter, A. L. (1969). Synthesis of nuclear and chromosomal proteins on light polyribosomes during cleavage in the sea urchin embryo. *J. Mol. Biol.* **45**, 337.

Keem, K., Smith, L. D., Wallace, R. A., and Wolf, D. (1979). Growth rate of oocytes in laboratory-maintained *Xenopus laevis*. *Gamete Res.* **2**, 125.

Keichline, L. D., and Wassarman, P. M. (1977). Developmental study of the structure of sea urchin embryo and sperm chromatin using micrococcal nuclease. *Biochim. Biophys. Acta* **475**, 139.

Keichline, L. D., and Wassarman, P. M. (1979). Structure of chromatin in sea urchin embryos, sperm, and adult somatic cells. *Biochemistry* **18**, 214.

Keller, R. E. (1975). Vital dye mapping of the gastrula and neurula of *Xenopus laevis*. I. Prospective areas and morphogenetic movements of the superficial layer. *Dev. Biol.* **42**, 222.

Keller, R. E. (1976). Vital dye mapping of the gastrula and neurula of *Xenopus laevis*. II. Prospective areas and morphogenetic movements of the deep layer. *Dev. Biol.* **51**, 118.

Kelly, S. J. (1975). Studies of the potency of the early cleavage blastomeres of the mouse. *In* "The Early Development of Mammals" (M. Balls and A. Wild, eds.), p. 97. Cambridge Univ. Press, London and New York.

Kelly, S. J. (1977). Studies of the developmental potential of 4- and 8-cell stage mouse blastomeres. *J. Exp. Zool.* **200**, 365.

Kelly, S. J., Mulnard, J. G., and Graham, C. F. (1978). Cell division and cell allocation in early mouse development. *J. Embryol. Exp. Morphol.* **48**, 37.

Kerr, J. B., and Dixon, K. E. (1974). An ultrastructural study of germ plasm in spermatogenesis of *Xenopus laevis*. *J. Embryol. Exp. Morphol.* **32**, 573.

Keshet, I., Yisraeli, J., and Cedar, H. (1985). Effect of regional DNA methylation on gene expression. *Proc. Natl. Acad. Sci. U.S.A.* **82**, 2560.

Kidd, S., Lockett, T. J., and Young, M. W. (1983). The *Notch* locus of *Drosophila melanogaster*. *Cell* **34**, 421.

Kidder, G. M. (1972). Gene transcription in mosaic embryos. II. Polyribosomes and messenger RNA in early development of the coot clam, *Mulinia lateralis*. *J. Exp. Zool.* **180**, 75.

Kidder, G. M. (1976). The ribosomal RNA cistrons in clam gametes. *Dev. Biol.* **49**, 132.

Kidder, G. M., and Pedersen, R. A. (1982). Turnover of embryonic messenger RNA in preimplantation mouse embryos. *J. Embryol. Exp. Morphol.* **67**, 37.

Kijima, S., and Wilt, F. H. (1969). Rate of nuclear ribonucleic acid turnover in sea urchin embryos. *J. Mol. Biol.* **40**, 235.

Killary, A. M., and Fournier, R. E. K. (1984). A genetic analysis of extinction: *Trans*-dominant loci regulate expression of liver-specific traits in hepatoma hybrid cells. *Cell* **38**, 523.

Kimble, J. (1981). Alterations in cell lineage following laser ablation of cells in the somatic gonad of *Caenorhabditis elegans*. *Dev. Biol.* **87**, 286.

Kimble, J., and Hirsh, D. (1979). The postembryonic cell lineages of the hermaphrodite and male gonads in *Caenorhabditis elegans*. *Dev. Biol.* **70**, 396.

Kinderman, N. B., and King, R. C. (1973). Oogenesis in *Drosophila virilis*. I. Interactions between the ring canal rims and the nucleus of the oocyte. *Biol. Bull. (Woods Hole, Mass.)* **144**, 331.

King, M. L., and Barklis, E. (1985). Regional distribution of maternal messenger RNA in the amphibian oocyte. *Dev. Biol.* **112**, 203.

King, R. C. (1970). "Ovarian Development in *Drosophila melanogaster*." Academic Press, New York.

King, R. C. (1979). Aberrant fusomes in the ovarian cystocytes of the *fs (1) 231* mutant of *Drosophila melanogaster*. Meigen (Diptera: Drosophilidae). *Int. J. Insect Morphol. Embryol.* **8**, 297.

King, R. C., and Aggarwal, S. K. (1965). Oogenesis in *Hyalophora cecropia*. *Growth* **29**, 17.

King, R. C., and Büning, J. (1985). The origin and functioning of insect oocytes and nurse cells. *In* "Comprehensive Insect Physiology, Biochemistry, and Pharmacology: Embryogenesis and Reproduction" (G. A. Kerkut, and L. I. Gilbert, eds.), Vol. 1, p. 37. Pergamon, Oxford.

King, R. C., and Mohler, J. D. (1975). The genetic analysis of oogenesis in *Drosophila melanogaster*. *In* "Handbook of Genetics" (R. C. King, ed.), Vol. 3, p. 757. Academic Press, New York.

King, R. C., and Riley, S. F. (1982). Ovarian pathologies generated by various alleles of the *otu* locus in *Drosophila melanogaster*. *Dev. Genet.* **3**, 69.

Kirschner, M., Gerhart, J. C., Hara, K., and Ubbels, G. A. (1980). Initiation of the cell cycle and establishment of bilateral symmetry in *Xenopus* eggs. *In* "The Cell Surface: Mediator of Developmental Processes" (S. Subtelny and N. K. Wessels, eds.), p. 187. Academic Press, New York.

Kitajima, T., and Okazaki, K. (1980). Spicule formation *in vitro* by the descendants of precocious micromere formed at the 8-cell stage of sea urchin embryo. *Dev., Growth Differ.* **22**, 265.

Klag, J. J., and Ubbels, G. A. (1975). Regional morphological and cytological differentiation of the fertilized egg of *Discoglossus pictus* (Anura). *Differentiation* **3**, 15.

Kleene, K. C., and Humphreys, T. (1977). Similarity of hnRNA sequences in blastula and pluteus stage sea urchin embryos. *Cell* **12**, 143.

Klein, W. H., Thomas, T. L., Lai, C., Scheller, R. H., Britten, R. J., and Davidson, E. H. (1978). Characteristics of individual repetitive sequence families in the sea urchin genome studied with cloned repeats. *Cell* **14**, 889.

Kleinschmidt, J. A., Fortkamp, E., Krohne, G., Zentgraf, H., and Franke, W. W. (1985). Coexistence of two different types of soluble histone complexes in nuclei of *Xenopus laevis* oocytes. *J. Biol. Chem.* **260**, 1166.

Knipple, D. C., Seifert, E., Rosenberg, U, B., Preiss, A., and Jäckle, H. (1985). Spatial and temporal patterns of *Krüpple* gene expression in early *Drosophila* embryos. *Nature (London)* **317**, 40.

Knöchel, W., and Bladauski, D. (1980). A comparison of sequence complexity of nuclear and polysomal poly(A)⁺ RNA from different developmental stages of *Xenopus laevis*. *Wilhelm Roux's Arch. Dev. Biol.* **188**, 187.

Knöchel, W., and Bladauski, D. (1981). Cloning of cDNA sequences derived from poly(A)⁺ nuclear RNA of *Xenopus laevis* at different developmental stages: Evidence for stage specific regulation. *Wilhelm Roux's Arch. Dev. Biol.* **190**, 97.

Knowland, J., and Graham, C. (1972). RNA synthesis at the two-cell stage of mouse development. *J. Embryol. Exp. Morphol.* **27**, 167.

Knowles, J. A., and Childs, G. J. (1984). Temporal expression of late histone messenger RNA in the sea urchin *Lytechinus pictus*. *Proc. Natl. Acad. Sci. U.S.A.* **81**, 2411.

Kobel, H. R., Brun, R. B., and Fischberg, M. (1973). Nuclear transplantation with melanophores, ciliated epidermal cells, and the established cell-line A-8 in *Xenopus laevis*. *J. Embryol. Exp. Morphol.* **29**, 539.

Koch, E. A., Smith, P. A., and King, R. C. (1967). The division and differentiation of *Drosophila* cystocytes. *J. Morphol.* **121**, 55.

Koebke, J. (1977). Über das Differenzierungsverhalten des mesodermalen Keimbezirks verscheiden alter Prägastrulationsstadien von *Ambystoma mexicanum*. Aufzucht von unbehandelten und mit Lithium behandelten Isolaten. *Z. Mikrosk. Anat. Forsch.* **91**, 215.

Kominami, T. (1984). Allocation of mesendodermal cells during early embryogenesis in the starfish, *Asterina pectinifera*. *J. Embryol. Exp. Morphol.* **84**, 177.

Kondoh, H., Yasuda, K., and Okada, T. S. (1983). Tissue-specific expression of a cloned chick δ-crystallin gene in mouse cells. *Nature (London)* **301**, 440.

Konieczny, S. F., and Emerson, C. P., Jr. (1984). 5-Azacytidine induction of stable mesodermal stem cell lineages from 10T1/2 cells: Evidence for regulatory genes controlling determination. *Cell* **38**, 791.

Konieczny, S. F., and Emerson, C. P. (1985). Differentiation, not determination, regulates muscle gene activation: Transfection of troponin I genes into multipotential and muscle lineages of 10T1/2 cells. *Mol. Cell. Biol.* **5**, 2423.

Konrad, K. D., Engstrom, L., Perrimon, N., and Mahowald, A. P. (1985). Genetic analysis of oogenesis and the role of maternal gene expression in early development. *In* "Developmental Biology. A Comprehensive Survey. Oogenesis" (L. W. Browder, ed.), Vol. 1, p. 577. Plenum, New York.

Korn, L. J., and Gurdon, J. B. (1981). The reactivation of developmentally inert 5S genes in somatic nuclei injected into *Xenopus* oocytes. *Nature (London)* **289**, 461.

Kornberg, T. (1981). *Engrailed*: A gene controlling compartment and segment formation in *Drosophila*. *Proc. Natl. Acad. Sci. U.S.A.* **78**, 1095.

Kornberg, T., Siden, I., O'Farrell, P., and Simon, M. (1985). The *engrailed* locus of *Drosophila: In situ* localization of transcripts reveals compartment-specific expression. *Cell* **40**, 45.

Koser, R. B., and Collier, J. R. (1976). An electrophoretic analysis of RNA synthesis in normal and lobeless *Ilyanassa* embryo. *Differentiaton* **6**, 47.

Kovesdi, I., and Smith, M. J. (1982). Sequence complexity in the maternal RNA of the starfish *Pisaster ochraceus* (Brant). *Dev. Biol.* **89**, 56.

Kovesdi, I., and Smith, M. J. (1985). Quantitative assessment of actin transcript number in eggs, embryos, and tube feet of the sea star *Pisaster ochraceus*. *Mol. Cell. Biol.* **5**, 3001.

Kowalevsky, A. (1867). Entwickelungsgeschichte des *Amphioxus lanceolatus*. *Mem. Acad. Imp. Sci., St. Petersbourg* **11**, No. 4.

Kowalevsky, A. (1871). Embryologische Studien an Würmern und Arthropoden. *Mem. Acad. Imp. Sci., St. Petersbourg*, **16**, No. 12.

Krieg, P. A., and Melton, D. A. (1984). Formation of the 3' end of histone mRNA by post-transcriptional processing. *Nature (London)* **308**, 203.

Kriegstein, H. J., and Hogness, D. S. (1974). Mechanism of DNA replication in *Drosophila* chromosomes: Structure of replication forks and evidence for bidirectionality. *Proc. Natl. Acad. Sci. U.S.A.* **71**, 135.

Krigsgaber, M. R., and Neyfakh, A. A. (1972). Investigation of the mode of nuclear control over protein synthesis in early development of loach and sea urchin. *J. Embryol. Exp. Morphol.* **28**, 491.

Krohne, G., and Franke, W. W. (1980). Immunological identification and localization of the predominant nuclear protein of the amphibian oocyte nucleus. *Proc. Natl. Acad. Sci. U.S.A.* **77**, 1034.

Kronenberg, L. H., and Humphreys, T. (1972). Double-stranded ribonucleic acid in sea urchin embryos. *Biochemistry* **11**, 2020.

Krumlauf, R., Hammer, R. E., Brinster, R., Chapman, V. M., and Tilghman, S. M. (1985a). Regulated expression of α-fetoprotein genes in transgenic mice. *Cold Spring Harbor Symp. Quant. Biol.* **50**, 371.

Krumlauf, R., Hammer, R. E., Tilghman, S. M., and Brinster, R. L. (1985b). Developmental regulation of α-fetoprotein genes in transgenic mice. *Mol. Cell. Biol.* **5**, 1639.

Kuner, J. M., Nakanishi, M., Ali, Z., Drees, B., Gustavson, E., Theis, J., Kauvar, L., Kornberg, T., and O'Farrell, P. H. (1985). Molecular cloning of *engrailed*: A gene involved in the development of pattern in *Drosophila melanogaster*. *Cell* **42**, 309.

Kung, C. S. (1974). On the size relationship between nuclear and cytoplasmic RNA in sea urchin embryos. *Dev. Biol.* **36**, 343.

Kunkel, N. S., and Weinberg, E. S. (1978). Histone gene transcripts in the cleavage and mesenchyme blastula embryo of the sea urchin, *S. purpuratus*. *Cell* **14**, 313.

Kunkel, N. S., Hemminki, K., and Weinberg, E. S. (1978). Size of histone gene transcripts in different embryonic stages of the sea urchin, *Strongylocentrotus purpuratus*. *Biochemistry* **17**, 2591.

Kunz, W. (1967a). Lampenbürstenchromosomen und multiple Nukleolen bei Orthopteren. *Chromosoma* **21**, 446.

Kunz, W. (1967b). Funktionsstrukturen im Oocytenkern von *Locusta migratoria*. *Chromosoma* **20**, 332.

Kunz, W., Trepte, H.-H., and Bier, K. (1970). On the function of the germ line chromosomes in the oogenesis of *Wachtliella persicariae* (Cecidomyiidae). *Chromosoma* **30**, 180.

Kunz, W., Grimm, C., and Franz, G. (1982). Amplification and synthesis of rDNA: *Drosophila*. *In* "The Cell Nucleus" (H. Busch, ed.), Vol. 12, p. 155. Academic Press, New York.

Kuo, C.-H., and Garen, A. (1978). Analysis of the coding activity and stability of messenger RNA in *Drosophila* oocytes. *Dev. Biol.* **67**, 237.

Kuroiwa, A., Hafen, E., and Gehring, W. H. (1984). Cloning and transcriptional analysis of the segmentation gene *fushi tarazu* of *Drosophila*. *Cell* **37**, 825.

Kurtz, D. T. (1981). Hormonal inducibility of rat α_{2u} globulin genes in transfected mouse cells. *Nature (London)* **291**, 629.

Kuwada, J. Y., and Goodman, C. S. (1985). Neuronal determination during embryonic development of the grasshopper nervous system. *Dev. Biol.* **110**, 114.

Labhart, P., and Reeder, R. H. (1984). Enhancer-like properties of the 60/81 bp elements in the ribosomal gene spacer of *Xenopus laevis*. *Cell* **37**, 285.

LaBonne, S. G., and Mahowald, A. P. (1985). Partial rescue of embryos from two maternal-effect neurogenic mutants by transplantation of wild-type ooplasm. *Dev. Biol.* **110**, 264.

Lacalli, T. C. (1981). Structure and development of the apical organ in trochophores of *Spirobranchus polycerus*, *Phyllodoce maculata* and *Phyllodoce mucosa* (Polychaeta). *Proc. R. Soc. London, Ser. B* **212**, 381.

Lacalli, T. C. (1984). Structure and organization of the nervous system trochophore larva of *Spirobranchus*. *Philos. Trans. R. Soc. London, Ser. B* **306**, 79.

Lai, E. C., Riser, M. E., and O'Malley, B. W. (1983). Regulated expression of the chicken ovalbumin gene in a human estrogen-responsive cell line. *J. Biol. Chem.* **258**, 12693.

Laird, C. D. (1971). Chromatid structure: Relationship between DNA content and nucleotide sequence diversity. *Chromosoma* **32**, 378.

Laird, C. D., and Chooi, W. Y. (1976). Morphology of transcription units in *Drosophila melanogaster*. *Chromosoma* **58**, 193.

Lallier, R. (1975). Animalization and vegetalization. *In* "The Sea Urchin Embryo: Biochemistry and Morphogenesis" (G. Czihak, ed.), p. 473. Springer-Verlag, Berlin and New York.

LaMarca, M. J., and Wassarman, P. M. (1979). Program of early development in the mammal: Changes in absolute rates of synthesis of ribosomal proteins during oogenesis and early embryogenesis in the mouse. *Dev. Biol.* **73**, 103.

LaMarca, M. J., and Wassarman, P. M. (1984). Relationship between rates of synthesis and intracellular distribution of ribosomal proteins during oogenesis in the mouse. *Dev. Biol.* **102**, 525.

Lamb, M. M., and Laird, C. D. (1976). The size of poly(A)-containing RNAs in *Drosophila melanogaster* embryos. *Biochem. Genet.* **14**, 357.

Lambert, C. C. (1971). Genetic transcription during the development and metamorphosis of the tunicate, *Ascidia callosa*. *Exp. Cell Res.* **66**, 401.

Lane, N. J. (1967). Spheroidal and ring nucleoli in amphibian oocytes. Patterns of uridine incorporation and fine structural features. *J. Cell Biol.* **35**, 421.

Langelan, R. E., and Whiteley, A. H. (1985). Unequal cleavage and the differentiation of echinoid primary mesenchyme. *Dev. Biol.* **109**, 464.

Lankester, E. R. (1877). Notes on the embryology and classification of the animal kingdom: Comprising a revision of speculations relative to the origin and significance of the germ layers. I. The planula theory. *Q. J. Microsc. Sci.* **17**, 399.

Laskey, R. A., and Gurdon, J. B. (1970). Genetic content of adult somatic cells tested by nuclear transplantation from cultured cells. *Nature (London)* **228**, 1332.

Laskey, R. A., and Harland, R. M. (1982). Replication origins in the *Xenopus* egg. *Cell* **31**, 503.

Laskey, R. A., Mills, A. D., Gurdon, J. B., and Partington, G. A. (1977). Protein synthesis in oocytes of *Xenopus laevis* is not regulated by the supply of messenger RNA. *Cell* **11**, 345.

Laskey, R. A., Gurdon, J. B., and Trendelenburg, M. (1979). Accumulation of materials involved in rapid chromosomal replication in early amphibian development. *Br. Soc. Dev. Biol. Symp.* **4**, 65.

Laskey, R. A., Kearsey, S. E., Mechali, M., Dingwall, C., Mills, A. D., Dilworth, S. M., and Kelinschmidt, J. (1985). Chromosome replication in early *Xenopus* embryos. *Cold Spring Harbor Symp. Quant. Biol.* **50**, 657.

Lasky, L. A., Lev, Z., Xin, J.-H., Britten, R. J., and Davidson, E. H. (1980). Messenger RNA prevalence in sea urchin embryos measured with cloned cDNAs. *Proc. Natl. Acad. Sci. U.S.A.* **77**, 5317.

Lassar, A. B., Martin, P. L., and Roeder, R. G. (1983). Transcription of class III genes: Formation of preinitiation complexes. *Science* **222**, 740.

Lau, J. T. Y., and Lennarz, W. J. (1983). Regulation of sea urchin glycoprotein mRNAs during embryonic development. *Proc. Natl. Acad. Sci. U.S.A.* **80**, 1028.

Laufer, J. S., Bazzicalupo, P., and Wood, W. B. (1980). Segregation of developmental potential in early embryos of *Caenorhabditis elegans*. *Cell* **19**, 569.

Laughon, A., and Scott, M. P. (1984). Sequence of a *Drosophila* segmentation gene: Protein structure homology with DNA-bindiong proteins. *Nature (London)* **310**, 25.

Lawrence, P. A. (1973). A clonal analysis of segment development in *Oncopeltus* (Hemiptera). *J. Embryol. Exp. Morphol.* **30**, 681.

Lawrence, P. A. (1981). The cellular basis of segmentation in insects. *Cell* **26**, 3.

Lawrence, P. A. (1982). Cell lineage of the thoracic muscles of *Drosophila*. *Cell* **29**, 493.

Lawrence, P. A., and Johnston, P. (1984a). On the role of the *engrailed*+ gene in the internal organs of *Drosophila*. *EMBO J.* **3**, 2839.

Lawrence, P. A., and Johnston, P. (1984b). The genetic specification of pattern in a *Drosophila* muscle. *Cell* **36**, 775.

Lawrence, P. A., and Morata, G. (1976). Compartments in the wing of *Drosophila*: A study of the *engrailed* gene. *Dev. Biol.* **50**, 321.

Lawrence, P. A., and Morata, G. (1983). The elements of the bithorax complex. *Cell* **35**, 595.

Lawrence, P. A., and Struhl, G. (1982). Further studies of the *engrailed* phenotype in *Drosophila*. *EMBO J.* **1**, 827.

Lawrence, P. A., Johnston, P., and Struhl, G. (1983). Different requirements for homeotic genes in the soma and germ line of *Drosophila*. *Cell* **35**, 27.

Lawson, K. A., Meneses, J. J., and Pedersen, R. A. (1986). Cell fate and cell lineage in the endoderm of the presomite mouse embyro, studied with an intracellular tracer. *Dev. Biol.* **115**, 325.

Lazarides, E., and Moon, R. T. (1984). Assembly and topogenesis of the spectrin-based membrane skeleton in erythroid development. *Cell* **37**, 354.

Leder, P. (1982). The genetics of antibody diversity. *Sci. Am.* **246**, 102.

Lee, C. S., and Pavan, C. (1974). Replicating DNA molecules from fertilizated eggs of *Cochliomyia hominivorax* (Diptera). *Chromosoma* **47**, 429.

Lee, D. C., McKnight, G. S., and Palmiter, R. D. (1980). The chicken transferrin gene. Restriction endonuclease analysis of gene sequences in liver and oviduct DNA. *J. Biol. Chem.* **255**, 1442.

Lee, G., Hynes, R., and Kirschner, M. (1984). Temporal and spatial regulation of fibronectin in early *Xenopus* development. *Cell* **36**, 729.

Lee, J. J., Shott, R. J., Rose, S. J., Thomas, T. L., Britten, R. J., and Davidson, E. H. (1984). Sea urchin actin gene subtypes: Gene number, linkage, and evolution. *J. Mol. Biol.* **172**, 149.

Lee, J. J., Calzone, F. J., Britten, R. J., Angerer, R. C., and Davidson, E. H. (1986a). Activation of sea urchin actin genes during embryogenesis: Measurement of transcript accumulation from five different genes in *Strongylocentrotus purpuratus. J. Mol. Biol.* **188**, 173.

Lee, J. J., Calzone, F. C., Britten, R. J., and Davidson, E. H. (1986). Activation of sea urchin actin genes during embryogenesis: Nuclear synthesis and decay rate measurements of transcripts from five different genes. Submitted for publication.

Lee, Y. R., and Whiteley, A. H. (1984). Gene transcription during oogenesis of *Schizobranchia insignis*, a tubiculous polychaete. *Fortschr. Zool.* **29**, 167.

Lehmann, R., Jiménez, F., Dietrich, U., and Campos-Ortega, J. A. (1983). On the phenotype and development of mutants of early neurogenesis in *Drosophila melanogaster. Wilhelm Roux's Arch. Dev. Biol.* **192**, 62.

Leivo, I., Vaheri, A., Timpl, R., and Wartiovaara, J. (1980). Appearance and distribution of collagens and laminin in the early mouse embryo. *Dev. Biol.* **76**, 100.

Lennarz, W. J. (1985). Regulation of glycoprotein synthesis in the developing sea urchin embryo. *Trends Biochem. Sci.* **10**, 248.

León, P. E. (1975). Function of lampbrush chromosomes: A hypothesis. *J. Theor. Biol.* **55**, 481.

Leonard, R. A., Hoffner, N. J., and DiBerardino, M. A. (1982). Induction of DNA synthesis in amphibian erythroid nuclei in *Rana* eggs following conditioning in meiotic oocytes. *Dev. Biol.* **92**, 343.

Lev, Z., Thomas, T. L., Lee, A. S., Angerer, R. C., Britten, R. J., and Davidson, E. H. (1980). Developmental expression of two cloned sequences coding for rare sea urchin embryo messages. *Dev. Biol.* **76**, 322.

Levenson, R. G., and Marcu, K. B. (1976). On the existence of polyadenylated histone mRNA in *Xenopus laevis* oocytes. *Cell* **9**, 311.

Levey, I. L., Stull, G. B., and Brinster, R. L. (1978). Poly(A) and synthesis of polyadenylated RNA in the preimplantation mouse embryo. *Dev. Biol.* **64**, 140.

Levine, M., Hafen, E., Garber, R. L., and Gehring, W. J. (1983). Spatial distribution of *Antennapedia* transcripts during *Drosophila* development. *EMBO J.* **2**, 2037.

Levine, M., Harding, K., Wedeen, C., Doyle, H., Hoey, T., and Radomska, H. (1985). Expression of the homeobox gene family in *Drosophila. Cold Spring Harbor Symp. Quant. Biol.* **50**, 209.

Levis, R., and Penman, S. (1977). The metabolism of poly(A)$^+$ and poly(A)$^-$ hnRNA in cultured *Drosophila* cells studied with a rapid uridine pulse-chase. *Cell* **11**, 105.

Levner, M. H. (1974). RNA transcription in mature sea urchin eggs. *Exp. Cell Res.* **85**, 296.

Levy W., B., and McCarthy, B. J. (1975). Messenger RNA complexity in *Drosophila melanogaster. Biochemistry* **14**, 2440.

Lewis, E. B. (1963). Genes and developmental pathways. *Am. Zool.* **3**, 33.

Lewis, E. B. (1978). A gene complex controlling segmentation in *Drosophila. Nature (London)* **276**, 565.

Lewis, E. B. (1982). Control of body segment differentiation in *Drosophila* by the bithorax gene complex. *In* "Embryonic Development. Part A: Genetic Aspects" (M. M. Burger and R. Weber, eds.), p. 269. Alan R. Liss, New York.

Lewis, E. B. (1985). Regulation of the genes of the bithorax complex in *Drosophila. Cold Spring Harbor Symp. Quant. Biol.* **50**, 155.

Lewis, J. C., and McMillan, D. B. (1965). The development of the ovary of the sea lamprey (*Petromyzon marinus*). *J. Morphol.* **117**, 425.

Lewis, R. A., Kaufman, T. C., Denell, R. E., and Tallerico, P. (1980a). Genetic analysis of the *Antennapedia* gene complex (ANT-C) of *Drosophila melanogaster*. I. Polytene chromosome segments 84B-D. *Genetics* **95**, 367.

Lewis, R. A., Wakimoto, B. T., Denell, R. E., and Kaufman, T. C. (1980b). Genetic analysis of the *Antennapedia* gene complex (ANT-C) of *Drosophila melanogaster*. II. Polytene chromosome segments 84A-B1,2. *Genetics* **95**, 383.

Leys, E. J., and Kellems, R. E. (1981). Control of dihydrofolate reductase messenger ribonucleic acid production. *Mol. Cell Biol.* **1**, 961.

Leys, E. J., Crouse, G. F., and Kellems, R. E. (1984). Dihydrofolate reductase gene expression in cultured mouse cells is regulated by transcript stabilization in the nucleus. *J. Cell Biol.* **99**, 180.

Lifton, R. P., Goldberg, M. L., Karp, R. W., and Hogness, D. S. (1978). The organization of the histone genes in *Drosophila melanogaster*: Functional and evolutionary implications. *Cold Spring Harbor Symp. Quant. Biol.* **42**, 1047.

Lillie, F. R. (1906). Observations and experiments concerning the elementary phenomena of embryonic development in *Chaetopterus*. *J. Exp. Zool.* **3**, 153.

Lillie, F. R. (1909). Polarity and bilaterality of the annelid egg. Experiments with centrifugal force. *Biol. Bull.* (*Woods Hole, Mass.*) **16**, 54.

Limbourg, B., and Zalokar, M. (1973). Permeabilization of *Drosophila* eggs. *Dev. Biol.* **35**, 382.

Lindsley, D. L., and Grell, E. H. (1968). Genetic variations of *Drosophila melanogaster*. *Carnegie Inst. Washington, Publ.* No. 627.

Lockshin, R. A. (1966). Insect embryogenesis: Macromolecular synthesis during early development. *Science* **154**, 775.

Lohs-Schardin, M. (1982). *Dicephalic*—a *Drosophila* mutant affecting polarity in follicle organization and embryonic patterning. *Wilhelm Roux's Arch. Dev. Biol.* **191**, 28.

Lohs-Schardin, M., Cremer, C., and Nüsslein-Volhard, C. (1979a). A fate map for the larval epidermis of *Drosophila melanogaster*: Localized cuticle defects following irradiation of the blastoderm with an ultraviolet laser microbeam. *Dev. Biol.* **73**, 239.

Lohs-Schardin, M., Sander, K., Cremer, C., Cremer, T., and Zorn, C. (1979b). Localized ultraviolet laser microbeam irradiation of early *Drosophila* embryos: Fate maps based on location and frequency of adult defects. *Dev. Biol.* **68**, 533.

Lorenz, L. J., and Richter, J. D. (1985). A cDNA clone for a poly(A) RNA binding protein of *Xenopus laevis* oocytes hybridizes to four developmentally regulated mRNAs. *Mol. Cell. Biol.* **5**, 2697.

Lovett, J. A., and Goldstein, E. S. (1977). The cytoplasmic distribution and characterization of poly(A)$^+$ RNA in oocytes and embryos of *Drosophila*. *Dev. Biol.* **61**, 70.

Loyd, J. E., Raff, E. C., and Raff, R. A. (1981). Site and timing of synthesis of tubulin and other proteins during oogenesis in *Drosophila melangoaster*. *Dev. Biol.* **86**, 272.

Loyez, M. (1905). Recherches sur le développement ovarien des oeufs méroblastiques. *Arch. Anat. Microsc. Morphol. Exp.* **8**, 69.

Lunan, K. D., and Mitchell, H. K. (1969). The metabolism of tyrosine-O-phosphate in *Drosophila*. *Arch. Biochem. Biophys.* **132**, 450.

Lund, E., Dahlberg, J. E., and Forbes, D. J. (1984). The two embryonic U1 small nuclear RNAs of *Xenopus laevis* are encoded by a major family of tandemly repeated genes. *Mol. Cell. Biol.* **4**, 2580.

Lynn, D. A., Angerer, L. M., Bruskin, A. M., Klein, W. H., and Angerer, R. C. (1983). Localization of a family of mRNAs in a single cell type and its precursor in sea urchin embryos. *Proc. Natl. Acad. Sci. U.S.A.* **80**, 2656.

Lyon, M. F. (1983). The use of Robertsonian translocations for studies of nondisjunction. *In* "Radiation-Induced Damage in Man" (T. Ishihara, ed.), p. 327. Alan R. Liss, New York.

Lyon, M. F., and Glenister, P. H. (1977). Factors affecting the observed number of young resulting from adjacent-2 disjunction in mice carrying a translation. *Genet. Res.* **29**, 83.

Mabuchi, I. (1976). Myosin from starfish egg: Properties and interaction with actin. *J. Mol. Biol.* **100**, 569.

Mabuchi, I., and Okuno, M. (1977). The effect of myosin antibody on the division of starfish blastomeres. *J. Cell Biol.* **74**, 251.

MacBride, E. W. (1914). "Textbook of Embryology: Invertebrata," Vol. 1. Macmillan, London.

McCarthy, R. A., and Spiegel, M. (1983a). Serum effects on the *in vitro* differentiation of sea urchin micromeres. *Exp. Cell Res.* **149**, 433.

McCarthy, R. A., and Spiegel, M. (1983b). Protein composition of the hyaline layer of sea urchin embryos and reaggregating cells. *Cell Differ.* **13**, 93.

McClay, D. R., and Chambers, A. F. (1978). Identification of four classes of cell surface antigens appearing at gastrulation in sea urchin embryos. *Dev. Biol.* **63**, 179.

McClay, D. R., and Fink, R. D. (1982). Sea urchin hyalin: Appearance and function in development. *Dev. Biol.* **92**, 285.

McClay, D. R., Cannon, G. W., Wessel, G. M., Fink, R. D., and Marchase, R. B. (1983). Patterns of antigenic expression in early sea urchin development. *In* "Time, Space, and Pattern in Embryonic Development" (W. R. Jeffery and R. A. Raff, eds.), p. 157. Alan R. Liss, New York.

McClendon, J. F. (1910). The development of isolated blastomeres of the frog's egg. *Am. J. Anat.* **10**, 425.

McColl, R. S., and Aronson, A. I. (1978). Changes in transcription patterns during early development of the sea urchin. *Dev. Biol.* **65**, 126.

McGhee, J. D. (1986). Isolation and characterization of an intestinal esterase from the nematode, *Caenorhabditis elegans*. In preparation.

McGinnis, W., Levine, M. S., Hafen, E., Kuroiwa, A., and Gehring, W. J. (1984a). A conserved DNA sequence in homeotic genes of the *Drosophila* Antennapedia and bithorax complexes. *Nature (London)* **308**, 428.

McGinnis, W., Garber, R. L., Wirz, J., Kuroiwa, A., and Gehring, W. J. (1984b). A homologous protein-coding sequence in *Drosophila* homeotic genes and its conservation in other metazoans. *Cell* **37**, 403.

McGinnis, W., Hart, C. P., Gehring, W. J., and Ruddle, F. H. (1984c). Molecular cloning and chromosome mapping of a mouse DNA sequence homologous to homeotic genes of *Drosophila*. *Cell* **38**, 675.

McGrath, J., and Solter, D. (1983). Nuclear transplantation in the mouse embryo by microsurgery and cell fusion. *Science* **220**, 1300.

McGrath, J., and Solter, D. (1984a). Inability of mouse blastomere nuclei transferred to enucleated zygotes to support development *in vitro*. *Science* **226**, 1317.

McGrath, J., and Solter, D. (1984b). Completion of mouse embryogenesis requires both the maternal and paternal genomes. *Cell* **37**, 179.

McGrath, J., and Solter, D. (1985). Nuclear transfer in mammalian embryos: Genomic requirements for successful development. *In* "Genetic Manipulation of the Mammalian Early Embryo" (F. Costantini and R. Jaenisch, eds.), Banbury Report 20, p. 31. Cold Spring Harbor Lab., Cold Spring Harbor, New York.

Macgregor, H. C. (1980). Recent developments in the study of lampbrush chromosomes. *Heredity* **44**, 3.

Macgregor, H. C., and Andrews, C. (1977). The arrangement and transcription of "middle repetitive" DNA sequences on lampbrush chromosomes of *Triturus*. *Chromosoma* **63**, 109.

Macgregor, H. C., and Horner, H. (1980). Heteromorphism for chromosome I, a requirement for normal development in crested newts. *Chromosoma* **76**, 111.

Macgregor, H. C., and Kezer, J. (1970). Gene amplification in oocytes with eight germinal vesicles from the tailed frog *Ascaphus truei* Stejneger. *Chromosoma* **29**, 189.

Macgregor, H., and Klosterman, L. (1979). Observations on the cytology of *Bipes* (Amphisbaenia) with special reference to its lampbrush chromosomes. *Chromosoma* **72**, 67.

Mackay, S., and Newrock, K. M. (1982). Histone subtypes and switches in synthesis of histone subtypes during *Ilyanassa* development. *Dev. Biol.* **93**, 430.

McKeon, C., Ohkubo, H., Pastan, I., and de Crombrugghe, B. (1982). Unusual methylation pattern of the $\alpha2(I)$ collagen gene. *Cell* **29**, 203.

McKnight, G. S., Hammer, R. E., Kuenzel, E. A., and Brinster, R. L. (1983). Expression of the chicken transferrin gene in transgenic mice. *Cell* **34**, 335.

McKnight, S. L., and Miller, O. L. (1976). Ultrastructural patterns of RNA synthesis during early embryogenesis of *Drosophilia melanogaster*. *Cell* **8**, 305.

McKnight, S. L., and Miller, O. L. (1977). Electron microscopic analysis of chromatin replication in the cellular blastoderm *Drosophilia melanogaster* embryo. *Cell* **12**, 795.

McLean, K. W., and Whiteley, A. H. (1974). RNA synthesis during the early development of the Pacific oyster, *Crassostraea gigas*. *Exp. Cell Res.* **87**, 132.

Macleod, D., and Bird, A. (1983). Transcription in oocytes of highly methylated rDNA from *Xenopus laevis* sperm. *Nature (London)* **306**, 200.

McMahon, A. P., Novak, T. J., Britten, R. J., and Davidson, E. H. (1984). Inducible expression of a cloned heat shock fusion gene in sea urchin embryos. *Proc. Natl. Acad. Sci. U.S.A.* **81**, 7490.

McMahon, A. P., Flytzanis, C. N., Hough-Evans, B. R., Katula, K. S., Britten, R. J., and Davidson, E. H. (1985). Introduction of cloned DNA into sea urchin egg cytoplasm: Replication and persistance during embryogenesis. *Dev. Biol.* **108**, 420.

Madhaven, M. M. (1973). The dual origin of the nurse chamber in the ovarioles of the gall midge, *Heteropeza pygmaea*. *Wilhelm Roux's Arch. Dev. Biol.* **173**, 164.

Maekawa, H., and Suzuki, Y. (1980). Repeated turn-off and turn-on of fibroin gene transcription during silk gland development of *Bombyx mori*. *Dev. Biol.* **78**, 394.

Maggio, R., Vittorelli, M. L., Rinaldi, A. M., and Monroy, A. (1964). *In vitro* incorporation of amino acids into proteins stimulated by RNA from unfertilized sea urchin eggs. *Biochem. Biophys. Res. Commun.* **15**, 436.

Magnuson, T., and Epstein, C. J. (1981). Characterization of concanavalin A precipitated proteins from early mouse embryos: A 2-dimensional gel electrophoresis study. *Dev. Biol.* **81**, 193.

Mahon, K. A., and Gall, J. G. (1984). The expression of repetitive sequences on amphibian lampbrush chromosomes. *In* "Molecular Biology of Development" (E. H. Davidson and R. A. Firtel, eds.), p. 227. Alan R. Liss, New York.

Mahowald, A. P. (1968). Polar granules of *Drosophila*. II. Ultrastructural changes during early embryogenesis. *J. Exp. Zool.* **167**, 237.

Mahowald, A. P. (1971a). Polar granules of *Drosophila*. III. The continuity of polar granules during the life cycle of *Drosophila*. *J. Exp. Zool.* **176**, 329.

Mahowald, A. P. (1971b). Polar granules of *Drosophila*. IV. Cytochemical studies showing loss of RNA from polar granules during early stages of embryogenesis. *J. Exp. Zool.* **176**, 345.

Mahowald, A. P. (1977). The germ plasm of *Drosophila*: A model system for the study of embryonic determination. *Am. Zool.* **17**, 551.

Mahowald, A. P. (1983). Genetic analysis of oogenesis and determination. *In* "Time, Space, and Pattern in Embryonic Development" (W. R. Jeffery and R. A. Raff, eds.), p. 349. Alan R. Liss, New York.

Mahowald, A. P., and Kambysellis, M. P. (1980). Oogenesis. *In* "The Genetics and Biology of *Drosophila*" (M. Ashburner and T. R. F. Wright, eds.), Vol. 2d, p. 141. Academic Press, London.

Mahowald, A. P., and Stoiber, D. (1974). The origin of the nurse chamber in ovaries of *Miastor* (Diptera: Cecidomyidae). *Wilhelm Roux's Arch. Dev. Biol.* **176**, 159.

Mahowald, A. P., and Tiefert, M. (1970). Fine structural changes in the *Drosophila* oocyte nucleus during a short period of RNA synthesis. An autoradiographic and ultrastructural study of RNA synthesis in the oocyte nucleus of *Drosophila*. *Wilhelm Roux's Arch. Dev. Biol.* **165**, 8.

Mahowald, A. P., Allis, C. D., Karrer, K. M., Underwood, E. M., and Waring, G. L. (1979a). Germ plasm and pole cells of *Drosophila*. *In* "Determinants of Spatial Organization" (S. Subtelny and I. R. Konigsberg, eds.), p. 127. Academic Press, New York.

Mahowald, A. P., Caulton, J. H., and Gehring, W. J. (1979b). Ultrastructural studies of oocytes and embryos derived from female flies carrying the *grandchildless* mutation in *Drosophila subobscura*. *Dev. Biol.* **69**, 118.

Mairy, M., and Denis, H. (1971). Recherches biochimiques sur l'oogenèse. I. Synthèse et accumulation du RNA pendant l'oogenèse du crapaud sud-africain *Xenopus laevis*. *Dev. Biol.* **24**, 143.

Mairy, M., and Denis, H. (1972). Recherches biochimiques sur l'oogenèse. 3. Assemblage des ribosomes pendant le grand accroissement des oocytes de *Xenopus laevis*. *Eur. J. Biochem.* **25**, 535.

Malacinski, G. M., and Spieth, J. (1979). Maternal effect genes in the Mexican axolotl (*Ambystoma mexicanum*). *In* "Maternal Effects in Development" (D. R. Newth and M. Balls, eds.), p. 241. Cambridge Univ. Press, London and New York.

Malacinski, G. M., Benford, H., and Chung, H.-M. (1974). Association of an ultraviolet irradiation sensitive cytoplasmic localization with the future dorsal side of the amphibian egg. *J. Exp. Zool.* **191**, 97.

Malacinski, G. M., Brothers, A. J., and Chung, H.-M. (1977). Destruction of components of the neural induction system of the amphibian egg with ultraviolet irradiation. *Dev. Biol.* **56**, 24.

Malacinski, G. M., Chung, H.-M., and Asashima, M. (1980). The association of primary embryonic organizer activity with the future dorsal side of amphibian eggs and early embryos. *Dev. Biol.* **77**, 449.

Malcolm, D. B., and Sommerville, J. (1974). The structure of chromosome-derived ribonucleoprotein in oocytes of *Triturus cristatus carnifex* (Laurenti). *Chromosoma* **48**, 137.

Malcolm, D. B., and Sommerville, J. (1977). The structure of nuclear ribonucleoprotein of amphibian oocytes. *J. Cell Sci.* **24**, 143.

Maller, J. L., and Krebs, E. G. (1980). Regulation of oocyte maturation. *Curr. Top. Cell. Regul.* **16**, 271.

Mancino, G., Ragghianti, M., and Bucci-Innocenti, S. (1977). Cytotaxonomy and cytogenetics in European newt species. *In* "The Reproductive Biology of Amphibians" (D. H. Taylor and S. I. Guttman, eds.), p. 411. Plenum, New York.

Manes, M. E., and Elinson, R. P. (1980). Ultraviolet light inhibits grey crescent formation on the frog egg. *Wilhelm Roux's Arch. Dev. Biol.* **189**, 73.

Manes, M. E., Elinson, R. P., and Barbieri, F. D. (1978). Formation of the amphibian grey crescent: Effects of colchicine and cytochalasin B. *Wilhelm Roux's Arch. Dev. Biol.* **185**, 99.

Mangold, O. (1920). Fragen der regulation und Determination an ungeordneten Furchungsstadien und verschmolzenen Keimen von *Triton*. *Arch. Entwicklungsmech. Org.* **47**, 249.

Manning, J. E., Schmid, C. W., and Davidson, N. (1975). Interspersion of repetitive and nonrepetitive DNA sequences in the *Drosophila melanogaster* genome. *Cell* **4**, 141.

Mantei, N., and Weissmann, C. (1982). Controlled transcription of a human α-interferon gene introduced into mouse L cells. *Nature (London)* **297**, 128.

Maréchal, J. (1907). Sur L'ovogénèse des sélaciens et de quelques autres chordates. Premier memoire: Morphologie de l'élément chromosomique dans l'ovocyte. I. chez les selaciens, les téléostéens, les tuniciers et l'amphioxus. *Cellule* **24**, 1.

Mariano, E. E., and Schram-Doumont, A. (1965). Rapidly labelled ribonucleic acid in *Xenopus laevis* embryonic cells. *Biochim. Biophys. Acta* **103**, 610.

Marinos, E. (1985). The number of mitochondria in *Xenopus laevis* ovulated oocytes. *Cell Differ.* **16**, 139.

Markert, C. L. (1982). Parthenogenesis, homozygosity, and cloning in mammals. *J. Hered.* **73**, 390.

Marmur, J., Rownd, R., and Schildkraut, C. L. (1963). Denaturation and renaturation of deoxyribonucleic acid. *Prog. Nucleic Acid Res.* **1**, 231.

Marsh, J. L., and Wieschaus, E. (1977). Germ line dependence of the *maroon-like* maternal effect in *Drosophila*. *Dev. Biol.* **60**, 396.

Marsh, J. L., Van Deusen, E. B., Wieschaus, E., and Gehring, W. J. (1977). Germ line dependence of the *deep orange* maternal effect in *Drosophila*. *Dev. Biol.* **56**, 195.

Marshall, J. A., and Dixon, K. E. (1977). Nuclear transplantation from intestinal epithelial cells of early and late *Xenopus laevis* tadpoles. *J. Embryol. Exp. Morphol.* **40**, 167.

Martin, K. A., and Miller, O. L. (1983). Polysome structure in sea urchin eggs and embryos: An electron microscopic analysis. *Dev. Biol.* **98**, 338.

Martin, K., Osheim, Y. N., Beyer, A. L., and Miller, O. L., Jr. (1980). Visualization of transcriptional activity during *Xenopus laevis* oogenesis. *Results Probl. Cell Differ.* **11**, 37.

Martindale, M. Q., and Brandhorst, B. P. (1984). Translational changes induced by 1-methyladenine in anucleate starfish oocytes. *Dev. Biol.* **101**, 512.

Martindale, M. Q., Doe, C. Q., and Morrill, J. B. (1985). The role of animal-vegetal interaction with respect to the determination of dorsoventral polarity in the equal-cleaving spiralian, *Lymnaea palustris*. *Wilhelm Roux's Arch. Dev. Biol.* **194**, 281.

Martinez-Arias, A. (1985). The development of *fused⁻* embryos of *Drosophila melanogaster. J. Embryol. Exp. Morphol.* **87**, 99.

Martinez-Arias, A., and Lawrence, P. A. (1985). Parasegments and compartments in the *Drosophila* embryo. *Nature (London)* **313**, 639.

Maruyama, Y. K., Nakaseko, Y., and Yagi, S. (1985). Localization of cytoplasmic determinants responsible for primary mesenchyme formation and gastrulation in the unfertilized egg of the sea urchin *Hemicentrotus pulcherrimus. J. Exp. Zool.* **236**, 155.

Marzluff, W. F., Jr., White, E. L., Benjamin, R., and Huang, R. C. C. (1975). Low molecular weight RNA species from chromatin. *Biochemistry* **14**, 3715.

Masuda, M. (1979). Species specific pattern of ciliogenesis in developing sea urchin embryos. *Dev., Growth Differ.* **21**, 545.

Masuda, M., and Sato, H. (1984). Asynchronization of cell division is concurrently related with ciliogenesis in sea urchin blastulae. *Dev., Growth & Differ.* **26**, 281.

Masui, Y., and Clarke, H. (1979). Oocyte maturation. *Int. Rev. Cytol.* **57**, 185.

Matsumoto, L., Kasamatsu, H., Pikó, L., and Vinograd, J. (1974). Mitochondrial DNA replication in sea urchin oocytes. *J. Cell Biol.* **63**, 146.

Mattaj, I. W., and De Robertis, E. M. (1985). Nuclear segregation of U2 snRNA requires binding of specific snRNP proteins. *Cell* **40**, 111.

Mauron, A., Levy, S., Childs, G., and Kedes, L. (1981). Monocistronic transcription is the physiological mechanism of sea urchin embryonic histone gene expression. *Mol. Cell. Biol.* **1**, 661.

Mauron, A., Kedes, L., Hough-Evans, B. R., and Davidson, E. H. (1982). Accumulation of individual histone mRNAs during embryogenesis of the sea urchin *Strongylocentrotus purpuratus. Dev. Biol.* **94**, 425.

Maxson, R. E., and Egrie, J. C. (1980). Expression of maternal and paternal histone genes during early cleavage stages of the echinoderm hybrid *Strongylocentrotus purpuratus* × *Lytechinus pictus. Dev. Biol.* **74**, 335.

Maxson, R. E., and Wilt, F. H. (1981). The rate of synthesis of histone mRNA during the development of sea urchin embryos (*Strongylocentrotus purpuratus*). *Dev. Biol.* **83**, 380.

Maxson, R. E., and Wilt, F. H. (1982). Accumulation of the early histone messenger RNAs during the development of *Strongylocentrotus purpuratus. Dev. Biol.* **94**, 435.

Maxson, R., Mohun, T., Gormezano, G., Childs, G., and Kedes, L. (1983). Distinct organizations and patterns of expression of early and late histone gene sets in the sea urchin. *Nature (London)* **301**, 120.

Maxson, R., Ito, M., Balcells, S., Thayer, M., and Etkin, L. (1986). Differential expression of sea urchin early and late histone genes in *Xenopus* oocytes in response to a *trans*-acting factor isolated from late stage sea urchin embryos. Submitted for publication.

Mayo, K. E., Warren, R., and Palmiter, R. D. (1982). The mouse metallothionein-I gene is transcriptionally regulated by cadmium following transfection into human or mouse cells. *Cell* **29**, 99.

Mazabraud, A., Wegnez, M., and Denis, H. (1975). Biochemical research on oogenesis: RNA accumulation in the oocytes of teleosts. *Dev. Biol.* **44**, 326.

Mazur, G. D., Regier, J. C., and Kafatos, F. C. (1980). The silkmoth chorion: Morphogenesis of surface structures and its relation to synthesis of specific proteins. *Dev. Biol.* **76**, 305.

Méchali, M., and Kearsey, S. (1984). Lack of specific sequence requirement for DNA replication in *Xenopus* eggs compared with high sequence specificity in yeast. *Cell* **38**, 55.

Meedel, T. H. (1983). Myosin expression in the developing ascidian embryo. *J. Exp. Zool.* **227**, 203.

Meedel, T. H., and Whittaker, J. R. (1978). Messenger RNA synthesis during early ascidian development. *Dev. Biol.* **66**, 410.

Meedel, T. H., and Whittaker, J. R. (1983). Development of translationally active mRNA for larval muscle acetylcholinesterase during ascidian embryogenesis. *Proc. Natl. Acad. Sci. U.S.A.* **80**, 4761.

Melli, M., Whitfield, C., Rao, K. V., Richardson, M., and Bishop, J. O. (1971). DNA-RNA hybridization in vast DNA excess. *Nature (London) New Biol.* **231**, 8.

Melton, D. A., De Robertis, E. M., and Cortese, R. (1980). Order and intracellular location of the events involved in the maturation of a spliced tRNA. *Nature (London)* **284**, 143.

Melton, D. A., Krieg, P. A., Rebagliati, M. R., Maniatis, T., Zinn, K., and Green, M. R. (1984). Efficient *in vitro* synthesis of biologically active RNA and hybridization probes from plasmids containing a bacteriophage SP6 promoter. *Nucleic Acids Res.* **12**, 7035.

Mercola, M., Wang, X.-F., Olsen, J., and Calame, K. (1983). Transcriptional enhancer elements in the mouse immunoglobulin heavy chain locus. *Science* **221**, 663.

Merlino, G. T., Water, R. D., Chamberlain, J. P., Jackson, D. A., El-Gewely, M. R., and Kleinsmith, L. J. (1980). Cloning of sea urchin actin gene sequences for use in studying the regulation of actin gene transcription. *Proc. Natl. Acad. Sci. U.S.A.* **77**, 765.

Mermod, J.-J., and Crippa, M. (1978). Variations in the amount of polysomes in mature oocytes of *Drosophila melanogaster*. *Dev. Biol.* **66**, 586.

Mermod, J.-J., Jacobs-Lorena, M., and Crippa, M. (1977). Changes in rate of RNA synthesis and ribosomal gene number during oogenesis of *Drosophila melanogaster*. *Dev. Biol.* **57**, 393.

Mermod, J.-J., Schatz, G., and Crippa, M. (1980). Specific control of messenger translation in *Drosophila* oocytes and embryos. *Dev. Biol.* **75**, 177.

Merriam, R. W., and Clark, T. G. (1978). Actin in *Xenopus* oocytes. II. Intracellular distribution and polymerizability. *J. Cell Biol.* **77**, 439.

Merriam, R. W., and Sauterer, R. A. (1983). Localization of a pigment-containing structure near the surface of *Xenopus* eggs which contracts in response to calcium. *J. Embryol. Exp. Morphol.* **76**, 51.

Merrill, G. F., Harland, R. M., Groudine, M., and McKnight, S. L. (1984). Genetic and physical analysis of the chicken *tk* gene. *Mol. Cell. Biol.* **4**, 1769.

Michael, P. (1984). Are the primordial germ cells (PGCs) in Urodela formed by the inductive action of the vegetal yolk mass? *Dev. Biol.* **103**, 109.

Miller, L. (1978). Relative amounts of newly synthesized poly(A)$^+$ and poly(A)$^-$ messenger RNA during development of *Xenopus laevis*. *Dev. Biol.* **64**, 118.

Miller, O. L., (1966). Structure and composition of peripheral nucleoli of salamander oocytes. *Natl. Cancer Inst. Monogr.* No. **23**, 53.

Miller, O. L., and Bakken, A. H. (1972). Morphological studies of transcription. *Acta Endocrinol. (Copenhagen)* **168**, Suppl., 155.

Miller, O. L., and Beatty, B. R. (1969a). Portrait of a gene. *J. Cell. Physiol.* **74**, Suppl. 1, 225.

Miller, O. L., and Beatty, B. R. (1969b). Extrachromosomal nucleolar genes in amphibian oocytes. *Genetics* **61**, Suppl., 1.

Miller, O. L., and Beatty, B. R. (1969c). Visualization of nucleolar genes. *Science* **164**, 955.

Miller, T. J., Stephens, D. L., and Mertz, J. E. (1982). Kinetics of accumulation and processing of simian virus 40 RNA in *Xenopus laevis* oocytes injected with simian virus 40 DNA. *Mol. Cell. Biol.* **2**, 1581.

Miñana, F. J., and Garcia-Bellido, A. (1982). Preblastoderm mosaics of mutants of the bithorax complex. *Wilhelm Roux's Arch. Dev. Biol.* **191**, 331.

Mintz, B. (1962). Formation of genotypically mosaic mouse embryos. *Am. Zool.* **2**, 432.

Mintz, B. (1965). Experimental genetic mosaicism in the mouse. *In* "Preimplantation Stages of Pregnancy" (C. E. W. Wolstenholme and M. O'Connor, eds.), Ciba Found. Symp., p. 194. Churchill, London.

Mirault, M.-E., Southgate, R., and Delwart, E. (1982). Regulation of heat-shock genes: A DNA sequence upstream of *Drosophila hsp70* genes is essential for their induction in monkey cells. *EMBO J.* **1**, 1279.

Mirkes, P. E. (1972). Polysomes and protein synthesis during development of *Ilyanassa obsoleta*. *Exp. Cell Res.* **74**, 503.

Mirsky, A. E. (1951). Some chemical aspects of the cell nucleus. *In* "Genetics of the 20th Century" (L. C. Dunn, ed.), p. 127. Macmillan, New York.

Mirsky, A. E. (1953). The chemistry of heredity. *Sci. Am.* **188**, 47.

Mirsky, A. E., and Ris, H. (1949). Variable and constant components of chromosomes. *Nature (London)* **163**, 666.

Mirsky, A. E., and Ris, H. (1951). The desoxyribonucleic acid content of animal cells and its evolutionary significance. *J. Gen. Physiol.* **34**, 451.

Mischke, D., and Pardue, M. L. (1982). Organization and expression of α-tubulin genes in *Drosophila melanogaster*. One member of the α-tubulin multigene family is transcribed in both oogenesis and later embryonic development. *J. Mol. Biol.* **156**, 449.

Mita-Miyazawa, I., Ikegami, S., and Satoh, N. (1985). Histospecific acetylcholinesterase development in the presumptive muscle cells isolated from 16-cell-stage ascidian embryos with respect to the number of DNA replications. *J. Embryol. Exp. Morphol.* **87**, 1.

Mitsunaga, K., Fujiwara, A., Yoshimi, T., and Yasumasu, I. (1983). Stage-specific effects on sea urchin embryogenesis of Zn^{2+}, Li^+, several inhibitors of cAMP-phosphodiesterase and inhibitors of protein synthesis. *Dev., Growth Differ.* **25**, 249.

Miwa, J., Schierenberg, E., Miwa, S., and von Ehrenstein, G. (1980). Genetics and mode of expression of temperature-sensitive mutations arresting embryonic development in *Caenorhabditis elegans*. *Dev. Biol.* **76**, 160.

Mizuno, S., Lee, Y. R., Whiteley, A. H., and Whiteley, H. R. (1974). Cellular distribution of RNA populations in 16-cell stage embryos of the sand dollar, *Dendraster excentricus*. *Dev. Biol.* **37**, 18.

Mlodzik, M., Fjose, A., and Gehring, W. J. (1985). Isolation of *caudal*, a *Drosophila* homeobox-containing gene with maternal expression, whose transcripts form a concentration gradient at the preblastoderm stage. *EMBO J.* **4**, 2961.

Moav, B., and Nemer, M. (1971). Histone synthesis. Assignment to a special class of polyribosomes in sea urchin embryos. *Biochemistry* **10**, 881.

Moen, T. L., and Namenwirth, M. (1977). The distribution of soluble proteins along the animal-vegetal axis of frog eggs. *Dev. Biol.* **58**, 1.

Mohan, J. (1971). Influence of the RNA content on oogenesis in the *bobbed* mutants of *Drosophila melanogaster*. *J. Embryol. Exp. Morphol.* **25**, 237.

Mohan, J., and Ritossa, F. M. (1970). Regulation of ribosomal RNA synthesis and its bearing on the *bobbed* phenotype of *Drosophila melanogaster*. *Dev. Biol.* **22**, 495.

Mohandas, T., Sparkes, R. S., and Shapiro, L. J. (1981). Reactivation of an inactive human X chromosome: Evidence for X inactivation by DNA methylation. *Science* **211**, 393.

Mohler, J. D. (1977). Developmental genetics of the *Drosophila* egg. I. Identification of 59 sex-linked cistrons with maternal effects on embryonic development. *Genetics* **85**, 259.

Mohler, J., and Wieschaus, E. F. (1985). *bicaudal* mutations of *Drosophila melanogaster*: Alteration of blastoderm cell fate. *Cold Spring Harbor Symp. Quant. Biol.* **50**, 105.

Mohun, T. J., Brennan, S., Dathan, N., Fairman, S., and Gurdon, J. B. (1984). Cell type-specific activation of actin genes in the early amphibian embryo. *Nature (London)* **311**, 716.

Mohun, T., Maxson, R., Gormezano, G., and Kedes, L. (1985). Differential regulation of individual late histone genes during development of the sea urchin (*Strongylocentrotus purpuratus*). *Dev. Biol.* **108**, 491.

Molloy, G. R., Thomas, W. L., and Darnell, J. E. (1972). Occurrence of uridylate-rich oligonucleotide regions in heterogeneous nuclear RNA of HeLa cells. *Proc. Natl. Acad. Sci. U.S.A.* **69**, 3684.

Monroy, A., and Tyler, A. (1963). Formation of active ribosomal aggregates (polysomes) upon fertilization and development of sea urchin eggs. *Arch. Biochem. Biophys.* **103**, 431.

Monroy, A., Maggio, R., and Rinaldi, A. M. (1965). Experimentally induced activation of the ribosomes of the unfertilized sea urchin egg. *Proc. Natl. Acad. Sci. U.S.A.* **54**, 107.

Moody, S. A., and Jacobson, M. (1983). Compartmental relationships between anuran primary spinal motoneurons and somatic muscle fibers that they first innervate. *J. Neurosci.* **3**, 1670.

Moon, R. T. (1983). Poly(A)-containing messenger ribonucleoprotein complexes from sea urchin eggs and embryos: Polypeptides associated with native and UV-crosslinked mRNPs. *Differentiation* **24**, 13.

Moon, R. T., Moe, K. D., and Hille, M. B. (1980). Polypeptides of nonpolyribosomal messenger ribonucleoprotein complexes of sea urchin eggs. *Biochemistry* **19**, 2723.

Moon, R. T., Danilchik, M. V., and Hille, M. B. (1982). An assessment of the masked message hypothesis: Sea urchin egg messenger ribonucleoprotein complexes are efficient templates for *in vitro* protein synthesis. *Dev. Biol.* **93**, 389.

Moor, R. (1983). Contact, signalling and cooperation between follicle cells and dictyate oocytes in mammals. *In* "Current Problems in Germ Cell Differentiation" (A. McLaren and C. C. Wylie, eds.), p. 307. Cambridge University Press, London and New York.

Moor, R. M., and Osborn, J. C. (1983). Somatic control of protein synthesis in mammalian oocytes during maturation. *In* "Molecular Biology of Egg Maturation" (R. Porter and J. Whelan, eds.), Ciba Found. Symp., Vol. **98**, p. 178. Pitman, London.

Moore, J. A. (1941). Developmental rate of hybrid frogs. *J. Exp. Zool.* **86**, 405.

Morata, G., and Lawrence, P. A. (1975). Control of compartment development by the *engrailed* gene in *Drosophila*. *Nature (London)* **255**, 614.

Morgan, G. T., Macgregor, H. C., and Colman, A. (1980). Multiple ribosomal gene sites revealed by *in situ* hybridization of *Xenopus* rDNA to *Triturus* lampbrush chromosomes. *Chromosoma* **80**, 309.

Morgan, G. T., Reeder, R. H., and Bakken, A. H. (1983). Transcription in cloned spacers of *Xenopus laevis* ribosomal DNA. *Proc. Natl. Acad. Sci. U.S.A.* **80**, 6490.

Morgan, T. H. (1910). Cytological studies of centrifuged eggs. *J. Exp. Zool.* **9**, 593.

Morgan, T. H. (1927). "Experimental Embryology." Columbia Univ. Press, New York.

Morgan, T. H. (1934). "Embryology and Genetics." Columbia Univ. Press, New York.

Morgan, T. H., and Boring, A. M. (1903). The relation of the first plane of cleavage and the grey crescent to the median plane of the embryo of the frog. *Arch. Entwicklungmech. Org.* **16**, 680.

Morgan, T. H., and Spooner, G. B. (1909). The polarity of centrifuged eggs. *Arch. Entwicklungmech. Org.* **28**, 104.

Morris, G. F., and Marzluff, W. F. (1985). Synthesis of U1 RNA in isolated nuclei from sea urchin embryos: U1 RNA is initiated at the first nucleotide of the RNA. *Mol. Cell. Biol.* **5**, 1143.

Morris, P. W., and Rutter, W. J. (1976). Nucleic acid polymerizing enzymes in developing *Strongylocentrotus franciscanis* embryos. *Biochemistry* **15**, 3006.

Morrow, J. F. (1974). Mapping the SV40 chromosome by use of restriction enzymes. Ph.D. Thesis, Stanford University, Stanford, California.

Mortensen, T. (1921). "Studies of the Development and Larval Forms of Echinoderms." G. E. C. Gad, Copenhagen.

Moss, T. (1982). Transcription of cloned *Xenopus laevis* ribosomal DNA microinjected into *Xenopus* oocytes, and the identification of an RNA polymerase I promoter. *Cell* **30**, 835.

Moss, T. (1983). A transcriptional function for the repetitive ribosomal spacer in *Xenopus laevis*. *Nature (London)* **302**, 223.

Mous, J., Stunnenberg, H., Georgiev, O., and Birnstiel, M. L. (1985). Stimulation of sea urchin H2b histone gene transcription by a chromatin-associated protein fraction depends on gene sequences downstream of the transcription start site. *Mol. Cell Biol.* **5**, 2764.

Muhlach, W. L., and Schwalm, F. E. (1977). Utilization of uridine in developing ovarioles of the kelp fly, *Coelopa*. *J. Insect Physiol.* **23**, 931.

Müller, M. M., Carrasco, A. E., and De Robertis, E. M. (1984). A homeobox-containing gene expressed during oogenesis in *Xenopus*. *Cell* **39**, 157.

Müller, W. P. (1974). The lampbrush chromosomes of *Xenopus laevis* (Daudin). *Chromosoma* **47**, 283.

Müller-Holtkamp, F., Knipple, D. C., Seifert, E., and Jäckle, H. (1985). An early role of maternal mRNA in establishing the dorsoventral pattern in *pelle* mutant *Drosophila* embryos. *Dev. Biol.* **110**, 238.

Mulnard, J. G. (1965). Studies of regulation of mouse ova *in vitro*. *In* "Preimplantation Stages of Pregnancy" (G. E. W. Wolstenholme and M. O'Connor, eds.), Ciba Found. Symp. p. 123. Churchill, London.

Nadjicka, M., and Hillman, N. (1974). Ultrastructural studies of the mouse blastocyst substages. *J. Embryol. Exp. Morphol.* **32**, 675.

Nägeli, C. (1884). "Mechanisch-physiologische Theorie der Abstammungslehre." R. Oldenbourg, Munich and Leipzig.

Nakahashi, T., and Yamana, K. (1976). Biochemical and cytological examination on the initiation of ribosomal RNA synthesis during gastrulation of *Xenopus laevis*. *Dev., Growth Differ.* **18**, 329.

Nakamura, O. (1978). Epigenetic formation of the organizer. *In* "Organizer—A Milestone of a Half-Century from Spemann" (O. Nakamura and S. Toivonen, eds.), p. 179. Elsevier/North-Holland, Amsterdam.

Nakamura, O., and Takasaki, H. (1970). Further studies on the differentiation capacity of the dorsal marginal zone in the morula of *Triturus pyrrhogaster*. *Proc. Jpn. Acad.* **46**, 546.

Nakamura, O., Takasaki, H., and Ishihara, M. (1970a). Formation of the organizer from combinations of presumptive ectoderm and endoderm. *Proc. Jpn. Acad.* **47**, 313.

Nakamura, O., Takasaki, H., and Mizohata, T. (1970b). Differentiation during cleavage in *Xenopus laevis*. I. Acquisition of self-differentiation capacity of the dorsal marginal zone. *Proc. Jpn. Acad.* **46**, 694.

Nasmyth, K. A. (1982). Molecular genetics of yeast mating type. *Annu. Rev. Genet.* **16**, 439.

Natzle, J. E., and McCarthy, B. J. (1984). Regulation of *Drosophila* α- and β-tubulin genes during development. *Dev. Biol.* **104**, 187.

Neff, A. W., Wkahara, M., Jurand, A., and Malacinski, G. M. (1984). Experimental analyses of cytoplasmic rearrangements which follow fertilization and accompany symmetrization of inverted *Xenopus* eggs. *J. Embryol. Exp. Morphol.* **80**, 197.

Nemer, M. (1962). Interrelation of messenger polyribonucleotides and ribosomes in the sea urchin egg during embryonic development. *Biochem. Biophys. Res. Commun.* **8**, 511.

Nemer, M. (1975). Developmental changes in the synthesis of sea urchin embryo messenger RNA containing and lacking polyadenylic acid. *Cell* **6**, 559.

Nemer, M. (1979). *S*-adenosylmethionine pool size and turnover rate in sea urchin embryos determined from simultaneous measurements of amounts and specific activities. *Dev. Biol.* **68**, 643.

Nemer, M. (1986). An altered series of ectodermal gene expressions accompanying reversible suspension of differentiation in the zinc-animalized sea urchin embryo. *Dev. Biol.* **114**, 214.

Nemer, M., and Bard, S. G. (1963). Polypeptide synthesis in sea urchin embryogenesis: An examination with synthetic polyribonucleotides. *Science* **140**, 664.

Nemer, M., and Infante, A. A. (1965). Messenger RNA in early sea urchin embryos: Size classes. *Science* **150**, 217.

Nemer, M., and Lindsay, D. T. (1969). Evidence that the s-polysomes of early sea urchin embryos may be responsible for the synthesis of chromosomal histones. *Biochem. Biophys. Res. Commun.* **35**, 156.

Nemer, M., Graham, M., and Dubroff, L. M. (1974). Coexistence of nonhistone messenger RNA species lacking and containing polyadenylic acid in sea urchin embryos. *J. Mol. Biol.* **89**, 435.

Nemer, M., Dubroff, L. M., and Graham, M. (1975). Properties of sea urchin embryo messenger RNA containing and lacking poly(A). *Cell* **6**, 171.

Nemer, M., Ginzburg, I., Surrey, S., and Litwin, S. (1979). Rates of synthesis and turnover of 5' cap structures of hnRNA and mRNA and their changes during sea urchin development. *Dev. Genet.* **1**, 151.

Nemer, M., Travaglini, E. C., Rondinelli, E., and D'Alonzo, J. (1984). Developmental regulation, induction, and embryonic tissue specificity of sea urchin metallothionein gene expression. *Dev. Biol.* **102**, 471.

Nemer, M., Wilkinson, D. G., and Travaglini, E. C. (1985). Primary differentiation and ectoderm-specific gene expression in the animalized sea urchin embryo. *Dev. Biol.* **109**, 418.

Newport, G. (1854). Researches on the impregnation of the ovum in the Amphibia; and on the early stages of development of the embryo. *Philos. Trans. R. Soc. London* **144**, 229.

Newport, J., and Kirschner, M. (1982a). A major developmental transition in early *Xenopus* embryos: I. Characterization and timing of cellular changes at the midblastula stage. *Cell* **30**, 675.

Newport, J., and Kirschner, M. (1982b). A major developmental transition in early *Xenopus* embryos: II. Control of the onset of transcription. *Cell* **30**, 687.

Newport, J. W., and Kirschner, M. W. (1984). Regulation of the cell cycle during early *Xenopus* development. *Cell* **37**, 731.

Newport, J., Spann, T., Kanki, J., and Forbes, D. (1985). The role of mitotic factors in regulating the timing of the midblastula transition in *Xenopus*. *Cold Spring Harbor Symp. Quant. Biol.* **50**, 651.

Newrock, K. M., and Raff, R. A. (1975). Polar lobe specific regulation of translation in embryos of *Ilyanassa obsoleta*. *Dev. Biol.* **42**, 242.

Newrock, K. M., Alfageme, C. R., Nardi, R. V., and Cohen, L. H. (1978a). Histone changes during chromatin remodeling in embryogenesis. *Cold Spring Harbor Symp. Quant. Biol.* **42**, 421.

Newrock, K. M., Cohen, L. H., Hendricks, M. B., Donnelly, R. J., and Weinberg, E. S. (1978b). Stage-specific mRNAs coding for subtypes of H2b and H2b histones in the sea urchin embryo. *Cell* **14**, 327.

Newrock, K. M., Freedman, N., Alfageme, C. R., and Cohen, L. H. (1982). Isolation of sea urchin embryo histone H2a$_{\alpha 1}$ and immunological identification of other stage-specific H2a proteins. *Dev. Biol.* **89**, 248.

Nguyen-Huu, M. C., Stratmann, M., Groner, B., Wurtz, T., Land, H., Giesecke, K., Sippel, A. E., and Schütz, G. (1979). Chicken lysozyme gene contains several intervening sequences. *Proc. Natl. Acad. Sci. U.S.A.* **76**, 76.

Nicholas, J. S., and Hall, B. V. (1942). Experiments on developing rats. II. The development of isolated blastomeres and fused eggs. *J. Exp. Zool.* **90**, 441.

Nieuwkoop, P. D. (1969). The formation of the mesoderm in urodelean amphibians. II. The origin of the dorsoventral polarity of the mesoderm. *Wilhelm Roux's Arch. Entwicklungsmech. Org.* **163**, 298.

Nieuwkoop, P. D. (1977). Origin and establishment of embryonic polar axes in amphibian development. *Curr. Top. Dev. Biol.* **11**, 115.

Nieuwkoop, P. D., and Faber, J. (1956). "Normal Table of *Xenopus laevis* (Daudin)." North-Holland Publ., Amsterdam.

Nijhawan, P., and Marzluff, W. F. (1979). Metabolism of low molecular weight ribonucleic acids in early sea urchin embryos. *Biochemistry* **18**, 1353.

Niki, Y. (1984). Developmental analysis of the *grandchildless* (*gs(1)N26*) mutation in *Drosophila melanogaster*: Abnormal cleavage patterns and defects in pole cell formation. *Dev. Biol.* **103**, 182.

Nilsson, M. O., and Hultin, T. (1974). Characteristics and intracellular distribution of messengerlike RNA in encysted embryos of *Artemia salina*. *Dev. Biol.* **38**, 138.

Nishida, H., and Satoh, N. (1983). Cell lineage analysis in ascidian embryos by intracellular injection of a tracer enzyme. I. Up to the eight-cell stage. *Dev. Biol.* **99**, 382.

Nishida, H., and Satoh, N. (1985). Cell lineage analysis in ascidian embryos by intracellular injection of a tracer enzyme. II. The 16- and 32-cell stages. *Dev. Biol.* **110**, 440.

Nishikura, K., and De Robertis, E. M. (1981). RNA processing in microinjected *Xenopus* oocytes. Sequential addition of base modifications in a spliced transfer RNA. *J. Mol. Biol.* **145**, 405.

Noble, J. S., Haight, M., and Pasternak, J. (1976). Relative ribosomal RNA cistron multiplicity in oocyte and postembryonic stages of the eutelic nematode *Panagrellus silusiae*. *Mol. Gen. Genet.* **147**, 343.

Nokkala, S., and Puro, J. (1976). Cytological evidence for a chromocenter in *Drosophila melanogaster* oocytes. *Hereditas* **83**, 265.

Noronha, J. M., Sheys, G. H., and Buchanan, J. M. (1972). Induction of a reductive pathway for deoxyribonucleotide synthesis during early embryogenesis of the sea urchin. *Proc. Natl. Acad. Sci. U.S.A.* **69**, 2006.

Novikoff, A. B. (1938). Embryonic determination in the annelid *Sabellaria vulgaris*. II. Transplantation of polar lobes and blastomeres as a test of their inducing capacities. *Biol. Bull. (Woods Hole, Mass.)* **74**, 211.

Nüsslein-Volhard, C. (1977). Genetic analysis of pattern formation in the embryo of *Drosophila melanogaster*. Characterization of the maternal effect mutant *bicaudal*. *Wilhelm Roux's Arch. Dev. Biol.* **183**, 249.

Nüsslein-Volhard, C. (1979). Maternal effect mutations that alter the spatial coordinates of the embryo of *Drosophila melanogaster*. *In* "Determinants of Spatial Organization" (S. Subtelny and I. R. Konigsberg, eds.), p. 185. Academic Press, New York.

Nüsslein-Volhard, C., and Wieschaus, E. (1980). Mutations affecting segment number and polarity in *Drosophila*. *Nature (London)* **287**, 795.

Nüsslein-Volhard, C., Lohs-Schardin, M., Sander, K. and Cremer, C. (1980). A dorsoventral shift of embryonic primordia in a new maternal-effect mutant of *Drosophila*. *Nature (London)* **283**, 474.

Nüsslein-Volhard, C., Wieschaus, E., and Kluding, H. (1984). Mutations affecting the pattern of the larval cuticle in *Drosophila melanogaster*. I. Zygotic loci on the second chromosome. *Wilhelm Roux's Arch. Dev. Biol.* **193**, 267.

Oi, V. T., Morrison, S. L., Herzenberg, L. A., and Berg, P. (1983). Immunoglobulin gene expression in transformed lymphoid cells. *Proc. Natl. Acad. Sci. U.S.A.* **80**, 825.

Okada, M., and Togashi, S. (1985). Isolation of a factor inducing pole cell formation from *Drosophila* embryos. *Int. J. Invertebr. Reprod. Dev.* **8**, 207.

Okada, M., Kleinman, I. A., and Schneiderman, H. A. (1974a). Repair of a genetically-caused defect in oogenesis in *Drosophila melanogaster* by transplantation of cytoplasm from wild-type eggs and by injection of pyrimidine nucleosides. *Dev. Biol.* **37**, 55.

Okada, M., Kleinman, I. A., and Schneiderman, H. A. (1974b). Restoration of fertility in sterilized *Drosophila* eggs by transplantation of polar cytoplasm. *Dev. Biol.* **37**, 43.

Okada, M., Kleinman, I. A. and Schneiderman, H. A. (1974c). Chimeric *Drosophila* adults produced by transplantation of nuclei into specific regions of fertilized eggs. *Dev. Biol.* **39**, 286.

Okazaki, K. (1975a). Normal development to metamorphosis. *In* "The Sea Urchin Embryo: Biochemistry and Morphogenesis" (G. Czihak, ed.), p. 177. Springer-Verlag, Berlin and New York.

Okazaki, K. (1975b). Spicule formation by isolated micromeres of the sea urchin embryo. *Am. Zool.* **15**, 567.

Okkelberg, P. (1921). The early history of the germ cells in the brook lamprey, *Entosphenus wilderi* (Gage), up to and including the period of sex differentiation. *J. Morphol.* **35**, 1.

Olds, P. J., Stern, S., and Biggers, J. D. (1973). Chemical estimates of the RNA and DNA contents of the early mouse embryo. *J. Exp. Zool.* **186**, 39.

O'Melia, A. F., and Villee, C. A. (1972). *De novo* synthesis of transfer and 5S RNA^{1f} in cleaving sea urchin embryos. *Nature (London) New Biol.* **239**, 51.

Ornitz, D. M., Palmiter, R. D., Hammer, R. E., Brinster, R. L., Swift, G. H., and MacDonald, R. J. (1985). Specific expression of an elastase-human growth hormone fusion gene in pancreatic acinar cells of transgenic mice. *Nature (London)* **313**, 600.

Osawa, S., and Hayashi, Y. (1953). Ribonucleic acid and protein in the growing oocytes of *Triturus pyrrhogaster*. *Science* **118**, 84.

Ott, M.-O., Sperling, L., Cassio, D., Levilliers, J., Sala-Trepat, J., and Weiss, M. C. (1982). Undermethylation at the 5' end of the albumin gene is necessary but not sufficient for albumin production by rat hepatoma cells in culture. *Cell* **30**, 825.

Otto, J. J., and Schroeder, T. E. (1984a). Microtubule arrays in the cortex and near the germinal vesicle of immature starfish oocytes. *Dev. Biol.* **101**, 274.

Otto, J. J., and Schroeder, T. E. (1984b). Assembly–disassembly of actin bundles in starfish oocytes: An analysis of actin-associated proteins in the isolated cortex. *Dev. Biol.* **101**, 263.

Ouwenweel, W. J. (1976). Developmental genetics of homeosis. *Adv. Genet.* **18**, 179.

Pachnis, V., Belayew, A., and Tilghman, S. M. (1984). Locus unlinked to α-fetoprotein under the control of the murine *raf* and *Rif* genes. *Proc. Natl. Acad. Sci. U.S.A.* **81**, 5523.

Paglia, L. M., Berry, S. J., and Kastern, W. H. (1976a). Messenger RNA synthesis, transport and storage in silkmoth ovarian follicles. *Dev. Biol.* **51**, 173.

Paglia, L. M., Kastern, W. H., and Berry, S. J. (1976b). Messenger ribonucleoprotein particles in silkmoth oogenesis. *Dev. Biol.* **51**, 182.

Palmiter, R. D. (1973). Rate of ovalbumin messenger ribonucleic acid synthesis in the oviduct of estrogen-primed chicks. *J. Biol. Chem.* **248**, 8260.

Palmiter, R. D., Norstedt, G., Gelinas, R. E., Hammer, R. E., and Brinster, R. L. (1983). Metallothionein-human GH fusion genes stimulate growth of mice. *Science* **222**, 809.

Pander, C. H. (1817). "Beiträge zur Entwickelungsgeschichte des Hühnchens im Ei." H. L. Brönner, Würzburg.

Pardue, M. L., and Gall, J. G. (1969). Molecular hybridization of radioactive DNA to the DNA of cytological preparation. *Proc. Natl. Acad. Sci. U.S.A.* **64**, 600.

Parker, C. S., and Roeder, R. G. (1977). Selective and accurate transcription of the *Xenopus laevis* 5S RNA genes in isolated chromatin by purified RNA polymerase III. *Proc. Natl. Acad. Sci. U.S.A.* **74**, 44.

Parker, C. S., and Topol, J. (1984). A *Drosophila* RNA polymerase II transcription factor binds to the regulatory site of an hsp 70 gene. *Cell* **37**, 273.

Pasteels, J. (1938). Recherches sur les facteurs initiaux de la morphogenèse chez les Amphibiens Anoures. I. Résultats de l'expérience de Schultze et leur interprétation. *Arch. Biol.* **49**, 629.

Pasteels, J. (1939). Recherches sur les facteurs initiaux de la morphogénèse chez les Amphibiens Anoures. II. Lèvres blastoporales successives dans un même oeuf. *Arch. Biol.* **50**, 291.

Pasteels, J. (1942). New observations concerning the maps of presumptive areas of the young amphibian gastrula (*Amblystoma* and *Discoglossus*). *J. Exp. Zool.* **89**, 255.

Pasteels, J. (1946). Sur la structure de l'oeuf insegmente d'axolotl et l'origine des prodromes morphogénétiques. *Acta Anat.* **2**, 1.

Patterson, J. B., and Stafford, D. W. (1971). Characterization of sea urchin ribosomal satellite deoxyribonucleic acid. *Biochemistry* **10**, 2775.

Payvar, F., DeFranco, D., Firestone, G. L., Edgar, B., Wrange, Ö., Okret, S., Gustafsson, J.-Å., and Yamamoto, K. R. (1983). Sequence-specific binding of glucocorticoid receptor to MTV DNA at sites within and upstream of the transcribed region. *Cell* **35**, 381.

Peacock, W. J. (1965). Chromosome replication. *Natl. Cancer Inst. Monogr.* No. 18, 101.

Pedersen, R. A., Meneses, J., Spindle, A., Wu, K., and Galloway, S. M. (1985). Cytochrome P-450 metabolic activity in embryonic and extraembryonic tissue lineages of mouse embryos. *Proc. Natl. Acad. Sci. U.S.A.* **82**, 3311.

Pederson, T. (1983). Nuclear RNA-protein interactions and messenger RNA processing. *J. Cell Biol.* **97**, 1321.

Pederson, T. and Munroe, S. H. (1981). Ribonucleoprotein organization of eukaryotic RNA. XV. Different nucleoprotein structures of globin messenger RNA sequences in nuclear and polyribosomal ribonucleoprotein particles. *J. Mol. Biol.* **150**, 509.

Pehrson, J. R., and Cohen, L. H. (1984). Embryonal histone H1 subtypes of the sea urchin *Strongylocentrotus purpuratus*: Purification, characterization, and immunological comparison with H1 subtypes of the adult. *Biochemistry* **23**, 6761.

Pehrson, J. R., and Cohen, L. H. (1986). The fate of the small micromeres in sea urchin development *Dev. Biol.* **113**, 522.

Pelham, H. R. B. (1982). A regulatory upstream promoter element in the *Drosophila* hsp 70 heat-shock gene. *Cell* **30**, 517.

Pelham, H. R. B., and Bienz, M. (1982). A synthetic heat-shock promoter element confers heat-inducibility on the herpes simplex virus thymidine kinase gene. *EMBO J.* **1**, 1473.

Pelham, H. R. B., and Brown, D. D. (1980). A specific transcription factor that can bind either the 5S RNA gene or 5S RNA. *Proc. Natl. Acad. Sci. U.S.A.* **77**, 4170.

Penners, A. (1926). Experimentelle Untersuchungen zum Determinationsproblem am Keim von *Tubifex rivulorum* Lam. II. Die Entwicklung teilweise abgetöteter Keime. *Z. Wiss. Zool.* **127**, 1.

Penners, A., and Schleip, W. (1928a). Die Entwicklung der Schultze'schen Doppelbildungen aus dem Ei von *Rana fusca*. Teil I–IV. *Z. Wiss. Zool.* **130**, 305.

Penners, A., and Schleip, W. (1928b). Die Entwicklung der Schultzes'chen Doppelbildungin aus dem Ei von *Rana fusca*. Teil V and VI. *Z. Wiss. Zool.* **131**, 1.

Pennock, D. G., and Reeder, R.H. (1984). *In vitro* methylation of *Hpa*III sites in *Xenopus laevis* rDNA does not affect its transcription in oocytes. *Nucleic Acids Res.* **12**, 2225.

Perkowska, E., Macgregor, H. C., and Birnstiel, M. L. (1968). Gene amplification in the oocyte nucleus of mutant and wild-type *Xenopus laevis*. *Nature (London)* **217**, 649.

Perlman, S., and Rosbash, M. (1978). Analysis of *Xenopus laevis* ovary and somatic cell polyadenylated RNA by molecular hybridization. *Dev. Biol.* **63**, 197.

Perlman, S. M., Ford, P. J., and Rosbash, M. M. (1977). Presence of tadpole and adult globin RNA sequences in oocytes of *Xenopus laevis*. *Proc. Natl. Acad. Sci. U.S.A.* **74**, 3835.

Perrimon, N., and Gans, M. (1983). Clonal analysis of the tissue specificity of recessive female sterile mutations of *Drosophila melanogaster* using a dominant female sterile mutation, *Fs(1)K1237*. *Dev. Biol.* **100**, 365.

Perrimon, N., Engstrom, L., and Mahowald, A. P. (1984). The effects of zygotic lethal mutations on female germ line functions in *Drosophila*. *Dev. Biol.* **105**, 404.

Perrimon, N., Engstrom, L., and Mahowald, A. P. (1985). A pupal lethal mutation with a paternally influenced maternal effect on embryonic development in *Drosophila melanogaster*. *Dev. Biol.* **110**, 480.

Perry, H. E., and Melton, D. A. (1983). A rapid increase in acetylcholinesterase mRNA during ascidian embryogenesis as demonstrated by microinjection into *Xenopus laevis* oocytes. *Cell Differ.* **13**, 233.

Perry, M., Thomsen, G. H., and Roeder, R. G. (1985). Genomic organization and nucleotide sequence of two distinct histone gene clusters from *Xenopus laevis*. Identification of novel conserved upstream sequence elements. *J. Mol. Biol.* **185**, 479.

Pestell, R. Q. W. (1975). Microtubule protein synthesis during oogenesis and early embryogenesis in *Xenopus laevis*. *Biochem. J.* **145**, 527.

Peterson, R. C., Doering, J. L., and Brown, D. D. (1980). Characterization of two *Xenopus* somatic 5S DNAs and one minor oocyte-specific 5S DNA. *Cell* **20**, 131.

Phillips, C. R. (1982). The regional distribution of poly(A) and total RNA concentrations during early *Xenopus* development. *J. Exp. Zool.* **223**, 265.

Phillips, C. R. (1985). Spatial changes in poly(A) concentrations during early embryogenesis in *Xenopus laevis*: Analysis by *in situ* hybridization. *Dev. Biol.* **109**, 299.

Piatigorsky, J. (1975). Gametogenesis. *In* "The Sea Urchin Embryo: Biochemistry and Morphogenesis" (G. Czihak, ed.), p. 42. Springer-Verlag, Berlin and New York.

Piatigorsky, J., Chepelinsky, A. B., Hejtmancik, J. F., Borrás, T., Das, G. C., Hawkins, J. W., Zelenka, P. S., King, C. R., Beebe, D. C., and Nickerson, J. M. (1984). Expression of crystallin gene families in the differentiating eye lens. *In* "Molecular Biology of Development" (E. H. Davidson and R. A. Firtel, eds.), p. 331. Alan R. Liss, New York.

Picard, B., and Wegnez, M. (1979). Isolation of a 7S particle from *Xenopus laevis* oocytes: A 5S RNA-protein complex. *Proc. Natl. Acad. Sci. U.S.A.* **76**, 241.

Picard, B., le Maire, M., Wegnez, M., and Denis, H. (1980). Biochemical research on oogenesis. Composition of the 42S storage particles of *Xenopus laevis* oocytes. *Eur. J. Biochem.* **109**, 359.

Pierandrei-Amaldi, P., Campioni, N., Beccari, E., Bozzoni, I., and Amaldi, F. (1982). Expression of ribosomal-protein genes in *Xenopus laevis* development. *Cell* **30**, 163.

Pierandrei-Amaldi, P., Beccari, E., Bozzoni, I., and Amaldi, F. (1985). Ribosomal protein production in normal and anucleate *Xenopus* embryos: Regulation at the posttranscriptional and translational levels. *Cell* **42**, 317.

Pietruschka, F., and Bier, K. (1972). Autoradiographische Untersuchengen zur RNS-und Protein-Synthesis in der fruehen Embryogenese von *Musca domestica*. *Wilhelm Roux's Arch. Entwicklungsmech. Org.* **169**, 56.

Pikó, L. (1970). Synthesis of macromolecules in early mouse embryos cultured *in vitro*: RNA, DNA, and a polysaccharide component. *Dev. Biol.* **21**, 257.

Pikó, L., and Chase, D. G. (1973). Role of the mitochondrial genome during early development in mice. *J. Cell Sci.* **58**, 357.

Pikó, L., and Clegg, K. B. (1982). Quantitative changes in total RNA, total poly(A), and ribosomes in early mouse embryos. *Dev. Biol.* **89**, 362.

Pikó, L., and Matsumoto, L. (1976). Number of mitochondria and some properties of mitochondrial DNA in the mouse egg. *Dev. Biol.* **49**, 1.

Piperno, G., Huang, B., and Luck, D. J. L. (1977). Two-dimensional analysis of flagellar proteins from wild-type and paralyzed mutants of *Chlamydomonas reinhardtii*. *Proc. Natl. Acad. Sci. U.S.A.* **74**, 1600.

Pittman, D., and Ernst, S. G. (1984). Developmental time, cell lineage, and environment regulate the newly synthesized proteins in sea urchin embryos. *Dev. Biol.* **106**, 236.

Poccia, D., Salik, J., and Krystal, G. (1981). Transitions in histone variants of the male pronucleus following fertilization and evidence for a maternal store of cleavage-stage histones in the sea urchin egg. *Dev. Biol.* **82**, 287.

Poccia, D., Wolff, R., Kragh, S., and Williamson, P. (1985). RNA synthesis in male pronuclei of the sea urchin. *Biochim. Biophys. Acta* **824**, 349.

Pollack, S. B., and Telfer, W. H. (1969). RNA in *Cecropia* moth ovaries: Sites of synthesis, transport, and storage. *J. Exp. Zool.* **170**, 1.

Poole, S. J., Kauvar, L. M., Drees, B., and Kornberg, T. (1985). The *engrailed* locus of *Drosophila*: Structural analysis of an embryonic transcript. *Cell* **40**, 37.

Porter, R., and Whelan, J., eds. (1983). "Molecular Biology of Egg Maturation," Ciba Found. Symp., Vol. 98, Pitman, London.

Posakony, J. W., Scheller, R. H., Anderson, D. M., Britten, R. J., and Davidson, E. H. (1981). Repetitive sequences of the sea urchin genome. Nucleotide sequences of cloned repeat elements. *J. Mol. Biol.* **149**, 41.

Posakony, J. W., Flytzanis, C. N., Britten, R. J., and Davidson, E. H. (1983). Interspersed sequence organization and developmental representation of cloned poly(A) RNAs from sea urchin eggs. *J. Mol. Biol.* **167**, 361.

Poulson, D. F. (1937). Chromosomal deficiencies and the embryonic development of *Drosophila melanogaster*. *Proc. Natl. Acad. Sci. U.S.A.* **23**, 133.

Poulson, D. F. (1950). Histogenesis, organogenesis, and differentiation in the embryo of *Drosophila melanogaster* Meigen. *In* "Biology of *Drosophila*" (M. Demerec, ed.), p. 168. Wiley, New York.

Pratt, H. P. M., Ziomek, C. A., Reeve, W. J. D., and Johnson, M. H. (1982). Compaction of the mouse embryo: An analysis of its components. *J. Embryol. Exp. Morphol.* **70**, 113.

Pratt, H. P. M., Bolton, V. N., and Gudgeon, K. A. (1983). The legacy from the oocyte and its role in controlling early development of the mouse embryo. *In* "Molecular Biology of Egg Maturation" (R. Porter and J. Whelan, eds.), Ciba Found. Symp. Vol. 98, p. 197. Pitman, London.

Preiss, A., Rosenberg, U. B., Kienlin, A., Seifert, E., and Jäckle, H. (1985). Molecular genetics of *Krüppel*, a gene required for segmentation of the *Drosophila* embryo. *Nature (London)* **313**, 27.

Probst, E., Kressman, A., and Birnstiel, M. L. (1979). Expression of sea urchin histone genes in the oocyte of *Xenopus laevis*. *J. Mol. Biol.* **135**, 709.

Pruitt, S. C., and Reeder, R. H. (1984a). Effect of topological constraint on transcription of ribosomal DNA in *Xenopus* oocytes. Comparison of plasmid and endogenous genes. *J. Mol. Biol.* **174**, 121.

Pruitt, S. C., and Reeder, R. H. (1984b). Effect of intercalating agents on RNA polymerase I promoter selection in *Xenopus laevis*. *Mol. Cell. Biol.* **4**, 2851.

Pukkila, P. J. (1975). Identification of the lampbrush chromosome loops which transcribe 5S ribosomal RNA in *Notophthalmus (Triturus) viridescens*. *Chromosoma* **53**, 71.

Rabagliati, M. R., Weeks, D. L., Harvey, R. P., and Melton, D. A. (1985). Identification and cloning of localized maternal RNAs from *Xenopus* eggs. *Cell* **42**, 769.

Raff, R. A. (1975). Regulation of microtubule synthesis and utilization during early embryonic development of the sea urchin. *Am. Zool.* **15**, 661.

Raff, R. A., and Kaufman, T. C. (1983). "Embryos, Genes, and Evolution. The Developmental–Genetic Basis of Evolutionary Change." Macmillan, New York.

Raff, R. A., and Kaumeyer, J. F. (1973). Soluble microtubule proteins of the sea urchin embryo: Partial characterization of the proteins and behavior of the pool in early development. *Dev. Biol.* **32**, 309.

Raff, R. A., Brandis, J. W., Green, L. H., Kaumeyer, J. F., and Raff, E. C. (1975). Microtubule protein pools in early development. *Ann. N.Y. Acad. Sci.* **253**, 304.

Raff, R. A., Anstrom, J. A., Huffman, C. J., Leaf, D. S., Loo, J.-H., Showman, R. M., and Wells, D. E. (1984). Origin of a gene regulatory mechanism in the evolution of echinoderms. *Nature (London)* **310**, 312.

Ragg, H., and Weissmann, C. (1983). Not more than 117 base pairs of 5'-flanking sequence are required for inducible expression of a human IFN-α gene. *Nature (London)* **303**, 439.

Rankin, J. K., and Darlington, G. J. (1979). Expression of human hepatic genes in mouse hepatoma-human amniocyte hybrids. *Somatic Cell Genet.* **5**, 1.

Rasch, E. M., Barr, H. J., and Rasch, R. W. (1971). The DNA content of sperm of *Drosophila melanogaster*. *Chromosoma* **33**, 1.

Rattenburg, J. C., and Berg, W. E. (1954). Embryonic segregation during early development of *Mytilus edulis*. *J. Morphol.* **95**, 393.

Rau, K.-G., and Kalthoff, K. (1980). Complete reversal of anteroposterior polarity in a centrifuged insect embryo. *Nature (London)* **287**, 635.

Raven, C. P. (1961). "Oogenesis: The Storage of Developmental Information." Pergamon, Oxford.

Rawls, J. M., and Fristrom, J. W. (1975). A complex genetic locus that controls the first three steps of pyrimidine biosynthesis in *Drosophila*. *Nature (London)* **255**, 738.

Razin, A., and Riggs, A. D. (1980). DNA methylation and gene function. *Science* **210**, 604.

Reeder, R. H., and Roan, J. G. (1985). The mechanism of nucleolar dominance in *Xenopus* hybrids. *Cell* **38**, 39.

Reeder, R. H., Roan, J. G., and Dunaway, M. (1983). Spacer regulation of *Xenopus* ribosomal gene transcription: Competition in oocytes. *Cell* **35**, 449.

Reeve, W. J. D. (1981). Cytoplasmic polarity develops at compaction in rat and mouse embryos. *J. Embryol. Exp. Morphol.* **62**, 351.

Regier, J. C., and Kafatos, F. C. (1977). Absolute rates of protein synthesis in sea urchins with specific activity measurements of radioactive leucine and leucyl-tRNA. *Dev. Biol.* **57**, 270.

Regier, J. C., Mazur, G. D., and Kafatos, F. C. (1980). The silkmoth chorion: Morphological and biochemical characterization of four surface regions. *Dev. Biol.* **76**, 286.

Regier, J. C., Mazur, G. D., Kafatos, F. C., and Paul, M. (1982). Morphogenesis of silkmoth chorion: Initial framework formation and its relation to synthesis of specific proteins. *Dev. Biol.* **92**, 159.

Render, J. A. (1983). The second polar lobe of the *Sabellaria cementarium* embryo plays an inhibitory role in apical tuft formation. *Wilhelm Roux's Arch. Dev. Biol.* **192**, 120.

Render, J. A., and Guerrier, P. (1984). Size regulation and morphogenetic localization in the *Dentalium* polar lobe. *J. Exp. Zool.* **232**, 79.

Renkawitz, R., and Kunz, W. (1975). Independent replication of the ribosomal RNA genes in the polytrophic-meroistic ovaries of *Calliphora erythrocephala*, *Drosophila hydei*, and *Sarcophaga barbata*. *Chromosoma* **53**, 131.

Renkawitz, R., Beug, H., Graf, T., Matthias, P., Grez, M., and Schütz, G. (1982). Expression of a chicken lysozyme recombinant gene is regulated by progesterone and dexamethasone after microinjection into oviduct cells. *Cell* **31**, 167.

Renkawitz, R., Schütz, G., von der Ahe, D., and Beato, M. (1984). Sequences in the promoter region of the chicken lysozyme gene required for steroid regulation and receptor binding. *Cell* **37**, 503.

Reverberi, G. (1971a). Ctenophores. *In* "Experimental Embryology of Marine and Fresh-Water Invertebrates" (G. Reverberi, ed.), p. 83. North-Holland Publ., Amsterdam.

Reverberi, G. (1971b). Ascidians. *In* "Experimental Embryology of Marine Fresh-Water Invertebrates" (G. Reverberi, ed.), p. 507. North-Holland Publ., Amsterdam.

Reverberi, G. (1971c). Annelids. *In* "Experimental Embryology of Marine Fresh-Water Invertebrates" (G. Reverberi, ed.), p. 126. North-Holland Publ., Amsterdam.

Reyer, R. W. (1956). Lens regeneration from homoplastic and heteroplastic implants of dorsal iris into the eye chamber of *Triturus viridescens* and *Amblystoma punctatum*. *J. Exp. Zool.* **133**, 145.

Ribbert, D. (1979). Chromomeres and puffing in experimentally induced polytene chromosomes of *Calliphora erythrocephala*. *Chromosoma* **74**, 269.

Ribbert, D., and Bier, K. (1969). Multiple nucleoli and enhanced nucleolar activity in the nurse cells of the insect ovary. *Chromosoma* **27**, 178.

Ribbert, D., and Kunz, W. (1969). Lampenbürstenchromosomen in den Oocytenkernen von *Sepia officinalis. Chromosoma* **28**, 93.

Rice, T. B., and Garen, A. (1975). Localized defects of blastoderm formation in maternal effect mutants of *Drosophila. Dev. Biol.* **43**, 277.

Richards, A. (1931). "Outline of Comparative Embryology." Wiley, New York.

Richards, G., Cassab, A., Bourouis, M., Jarry, B., and Dissous, C. (1983). The normal developmental regulation of a cloned *sgs3* 'glue' gene chromosomally integrated in *Drosophila melanogaster* by P element transformation. *EMBO J.* **2**, 2137.

Richter, J. D., and Smith, L. D. (1981). Differential capacity for translation and lack of competition between mRNAs that segregate to free and membrane-bound polysomes. *Cell* **27**, 183.

Richter, J. D., and Smith, L. D. (1983). Developmentally regulated RNA binding proteins during oogenesis in *Xenopus laevis. J. Biol. Chem.* **258**, 4864.

Richter, J. D., and Smith, L. D. (1984). Reversible inhibition of translation by *Xenopus* oocyte-specific proteins. *Nature (London)* **309**, 378.

Richter, J. D., Wasserman, W. J., and Smith, L. D. (1982). The mechanism for increased protein synthesis during *Xenopus* oocyte maturation. *Dev. Biol.* **89**, 159.

Richter, J. D., Evers, D. C., and Smith, L. D. (1983). The recruitment of membrane-bound mRNAs for translation in microinjected *Xenopus* oocytes. *J. Biol. Chem.* **258**, 2614.

Richter, J. D., Anderson, D. M., Davidson, E. H., and Smith, L. D. (1984). Interspersed poly(A) RNAs of amphibian oocytes are not translatable. *J. Mol. Biol.* **173**, 227.

Rinaldi, A. M., and Giudice, G. (1985). Nuclear-cytoplasmic interactions in early development. *In* "Biology of Fertilization," (C. B. Metz and A. Monroy, eds.), Vol. 3, p. 367. Academic Press, Orlando, Florida.

Rinaldi, A. M., and Monroy, A. (1969). Polyribosome formation and RNA synthesis in the early postfertilization stages of the sea urchin egg. *Dev. Biol.* **19**, 73.

Rinaldi, A. M., De Leo, G., Arzone, A., Salcher, I., Storace, A., and Mutolo, V. (1979). Biochemical and electron microscope evidence that cell nucleus negatively controls mitochondrial genomic activity in early sea urchin development. *Proc. Natl. Acad. Sci. U.S.A.* **76**, 1916.

Ripley, S., and Kalthoff, K. (1983). Changes in the apparent localization of anterior determinants during early embryogenesis (*Smittia* spec., Chironomidae, Diptera). *Wilhelm Roux's Arch. Dev. Biol.* **192**, 353.

Ripoll, P. (1977). Behavior of somatic cells homozygous for zygotic lethals in *Drosophila melanogaster. Genetics* **86**, 357.

Ritossa, F. (1976). The *bobbed* locus. *In* "The Genetics and Biology of *Drosophila*" (M. Ashburner and E. Novitski, eds.), Vol. 1b, p. 801. Academic Press, London.

Roark, M., Mahoney, P. A., Graham, M. L., and Lengyel, J. A. (1985). Blastoderm-differential and blastoderm-specific genes of *Drosophila melanogaster. Dev. Biol.* **109**, 476.

Robbins, L. G. (1980). Maternal-zygotic lethal interactions in *Drosophila* melanogaster: The effects of deficiencies in the *zeste-white* region of the X chromosome. *Genetics* **96**, 187.

Robbins, L. G. (1983). Maternal-zygotic lethal interactions in *Drosophila* melanogaster: *zeste-white* region single-cistron mutations. *Genetics* **103**, 633.

Roberts, S. B., Weisser, K. E., and Childs, G. (1984). Sequence comparisons of nonallelic late histone genes and their early stage counterparts. Evidence for gene conversion within the sea urchin late stage gene family. *J. Mol. Biol.* **174**, 647.

Roccheri, M. C., Di Bernardo, M. G., and Giudice, G. (1981). Synthesis of heat shock proteins in developing sea urchins. *Dev. Biol.* **83**, 173.

Roccheri, M. C., Sconzo, G., La Rosa, M., Oliva, D., Abrignani, A., and Giudice, G. (1986). Response to heat shock of different sea urchin species. *Cell Differ.* **18**, 131.

Rochaix, J.-D., Bird, A., and Bakken, A. (1974). Ribosomal RNA gene amplification by rolling circles. *J. Mol. Biol.* **87**, 473.

Rodgers, W. H., and Gross, P. R. (1978). Inhomogeneous distribution of egg RNA sequences in the early embryo. *Cell* **14**, 279.

Roeder, R. G. (1974). Multiple forms of deoxyribonucleic acid-dependent ribonucleic acid polymerase in *Xenopus laevis*. Levels of activity during oocyte and embryonic development. *J. Biol. Chem.* **249**, 249.

Roeder, R. G., and Rutter, W. J. (1970). Multiple RNA polymerases and ribonucleic acid synthesis during sea urchin development. *Biochemistry* **9**, 2543.

Romanoff, A. L. (1960). "The Avian Embryo," p. 13. Macmillan, New York.

Rosbash, M. (1981). A comparison of *Xenopus laevis* oocyte and embryo mRNA. *Dev. Biol.* **87**, 319.

Rosbash, M., and Ford, P. J. (1974). Polyadenylic acid-containing RNA in *Xenopus laevis* oocytes. *J. Mol. Biol.* **85**, 87.

Rosbash, M., Ford, P. J., and Bishop, J. O. (1974). Analysis of the C-value paradox by molecular hybridization. *Proc. Natl. Acad. Sci. U.S.A.* **71**, 3746.

Rosenberg, U. B., Preiss, A., Seifert, E., Jäckle, H., and Knipple, D. C. (1985). Production of phenocopies by *Krüppel* antisense RNA injection into *Drosophila* embryos. *Nature (London)* **313**, 703.

Rosenthal, E. T., and Wilt, F. H. (1986). Patterns of maternal messenger RNA accumulation and adenylation during oogenesis in *Urechis caupo*. *Dev. Biol.* In press.

Rosenthal, E. T., Hunt, T., and Ruderman, J. V. (1980). Selective translation of mRNA controls the pattern of protein synthesis during early development of the surf clam *Spisula solidissima*. *Cell* **20**, 487.

Rosenthal, E. T., Brandhorst, B. P., and Ruderman, J. V. (1982). Translationally mediated changes in patterns of protein synthesis during maturation of starfish oocytes. *Dev. Biol.* **91**, 215.

Rosenthal, E. T., Tansey, T. R., and Ruderman, J. V. (1983). Sequence-specific adenylations and deadenylations accompany changes in the translation of maternal messenger RNA after fertilization of *Spisula* oocytes. *J. Mol. Biol.* **166**, 309.

Rossant, J., and Lis, W. T. (1979). Potential of isolated mouse inner cell masses to form trophectoderm derivatives in vivo. *Dev. Biol.* **70**, 255.

Rossant, J., and Vijh, K. M. (1980). Ability of outside cells from preimplantation mouse embryos to form inner cell mass derivatives. *Dev. Biol.* **76**, 475.

Roux, W. (1883). "Über die Bedeutung der Kerntheilungsfiguren. Eine hypothetische Erörterung." Engelmann, Leipzig.

Roux, W. (1887). Beiträge zur Entwickelungsmechanik des Embryo. 4. Die Richtungsbestimmung der Medianebene des Froschembryo durch die Copulationsrichtung des Eikernes und Spermakernes. *Arch. Mikrosk. Anat.* **29**, 344.

Roux, W. (1888). Beiträge zur Entwickelungsmechanik des Embryo. 5. Über die künstliche Hervorbringung halber Embryonen durch Zerstörung einer der beiden ersten Furchungskugeln, sowie über die Nachentwickelung (Postgeneration) der fehlenden Körperhälfte. *Arch. Pathol. Anat. Physiol. Klin. Med.* **114**, 113.

Roux, W. (1903). Über die Ursachen der Bestimmung der Hauptrichtungen des Embryo im Froschei. *Anat. Anz.* **23**, 65.

Rozek, C. E., and Davidson, N. (1983). *Drosophila* has one myosin heavy-chain gene with three developmentally regulated transcripts. *Cell* **32**, 23.

Rubin, G. M., and Spradling, A. C. (1982). Genetic transformation of *Drosophila* with transposable element vectors. *Science* **218**, 348.

Rückert, J. (1892). Zur Entwickelungsgeschichte des Ovarialeies bei Selachiern. *Anat. Anz.* **7**, 107.

Ruddell, A., and Jacobs-Lorena, M. (1983). Abrupt decline in the rate of accumulation of total protein and yolk in postvitellogenic egg chambers of *Drosophila*. *Wilhelm Roux's Arch. Dev. Biol.* **192**, 189.

Ruddell, A., and Jacobs-Lorena, M. (1984). Preferential expression of actin genes during oogenesis of *Drosophila*. *Dev. Biol.* **105**, 115.

Ruddell, A., and Jacobs-Lorena, M. (1985). Biphasic pattern of histone gene expression during *Drosophila* oogenesis. *Proc. Natl. Acad. Sci. U.S.A.* **82**, 3316.

Ruderman, J. V., and Gross, P. R. (1974). Histones and histone synthesis in sea urchin development. *Dev. Biol.* **36**, 286.

Ruderman, J. V., and Pardue, M. L. (1977). Cell free translation analysis of messenger RNA in echinoderm and amphibian early development. *Dev. Biol.* **60**, 48.

Ruderman, J. V., and Schmidt, M. R. (1981). RNA transcription and translation in sea urchin oocytes and eggs. *Dev. Biol.* **81**, 220.

Ruderman, J. V., Woodland, H. R., and Sturgess, E. A. (1979). Modulations of histone messenger RNA during the early development of *Xenopus laevis*. *Dev. Biol.* **71**, 71.

Rungger, D., and Türler, H. (1978). DNAs of simian virus 40 and polyoma direct the synthesis of viral tumor antigens and capsid proteins in *Xenopus* oocytes. *Proc. Natl. Acad. Sci. U.S.A.* **75**, 6073.

Runnström, J., and Immers, J. (1971). Treatment with lithium as a tool for the study of animal-vegetal interactions in sea urchin embryos. *Wilhelm Roux's Arch. Entwicklungsmech. Org.* **167**, 222.

Rustad, R. C. (1960). Dissociation of the mitotic time-schedule from the micromere "clock" with X-rays. *Acta Embryol. Morphol. Exp.* **3**, 155.

Ruud, G. (1925). Die Entwicklung isolierter Keimfragmente frühester Stadien von *Triton taeniatus*. *Wilhelm Roux's Arch. Entwicklungsmech. Org.* **105**, 1.

Sagata, N., Nakahashi, T., Shiokawa, K., and Yamana, K. (1978). Poly(A)-containing RNA synthesis in *Xenopus laevis* embryos. *Cell Struct. Funct.* **3**, 71.

Sagata, N., Shiokawa, K., and Yamana, K. (1980). A study on the steady-state population of poly(A)$^+$ RNA during early development of *Xenopus laevis*. *Dev. Biol.* **77**, 431.

Sakai, M., and Kubota, H. Y. (1981). Cyclic surface changes in the nonnucleate egg fragment of *Xenopus laevis*. *Dev., Growth Differ.* **23**, 41.

Sakata, S., and Kotani, M. (1985). Decrease in the number of primordial germ cells following injection of the animal pole cytoplasm into the vegetal pole region of *Xenopus* eggs. *J. Exp. Zool.* **233**, 327.

Sakonju, S., and Brown, D. D. (1982). Contact points between a positive transcription factor and the *Xenopus* 5S RNA gene. *Cell* **31**, 395.

Sakonju, S., Bogenhagen, D. F., and Brown, D. D. (1980). A control region in the center of the 5S RNA gene directs specific initiation of transcription: I. The 5' border of the region. *Cell* **19**, 13.

Sakoyama, Y., and Okubo, S. (1981). Two-dimensional gel patterns of protein species during development of *Drosophila* embryos. *Dev. Biol.* **81**, 361.

Salditt-Georgieff, M., and Darnell, J. E. (1982). Further evidence that the majority of primary nuclear RNA transcripts in mammalian cells do not contribute to mRNA. *Mol. Cell. Biol.* **2**, 701.

Salditt-Georgieff, M., Harpold, M. M., Wilson, M. C., and Darnell, J. E. (1981). Large heterogeneous nuclear ribonucleic acid has three times as many 5' caps as polyadenylic acid segments, and most caps do not enter polyribosomes. *Mol. Cell. Biol.* **1**, 179.

Salik, J., Herlands, L., Hoffman, H. P., and Poccia, D. (1981). Electrophoretic analysis of the stored histone pool in unfertilized sea urchin eggs: Quantification and identification by antibody binding. *J. Cell Biol.* **90**, 385.

Sanchez, F., Tobin, S. L., Rdest, U., Zulauf, E., and McCarthy, B. J. (1983). Two *Drosophila* actin genes in detail. Gene structure, protein structure, and transcription during development. *J. Mol. Biol.* **163**, 533.

Sánchez-Herrero, E., Verrós, I., Marco, R., and Morata, G. (1985). Genetic organization of *Drosophila* bithorax complex. *Nature (London)* **313**, 108.

Sander, K. (1976). Specification of the basic body pattern in insect embryogenesis. *Adv. Insect Physiol.* **12**, 125.

Sander, K. (1981). Pattern generation and conservation in insect ontogenesis: Problems, data, and models. *Fortschr. Zool.* **26**, 101.

Santamaria, P., and Nüsslein-Volhard, C. (1983). Partial rescue of *dorsal*, a maternal effect mutation affecting the dorsoventral pattern of the *Drosophila* embryo, by the injection of wild-type cytoplasm. *EMBO J.* **2**, 1695.

Santon, J. B., and Pellegrini, M. (1980). Expression of ribosomal proteins during *Drosophila* early development. *Proc. Natl. Acad. Sci. U.S.A.* **77**, 5649.

Santon, J. B., and Pellegrini, M. (1981). Rates of ribosomal protein and total protein synthesis during *Drosophila* early embryogenesis. *Dev. Biol.* **85**, 252.

Sardet, C., and Chang, P. (1985). A marker of animal-vegetal polarity in the egg of the sea urchin *Paracentrotus lividus*. The pigment band. *Exp. Cell Res.* **160**, 73.

Sargent, T. D., and Dawid, I. B. (1983). Differential gene expression in the gastrula of *Xenopus laevis*. *Science* **222**, 135.

Sargent, T. D., and Raff, R. A. (1976). Protein synthesis and messenger RNA stability in activated enucleate sea urchin eggs are not affected by actinomycin D. *Dev. Biol.* **48**, 327.

Sarmiento, L. A., and Mitchell, H. K. (1982). *Drosophila* salivary gland proteins and pupation. *Dev. Genet.* **3**, 255.

Sato, T., Hayes, P. H., and Denell, R. E. (1985). Homeosis in *Drosophila*: Roles and spatial patterns of expression of the *Antennapedia* and *Sex combs reduced* loci in embryogenesis. *Dev. Biol.* **111**, 171.

Satoh, N. (1977). 'Metachronous' cleavage and initiation of gastrulation in amphibian embryos. *Dev., Growth Differ.* **19**, 111.

Satoh, N., and Ikegami, S. (1981a). On the 'clock' mechanism determining the time of tissue-specific enzyme development during ascidian embryogenesis. II. Evidence for association of the clock with the cycle of DNA replication. *J. Embryol. Exp. Morphol.* **64**, 61.

Satoh, N., and Ikegami, S. (1981b). A definite number of aphidicolin-sensitive cell-cyclic events are required for acetylcholinesterase development in the presumptive muscle cells of the ascidian embryos. *J. Embryol. Exp. Morphol.* **61**, 1.

Satoh, N., Kageyama, T., and Sirakami, K. I. (1976). Mobility of dissociated embryonic cells in *Xenopus laevis*: Its significance to morphogenetic movements. *Dev., Growth Differ.* **18**, 55.

Savić A., Richman, P., Williamson, P., and Poccia, D. (1981). Alterations in chromatin structure during early sea urchin embryogenesis. *Proc. Natl. Acad. Sci. U.S.A.* **78**, 3706.

Savoini, A., Micali, F., Marzari, R., de Cristini, F., and Graziosi, G. (1981). Low variability of the protein species synthesized by *Drosophila melanogaster* embryos. *Wilhelm Roux's Arch. Dev. Biol.* **190**, 161.

Sawada, T., and Osanai, K. (1981). The cortical contraction related to the ooplasmic segregation in *Ciona intestinalis* eggs. *Wilhelm Roux's Arch. Dev. Biol.* **190**, 208.

Sawada, T., and Osanai, K. (1985). Distribution of actin filaments in fertilized egg of the ascidian *Ciona intestinalis*. *Dev. Biol.* **111**, 260.

Sawai, T. (1979). Cyclic changes in the cortical layer of nonnucleated fragments of the newt's egg. *J. Embryol. Exp. Morphol.* **51**, 183.

Schaefer, U., Golden, L., Hyman, L. E., Colot, H. V., and Rosbash, M. (1982). Some somatic sequences are absent or exceedingly rare in *Xenopus* oocyte RNA. *Dev. Biol.* **94**, 87.

Schaffner, W., Kunz, G., Daetwyler, H., Telford, J., Smith, H. O., and Birnstiel, M. L. (1978). Genes and spacers of cloned sea urchin histone DNA analyzed by sequencing. *Cell* **14**, 655.

Scharf, S. R., and Gerhart, J. C. (1980). Determination of the dorsal–ventral axis in eggs of *Xenopus laevis*: Complete rescue of UV-impaired eggs by oblique orientation before first cleavage. *Dev. Biol.* **79**, 181.

Scharf, S. R., and Gerhart, J. C. (1983). Axis determination in eggs of *Xenopus laevis*: A critical period before first cleavage, identified by the common effects of cold, pressure and ultraviolet irradiation. *Dev. Biol.* **99**, 75.

Schatten, G. (1982). Motility during fertilization. *Int. Rev. Cytol.* **79**, 59.

Schatten, G., and Schatten, H. (1981). Effects of motility inhibitors during sea urchin fertilization. *Exp. Cell Res.* **135**, 311.

Scheer, U. (1973). Nuclear pore flow rate of ribosomal RNA and chain growth rate of its precursor during oogenesis of *Xenopus laevis*. *Dev. Biol.* **30**, 13.

Scheer, U. (1978). Changes of nucleosome frequency in nucleolar and nonnucleolar chromatin as a function of transcription: An electron microscopic study. *Cell* **13**, 535.

Scheer, U. (1981). Identification of a novel class of tandemly repeated genes transcribed on lampbrush chromosomes of *Pleurodeles waltlii*. *J. Cell Biol.* **88**, 599.

Scheer, U., and Sommerville, J. (1982). Size of chromosome loops and hnRNA molecules in oocytes of Amphibia of different genome sizes. *Exp. Cell Res.* **139**, 410.

Scheer, U., Franke, W. W., Trendelenburg, M. F., and Spring, H. (1976a). Classification of loops of lampbrush chromosomes according to the arrangement of transcriptional complexes. *J. Cell Sci.* **22**, 503.

Scheer, U., Trendelenburg, M. F., and Franke, W. W. (1976b). Regulation of transcription of genes of ribosomal RNA during amphibian oogenesis. A biochemical and morphological study. *J. Cell Biol.* **69**, 465.

Scheer, U., Trendelenburg, M. F., Krohne, G., and Franke, W. W. (1977). Lengths and patterns of transcriptional units in the amplified nucleoli of oocytes of *Xenopus laevis*. *Chromosoma* **60**, 147.

Scheller, R. H., Thomas, T. L., Lee, A. S., Klein, W. H., Niles, W. D., Britten, R. J., and Davidson, E. H. (1977). Clones of individual repetitive sequences from sea urchin DNA constructed with synthetic *Eco*RI sites. *Science* **196**, 197.

Scheller, R. H., Costantini, F. D., Kozlowski, M. R., Britten, R. J., and Davidson, E. H. (1978). Representation of cloned interspersed repetitive sequences in sea urchin RNAs. *Cell* **15**, 189.

Scheller, R. H., McAllister, L. B., Crain, W. R., Durica, D. S., Posakony, J. W., Thomas, T. L., Britten, R. J., and Davidson, E. H. (1981a). Organization and expression of multiple actin genes in the sea urchin. *Mol. Cell Biol.* **1**, 609.

Scheller, R. H., Anderson, D. M., Posakony, J. W., McAllister, L. B., Britten, R. J., and Davidson, E. H. (1981b). Repetitive sequences of the sea urchin genome. Subfamily structure and evolutionary conservation. *J. Mol. Biol.* **149**, 41.

Schenkel, H., and Schnetter, W. (1979). Transcription during early embryogenesis of *Leptinotarsa* (Coleoptera). *Wilhelm Roux's Arch. Dev. Biol.* **186**, 179.

Scherer, G., Telford, J., Baldari, C., and Pirrotta, V. (1981). Isolation of cloned genes differentially expressed at early and late stages of *Drosophila* embryonic development. *Dev. Biol.* **86**, 438.

Schibler, U., Pittet, A.-C., Young, R. A., Hagenbüchle, O., Tosi, M., Gellman, S., and Wellauer, P. K. (1982). The mouse α-amylase multigene family. Sequence organization of members expressed in the pancreas, salivary gland and liver. *J. Mol. Biol.* **155**, 247.

Schibler, U., Hagenbüchle, O., Wellauer, P. K., and Pittet, A. C. (1983). Two promoters of different strengths control the transcription of the mouse α-amylase gene *Amy-1ᵅ* in the parotid gland and the liver. *Cell* **33**, 501.

Schierenberg, E., Miwa, J., and von Ehrenstein, G. (1980). Cell lineages and developmental defects of temperature-sensitive embryonic arrest mutants in *Caenorhabditis elegans*. *Dev. Biol.* **76**, 141.

Schierenberg, E., Carlson, C., and Sidio, W. (1984). Cellular development of a nematode: 3-D computer reconstruction of living embryos. *Wilhelm Roux's Arch. Dev. Biol.* **194**, 61.

Schlesinger, M. J., Ashburner, M., and Tissieres, A., eds. (1982). "Heat Shock: From Bacteria to Man." Cold Spring Harbor Lab., Cold Spring Harbor, New York.

Schlissel, M. S., and Brown, D. D. (1984). The transcriptional regulation of *Xenopus* 5S RNA genes in chromatin: The roles of active stable transcription complexes and histone H1. *Cell* **37**, 903.

Schmid, V., and Alder, H. (1984). Isolated, mononucleated, striated muscle can undergo pluripotent transdifferentiation and form a complex regenerate. *Cell* **38**, 801.

Schmidt, O., and Jäckle, H. (1978). RNA synthesized during oogenesis and early embryogenesis in an insect egg (*Euscelis plebejus*). *Wilhelm Roux's Arch. Dev. Biol.* **184**, 143.

Schmidt, O., Zissler, D., Sander, K., and Kalthoff, K. (1975). Switch in pattern formation after puncturing the anterior pole of *Smittia* eggs (Chironomidae, Diptera). *Dev. Biol.* **46**, 216.

Scholnick, S. B., Morgan, B. A., and Hirsh, J. (1983). The cloned dopa decarboxylase gene is developmentally regulated when reintegrated into the *Drosophila* genome. *Cell* **34**, 37.

Schroeder, T. E. (1980a). Expressions of the prefertilization polar axis in sea urchin eggs. *Dev. Biol.* **79**, 428.

Schroeder, T. E. (1980b). The jelly canal: Marker of polarity for sea urchin oocytes, eggs, and embryos. *Exp. Cell Res.* **128**, 490.

Schroeder, T. E. (1981). Development of a "primitive" sea urchin (*Eucidaris tribuloides*): Irregularities in the hyaline layer, micromeres and primary mesenchyme. *Biol. Bull. (Woods Hole, Mass.)* **161**, 141.

Schroeder, T. E. (1982). Distinctive features of the cortex and cell surface of micromeres: Observations and cautions. *Cell Differ.* **11**, 289.

Schroeder, T. E. (1985). Cortical expression of polarity in the starfish oocyte. *Dev., Growth Differ.* **27**, 311.

Schroeder, T. E., and Stricker, S. A. (1983). Morphological changes during maturation of starfish oocytes: Surface ultrastructure and cortical actin. *Dev. Biol.* **98**, 373.

Schubiger, G. (1976). Adult differentiation from partial *Drosophila* embryos after egg ligation during stages of nuclear multiplication and cellular blastoderm. *Dev. Biol.* **50**, 476.

Schubiger, G., Moseley, R. C., and Wood, W. J. (1977). Interaction of different egg parts in determination of various body regions in *Drosophila melanogaster*. *Proc. Natl. Acad. Sci. U.S.A.* **74**, 2050.

Schuler, M. A., McUsker, P., and Keller, E. B. (1983). DNA sequence of two linked actin genes of sea urchin. *Mol. Cell. Biol.* **3**, 448.

Schultz, G. A., Clough, J. R., Braude, P. R., Pelham, H. R. B., and Johnson, M. H. (1981a) A reexamination of messenger RNA populations in the preimplantation mouse embryo. *In* "Cellular and Molecular Aspects of Implantation" (S. R. Glasser and D. W. Bullock, eds.), p. 137. Plenum, New York.

Schultz, G. A., Kaye, P. L., McKay, D. J., and Johnson, M. H. (1981b). Endogenous amino acid pool sizes in mouse eggs and preimplantation embryos. *J. Reprod. Fertil.* **61**, 387.

Schultz, R. M., and Wassarman, P. M. (1977). Specific changes in the pattern of protein synthesis during meiotic maturation of mammalian oocytes *in vitro*. *Proc. Natl. Acad. Sci. U.S.A.* **74**, 538.

Schultz, R. M., La Marca, M. J., and Wassarman, P. M. (1978a). Absolute rates of protein synthesis during meiotic maturation of mammalian oocytes *in vitro*. *Proc. Natl. Acad. Sci. U.S.A.* **75**, 4160.

Schultz, R. M., Letourneau, G. E., and Wassarman, P. M. (1978b). Meiotic maturation of mouse oocytes *in vitro*: Protein synthesis in nucleate and anucleate oocyte fragments. *J. Cell Sci.* **30**, 251.

Schultz, R. M., Letourneau, G. E., and Wassarman, P. M. (1979a). Program of early development in the mammal: Changes in patterns and absolute rates of tubulin and total protein synthesis during oogenesis and early embryogenesis in the mouse. *Dev. Biol.* **68**, 341.

Schultz, R. M., Letourneau, G. E., and Wassarman, P. M. (1979b). Program of early development in the mammal: Changes in the patterns and absolute rates of tubulin and total protein synthesis during oocyte growth in the mouse. *Dev. Biol.* **73**, 120.

Schultze, O. (1894). Die künstliche Erzeugung von Doppelbildungen bei Froschlarven mit Hilfe abnormer Gravitationswirkung. *Arch. Entwicklungsmech. Org.* **1**, 269.

Schultze, O. (1900). Zur Frage von der Bedeutung der Schwerkraft für die Entwicklung des thierischen Embryo. *Arch. Mikrosk. Anat.* **56**, 309.

Schupbach, T., and Wieschaus, E. (1986a). Germ line autonomy of maternal effect mutations altering the embryonic body pattern of *Drosophila. Dev. Biol.* **113**, 443.

Schupbach, T., and Wieschaus, E. (1986b). Maternal effect mutations altering the anterior-posterior pattern of the *Drosophila* embryo. *Wilhelm Roux's Arch. Dev. Biol.* In press.

Sconzo, G., Bono, A., Albanese, I., and Giudice, G. (1972). Studies on sea urchin oocytes. II. Synthesis of RNA during oogenesis. *Exp. Cell Res.* **72**, 95.

Sconzo, G., Albanese, I., Rinaldi, A. M., Lo Presti, G., and Giudice, G. (1974). Cytoplasmic giant RNA in sea urchin embryos. II. Physicochemical characterization. *Cell Differ.* **3**, 297.

Sconzo, G., Roccheri, M. C., Di Carlo, M., Di Bernardo, M. G., and Giudice, G. (1983). Synthesis of heat shock proteins in dissociated sea urchin embryonic cells. *Cell Differ.* **12**, 317.

Scott, M. P., and Weiner, A. J. (1984). Structural relationships among genes that control development: Sequence homology between the *Antennapedia, Ultrabithorax,* and *fushi tarazu* loci of *Drosophila. Proc. Natl. Acad. Sci. U.S.A.* **81**, 4115.

Scott, M. P., Weiner, A. J., Hazelrigg, T. I., Polisky, B. A., Pirrotta, V., Scalenghe, F., and Kaufman, T. C. (1983). The molecular organization of the *Antennapedia* locus of *Drosophila. Cell* **35**, 763.

Scott, R. W., Vogt, T. F., Croke, M. E., and Tilghman, S. M. (1984). Tissue-specific activation of a cloned α-fetoprotein gene during differentiation of a transfected embryonal carcinoma cell line. *Nature (London)* **310**, 562.

Seale, R. L., and Aronson, A. I. (1973). Chromatin-associated proteins of the developing sea urchin embryo. I. Kinetics of synthesis and characterization of nonhistone proteins. *J. Mol. Biol.* **75**, 633.

Searle, A. G., and Beechey, C. V. (1978). Complementation studies with mouse translocations. *Cytogenet. Cell Genet.* **20**, 282.

Segraves, W. A., Louis, C., Schedl, P., and Jarry, B. P. (1983). Isolation of the *rudimentary* locus of *Drosophila melanogaster. Mol. Gen. Genet.* **189**, 34.

Segraves, W. A., Louis, C., Tsubota, S., Schedl, P., Rawls, J. M., and Jarry, B. P. (1984). The *rudimentary* locus of *Drosophila melanogaster. J. Mol. Biol.* **175**, 1.

Seidel, F. (1960). Die Entwicklungsfähigkeiten isolierter Furchungszellen aus dem Ei des Kaninchens *Oryctolagus cuniculus. Wilhelm Roux's Arch. Entwicklungsmech. Org.* **152**, 43.

Selman, K., and Kafatos, F. C. (1974). Transdifferentiation in the labial gland of silk moths: Is DNA required for cellular metamorphosis? *Cell Differ.* **3**, 81.

Senger, D. R., and Gross, P. R. (1978). Macromolecule synthesis and determination in sea urchin blastomeres at the sixteen-cell stage. *Dev. Biol.* **65**, 404.

Serfling, E., Jasin, M., and Schaffner, W. (1985). Enhancers and eukaryotic gene transcription. *Trends Genet.* **1**, 224.

Sexsmith, E. (1968). DNA values and karyotypes of amphibia. Ph.D. Thesis, Univ. of Toronto, Toronto.

Shani, M. (1985). Tissue-specific expression of the rat myosin light-chain 2 gene in transgenic mice. *Nature (London)* **314**, 283.

Shannon, M. P. (1972). Characterization of the female sterile mutant *almondex* of *Drosophila melanogaster. Genetics* **43**, 244.

Shannon, M. P., Kaufman, T. C., Shen, M. W., and Judd, B. H. (1972). Lethality patterns and morphology of selected lethal and semilethal mutations in the *zeste-white* region of *Drosophila melanogaster. Genetics* **72**, 615.

Shapiro, B. M., Schackmann, R. W., and Gabel, C. A. (1981). Molecular approaches to the study of fertilization. *Annu. Rev. Biochem.* **50**, 815.

Shastry, B. S., Honda, B. M., and Roeder, R. G. (1984). Altered levels of a 5S gene-specific transcription factor (TFIIIA) during oogenesis and embryonic development of *Xenopus laevis*. *J. Biol. Chem.* **259**, 11373.

Sheffery, M., Rifkind, R. A., and Marks, P. A. (1982). Murine erythroleukemia cell differentiation: DNase I hypersensitivity and DNA methylation near the globin genes. *Proc. Natl. Acad. Sci. U.S.A.* **79**, 1180.

Shen, S. S., and Steinhardt, R. A. (1979). Direct measurement of intracellular pH during metabolic derepression at fertilization and ammonia activation of the sea urchin egg. *Nature (London)* **272**, 253.

Shepherd, G. W., and Nemer, M. (1980). Developmental shifts in frequency distribution of polysomal mRNA and their posttranscriptional regulation in the sea urchin embryo. *Proc. Natl. Acad. Sci. U.S.A.* **77**, 4653.

Shepherd, J. C. W., McGinnis, W., Carrasco, A. E., De Robertis, E. M., and Gehring, W. J. (1984). Fly and frog homeo domains show homologies with yeast mating type regulatory proteins. *Nature (London)* **310**, 70.

Sherman, M. I. (1979). Developmental biochemistry of preimplantation mammalian embryos. *Annu. Rev. Biochem.* **48**, 443.

Shih, R. J., O'Connor, C. M., Keem, K., and Smith, L. D. (1978). Kinetic analysis of amino acid pools and protein synthesis in amphibian oocytes and embryos. *Dev. Biol.* **66**, 172.

Shih, R. J., Smith, L. D., and Keem, K. (1980). Rates of histone synthesis during early development of *Rana pipiens*. *Dev. Biol.* **75**, 329.

Shiokawa, K., and Yamana, K. (1968). Ribonucleic acid (RNA) synthesis in dissociated embryonic cells of *Xenopus laevis*. IV. Synthesis of messenger RNA in the presence of an inhibitory factor of ribosomal RNA synthesis. *Proc. Jpn. Acad.* **44**, 379.

Shiokawa, K., Misumi, Y., Yasuda, Y., Nishio, Y., Kurata, S., Sameshima, M., and Yamana, K. (1979). Synthesis and transport of various RNA species in developing embryos of *Xenopus laevis*. *Dev. Biol.* **68**, 503.

Shiokawa, K., Misumi, Y., and Yamana, K. (1981a). Mobilization of newly synthesized RNAs into polysomes in *Xenopus laevis* embryos. *Wilhelm Roux's Arch. Dev. Biol.* **190**, 103.

Shiokawa, K., Misumi, Y., and Yamana, K. (1981b). Demonstration of rRNA synthesis in pregastrular embryos of *Xenopus laevis*. *Dev., Growth Differ.* **23**, 579.

Shiokawa, K., Tashiro, K., Misumi, Y., and Yamana, K. (1981c). Noncoordinated synthesis of RNAs in pregastrular embryos of *Xenopus laevis*. *Dev., Growth Differ.* **23**, 589.

Shiokawa, K., Saito, A., Kageura, H., Higuchi, K., Koga, K., and Yamana, K. (1984). Protein synthesis in dorsal, ventral, animal, and vegetal half-embryos of *Xenopus laevis* isolated at the 8-cell stage. *Cell Struct. Funct.* **9**, 369.

Shiokawa, K., Takeichi, T., Miyata, S., Tashiro, K., and Matsuda, K. (1986). Timing of the initiation of rRNA gene expression and nucleolar formation in cleavage embryos arrested by cytochalasin B and podophyllotoxin and in cytoplasm-extracted embryos of *Xenopus laevis*. *Cytobios*. In press.

Shott, R. J., Lee, J. J., Britten, R. J., and Davidson, E. H. (1984). Differential expression of the actin gene family of *Strongylocentrotus purpuratus*. *Dev. Biol.* **101**, 295.

Showman, R. M., Wells, D. E., Anstrom, J., Hursh, D. A., and Raff, R. A. (1982). Message-specific sequestration of maternal histone mRNA in the sea urchin egg. *Proc. Natl. Acad. Sci. U.S.A.* **79**, 5944.

Signoret, J., and Lefresne, J. (1971). Contribution a l'étude de la segmentation de l'oeuf d' axolotl: I. Définition de la transition blastuléenne. *Ann. Embryol. Morphog.* **4**, 113.

Signoret, J., Lefresne, J., Vinson, D., and David, J. C. (1981). Enzymes involved in DNA replication in the axolotl. II. Control of DNA ligase activity during very early development. *Dev. Biol.* **87**, 126.

Sina, B. J., and Pellegrini, M. (1982). Genomic clones coding for some of the initial genes expressed during *Drosophila* development. *Proc. Natl. Acad. Sci. U.S.A.* **79**, 7351.

Sirotkin, K., and Davidson, N. (1982). Developmentally regulated transcription from *Drosophila melanogaster* chromosomal site 67B. *Dev. Biol.* **89**, 196.

Skoultchi, A., and Gross, P. R. (1973). Maternal histone messenger RNA: Detection by molecular hybridization. *Proc. Natl. Acad. Sci. U.S.A.* **70**, 2840.

Slack, J. M. W. (1983). "From Egg to Embryo: Determinative Events in Early Development." Cambridge Univ. Press, London and New York.

Slack, J. M. W. (1984a). *In vitro* development of isolated ectoderm from axolotl gastrulae. *J. Embryol. Exp. Morphol.* **80**, 321.

Slack, J. M. W. (1984b). Regional biosynthetic markers in the early amphibian embryo. *J. Embryol. Exp. Morphol.* **80**, 289.

Slack, J. M. W., and Forman, D. (1980). An interaction between dorsal and ventral regions of the marginal zone in early amphibian embryos. *J. Embryol. Exp. Morphol.* **56**, 283.

Slater, D. W., and Spiegelman, S. (1966). An estimation of genetic messages in the unfertilized echinoid egg. *Proc. Natl. Acad. Sci. U.S.A.* **56**, 164.

Slater, D. W., Slater, I., and Gillespie, D. (1972). Postfertilization synthesis of polyadenylic acid in sea urchin embryos. *Nature (London)* **240**, 333.

Slater, D. W., Slater, I., and Bollum, F. J. (1978). Cytoplasmic poly(A) polymerase from sea urchin eggs, merogons, and embryos. *Dev. Biol.* **63**, 94.

Slater, I., and Slater, D. W. (1974). Polyadenylation and transcription following fertilization. *Proc. Natl. Acad. Sci. U.S.A.* **71**, 1103.

Slater, I., Gillespie, D., and Slater, D. W. (1973). Cytoplasmic adenylation and processing of maternal RNA. *Proc. Natl. Acad. Sci.* **70**, 406.

Slegers, H., and Kondo, M. (1977). Messenger ribonucleoprotein complexes of cryptobiotic embryos of *Artemia salina*. *Nucleic Acids Res.* **4**, 625.

Smith, D. R., Jackson, I. J., and Brown, D. D. (1984). Domains of the positive transcription factor specific for the *Xenopus* 5S RNA gene. *Cell* **37**, 645.

Smith, J. C., and Slack, J. M. W. (1983). Dorsalization and neural induction: Properties of the organizer in *Xenopus laevis*. *J. Embryol. Exp. Morphol.* **78**, 299.

Smith, L. D. (1965). Transplantation of the nuclei of primordial germ cells into enucleated eggs of *Rana pipiens*. *Proc. Natl. Acad. Sci. U.S.A.* **54**, 101.

Smith, L. D. (1966). Role of a "germinal plasm" in the formation of primordial germ cells in *Rana pipiens*. *Dev. Biol.* **14**, 330.

Smith, L. D. (1975). Molecular events during oocyte maturation. *In* "Biochemistry of Animal Development" (R. Weber, ed.), Vol. 3, p. 1. Academic Press, New York.

Smith, L. D., and Ecker, R. E. (1965). Protein synthesis in enucleated eggs of *Rana pipiens*. *Science* **150**, 777.

Smith, L. D., and Ecker, R. E. (1969). Role of the oocyte nucleus in physiological maturation in *Rana pipiens*. *Dev. Biol.* **19**, 281.

Smith, L. D., and Richter, J. D. (1985). Synthesis, accumulation, and utilization of maternal macromolecules during oogenesis and oocyte maturation. *In* "The Biology of Fertilization" (C. B. Mertz and A. Monroy, eds.), Vol. 1, p. 141. Academic Press, Orlando, Florida.

Smith, L. D., and Williams, M. A. (1975). Germinal plasm and determination of the primordial germ cells. *In* "The Developmental Biology of Reproduction" (C. L. Markert and J. Papaconstantinou, eds.), p. 3. Academic Press, New York.

Smith, L. D., and Williams, M. (1979). Germinal plasm and germ cell determinants in anuran amphibians. *In* "Maternal Effects in Development" (D. R. Newth and M. Balls, eds.), p. 167. Cambridge University Press, London and New York.

Smith, L. D., Michael, P., and Williams, M. A. (1983). Does a predetermined germ line exist in amphibians? *In* "Current Problems in Germ Cell Differentiation" (A. McLaren and C. C. Wylie, eds.), p. 19. Cambridge Univ. Press, London and New York.

Smith, L. D., Richter, J. D., and Taylor, M. A. (1984). Regulation of translation during oogenesis. *In* "Molecular Biology of Development" (E. H. Davidson and R. A. Firtel, eds.), p. 129. Alan R. Liss, New York.

Smith, K. D. (1967). Genetic control of macromolecular synthesis during development of an ascidian: *Ascidia nigra. J. Exp. Zool.* **164**, 393.

Smith, M. J., Hough, B. R., Chamberlin, M. E., and Davidson, E. H. (1974). Repetitive and non-repetitive sequence in sea urchin heterogeneous nuclear RNA. *J. Mol. Biol.* **85**, 103.

Smith, M. J., Britten, R. J., and Davidson, E. H. (1975). Studies on nucleic acid reassociation kinetics: Reactivity of single stranded tails in DNA–DNA renaturation. *Proc. Natl. Acad. Sci. U.S.A.* **72**, 4805.

Smith, R., and McLaren, A. (1977). Factors affecting the time of formation of the mouse blastocoel. *J. Embryol. Exp. Morphol.* **41**, 79.

Smith, R. C., and Knowland, J. (1984). Protein synthesis in dorsal and ventral regions of *Xenopus laevis* embryos in relation to dorsal and ventral differentiation. *Dev. Biol.* **103**, 355.

Sobel, J. S., and Alliegro, M. A. (1985). Changes in the distribution of a spectrin-like protein during development of the preimplantation mouse embryo. *J. Cell Biol.* **100**, 333.

Sodja, A., Arking, R., and Zafar, R. S. (1982). Actin gene expression during embryogenesis of *Drosophila melanogaster. Dev. Biol.* **90**, 363.

Sollner-Webb, B., Wilkinson, J. A. K., Roan, J., and Reeder, R. H. (1983). Nested control regions promote *Xenopus* ribosomal RNA synthesis by RNA polymerase I. *Cell* **35**, 199.

Solursh, M., and Katow, H. (1982). Initial characterization of sulfated macromolecules in the blastocoels of mesenchyme blastulae of *Strongylocentrotus purpuratus* and *Lytechinus pictus. Dev. Biol.* **94**, 326.

Sommerville, J. (1973). Ribonucleoprotein particles derived from the lampbrush chromosomes of newt oocytes. *J. Mol. Biol.* **78**, 487.

Sommerville, J., and Malcolm, D. B. (1976). Transcription of genetic information in amphibian oocytes. *Chromosoma* **55**, 183.

Sommerville, J., and Scheer, U. (1982). Transcription of complementary repeat sequences in amphibian oocytes. *Chromosoma* **86**, 95.

Sommerville, J., Malcolm, D. B., and Callan, H. G. (1978a). The organization of transcription on lampbrush chromosomes. *Philos. Trans. R. Soc. London, Ser. B* **283**, 359.

Sommerville, J., Crichton, C., and Malcolm, D. (1978b). Immunofluorescent localization of transcriptional activity on lampbrush chromosomes. *Chromosoma* **66**, 99.

Sonneborn, T. M. (1950). The cytoplasm in heredity. *Heredity* **4**, 11.

Sorenson, R. A., and Wassarman, P. M. (1976). Relationship between growth and meiotic maturation of the mouse oocyte. *Dev. Biol.* **50**, 531.

Spadafora, C., Bellard, M., Compton, J. L., and Chambon, P. (1976). The DNA repeat lengths in chromatins from sea urchin sperm and gastrula cells are markedly different. *FEBS Lett.* **69**, 281.

Speksnijder, J. E., Mulder, M. M., Dohmen, M. R., Hage, W. J., and Bluemink, J. G. (1985). Animal–vegetal polarity in the plasma membrane of a molluscan egg: A quantitative freeze–fracture study. *Dev. Biol.* **108**, 38.

Spemann, H. (1903). Entwickelungsphysiologische Studien am *Triton*-Ei. II. *Arch. Entwicklungsmech. Org.* **15**, 448.

Spemann, H. (1938). "Embryonic Development and Induction." Yale University Press, New Haven, Connecticut.

Spemann, H., and Mangold, H. (1924). Über Induktion von Embryonalanlagen durch Implantation artfremder Organisatoren. *Wilhelm Roux's Arch. Dev. Biol.* **100**, 599.

Spiegel, E., Burger, M. M., and Spiegel, M. (1983). Fibronectin and laminin in the extracellular matrix and basement membrane of sea urchin embryos. *Exp. Cell Res.* **144**, 47.

Spierer, P. (1984). A molecular approach to chromosome organization. *Dev. Genet.* **4**, 333.

Spierer, P., Spierer, A., Bender, W., and Hogness, D. S. (1983). Molecular mapping of genetic and chromomeric units in *Drosophila melanogaster*. *J. Mol. Biol.* **168**, 35.

Spieth, J., and Whiteley, A. H. (1981). Polyribosome formation and poly(A)-containing RNA in embryos of the sand dollar, *Dendraster excentricus*. *Wilhelm Roux's Arch. Dev. Biol.* **190**, 111.

Spindle, A. I. (1978). Trophoblast regeneration by inner cell masses isolated from cultured mouse embryos. *J. Exp. Zool.* **203**, 483.

Spirin, A. S. (1966). On 'masked' forms of messenger RNA in early embryogenesis and in other differentiating systems. *Curr. Top. Dev. Biol.* **1**, 1.

Spooner, G. B. (1911). Embryological studies with the centrifuge. *J. Exp. Zool.* **10**, 23.

Spradling, A. C. (1981). The organization and amplification of two chromosomal domains containing *Drosophila* chorion genes. *Cell* **27**, 193.

Spradling, A. C., and Mahowald, A. P. (1979). Identification and genetic localization of mRNAs from ovarian follicle cells of *Drosophila melanogaster*. *Cell* **16**, 589.

Spradling, A. C., and Mahowald, A. P. (1981). A chromosome inversion alters the pattern of specific DNA replication in *Drosophila* follicle cells. *Cell* **27**, 203.

Spradling, A. C., and Rubin, G. M. (1982). Transposition of cloned P elements into *Drosophila* germ line chromosomes. *Science* **218**, 341.

Spradling, A. C., and Rubin, G. M. (1983). The effect of chromosomal position on the expression of the *Drosophila* xanthine dehydrogenase gene. *Cell* **34**, 47.

Stafford, D. W., Sofer, W. H., and Iverson, R. M. (1964). Demonstration of polyribosomes after fertilization of the sea urchin egg. *Proc. Natl. Acad. Sci. U.S.A.* **52**, 313.

Stalder, J., Groudine, M., Dodgson, J. B., Engel, J. D., and Weintraub, H. (1980). Hb switching in chickens. *Cell* **19**, 973.

Standart, N. M., Bray, S. J., George, E. L., Hunt, T., and Ruderman, J. V. (1985). The small subunit of ribonucleotide reductase is encoded by one of the most abundant translationally regulated maternal mRNAs in clam and sea urchin eggs. *J. Cell Biol.* **100**, 1968.

Stedman, E., and Stedman, E. (1950). Cell specificity of histones. *Nature (London)* **166**, 780.

Stein, R., Gruenbaum, Y., Pollack, Y., Razin, A., and Cedar, H. (1982). Clonal inheritance of the pattern of DNA methylation in mouse cells. *Proc. Natl. Acad. Sci. U.S.A.* **79**, 61.

Stein, R., Sciaky-Gallili, N., Razin, A., and Cedar, H. (1983). Pattern of methylation of two genes coding for housekeeping functions. *Proc. Natl. Acad. Sci. U.S.A.* **80**, 2422.

Steinmetz, M., Frelinger, J. G., Fisher, D., Hunkapiller, T., Pereira, D., Weissman, S. M., Uehara, H., Nathenson, S., and Hood, L. (1981). Three cDNA clones encoding mouse transplantation antigens: Homology to immunoglobulin genes. *Cell* **24**, 125.

Steller, H., and Pirrotta, V. (1984). Regulated expression of genes injected into early *Drosophila* embryos. *EMBO J.* **3**, 165.

Stephens, R. E. (1972). Studies on the development of the sea urchin *Strongylocentrotus dröebachiensis*. III. Embryonic synthesis of ciliary proteins. *Biol. Bull. (Woods Hole, Mass.)* **142**, 489.

Stephens, R. E. (1977). Differential protein synthesis and utilization during cilia formation in sea urchin embryos. *Dev. Biol.* **61**, 311.

Stephens, R. E. (1978). Primary structural differences among tubulin subunits from flagella, cilia, and the cytoplasm. *Biochemistry* **17**, 2882.

Stephens, R. E., and Kane, R. E. (1970). Some properties of hyalin. *J. Cell Biol.* **44**, 611.

Stephenson, E. C., Erba, H. P., and Gall, J. G. (1981). Histone gene clusters of the newt *Notophthalmus* are separated by long tracts of satellite DNA. *Cell* **24**, 639.

Stern, M. S. (1973). Development of cleaving mouse embryos under pressure. *Differentiation* **1**, 407.

Stern, M. S., and Wilson, I. B. (1972). Experimental studies on the organization of the preimplantation mouse embryo. I. Fusion of asynchronously cleaving eggs. *J. Embryol. Exp. Morphol.* **28**, 247.

Sternberg, P. W., and Horvitz, H. R. (1982). Postembryonic nongonadal cell lineages of the nematode *Panagrellus redivivus*: Description and comparison with those of *Caenorhabditis elegans*. *Dev. Biol.* **93**, 181.

Sternberg, P. W., and Horvitz, H. R. (1984). The genetic control of cell lineage during nematode development. *Annu. Rev. Genet.* **18**, 489.

Sternlicht, A. L., and Schultz, R. M. (1981). Biochemical studies of mammalian oogenesis: Kinetics of accumulation of total and poly(A)-containing RNA during growth of the mouse oocyte. *J. Exp. Zool.* **215**, 191.

Stevens, N. M. (1909). The effect of ultra-violet light upon the developing eggs of *Ascaris megalocephala*. *Arch. Entwicklungsmech. Org.* **27**, 622.

Steward, R., McNally, F. J., and Schedl, P. (1984). Isolation of the dorsal locus of *Drosophila*. *Nature (London)* **311**, 262.

Steward, R., Ambrosio, L., and Schedl, P. (1985). Expression of the *dorsal* gene. *Cold Spring Harbor Symp. Quant. Biol.* **50**, 223.

Stewart, C. L., Stuhlmann, H., Jähner, D., and Jaenisch, R. (1982). *De novo* methylation, expression, and infectivity of retroviral genomes introduced into embryonal carcinoma cells. *Proc. Natl. Acad. Sci. U.S.A.* **79**, 4098.

Stewart, T. A., Pattengale, P. K., and Leder, P. (1984). Spontaneous mammary adenocarcinomas in transgenic mice that carry and express MTV/*myc* fusion genes. *Cell* **38**, 627.

Stick, R., and Hausen, P. (1985). Changes in the nuclear lamina composition during early development of *Xenopus laevis*. *Cell* **41**, 191.

Stinchcomb, D. T., Shaw, J., Carr, S., and Hirsh, D. (1985). Extrachromosomal DNA transformation of *C. elegans*. *In* "Genetic Manipulation of the Early Mammalian Embryo" (F. Costantini and R. Jaenisch, eds.), Banbury Report 20, p. 251. Cold Spring Harbor Lab., Cold Spring Harbor, New York.

Storb, U., O'Brien, R. L., McMullen, M. D., Gollahon, K. A., and Brinster, R. L. (1984). High expression of cloned immunoglobulin κ gene in transgenic mice is restricted to B lymphocytes. *Nature (London)* **310**, 238.

Strasburger, E. (1884). New investigations on the course of fertilization in the phanerograms as basis for a theory of heredity. (Engl. trans.) *In* Voeller, B. R., ed. (1968). "The Chromosome Theory of Inheritance: Classic Papers in Development and Heredity." Appleton, New York.

Strathmann, R. R. (1975). Larval feeding in echinoderms. *Am. Zool.* **15**, 717.

Strome, S., and Wood, W. B. (1982). Immunofluorescence visualization of germ line-specific cytoplasmic granules in embryos, larvae, and adults of *Caenorhabditis elegans*. *Proc. Natl. Acad. Sci. U.S.A.* **79**, 1558.

Strome, S., and Wood, W. B. (1983). Generation of asymmetry and segregation of germ line granules in early *C. elegans* embryos. *Cell* **35**, 15.

Strub, K., Galli, G., Busslinger, M., and Birnstiel, M. L. (1984). The cDNA sequences of the sea urchin U7 small nuclear RNA suggest specific contacts between histone mRNA precursor and U7 RNA during RNA processing. *EMBO J.* **3**, 2801.

Struhl, G. (1981). A homeotic mutation transforming leg to antenna in *Drosophila*. *Nature (London)* **292**, 635.

Struhl, G. (1982). Genes controlling segmental specification in the *Drosophila* thorax. *Proc. Natl. Acad. Sci. U.S.A.* **79**, 7380.

Struhl, G. (1983). Role of the *esc*⁺ gene product in ensuring the selective expression of segment-specific homeotic genes in *Drosophila*. *J. Embryol. Exp. Morphol.* **76**, 297.

Struhl, G. (1984). Splitting the bithorax complex of *Drosophila*. *Nature (London)* **308**, 454.

Struhl, G. (1985). Near reciprocal phenotypes caused by inactivation or indiscriminate expression of the *Drosophila* segmentation gene *ftz*. *Nature (London)* **318**, 677.

Struhl, G., and Akam, M. (1985). Altered distributions of *Ultrabithorax* transcripts in *extra sex combs* mutant embryos of *Drosophila*. *EMBO J.* **4**, 3259.

Struhl, G., and Brower, D. (1982). Early role of the *esc*⁺ gene product in the determination of segments in *Drosophila*. *Cell* **31**, 285.

Struhl, G., and White, R. A. H. (1985). Regulation of the *Ultrabithorax* gene of *Drosophila* by other bithorax complex genes. *Cell* **43**, 507.

Stuhlmann, H., Jähner, D., and Jaenisch, R. (1981). Infectivity and methylation of retroviral genomes is correlated with expression in the animal. *Cell* **26**, 221.

Sturgess, E. A., Ballantine, J. E. M., Woodland, H. R., Mohun, P. R., Lane, C. D., and Dimitriadis, G. J. (1980). Actin synthesis during the early development of *Xenopus laevis*. *J. Embryol. Exp. Morphol.* **58**, 303.

Sturtevant, A. H. (1923). Inheritance of direction of coiling in *Limnaea*. *Science* **58**, 269.

Subtelny, S. (1965). On the nature of the restricted differentiation-promoting ability of transplanted *Rana pipiens* nuclei from differentiating endoderm cells. *J. Exp. Zool.* **159**, 59.

Subtelny, S. (1980). Migration and replication of the germ cell line in *Rana pipiens*. *In* "Differentiation and Neoplasia" (R. D. McKinnell, M. A. DiBernardino, M. Blumenfeld, and R. D. Bergad, eds.), p. 157. Springer-Verlag, Berlin and New York.

Sucov, H. M., Benson, S. C., Robinson, J. J., Britten, R. J., Wilt, F., and Davidson, E. H. (1986). A lineage specific gene encoding a major matrix protein of the sea urchin embryo spicule. II. Structure of the gene and derived sequence of the protein. Submitted for publication.

Sudarwati, S., and Nieuwkoop, P. D. (1971). Mesoderm formation in the anuran *Xenopus laevis* (Daudin). *Wilhelm Roux's Arch. Dev. Biol.* **166**, 189.

Sulston, J. E. (1983). Neuronal cell lineages in the nematode *Caenorhabditis elegans*. *Cold Spring Harbor Symp. Quant. Biol.* **48**, 443.

Sulston, J. E., and Horvitz, H. R. (1977). Postembryonic cell lineages of the nematode, *Caenorhabditis elegans*. *Dev. Biol.* **56**, 110.

Sulston, J. E., and Horvitz, H. R. (1981). Abnormal cell lineages in mutants of the nematode *Caenorhabditis elegans*. *Dev. Biol.* **82**, 41.

Sulston, J. E., Albertson, D. G., and Thomson, J. N. (1980). The *Caenorhabditis elegans* male: Postembryonic development of nongonadal structures. *Dev. Biol.* **78**, 542.

Sulston, J. E., Schierenberg, E., White, J. G., and Thomson, J. N. (1983). The embryonic cell lineage of the nematode *Caenorhabditis elegans*. *Dev. Biol.* **100**, 64.

Summers, M. C., Bedian, V., and Kauffman, S. A. (1986). An analysis of stage-specific protein synthesis in the early *Drosophila* embryo using high-resolution, two-dimensional gel electrophoresis. *Dev. Biol.* **113**, 49.

Surani, M. A. H., and Barton, S. C. (1983). Development of gynogenetic eggs in the mouse: Implications for parthenogenetic embryos. *Science* **222**, 1034.

Sures, I., Lowry, J., and Kedes, L. H. (1978). The DNA sequences of sea urchin (*S. purpuratus*) H2a, H2b, and H3 histone coding and spacer regions. *Cell* **15**, 1033.

Sures, I., Levy, S., and Kedes, L. H. (1980). Leader sequences of *Strongylocentrotus purpuratus* histone mRNAs start at a unique heptanucleotide common to all five histone genes. *Proc. Natl. Acad. Sci. U.S.A.* **77**, 1265.

Surrey, S., and Nemer, M. (1976). Methylated blocked 5' terminal sequences of sea urchin embryo messenger RNA classes containing and lacking poly(A). *Cell* **9**, 589.

Surrey, S., Ginzburg, I., and Nemer, M. (1979). Ribosomal RNA synthesis in pre- and post-gastrula-stage sea urchin embryos. *Dev. Biol.* **71**, 83.

Sutasurya, L. A., and Nieuwkoop, P. D. (1974). The induction of the primordial germ cells in the urodeles. *Wilhelm Roux's Arch. Dev. Biol.* **175**, 199.

Sutton, W. S. (1903). The chromosomes in heredity. *Biol. Bull. (Woods Hole, Mass.)* **4**, 231.

Suzuki, Y., and Adachi, S. (1984). Signal sequences associated with fibroin gene expression are identical in fibroin-producer and -nonproducer tissues. *Dev., Growth Differ.* **26**, 139.

Suzuki, Y., Gage, L. P., and Brown, D. D. (1972). The genes for silk fibroin in *Bombyx mori*. *J. Mol. Biol.* **70**, 637.

Swift, G. H., Hammer, R. E., MacDonald, R. J., and Brinster, R. L. (1984). Tissue-specific expression of the rat pancreatic elastase I gene in transgenic mice. *Cell* **38**, 639.

Szabad, J., Schüpbach, T., and Wieschaus, E. (1979). Cell lineage and development in the larval epidermis of *Drosophila melanogaster*. *Dev. Biol.* **73**, 256.

Takahashi, K., and Yoshii, M. (1981). Development of sodium, calcium and potassium channels in the cleavage-arrested embryo of an ascidian. *J. Physiol. (London)* **315**, 515.

Takahashi, M. M., and Okazaki, K. (1979). Total cell number and number of the primary mesenchyme cells in whole, 1/2 and 1/4 larvae of *Clypeaster japonicus*. *Dev., Growth Differ.* **21**, 553.

Takaya, H. (1978). Dynamics of the organizer. A. Morphogenetic movements and specificities in induction and differentiation of the organizer. *In* "Organizer—A Milestone of a Half-Century from Spemann" (O. Nakamura and S. Toivonen, eds.), p. 49. Elsevier/North-Holland, Amsterdam.

Takeichi, T., Satoh, N., Tashiro, K., and Shiokawa, K. (1985). Temporal control of rRNA synthesis in cleavage-arrested embryos of *Xenopus laevis*. *Dev. Biol.* **112**, 443.

Takeshima, K., and Nakano, E. (1983). Modification of ribosomal proteins in sea urchin eggs following fertilization. *Eur. J. Biochem.* **137**, 437.

Tanabe, K., and Kotani, M. (1974). Relationship between the amount of the 'germinal plasm' and the number of primordial germ cells in *Xenopus laevis*. *J. Embryol. Exp. Morphol.* **31**, 89.

Tanaka, Y. (1976). Effects of surfactants on the cleavage and further development of the sea urchin embryo. I. The inhibition of micromere formation at the fourth cleavage. *Dev., Growth Differ.* **18**, 113.

Tansey, T. R., and Ruderman, J. V. (1983). Differential mRNA accumulation and translation during *Spisula* development. *Dev. Biol.* **99**, 338.

Tarkowski, A. K. (1959a). Experiments on the development of isolated blastomeres of mouse eggs. *Nature (London)* **184**, 1286.

Tarkowski, A. K. (1959b). Experimental studies on regulation in the development of isolated blastomeres of mouse eggs. *Acta Theriol.* **3**, 191.

Tarkowski, A. K. (1961). Mouse chimaeras developed from fused eggs. *Nature (London)* **190**, 857.

Tarkowski, A. K. (1963). Studies on mouse chimaeras developed from eggs fused *in vitro*. *Natl. Cancer Inst. Monogr.* No. 11, 37.

Tarkowski, A. K., and Wróblewska, J. (1967). Development of blastomeres of mouse eggs isolated at the 4- and 8-cell stage. *J. Embryol. Exp. Morphol.* **18**, 155.

Tartof, K. D. (1973). Regulation of ribosomal RNA gene multiplicity in *Drosophila melanogaster*. *Genetics* **73**, 57.

Tashima, M., Calabretta, B., Torelli, G., Scofield, M., Maizel, A., and Saunders, G. F. (1981). Presence of a highly repetitive and widely dispersed DNA sequence in the human genome. *Proc. Natl. Acad. Sci. U.S.A.* **78**, 1508.

Taylor, M. A., and Smith, L. D. (1985). Quantitative changes in protein synthesis during oogenesis in *Xenopus laevis*. *Dev. Biol.*, **110**, 230.

Taylor, M. A., Johnson, A. D., and Smith, L. D. (1985). Growing *Xenopus* oocytes have spare translational capacity. *Proc. Natl. Acad. Sci. U.S.A.* **82**, 6586.

Technau, G. M., and Campos-Ortega, J. A. (1985). Fate-mapping in wild-type *Drosophila melanogaster*. II. Injections of horseradish peroxidase in cells of the early gastrula stage. *Wilhelm Roux's Arch. Dev. Biol.* **194**, 196.

Telfer, W. H. (1975). Development and physiology of the oöcyte-nurse cell syncytium. *Adv. Insect Physiol.* **11**, 223.

Telfer, W. H., Woodruff, R. I., and Huebner, E. (1981). Electrical polarity and cellular differentiation in meroistic ovaries. *Am. Zool.* **21**, 675.

Teugels, E., and Ghysen, A. (1985). Domains of action of bithorax genes in *Drosophila* central nervous system. *Nature (London)* **314**, 558.

Thiébaud, C. H. (1979). Quantitative determination of amplified rDNA and its distribution during oogenesis in *Xenopus laevis*. *Chromosoma* **73**, 37.

Thiebaud, P., Lefresne, J., Signoret, J., and David, J. C. (1983). Isolation of the messenger RNA for 8S DNA ligase in early developing axolotl egg and its cell free translation. *Nucleic Acids Res.* **11**, 2563.

Thomas, C. (1974). RNA metabolism in previtellogenic oocytes of *Xenopus laevis*. *Dev. Biol.* **39**, 191.

Thomas, J. B., Bastiani, M. J., Bate, M., and Goodman, C. S. (1984). From grasshopper to *Drosophila*: A common plan for neuronal development. *Nature (London)* **310**, 203.

Thomas, T. L., Britten, R. J., and Davidson, E. H. (1982). An interspersed region of the sea urchin genome represented in both maternal poly(A) RNA and embryo nuclear RNA. *Dev. Biol.* **94**, 230.

Thomas, V., Heasman, J., Ford, C., Nagajski, D., and Wylie, C. C. (1983). Further analysis of the effect of ultraviolet irradiation on the formation of the germ line in *Xenopus laevis*. *J. Embryol. Exp. Morphol.* **76**, 67.

Togashi, S., and Okada, M. (1983). Effects of UV-irradiation at various wavelengths on sterilizing *Drosophila* embryos. *Dev., Growth Differ.* **25**, 133.

Topol, J., Ruden, D. M., and Parker, C. S. (1985). Sequences required for *in vitro* transcriptional activation of a *Drosophila* hsp70 gene. *Cell* **42**, 527.

Townes, T. M., Lingrel, J. B., Chen, H. Y., Brinster, R. L., and Palmiter, R. D. (1985). Erythroid-specific expression of human β-globin genes in transgenic mice. *EMBO J.* **4**, 1715.

Treisman, R., Green, M. R., and Maniatis, T. (1983). *cis* and *trans* activation of globin gene transcription in transient assays. *Proc. Natl. Acad. Sci. U.S.A.* **80**, 7428.

Trendelenburg, M. F., and McKinnell, R. G. (1979). Transcriptionally active and inactive regions of nucleolar chromatin in amplified nucleoli of fully grown oocytes of hibernating frogs, *Rana pipiens* (Amphibia, Anura). A quantitative electron microscopic study. *Differentiation* **15**, 73.

Trendelenburg, M. F., Scheer, U., Zentgraft, H., and Franke, W. W. (1976). Heterogeneity of spacer lengths in circles of amplified ribosomal DNA of two insect species, *Dytiscus marginalis* and *Acheta domesticus*. *J. Mol. Biol.* **108**, 453.

Trendelenburg, M. F., Franke, W. W., and Scheer, U. (1977). Frequencies of circular units of nucleolar DNA in oocytes of two insects, *Acheta domesticus* and *Dytiscus marginalis*, and changes of nucleolar morphology during oogenesis. *Differentiation* **7**, 133.

Trinkaus, J. P. (1984). "Cells into Organs. The Forces that Shape the Embryo," 2nd ed. Prentice-Hall, Engelwood Cliffs, New Jersey.

Trumbly, R. J., and Jarry, B. (1983). Stage-specific protein synthesis during early embryogenesis in *Drosophila melanogaster*. *EMBO J.* **2**, 1281.

Tsafriri, A. (1978). Oocyte maturation in mammals. *In* "The Vertebrate Ovary: Comparative Biology and Evolution" (R. E. Jones, ed.), p. 409. Plenum, New York.

Tsujimoto, Y., and Suzuki, Y. (1984). Natural fibroin genes purified without using cloning procedures from fibroin-producing and -nonproducing tissues reveal indistinguishable structure and function. *Proc. Natl. Acad. Sci. U.S.A.* **81**, 1644.

Tufaro, F., and Brandhorst, B. P. (1979). Similarity of proteins synthesized by isolated blastomeres of early sea urchin embryos. *Dev. Biol.* **72**, 390.

Tung, T. C., Wu, S. C., Yeh, Y. F., Li, K. S., and Hsu, M. C. (1977). Cell differentiation in ascidian studied by nuclear transplantation. *Sci. Sin. (Engl. Ed.)* **20**, 222.

Turner, S. H., and Laird, C. D. (1973). Diversity of RNA sequences in *Drosophila melanogaster*. *Biochem. Genet.* **10**, 263.

Tyler, A. (1963). The manipulations of macromolecular substances during fertilization and early development of animal eggs. *Am. Zool.* **3**, 109.

Tyler, A. (1965). The biology and chemistry of fertilization. *Am. Nat.* **99**, 309.

Ubbels, G. A., Hara, K., Koster, C. H., and Kirschner, M. W. (1983). Evidence for a functional role of the cytoskeleton in determination of the dorsoventral axis in *Xenopus laevis* eggs. *J. Embryol. Exp. Morphol.* **77**, 15.

Ueda, R., and Okada, M. (1982). Induction of pole cells in sterilized *Drosophila* embryos by injection of subcellular fraction from eggs. *Proc. Natl. Acad. Sci. U.S.A.* **79**, 6946.

Underwood, E. M., Turner, F. R., and Mahowald, A. P. (1980). Analysis of cell movements and fate mapping during early embryogenesis in *Drosophila melanogaster*. *Dev. Biol.* **74**, 286.

Urieli-Shoval, S., Gruenbaum, Y., Sedat, J., and Razin, A. (1982). The absence of detectable methylated bases in *Drosophila melanogaster* DNA. *FEBS Lett.* **146**, 148.

Uzman, J. A., and Wilt, F. H. (1984). The role of RNA polymerase initiation and elongation in control of total RNA and histone mRNA synthesis in sea urchin embryos. *Dev. Biol.* **106**, 174.

Vacquier, V. D. (1980). The adhesion of sperm to sea urchin eggs. *In* "The Cell Surface, Mediator of Developmental Processes" (N. K. Wessells, ed.), p. 151. Academic Press, New York.

Van Beneden, E. (1883). Recherches sur la maturation de l'oeuf et la fécondation et la division cellulaire. *Arch. Biol.* **4**, 265.

Van Blerkom, J. (1979). Molecular differentiation of the rabbit ovum. III. Fertilization-autonomous polypeptide synthesis. *Dev. Biol.* **72**, 188.

Van Blerkom, J. (1981). Structural relationship and posttranslational modification of stage-specific proteins synthesized during early preimplantation development in the mouse. *Proc. Natl. Acad. Sci. U.S.A.* **78**, 7629.

Van Blerkom, J., and McGaughey, R. W. (1978). Molecular differentiation of the rabbit ovum. I. During oocyte maturation *in vivo* and *in vitro*. *Dev. Biol.* **63**, 139.

Van Blerkom, J., Barton, S. C., and Johnson, M. H. (1976). Molecular differentiation in the preimplantation mouse embryo. *Nature (London)* **259**, 319.

Van Dam, W. I., and Verdonk, N. H. (1982). The morphogenetic significance of the first quartet micromeres for the development of the snail *Bithynia tentaculata*. *Wilhelm Roux's Arch. Dev. Biol.* **191**, 112.

Van Dam, W. I., Dohmen, M. R., and Verdonk, N. H. (1982). Localization of morphogenetic determinants in a special cytoplasm present in the polar lobe of *Bithynia tentaculata* (Gastropoda). *Wilhelm Roux's Arch. Dev. Biol.* **191**, 371.

van den Biggelaar, J. A. M., and Guerrier, P. (1979). Dorsoventral polarity and mesentoblast determination as concomitant results of cellular interactions in the mollusk *Patella vulgata*. *Dev. Biol.* **68**, 462.

van den Biggelaar, J. A. M., Dorresteijn, A, W. C., de Laat, S. W., and Bluemink, J. G. (1981). The role of topographical factors in cell interaction and determination of cell lines in molluscan development *In* "International Cell Biology 1980–1981" (H. G. Schweiger, ed.), p. 526. Springer-Verlag, Berlin and New York.

van der Ploeg, L. H. T., and Flavell, R. A. (1980). DNA methylation in the human γδβ-globin locus in erythroid and nonerythroid tissues. *Cell* **19**, 947.

Van Dongen, C. A. M. (1976). The development of *Dentalium* with special reference to the significance of the polar lobe. VII. Organogenesis and histogenesis in lobeless embryos of *Dentalium vulgare* (da Costa) as compared to normal development. *Proc. K. Ned. Akad. Wet., Ser. C* **79**, 421.

Van Dongen, C. A. M., and Geilenkirchen, W. L. M. (1974). The development of *Dentalium* with special reference to the significance of the polar lobe. I. Division chronology and development of the cell pattern in *Dentalium dentale* (Scaphopoda). *Proc. K. Ned. Akad. Wet., Ser. C* **77**, 57.

Van Dongen, C. A. M., and Geilenkirchen, W. L. M. (1975). The development of *Dentalium* with special reference to the significance of the polar lobe. IV. Division chronology and development of the cell pattern in *Dentalium dentale* after removal of the polar lobe at first cleavage. *Proc. K. Ned. Akad. Wet., Ser. C* **78**, 358.

Van Dongen, W., Zaal, R., Moorman, A., and Destrée, O. (1981). Quantitation of the accumulation of histone messenger RNA during oogenesis in *Xenopus laevis*. *Dev. Biol.* **86**, 303.

Van Dongen, W. M. A. M., Moorman, A. F. M., and Destrée, O. H. J. (1983a). Histone gene expression in early development of *Xenopus laevis*. Analysis of histone mRNA in oocytes and embryos by blot-hybridization and cell-free translation. *Differentiation* **24**, 226.

Van Dongen, W. M. A. M., Moorman, A. F. M., and Destrée, O. H. J. (1983b). The accumulation of the maternal pool of histone H1a during oogenesis in *Xenopus laevis*. *Cell Differ.* **12**, 257.

Van Gansen, P., and Schram, A. (1974). Incorporation of [³H]uridine and [³H]thymidine during the phase of nucleolar multiplication in *Xenopus laevis* oögenesis: A high-resolution autoradiographic study. *J. Cell Sci.* **14**, 85.

Van Ness, J., and Hahn, W. E. (1982). Physical parameters affecting the rate and completion of RNA driven hybridization of DNA: New measurements relevant to quantitation based on kinetics. *Nucleic Acids Res.* **10**, 8061.

Vardimon, L., Kressmann, A., Cedar, H., Maechler, M., and Doerfler, W. (1982). Expression of a cloned adenovirus gene is inhibited by *in vitro* methylation. *Proc. Natl. Acad. Sci. U.S.A.* **79**, 1073.

Varley, J. M., Macgregor, H. C., and Erba, H. P. (1980a). Satellite DNA is transcribed on lampbrush chromosomes. *Nature (London)* **283**, 686.

Varley, J. M., Macgregor, H. C., Nardi, I., Andrews, C., and Erba, H. P. (1980b). Cytological evidence of transcription of highly repeated DNA sequences during the lampbrush stage in *Triturus cristatus carnifex*. *Chromosoma* **80**, 289.

Venezky, D. L., Angerer, L. M., and Angerer, R. C. (1981). Accumulation of histone repeat transcripts in the sea urchin egg pronucleus. *Cell* **24**, 385.

Verdonk, N. H. (1968). The effect of removing the polar lobe in centrifuged eggs of *Dentalium*. *J. Embryol. Exp. Morphol.* **19**, 33.

Verdonk, N. H., and Cather, J. N. (1973). The development of isolated blastomeres in *Bithynia tentaculata* (Prosobranchia, Gastropoda). *J. Exp. Zool.* **186**, 47.

Vincent, A., O'Connell, P., Gray, M. R., and Rosbash, M. (1984). *Drosophila* maternal and embryo mRNAs transcribed from a single transcription unit use alternate combinations of exons. *EMBO J.* **3**, 1003.

Vincent, J.-P., Oster, G. F., and Gerhart, J. C. (1986). Kinematics of gray crescent formation in *Xenopus* eggs: The displacement of subcortical cytoplasm relative to the egg surface. *Dev. Biol.* **113**, 484.

Vincent, W., Halvorsen, H., Chen, H., and Shin, D. (1969). A comparison of the ribosomal gene amplification in uni- and multinucleate oocytes. *Exp. Cell Res.* **57**, 240.

Vitelli, L., Kemler, I., Busslinger, M., and Birnstiel, M. L. (1986). Developmental regulation of early histone genes injected into sea urchin eggs depends on 5' gene sequences. In preparation.

Vlad, M., and Macgregor, H. C. (1975). Chromomere number and its genetic significance in lampbrush chromosomes. *Chromosoma* **50**, 327.

Vogel, O. (1978). Pattern formation in the egg of the leafhopper *Euscelis plebejus* Fall. (Homoptera): Developmental capacities of fragments isolated from the polar egg regions. *Dev. Biol.* **67**, 357.

Vogel, O. (1982a). Development of complete embryos in drastically deformed leafhopper eggs. *Wilhelm Roux's Arch. Dev. Biol.* **191**, 134.

Vogel, O. (1982b). Experimental test fails to confirm gradient interpretation of embryonic patterning in leafhopper eggs. *Dev. Biol.* **90**, 160.

Vogel, O. (1983). Pattern formation by interaction of three cytoplasmic factors in the egg of the leafhopper *Euscelis plebejus*. *Dev. Biol.* **99**, 166.

Vogt, W. (1929). Gestaltungsanalyse am Amphibienkeim mit örtlicher Vitalfärbung. II. Teil. Gastrulation und Mesodermbildung bei Urodelen und Anuren. *Wilhelm Roux's Arch. Entwicklungsmech. Org.* **120**, 384.

von Baer, K. E. (1828, 1837). Über Entwicklungsgeschichte der Tiere: Beobachtung und Reflexion. 1st part, Königsberg, 1828; 2nd part, Königsberg, 1837; 3rd part (notes), Stieda, Königsberg, 1888.

von Beroldingen, C. H. (1981). The developmental potential of synchronized amphibian cell nuclei. *Dev. Biol.* **81**, 115.

von Brunn, A., and Kalthoff, K. (1983). Photoreversible inhibition by ultraviolet light of germ line development in *Smittia* sp. (Chironomidae, Diptera). *Dev. Biol.* **100**, 426.

von Holt, C., De Groot, P., Schwager, S., and Brandt, W. F. (1984). The structure of sea urchin histones and considerations on their function. *In* "Histone Genes: Structure, Organization, and Regulation" (G. S. Stein, J. L. Stein, and W. F. Marzluff, eds.), p. 65. Wiley, New York.

Wabl, M. R., Brun, R. B., and Du Pasquier, L. (1975). Lymphocytes of the toad *Xenopus laevis* have the gene set for promoting tadpole development. *Science* **190**, 1310.

Wadsworth, S. C., Madhavan, K., and Bilodeau-Wentworth, D. (1985). Maternal inheritance of transcripts from three *Drosophila src*-related genes. *Nucleic Acid Res.* **13**, 2153.

Wakahara, M. (1977). Partial characterization of 'primordial germ cell-forming activity' localized in vegetal pole cytoplasm in anuran eggs. *J. Embryol. Exp. Morphol.* **39**, 221.

Wakahara, M. (1978). Induction of supernumerary primordial germ cells by injecting vegetal pole cytoplasm into *Xenopus* eggs. *J. Exp. Zool.* **203**, 159.

Wakahara, M., Neff, A. W., and Malacinski, G. M. (1984). Topology of the germ plasm and development of primordial germ cells in inverted amphibian eggs. *Differentiation* **26**, 203.

Wakefield, L., and Gurdon, J. B. (1983). Cytoplasmic regulation of 5S RNA genes in nuclear transplant embryos. *EMBO J.* **2**, 1613.

Wakefield, L., Ackerman, E., and Gurdon, J. B. (1983). The activation of RNA synthesis by somatic nuclei injected into amphibian oocytes. *Dev. Biol.* **95**, 468.

Wakimoto, B. T., and Kaufman, T. C. (1981). Analysis of larval segmentation in lethal genotypes associated with the *Antennapedia* gene complex in *Drosophila melanogaster*. *Dev. Biol.* **81**, 51.

Wakimoto, B. T., Turner, F. R., and Kaufman, T. C. (1984). Defects in embryogenesis in mutants associated with the *Antennapedia* gene complex of *Drosophila melanogaster*. *Dev. Biol.* **102**, 147.

Walker, M. D., Edlund, T., Boulet, A. M., and Rutter, W. J. (1983). Cell-specific expression controlled by the 5'-flanking region of insulin and chymotrypsin genes. *Nature (London)* **306**, 557.

Wallace, H., and Elsdale, T. R. (1963). Effects of actinomycin D on amphibian development. *Acta Embryol. Morphol. Exp.* **6**, 275.

Wallace, R. A. (1983). Interactions between somatic cells and the growing oocyte of *Xenopus laevis*. *In* "Current Problems in Germ Cell Differentiation" (A. McLaren and C. C. Wylie, eds.), p. 285. Cambridge Univ. Press, London and New York.

Wallace, R. A., and Misulovin, Z. (1978). Long-term growth and differentiation of *Xenopus* oocytes in a defined medium. *Proc. Natl. Acad. Sci. U.S.A.* **75**, 5534.

Wallace, R. B., Dube, S. K., and Bonner, J. (1977). Localization of the globin gene in the template active fraction of chromatin of Friend leukemia cells. *Science* **198**, 1166.

Ward, G. E., Vacquier, V. D., and Michel, S. (1983). The increased phosphorylation of ribosomal protein S6 in *Arbacia punctulata* is not a universal event in the activation of sea urchin eggs. *Dev. Biol.* **95**, 360.

Waring, G. L., and Mahowald, A. P. (1979). Identification and time of synthesis of chorion proteins in *Drosophila melanogaster*. *Cell* **16**, 599.

Waring, G. L., Allis, C. D., and Mahowald, A. P. (1978). Isolation of polar granules and the identification of polar granule-specific protein. *Dev. Biol.* **66**, 197.

Waring, G. L., DiOrio, J. P., and Hennen, S. (1983). Isolation of germ line-dependent female sterile mutation that effects yolk specific sequestration and chorion formation in *Drosophila*. *Dev. Biol.* **100**, 452.

Waring, M., and Britten, R. J. (1966). Nucleotide sequence repetition: A rapidly reassociating fraction of mouse DNA. *Science* **154**, 791.

Warn, R. (1975). Restoration of the capacity to form pole cells in UV-irradiated *Drosophila* embryos. *J. Embryol. Exp. Morphol.* **33**, 1003.

Warn, R. M., Gutzeit, H. O., Smith, L., and Warn, A. (1985). F-actin rings are associated with the ring canals of the *Drosophila* egg chamber. *Exp. Cell Res.* **157**, 355.

Warren, T. G. and Mahowald, A. P. (1979). Isolation and partial chemical characterization of the three major yolk polypeptides from *Drosophila melanogaster*. *Dev. Biol.* **68**, 130.

Wassarman, P. (1983). Oogenesis: Synthetic events in the developing mammalian egg. *In* "Mechanism and Control of Animal Fertilization" (J. F. Hartmann, ed.), p. 1. Academic Press, New York.

Wassarman, P. M., and Bleil, J. D. (1982). The role of zona pellucida glycoproteins as regulators of sperm-egg interactions in the mouse. *In* "Cellular Recognition" (W. A. Frazier and L. Glaser, eds.), p. 845. Alan R. Liss, New York.

Wassarman, P. M., and Mrozak, S. C. (1981). Program of early development in the mammal: Synthesis and intracellular migration of histone H4 during oogenesis in the mouse. *Dev. Biol.* **84**, 364.

Wassarman, P. M., Schultz, R. M., Letourneau, G. E., LaMarca, M. J., Josefowicz, W. J., and Bleil, J. D. (1979). Meiotic maturation of mouse oocytes *in vitro*. *In* "Ovarian Follicular and Corpus Luteum Function" (C. P. Channing, J. M. Marsh, and W. A. Sadler, eds.), p. 251. Plenum, New York.

Wassarman, W. J., and Smith, L. D. (1978). Oocyte maturation in nonmammalian vertebrates. *In* "The Vertebrate Ovary: Comparative Biology and Evolution" (R. E. Jones, ed.), p. 443. Plenum, New York.

Wassarman, W. J., Richter, J. D., and Smith, L. D. (1982). Protein synthesis during maturation promoting factor- and progesterase-induced maturation in *Xenopus* oocytes. *Dev. Biol.* **89**, 152.

Webb, A. C., and Smith, L. D. (1977). Accumulation of mitochondrial DNA during oogenesis in *Xenopus laevis*. *Dev. Biol.* **56**, 219.

Webb, A. C., LaMarca, M. J., and Smith, L. D. (1975). Synthesis of mitochondrial RNA by full-grown and maturing oocytes of *Rana pipiens* and *Xenopus laevis*. *Dev. Biol.* **45**, 44.

Wedeen, C., Harding, K., and Levine, M. (1986). Spatial regulation of Antennapedia and bithorax gene expression by the *Polycomb* locus in *Drosophila*. *Cell* **44**, 739.

Wedlich, D., Dreyer, C., and Hausen, P. (1985). Occurrence of a species-specific nuclear antigen in the germ line of *Xenopus* and its expression from paternal genes in hybrid frogs. *Dev. Biol.* **108**, 220.

Weinberg, E. S., Overton, G. C., Shutt, R. H., and Reeder, R. H. (1975). Histone gene arrangement in the sea urchin, *Strongylocentrotus purpuratus*. *Proc. Natl. Acad. Sci. U.S.A.* **72**, 4815.

Weinberg, E. S., Hendricks, M. B., Hemminki, K., Kuwabara, P. E., and Farrelly, L. A. (1983). Timing and rates of synthesis of early histone mRNA in the embryo of *Strongylocentrotus purpuratus*. *Dev. Biol.* **98**, 117.

Weiner, A. J., Scott, M. P., and Kaufman, T. C. (1984). A molecular analysis of *fushi tarazu*, a gene in *Drosophila melanogaster* that encodes a product affecting embryonic segment number and cell fate. *Cell* **37**, 843.

Weinstock, R., Sweet, R., Weiss, M., Cedar, H., and Axel, R. (1978). Intragenic DNA spacers interrupt the ovalbumin gene. *Proc. Natl. Acad. Sci. U.S.A.* **75**, 1299.

Weintraub, H. (1985a). Tissue-specific gene expression and chromatin structure. *Harvey Lect.* **79**, 217.

Weintraub, H. (1985b). Assembly and propagation of repressed and derepressed chromosomal states. *Cell* **42**, 705.

Weintraub, H., and Groudine, M. (1976). Chromosomal subunits in active genes have an altered conformation. Globin genes are digested by deoxyribonuclease I in red blood cell nuclei but not in fibroblast nuclei. *Science* **193**, 848.

Weintraub, H., Beug, H., Groudine, M., and Graf, T. (1982). Temperature-sensitive changes in the structure of globin chromatin in lines of red cell precursors transformed by ts-AEV. *Cell* **28**, 931.

Weir, M. P., and Kornberg, T. C. (1985). Patterns of *engrailed* and *fushi tarazu* transcripts reveal novel intermediate stages in *Drosophila* segmentation. *Nature (London)* **318**, 433.

Weisbrod, S. (1982). Active chromatin. *Nature (London)* **297**, 289.

Weismann, A. (1885). "Die Continuität des Keimplasmas als Grundlage einer Theorie de Vererbung." Fischer, Jena. [The continuity of the germ plasm as the foundation of a theory of heredity. (Engl. transl.). *In* "Essays upon Heredity and Kindred Biological Problems," Poulton, E. B., Schönland S., and Shipley, A. E., eds. (1891). Vol. 1, p. 162. Clarendon Press, Oxford.]

Weismann, A. (1892). "Das Keimplasma, eine Theorie der Vererbung." Fischer, Jena. [Engl. transl. *In* "The Germ-Plasm, a Theory of Heredity." Parker, W. N., and Rönnfeldt, H., eds. (1893). Walter Scott, London.]

Weiss, Y. C., Vaslet, C. A., and Rosbash, M. (1981). Ribosomal protein mRNAs increase dramatically during *Xenopus* development. *Dev. Biol.* **87**, 330.

Wellauer, P. K., Dawid, I. B., Brown, D. D., and Reeder, R. H. (1976a). The molecular basis for length heterogeneity in ribosomal DNA from *Xenopus laevis. J. Mol. Biol.* **105**, 461.

Wellauer, P. K., Reeder, R. H., Dawid, I. B., and Brown, D. D. (1976b). The arrangement of length heterogeneity in repeating units of amplified and chromosomal ribosomal DNA from *Xenopus laevis. J. Mol. Biol.* **105**, 487.

Wells, D. E., Showman, R. M., Klein, W. H., and Raff, R. A. (1981). Delayed recruitment of maternal histone H3 mRNA in sea urchin embryos. *Nature (London)* **292**, 477.

Wells, D. E., Bruskin, A. M., O'Brochta, D. A., and Raff, R. A. (1982). Prevalent RNA sequences of mitochondrial origin in sea urchin embryos. *Dev. Biol.* **92**, 557.

Welply, J. K., Lau, J. T., and Lennarz, W. J. (1985). Developmental regulation of glycosyltransferases involved in synthesis of *N*-linked glycoproteins in sea urchin embryos. *Dev. Biol.* **107**, 252.

Wessel, G. M., and McClay, D. R. (1985). Sequential expression of germ layer specific molecules in the sea urchin embryo. *Dev. Biol.* **111**, 451.

Wessel, G. M., Marchase, R. B., and McClay, D. R. (1984). Ontogeny of the basal lamina in the sea urchin embryo. *Dev. Biol.* **103**, 235.

Wetmur, J. G. (1971). Excluded volume effects on the rate of renaturation of DNA. *Biopolymers* **10**, 601.

Wetmur, J. G. (1976). Hybridization and renaturation kinetics of nucleic acids. *Annu. Rev. Biophys. Bioeng.* **5**, 337.

Wetmur, J. G., and Davidson, N. (1968). Kinetics of renaturation of DNA. *J. Mol. Biol.* **31**, 349.

Wharton, K. A., Johansen, K. M., Xu, T., and Artavanis-Tsakonas, S. (1985). Nucleotide sequence from the neurogenic locus *Notch* implies a gene product which shares homology with proteins containing EGF-like repeats. *Cell* **43**, 567.

White, M. J. D. (1954). "Animal Cytology and Evolution," 2nd ed. Cambridge Univ. Press, London and New York.

White, M. J. D. (1978). "Modes of Speciation." Freeman, San Francisco, California.

White, R. A. H., and Akam, M. E. (1985). *Contrabithorax* mutations cause inappropriate expression of *Ultrabithorax* products in *Drosophila. Nature (London)* **318**, 567.

White, R. A. H., and Wilcox, M. (1984). Protein products of the bithorax complex in *Drosophila. Cell* **39**, 163.

White, R. A. H., and Wilcox, M. (1985). Regulation of the distribution of *Ultrabithorax* proteins in *Drosophila. Nature (London)* **318**, 563.

Whiteley, A. H. (1949). The phosphorus compounds of sea urchin eggs and the uptake of radiophosphate upon fertilization. *Am. Nat.* **83**, 249.

Whiteley, A. H., McCarthy, B. J., and Whiteley, H. R. (1966). Changing populations of messenger RNA during sea urchin development. *Proc. Natl. Acad Sci. U.S.A.* **55**, 285.

Whiteley, H. R., McCarthy, B. J., and Whiteley, A. H. (1970). Conservatism of base sequences in RNA for early development of echinoderms. *Dev. Biol.* **21**, 216.

Whitington, P. McD., and Dixon, K. E. (1975). Quantitative studies of germ plasm and germ cells during early embryogenesis of *Xenopus laevis. J. Embryol. Exp. Morphol.* **33**, 57.

Whitman, C. O. (1878). The embryology of *Clepsine. Q. J. Microsc. Sci.* **18**, 215.

Whitman, C. O. (1893). The inadequacy of the cell-theory of development. *J. Morphol.* **8**, 639.

Whitman, C. O. (1895a). Evolution or epigenesis. *In* "Biological Lectures Delivered at the

Marine Biological Laboratory of Woods Hole," 10th Lecture, p. 205. Ginn, Boston, Massachusetts.

Whitman, C. O. (1895b). Bonnet's theory of evolution. A system of negations. *In* "Biological Lectures Delivered at the Marine Biological Laboratory of Woods Hole," 12th Lecture, p. 225. Ginn, Boston, Massachusetts.

Whittaker, J. R. (1973). Segregation during ascidian embryogenesis of egg cytoplasmic information for tissue-specific enzyme development. *Proc. Natl. Acad. Sci. U.S.A.* **70**, 2096.

Whittaker, J. R. (1977). Segregation during cleavage of a factor determining endodermal alkaline phosphatase development in ascidian embryos. *J. Exp. Zool.* **202**, 139.

Whittaker, J. R. (1979a). Cytoplasmic determinants of tissue differentiation in the ascidian egg. *In* "Determinants of Spatial Organization" (S. Subtelny and I. R. Konigsberg, eds.), p. 29. Academic Press, New York.

Whittaker, J. R. (1979b). Quantitative control of end products in the melanocyte lineage of the ascidian embryo. *Dev. Biol.* **73**, 76.

Whittaker, J. R. (1980). Acetylcholinesterase development in extra cells caused by changing the distribution of myoplasm in ascidian embryos. *J. Embryol. Exp. Morphol.* **55**, 343.

Whittaker, J. R. (1982). Muscle lineage cytoplasm can change the developmental expression in epidermal lineage cells of ascidian embryos. *Dev. Biol.* **93**, 463.

Whittaker, J. R. (1983). Quantitative regulation of acetylcholinesterase development in the muscle lineage cells of cleavage-arrested ascidian embryos. *J. Embryol. Exp. Morphol.* **76**, 235.

Whittaker, J. R., Ortolani, G., and Farinella-Ferruzza, N. (1977). Autonomy of acetylcholinesterase differentiation in muscle lineage cells of ascidian embryos. *Dev. Biol.* **55**, 196.

Wickens, M. P., and Gurdon, J. B. (1983). Posttranslational processing of simian virus 40 late transcripts in injected frog oocytes. *J. Mol. Biol.* **163**, 1.

Wickens, M. P., Woo, S., O'Malley, B. W., and Gurdon, J. B. (1980). Expression of a chicken chromosomal ovalbumin gene injected into frog oocyte nuclei. *Nature (London)* **285**, 628.

Wieschaus, E. (1979). *fs(1)K110*, a female sterile mutation altering the pattern of both the egg coverings and the resultant embryos in *Drosophila*. *In* "Cell Lineage, Stem Cells, and Cell Determination" (N. Le Douarin, ed.), INSERM Symp. 10, p. 291. Elsevier/North-Holland, Amsterdam.

Wieschaus, E., and Gehring, W. (1976). Clonal analysis of primordial disc cells in the early embryo of *Drosophila melanogaster*. *Dev. Biol.* **50**, 249.

Wieschaus, E., Marsh, J. L., and Gehring, W. (1978). *Fs(1)K10*, a germ line-dependent female sterile mutation causing abnormal chorion morphology in *Drosophila melanogaster*. *Wilhelm Roux's Arch. Dev. Biol.* **184**, 75.

Wieschaus, E., Nüsslein-Volhard, C., and Jürgens, G. (1984a). Mutations affecting the pattern of the larval cuticle in *Drosophila melanogaster*. III. Zygotic loci on the X-chromosome and fourth chromosome. *Wilhelm Roux's Arch. Dev. Biol.* **193**, 296.

Wieschaus, E., Nüsslein-Volhard, C., and Kluding, H. (1984b). *Krüppel*, a gene whose activity is required early in the zygotic genome for normal embryonic segmentation. *Dev. Biol.* **104**, 172.

Wilkinson, D. G., and Nemer, M. (1986a). Metallothionein mRNAs expressed under distinct quantitative and tissue-specific regulation in sea urchin embryos. Submitted for publication.

Wilkinson, D. G., and Nemer, M. (1986b). Different sea urchin metallothionein genes encode tissue-specific mRNAs in isotypic proteins with potentially distinct structures. Submitted for publication.

Williams, D. H. C., and Anderson, D. T. (1975). The reproductive system, embryonic development, larval development, and metamorphosis of the sea urchin *Heliocidaris erythrogramma* (Val.) (Echinoidea: Echinometridae). *Aust. J. Zool.* **23**, 371.

Williams, M. A., and Smith, L. D. (1971). Ultrastructure of the 'germinal plasm' during maturation and early cleavage in *Rana pipiens*. *Dev. Biol.* **25**, 568.

Williams, M. A., and Smith, L. D. (1984). Ultraviolet irradiation of *Rana pipiens* embryos delays the migration of primordial germ cells into the genital ridges. *Differentiation* **26**, 220.

Willing, M. C., Nienhuis, A. W., and Anderson, W. F. (1979). Selective activation of human β- but not γ-globin gene in human fibroblast X mouse erythroleukaemia cell hybrids. *Nature (London)* **277**, 534.

Wilson, E. B. (1896a). On cleavage and mosaic work. See Crampton (1896), Appendix, p. 19.

Wilson, E. B. (1896b). "The Cell in Development and Inheritance." Macmillan, New York.

Wilson, E. B. (1898). Considerations on cell-lineage and ancestral reminiscence, based on a re-examination of some points in the early development of annelids and polyclades. *Ann. N.Y. Acad. Sci.* **11**, 1.

Wilson, E. B. (1903). Experiments on the cleavage and localization in the nemertine-egg. *Arch. Entwicklungsmech. Org.* **16**, 411.

Wilson, E. B. (1904a). Experimental studies in germinal localization. II. Experiments on the cleavage-mosaic in *Patella* and *Dentalium*. *J. Exp. Zool.* **1**, 197.

Wilson, E. B. (1904b). Experimental studies in germinal localization. I. The germ-regions in the egg of *Dentalium*. *J. Exp. Zool.* **1**, 1.

Wilson, E. B. (1925). "The Cell in Development and Heredity," 3rd ed. Macmillan, New York.

Wilson, I. B., Bolton, F., and Cuttler, R. H. (1972). Preimplantation differentiation in the mouse egg as revealed by microinjection of vital markers. *J. Embryol. Exp. Morphol.* **27**, 467.

Wilt, F. H. (1963). The synthesis of ribonucleic acid in sea urchin embryos. *Biochem. Biophys. Res. Commun.* **11**, 447.

Wilt, F. H. (1964). Ribonucleic acid synthesis during sea urchin embryogenesis. *Dev. Biol.* **9**, 299.

Wilt, F. H. (1965). Regulation of the initiation of chick embryo hemoglobin synthesis. *J. Mol. Biol.* **12**, 331.

Wilt, F. H. (1970). The acceleration of ribonucleic acid synthesis in cleaving sea urchin embryos. *Dev. Biol.* **23**, 444.

Wilt, F. H. (1973). Polyadenylation of maternal RNA of sea urchin eggs after fertilization. *Proc. Natl. Acad. Sci. U.S.A.* **70**, 2345.

Wilt, F. H. (1977). The dynamics of maternal poly(A)-containing mRNA in fertilized sea urchin eggs. *Cell* **11**, 673.

Wilt, F. H., and Hultin, T. (1962). Stimulation of phenylalanine incorporation by polyuridylic acid in homogenates of sea urchin eggs. *Biochem. Biophys. Res. Commun.* **9**, 313.

Wilt, F. H., Aronson, A. I., and Wartiovaara, J. (1969). Function of the nuclear RNA of sea urchin embryos. *In* "Problems in Biology: RNA in Development" (E. W. Hanly, ed.), p. 331. Univ. of Utah Press, Salt Lake City.

Wilt, F. H., Anderson, M., and Ekenberg, E. (1973). Centrifugation of nuclear ribonucleoprotein particles of sea urchin embryos in cesium sulfate. *Biochemistry* **12**, 959.

Winkler, M. M., and Steinhardt, R. A. (1981). Activation of protein synthesis in a sea urchin cell-free system. *Dev. Biol.* **84**, 432.

Winkler, M. M., Bruening, G., and Hershey, J. W. B. (1983). An absolute requirement for the 5' cap structure for mRNA translation in sea urchin eggs. *Eur. J. Biochem.* **137**, 227.

Winkler, M. M., Nelson, E. M., Lashbrook, C., and Hershey, J. W. B. (1985). Multiple levels of regulation of protein synthesis at fertilization in sea urchin eggs. *Dev. Biol.* **107**, 290.

Winkles, J. A., and Grainger, R. M. (1985a). Differential stability of *Drosophila* embryonic mRNAs during subsequent larval development. *J. Cell Biol.* **101**, 1808.

Winkles, J. A., and Grainger, R. M. (1985b). Polyadenylation state of abundant mRNAs during *Drosophila* development. *Dev. Biol.* **110**, 259.

Winkles, J. A., Sargent, T. D., Parry, D. A. D., Jonas, E., and Dawid, I. B. (1985). A developmentally regulated cytokeratin gene in *Xenopus laevis*. *Mol. Cell. Biol.* **5**, 2575.

Winter, H., Wiemann-Weiss, D., and Duspiva, F. (1977). Endogene synthese kurtzlebiger Messenger RNS in der Oocyte von *Dysdercus intermedius* Dist. nach Anschluss der Vitellogenese. *Wilhelm Roux's Arch. Dev. Biol.* **182**, 39.

Wold, B. J. (1978). Studies of structural gene transcripts in sea urchin embryos and adult tissues. Ph.D. Thesis, California Inst. Technol., Pasadena.

Wold, B. J., Klein, W. H., Hough-Evans, B. R., Britten, R. J., and Davidson, E. H. (1978). Sea urchin blastula mRNA sequences expressed in the nuclear RNAs of adult tissues. *Cell* **14**, 941.

Wolf, N., Priess, J., and Hirsh, D. (1983). Segregation of germ line granules in early embryos of *Caenorhabditis elegans*: An electron microscopic analysis. *J. Embryol. Exp. Morphol.* **73**, 297.

Wolff, C. F. (1759). "Theoria Generationis." Halle.

Wolff, C. F. (1768). De formatione intestinorum praecipue, tum et de amnio spurio, aliisque partibus embryonis gallinacei nondum visis. *Novi Comment. Acad. Sci. Imp. Petropol.* **12**. ["Über die Bildung des Darmkanals im befruchteten Hühnchen" (J. F. Meckel, trans.). Halle, 1812.]

Wolff, G. (1895). Entwickelungsphysiologische Studien. I. Die Regeneration der Urodelenlinse. *Arch. Entwicklungsmech. Org.* **1**, 380.

Wolgemuth, D. J., Jagiello, G. M., and Henderson, A. S. (1979). Quantitation of ribosomal RNA genes in fetal human oocyte nuclei using rRNA:DNA hybridization *in situ*. *Exp. Cell Res.* **118**, 181.

Wolgemuth, D. J., Jagiello, G. M., and Henderson, A. S. (1980). Baboon late diplotene oocytes contain micronucleoli and a low label of extra rDNA templates. *Dev. Biol.* **78**, 598.

Wolpert, L., and Mercer, E. H. (1963). An electron microscope study of the development of the blastula of the sea urchin embryo and its radial polarity. *Exp. Cell Res.* **30**, 280.

Wolstenholme, D. R. (1973). Replicating DNA molecules from eggs of *Drosophila melanogaster*. *Chromosoma* **43**, 1.

Wood, W. B., and Revel, H. R. (1976). The genome of bacteriophage T4. *Bacteriol. Rev.* **40**, 847.

Wood, W. B., Hecht, R., Carr, S., Vanderslice, R., Wolf, N., and Hirsh, D. (1980). Parental effects and phenotypic characterization of mutations that affect early development in *Caenorhabditis elegans*. *Dev. Biol.* **74**, 446.

Wood, W. B., Strome, S. and Laufer, J. S. (1983). Localization and determination in embryos of *Caenorhabditis elegans*. *In* "Time, Space, and Pattern in Embryonic Development" (W. R. Jeffery and R. A. Raff, eds.), p. 221. Alan R. Liss, New York.

Wood, W. B., Schierenberg, E., and Strome, S. (1984). Localization and determination in early embryos of *Caenorhabditis elegans*. *In* "Molecular Biology of Development" (E. H. Davidson and R. A. Firtel, eds.), p. 37. Alan R. Liss, New York.

Woodland, H. R. (1974). Changes in the polysome content of developing *Xenopus laevis* embryos. *Dev. Biol.* **40**, 90.

Woodland, H. R. (1980). Histone synthesis during the development of *Xenopus*. *FEBS Lett.* **121**, 1.

Woodland, H. R., and Adamson, E. D. (1977). The synthesis and storage of histones during the oogenesis of *Xenopus laevis*. *Dev. Biol.* **57**, 118.

Woodland, H. R., and Ballantine, J. E. M. (1980). Paternal gene expression in developing hybrid embryos of *Xenopus laevis* and *Xenopus borealis*. *J. Embryol. Exp. Morphol.* **60**, 359.

Woodland, H. R., and Gurdon, J. B. (1968). The relative rates of synthesis of DNA, sRNA and rRNA in the endodermal region and other parts of *Xenopus laevis* embryos. *J. Embryol. Exp. Morphol.* **19**, 363.

Woodland, H. R., and Wilt, F. H. (1980a). The stability and translation of sea urchin histone messenger RNA molecules injected into *Xenopus laevis* eggs and developing embryos. *Dev. Biol.* **75**, 214.

Woodland, H. R., and Wilt, F. H. (1980b). The functional stability of sea urchin histone mRNA injected into oocytes of *Xenopus laevis*. *Dev. Biol.* **75**, 199.

Woodland, H. R., Flynn, J. M., and Wylie, A. J. (1979). Utilization of stored mRNA in *Xenopus* embryos and its replacement by newly synthesized transcripts: Histone H1 synthesis using interspecies hybrids. *Cell* **18**, 165.

Woodland, H. R., Old, R. W., Sturgess, E. A., Ballantine, J. E. M., Aldridge, T. C., and Turner, P. C. (1983). The strategy of histone gene expression in the development of *Xenopus*. *In* "Current Problems in Germ Cell Differentiation" (A. McLaren and C. C. Wylie, eds.), p. 353. Cambridge Univ. Press, London and New York.

Woodland, H. R., Warmington, J. R., Ballantine, J. E. M., and Turner, P. C. (1984). Are there major developmentally regulated H4 gene classes in *Xenopus*? *Nucleic Acids Res.* **12**, 4939.

Woodruff, R. I., and Telfer, W. H. (1980b). Electrophoresis of proteins in intercellular bridges. *Nature (London)* **286**, 84.

Woods, D. E., and Fitschen, W. (1978). The mobilization of maternal histone messenger RNA after fertilization of the sea urchin egg. *Cell Differ.* **7**, 103.

Wormington, W. M., and Brown, D. D. (1983). Onset of 5S RNA gene regulation during *Xenopus* embryogenesis. *Dev. Biol.* **99**, 248.

Wormington, W. M., Bogenhagen, D. F., Jordan, E., and Brown, D. D. (1981). A quantitative assay for *Xenopus* 5S RNA gene transcription *in vitro*. *Cell* **24**, 809.

Wright, S., de Boer, E., Grosveld, F. G., and Flavell, R. A. (1983). Regulated expression of the human β-globin gene family in murine erythroleukaemia cells. *Nature (London)* **305**, 333.

Wright, T. R. F. (1970). The genetics of embryogenesis in *Drosophila*. *Adv. Genet.* **15**, 261.

Wright, T. R. F., Beermann, W., Marsh, J. L., Bishop, C. P., Steward, R., Black, B. C., Tomsett, A. D., and Wright, E. (1981). The genetics of dopa decarboxylase in *Drosophila melanogaster*. IV. The genetics and cytology of the 37B10-37D1 region. *Chromosoma* **83**, 45.

Wu, C. (1984). Activating protein factor binds *in vitro* to upstream control sequences in heat shock gene chromatin. *Nature (London)* **311**, 81.

Wu, R. S., and Wilt, F. H. (1974). The synthesis and degradation of RNA containing polyriboadenylate during sea urchin embryogeny. *Dev. Biol.* **41**, 352.

Wylie, C. C., Heasman, J., Snape, A., O'Driscoll, M., and Holwill, S. (1985). Primordial germ cells of *Xenopus laevis* are not irreversibly determined early in development. *Dev. Biol.* **112**, 66.

Xin, J.-H., Brandhorst, B. P., Britten, R. J., and Davidson, E. H. (1982). Cloned embryo mRNAs not detectably expressed in adult sea urchin coelomocytes. *Dev. Biol.* **89**, 527.

Xu, W.-L., and Jacobson, M. (1986). Fate maps in *Xenopus* embryos. III. States of determination of single cell and multicellular transplants at blastula, gastrula, and neurula stages. *Dev. Biol.* In press.

Yablonka-Reuveni, Z., and Hille, M. B. (1983). Isolation and distribution of elongation factor 2 in eggs and embryos of sea urchins. *Biochemistry* **22**, 5205.

Yajima, H. (1960). Studies on embryonic determination of the harlequin-fly, *Chironomus dorsalis*. I. Effects of centrifugation and its combination with constriction and puncturing. *J. Embryol. Exp. Morphol.* **8**, 198.

Yamada, T., and McDevitt, D. S. (1974). Direct evidence for transformation of differentiated iris epithelial cells into lens cells. *Dev. Biol.* **38**, 104.

Yamamoto, K. R. (1985). Steroid receptor regulated transcription of specific genes and gene networks. *Annu. Rev. Genet.* **19**, 209.

Yatsu, N. (1904). Experiments on the development of egg fragments in *Cerebratulus*. *Biol. Bull. (Woods Hole, Mass.)* **6**, 123.

Yatsu, N. (1910). Experiments on germinal localization in the egg of *Cerebratulus*. *J. Coll. Sci., Imp. Univ. Tokyo* **27**, No. 17.

Yatsu, N. (1912). Observations and experiments on the Ctenophore egg. III. Experiments on germinal localization of the egg of *Beroe ovata*. *Annot. Zool. Jpn.* **8**, 5.

Yedvobnick, B., Muskavitch, M. A. T., Wharton, K. A., Halpern, M. E., Paul, E., Grimwade, B. G., and Artavanis-Tsakonas, S. (1985). Molecular genetics of *Drosophila* neurogenesis. *Cold Spring Harbor Symp. Quant. Biol.* **50**, 841.

Youn, B. W., and Malacinski, G. M. (1980). Action spectrum for ultraviolet irradiation inactivation of a cytoplasmic component(s) required for neural induction in the amphibian egg. *J. Exp. Zool.* **211**, 369.

Young, E. M., and Raff, R. A. (1979). Messenger ribonucleoprotein particles in developing sea urchin embryos. *Dev. Biol.* **72**, 24.

Young, R. J., and Sweeney, K. (1979). Adenylation and ADP-ribosylation in the mouse 1-cell embryo. *J. Embryol. Exp. Morphol.* **49**, 139.

Zalokar, M. (1973). Transplantation of nuclei into the polar plasm of *Drosophila* eggs. *Dev. Biol.* **32**, 189.

Zalokar, M. (1976). Autoradiographic study of protein and RNA formation during early development of *Drosophila* eggs. *Dev. Biol.* **49**, 425.

Zalokar, M., and Erk, I. (1976). Division and migration of nuclei during early embryogenesis of *Drosophila melanogaster*. *J. Microsc. Biol. Cell.* **25**, 97.

Zalokar, M., and Sardet, C. (1984). Tracing of cell lineage in embryonic development of *Phallusia mammillata* (Ascidia) by vital staining of mitochondria. *Dev. Biol.* **102**, 195.

Zalokar, M., Audit, C., and Erk, I. (1975). Developmental defects of female-sterile mutants of *Drosophila melanogaster*. *Dev. Biol.* **47**, 419.

Zeleny, C. (1904). Experiments on the localization of developmental factors in the nemertine egg. *J. Exp. Zool.* **1**, 293.

Zeller, R., Nyffenegger, T., and De Robertis, E. M. (1983). Nucleocytoplasmic distribution of snRNPs and stockpiled snRNA-binding proteins during oogenesis and early development in *Xenopus laevis*. *Cell* **32**, 425.

Zengel, J. M., and Epstein, H. F. (1980). Identification of genetic elements associated with muscle structure in the nematode *Caenorhabditis elegans*. *Cell Motil.* **1**, 73.

Zierler, M. K., Marini, N. J., Stowers, D. J., and Benbow, R. M. (1985). Stockpiling of DNA polymerases during oogenesis and embryogenesis in the frog, *Xenopus laevis*. *J. Biol. Chem.* **260**, 974.

Zimmerman, J. L., Fouts, D. L., and Manning, J. E. (1980). Evidence for a complex class of nonadenylated mRNA in *Drosophila*. *Genetics* **95**, 673.

Zimmerman, J. L., Fouts, D. L., Levy, L. S., and Manning, J. E. (1982). Nonadenylylated mRNA is present as polyadenylylated RNA in nuclei of *Drosophila*. *Proc. Natl. Acad. Sci. U.S.A.* **79**, 3148.

Zimmerman, J. L., Petri, W., and Meselson, M. (1983). Accumulation of a specific subset of *Drosophila melanogaster* heat shock mRNAs in normal development without heat shock. *Cell* **32**, 1161.

Ziomek, C. A., and Johnson, M. H. (1980). Cell surface interaction induces polarization of mouse 8-cell blastomeres at compaction. *Cell* **21**, 935.

Ziomek, C. A., and Johnson, M. H. (1982). The roles of phenotype and position in guiding the fate of 16-cell mouse blastomeres. *Dev. Biol.* **91**, 440.

Ziomek, C. A., Johnson, M. H., and Handyside, A. H. (1982a). The developmental potential of mouse 16-cell blastomeres. *J. Exp. Zool.* **221**, 345.

Ziomek, C. A., Pratt, H. P. M., and Johnson, M. H. (1982b). The origins of cell diversity in the early mouse embryo. *In* "Functional Integration of Cells in Animal Tissues" (J. D. Pitts and M. E. Finbow, eds.), p. 149. Cambridge University Press, London and New York.

Zusman, S. B., and Wieschaus, E. F. (1985). Requirements for zygotic gene activity during gastrulation in *Drosophila melanogaster*. *Dev. Biol.* **111**, 359.

Züst, B., and Dixon, K. E. (1977). Events in the germ cell lineage after entry of the primordial germ cells into the genital ridges in normal and UV-irradiated *Xenopus laevis*. *J. Embryol. Exp. Morphol.* **41**, 33.

Subject Index

A

α, fraction, 531, 541
Ablation, *see also* Cautery; Laser ablation;
  Pricking; UV irradiation
  Dentalium, 458–459
  gastropod, 463
  silk moth, 317
Accessory cells, in oogenesis, 309
Accumulation kinetics, 140–184
Acetylation, histone, 186
Acetylcholinestarase (ACE), ascidian
  marker, 431–436
Acheta domestica, oogenesis, 315
Acmaea scotum, repetitive sequences, 56
Actias luna, nurse cell RNA, 317
Actin, 116, 172
  ascidian, lamina associated, 489–490
  Caenorhabditis, in situ hybridization, 211
  Drosophila, 114–115, 270
    oogenesis, 321–323
  mouse, 519
    mRNA, 105–106, 197, 400, 403
    protein, 403
  sea urchin, 160
    gene family 245–246
    lineage-specific expression, 224–225,
      230, 233–236, 239–240, 500
    maternal RNA, 68, 85–86, 88, 168
  sheep, maternal RNA, 342
  Smittia, 114–115
  starfish, 85, 496–497
  Xenopus
    cell lineage expression, 42, 97–98, 262–
      266
    maternal RNA, 370, 374
Actinomycin, transcription inhibitor, 82
  anurans, 49, 93, 260
  ascidians, 431, 436

Drosophila, 113
Ilyanassa, 467–468
Pleurodeles, 99
sea urchin, 77, 87
silk moth, 317
Adaptive value of
  developmental strategies, 3
  lampbrush chromosomes, 374
  localization, 8
  meroistic oogenesis, 315, 325
  stress protection, 191
S-adenosylmethionine pool, 546–547
Adenosyl phosphoribosyl transferase
  (APRT), methylation of
  hamster gene, 41–42
  mouse gene, 34
Adenovirus, methylation, 34
Adhesion, changes in *Xenopus* development,
  221–224
Albumin, human, 17
  methylation, 34
Alcohol dehydrogenase (ADH)
  Ambystoma activation, 38
  Drosophila transgenic studies, 29
Aldehyde oxidase, *Drosophila* mutant, 108
Alkaline phosphatase, lineage tracer in
  ascidian gut, 435
  mouse ICM, 518
  Ilyanassa polar lobe isozyme variant,
    467
Allolophora, germ bands, 412
Alternate processing, 112
Alu sequences, maternal RNA, 100
Amanitin (α-), transcriptional inhibitor
  Caenorhabditis, 210
  Drosophila, 114, 115
  mouse, 103, 105, 160
  Pleurodeles, 38
  Xenopus, 99, 134

Index to Figures by Author